Ziegler–Natta Catalysts
and Polymerizations

John Boor, Jr. (*1930–1974*)

Ziegler–Natta Catalysts and Polymerizations

John Boor, Jr.

Shell Development Company
Houston, Texas

ACADEMIC PRESS New York San Francisco London 1979
A Subsidiary of Harcourt Brace Jovanovich, Publishers

ACADEMIC PRESS, INC.
111 Fifth Avenue, New York, New York 10003

United Kingdom Edition published by
ACADEMIC PRESS, INC. (LONDON) LTD.
24/28 Oval Road, London NW1 7DX

Library of Congress Cataloging in Publication Data

Boor, John, 1930–1974.
    Ziegler–Natta catalysts and polymerizations.

    Includes bibliographical references and index.
    1. Ziegler–Natta catalysts. 2. Polymers and
polymerization. I. Title.
TP159.C3B66 1978   660.2'9'95   77–92237
ISBN 0–12–115550–1

PRINTED IN THE UNITED STATES OF AMERICA
79 80 81 82   9 8 7 6 5 4 3 2 1

To my wife Janet
and to my sons Michael and Colin

Contents

1

HIGHLIGHTS OF ZIEGLER–NATTA CATALYSTS AND POLYMERIZATIONS

2

GENESIS OF ZIEGLER–NATTA CATALYSTS

3

DEFINITIONS, STEREOCHEMISTRY, EXPERIMENTAL METHODS, AND COMMERCIAL POLYMERS

4

CHEMICAL DESCRIPTION OF ZIEGLER–NATTA CATALYSTS FOR OLEFINS

5

THE CHEMICAL DESCRIPTION OF CATALYSTS FOR CONJUGATED DIENES

6

INITIAL PHYSICAL STATE OF THE CATALYST

7

PHYSICAL STATE OF THE POLYMER DURING POLYMERIZATION

8

GROWTH OF THE POLYMER PARTICLE

9

MODIFICATION OF ZIEGLER–NATTA CATALYSTS BY THIRD COMPONENTS

10

TERMINATION OF POLYMER CHAIN GROWTH

11

OXIDATION STATE OF CATALYSTS AND ACTIVE CENTERS

16

MECHANISMS FOR STEREOCHEMICAL CONTROL OF CONJUGATED AND NONCONJUGATED DIENES

17

MECHANISMS FOR STEREOCHEMICAL CONTROL OF STEREOSELECTIVE AND STEREOELECTIVE ISOTACTIC PROPAGATIONS

18

KINETICS

19

POLYMERIZATION OF MONOMERS

20
COPOLYMERIZATIONS

21
BLOCK POLYMERIZATIONS

22
OTHER USES OF THE ZIEGLER–NATTA CATALYST

23
FINAL COMMENTS AND OUTLOOK   611

Foreword

In publishing this book posthumously, I wish to express my appreciation to Shell Development Company for its assistance. In particular, I would like to thank Kenneth Lackey and his staff for their work in drawing and photographing figures; Aphrodite Mamoulides and Shirley Thompson for library work; Marjorie Wallace for her superb job in retyping the manuscript and tables; and Ronald S. Bauer, R. William Glass, I. Gerry Burstain, Arthur L. Otermat, Edgar J. Smutny, Simone Mostert, G. Bruce Klingensmith, and Harold E. De La Mare for their final reading of the manuscript. I also wish to recognize the splendid work that G. Lynn Wiesehuegel has done in editing the manuscript.

Finally, my deep gratitude to Harold De La Mare for his faithful and unrelenting effort both in coordinating the work at Shell and for the many hours of long, hard work put into the book at home. Without his tremendous help, this book could never have been published.

Janet Boor

Preface

Since 1955 there has been an explosion of literature on the Ziegler–Natta catalyst and polymerization. There have appeared probably well over ten thousand distinct journal papers and patents covering many aspects of this catalyst. How does an individual cope with this voluminous literature? Review papers on specific and general aspects of this catalysis have helped immensely. A book on Ziegler–Natta catalysts and polymerizations is timely now that two decades have passed since the Ziegler–Natta discoveries were first announced, and the rate at which new publications appear has lessened to a manageable level.

In writing this book I had a number of objectives. Foremost, I wished to tell the Ziegler–Natta story in some depth and in a way that would "put it all together" for the reader who wants to learn something about it. It is a unique and beautiful catalyst. It is commercially very important, and it deserves the attention of the scientific and engineering community. This book was written especially for the chemistry and engineering graduate student as a source of literature on this catalyst, and I hope it will be beneficial in courses and graduate seminars. It was also written for the industrial chemist, engineer, and manager who may become involved in a Ziegler–Natta problem. Whether the problem is short term or long term, this book should help obtain the necessary background and understanding and direct him to proper references for more detailed help. Some of the chapters will be useful for general information to those whose assignment places them between the research–process laboratory and the customers of their products.

I have attempted at all times to present the literature findings as accurately and fairly as possible, always striving to give the view of the worker being cited. Where possible, I added my personal view.

This book is not a catchall for everything that has been published on Ziegler–Natta polymerizations. Indeed, the references cited in this book represent only a small fraction of the total. Some publications are discussed in more detail than others, depending on my assessment of their significance and relevance. I believe that the manner in which the subject matter has been written will be beneficial to those institutions having limited access to patent and journal literature. A compilation of earlier reviews was included to help the reader obtain additional background material.

As this book attests, polymer chemists have freely utilized knowledge of catalysis and organometallic chemistry to solve their problems. They have, in turn, made many significant contributions to the field which should be of interest to catalyst and organometallic chemists. I hope this book is successful in communicating this chemistry to them.

Each chapter was written to be independently coherent and useful in describing a particular subject area. After an initial reading of Chapters 1 and 3 (which cover the subject matter in a broad way), it should be possible to read any chapter without having to read the others. For the reader who seeks only an overview of Ziegler–Natta polymerizations, Chapter 1 may suffice.

So far as I am aware, this is the first single-author book on the general aspects of Ziegler–Natta catalysts and polymerizations. In 1959, Professors Gaylord and Mark devoted a major portion of their book "Linear and Stereoregular Addition Polymers" to this catalyst. In 1972, Professor Keii focused largely on the kinetic features of the catalyst in his book "Kinetics of Ziegler–Natta Polymerization." My book covers seventeen important subjects of Ziegler–Natta catalysts and polymerizations. The polymerizations of olefins, dienes, and many other types of monomers are described. Each of these areas of Ziegler–Natta polymerizations holds a certain fascination which attracted the efforts of individual workers from many worldwide laboratories.

This book could not have been written without the assistance of my wife, Janet, who typed the original manuscript and continually encouraged me to complete it. I am grateful to the Shell Development Company for its help in bringing this manuscript to fruition. I wish to thank Ken Lackey for photographic and reproduction work, Joan Wilkinson and Marjorie Wallace for stenographic assistance, Aphrodite Mamoulides and Shirley Thompson for library assistance, and Laurie Doerr and Martin Baer for patent assistance.

Finally, I want to take this opportunity to thank my colleagues at the Shell Laboratories who were helpful to me in my own experimental work in the Ziegler–Natta and other areas of chemistry. I cite especially D. W. Penhale, who assisted me technically for many years; Dr. E. A. Youngman, who was my immediate manager and with whom I had the pleasure of collaborating on a number of papers; and R. M. Cole, who was head of the Plastics Department at that time and is now retired. As this book attests, the Shell Laboratories in England, The Netherlands, and United States have made notable contributions to Ziegler–Natta chemistry. I have found it personally profitable to be associated with the many people who were responsible for these contributions. My assignment in the Shell Research Laboratory in Amsterdam in 1967–1968 as an exchange scientist gave me the opportunity to work directly with my European colleagues.

John Boor, Jr.

John Boor, Jr.
(1930–1974)

John Boor, Jr., was born in Czechoslovakia on April 1, 1930. He came to the United States in 1938. As a teenager he lived at the Bonnie Brae Farm for Boys in West Millington, New Jersey, and attended the public schools in that area. He graduated from Bernards High School, Bernardsville, New Jersey and then enrolled in Rutgers University where he had received a State Scholarship. He graduated from Rutgers University in 1953 with a B.S. in chemistry. He did his graduate research work with Professor T. L. Jacobs at UCLA, and worked on the polymerization of chloroallene. He received his Ph.D. in 1957, and was employed by the Shell Development Company in July of that year.

From 1967 through 1968 he was on a special exchange scientist assignment with the Shell Laboratory in Amsterdam. Dr. Boor was widely known in scientific circles for his work in Ziegler–Natta catalyzed polymerizations of olefins, and was often invited to give lectures on that subject. He has authored and coauthored a number of very significant papers, particularly in that field. He was also the sole inventor or coinventor of many U.S. patents assigned to the Shell Development Company. He was a member of the American Chemical Society and the New York Academy of Sciences.

Dr. Boor died in a tragic automobile accident December 31, 1974.

Janet Boor

Highlights of Ziegler–Natta Catalysts and Polymerizations

I. Highlights

The Ziegler–Natta catalyst is unique and marvelous, and it is doubtful that it can be challenged by any other catalyst for its versatility. It was discovered in the laboratories of K. Ziegler in the fall of 1953 and G. Natta in spring of 1954. The Ziegler school revealed that high density polyethylene was easily made at low pressures with binary mixtures of metal alkyls and transition metal salts, such as $AlEt_3 + TiCl_4$. The Natta school then demonstrated the ability of the same catalysts, and to a greater extent, catalysts containing lower valent transition metal chloride salts such as $TiCl_3$, to form isotactic polymers from α-olefins. These discoveries provoked a worldwide research and development effort that culminated in many new commercial plastics and elastomers. The Ziegler–Natta catalyst has now joined ranks of conventional cationic, anionic, and radical initiators as one of the major methods that we have to initiate polymerizations. The Ziegler–Natta catalysts became prominent in a special period in the history of polymer science, a period that produced not only many new commercial polymers but also enhanced our basic knowledge of polymers and polymerization. Chapter 2 describes the genesis of the Ziegler–Natta catalyst and briefly recalls other important contemporary discoveries.

According to the broad patent definition, the Ziegler–Natta catalyst is a mixture of a metal alkyl of base metal alkyls of group I to III and a transition metal salt of metals of groups IV to VIII. Not all of the possible combinations are effective, and many of these combinations are active only for certain monomers or under certain conditions.

TABLE 1–1

Selected Example of Olefin Polymerizations[a]

Olefin	Prevalent structural units	Catalyst	Product
Ethylene	$-CH_2-CH-$ Linear	$AlEt_3 + TiCl_4$	Crystalline, MP $\simeq$ 135°C
Propylene and higher α-olefins	(see structure) Isotactic	$AlEt_2Cl + TiCl_3$	Crystalline, MP $\simeq$ 170°C
Propylene only	(see structure) Syndiotactic	$AlEt_2Cl + VCl_4$ ($< -45°C$)	Crystalline, MP $\simeq$ 130°C
(RS)-3-Methyl-1-pentene and similar racemic α-olefins	Equal mixture of polymer chains, each of which contains prevailingly either (R)- or (S)-α-olefin	$Al\text{-}i\text{-}Bu_3 + TiCl_4$	Optically inactive

Isotactic structure:

$$-CH_2-\underset{H}{\overset{CH_3}{C}}-CH_2-\underset{H}{\overset{CH_3}{C}}-CH_2-\underset{H}{\overset{CH_3}{C}}-$$

Syndiotactic structure:

$$-CH_2-\underset{H}{\overset{CH_3}{C}}-CH_2-\underset{CH_3}{\overset{H}{C}}-CH_2-\underset{H}{\overset{CH_3}{C}}-$$

Vinylcyclopropane	$-CH_2-CH->-(CH_2-)_2CH=CH-CH_2-$ $\triangle$	$AlEt_2Cl + VCl_3$	—
Cyclobutene	$\square$ or	$AlEt_2Cl + V(acac)_3$	Crystalline
	$-CH=CH-(CH_2-)_2-$	$AlEt_3 + VCl_4$	Elastomer
2-Butene	$-CH-CH-CH_2-CH_2-$ $\quad | \quad |$ $\quad CH_3 \quad CH_3$	$AlEt_2Cl + V(acac)_3$ $(\leq -30°C)$	Crystalline
	2-Butene polymerizes only if ethylene is present		

^{a} Most isotactic polymerizations are made between 25° and 100°C; however, some catalysts require specific temperatures.

TABLE 1-2

Selected Examples of Diene Polymerizations

Diene	Prevalent structural units	Catalyst	Product
Butadiene	cis-1,4	AlEt₂Cl + soluble cobalt salt	Elastomeric
Butadiene	trans-1,4	AlEt₃ + VCl₃	Crystalline, MP ≈ 145°C
Butadiene	1,2-isotactic	AlEt₃ + Ti(O-n-Bu)₄	Crystalline, MP ≈ 128°C
Butadiene	1,2-syndiotactic	AlEt₃ + Cr(acac)₃	Crystalline, MP ≈ 150°C

Monomer	Structure	Catalyst	Property
Isoprene	CH_3 $-H_2C$ $C=C$ CH_2- *cis*-1,4	$AlEt_3 + TiCl_4$ $Al/Ti \simeq 1$	Elastomeric
Piperylene	H CH_2- $C=C$ $-HC$ CH_3 H isotactic *trans*-1,4	$AlEt_3 + VCl_3$	—
1,5-Hexadiene	$-CH_2-CH-$ $(CH_2)_2$ CH $=CH_2$	$AlEt_2Cl + V(acac)_3$	Elastomeric
	CH_2- CH_2 $+$ $-CH_2-C$ $CH-$ CH_2-CH-	$Al\text{-}i\text{-}Bu_3 + TiCl_3$	Crystalline, MP $\simeq$ 135°C
1,3-Cyclohexadiene		$AlEt_3 + VCl_3$	—

5

Stereochemical control is one of the most important attributes of the Ziegler–Natta catalyst. Tables 1–1 and 1–2 collect some of the more important examples and show the different stereochemical structures that have been isolated in high concentrations. Of the many polymers that have been investigated in the laboratory, about ten (shown in the tabulation below) have been or are about to be commercially produced.

Plastics	Elastomers
Intermediate and high density polyethylenes	cis-1,4-Polybutadiene
Isotactic polypropylene	trans-1,4-Polybutadiene
Isotactic poly-1-butene	trans-1,5-Polypentenamer
Isotactic poly-4-methyl-1-pentene	Random copolymers of ethylene and propylene
Block copolymers of ethylene and propylene	
trans-1,4-Polyisoprene	

Chapter 3 elaborates the definition of the catalyst in detail and the stereochemistry of these important polymers. It also describes both the synthetic methods used to prepare them and analytical methods used to identify and evaluate the stereochemical content.

The versatility of the Ziegler-Natta catalyst is the outcome of its inherent sensitivity to a number of factors that influence its kinetic and stereochemical behavior. This sensitivity has provided a powerful handle by which workers can modify the behavior of the catalyst. Several important factors have been identified: (a) metal alkyl structure—choice of metal and nature of ligands; (b) transition metal structure—choice of metal, nature of ligands, crystal structure, valence; and (c) the way the components are brought together and used in polymerizations—absolute and relative concentrations, aging, temperature, and time. As an illustration, compare examples 2 and 3 in Table 1–1.

While many different catalysts have been investigated, a number have emerged that we can cite as more important on the basis of mechanistic studies or as probable commercial catalysts. Both the dominating factors listed above and these important catalysts are discussed in Chapter 4 (olefins) and Chapter 5 (dienes).

The Ziegler–Natta catalyst has been used in a soluble, colloidal, or heterogeneous form for both olefin and diene polymerizations. Often there is a distinct advantage obtained for each type, such as high activity of catalyst, a desired morphology of the formed polymer particles, comonomer content in copolymer, and type and degree of stereochemical structures. As a specific illustration, crystalline isotactic polyolefins are prepared only with heterogeneous systems, e.g., $AlEt_2Cl + TiCl_3$. To prepare a highly crystalline syndiotactic polypropylene, not only is a soluble catalyst required, but it must be used below $-45°C$. Highly stereoregular polybutadienes, on the other hand, have been synthesized with both heterogeneous and soluble vanadium-based salts. Enhancements of activities up to ten times and higher have been found for ethylene polymerizations if the transition metal salt is chemically fixed on high surface supports, e.g., $TiCl_4$ on $Mg(OH)Cl \rightarrow Mg(OTiCl_3)Cl$, before the metal alkyl component is added. Chapter 6 describes other examples and elaborates on the significance of the physical state.

The literature indicates that, while slurry processes have been more widely used for olefin polymerizations, solution processes have been favored for dienes. The advantage of a slurry process over a solution one for a catalyst of comparable activity is a higher throughput of polymer in the reactor system and a lower solvent recovery. In recent years, vapor phase polymerizations (no solvent added) have been reported for polyethylene synthesis. These can offer a considerable economic advantage since capital expenditure for solvent storage and recovery are eliminated. The main features of these three types of processes and their advantage–disadvantage balance are discussed in Chapter 7.

One of the most fascinating aspects of Ziegler–Natta catalysts is the growth of the polymer particle from the catalyst particle (Chapter 8). A dominating feature of the catalyst particle is the ability to replicate its morphology in the morphology of the progeny polymer particle. The mechanistic path by which this occurs is partially understood. The secondary catalyst $TiCl_3$ particle (about 30 μm) is actually made up of much smaller (primary) subparticles (0.01 to 0.1 μm). Growth of the polymer takes place on these subparticles and they remain within the same polymer particle because the formed polymer acts to cement them. Just how individual chains become assembled to form the optically visible polymer particle has not yet been completely elucidated. Several intermediate substructures have been identified, and a fiber with a 0.5 μm diameter may be a basic building unit. This chemistry has important economic consequences because the morphology of the polymer particles can be largely controlled through the synthesis of the $TiCl_3$ catalyst. The use of dense and highly regular catalyst particles in a slurry process offers obvious advantages in the production of a polyolefin.

Not long after the Ziegler–Natta catalysts were first discovered, many different workers reported how the influence of electron donors greatly affected their kinetic and stereochemical behavior. Typically, such organic molecules as amines, phosphines, esters, and ketones were added in controlled amounts. Changes in catalyst activity, or the molecular weight and stereochemical character of the formed polymer, were usually cited as the outcome of this addition. It has become obvious from many years of literature that these modifications require precise or narrow experimental conditions to achieve the desired change. Studies involving electron-donors have aided mechanistic studies. The exact mechanistic paths by which these donors work depend largely on the catalyst–monomer system investigated and, to a degree, on the experimental conditions. Chapter 9 describes the various systems and the mechanistic explanations offered by individual authors.

Some third components can terminate chain growth and, in this way, the molecular weight of a polymer product can be controlled. Because specific applications require limited molecular weights, molecular weight control is practically very important in a commercial process. While many of the organic and inorganic reactants can terminate chain growth, only a few do so without simultaneously killing the active center. Examples of true transfer agents are molecular hydrogen and $ZnEt_2$, and these are used in olefin polymerizations. Olefins in small concentrations have been found effective in butadiene polymerizations with cobalt-based catalysts. Chapter 10 describes six different methods by which chain termination has been effected.

Knowledge of the oxidation state of the active center is needed to formulate its structure and describe its electronic and steric character. Because we are dealing with such a small number of active centers, it is not always clear whether the measured oxidation state of the whole catalyst reveals the true oxidation state of the active center. Indirect evidence indicates that for many transition metal salts more than one oxidation state is active, while for others only one appears to be active. There appears to be no doubt that the ligand environment of the center is playing a dominant role in deciding not only which oxidation state of the transition metal is active but also the activity of the center. Chapter 11 offers many examples for transition metals from Sc to Co and touches on the theoretical aspects.

At about the time Ziegler–Natta catalysts were discovered, workers at Phillips and Standard of Indiana companies disclosed the synthesis of high density polyethylene catalysts consisting of transition metal oxides on SiO_2, $SiO_2 \cdot Al_2O_3$, and other supports. These were some of the earliest disclosed metal alkyl-free catalysts (no added metal alkyls of group I to III base metals). In the interim, many other different examples have been disclosed which polymerize ethylene, propylene and α-olefins, and butadiene and a few other monomers. Many of these are active only for specific monomers. The importance of these metal alkyl-free catalysts is twofold. First, they are potential commercial catalysts and, in fact, one of these has reached this status. Secondly, their very existence demonstrates that the base metal alkyl component, $AlEt_3$, is not essential for the stereochemical process to occur. Typical examples are shown in Table 1–3, while others are described in detail in Chapter 12.

TABLE 1–3

Selected Examples of Metal Alkyl-Free Catalysts

Catalyst	Olefin	Product
$TiCl_2$ (ball-milled)	Ethylene	Highly linear
$TiCl_3$ + amine	Propylene	Highly isotactic
$ZrBz_3Cl$ (Bz = benzyl)	4-Methyl-1-pentene	Isotactic
$Cr(\pi\text{-allyl})_3$	Ethylene	Highly linear
π-allyl NiX	Butadiene	X = Cl, *cis*-1,4
		X = Br, *trans*-1,4
$(C_5H_5)_2Cr$ on SiO_2 (supported)	Ethylene	Highly linear

The structure of the active center in the polymerization of olefins has been the subject of much discussion (Chapter 13). While many proposals have been made, two views of the active center have predominated, and these are represented by structures I and II.

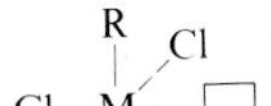

I
Part of crystal surface

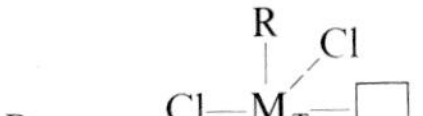

II
Soluble species

M_T = transition metal, e.g., Ti
M_B = base metal, e.g., Al.

The primary role of the base metal alkyl in the Ziegler–Natta catalyst is to alkylate the transition metal salt. In this way, centers of types I and II are formed. Structure II is believed to exist for certain soluble alkylated transition metal centers, which would decompose if not stabilized by complexing with the metal alkyl. Structure I is believed to be the active center in heterogeneous catalysts. In both structures, growth occurs at the transition metal–carbon bond. The view that propagation takes place at a base metal–carbon bond is held by only a few workers today.

It is generally agreed that propagation in an isotactic polymerization involves several steps, as illustrated for complex I in Eq. 1–1.

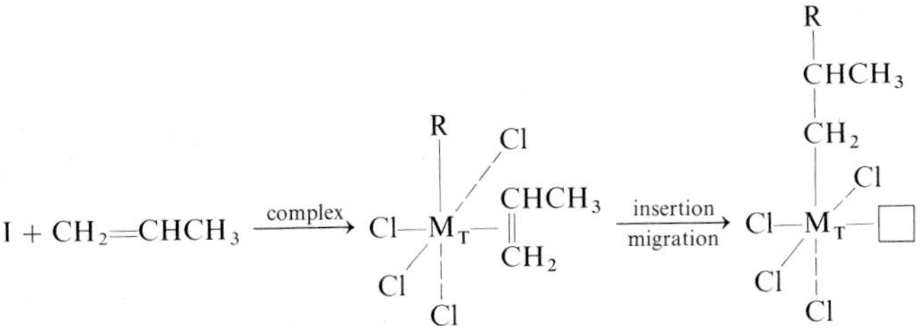

$$I + CH_2=CHCH_3 \xrightarrow{complex} \quad \xrightarrow{\substack{insertion \\ migration}} \quad (1\text{--}1)$$

In the polymerization of dienes, the growth center is also believed to be a transition metal–carbon bond, and both σ- and π-allyl-type bonds have been suggested, depending on the nature of transition metal, as shown in structures III and IV.

III	IV
σ Bond center	π-Allyl bond center

Chapters 4 and 5 discuss the many other proposals that have been made for the structure of the active centers, and they present experimental evidence for the basic views.

In both isotactic and syndiotactic propagations, the growth step consists of a cis insertion (cis-ligand migration) of the olefin into the transition metal bond. Recent evidence suggests that propylene adds to form a primary metal alkyl (primary addition) in an isotactic propagation (see Eq. 1–1) and a secondary metal alkyl (secondary addition) in a syndiotactic propagation. Steric interactions between the complexing propylene and the ligands of the asymmetric center of the heterogeneous catalyst continuously force a complexing mode that has the same configurations (isotactic propagation). In a syndiotactic polymerization with a soluble catalyst, this driving force is absent, and steric interactions between the complexing propylene and the last-added propylene unit force complexing modes in which opposite configurations alternate. While several cogent mechanistic schemes have been proposed to explain the obtained stereochemical structures in diene polymerization, none are generally accepted. Chapters 15 and 16 describe the different mechanistic views that have been published for olefins and dienes and cite the important experimental evidence for the more prominent schemes. Chapter 17 describes a special case of isotactic polymerization involving the polymerization of racemic α-olefins with conventional and optically active Ziegler–Natta catalysts. The active centers of the heterogeneous forms also have the ability to select and polymerize one or the other enantiomers of the racemic mixture in one chain.

The polymerization kinetic data follow mainly two types of rate curves, which are characterized by constant or decaying rates, respectively. The origin of these rate curves, as well as the nature of the polymer products, has been the object of many kinetic studies. Operational contributions, such as nature of catalyst, stirring speeds, temperature, and time of polymerization have been identified fairly well for many systems. The relative importance of basic parameters, however, has been the object of much debate. General agreement does not exist when it comes to identifying and evaluating such factors as the rate-determining step, nature and concentration of centers as a function of time or yield of polymer, living character of polymer chain, and the importance of polymer itself as a barrier to diffusion of olefin to centers. The use of Rideal and Langmuir–Hinshelwood models has not given us greater insight of the mechanistic features of the catalyst. Chapter 18 identifies the kinetic rate curves and the polymer product and then discusses the kinetic schemes and operational and basic factors which have been reported.

Ziegler–Natta catalysts are primarily used for the polymerization of olefins and dienes. Any olefin is polymerizable as long as the 3-position is not fully substituted. The activity of an olefin decreases as the size of the pendant group increases, the branched carbon moves closer to the double bond, and the size of the group attached to the branched point increases. Consider, for example, the following:

$$CH_2{=}CH_2 \gg CH_2{=}CHCH_3 > CH_2{=}CH{-}C_2H_5 > CH_2{=}CH{-}CH_2{-}\overset{\overset{\displaystyle CH_3}{|}}{CH}{-}CH_3 >$$

$$CH_2{=}CH{-}\overset{\overset{\displaystyle CH_3}{|}}{CH}{-}CH_2{-}CH_3 > CH_2{=}CH{-}\overset{\overset{\displaystyle C_2H_5}{|}}{CH}{-}CH_2CH_3 \gg CH_2{=}CH{-}\overset{\overset{\displaystyle CH_3}{|}}{\underset{\underset{\displaystyle CH_3}{|}}{C}}{-}CH_3$$

Alkenelike polar monomers can be homopolymerized and copolymerized if the double bond is insulated by methylene units and/or the heteroatom is shielded by bulky substituents, as shown in structure V.

$$CH_2{=}CH{-}(CH_2)_5\,N\overset{\displaystyle \diagup C_6H_5}{\diagdown C_6H_5}$$

V

Conjugated and nonconjugated dienes are easily polymerized, especially the lower homologs, i.e., butadiene, isoprene, and 1,5-hexadiene. Many other monomers have been reported to be polymerized—allenes, acetylenes, 1,1- and 1,2-disubstituted ethylene, and cycloolefins. Chapter 19 describes the reported polymerizations of eleven classifications of monomers.

Random, alternating, and block copolymers also have been synthesized with selected Ziegler–Natta catalysts. Here, also the composition of the copolymer has been controlled by selection of a proper catalyst and adherence to specific conditions of polymerization. Chapter 20 discusses these operational factors and describes briefly the copolymerization of nine selected comonomer pairs.

While many patents have claimed the synthesis of block polymers with Ziegler–Natta catalysts, there is no evidence to confirm the reported structures. At best, the products probably contain only low contents of the claimed block polymers as contaminants with homopolymer and copolymer. Chapter 21 briefly presents the historical developments and shows why the Ziegler–Natta catalyst is not suited for block polymerizations.

Other uses have also been found for certain binary mixtures of metal alkyl and transition metal salts other than polymerizations described above. These include radical or cationic polymerizations, metathesis, oligomerization, isomerization, hydrogenation, and alkylation. Chapter 22 gives a few examples of each type.

II. Collected Reviews

Throughout the last twenty years, many reviews have appeared covering either general or specific aspects of Ziegler–Natta catalysts and polymerizations. Many were timely "progress reports," especially those which emerged from the Natta school. Naturally, some of the older ones are partly outdated, but they still have an important historical value. Because the publication rate has waned considerably in the last ten years, more recent reviews still have an active value, and they should be consulted by those seeking more detailed data or subject matter which is outside the scope of this book. They will serve as a valuable supplement to this book (and vice versa). Not all of the following reviews were available to the author, however, but are included since they may be accessible to many readers.

As the reader will discover from this book, many aspects of the Ziegler–Natta polymerization are controversial and have been subjected to much personal interpretation. In this respect, some of these reviews, especially 1970 (Ref. 1), 1967 (Ref. 2c), 1974 (Ref. 1), and 1971 (Ref. 3), are invaluable because they were written from a vantage point unique to the respective authors who were able to focus on special topics in great depth. Such reviews will continue to be important in the future as new data provokes new interpretations and mechanistic views.

References

1955

1. G. Natta, Isotactic polymers. *Makromol. Chem.* **16**, No. 3, 213–237 (1955).

1956

1. G. Natta, Stereospecific polymerization of olefins. *Chim. Ind. (Milan)* **38**, 124–127 (1956).
2. G. Natta, Stereospecific catalysis and isotactic polymers. *Chim. Ind. (Milan)* **38**, 751–765 (1956); see also *Angew. Chem.* **68**, 393–403 (1956).

factors as the rate-determining step, nature and concentration of centers as a function of time or yield of polymer, living character of polymer chain, and the importance of polymer itself as a barrier to diffusion of olefin to centers. The use of Rideal and Langmuir–Hinshelwood models has not given us greater insight of the mechanistic features of the catalyst. Chapter 18 identifies the kinetic rate curves and the polymer product and then discusses the kinetic schemes and operational and basic factors which have been reported.

Ziegler–Natta catalysts are primarily used for the polymerization of olefins and dienes. Any olefin is polymerizable as long as the 3-position is not fully substituted. The activity of an olefin decreases as the size of the pendant group increases, the branched carbon moves closer to the double bond, and the size of the group attached to the branched point increases. Consider, for example, the following:

$$CH_2{=}CH_2 \gg CH_2{=}CHCH_3 > CH_2{=}CH{-}C_2H_5 > CH_2{=}CH{-}CH_2{-}\underset{\underset{CH_3}{|}}{CH}{-}CH_3 >$$

$$CH_2{=}CH{-}\underset{\underset{CH_3}{|}}{CH}{-}CH_2{-}CH_3 > CH_2{=}CH{-}\underset{\underset{C_2H_5}{|}}{CH}{-}CH_2CH_3 \ggg CH_2{=}CH{-}\underset{\underset{CH_3}{|}}{\overset{\overset{CH_3}{|}}{C}}{-}CH_3$$

Alkenelike polar monomers can be homopolymerized and copolymerized if the double bond is insulated by methylene units and/or the heteroatom is shielded by bulky substituents, as shown in structure V.

$$CH_2{=}CH{-}(CH_2)_5\ N \overset{\displaystyle C_6H_5}{\underset{\displaystyle C_6H_5}{\big\backslash}}$$

V

Conjugated and nonconjugated dienes are easily polymerized, especially the lower homologs, i.e., butadiene, isoprene, and 1,5-hexadiene. Many other monomers have been reported to be polymerized—allenes, acetylenes, 1,1- and 1,2-disubstituted ethylene, and cycloolefins. Chapter 19 describes the reported polymerizations of eleven classifications of monomers.

Random, alternating, and block copolymers also have been synthesized with selected Ziegler–Natta catalysts. Here, also the composition of the copolymer has been controlled by selection of a proper catalyst and adherence to specific conditions of polymerization. Chapter 20 discusses these operational factors and describes briefly the copolymerization of nine selected comonomer pairs.

While many patents have claimed the synthesis of block polymers with Ziegler–Natta catalysts, there is no evidence to confirm the reported structures. At best, the products probably contain only low contents of the

claimed block polymers as contaminants with homopolymer and copolymer. Chapter 21 briefly presents the historical developments and shows why the Ziegler–Natta catalyst is not suited for block polymerizations.

Other uses have also been found for certain binary mixtures of metal alkyl and transition metal salts other than polymerizations described above. These include radical or cationic polymerizations, metathesis, oligomerization, isomerization, hydrogenation, and alkylation. Chapter 22 gives a few examples of each type.

II. Collected Reviews

Throughout the last twenty years, many reviews have appeared covering either general or specific aspects of Ziegler–Natta catalysts and polymerizations. Many were timely "progress reports," especially those which emerged from the Natta school. Naturally, some of the older ones are partly outdated, but they still have an important historical value. Because the publication rate has waned considerably in the last ten years, more recent reviews still have an active value, and they should be consulted by those seeking more detailed data or subject matter which is outside the scope of this book. They will serve as a valuable supplement to this book (and vice versa). Not all of the following reviews were available to the author, however, but are included since they may be accessible to many readers.

As the reader will discover from this book, many aspects of the Ziegler–Natta polymerization are controversial and have been subjected to much personal interpretation. In this respect, some of these reviews, especially 1970 (Ref. 1), 1967 (Ref. 2c), 1974 (Ref. 1), and 1971 (Ref. 3), are invaluable because they were written from a vantage point unique to the respective authors who were able to focus on special topics in great depth. Such reviews will continue to be important in the future as new data provokes new interpretations and mechanistic views.

References

1955

1. G. Natta, Isotactic polymers. *Makromol. Chem.* **16**, No. 3, 213–237 (1955).

1956

1. G. Natta, Stereospecific polymerization of olefins. *Chim. Ind.* (*Milan*) **38**, 124–127 (1956).
2. G. Natta, Stereospecific catalysis and isotactic polymers. *Chim. Ind.* (*Milan*) **38**, 751–765 (1956); see also *Angew. Chem.* **68**, 393–403 (1956).

1957

1. G. Natta, Stereospecific catalysis and stereoisomeric polymers. Preparation of new fibers, plastics, and elastomers. *Experientia, Suppl.* **7**, 21–59 (1957); *Mater. Plast. (Milan)* **24**, No. 1, 3–23 (1958).
2. G. Natta, How giant molecules are made. *Sci. Am.* **197**, No. 3, 98–104 (1957).

1958

1. J. K. Stille, Polymerization of olefins by complex metal catalysts. *Chem. Rev.* **58**, 541–580 (1958).
2. D. C. Costescu, Stereoisomeric polymers. *Rev. Chim. (Bucharest)* **9**, 193–201 (1958).
3. G. Natta, Stereospecific polymerization and catalysis. VIII. *Nat. Congr. Soc. Chim. Ital. Soc. Chem. Ind., 1958; Gazz. Chim. Ital.* **89**, 89–125 (1959).
4. G. Natta, Stereospecific polymerization by means of coordinated-anionic catalysts. *J. Inorg. Nucl. Chem.* **4**, 589–611 (1958); *Int. Symp. Chem. Coord. Compounds, 1957* Supplement (1957).

1959

1. N. G. Gaylord and H. F. Mark, "Linear and Stereoregular Addition Polymers." Wiley (Interscience), New York, 1959.
2. G. Natta and I. Pasquon, The kinetics of the stereospecific polymerization of olefin. *Adv. Catal.* **11**, 1–65 (1959).
3. G. Natta, Progress in five years of research in stereospecific polymerization. *SPE J.* **15**, 373–382 (1959).
4. J. Furukawa, Catalytic reactivity and stereospecificity of organometallic compounds in olefin polymerization. *J. Polym. Sci.* **36**, 275–286 (1959).
5. G. Natta, A new Italian assertion in the textile field: Polypropylene fibers. *Chim. Ind. (Milan)* **41**, No. 7, 647–652 (1959).
6. G. Natta and F. Danusso, Nomenclature relating to polymers having sterically ordered structure. *J. Polym. Sci.* **34**, 3–11 (1959); *Chim. Ind. (Milan)* **40**, 743 (1958).

1960

1. K. Ziegler, Organo-aluminum compounds. In "Organometallic Chemistry" (H. Zeiss, ed.), Monogr. Ser. No. 147. Van Nostrand: Reinhold, Princeton, New Jersey, 1960.
2. C. E. H. Bawn, Recent developments in polyolefins. *J. Inst. Pet., London* **46**, 374–381 (1960).
3. M. Sittig, Polyolefin processes today. *Pet. Refiner* **39**, 162–222 (1960).
4. M. Roha, The chemistry of coordinate polymerization of dienes. *Fortschr. Hochpolym.-Forsch.* **1**, No. 4, 512–539 (1960).
5. G. Bier, A. Gumboldt, and G. Lehmann, A new polymerization technique with Ziegler catalysts. *Plast. Inst., Trans. J.* **28**, No. 75, 98–110 (1960).
6. K. S. Minsker and V. S. Etlis, Mechanism of stereospecific polymerization. *Usp. Khim. Tekhnol. Polim.* **3**, 14–38 (1960).
7. G. Lefebvre, Organometallic compounds and stereospecific polymerization. *Rev. Inst. Fr. Pet. Ann. Combust. Liq.* **15**, No. 4, 730 (1960).
8. G. Natta, Stereospecific polymerizations. *J. Polym. Sci.* **48**, 219–239 (1960).
9. A. V. Subramanian and S. L. Kapur, Stereospecific polymerization. *J. Sci. Ind. Res., Sect. A* **19**, No. 5, 200–212 (1960).
10. G. Natta, Progress in the stereospecific polymerization. *Makromol. Chem.* **35**, 94–131 (1960).

11. G. Natta, Catalytic aspects of stereospecific polymerization. *Chim. Ind. (Milan)* **42**, 1207 (1960).

1961

1. W. Cooper, Stereospecific polymerization (J. C. Robb and F. W. Peaker, eds.) *Prog. High Polym.*, Academic Press, New York, **1**, 279–340, (1961).
2. H. Hopff and H. G. Elias, Stereospecific polymerization. *Chimia* **15**, 479–492 (1961) (in German).
3. M. Wajnryb, Stereoregular structure of polymers. *Tworzywa Wielkoczasteczkowe* **6**, No. 5, 144–149 (1961).
4. O. Solomon, Stereospecific polymerization of isoprene with triethyl aluminium and titanium tetrachloride. *Rev. Chim. (Bucharest)* **12**, 284–287 (1961).
5. N. Gaylord, The mechanism of stereospecific polymerization. *Makromol. Chem.* **44–46**, 448–460 (1961).

1962

1. C. E. H. Bawn and A. Ledwith, Stereoregular addition polymerization. *Q. Rev. (London)* **16**, No. 4, 361–434 (1962).
2. F. Danusso, Stereoregular polymers and polymerization. *Polymer* **3**, 423–448 (1962).
3. F. Dawans and G. Lefebvre, The stereospecific polymerization of isoprene. *Rev. Inst. Fr. Pet. Ann. Combust. Liq.* **17**, 110–163 (1962).
4. C. E. H. Bawn, Stereospecific polymerization. *Proc. Chem. Soc. London* No. 5, pp. 165–175 (1962).
5. G. Natta, Stereospecific polymerization and asymmetric synthesis of macromolecules. *Pure Appl. Chem.* **4**, 363–385 (1962).
6. F. Danusso, The mechanism of stereospecific polymerization and copolymerization of vinyl aromatic monomers. II. Kinetic and chemical mechanism. *Chim. Ind. (Milan)* **44**, 611–620 (1962); also pp. 474–482.
7. G. P. Belov, Stereoregular butadiene rubber. *Kauch. Rezina* **21**, No. 11, 34–38 (1962).
8. E. Guccione, Stereospecific catalysis: A new dimension in polymerization. *Chem. Eng. (N.Y.)* **69**, No. 7, 93–106 (1962).
9. G. Natta, Progress in stereospecific polymerization and asymmetric synthesis of macro-molecules. *Pure Appl. Chem.* **4**, 363–385 (1962).

1963

1. W. Kerner and D. Braun, Stereospecific polymerization in a homogeneous medium. *Chem.-Ztg., Chem. Appar.* **87**, No. 22, 799–807 (1963).
2. G. Natta, Crystallinity in high polymers and copolymers. *SPE Trans.* 99–110 (1963).

1964

1. J. P. Kennedy and A. W. Langer, Jr., Recent advances in cationic polymerization. *Fortschr. Hochpolym.-Forsch.* **3**, 508–580 (1964).
2. G. E. Ham, ed., "Copolymerization." Wiley (Interscience), New York, 1964, see individual contributions:
   a. C. A. Lukach and H. M. Spurlin, Chapter IVA. Copolymers of α-olefins.
   b. G. Bier and G. Lehmann, Chapter IVB. Block copolymers with ethylene as one of the components.
   c. G. Crespi, A. Valvassori, and G. Sartori, Chapter IVC. Ethylene-propylene copolymers as rubbers.
   d. G. Crespi, A. Valvassori, and G. Sartori, Chapter IVD. Other α-olefin copolymers.

1965

1. R. A. Raff and K. W. Doak, eds., "Crystalline Olefin Polymers." Wiley (Interscience), New York, 1965; see following individual chapters:
   a. R. A. Raff and E. Lyle, Chapter 1. Historical developments.
   b. F. W. Breuer, L. E. Geipel, and A. B. Loebel, Chapter 3. Catalyst Systems.
   c. M. Goodman, J. Brandrup, and H. F. Mark, Chapter 4. Stereospecific polymerizations.
   d. A. Schindler, Chapter 5. Mechanism of anionic (low pressure) polymerization.
   e. K. W. Doak and A. Schrage, Chapter 8. Polymerization and copolymerization processes.
   f. R. L. Miller, Chapter 12. Crystalline and spherulitic properties.
   g. J. A. Faucher and F. P. Reding, Chapter 13. Relationship between structure and fundamental properties.
2. M. Roha, Ionic factors in steric control. *Adv. Polym. Sci.* **4**, 353–392 (1965).
3. G. Natta, Macromolecular chemistry—from the stereospecific polymerization to the asymmetric autocatalytic synthesis of macromolecules. *Science* **147**, 261–272 (1965).
4. P. Pino, Optically active addition polymers. *Adv. Polym. Sci.* **4**, 393–456 (1965).

1966

1. L. Reich and A. Schindler, "Polymerization by Organometallic Compounds," Chapters III-1 and IV. Wiley (Interscience), New York, 1966.
2. H. W. Coover, Jr., R. L. McConnell, and F. B. Joyner, Relationship of catalyst composition to catalytic activity for the polymerization of α-olefins. *Macromol. Rev.* **1**, 91–118 (1966).

1967

1. W. Cooper and G. Vaughan, Recent developments in the polymerization of conjugated dienes. *Prog. Polym. Sci.* **1**, 93–160 (1967).
2. A. D. Ketley, ed., "The Stereochemistry of Macromolecules." Dekker, New York, 1967; see following chapters:
   a. D. O. Jordan, Chapter 1. Ziegler–Natta polymerization: Catalysts, monomers and polymerization procedures.
   b. D. F. Hoeg, Chapter 2. The mechanism of Ziegler–Natta catalysis. I. Experimental foundations.
   c. P. Cossee, Chapter 3. The mechanism of Ziegler–Natta polymerization. II. Quantum-chemical and crystal-chemical aspects.
   d. I. Pasquon, A. Valvassori, and G. Sartori, Chapter 4. The copolymerization of olefins by Ziegler–Natta catalysts.
   e. W. Marconi, Chapter 5. The polymerization of dienes by Ziegler–Natta catalysts.
   f. M. Compostella, Chapter 6. The manufacture and commercial applications of stereoregular polymers.
3. A. Peterlin, M. Goodman, S. Okamura, B. H. Zimm, and H. F. Mark, eds., "Macromolecular Reviews," Vol. 2. Wiley (Interscience), New York, 1967; see following individual papers:
   a. E. A. Youngman and J. Boor, Syndiotactic polypropylene. pp. 33–69.
   b. J. Boor, The nature of the active site in the Ziegler–type catalyst. pp. 115–268.
4. F. Schué, Optically active polymers. *Rev. Gen. Caoutch. Plast. Ed. Plast.* **4**, 261–267 (1967).
5. D. C. Feay, Heterogeneous polymerization of unsaturated monomers. *In* "Organic Chemistry of Synthetic High Polymers" (R. W. Lenz, ed.) pp. 579–654. Wiley (Interscience), New York, 1967.
6. I. Pasquon, Some aspects of the mechanism of the stereospecific polymerization of α-olefins. *Pure Appl. Chem.* **15**, 465–480 (1967).
7. A. D. Ketley, ed., "The Stereochemistry of Macromolecules," Vol. 3. Dekker, New York, 1968, discusses the way the steric structure of polymers is determined and the way such structural features affect the physical and mechanical properties of the polymers.

1968

1. G. Natta and L. Porri, Diene elastomers. *High Polym.* **23**, Part 2, 597–678 (1968).
2. P. Pino, P. Salvadori, E. Chiellini, and P. L. Luisi, Optical activity and optical rotatory dispersion in synthetic polymers. *Pure Appl. Chem.* **16**, No. 2–3, 496–590 (1968).
3. E. I. Klabunovskii, Stereospecific catalysis in the field of optically active polymers. *Russ. Chem. Rev. (Engl. Transl.)* **37**, 969–983 (1968).
4. S. S. Medvedev, Active centers in anionic-coordination polymerization reactions of hydrocarbon monomers. *Usp. Khim.* **37**, 1923–1945 (1968) (in Russian).
5. G. Alliger and F. C. Weissert, Elastomers. *Indian Rubber Bull.* **238**, 7–15 and 18–23; **239**, 20–25 (1968).
6. I. D. Rubin, Poly(1-butene): Its preparation and properties. "Polymer Monographs." Gordon & Breach, New York, 1968.
7. *Corsi Semin. Chim.* **8**; see following papers:
   a. F. Ciardelli, C. Carlini, G. Montagnoli, and P. Pino, Stereoselective copolymerization of racemic alpha-olefins with optically active alpha-olefins. pp. 104–106.
   b. I. Pasquon, Polymerization of alpha-olefins in the presence of anionic coordination catalysts. pp. 74–76.
   c. L. Porri, Stereospecific polymerization of conjugated diolefins. pp. 86–88.
   d. M. Farina, Optically active polymers: Asymmetric polymerization and the synthesis of configurational models of polymers. pp. 272–274.
   e. R. Palumbo, Model compounds for Stereospecific polymerization catalysis; molecular asymmetry of prochiral olefin-transition metal complexes. pp. 95–98.
   f. P. Pino, F. Ciardelli, G. Montagnoli, and O. Pieroni, Stereospecific polymerization of olefins containing an asymmetric carbon and having various degrees of optical purity. pp. 197–199.
8. B. A. Dolgoplosk and E. I. Tinyakova, Mechanism of diene polymerization and microstructure of the polymer chain. *Kinet. Mekh. Obraz. Prevrashch. Makromol.* pp. 69–112 (1968).

1969

1. J. P. Kennedy and E. G. M. T. Törnqvist, "Polymer Chemistry of Synthetic Elastomers," High Polym. Ser. No. 23. Wiley (Interscience), New York, 1969; see following individual papers:
   a. G. Natta and L. Porri, Diene elastomers.
   b. G. Natta, A. Valvassori, and G. Sartori, Ethylene-propylene rubbers.
   c. G. Natta and G. Dall'Asta, Elastomers from cyclic olefins.
2. M. N. Berger, G. Boocock, and R. N. Haward, The polymerization of olefins by Ziegler catalysts. *Adv. Catal.* **19**, 211–240 (1969).
3. N. D. Zavorokhin, Anionic-coordinated polymerization of olefins with soluble complexes. *Tr. Inst. Khim. Nauk, Akad. Nauk Kaz. SSR* **23**, 3–35 (1969).
4. S. Pasynkiewicz, Mechanism of polymerization with organometallic catalysts. *Wiad. Chem.* **23**, 639–652 (1969).
5. J. L. Jezl, Propylene polymers. *Encycl. Polym. Sci. Technol.* **11**, 597–619 (1969).
6. W. E. Smith, The mechanism of stereospecific polymerization of propylene. Vinyl Polymerization, Part II; G. E. Ham, ed, "Kinetics and Mechanism of Polymerization," Marcel Dekker, **1**, 187–209 (1969).
7. G. Henrici-Olive and S. Olive, Coordinative polymerization with soluble transition metal catalysts. *Fortschr. Hochpolym.-Forsch.* **6**, 421–472 (1969).

8. E. W. Duck and B. J. Ridgewell, Synthetic rubbers. Part II. *Chem. Ind.* (*London*) **9**, 254–257 (1969).
9. A. Valvassori and N. Cameli, 3. Ethylene-propylene copolymers. *Ethylene Its Ind. Deriv.* pp. 413–425 (1969).
10. G. Mazzanti, Polyolefins in Italy. *Chem. Ind.* (*London*) **35**, 1204–1212 (1969).
11. H. Mori and F. Imaizumi, Mechanism and progress of diene polymerization catalysts. *Shokubai* **11**, 196–209 (1969) (in Japanese).
12. K. W. Scott, N. Calderon, E. A. Ofstead, W. A. Judy, and J. P. Ward, Ring-opening polymerization of cycloolefins—some mechanistic aspects. *Adv. Chem. Ser.* **91**, 399–418 (1969).
13. A. J. Foglia, Polybutylene, its chemistry, properties and applications. *Appl. Polym. Symp.* **11**, 1–18 (1969).

1970

1. W. Cooper, Aspects of the mechanism of coordination polymerization of conjugated dienes. *Ind. Eng. Chem., Prod. Res. Dev.* **9**, No. 4, 457–466 (1970).
2. P. Pino, F. Ciardelli, and M. Zandomeneghi, Optical activity in stereoregular synthetic polymers. *Annu. Rev. Phys. Chem.* **21**, 561–608 (1970).
3. I. Pasquon, Stereoregular linear polymers. Preparation. *Encyl. Polym. Sci. Technol.* **13**, 13–86 (1970).
4. M. Farina, Stereospecific polymerization. *Nuova Chim.* **46**, 30–32 and 55–59 (1970).
5. J. Boor, Review of recent literature on Ziegler–type catalysts. *Ind. Eng. Chem., Prod. Res. Dev.* **9**, No. 4, 437–456 (1970).
6. A. Yamamoto, Polymerization by transition metal alkyl and hydride complexes. *Kobunshi* **19**, 765–774 (1970) (in Japanese).
7. B. A. Dolgoplosk and E. I. Tinyakova, Stereospecific catalysis during diene polymerization under the effect of π-allyl complexes of transition metals. *Mekh. Kinet. Slozhnykh Katal. Reakts., Lektsii, Simp., 1st, 1968* pp. 142–162 (1970) (in Russian).
8. P. E. Matkovskii, V. P. Konovalov, N. M. Chirkov, A. D. Pomogailo, and G. A. Beikhol'd, Mechanism of the deactivation of complex catalysts and processes for their stabilization during olefin polymerization. *Tr. Inst. Khim. Nefti Prir. Solei, Akad. Nauk Kaz. SSR* **1**, 197–208 (1970) (in Russian).
9. A. D. Pomogailo and P. E. Matkovskii. Complexing of organoaluminum compounds with electron donor ligands and its effect on the polymerization of some monomers. *Tr. Inst. Khim. Nefti Prir. Solei. Akad. Nauk Kaz. SSR* **1**, 214–224 (1970).
10. E. Pajda, M. Uhniat, and A. Pajda, Effect of impurities on the polymerization process of ethylene catalyzed by Ziegler–Natta catalysts and on the polymer properties. *Chemik* **23**, 86 (1970 (in Polish).
11. H. Schnecko and W. Kern, End group determinations and number of active centers in heterogeneous ionic (coordinative) polymerization. *Chem.-Ztg.* **94**, 229–239 (1970).
12. K. Azuma, Complexes in catalytic polymerization. *Kagaku Kogyo* **21**, 477–482 (1970).
13. M. A. Golub, Polymers containing C=C bonds. *Chem. Alkenes* **2**, 411–509 (1970).
14. I. Pasquon, Stereoregular linear polymers. Preparation. *Encycl. Polym. Sci. Technol.* **13**, 13–86 (1970).
15. F. Dawans and Ph. Teyssie, π-Allyl-type polymerization. Presented in Toronto at joint meeting of American Chemical Society and Chemical Institute of Canada. *Am. Chem. Soc., Div. Org. Coat. Plast. Chem., Pap.* **30**, No. 1, 208–219 (1970).

1971

1. G. Henrici-Olive and S. Olive, Vanadium and chromium catalysts for polymerization of ethylene. *Angew. Chem., Int. Ed. Engl.* **10**, 776–786 (1971).

2. B. A. Dolgoplosk, Stereospecific catalysis of diene polymerizations and copolymerization, and the mechanism of stereoregulation. *Polym. Sci. USSR (Engl. Transl.)* **13**, 367–393 (1971); *Vysokomol. Soedin., Ser. A* **13**, 325–347 (1971).

3. G. Henrici-Olive and S. Olive, Influence of ligands on the activity and specificity of soluble transition metal catalysts. *Angew. Chem. Int. Ed. Engl.* **10**, No. 2, 105–115 (1971).

4. P. Teyssie and F. Dawans. π-Allyl-type polymerizations. *Ind. Eng. Chem., Prod. Res. Dev.* **10**, 261–269; also P. Teyssie, *Macromol. Prepr., Int. Cong. Pure Appl. Chem., 23rd, 1971,* Vol. I, pp. 118–128 (1971).

5. H. Weber, Organometallic compounds as catalysts in stereorubber manufacturing. *Fortschr. Chem. Forsch.* **16**, 329–363 (1971).

6. H. E. Schroeder, Recent developments in synthetic elastomers. *Angew. Makromol. Chem.* **16/17**, 1–25 (1971).

7. V. P. Mardykin, A. M. Antipova, and P. N. Gaponik, Three-component complex organometallic catalysts. *Usp. Khim.* **40**, 24–42 (1971).

8. R. E. Haney, Ethylene, propylene and 1-butene. *High Polym.* **24**, Part 2, 577–689 (1971).

9. A. Gumboldt, Metal organic compounds as olefin polymerization catalysts. *Fortschr. Chem. Forsch.* **16**, No. 3–4, 299–328 (1971).

10. S. Cuicinella, A. Mazzei, and W. Marconi, Synthesis and reactions of aluminum hydride Derivatives. *Inorg. Chim. Acta* **4**, 51–72 (1971).

1972

1. F. P. Baldwin and G. Ver Strate, Polyolefin elastomers based on ethylene and propylene. *Rubber Chem. Technol.* **45**, No. 3, 709–881 (1972).

2. T. Mole and E. A. Jeffery, "Organoaluminum Compounds." Elsevier, Amsterdam, 1972.

3. T. Keii, "Kinetics of Ziegler–Natta Polymerization." Kodansha Ltd., Tokyo, 1972 [Japan Kodansha EDP No. 3043-247179-2253(0)].

1973

1. D. G. H. Ballard, Pi and sigma transition metal-carbon compounds as catalysts for the polymerization of vinyl monomers and olefins. *Adv. Catal.* **23**, 263–325 (1973).

2. T. G. Heggs, Block copolymers made with Ziegler catalysts. Physical properties of block copolymers made with Ziegler catalysts. *In* "Block Copolymers" (D. C. Allport and W. H. Janes, eds.), Chapters 4 and 8D. Wiley, New York, 1973.

3. W. L. Carrick, The mechanism of olefin polymerization by Ziegler–Natta catalysts. *Adv. Polym. Sci.* **12**, 65–85 (1973).

1974

1. A. Zambelli and C. Tosi, Stereospecific polymerization of propylene. *Adv. in Polym. Sci.* **15**, 32–60 (1974); a preprint was kindly given to me by Professor Zambelli before its publication.

2. H. Chanzy, B. Fisa, and R. H. Marchessault, Nascent morphology of polymers. *Crit. Rev. Macromol. Chem.* (to be published); a preprint was kindly given to me by Professor Marchessault.

1975

1. F. J. Karol, Supported catalysts for olefin polymerization. *Encycl. Polym. Sci. Technol.* (in press); a preprint was kindly given to me by Dr. Karol.

2. E. J. Vandenberg and B. C. Repka, Ziegler type polymerizations. *In* "Polymer Processes" (C. Schildknecht, ed.). Wiley (Interscience), New York, 1975 (to be published) (a preprint was kindly given to me by Mr. Repka).

Genesis of Ziegler–Natta Catalysts

I. Scientific and Commercial Importance

The Ziegler–Natta catalyst is unique, versatile, and truly marvelous. Consider being asked at the beginning of the 1950's to invent a catalyst capable of polymerizing olefins and dienes to the different stereochemical structures shown in Tables 1–1 and 1–2. Furthermore, suppose the catalyst had to be active in any one of the several possible physical states such as soluble, colloidal, or heterogeneous or had to be active in the temperature range of $-78°$ to $200°C$. A further constraint was that the olefin pressure had to be relatively low, only about 1 to 30 atm. In order to be versatile, the polymerizations could be done in solution, slurry, or vapor phase type processes.

Up to this time, it was not at all possible to polymerize α-olefins to very high molecular weight chains, let alone to stereoregular polymers (1, 2). Low density (highly branched) polyethylene ($d = 0.92$ g/cm^3 and melting point about 115° to 120°C) was available commercially via high pressure radical processes. The Phillips and Standard of Indiana medium pressure processes were about to make their debut. High cis-1,4- and trans-1,4-polyisoprenes were obtained as the natural products *Hevea* and gutta-percha, respectively (3). The known methods of initiation and catalysis in addition-type polymerizations were cationic, anionic, and free radical, and these were used to synthesize polymers such as polyisobutylene, polyvinyl ethers, styrene–butadiene rubbers, styrene–butadiene–acrylonitrile plastics, and polyacrylates. (4, 5).

Stereoregular poly-1-olefins and polydienes with high cis-1,4 or trans-1,4 microstructure similarly could not be synthesized. Radical initiation of propylene (2,800 atm at 80°C) produced atactic product of a low degree of polymerization (6). Cationic initiation of propylene (Al_2Br_6 + HBr at $-78°C$ in butane) produced products that contain ethyl and propyl groups resulting from intramolecular and intermolecular hydride transfer (7).

Since olefins and diolefins were major products that could be derived from cracking of oils or as first derivatives of hydrocarbons, the petroleum industry was undoubtedly most anxious to find new uses for them. Propylene had not yet found a wide volume application in high polymers.

But alas, most of us would have rejected the notion that such a versatile single polymerization catalyst was possible. However, within a few years such a catalyst was discovered in the laboratories of Ziegler and Natta, and subsequent investigations by the Natta group and by many industrial and academic laboratories throughout the world showed it to be even more versatile.

The author recalls the keen interest which this new catalyst generated in the late 1950's and the wide attention that the subject attracted at scientific meetings. One of the most exciting events of this period and for many years afterward was the annual lecture at the industrial laboratories and at scientific meetings of the international ambassador of science, Professor H. F. Mark of the Brooklyn Polytechnic Institute, who informed packed auditoriums about the current advances made in this rapidly emerging field.

A gigantic experimental and developmental effort by teams of chemists and engineers was made into commercializing the synthesis of plastics and elastomers with Ziegler–Natta catalysts. Since 1954 an endless stream of patents has continued to flow; it seems that the number of patents that have been issued is beyond count.

This effort came into fruition at the end of the 1950's when pilot and commercial plants began producing the new polymers. Eventually, the plastics (high density polyethylene, isotactic polypropylene (8), isotactic poly-1-butene, isotactic poly-4-methyl-1-pentene, *trans*-1,4-polyisoprene) and the elastomers (ethylene–propylene copolymers, *cis*-1,4-polybutadiene, *cis*-1,4-polyisoprene, and polypentenamers) became commercially available via Ziegler–Natta catalysis.

By 1971, there were nine producers of polypropylene in the United States, and they made about 1.2 billion pounds of this polymer, which was about one-third of the world production (9). Except for a bad year in 1970, total sales of polypropylene increased from 25 to 31% per year in the period 1968–1973 (10). Polypropylene in 1974 was considered a "tiger on a leash." The 1975 planned capacity expansions were expected to be sold out (10).

How did the discovery of this unique catalyst come about? The story began in the laboratory of Professor K. Ziegler at the Max Planck Institute for Coal Research in Mülheim, Germany.

II. Historical Origins

The discovery of the first examples of this catalyst was the culmination of many years of basic research in the area of organometallic chemistry by Professor Ziegler and his students (11). The overt experiment was made at the end of 1953 when ethylene was reacted with $AlEt_3$ and zirconium acetylacetonate (12, 13). The white powder which filled the autoclave was a high molecular weight, linear polyethylene. What prompted Ziegler and his co-workers to react ethylene in the presence of these two particular components?

The story can go back a few or many years, depending on where one wants to place the seed of this great discovery. During the period of 1900–1953, a number of research groups were actively or sporadically investigating organometallic chemistry, particularly the reactions of dienes, ethylene, and substituted ethylenes.

In 1930 Friedrich and Marvel discovered that ethylene was polymerized to low molecular weight products in presence of lithium alkyls (14). Ziegler and co-workers extended this finding by investigating the mechanism by which alkali metal alkyls initiated the polymerization of conjugated dienes (15). No work was done on ethylene until after 1945. In the interim, Du Pont workers reported that ethylene, at high pressures, could be polymerized with a lithium–copper powder catalyst (16) and with lithium alkyl and aryl compounds (17). Only low molecular weight compounds were obtained.

Following World War II, Ziegler, who was supported by German industries, reinitiated a program on the polymerization of ethylene with lithium alkyls with the goal of synthesizing higher molecular weight polymer chains. This, however, proved to be an unsuitable initiator because the growing alkyl polymer was prematurely eliminated (see Eq. 2–1, where n is a small number)

$$LiR + nC_2H_4 \longrightarrow Li(CH_2-CH_2-)_nR \longrightarrow$$
$$LiH\downarrow + CH_2{=}CH-(CH_2-CH_2-)_{n-1}R \qquad (2-1)$$

with precipitation of lithium hydride (18). Once out of solution, LiH was at best negligibly active. This problem became the source of the next important innovation. While attempting to improve this reaction, they found that it was promoted by ether. Ziegler and co-workers, however, speculated that the ether-soluble $LiAlH_4$ (which became known at the time) could also serve as a source of lithium alkyl, and this led them to the next advance (19). They demonstrated that $LiAlH_4$ reacted with ethylene to form $LiAlEt_4$. At first they thought that ethylene would polymerize only on the LiEt portion and not on $AlEt_3$; this turned out to be wrong—$AlEt_3$ polymerized ethylene even more efficiently. Ziegler and Gellert switched to $AlEt_3$ because of experimental advantages. The equilibrium between metal alkyl and metal alkyl hydride plus olefin components could now be established since all of the metal alkyl and metal alkyl hydride components remain soluble. This feature made it possible for Ziegler and Gellert to study the organometallic synthesis of olefins by the following steps (which Ziegler called the "Aufbau reaction"), shown in Eq. 2–2.

$$AlH_3 + 3\,CH_2\!\!=\!\!CH_2 \longrightarrow Al\!\!\begin{array}{l}{-CH_2CH_3}\\{-CH_2CH_3}\\{-CH_2CH_3}\end{array} \xrightarrow{CH_2=CH_2} Al\!\!\begin{array}{l}{-CH_2CH_2CH_2CH_3}\\{-CH_2CH_3}\\{-CH_2CH_3}\end{array} \xrightarrow{CH_2=CH_2}$$

$$Al\!\!\begin{array}{l}{-CH_2CH_2CH_2CH_3}\\{-CH_2CH_2CH_2CH_3}\\{-CH_2CH_3}\end{array} \xrightarrow{CH_2=CH_2} etc. \longrightarrow Al\!\!\begin{array}{l}{-(CH_2CH_2)_mC_2H_5}\\{-(CH_2CH_2)_nC_2H_5}\\{-(CH_2CH_2)_oC_2H_5}\end{array}$$

$$(2\text{--}2)$$

where m, n, and o are small numbers.

The growth step is favored at the Al–C bond at 100° to 120°C; molecular weights in the range 3,000 to 30,000 were claimed depending on conditions used, especially the concentration of ethylene. Higher molecular weight polymers were not formed because a displacement reaction also took place at these high temperatures, as in Eq. 2–3.

$$\diagdown\!\!Al\!\!-\!\!(CH_2CH_2)_m\!\!-\!\!C_2H_5 \diagup \longrightarrow \diagdown\!\!Al\!\!-\!\!H \diagup + CH_2\!\!=\!\!CH\!\!-\!\!(CH_2CH_2)_{m-1}\!\!-\!\!C_2H_5 \quad (2\text{-}3)$$

The Al–H bond reacts with the remaining ethylene to form polymer chains of comparable molecular weight. This termination step removes the growing alkyl from the aluminum before it has a chance to become very long. This was typical of this reaction, except one day an anomolous finding was made by Ziegler and Holzkamp.

Instead of polyethylene waxes, 1-butene was predominantly recovered from the reactor (20, 21). They speculated that the displacement reaction had become catalyzed by a contaminant. As it turned out, this contaminant was shown to be colloidal nickel remaining in the reactor from a previous hydrogenation experiment. This nickel effect was a lucky find. They now had a handle with which to modify the "Aufbau reaction."

A series of transition metal salts were next examined by Ziegler and Breil in combination with $AlEt_3$ in an effort to find other displacement catalysts. Meanwhile, Ziegler and Holzkamp continued to exclusively study the nickel effect in order to learn how to control it. According to Ziegler's account, Holzkamp had actually already performed the "discovery experiment" without the group realizing it. He had made two experiments with a combination of AlEt$_3$ and zirconium acetylacetonate, one which produced results analogous to the "nickel effect" and the other which produced a certain amount of a solid white precipitate. But Ziegler said that because they often previously found a little polyethylene on the reactor wall in their Aufbau studies, this observation was not taken seriously enough to alter the course of their program. It was left to Breil on a second trial to make the overt experiment.

Cobalt and platinum salts behaved as nickel, but zirconium acetylacetonate did not. Instead of 1-butene or waxes, they now synthesized high molecular weight polyethylene at low olefin pressures (12, 13).

Transition metal salts (groups IV to VI) were also active, but the most active catalyst was made from TiCl$_4$ and AlEt$_3$, and this was developed for large scale production of high density polyethylene plastics (and designated by Ziegler as the "Mülheim Atmospheric Polyethylene Process"). The giant leap was thus made from the limited "Aufbau reaction" to the versatile Mülheim Atmospheric Polyethylene Process.

Since Ziegler was concerned with ethylene at the Max Planck Institute for Coal Research, much of his initial effort with this new catalyst was apparently devoted to polyethylene and copolymers of ethylene and propylene. Because the polymerizations of propylene, higher α-olefins, and dienes were left to others, the full potential of this new catalyst was not realized in Ziegler's laboratory.

Before he informed the scientific community of his discovery, Ziegler disclosed the catalyst to Montecatini Company (Italy) and Goodrich–Gulf Chemical Company (USA) (22). Professor Natta was a consultant to Montecatini at that time and he brought Ziegler's discovery to its attention. He undertook the research investigations of this new catalyst within the terms of an agreement between Ziegler and Montecatini.

Natta and his school were in a very good position to explore the catalyst (23). Being already active in studying the kinetics of ethylene addition to aluminum alkyls, they were experimentally ready to study the new catalyst (24). The Ziegler and Natta schools were aware of each other's activities long before they were made public. Whether or not Natta had hopes that the new catalyst would produce stereospecific (crystalline) polymers from α-olefins is not known to the author. One cannot help but feel that he had great expectations for this novel catalyst.

Early in 1954, the Natta group succeeded in synthesizing crystalline polymers (25–29). Their first experiments were with the Mülheim catalyst (AlEt$_3$ + TiCl$_4$) which they found to produce a mixture of amorphous and crystalline fraction polypropylenes. The crystalline portion was isolated by solvent extraction. When other titanium chlorides (especially αTiCl$_3$, made by reducing TiCl$_4$ with H$_2$ at high temperatures) were used in place of TiCl$_4$, the polymers obtained were much more crystalline. In this way, highly crystalline polymers from propylene, 1-butene, and styrene were easily synthesized for the first time. Natta and Corradini established that the polymer chains were made up of long sequences of monomeric units having the same configuration. Natta designated these polymers as "isotactic".

Natta next assembled a large team of talented scientists who were specialists in various disciplines. Under his direction, this group continued to lead and dominate the investigation of these Ziegler–Natta catalysts and to open new areas of research in polymerization chemistry. Other novel polymers soon became available which offered new opportunities to enhance our knowledge of polymer physics.

As a result of the extensive, highly original, and profound synthetic and structural investigations of Natta and his prestigious school of co-workers, the basic principles of stereoregularity and stereoregulation in polymerization of α-olefins were clearly established. What was known for a long time for low molecular weight organic compounds was now accomplished for synthetic high polymer chains.

As mentioned above, the Goodrich–Gulf Chemical Company was also informed by Ziegler of his discovery before it became public, and this company examined it for polymerization of dienes. Using the same catalyst ($AlEt_3 + TiCl_4$), but in specific ratios, Horne and co-workers were able to synthesize high cis-1,4-polyisoprene, the basic component of natural *Hevea* rubber (30). The Natta school, and subsequently workers in other laboratories, independently discovered that the new catalyst could also be used for the polymerization of diolefins to polymers which had high cis-1,4 or trans-1,4 microstructures. It appears, however, that the first public disclosures for isoprene were made by Goodrich–Gulf.

Once the Ziegler–Natta discoveries were made public, worldwide investigations were immediately undertaken to elucidate the mechanisms of these catalysts and to develop them for commercial application (31). A number of significant discoveries were made in the following years which deserve comment here.

In 1955 Vandenberg at Hercules (32) and Ettore and Luciano at Montecatini (32a) applied for patents which disclosed that molecular hydrogen ˒ acted as a true transfer agent for the Ziegler–Natta catalyst. This was a very important discovery because many of the Ziegler–Natta catalysts produce polymers whose molecular weights are too high for facile processing and certain product applications. The desired molecular weight is easily obtained in the presence of hydrogen.

The other discoveries came from the Natta school. In addition to the many contributions made in the synthesis of new stereospecific polymers (33), this group elucidated the physical chemistry of these stereospecific polymers by means of sophisticated X-ray diffraction studies, dilatometry, and infrared spectroscopy (34–36).

Professor Natta cites the 1954 contributors as P. Pino, G. Mazzanti, E. Giachetti, P. Chini, P. Corradini, F. Danusso, G. Moraglio, E. Mantica, M. Peraldo, L. Bicelli, and G. Lutzu (36a). For years afterward many of these co-workers continued in the Natta group. Other co-workers whose names appear on major papers during the next several years include: G. Dall'Asta, L. Porri, M. Farina, G. Allegra, I. Pasquon, I. W. Bassi, U. Giannini, G. Sartori, D. Sianesi, A. Valvassori, and A. Zambelli.

Three very important synthetic discoveries were dislcosed in 1962 which enhanced the versatility of this catalyst. Zambelli, Natta, and co-workers (37) announced that the synthesis of highly syndiotactic polypropylene at $-78°C$ when the catalyst $AlEt_2Cl + VCl_4$ was specifically used (titanium chlorides and aluminum trialkyls were not effective). Pino, Natta, and co-workers (38) disclosed that when racemic α-olefins such as (RS)-4-methyl-1-hexene were polymerized with the ordinary Ziegler–Natta catalyst (Al-i-$Bu_3 + TiCl_4$), the individual polymer chains of the product were made predominantly from the (R)- or (S)- enantiomer. Dall'Asta, Natta, and co-workers (39) disclosed that while certain cycloolefins were only sluggishly or not at all polymerizable, they could copolymerize when ethylene was also present. They also disclosed (40, 41) that cis-1,5- or $trans$-1,5-polypentenamers could be made from cyclopentene with the following specific catalysts:

$$
\begin{array}{l}
\text{AlEt}_3 + \text{MoCl}_5 \quad \overset{\underset{\text{H}}{|}\;\;\overset{\underset{\text{H}}{|}}{}}{-\text{C}=\text{C}-\text{CH}_2-\text{CH}_2-\text{CH}_2-} \\
\qquad\qquad\qquad\qquad\qquad cis\text{-}1,5 \\[2mm]
\text{AlEt}_3 + \text{WCl}_6 \quad -\text{C}=\text{C}-\text{CH}_2-\text{CH}_2-\text{CH}_2- \\
\end{array}
\qquad (2\text{--}4)
$$

trans-1,5

The $trans$-1,5-polymer is currently being developed as a new elastomer with properties superior to poly-cis-1,4-butadiene (see Chapters 3 and 19).

The Royal Academy of Science of Sweden awarded the 1963 Nobel Prize for Chemistry to Professors K. Ziegler and G. Natta. The polymer and catalysis fields of science became profoundly enriched through the achievements of these two giants and their co-workers. The other scientific achievements and personal side of these two men are found in brief biographies; H. Mark (41a) and G. Wilke (41b) write about Ziegler, and W. Kern and P. Pino (41c) describe Natta.

III. Anticipation of the Catalyst and Contemporary Discoveries

Polymer research in the period 1943–1950 was directed toward discovering new types of catalysts and ways to control stereoisomerism and geometric isomerism in the growth step only to a small extent. A brief review of these somewhat isolated events in various parts of the world is presented here with the objective of setting the stage for the Ziegler–Natta discoveries and to review for the reader contemporary discoveries in related areas.

In 1943, Max Fischer of BASF (Germany) discovered solid polyethylene was made in small amounts as a by-product with low molecular weight oils when ethylene was reacted (about 150°C and 50 atm) with a mixture of $AlCl_3$, aluminum powder, and $TiCl_4$ (42). A patent was filed in 1943 but was not issued until April 20, 1953. It was not declared public property by Allied technical groups who surveyed German scientific information after World War II. The stated purpose of the aluminum powder in the ternary mixture was to remove HCl. If Fischer had been aware that Hall and Nash (42a) had discovered in 1937 that reacting aluminum powder, $AlCl_3$, and ethylene at 50–100 atm and 100° to 200°C produced a mixture of aluminum alkyls, he probably would have realized that the polyethylene was produced by a different catalyst (aluminum alkyl + $TiCl_4$) rather than a cationic one ($AlCl_3$ + $TiCl_4$). One must remember, however, that his primary objective was to find inexpensive alternate routes to lubricating oils. The concurrently formed polyethylene may have been a nuisance.

About the same time that the Fischer patent appeared, a patent was also issued to Du Pont (December 11, 1952) which described a catalyst comprised of LiBu and Ni on kieselguhr in benzene solvent under anhydrous conditions (43). Polyethylene was formed that had a melting point of 128°C as determined by disappearance of spherulites on a hot-stage microscope. This was only 5°C higher than the highest obtained in the other 23 samples of the patent, which clearly described radicallike polymerizations. While anhydrous conditions were used, it is not obvious whether oxygen was excluded in the entire operation. If not, the catalyst could have served as a radical source, as the patent implies.

Other Du Pont workers [Gresham and Merckling (44)] applied for a U.S. Patent on August 30, 1954, which definitely describes the synthesis of a high density polyethylene. The catalyst consisted of $MoCl_5$ + Bu_4Sn or C_6H_5MgBr and experimental conditions called for the absence of oxygen and moisture. Ethylene was used at pressures of 1000 and 2550 psig and the product had a density of 0.96 g/cm.[3] This patent, however, was not issued until August 18, 1959. British Patent 777,538 was also issued to Du Pont on June 26, 1957, which cites an August 16, 1954 U.S. application date. This patent describes $LiAlR_4$–$TiCl_4$ as a catalyst for polymerization of propylene and mixtures of olefins. These Du Pont patents show that metal alkyl–transition metal catalysts were also emerging from the Du Pont laboratory at about the same time.

It is worthwhile noting that binary mixtures of a metal alkyl ($ZnEt_2$, C_6H_5, Li, n-BuLi, and EtMgX) and transition metal salts ($TiCl_4$ + $ZrCl_4$) were not novel at the time of Ziegler's discovery. Gilman and Jones (44a) describe such combinations in a 1945 publication that deals with the attempted synthesis of organometallic compounds of titanium by this route. However, no polymerizations were done at that time or earlier. This paper cites similar attempts by other workers, which were made as far back as 1862 to 1887.

Two very important disclosures were made in the late 1940's which demonstrated that some stereochemical control was possible in the polymerization of vinyl and diene monomers.

Morton demonstrated that a mixture of three sodium salts: allyl sodium, sodium isopropoxide, and sodium chloride (the Alfin catalyst) polymerized butadiene to a *trans*-1,4-polymer (45). Prior to this, butadiene was polymerized by organosodium compounds, primarily by 1,2-addition.

About the same time, Schildknecht described the first successful stereoregular polymerization of a vinyl monomer using vinyl isobutyl ether and BF_3 etherates (catalyst) at $-78°C$ in propane solvents (46). From the crude product, he was able to fractionate, by solvent extraction, a crystalline portion. He erroneously assigned a syndiotactic structure (using present terminology) to this crystalline material. Later, Natta showed it to be isotactic. Schildknecht's contribution is a milestone because it was the first laboratory demonstration and recognition of the stereoregulation of a vinyl monomer.

About the same time that the Ziegler–Natta discoveries were made public, two supported catalyst systems based on transition metal oxides were reported active for synthesizing highly linear polyethylene. Hogan and Banks of the Phillips Petroleum Company disclosed the use of chromium oxide or a mixture of chromium oxide and strontium oxides supported on a silica, alumina, silica–alumina, zirconium oxide, or thorium oxide bed (47). Peters and Evering of the Standard Oil Company of Indiana described two different catalysts: (1) nickel or cobalt metal on charcoal, and (2) molybdenum on alumina. The chromium oxide catalyst was activated by treatment with dry air or air containing 3–10% steam at elevated temperatures (48). The molybdenum oxide and nickel catalyst were activated by heating in the presence of H_2 or various hydrides (borohydrides or aluminohydrides). As will be shown later for these metal oxides and the Ziegler–Natta catalyst, the growth center was shown to be a transition metal–carbon bond. The catalysts differ in nature of the initiation step—that is, the way in which active centers are formed.

These catalysts were also found by Hogan and Banks (47) and Zletz (48a) to polymerize propylene. Because the yields were very low and the crude products were described as tacky semisolids, and as having high molecular weight, rubberlike polymers, respectively, these findings did not attract much attention at first. They later became important when the U.S. patent for the composition of matter for isotactic polypropylene was being decided in a five party patent interference. Zletz had already received such a composition of matter patent in Canada. This interference, however, was resolved in favor of Montecatini with the awarding of U.S. Patent 3,715,344 to Natta, Pino, and Mazzanti (48b). An appeal to this decision is pending.

Two other significant discoveries of this period also deserve mention. A research group at Firestone Tire and Rubber Company discovered that using lithium or lithium alkyls under controlled conditions, high *cis*-1,4-polyisoprene was synthesized (49). Thus, two totally different means became available in a short period for a source of a substitute of natural (*Hevea*) rubber, which until this time had been obtained from rubber trees.

Also, in 1955 Pruitt and Baggett reported the first stereospecific polymerization of propylene oxide with a catalyst made by reacting specific amounts of $FeCl_3$ and propylene oxide (50). The crude product contained a rubbery fraction and crystalline fraction with a melting point of 70°C. This crystalline portion had the same X-ray diffraction pattern as a product made a year later by C. Price, who polymerized optically active propylene oxide with KOH (51). The product made by Pruitt and Baggett was a mixture of polymer chains, each made predominantly of *d* or *l* structural units.

IV. The Golden Age of Polymer Science

The Ziegler–Natta discoveries played a major role in making the two decades following 1950 a "golden age" for polymer science. Not only were many new polymers synthesized by means of this catalyst, but the use of it also made alternate routes possible for others. Independent discoveries were also made, and new polymers were synthesized by other means which also made this period so significant.

Because polymer science was taught in only a few colleges and universities, the number of polymer chemists and physicists in the 1950's was relatively small. It was necessary to draw upon chemists, physicists, and engineers from other disciplines to satisfy this great demand.

As already mentioned, the Phillips and Standard of Indiana catalysts provided alternate attractive commercial routes to high density polyethylene.

Major progress was made in anionic polymerization, starting with the discovery of anionic "living polymerizations" by M. Szwarc (52) and highlighted by the discovery of self-vulcanizing thermoplastic elastomers (KRATON®) at the Shell Laboratories (53). There was a revival in cationic polymerizations with the discovery of isomerization–polymerizations, of living initiators for cyclic ethers, and of metal alkyl coreactant initiators (54). Many new heat-resistant organic polymers (the superpolymers) emerged during this period (55, 56), notably the polybenzimidazoles discovered by C. Marvel and co-workers. Many new engineering thermoplastics were introduced (57), especially polyacetals, polycarbonates, polysulfones, polyimides, polyphenylene oxide, and chlorinated polyethers. High performance elastomers based on epichlorohydrin appeared on the market during the early 1960's (58). Some polymers emerged that required sophisticated application along with their synthesis; for example, poly-p-xylenes were deposited as thin films on substrates as they were synthesized (59).

This is only a partial list. Polymer science, which was born in the early 1930's and which was given great impetus by the brilliant pioneering investigations of Staudinger, Carothers, Flory, and other eminent scientists, came into maturity in the 1950's and exploded into a "golden age." The Ziegler–Natta discoveries were an integral part of this special period.

With the death of Karl Ziegler on August 11, 1973, science lost one of its most eminent chemists (60). A special symposium in memorial to Professor Ziegler was presented at the American Chemical Society Meeting held in Los Angeles, California, in April 1 to 6, 1974.

References

1. C. E. H. Bawn, "The Chemistry of High Polymers." Wiley (Interscience), New York, 1948.
2. P. J. Flory, "Principles of Polymer Chemistry." Cornell Univ. Press, Ithaca, New York, 1953.
3. G. S. Whitby, "Synthetic Rubber." Wiley, New York, 1954.
4. C. E. Schildknecht, "Vinyl and Related Polymers." Wiley, New York, 1952.
5. G. F. D'Alelio, "Fundamental Principles of Polymerizations." Wiley, New York, 1952.
6. J. Osugi, K. Hamanowne, and T. Tachibana, *Rev. Phys. Chem. Jpn.* **38**, 96 (1968).
7. A. D. Ketley and M. C. Harvey, *J. Org. Chem.* **26**, 4649 (1961).
8. Montecatini first made isotactic polypropylene available commercially in 1957.
9. *Modern Plastics*, Plastiscope 1, Feb. 1972.
10. *Modern Plastics*, Plastiscope 1, Jan. 1974.
11. Starting with German Patent 487,727 (issued January 7, 1930) and ending with Belgian Patent 527,736 (issued October 1, 1954), K. Ziegler and his co-workers published approximately 32 papers and patents from 1927–1953.
12. K. Ziegler, Belgian Patent 534,792, issued May 1, 1955; also Belgian Patent 534,888, issued May 1, 1955; also Belgian Patent 540,459, issued February 9, 1956.
13. K. Ziegler, E. Holzkamp, H. Breil, and H. Martin, *Angew. Chem.* **67**, 541 (1955).

14. M. E. P. Friedrich and C. S. Marvel, *J. Am. Chem. Soc.* **52**, 376 (1930).
15. K. Ziegler, F. Dersch, and H. Wollthan, *Justus Liebigs Ann. Chem.* **511**, 13 (1934); K. Ziegler and L. Jakob, *ibid.* p. 45; K. Ziegler, L. Jakob, H. Wollthan, and A. Wenz, *ibid.* p. 64; K. Ziegler, *Angew. Chem.* **49**, 499 (1936); K. Ziegler, H. Grimm, and R. Willer, *Justus Liebigs Ann. Chem.* **542**, 90 (1940).
16. L. M. Ellis, U.S. Patent 2,212,155, issued August 20, 1940, E. I. Du Pont de Nemours & Company.
17. W. E. Hanford, J. R. Roland, and H. S. Young, U.S. Patent 2,377,779, June 5, 1945, E. I. Du Pont de Nemours & Company.
18. K. Ziegler and H. G. Gellert, *Justus Liebigs Ann. Chem.* **567**, 195 (1950).
19. K. Ziegler, *Brennst-Chem.* **33**, 193 (1952); *Angew. Chem.* **64**, 323 (1952); *Chim. Ind.* (*Milan*) **34**, 520 (1952); *Glueckauf* **88**, 380 (1952); K. Ziegler and H. G. Gellert, German Patent 878,560, issued April 16, 1953; K. Ziegler and H. G. Gellert, German Patent 917,006, issued July 15, 1954; K. Ziegler, Belgian Patent 512,267, issued December 20, 1952.
20. K. Ziegler, *Brennst.-Chem.* **35**, 321 (1954); Belgian Patent 527,736, issued October 1, 1954.
21. E. Holzkamp, Thesis, Technische Hochschule, Aachen (May 21,1954); student of Ziegler.
22. K. Ziegler, Metallorganische verbindungen vom Standpunkte der Makromolecularen Synthese, *in* "International Symposium on Macromolecular Chemistry Proc," Prague, 1957, Pergamon Press, 1957, pp. 295–306.
23. See comments made later by Professor Natta in *Science* **147**, 261 (1965).
24. G. Natta, P. Pino, M. Farina, Cinetica della polimerizzazione dell'etilene catalizzata da composti allumininio-alchilici, *in* International Symposium of Macromolecular Chemistry, Milan, 1954, Supplement A of La Ricerca Scientifica **25**, 1955, p. 120–133.
25. G. Natta, *Atti Accad. Naz. Lincei., Mem., Cl. Sci. Fis., Mat. Nat., Sez 3a* [8] **4**, 61 (1955). June 8, 1954 is given as the Italian priority date in U.S. Patents 3,112,300 and 3,112,301, which issued on November 26, 1963 to G. Natta, P. Pino, and G. Mazzanti.
26. G. Natta, P. Pino, P. Corradini, F. Danusso, E. Mantica, G. Mazzanti, and G. Moraglio, *J. Am. Chem. Soc.* **77**, 1708 (1955).
27. G. Natta, *Angew. Chem.* **67**, 430 (1955).
28. G. Natta, *Chim. Ind.* (*Milan*) **37**, 88 (1955).
29. G. Natta, *J. Polym. Sci.* **16**, 143 (1955).
30. "First Synthesis of Natural Rubber Molecules," in *Chem. Eng. News* **32**, 4913 (1954); see Horne *et al.* (6) in Chapter 3; also S. E. Horne, *Ind. Eng. Chem.* **48**, 784 (1956); Great Britain Patent 827,365, February 3, 1960 (S. E. Horne, F. Gibbs, and E. J. Carlson).
31. The explosive early literature was captured in a comprehensive book by N. G. Gaylord and H. F. Mark entitled, "Linear and Stereoregular Addition Polymers." Wiley (Interscience), New York, 1959.
32. E. J. Vandenberg, U.S. Patent 3,051,690, August 28, 1962, Hercules Powder Company.
32a. B. Ettore and L. Luciano, Italian Patent 554,013, published January 5, 1957, Montecatini.
33. The early kinetic work up to 1959 was reviewed by G. Natta and I. Pasquon, in *Adv. Catal.* **11**, 1 (1959).
34. G. Natta, *Gazz. Chim. Ital.* **89**, 89 (1959); F. Danusso, *Polymer* **3**, 423 (1962); R. L. Miller, *in* "Crystalline Olefin Polymers" (R. A. Raff and K. W. Doak, eds.), Chapter 12. Wiley (Interscience), New York, 1965.
35. G. Natta, P. Corradini, I. W. Bassi, and L. Porri, *Atti Accad. Naz. Lincei., Cl. Sci. Fis., Mat. Nat., Rend.* [8] **24**, 121 (1958).
36. G. Natta, P. Corradini, and G. Allegra, *J. Polym. Sci.* **51**, 399 (1961).
36a. G. Natta, *Atti Accad. Naz. Lincei., Mem., Cl. Sci. Fis., Mat. Nat., Sez. 3a* [8] **4**, 61 (1955); *J. Polym. Sci.* **16**, 143 (1955).
37. G. Natta, I. Pasquon, and A. Zambelli, *J. Am. Chem. Soc.* **84**, 1488 (1962).

38. P. Pino, F. Ciardelli, G. P. Lorenzi, and G. Natta, *J. Am. Chem. Soc.* **84**, 1487 (1962).
39. G. Dall'Asta, G. Mazzanti, G. Natta, and L. Porri, *Makromol. Chem.* **56**, 224 (1962).
40. G. Dall'Asta, G. Natta, and G. Mazzanti, *Pap. Macromol. Colloq.*, (1964).
41. G. Natta, G. Dall'Asta, and G. Mazzanti, *Angew. Chem.* **76**, 765 (1964).
41a. H. Mark, *J. Polym. Sci.* **12**, see introduction p. i (1974).
41b. G. Wilke, *Polym. Prepr., Am. Chem. Soc., Div. Polym. Chem.* **15**, 277 (1974).
41c. W. Kern and P. Pino, *Makromol. Chem.* **164**, 7 (1973).
42. M. Fischer, German Patent 874,215, published June 19, 1952, issued March 12, 1953, to Badische Anilin & Soda Fabrik.
42a. F. C. Hall and A. W. Nash, *J. Inst. Pet. Technol.* **23**, 679 (1937).
43. Great Britain Patent 682,420, published November 12, 1952, E. I. Du Pont de Nemours and Company.
44. W. F. Gresham and N. G. Merckling, U.S. Patent 2,900,372, August 18, 1959, E. I. Du Pont de Nemours and Company; see also U.S. Patent 2,872,439 which issued on February 3, 1959 and was applied for on August 30, 1954; Great Britain Patent 777,538 (U.S. application on August 16, 1954), authors: A. W. Anderson, W. L. Truett, W. N. Baxter, N. G. Merckling, I. M. Robinson, G. S. Stamatoff, and D. H. Payne.
44a. H. Gilman and R. G. Jones, *J. Org. Chem.* **10**, 505 (1945).
45. A. A. Morton, E. E. Magat, and R. L. Letsinger, *J. Am. Chem. Soc.* **69**, 950 (1947); A. A. Morton, *Ind. Eng. Chem.* **42**, 1488 (1950); A. A. Morton, F. H. Bolton, F. W. Collins, and E. F. Cluff, *ibid.* **44**, 2876 (1952); for a recent account and interim status report, see A. A. Morton and E. J. Lanpher, *J. Elastomers Plast.* **6**, 73 (1974).
46. C. E. Schildknecht, A. O. Zoss, and C. McKinley, *Ind. Eng. Chem.* **39**, 180 (1947); C. E. Schildknecht, C. E. S. T. Gross, H. R. Davidson, J. M. Lambert, and A. O. Zoss, *ibid.* **40**, 2104 (1948).
47. J. P. Hogan and R. L. Banks, Belgian Patent 530,617, July 23, 1954, Phillips Petroleum Company; also U.S. Patent 2,825,721, March 4, 1958, Phillips Petroleum Company.
48. E. F. Peters and B. L. Evering, U.S. Patent 2,692,261, October 19, 1954, Standard Oil of Indiana; for a comprehensive compilation of early Phillips and Standard Oil of Indiana references, see Chapter 12.
48a. A. Zletz, U.S. Patent 2,692,257, October 19, 1954, Standard Oil Company of Indiana.
48b. G. Natta, P. Pino, and G. Mazzanti, U.S. Patent 3,715,344, February 6, 1973, Montecatini-Edison; see also U.S. Patents 3,112,300 and 3,112,301 by same authors for related patents.
49. F. W. Stavely, *Ind. Eng. Chem.* **48**, 778 (1956).
50. M. E. Pruitt and J. M. Baggett, U.S. Patent 2,706,181, April 12, 1955, Dow Chemical Company; see also U.S. Patent 2,706,189, April 12, 1955.
51. C. C. Price and M. Osgan, *J. Am. Chem. Soc.* **78**, 4787 (1956).
52. M. Szwarc, *Nature (London)* **178**, 1168 (1956); see also *Fortschr. Hochpolym.-Forsch.* **2**, 275 (1960).
53. G. Holden and R. Milkovich, U.S. Patent 3,265,765, August 9, 1966, Shell Oil Company; G. Holden, E. T. Bishop, and N. R. Legge, *J. Polym. Sci., Part C* **26**, 37 (1969).
54. A. Tsukamoto and O. Vogl, *Prog. Polym. Sci.* **3**, 199–279, (1971); *Phys. Chem., Ser. One* **8**, 49–104 (1972).
55. See review by W. R. Dunnavant, *Plast. Des. Process.*, 11 (1966).
56. R. E. Funer and R. S. Dudinyak, *Mach. Des.* Nov 26, 1970 **70** p. 72 (1970).
57. See review "Engineering Thermoplastics; the Best is Yet to Come," *Mod. Plast.* (Feb) **48**, p. 51 (1971).
58. E. J. Vandenberg, *J. Polym. Sci.* **47**, 486 (1960).
59. W. F. Gorham, Am. Chem. Soc. Div. Polymer Chemistry, Polymer Preprints, **6** (1), 73 (1965).
60. *New York Times*, August 13, 1973.

3

Definitions, Stereochemistry, Experimental Methods, and Commercial Polymers

I. Introduction

The broad patent definition of a Ziegler–Natta catalyst describes it as a mixture of a base metal alkyl of the group I to III metals and a transition metal salt of groups IV to VIII metals. Not all of the possible combinations, however, that have been tried are active.

Stereochemical control is one of the most significant properties of this catalyst. Depending on the structure of the monomer and catalyst, as well as the experimental conditions, it is possible to obtain one of many different stereoregular structures from α-olefins and diene monomers in high purity, such as isotactic and syndiotactic configurations in 1,2-addition polymerizations, or the cis-1,4 and trans-1,4 geometric units in 1,4-addition polymerizations.

A variety of analytical methods have been worked out to identify and quantitatively measure each of these structures, including X-ray, infrared, ^{13}C-, and ^{1}H-NMR spectroscopy. The synthesis of these polymers has been accomplished in a simple glass apparatus or in more sophisticated pressure reactors. Care must always be taken to control the level of impurities, especially water and oxygen.

About ten polymers are produced commercially with Ziegler–Natta catalysts. High density polyethylene, isotactic polypropylene, isotactic poly-1-butene and isotactic poly-4-methylpentene, block copolymers of ethylene and propylene, and *trans*-1,4-polyisoprene are made for plastics applications. The following elastomers are available: *cis*-1,4-polybutadiene, *cis*-1,4-polyisoprene, *trans*-1,5-polypentenamer, and random copolymers of ethylene and propylene (with and without some diene).

II. Definition of Ziegler–Natta Catalysts

Ziegler–Natta catalysts are formed by reacting a metal alkyl (or hydride) and a transition metal salt, such as $AlEt_3 + VCl_3$, under an inert atmosphere (1). Binary mixtures have usually been used; however, more than one metal alkyl or transition metal salt can be present. It has been demonstrated that there is a definite advantage to having a ternary or quaternary system in only a few cases (2, 3). Organic and inorganic molecules are sometimes added to modify some aspects of the polymerization; hydrogen, for example, is added to terminate chain growth (see Chapter 10, Section IV).

The patent literature usually claims all metal alkyls (hydrides) of group I to III base metals and transition metal salts of group IV to VIII transition metals, including Sc, Th, and U. [For a more detailed treatment, see Gaylord and Mark (reference 1, pp. 90–106).] This definition is too broad when applied in the laboratory. In practice, only a few group I to III metal alkyls are effective. Aluminum alkyls have been overwhelmingly preferred; zinc, magnesium, beryllium (caution, very toxic), and lithium alkyls have been relatively less investigated. The most studied transition metal salts were based on Ti, V, Cr, Co, and Ni metals.

Just because a metal alkyl or a transition metal salt forms an active catalyst for one type of monomer does not mean it will be active for all monomers. In practice, the choice of a particular combination of a metal alkyl and a transition metal salt is largely governed by the structure of the monomer polymerized. For example, Ziegler–Natta catalysts based on group VIII transition metal salts, such as $AlEt_2Cl + CoCl_2$, readily polymerize dienes (butadiene) but not ethylene or α-olefins. On the other hand, catalysts based on group IV, V, and VI transition metal are active for both dienes and α-olefins, such as Ti-, V-, and Cr-based catalysts. Another example will illustrate the specificity of a particular catalyst: $(C_5H_5)_2TiCl_2$, in combination with an aluminum alkyl, polymerizes ethylene but not propylene (4, 5). Ethylene is polymerized by a substantially larger number of catalysts than are propylene and higher α-olefins. While all catalysts which are active for polymerization of α-olefins are also active for polymerization of ethylene, the reverse is not true.

The particular experimental conditions employed can also determine whether a particular combination of a metal alkyl and a transition metal salt will produce the desired product. For example, the mixture of $AlEt_3$ + $TiCl_4$ polymerizes isoprene to a 95% cis-1,4 product when the Al/Ti ratio is equal to about one, and to a predominantly trans structure if the ratio is lower than one (6). Additionally, the $AlEt_2Cl$–VCl_4 mixture polymerizes propylene to a syndiotactic polymer only if the polymerization temperature is kept below $-45°C$ (7).

Sometimes an exchange reaction takes place between the transition metal salt and the metal alkyl to generate a new transition metal salt. When $Ti(OR)_4$ and $AlEtCl_2$ are reacted, for example, the exchange shown in Eq. 3–1 takes place (8).

$$\text{>Al—Cl} + \text{>Ti—OR} \longrightarrow \text{>Al—OR} + \text{>Ti—Cl} \qquad (3\text{–}1)$$

The real catalyst may be the end product of this exchange rather than the starting mixture; at an Al/Ti ratio of 4, $\beta TiCl_3$ is formed.

Similarly, a new metal alkyl may be generated. Natta and co-workers (8a) found that $AlEt(OEt)Cl$ in combination with $TiCl_3$ is active only if some $AlCl_3$ is present (added directly or present as an alloy in $TiCl_3AA$). They showed that $AlEt(OEt)Cl$ reacts with $AlCl_3$ to form $AlEt_2Cl$ and suggested that the latter reacts with $TiCl_3AA$ to form the active catalyst. $TiCl_3AA$ has the composition $TiCl_3 \cdot \frac{1}{3} AlCl_3$.

The notation "Ziegler–Natta catalyst" has been adopted in this book. The reader may have noted that the literature contains a variety of other designations for this catalyst, including Ziegler catalyst, Ziegler-type catalyst, Natta catalyst, coordinated-anionic catalyst, mixed metal complex catalyst, etc. In the initial period following Ziegler's discovery, the catalysts reported by his school to polymerize ethylene (and copolymerize ethylene and propylene) contained the transition metal in the highest valence state, e.g., $TiCl_4$ and VCl_4 (9). These were designated as Ziegler catalysts. Later, Natta discovered that polypropylene of greater isotacticity was made if preformed lower valence state transition metal salts were used, such as $TiCl_3$ and VCl_3. While he referred to these as modified Ziegler catalysts or Ziegler-type catalysts (10), others, in order to differentiate them from the higher valence state Ziegler catalysts, designated them as Natta catalysts. Particular catalysts developed by other industrial laboratories sometimes assumed the names of that company to denote a special modification of the catalyst.

While some workers chose to refer to all of the above as Ziegler-type catalysts, others adopted the name Ziegler–Natta catalysts so as to include the large number of catalyst types discovered and elucidated later, not only by the Natta school, but also by other workers worldwide. The latter nomenclature has been favored in more recent years. It is used throughout this book when the active catalyst contains a metal alkyl (or hydride) and a transition metal salt regardless of additional modifications, such as the presence of third components, support of the catalyst, and *in situ* synthesis of catalyst components.

At Natta's suggestion (11), these catalysts have also been designated as "coordinated-anionic," to emphasize that the monomer is first coordinated to the metal center before being inserted into the growing chain and that the polarized metal–carbon bond, while covalent, has the metal partially positive and the attached carbon partially negative. A free anion mechanism was not intended by this nomenclature.

The special group of catalysts containing only a transition metal component and sometimes a modifier, but no added base metal alkyl from the group I to III metals, are not considered here to be Ziegler–Natta catalysts. These have been designated as metal alkyl-free catalysts (see Chapter 12).

No attempt is made here or desired to delineate (nor bias) the patent aspects of any catalyst; such matters involve legal considerations and are best settled in the patent office of each country in accordance with the laws, customs, and regulations of that country.

III. Stereochemical Structures of Polymers and Methods of Characterization

As will be described in Chapters 4 and 5 and 18 to 22, Ziegler–Natta catalysts are most effective for the homopolymerization and copolymerization of ethylene, α-olefins (propylene) and dienes (butadiene and isoprene), and closely related monomers. This section identifies, for the simpler members of these monomer types, the various stereochemical structures that have been synthesized in high purity and the analytical methods that were used to evaluate their content.

A. OLEFINS

Ethylene is polymerized to a highly linear polyethylene,

$$n\mathrm{CH_2{=}CH_2} \longrightarrow \mathrm{{\sim}{\sim}CH_2{-}CH_2{-}CH_2{-}CH_2{-}CH_2{-}CH_2{\sim}{\sim}} \qquad (3\text{--}2)$$

where $\sim\!\sim$ denotes a long polymer chain. In the absence of secondary reactions, the polymer chain would exclusively consist of methylene units

TABLE 3–1

Densities, Crystallinities, and Branching Values for Four Types of Polyethylenes (12)

Samples[a]	Density (20°C)	Crystallinity (%)	CH_3 1000 C[d]	$RCH=CH_2$ 1000 C	$RR'C=CH_2$[b] 1000 C	$RCH=CHR$[b] 1000 C	$RCH=CH_2$[b] / $RR'C=CH_2$
H.d. polyethylene produced by l.p. process (Phillips)	0.9671	76.72	0.54	1.579	0.0299	—	52.81
H.d. polyethylene produced by l.p. process (Standard Oil, Ind.)	0.9577	72.37	6.71	0.240	0.067	2.146	3.582
H.d. polyethylene produced by l.p. process (Ziegler)	0.9548	66.11	3.09	0.430	0.090	1.121	4.778
L.d. polyethylene produced by h.p. process	0.9257	45.36	17.62	0.045	0.195	0.053	0.231

[a] H.d., high density; l.p., low pressure; L.d., low density; h.p., high pressure.
[b] Types of unsaturation.

	$v(cm^{-1})$	ε^c	$12/\varepsilon$
$RCH=CHR'$	964	85.4	0.140
$RCH=CH_2$	908	121.0	0.099
$RR'C=CH_2$	888	103.4	0.116

[c] Figures obtained by L. H. Cross, R. B. Richards, and H. A. Willis, *Trans. Faraday Soc.* **9**, 235 (1950). The values of the branching shown above were obtained by the method of Willbourn, *J. Polym. Sci.* **34**, 569 (1959).
[d] Notation represents 1000 carbons or carbon atoms.

as shown above, and these products would have a density near 0.98 g/cm^3. Such a polymer had been made earlier by a catalyzed decomposition of diazomethane.

In practice, however, some of the ethylene molecules dimerize or oligo-merize to form 1-butene or higher α-olefins. These α-olefin molecules copolymerize with ethylene and the effect is to interrupt the long sequences of methylene units by methyl or higher substituent branches. The crystal-linity of the partially branched polyethylene is subsequently decreased, and the polymer has a lower density (about 0.95 to 0.96 g/cm^3). For many applications, polyethylenes of intermediate densities are preferred (see Section V); in fact, in the commercial production of polyethylene, some 1-butene or 1-hexene is intentionally added.

Relative to the low density polyethylene synthesized with radical initiators and at high ethylene pressures, the Ziegler–Natta polyethylenes have fewer branches and are more crystalline. Table 3–1 (12) compares the densities, crystallinities, and branching values for polyethylenes synthesized with the Phillips, Standard of Indiana, Ziegler–Natta, and radical systems. Polyeth-ylenes formed by the first three systems are comparable and are designated as high density (low pressure) polyethylenes, while the radical polymer is designated as low density (high pressure) polyethylene. The latter requires ethylene pressures above 3000 atm, while the former is done between 1 to 30 atm.

Melting points of about 100°, 125°, and 135°C have been measured for low, intermediate, and high density polyethylenes, respectively. Poly-ethylene chains crystallize in a lamellar form. The mechanism by which branches are formed in radical processes is different from that described above for Ziegler–Natta polymerizations. In the radical process, back-biting transfers the radical center to an internal carbon and a hydrogen to the carbon which bore the radical. Ethyl and butyl groups are formed in the greatest amount.

The degree of branching has been directly measured by means of infrared, ^{1}H-, and ^{13}C-NMR spectroscopy. In addition, physical methods that mea-sure a property related to crystallinity have often been used such as X-ray diffraction, differential scanning calorimetry, optical microscopy (bire-fringence), and density gradient measurements.

Figures 3–1 and 3–2 compare the infrared spectra of low and high density polyethylenes. The absorption band at 13.7 μ (760 cm^{-1}) is related to crys-tallinity, and it is used to evaluate the relative degree of crystallinity by measuring its intensity relative to the reference 13.9 μ band. Absolute crys-tallinities are obtained from measurements of densities and from X-ray diffraction spectra.

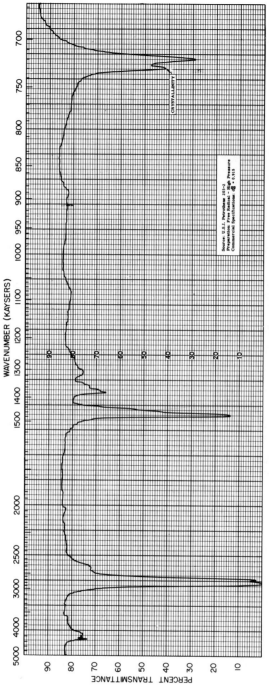

Fig. 3–1. Polyethylene, low density. Source: U.S.I. Petrothene, 203–2; preparation: free radical, high pressure; commercial specifications: $d_{23}^{23} = 0.915$.

38

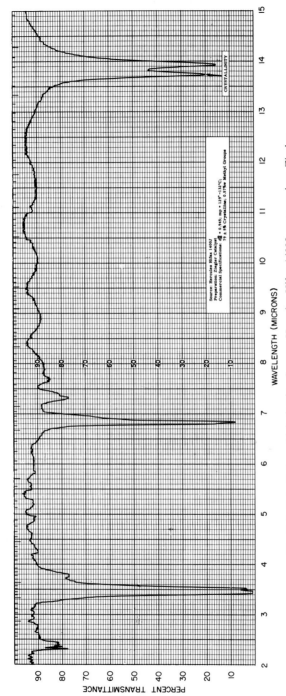

Fig. 3–2. Polyethylene, high density. Source: Hercules Hifax 1400J; preparation: Ziegler catalyst; commercial specifications: $d_{23}^{23} = 0.945$, melting point $= 129°C–132°C$, $79 \pm 5\%$ crystalline, 0.5 wt% methyl groups.

39

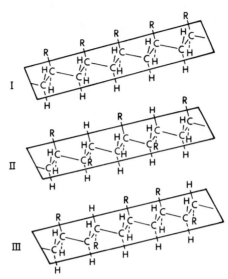

Fig. 3–3.   Planar presentation of the chains of polymeric α-olefins. (I) isotactic, (II) syndiotactic, and (III) atactic (13).

Propylene has been polymerized to isotactic and syndiotactic structures in a high purity, as shown in the planar presentation in Fig. 3–3 (13). In the perfect isotactic chain, every tertiary carbon would have the same configuration, while in the perfect syndiotactic chain, these tertiary carbons have opposite configurations. A chain is designated as atactic if a random distribution of the two possible configurations occurs.

The isotactic and syndiotactic polymer chains form helices which aggregate into crystallites. Figures 3–4 (14), 3–5 (14, 15), and 3–6 (15) show

Fig. 3–4.   Chain model of isotactic polypropylene (14). Melting point of highly crystalline sample (density $= 0.91$ g/cm^3 is 170°C.

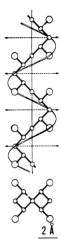

Fig. 3–5. Chain model of syndiotactic polypropylene (14). Melting point of highly crystalline sample (density = 0.885 g/cm^3) is about 130°C (15).

2 Å

(a)

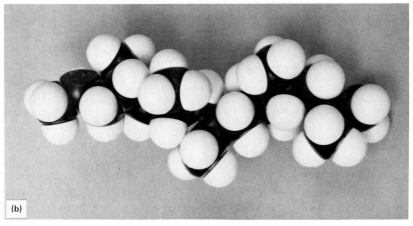

(b)

Fig. 3–6. Molecular models of helices of isotactic and syndiotactic polypropylenes (15).

different representations of helices of isotactic and syndiotactic polypropylenes. The number of repeating units of the α-olefin molecule determines the type of helix that is formed.

The practical benefit of stereoisomerism is the ability of highly tactic chains to crystallize into larger structures having beneficial properties such as high melting temperature, high crystallinity, high heat distortion temperature, and high tensile strength. Atactic polymers do not crystallize (they are said to be amorphous), and they lack the above properties.

Isotactic polypropylene has a melting point of 165° to 175°C and has gained important commercial significance. Syndiotactic polypropylene has a melting point near 130° to 159°C and does not appear to have practical applications of significance.

The isotactic content of a polymer can be determined in a number of ways. The most widely used method in the early days of Ziegler–Natta chronology involved extracting the polymer in a Kumagawa or a Soxhlet-type extractor

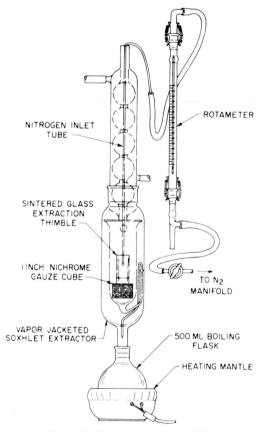

Fig. 3–7. Extraction apparatus (17).

with a boiling solvent (16). Figure 3–7 (17) shows a typical extraction apparatus (18). The solvent and temperature of extraction varies with the polymer. For isotactic polypropylene, boiling heptane or isooctane was commonly used. For poly-1-butenes, boiling ether was used. The fraction of the whole polymer unextracted × 100 was taken as a percent index of isotacticity. X-ray diffraction spectroscopy was used to calibrate this method and show that the fractionation differentiated isotacticity and not molecular weight.

This method was very valuable for it allowed a facile determination of an index related to isotacticity; it, however, possessed certain deficiencies. For polymers that had similar molecular weights (and molecular weight distributions), the relative comparison was valid. But polymers that had very low or very high molecular weights gave false comparisons. A polymer that had very low molecular weights, such as $[\eta] = 0.1$–0.8 dl/g, would be considerably more soluble in hot solvent than one that had $[\eta] = 2$–4 dl/g, even though both had identical isotacticity (see Chapter 4, Section II,C). High molecular weight polymers might show lower solubility due to higher solution viscosity.

Figure 3–8 and Table 3–2 correlate the amount of insoluble residue with density for polypropylenes of different intrinsic viscosity (19).

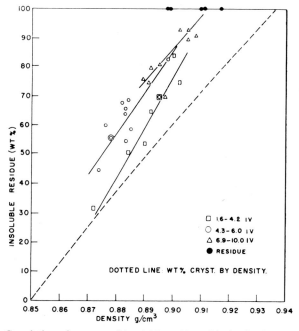

Fig. 3–8. Correlation of amount of insoluble residue with density for polypropylenes of different intrinsic viscosities (IV) (19).

TABLE 3–2

Properties of Insoluble Residues Compared to Properties of the Original Polymer (19)

Original polymer	Residue (wt %)	Original $[\eta]$ (dl/g)	Density of original polymer (g/cm³)	Density of residue (g/cm³)	Crystallinity of residue (wt %)
1	26[a]	1.6	0.872[b]	0.898	58
7	56	4.3	0.878	0.899	59
15	70[a]	3.0	0.895	0.911	73
16	70[a]	6.0	0.895	0.910	72
17	70[a]	9.6	0.897	0.917	80

[a] Extracted with heptane only.
[b] Not measured; estimated from infrared data.

Infrared methods later became popular and were used by many workers on a routine basis. ^{1}H- and ^{13}C-NMR spectroscopy have received considerable attention in recent years but more so as specialized tools rather than for routine analysis. Mechanistic studies, involving, for example, the determination of addition mode and stereochemistry, have greatly profited by their use, especially when labeled monomers were used (see Chapter 15, Section V).

Figure 3–9 compares the infrared spectra of isotactic and atactic polypropylenes (18). The absorption band at 99 cm^{-1} (10.02 μ) is due to isotactic helices. The band at 974 cm^{-1} (10.28 μ) is independent of isotacticity and has been used as a reference band. The isotactic index is taken as the ratio: absorbance at 995 cm^{-1}/absorbance at 974 cm^{-1}.

Infrared techniques also have been used to evaluate the syndiotacticity of polypropylene samples. The absorption band at 11.53 μ has been most

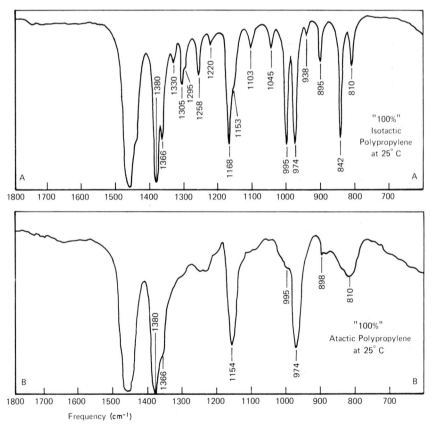

Fig. 3–9. Spectra of 100% isotactic (A) and 100% atactic (B) polypropylene, both at 25°C (18).

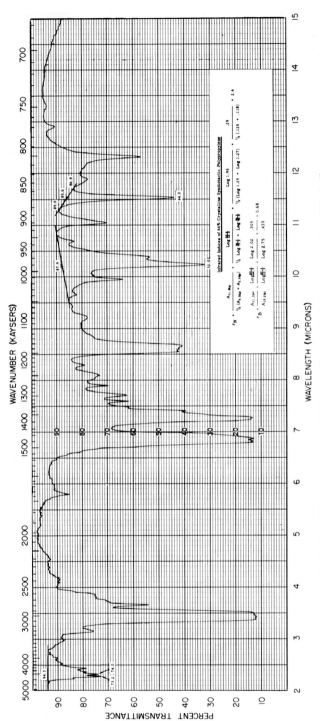

Fig. 3–10. Infrared spectrum of 66% crystalline syndiotactic polypropylene (20). Indices:

$$r_N = A_{11.53\mu}/\tfrac{1}{2}(A_{2.32\mu} + A_{2.35\mu})$$
$$= \log(85.6/44.0)/\tfrac{1}{2}\cdot[\log(94.3/73.2) + \log(94.3/74.0)]$$
$$= \log 1.95 \, \tfrac{1}{2}(\log 1.29 + \log 1.27)$$
$$= 0.29\tfrac{1}{2}(0.125 + 0.118) = 24;$$
$$r_B = A_{11.53\mu}/A_{10.25\mu}$$
$$= \log(89.0/44.0)/\log(87.9/32.0)$$
$$= \log 2.02/\log 2.75 = 0.305/0.439 = 0.68.$$

widely used. Figure 3–10 shows a spectra of a highly syndiotactic sample with an illustration of a calculation (20). Because the base lines are arbitrarily drawn, these indices have only a relative significance.

X-ray diffraction methods were widely used by the Natta school to identify and evaluate the stereochemical structure of the poly-α-olefin. Figures 3–11 and 3–12 show the X-ray diffraction spectra of isotactic (20a, 21) and

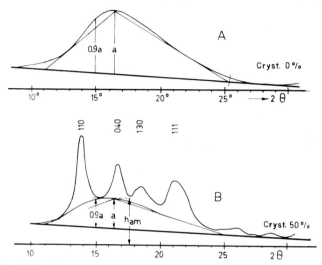

Fig. 3–11. Diffraction curves of polypropylene after Natta *et al.* (A) Entirely amorphous sample; (B) partially crystalline sample (20a).

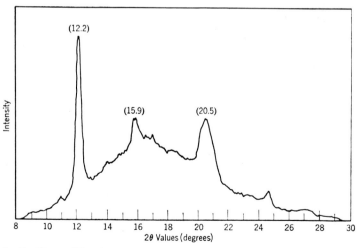

Fig. 3–12. X-ray diffraction trace of syndiotactic polypropylene at room temperature. Helical form; sample density = 0.855 g/cm^3 (20).

syndiotactic (20) polypropylenes, respectively. Weidinger and Hermans (20a) used the area method shown in Fig. 3–11 to evaluate the relative contents of amorphous and isotactic crystalline polypropylene.

An absolute X-ray method was later devised by Corradini and co-workers (21a) in which polypropylene mixtures of urotropine and adamantine were examined. Very good agreement was found between the two methods.

Figures 3–13 and 3–14 correlate density and specific volume of poly-propylenes with their corresponding X-ray crystallinity or infrared index.

^{1}H- and ^{13}C-NMR resonance spectroscopy have become very powerful tools to study tacticity. Figure 3–15 shows a ^{1}H-NMR spectra of isotactic polypropylene.

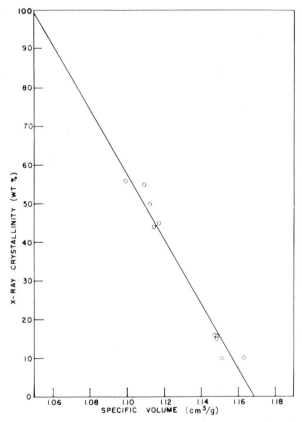

Fig. 3–13. Crystallinity as determined by X-ray analysis versus specific volume for polypro-pylene (19).

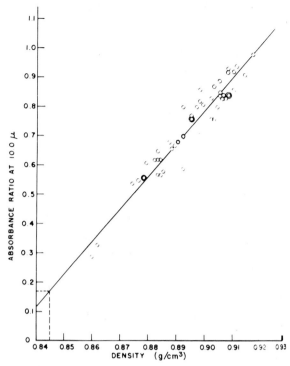

Fig. 3–14. Correlation of infrared absorption at 10.03 μ with density for polypropylene (19).

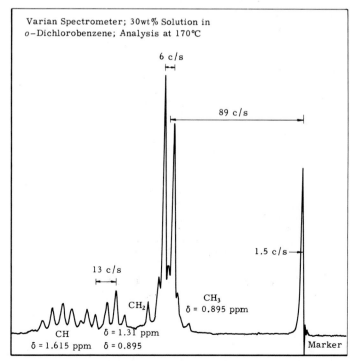

Fig. 3–15. NMR spectrum (100 Mc/sec) of 65% crystalline isotactic polypropylene.

TABLE 3–3

Carbon-13 NMR Chemical Shifts for Isotactic and Syndiotactic Polypropylenes (22)

Structure	ppm for each type carbon		
	CH_2	CH	CH_3
Isotactic	46.44	28.86	21.85
Syndiotactic	47.09	28.32	20.34

^{13}C-NMR spectroscopy is proving to be an ever more powerful method for differentiating between isotactic and syndiotactic structures. The chemical shifts due to CH, CH_2, and CH_3 carbons are widely separated. Table 3–3 compares the chemical shifts due to the isotactic and syndiotactic structures (22).

Higher α-olefins such as 1-butene, 4-methyl-1-pentene, and styrene have been polymerized only in the isotactic structure. Similar methods have been

TABLE 3–4

Melting Points of Selected Linear and Branched Polymers of Ethylene and α-Olefins (23)

Melting point^a (°C)	Polymer
137	Polyethylene (high density)
170–175	Polypropylene (isotactic)
130–135	Polypropylene (syndiotactic)
137	Poly-1-butene (modification 1, isotactic)
125	Poly-1-butene (modification 2, isotactic)
70	Poly-1-pentene
45	Poly-1-decene
70	Polyoctadecene
310	Poly-3-methyl-1-butene
235	Poly-4-methyl-1-pentene
110	Poly-5-methyl-1-hexene
amorphous	Poly-6-methyl-1-heptene
250	Polystyrene
372	Polyvinylcyclohexane
200	Poly-3-methyl-1-pentene

^{a} Polymers from 1-pentene and higher olefins have melting points derived from isotactic crystallinity.

used to characterize them. Table 3–4 (23) compares the crystalline melting points for selected linear and branched polyolefins.

Many of these polyolefins can exist in more than one isomorphic form, and it is essential to control the thermal history of a test sample if the obtained result is to be meaningful. Figure 3–16 shows the X-ray diffraction diagrams of the possible crystalline modifications of isotactic polypropylene (23a). Poly-1-butene can exist in four different crystalline modifications, and these exhibit different infrared spectra (Fig. 3–17), melting points, and X-ray diffraction spectra (23b). Modifications 1 and 2 differ in the structure of the helix which forms from the isotactic chain. The less stable form 2 contains about 11 monomers per 3 helical turns, while the stable form 1 contains 3 monomers per helical turn. Relative to the unstable form 2, form 1 is harder, more crystalline, more rigid, and has a greater tensile strength and a higher melting point.

When two or more olefins are copolymerized—ethylene and propylene (with and without diene)—the analysis can become more complicated.

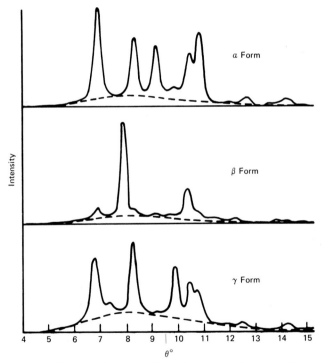

Fig. 3–16. X-ray diffraction diagrams of isotactic polypropylene crystalline modifications (23a).

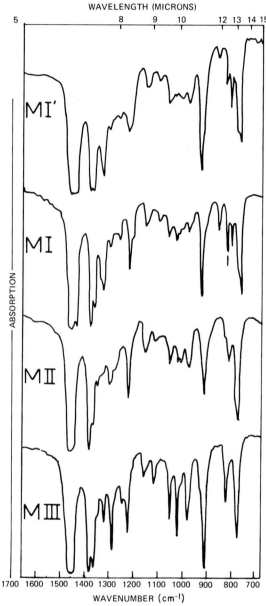

Fig. 3–17. Infrared spectra of poly-1-butene polymorphs. The degree of isotacticity is measured by the ratio of optical density at 1221 cm^{-1} (band of isotacticity) and 1050 cm^{-1} (internal standard) for a sample that is in the stable modification I (23b).

Because of the importance of this copolymer system, much work involving different characterization techniques has been employed, such as infrared, pyrolysis, and NMR (^{1}H- and ^{13}C-). Baldwin and Ver Strate discuss these methods in a comprehensive review (23c). Recent reviews by Tosi and Ciampelli (23d) and by Kissin (23e) also offer excellent discussions on structural analysis.

B. DIENES

What makes the polymerizations of dienes with Ziegler–Natta catalysts so fascinating is their ability to form polymers that have almost exclusively either the 1,2- or the 1,4-addition structures. Often mixtures of both are formed. The practical outcome is the synthesis of elastomeric or plastic polymers, depending on the diene used and the particular isomerism prevailing in the chain (Table 3–5). Of great commercial importance as elas-

TABLE 3–5

Physical Properties of Butadiene, Isoprene, and Pentadiene Polymers Obtained with Stereospecific Catalysts (24)

Polymer	Melting temp. (°C)	Identity period (Å)	Density
trans-1,4-Polybutadiene[a]	70–75 (transition temp.)	4.85, mod. 1	1.02
	145 (melting temp.)	4.7, mod. 2	0.93
cis-1,4-Polybutadiene	3	8.60	1.01
Isotactic 1,2-polybutadiene	128	6.5	0.96
Syndiotactic 1,2-polybutadiene	156	5.14	0.96
cis-1,4-Polyisoprene	20–30	8.10	1.00
trans-1,4-Polyisoprene	65	4.77	1.04
		8.75	
		9.2	
3,4-Polyisoprene	amorphous	—	—
trans-1,4-Polypentadiene	95	4.82	0.98
Isotactic cis-1,4-polypentadiene	42	8.15	n.d.[b]
Syndiotactic cis-1,4-polypentadiene	52–3	8.5	n.d.
1,2-Polypentadiene	10–20	515	n.d.

[a] trans-1,4-Polybutadiene can exist in two crystalline modifications; one (mod. 1) is stable below about 75°C, the other from 75° to 145°C. The two modifications have different identity periods and different equatorial encumberment per each chain (the latter being greater for mod. 2). Only the equatorial encumberment is greater for mod. 2. Identity period is greater for mod. 1.

[b] n.d., no data.

tomers are *cis*-1,4-polybutadiene and *cis*-1,4-polyisoprene. For a detailed discussion of isomerism, consult Natta and Porri (24). Only a brief summary is made here.

The simplest 1,3-diene is butadiene, and it can be polymerized exclusively or predominantly into one of the following four possible isomeric structures I to IV in Fig. 3–18.

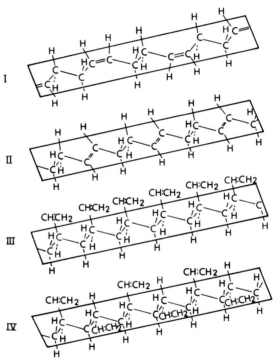

Fig. 3–18. Planar presentation of the chains of various polymers of butadiene: (I) *cis*-1,4-polybutadiene; (II) *trans*-1,4-polybutadiene; (III) isotactic 1,2-polybutadiene; (IV) syndiotactic 1,2-polybutadiene (13).

When butadiene is substituted in the 3- or 4-position as in isoprene (2-methyl-1,3-butadiene) and piperylene (1,3-pentadiene), additional isomeric structures are possible.

Polyisoprene can, in principle, exist as six possible isomers. Two of these are geometric, namely the cis-1,4 and trans-1,4 structures (V and VI). The other four structures have 1,2 (syndio- or isotactic, VII) or 3,4 (syndio- or isotactic, VII) isomerism.

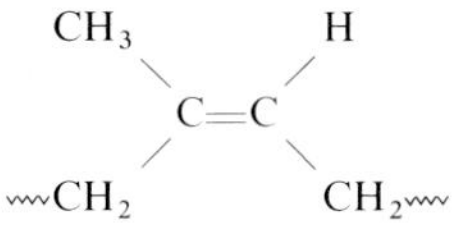

cis-1,4

V

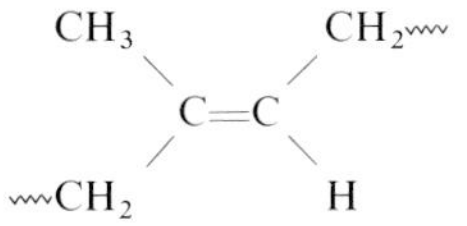

trans-1,4

VI

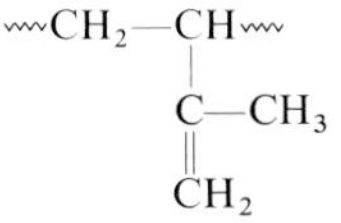

3,4 (isotactic, syndiotactic)

VII

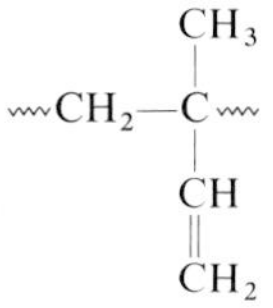

1,2 (isotactic, syndiotactic)

VIII

In practice, only the cis-1,4, trans-1,4 and 3,4 types have assumed importance.

^{13}C-NMR spectroscopy has been successfully used to distinguish between the cis-1,4 and trans-1,4 structures in polyisoprene samples (Fig. 3–19 and Table 3–6) (25).

Infrared spectroscopy has been most widely used to identify the different butadiene structures (Figs. 3–20, 3–21, and 3–22) (26).

The ^{13}C-NMR spectrum of 1,4-polybutadiene shows two resonance signals in the aliphatic region at 27.9 and 33.2 ppm (relative to TMS) corresponding to *cis*- and *trans*-methylene carbons (27). Mochel and co-workers (28) concluded from ^{13}C-NMR spectra of several polybutadienes that cis-1,4 and trans-1,4 units must always be separated by 1,2-vinyl units, but this has been disputed by others (29). Samples having variable contents of the cis-1,4 and trans-1,4, and 1,2-vinyl units have been examined recently by Alaki and co-workers (30) for configurational sequences using ^{13}C–^{1}H decoupled Fourier transform NMR spectroscopy. Without doubt, ^{13}C-NMR spectroscopy will be refined and developed further to be capable of the solution of problems as complex as this.

A different structural unit is possible for poly-1,3-pentadiene (IX). When cis-1,4- or trans-1,4-addition occurs, carbon-4 can exhibit isotactic or syndiotactic stereoisomerism.

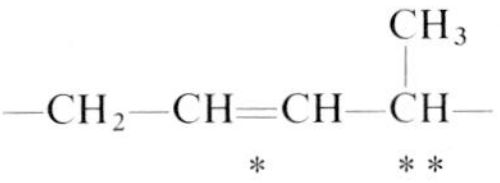

IX

* cis-1,4 or trans-1,4 structures possible

* * isotactic or syndiotactic structures possible at carbon-4

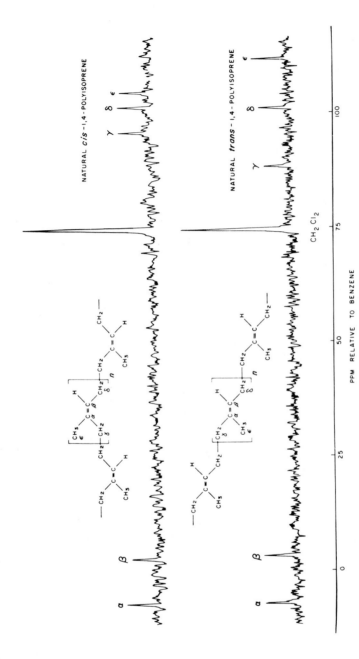

Fig. 3–19. Random noise, proton-decoupled 15.08 MHz carbon-13 nuclear magnetic resonance spectra of approximately 10% (wt/vol) solutions of natural *cis*-1,4-polyisoprene and natural *trans*-1,4-polyisoprene in dichloromethane. The spectrum of *cis*-1,4-polyisoprene consists of the time-averaged accumulation of 348 scans and the spectrum of *trans*-1,4-polyisoprene is the accumulation of 183 scans. Each scan was swept 2156 Hz in 250 seconds. Both sweep scales are equal and reported in parts per million with respect to benzene (25).

TABLE 3–6

Carbon-13 NMR Chemical Shift Data for the 1,4-Polyisoprenes[a] (25)

	Carbon position[b]									
	α		β		γ		δ		ε	
	cis	trans	cis	trans	cis	trans	cis	trans	cis	trans
cis-1,4-Polyisoprene										
Natural	−6.8		3.4		96.2		101.7		105.1	
Synthetic	−6.9		3.2		96.2		101.9		105.3	
trans-1,4-Polyisoprene										
Natural		−6.5		4.1		88.7		101.6		112.7
Synthetic		−6.5		4.2		88.7		101.5		112.8
Mixed synthetic	−6.5 band[c]		3.1	3.9	96.5	88.5	101.7 (band)		105.1	112.5

[a] Chemical shift values in parts per million ±0.2, are relative to benzene.
[b] Carbon positions as designated in Fig. 3–19.
[c] Unresolved broad peaks.

57

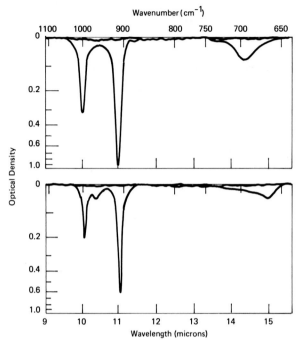

Fig. 3–20. IR spectra of 1,2-polybutadienes in solution: Above: isotactic. Below: syndiotactic. Resolution 935, corresponding to a slit width of 142 μ at 11 μ; scale, 5 cm/μ (26).

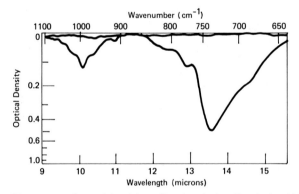

Fig. 3–21. IR spectrum of a *cis*-1,4-polybutadiene in solution. Resolution 935, corresponding to a slit width of 266 μ at 13.5 μ; scale, 5 cm/μ (26).

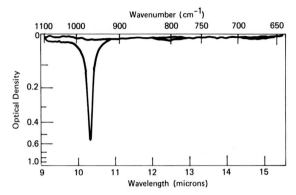

Fig. 3–22. IR spectrum of a *trans*-1,4-polybutadiene in solution. Resolution 935, corresponding to a slit width of 127 μ at 10.34 μ; scale, 5 cm/μ (26).

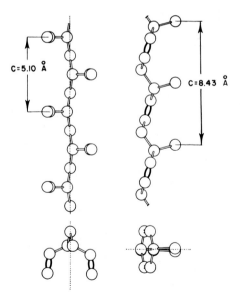

Fig. 3–23. Chain models of syndiotactic 1,2-polybutadiene and syndiotactic *cis*-1,4-polypentadiene (14).

Thus, each repeating unit can contain two types of isomeric structures. When both 1,2- or 3,4-addition occur, the structures shown in the following tabulation are possible.

1,2-Addition	3,4-Addition
Isotactic, cis	Disyndiotactic
Isotactic, trans	Erythrodiisotactic
Syndiotactic, cis	Threodiisotactic
Syndiotactic, trans	

Figure 3–23 shows chain models of syndiotactic 1,2-polybutadiene and syndiotactic cis-1,4-polypentadiene. Polymers containing the isotactic trans-1,4, isotactic cis-1,4, syndiotactic cis-1,4, or syndiotactic 1,2 structures in high purity or as the predominant structure have been synthesized (see Chapter 5).

Table 3–5 [compiled by Natta and Porri (24)] compares the melting temperatures, identity periods, and densities of polybutadienes, polyisoprenes, and polypentadienes that have been synthesized to date.

The reader should recognize that additional irregularities will occur in the chain when 1,2-additions or 1,2-additions followed by 1,4-additions take place and head-to-tail additions are accompanied by head-to-head and tail-to-tail additions. The detection of such irregularities has not been elucidated unequivocally.

C. COMMENTS

The analytical methods that were described in Sections III,A and B were generally used in the exploratory work from which new and improved catalysts evolved. When a promising catalyst was examined for larger scale development, it was necessary to examine the product by many other tests (tensile strengths, percent elongation, etc.) to establish its potential commercial application.

In the early period of Ziegler–Natta chronology, when much of this exploratory work was done, emphasis was placed on infrared and extraction methods as routine tools to measure tacticity. In more recent years, another order of sophistication was achieved through the powerful tools of ^{1}H- and ^{13}C-NMR spectroscopy. The detailed structure of the polymer chain is now within grasp of the worker who is willing to exploit it. There is no doubt that continued use of these powerful tools, especially when used in concert with the earlier methods, will enhance our capability and understanding of the mechanism by which tactic polymers are formed.

IV. Experimental Methods

A. REACTORS AND PROCEDURES

Ziegler–Natta polymerizations have been done in low pressure glass vessels and in high pressure metal reactors.

A variety of glass vessels were found useful. The simplest consisted of small bottles of 100 to 300 cm³ capacity, which were designed for holding such commercial products as hydrogen peroxide or soda drinks (31). Multineck, ground glass round bottom flasks commonly used in organic chemistry preparations have also been very popular (see Fig. 3–24) (32). Specially designed glass vessels, each bearing a unique operational feature, have also been described (32a–32c) (See Figs. 3–25, 3–26, and 3–27.). Polymerizations have been carried out in a special dilatometer (32d) or in a simple flask that can be attached to a high vacuum line (32e). Properly designed glass vessels are often safe to use up to 10 atm; the exact limit obviously depends on its construction and the safety shields protecting the worker.

Pressure reactors must be used when higher pressures are obtained during the polymerization. The Natta group used a rocking autoclave (33), shown in Fig. 3–28, for much of its early kinetic and exploratory polymerizations of olefins (ethylene, propylene, and 1-butene). Because stainless steel autoclaves were commercially available, in which agitation was accomplished by a motor driven paddle, much work has been done with these reactors.

Batch-type processes are done in these reactors; that is, the polymerization is carried to a desired percent conversion and the entire reaction is terminated. At advanced levels of development, continuous processes are often done. In these, the catalyst and monomer are continuously added while simultaneously, the polymer product is withdrawn. Only the withdrawn polymer is terminated.

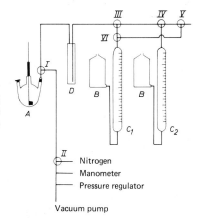

Fig. 3–24. Apparatus design for the polymerization of α-olefins (32). (A) Reaction flask; (B) pressure equilizing flask; (C₁, C₂) 1000 ml gas burettes; (D) bubble counter; (I–VI) three-way stopcocks.

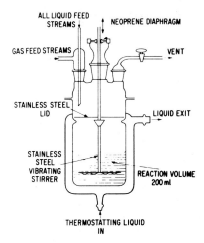

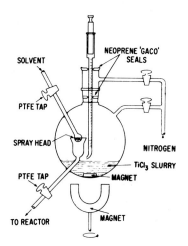

Fig. 3–25. Polymerization reactor and vessel for storing and sampling TiCl₃ (32e).

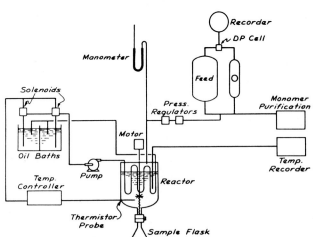

Fig. 3–26. Reactor assembly (schematic) (32b).

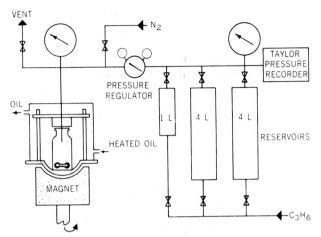

Fig. 3–27. Magnetically stirred kinetic apparatus (32c).

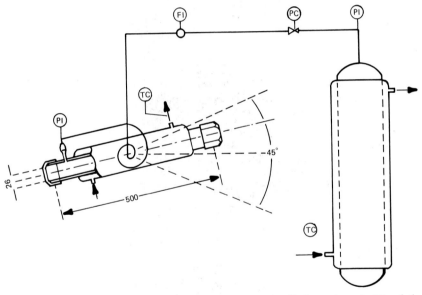

Fig. 3–28. Apparatus used for the kinetic measures (oscillations, through 45°, of the reactor numbered 45 per minute). (PI) pressure gauge; (PC) pressure regulator; (FI) flow indicator; (TC) thermoregulator (33).

Once the basic process and evaluations are done in small bench scale and pilot plant reactors, large manufacturing facilities are built which can produce polymer in volumes of 100 million pounds per year and often many times this value. Figure 3–29 shows the Shell Chemical Company manufacturing plant for isotactic polypropylene.

Fig. 3–29. Shell Chemical Company plant for manufacturing isotactic polypropylene.

The polymerization procedure closely depends on the particular catalyst used and monomer polymerized. Some monomer–catalyst systems are rugged while others are delicate. Some must be used under narrow operating conditions while others can be used over a wide range. There are no simple rules to follow. Nevertheless, all polymerizations require an inert atmosphere such as N_2 or Ar and the absence of polar reactive impurities, or at least require that they be present in very small levels. Compounds containing oxygen, water, and sulfur can be deleterious to catalyst activity, and their concentration must be controlled to a level at which they do not adversely affect the polymerization. Chapter 9 shows, however, that a beneficial effect can be gained from specific reactants and electron-donors under controlled conditions.

It one wants to prepare a small sample of a polymer without regard to the level of conversion or stereoregularity, minimal precautions often suffice. However, a detailed kinetic study or an investigation aimed at optimizing the performance of a catalyst for potential industrial use, or a synthesis of a polymer of highest steric purity, require maximum care in purification. This is illustrated by Wisseroth and co-workers, who showed vast improvements in the activity of the catalyst if the propylene was purified by passing it through Al_2O_3 (34).

Sometimes only after the polymer is characterized in advanced stages of development work by practical tests, such as processing behavior, do subtle differences in the method of synthesis appear. The patent literature contains many examples of claims that an improvement is achieved only if a specific procedure is followed. The worker must be aware that many factors may affect the efficiency of his polymerization and also the architecture of his polymer product. Chapters 6 and 7 discuss the types of catalysts and processes that can be used in laboratory and large scale synthesis.

B. SOLVENTS, CONCENTRATIONS, AND TEMPERATURE OF POLYMERIZATION

Only the scope of these experimental factors is presented here. Specific examples will be cited in Chapters 4, 5, and 18.

1. Solvents

Any solvent can be used as long as it does not react with the catalyst components in a deleterious way. In general, aliphatic and aromatic hydrocarbon solvents have been preferred in laboratory studies and appear to be used in most industrial processes if patents are an indication. The choice of a particular solvent depends on the catalyst and the conditions of polymerization. In an industrial process, its availability in a sufficiently pure state can be important.

In general, monoolefins are preferentially polymerized in aliphatic hydrocarbon solvents ranging from propane to octane and higher alkanes as well as in cyclohexane. While propane can be used in high pressure reactors, glass laboratory reactors require higher boiling solvents. While aromatic solvents (toluene or benzene) can also be used, they have not received as wide an application. As will be described below, polyolefins are usually synthesized in a particulate form and aliphatic solvents are most favorable.

In contrast, aromatic solvents and cyclohexane are preferred solvents for the polymerization of dienes such as butadiene and isoprene. The desired products (high cis-1,4- and trans-1,4-polymers) are soluble in these solvents, and a higher yield based on catalyst is obtained when soluble catalysts are used. The catalyst metal centers do not become entrapped in agglomerates of swollen impenetrable polymer.

Halogenated solvents (such as perchloroethylene, chloromethane, chloroethane, dichloroethane, and chlorobenzene) have been used in specific polymerizations. But if the chlorine is weakly held, such as in benzyl chloride, *tert*-BuCl, or *sec*-BuCl, the halogenated solvent will react with the aluminum alkyl to generate secondary metal alkyls. The addition of certain halogenated molecules in controlled amounts, however, can lead to increased catalyst activities (Chapter 9).

2. Concentration of Monomer

The monomer can be present in high or low concentrations, depending on several factors. Highly active monomers are usually diluted in the laboratory synthesis. Sluggishly active monomers are polymerized in higher concentrations; often these monomers serve as their own solvents, such as styrene and 4-methylpentene. But even more active olefins such as propylene and 1-butene can be polymerized, whereby the liquid olefin is the solvent for the polymerization. Only small amounts of alkane solvents are present, these being added only as carriers for the catalyst. Chapter 7 describes different processes that have been used.

3. Concentration of Catalyst Components

Many examples are cited throughout this book to show that both absolute and relative concentrations of the metal alkyl and the transition metal salt can influence catalyst behavior. When electron-donors are also added, their effect also depends on both absolute and relative concentrations.

4. Temperature of Polymerization

Specific Ziegler–Natta catalysts have been reported that are active as low as $-90°C$ and as high as 200°C. Syndiotactic polymerizations of propylene occur most favorably between $-45°$ and $-78°C$. Isotactic polymerizations of propylene are done between 25° and 90°C if the polymer maintains a particle form (slurry process) and between 110° and 150°C if it is dissolved (solution process). See Chapter 7 for discussion of processes. Polydienes (such as from butadiene and isoprene) have been synthesized between 0° and 80°C, most favorably at the lower temperatures.

The temperature of polymerization can affect stereoregularity, molecular weight, and polymerization rate. This effect is not always predictable, and one must establish it experimentally. Examples will be given later, especially in Chapter 18.

V. Commercial Polymers

Isotactic polypylene first became available commercially from Montecatini in 1957. Since then, many other polymers have been synthesized commercially with Ziegler–Natta catalysts for application as plastics or elastomers. Tables 3–7 and 3–8 list these polymers. Some can be synthesized only with Ziegler–Natta catalysts, while others can be made commercially by alternate means of initiation or catalysis.

TABLE 3–7

Commercially Available Plastics Synthesized by Ziegler–Natta Catalysts

Polymer	Alternate commercial synthetic routes[a]
High density polyethylene	Phillips, Standard of Indiana and Union Carbide catalysts[b]
Isotactic polypropylene	None
Isotactic poly-1-butene	None
Isotactic poly-4-methyl-1-pentene	None
Ethylene propylene block copolymers	None
trans-1,4-Polyisoprene	Nature (gutta-percha and balata)

[a] Some of these polymers can be made in low yields with metal alkyl-free catalysts, but these are not yet sufficiently active for commercial production.

[b] See Chapter 12 for descriptions of some of these metal alkyl-free catalysts.

TABLE 3–8

Commercially Available Elastomers Synthesized by Ziegler–Natta Catalysts

Polymer	Alternate commercial synthetic routes[a]
cis-1,4-Polybutadiene (90–98%)	None[b]
cis-1,4-Polyisoprene (90–98%)	Li-alkyl (ca 94% cis-1,4) nature (*Hevea*)
Ethylene–propylene random copolymers	None
trans-1,5-Polypentenamer	None

[a] Some of these polymers can be made in low yields with metal alkyl-free catalysts, but these are not yet sufficiently active for commercial production.

[b] Commercial scale synthesis may, however, become possible with some of metal alkyl-free catalysts in the near future if the reported high activities also prevail under scaled-up conditions.

A. PLASTICS

Table 3–9 (34a) compares the more important properties of polymers used in plastics application. Unquestionably, the most important plastic polymer to emerge from the Ziegler–Natta discoveries is isotactic polypropylene. By 1973 there were nine United States producers of this polymer, accounting for about 2.13 billion pounds of polypropylene. This was an increase of almost a billion pounds over the production reported in the

TABLE 3–9

Properties of Polyolefin Plastics[a]

| Property | Polyethylene | | |
	Low density	Medium density	High density
Melting point (°C)	108	125	135
Compression molding temp. (°C)	275–350	300–375	300–450
Injection molding temp. (°C)	300–500	300–500	300–600
Density (g/cm³)	0.910–0.925	0.926–0.940	0.941–0.965
Tensile strength (psi)	600–2300	1200–3500	3100–5500
Elongation (%)	40–800	50–600	20–1000
Tensile modulus (10^5 psi)	0.14–0.38	0.25–0.55	0.6–1.8
Impact strength (ft lb)/in (notched)	No break	0.5–16	0.5–20
Deflection temp. (°F)[b]			
264 psi fiber stress	90–105	105–120	110–130
66 psi fiber stress	100–121	120–165	140–190
Hardness, Rockwell	D41–D46 (Shore) R10	D50–D60 (Shore) R15	D60–D70 (Shore) —

[a] Data compiled from "Modern Plastics Encyclopedia" (34a) and other sources.

[b] Also known as heat distribution distortion temperature.

United States in 1971. The United States producers, who account for about one third of the world production of this polymer, were expected to have an annual capacity of 2.82 billion pounds by 1975.

High density polyethylene is next in production volume, but the Ziegler–Natta processes have to compete with the Phillips and Standard of Indiana processes for the world markets. The other polyolefins are produced in much lower volumes.

Most of these polymers are sold in different grades, each of which has been synthesized and blended with additives with a specific end use in mind.

A comprehensive review by Vandenberg and Repka (34b) contains excellent accounts of the characterization, fabrication properties, and uses of the major commercial plastics. It is highly recommended to the reader whose interests lie in this aspect of Ziegler–Natta polymerizations.

1. Polyethylene (35–37)

The main structural features that determine the properties of polyethylene are percent crystallinity, average molecular weight, and molecular weight distribution (35).

There are six properties that make polyethylene attractive: (1) toughness, (2) excellent chemical resistance to solvents, acids, and alkalis, (3) good

Polypropylene		Poly-1-butene	Poly-4-methyl-1-pentene	Polyallomer	trans-1,4-Polyisoprene
Unmodified	Rubber-filled				
165–170	—	132–138	235	—	145
340–450	300–450	300–350	540–555	—	—
400–550	400–500	290–380	500–570	430–445	—
0.901–0.906	0.90–0.91	0.910–0.915	0.83	0.896–0.899	0.95
4300–5500	2800–4400	3300–4400	4000	3050–3850	4700
200–700	350–500	300–380	15	400–500	—
1.6–2.25	1.0–1.7	0.26	1.6	—	—
0.5–2.0	1.0–15				—
(73°F)	(73°F)	No break	0.8	1.7C–6.0P	—
125–140	120–135	130–140	—	124–133	—
200–230	160–200	215–235	—	165–192	—
R85–110	R50–R85	65 (Shore D)	L67–74	R50–85	Shore D 73–78

barrier to gases, (4) outstanding dielectric characteristics, (5) adaptability to various fabrication techniques, and (6) facile adjustment of properties by tailoring branching and molecular weight and by incorporating additives.

As crystallinity is increased (decreasing the degree of branching), the following properties increase: rigidity, tensile strength, hardness, chemical resistance, and opacity. Permeability to gases simultaneously decreases.

In general, high molecular weight polyethylenes have better properties than do low molecular weight polymers, such as greater impact resistance (toughness) and chemical resistance. Processing, however, becomes more difficult as their molecular weight increases. Polyethylenes that have a wide molecular weight distribution are easier to process, which is important in molding and extrusion processes. In contrast, polymers that have a narrow molecular weight distribution have a higher impact resistance, better low temperature toughness, and better resistance to environmental stress cracking.

For many applications, polyethylenes must be stabilized against oxidation and ultraviolet light initiated degradation (antioxidants and carbon black have been added). Polyethylenes offer great flexibility and versatility through the ease with which they can be blended and modified by using fillers. For example, carbon black, glass fibers, and clay act as reinforcing fillers to improve tensile strength and rigidity. High density polyethylene (HDPE) can be foam molded into thick wall structures containing a gas filler.

By means of irradiation with high energy electron beams or chemical reagents, cross-linked (nonfusible) polyethylene products are formed.

The excellent balance between cost and end-use properties accounts for the large volume uses of high density polyethylene. The largest use of HDPE is in blow molding, by which bottles and containers for bleaches, liquid detergents, milk, and antifreeze are made (rigidity, toughness, and chemical resistance are required). High density polyethylene is used for injection molding for such items as pails, lids, seats, dish pans, and waste receptacles. Its excellent electrical properties have been applied in insulation and in the jacketing of wire. Sometimes the actual volume used is very small, yet the use is very important. According to a recent report (37a), thousands of victims of osteoarthritis are being helped by an operation in which the ball and socket of the hip joint is replaced with a high density polyethylene and a metal substitute.

2. Isotactic Polypropylene (37, 38–40a)

The main structural features that determine the properties of polypropylene include the percent isotactic crystallinity, molecular weight, and a very high degree of head-to-tail 1,2-addition.

The properties that make isotactic polypropylene attractive include: (1) high melting temperature (ca. 170°C); (2) ability to withstand heat sterilization without deformation; (3) ability to withstand constant exposure as high as 225°F under light or no-load conditions; (4) high stiffness and high tensile strength; (5) low density; (6) excellent chemical resistance; (7) excellent stress-cracking resistance; (8) excellent electrical properties.

Isotactic polypropylene processes well in both injection molding and extrusion operations. It can be molded into thin walled parts, such as disposable cups and food containers, at a faster rate than any other thermoplastic. In heavier walled items, such as bread trays and tote boxes, warpage has been practically eliminated. Polypropylene has been molded in a wide range of pastel, dark, and metallic colors. Molded parts have been decorated by hot stamping, silk screening, and electroplating methods.

Isotactic polypropylene has several deficiencies. It must be stabilized against oxidation and degradation by ultraviolet radiation. For this reason, the commercial products contain one or more additives. A second deficiency is the pronounced brittleness at temperatures near or below 0°C. This deficiency was solved by blending ethylene–propylene elastomers into the polymer (see Section V,B,4).

The cited properties have enabled isotactic polypropylene to capture a wide range of markets, including automotive industry, hospital ware, luggage, carpeting, upholstery fabrics, cordage and tying twine, and film. Most commercial producers make a variety of grades available, each of which is directed to a specific end use. Isotactic polypropylene has also found many novel uses, including artificial seaweeds.

Polypropylene markets break down as follows: injection molding, 35%; fibers, 25%; film, 8%; all others (blow molding, extrusion, pipe, etc.), 32% (40). Single items such as injection molded battery cases consumed about 30 to 40 million pounds in 1972 and expectations are high, about 100 million pounds in 1976 (40). The prediction was made that, by 1980, as much as 45 pounds per car of polypropylene could be used in assembly line automobiles. This contrasts with about 23 pounds per car in 1973.

3. Isotactic Poly-1-butene (41–49)

Because it can exist in four different crystalline modifications, isotactic poly-1-butene has been of considerable scientific interest since about 1957. By the early 1960's, there were indications that it was going to take its place after HDPE and isotactic polypropylene as the next largest volume polyolefin. In fact, a projection was made in 1973 that if the price of poly-1-butene could be lowered to the levels of isotactic polypropylene, it could reach the billion pound level (41).

Poly-1-butene has a unique combination of properties that differentiate it from HDPE and isotactic polypropylene. It crystallizes quickly from a melt into an unstable form (modification 2), which then is slowly converted to a stable form (modification 1). When a tensile bar is pulled in an Instron, the sample elongates with increasing applied stress after the yield point is reached. Below the tensile yield point, poly-1-butene has a high resistance to cold flow (does not creep under pressure). The tensile and yield strengths of poly-1-butene film are approximately $3\frac{1}{4}$ to 4 times higher at 70°C in comparison to polyethylene films. Poly-1-butene films are said to be unusually tough and have high tensile strength and high tear resistance. Another favorable property is its resistance to stress cracking.

These properties have encouraged its use in the manufacture of film and pipe.

4. Poly-4-methyl-1-pentene (50, 51)

Because its melting point is 70°C higher than that of isotactic polypropylene, isotactic poly-4-methyl-1-pentene has received attention for higher temperature applications. Because of its inherent higher cost, however, it has been reserved for special applications rather than replacement of the less expensive polyolefins. Commercial products are actually copolymers containing small amounts of 1-pentene comonomer.

The balance of properties that differentiate these copolymers from the lower polyolefins are higher melting point, high transparency, low density, and high dielectric properties. Like the other polyolefins, it has good resistance to chemical reagents, reasonable stiffness, tensile strength, and toughness.

Some applications examined for these 4-methyl-1-pentene copolymer plastics are electrical and lighting, medical items such as syringes, and packaging materials.

5. Block Copolymers of Propylene and Ethylene (52)

These products have become available by Eastman Kodak under the trade name Polyallomers. They have a density of about 0.876 g/cm^3, and the polymer chain contains segments of each olefin. The polymer shows crystallinity characteristics of isotactic polypropylene and linear polyethylene.

They are said to have good flow properties, high impact strength (1.7 to 6 ft/lb), excellent molding ability, and low susceptibility to blush marks.

A range of products of varying stiffness and impact resistance has been synthesized by control of synthesis parameters. Suggested applications include packages for frozen food and heat-sterilizable containers, automotive trim parts, toys, appliance parts, blow-molded bottles, and high clarity film.

6. *trans*-1,4-Polyisoprene (53–55)

This polymer has a melting point of 145°C. Some of its more favorable properties are high hardness values, high tensile strength, and excellent resilience and abrasion resistance. It has properties of both rubber and plastic. It can be molded like a plastic and vulcanized like a rubber with conventional cross-linking agents.

A major application is in the manufacture of golf ball covers. Also, *trans*-1,4-polyisoprene has found application in orthopedic and rehabilitation medicine, such as construction of spinal braces, arch supports, and splints and sockets for artificial limbs (54). Its properties allow it to be molded directly against the skin. It does not require a model or replica of the affected part of the patient.

B. ELASTOMERS (55a)

Table 3–10 compares some of the more important properties of elastomers that have been vulcanized, reinforced, and compounded. In describing and comparing the properties of these commercial rubbers, one should remember that their properties are dependent not only on the properties of the neat elastomer used but also on the method of vulcanization, reinforcement, and compounding. Each manufacturer has various grades intended for different end uses, each with its special set of properties. A listing and description of the various grades can be found in *Rubber World* (55a).

One of the most important properties of elastomers is their second-order transition temperature, frequently called glass transition temperature (T_g). In order for the elastomer to be useful at low temperature, its T_g value must be low. Table 3–11 compares the T_g values of some of the most important elastomers.

1. *cis*-1,4-Polyisoprene (55, 55a)

This polymer has properties comparable to natural rubber (*Hevea*). High cis-1,4-polymer has been synthesized via LiR initiators (gives 92 to 94% cis-1,4) and via Ziegler–Natta catalysts (gives 96 to 98% cis-1,4). Its outstanding properties are tear resistance and elasticity. Some of the products made from this elastomer include: tires, mechanical goods, injection-molded items, boots, footwear, adhesives, coated fabrics, and rubber bands.

2. *cis*-1,4-Polybutadiene (55–57)

Commercially produced polymer having 96 to 99% cis-1,4 content have been produced with Ziegler–Natta catalysts. The largest use of this elastomer has been in the passenger car and truck tire (retreads). It is usually blended

TABLE 3-10

Properties of Various Elastomers in Their Typical Fields of Application (55a)

Classification: Type of rubber:	Natural rubber	Styrene–butadiene rubber	Non oil-resistant		
			Polybutadiene rubber	Butyl rubber	Ethylene–propylene copolymer C-23
Spec. gravity	0.93	0.94	0.94	0.93	0.85–0.86
Tensile strength, vulcanized, unreinforced (kg/cm^2)	230	30	20–70	180	120[a]
Tensile strength, vulcanized, reinforced (kg/cm^2)	300	250	110–200	210	250–280
Elongation (%)	600	450	300–600	600	450–500
Wear resistance	Very good	Outstanding	Very good	Good	Good[b]
Tear resistance	Outstanding	Fair	Fair	Good	—
Elasticity	Outstanding	Good	Outstanding	Very poor	Outstanding[c]
Crack resistance	Good	Good	Good	Very good	
Maximum service temperature range	−60° to 160°C	−40° to 170°C	−90°C	−30° to 190°C	−35° to 200°C[d]
Light resistance	Poor	Fair	Poor	Outstanding	Very good
Oxidation resistance	Fair	Good	Very good	Very good	Very good
Oil resistance	Very poor	Very poor	Very poor	Very poor	Poor[e]
Swelling resistance in contact with aliphatic, aromatic, or halogenated solvents	Very poor	Very poor	Very poor	Very poor	Very poor
Chemical resistance in contact with					
Water	Good	Very good	—	Very good	Good
Alkali	Good	Good	—	Very good	Good
Acid	Good	Good	—	Very good	Good

[a] According to Montecatini, Great Britain Patent 563,834 vulcanized with benzoyl peroxide, maleic anhydride, and zinc oxide.
[b] According to Montecatini.
[c] Determined at Chemische Werke Hüls A.G.
[d] Estimated; exact data not available.
[e] According to Montecatini; good resistance to hydraulic liquids.

TABLE 3–11

Second-Order Transition Temperatures (T_g) of
Some Elastomers[a]

Elastomer	T_g (°C)
trans-1,4-Polychloroprene	−45
Ethylene–propylene copolymer	−50
Styrene–butadiene rubber	−65
Natural rubber	−70
Butyl rubber	−70
Propyleneoxide rubber	−75
trans-Polypentenamer	−95
cis-1,4-Polybutadiene	−115
Silicon rubber	−120
cis-Polypentenamer	−140

[a] Taken from G. Natta and G. Dall'Asta, *in* "Polymer Chemistry of Synthetic Elastomers, Part 2," edited by J. P. Kennedy and G. M. Törnqvist, Wiley Interscience, New York, 1969, p. 713.

with other elastomers. Some applications have been in the manufacture of industrial conveyer belts, wire and cable insulation, and shoe soles and heels.

cis-1,4-Polybutadiene has certain advantages over the other elastomers in tire applications. Increased tread life is one of the most important. One manufacturer stated that tests showed that tire wear improved almost 1% for each percent of polybutadiene blended with conventional styrene–butadiene rubber in the tread (56). Good cold weather performance of polybutadiene allowed its use in snow tires.

Garlanda (57) found that the properties of carbon black reinforced *cis*-1,4-polybutadiene were between those of styrene–butadiene elastomer and natural rubber. The elastomer had excellent abrasion resistance, and its elastic properties were superior to those of natural rubber, especially at low temperatures. The polymer accepts high filler and oil loadings.

3. *trans*-1,5-Polypentenamer (58–60)

This polymer, the latest to be developed, is synthesized from cyclopentene by a ring-opening mechanism that is strictly different than that by which propylene and other α-olefins polymerize (see Chapter 19, Section V,I).

This rubber is said to offer the desirable processing properties of natural rubber and the abrasion and aging resistance of various synthetic rubbers.

It has good low temperature properties and is inexpensive to compound. Small amounts of cross-linking agents are said to produce a high cross-linking yield.

Polypentenamer has been referred to as a general purpose synthetic rubber which will be suitable for tire manufacture (59, 60). However, it has not yet become commercially available, although it was scheduled at one time to make its debut by early 1974 (60).

4. Random Copolymers and Terpolymers of Ethylene and Propylene (55a)

Two different applications have been made:

(1) The copolymer (EPM, ethylene–propylene copolymer) has been blended into highly isotactic polypropylene in order to toughen it, that is, to increase its impact strength. Polypropylene has poor service performance near $0°C$ (refrigerator temperature). Because it becomes brittle, it cannot be used for many applications that require refrigeration, such as milk bottles. Addition of 5 to 20% of an ethylene–propylene elastomer ($T_g = -50°C$) overcomes this deficiency of isotactic polypropylene. Various grades of rubber-modified polypropylene are sold commercially. Another application is for bumper guards on automobiles.

(2) For applications requiring vulcanization, a nonconjugated diene is also copolymerized with ethylene and propylene to form EPDM (ethylene–propylene-diene "monomer" terpolymer) elastomers. The unreactive double bond acts as a promoter of cross-linking during the vulcanization process. Because these vulcanizates were anticipated to be inexpensive and because the largely hydrocarbon elastomer was shown to have excellent oxygen and ozone resistance, much interest was generated among rubber manufacturers. The mechanical properties are said to be between natural rubber and synthetic styrene–butadiene rubber (SBR). There was great expectation that EPDM rubbers would be used to manufacture tires and tire products. However, many problems were encountered (for example, it lacked building tack), and, with other suitable elastomers on hand, it lost the early battle. The consumption of EPDM elastomers rose from about several thousand long tons in 1963 to over 95,000 long tons in 1973.

The largest application of EPDM rubbers has been in the automotive field, for manufacture of sponges, seals, hose, weather strips, brake components, side walls of tires, and bumpers.

Other uses of EPDM rubbers take advantage of the chemical, heat, and abrasion resistance. These uses include wire and cable and many types of rubber products (molded and extrusion applications).

An excellent and comprehensive review on the synthesis, characterization, and use of ethylene–propylene-based rubbers has been recently published by Baldwin and Ver Strate (23c).

References*

1. N. G. Gaylord and H. F. Mark, "Linear and Stereoregular Addition Polymers." Wiley (Interscience), New York, 1959.
2. Great Britain Patent 884,583, December 13, 1961, Farbwerke Hoechst A.G.
3. M. Gippin, *Ind. Eng. Chem., Prod. Res. Dev.* **4**, 160 (1965); *Rubber Chem. Technol.* **39**, 508 (1966); see also Chapter 5, Marconi *et al.* (12).
4. G. Natta, P. Pino, G. Mazzanti, and R. Lanzo, *Chim. Ind. (Milan)* **39**, 1032 (1957).
5. D. S. Breslow and N. R. Newburg, *J. Am. Chem. Soc.* **79**, 5073 (1957).
6. S. E. Horne, Jr., C. F. Gibbs, V. L. Folt, and E. J. Carlson, Belgian Patent 543,292, June 2, 1956, Goodrich-Gulf Chemicals, Inc. (equivalent to U.S. Patent 3,114,743 December 17, 1963).
7. G. Natta, I. Pasquon, and A. Zambelli, *J. Am. Chem. Soc.* **84**, 1488 (1962).
8. S. Cucinella, A. Mazzei, W. Marconi, and C. Busetto, *J. Macromol. Sci. Chem.* **7**, 1549 (1970).
8a. G. Natta, U. Giannini, E. Pellino, and D. DeLuca, *Eur. Polym. J.* **3**, 391 (1967).
9. K. Ziegler, E. Holzkamp, H. Breil, and H. Martin, *Angew. Chem.* **67**, 541 (1955).
10. G. Natta and G. Mazzanti, *Tetrahedron* **8**, 86 (1960).
11. G. Natta and L. Porri *in* "Polymer Chemistry of Synthetic Elastomers, Pt. 2," edited by J. P. Kennedy and G. M. Törnqvist, Wiley (Interscience), New York, 1969 p. 597.
12. E. Cernia, C. Mancini, and G. Montaudo, *J. Polym. Sci., Part B* **1**, 371 (1963).
13. G. Natta, *Angew. Chem.* **68**, 393 (1956).
14. G. Natta and P. Corradini, *Chim. Ind. (Milan)* **45**, 299 (1963).
15. J. Boor, *Macromol. Rev.* **2**, 115–268 (1967).
16. I. Pasquon, *Chim. Ind. (Milan)* **41**, No. 6, 534 (1959).
17. C. A. Russell, *J. Appl. Polym. Sci.* **4**, 219 (1960).
18. J. P. Luongo, *J. Appl. Polym. Sci.* **3**, 302 (1960).
19. R. G. Quynn, J. L. Riley, D. A. Young, and H. D. Noether, *J. Appl. Polym. Sci.* **2**, 166 (1959).
20. E. A. Youngman and J. Boor, *Macromol. Rev.* **2**, 33–69 (1967).
20a. A. Weidinger and P. H. Hermans, *Makromol. Chem.* **50**, 98 (1961).
21. G. Natta, P. Corradini, and M. Cesari, *Atti Accad. Naz. Lincei., Cl. Sci. Fis., Mat. Nat., Rend.* [8] **22**, 11 (1957).
21a. P. Corradini, E. Martuscelli, and M. A. Martynov, *Makromol. Chem.* **108**, 285 (1967).
22. C. A. Reilly, Shell Development Co., private communication.
23. J. A. Faucher and F. P. Reding, *in* "Crystalline Olefin Polymers" (R. Raff and K. W. Doak, eds.), p. 697. Wiley (Interscience), New York, 1965.
23a. F. Danusso, *Polymer* **8**, 281 (1967).
23b. G. Goldbach and G. Peitscher, *J. Polym. Sci., Part B* **6**, 783 (1968).
23c. F. P. Baldwin and G. Ver Strate, *Rubber Chem. Technol.* **45**, No. 3, 709–881 (1972).
23d. C. Tosi and F. Ciampelli, *Adv. Polym. Sci.* **12**, 88–125 (1973).
23e. Y. Kissin, *Adv. Polym. Sci.* **12**, 92–151 (1973).
24. G. Natta and L. Porri *in* "Polymer Chemistry of Synthetic Elastomers, Pt. 2," edited by J. P. Kennedy and G. M. Törnqvist, Wiley (Interscience), New York, 1969 p. 597.
25. M. W. Duch and D. M. Grant, *Macromolecules* **3**, 165 (1970).
26. D. Morero, A. Santambrogio, L. Porri, and F. Ciampelli, *Chim. Ind. (Milan)* **41**, 758 (1959).
27. A. D. H. Clague, J. A. M. Van Broekhoven, and J. W. de Haan, *J. Polym. Sci., Polym. Lett. Ed.* **11**, 299 (1973).

* The dates cited in the patent references throughout this volume are disclosure or publication dates.

28. For comprehensive references and discussions of the experimental data on which this was based, see V. D. Mochel, *J. Macromol. Sci., Rev. Macromol. Chem.* **8**, 289–347 (1972); see also *J. Polym. Sci., Polym. Chem. Ed.* **10**, 1009 (1972).
29. F. Conti, A. Segré, P. Pini, and L. Porri, *Polymer* **15**, 5 (1974).
30. Y. Alaki, T. Yoshimoto, M. Imanari, and M. Takeuchi, *Rubber Chem. Technol.* **46**, 350 (1973).
31. J. Boor, *J. Polym. Sci., Part C* **1**, 237 (1963).
32. H. M. Schnecko, M. Reinmoller, K. Weirauch, W. Lintz, and K. Kern, *Makromol. Chem.* **69**, 105 (1963).
32a. M. N. Berger and B. M. Grieveson, *Makromol. Chem.* **83**, 80 (1965).
32b. E. Kohn, H. J. L. Schuurmans, J. V. Cavender, and R. A. Mendelson, *J. Polym. Sci.* **58**, 681 (1962).
32c. H. W. Coover, J. E. Guillet, R. L. Combs, and F. B. Joyner, *J. Polym. Sci., Part A-1* **4**, 2583 (1966).
32d. I. H. Anderson, G. M. Burnett, and P. J. T. Tait, *J. Polym. Sci.* **56**, 391 (1962).
32e. D. R. Burfield and P. J. Tait, *Polymer* **13**, 315 (1972).
33. G. Natta, *Chim. Ind. (Milan)* **39**, 993 (1957).
34. K. Wisseroth, H. Mohr, L. Reuter, and R. Scholl, Great Britain Patent 1,006,939, October 6, 1965, BASF.
34a. *Mod. Plast. Ency.* **49**, No. 2, 14 (1972); *Chem. & Eng. News* **51**, No. 27, 5 (1973).
34b. E. J. Vandenberg and B. C. Repka, "Ziegler type polymerizations, *in* Polymerization Processes," edited by C. Schildknecht and I. Skeist, Wiley, 1977, p. 337–423. Wiley (Interscience), New York, 1975.
35. R. B. Staub, *Mod. Plast. Ency.* **47**, No. 10A, (1970–1971).
36. J. A. Brydson, "Plastics Materials," Chapters 7 and 8, pp. 98–132 and 133–155 resp. Iliffe, London, 1966.
37. F. W. Billmeyer, "Textbook of Polymer Science," Chapters 13 and 14. Wiley (Interscience), New York, 1965.
37a. *Newsweek* p. 82, November 25 (1974).
38. M. Compostella, *in* "The Stereochemistry of Macromolecules" (A. D. Ketley, ed.), Chapter 6. Dekker, New York, 1967.
39. P. Junghans, *Mod. Plast. Ency.* Vol. **47**, No. 10A, 213, (1970–1971).
40. *Mod. Plast.* **50**, No. 8, 12 (1973).
40a. *Mod. Plast.* **50**, No. 12, 96 (1973).
41. "Tough, new polybutylene gets its chance," *Business Week* February 24, 1973.
42. C. L. Rohn and H. G. Tinger, *Mod. Plast.* **48**, No. 3, 78–87 (1971).
43. T. Yamamoto, *Sekiyu To Sekiyu Kagaku* **15**, 69–74 (1971).
44. B. Frost-Smith, *Chem. Process. (London)* Oct. 1970, 13–17 (1970).
45. R. C. Lindengren, *Polym. Eng. Sci.* **10**, 163 (1970).
46. A. J. Foglia, *Appl. Polym. Symp.* **11**, 1–18 (1969).
47. R. C. Lindengren, *Plast. Packag., Natl. Tech. Conf., Soc. Plast. Eng., 1969* 108–15 (1969).
48. "New Thermoplastic Resin," *Mod. Plast. Ency.* **43**, No. 2, 94–95 and 175 (1965).
49. I. D. Rubin, "Poly(1-butene)—Its Preparation and Properties," Gordon and Breach, New York, 1968.
50. H. C. Raine, *Appl. Poly. Symp.* **11**, 39–45 (1969).
51. J. Westall, *Mod. Plast. Ency.* **47**, No. 10A, 152 (1970–1971).
52. W. P. Gideon, *Mod. Plast. Ency.* **47**, No. 10A, 175 (1970–1971).
53. *Chem. & Eng. News*, April 11, 1960, 37 (1960).
54. R. H. Jones and Y. K. Wei, *Am. Chem. Soc., Div. Org. Coat. Plast. Chem. Pap.* **30**, No. 1, 10–17 (1970).
55. P. Baumann, *Chem. Ind. (London)* 1498 (1959).

55a. J. V. Del Gatto and S. R. Hague, eds., "Rubber World: Materials and Compounding Ingredients for Rubber." Automotive Publications, Inc., New York, 1970.
56. "Polybutadiene Gets off to Fast Start," *Chem. & Eng. News* **40**, No. 6, 32 (1962).
57. T. Garlanda, *Chim. Ind. (Milan)* **43**, 368 (1961).
58. "Engineering Applications of Thermoplastics," *Polym. News* **1**, Nos. 11–12, 23 (1974).
59. *Chem. & Eng. News*, Sept. 27, 1971, 25 (1971).
60. *Chem. Eng.* (N.Y.) **78**, No. 2, 25 (1971).

Chemical Description of
Ziegler–Natta Catalysts
for Olefins

I. Introduction

Many binary combinations of metal alkyls and transition metal salts have been found active and stereoregulating for polymerization of ethylene and α-olefins. The polymerization behavior of any given catalyst, however, depends on how the two particular components interact before and after the addition of monomer (for earlier reviews, Refs. 1 to 8 are recommended). There is no such thing as a universally "most stereoregulating" or "most active" metal alkyl or transition metal salt. Of the many combinations of components examined, zinc and aluminum alkyls and titanium and vanadium transition metal salts have led to the most attractive catalysts. For example, in the polymerization of propylene to isotactic polymer, VCl_3 combined with $AlEt_3$ produces a more isotactic-specific catalyst than when it is combined with $AlEt_2Cl$. In contrast, when $TiCl_3$ is used, $AlEt_2Cl$ produces the more specific binary catalyst (9, 10).

Besides the choice of metal for each component, other factors such as the nature of the ligand can be significant, and it is the sensitivity of the final catalyst to the interplay of these factors that makes these catalysts so versatile, and yet so difficult to elucidate. It is not possible, however, to make broad generalizations and extensions in going from one type of catalyst to another or from one type of monomer to another.

This chapter will describe the chemical nature of the catalyst, that is, the choice of metal alkyls and transition metal salts of different chemical composition for the polymerization of olefins. Chapter 5 will do the same for dienes. Particular emphasis will be placed on the experimental factors that affect the behavior of the catalysts in varying degrees, and then on selected catalysts that appear to be the most important.

As will be discussed in subsequent chapters, the metal alkyl and transition metal salt have a specific function. Alkylation of the transition metal salt by the base metal alkyl leads to the formation of active centers. The new transition metal–carbon bonds are these active centers where continuous insertion of monomer into the polarized metal–carbon bond constitutes the growth step of the polymerization. The additional role which the metal alkyl can play will also be elaborated upon later.

II. Factors Determining Behavior of Catalysts for Olefins

By behavior we mean the activity, stereoregulating ability, and tendency to form random, alternating, or blocky copolymers. The factors described are related to the chemical structure of the metal alkyl and the transition metal salt.

A. THE METAL ALKYL

1. Group of Metal

Not every metal in groups I to III forms an active metal alkyl as is claimed in many of the patents relating to Ziegler–Natta catalysts. Active catalysts from the following metals have been established for ethylene and/or propylene.

Group I	Group II	Group III
Li	Be	Al
Na	Mg	Ga
K	Zn	
	Cd	

Tables 4–1 to 4–3 (9, 11–25) show selected examples of typical catalysts for group I, group II, and group III metal alkyls. Of these, aluminum alkyls have been the most extensively used, and the reason for this is partly scientific and partly economic. By the end of the 1950's, aluminum alkyls such as $AlEt_3$ and $Al\text{-}i\text{-}Bu_3$, became available in research and developmental quantities in the pure state or in hydrocarbon solutions. Facile synthetic methods were developed using Al, H_2, and olefin. They were much safer to use in solution because, once properly diluted, they were less pyrophoric or not flammable unless contacted with a combustible material. Simple syringing techniques, using N_2 or A as an inert atmosphere, became widely

TABLE 4–1

Active Catalysts from Group I Metal Alkyls

Catalyst	Olefin	Product	Ref.
Isoamyl Li + TiCl$_4$	Ethylene	$T_m = 125°$ to $135°$C	11
Amyl Na + TiCl$_4$[a]	Propylene	10 to 80% isotactic[a]	12
n-BuLi + TiCl$_4$[b]	Propylene	10 to 60% isotactic[a]	12
n-Octyl Na + TiCl$_4$[c]	Propylene	37 to 51% isotactic[a]	13
Octyl Na, TiCl$_4$ + dimethyl ether of glycol	Propylene	79.1% isotactic[a]	13
n-BuK + TiCl$_4$[d,e]	Ethylene	—	14
Amyl Na + TiCl$_4$[b]	Ethylene	—	15
(CH$_2$)$_5$Li$_2$ + TiCl$_4$	Ethylene	Linear, high MW	16
Octyl Na + TiCl$_3$	Propylene	Isotactic $T_m = 165°$ to $170°$C[f]	17

[a] As measured approximately by insolubility in boiling hydrocarbon solvent.

[b] Highest isotactic values obtained when LiR/Ti > 1.8 to 4.5 and when NaR/Ti < 4.4.

[c] Highest yields were obtained at 10° to 20°C and at a Na/Ti ratio of 5.

[d] Optimum activities obtained when BuK/TiCl$_4$ was between 2.8 and 6.1 and for amyl K/TiCl$_4$, between 1.1 and 3.7.

[e] Inactive for propylene.

[f] 12.5% removed by ether extraction compared to 10.9% from product prepared with AlEt$_3$–TiCl$_3$.

TABLE 4–2

Active Catalysts from Group II Metal Alkyls

Catalyst	Olefin	Product	Ref.
R$_2$Cd or RCdI + TiCl$_4$[a]	Ethylene	Linear	18
n-Pr$_2$Cd, n-PrCdX(X = Cl, I)[b] + TiCl$_4$	Ethylene	Linear	19
Et$_2$Cd + TiCl$_4$	Propylene	MW = 355,000 % crystalline = 74	20
BeEt$_2$ + TiCl$_3$	Propylene	$[\eta]$ = 3.1 % Isotactic = 96 T_m = 174°C	21
ZnEt$_2$ + αTiCl$_3$	Propylene	58 to 65.6% Crystallinic	22
ZnEt$_2$ + β, α, and δTiCl$_3$ · xAlCl$_3$	Propylene	60 to 95% Isotactic	9

[a] R = Me, Et, Bu, and phenyl.

[b] Prepared and used in polymerization at −60° to −50°C.

TABLE 4–3

Active Catalysts from Group III Metal Alkyls

Catalyst	Olefin	Product	Ref.
$LiAlH_4$ or $NaAlH_4$ + $AlTi_2Cl_8$	Propylene	Isotactic	22a
$AlEt_3$ + $TiCl_3 \cdot xAlCl_3$	Propylene	Isotactic	23
$GaEt_3$ + $TiCl_3$	Propylene	Isotactic	24
GaR_3 + $TiCl_4$	Ethylene	Linear	25

used to dispense metal alkyl solutions. Eventually, many different aluminum alkyls became commercially available, and with modest precaution they could be used safely and routinely in research work by persons with little experience.

The most important impetus, however, was discovering that all of the desired stereoregular polymers could be made by a proper selection of an aluminum alkyl and a transition metal salt. The marriage of these factors has probably made aluminum alkyls the choice metal alkyl component in most, if not all, commercial processes. The most widely used aluminum alkyls have been $AlEt_3$, $Al\text{-}i\text{-}Bu_3$, $AlEt_2Cl$, $AlEtCl_2$, and $AlEt_2OR$. Binary complexes of lithium alkyls and aluminum alkyls, such as $LiAlR_4$, have also been explored.

Of the other metals of group III, only gallium is attractive, but its high cost has prevented extensive examination. Tanaka and co-workers have found that the catalytic activity of GaR_3–$TiCl_4$ catalysts in the low pressure polymerization of ethylene was lower compared with catalysts containing analogous aluminum alkyls (25). The order: $GaPr_3 > GaMe_3 > GaEt_3$ was found. There is very little data on metal alkyls of indium and thalium to draw clear-cut conclusions about their relative effectiveness.

In general, attempts to form active catalysts with BR_3 type alkyls and titanium chloride have been largely unsuccessful. A claim is made that a mixture of triethylamineborazane ($BH_3 \cdot NEt_3$) and $TiCl_4$ polymerized ethylene to a high molecular weight (MW $\simeq$ 200,000) and high melting (T $\simeq$ 130°C) polymer (26). Sometimes, the boron hydride is present in a lithium derivative, such as $LiBH_4$, and mixtures of the latter and Ti, Zr, or V chlorides have been reported to form solid crystalline polymers from propylene and 1-butene (27). The author is not aware of examples where mercury alkyls formed active catalysts.

Of the group I metals, metal alkyls from lithium, sodium, and potassium have received the most attention. Relative to group II and III metal alkyls, these are more ionic and, with the exception of some lithium alkyls, they have only limited solubility in the hydrocarbon solvents normally used in olefin polymerizations (28).

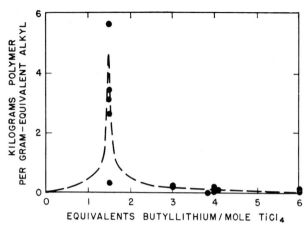

Fig. 4–1. Polymerization of ethylene with ethylene absent during interaction (29).

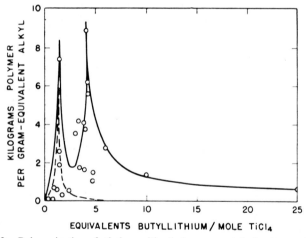

Fig. 4–2. Polymerization of ethylene with ethylene present during interaction (29).

The activity of the n-BuLi–TiCl$_4$ catalyst for polymerization of ethylene was found very sensitive to Li/Ti ratio and presence of ethylene when the catalyst components were mixed (Figs. 4–1 and 4–2) (29). This sensitivity was later found for many other combinations of metal alkyl and transition metal salts, especially when the latter was added in its highest valence state.

Of the group II metal alkyls, those based on Zn have been most widely examined, and examples will be presented throughout this book (see Tables 4–13 and 4–14). If it were not for their toxicity, beryllium alkyls probably would have been more extensively investigated. In combination with α-, γ-, or δTiCl$_3$, BeEt$_2$ formed highly isotactic-specific catalysts (see Table 4–16).

Some metal alkyls that fall outside of the group I to III patent definition form active catalysts for ethylene and occasionally for propylene. An alkylated transition metal, such as CH_3TiCl_3 or $Cp_2Ti(CH_3)_2$ (where Cp = Cyclopentadienyl), forms a stereoregulating catalyst for propylene when combined with $TiCl_3$. Also, $SiEt_4$ and $SnEt_4$ combined with $TiCl_4$, polymerize ethylene to high molecular weight linear polymer. These related catalysts, which do not contain added base metal alkyls of group I to III metals, are described separately in Chapter 12 under the designation of Metal Alkyl-Free Catalysts. Catalysts formed by mixing sodium or aluminum metal with $AlCl_3$ and $TiCl_3$ (or $TiCl_4$) will also be treated in that chapter.

2. Ligands

Tables 4–4 through 4–8 present some examples of polymerizations in which the structure of the metal alkyl was varied. Two types of ligands are compared: (1) an all-hydrocarbon alkyl or an aryl, such as ethyl or phenyl; and (2) a heteroatom or a radical containing a heteroatom, such as Cl or $-OC_6H_5$.

Both of these types of ligands have the potential of undergoing exchange reactions with the ligands of the transition metal. When the alkyl or aryl group is exchanged, a transition metal–carbon bond is formed. As shown in Chapter 13, the active center is a transition metal–carbon bond. Should the heteroatom be exchanged for a ligand of the transition metal, an active center is not formed because the transition metal–heteroatom bond is not active. However, the electronic and steric environment of the metal atom is altered. If this metal atom later becomes alkylated, then this center will probably show a different stereoregulating ability relative to the other centers not bearing the heteroatom.

The chemical behavior of a metal alkyl is strongly influenced by the nature and number of these two types of ligands. The following tabulation illustrates the variation in ligand structure in these catalysts:

Hydrocarbon	Halogen- or alkyl-bearing heteroatom
Saturated aliphatic R (Me, Et, Octyl)	Halogen (Cl, Br, I, F)
Cyclic R (cyclohexyl)	OC_6H_5, OR (R = alkyl, aryl)
Aromatic R (phenyl)	SC_6H_5
Unsaturated aliphatic R (isopropenyl)	SeC_6H_5
(see Table 4–8 for exotic metal alkyls)	NC_5H_{10}, NR (R = alkyl, aryl)

The changes in catalyst behavior due to changes in type of ligand are now illustrated.

a. Size of Alkyl Group in Metal Alkyl.   The preferred metal alkyls possess ethyl and isobutyl ligands. Typical examples are $AlEt_3$, $Al\text{-}i\text{-}Bu_3$, $AlEt_2Cl$, $Al\text{-}i\text{-}Bu_2Cl$, $AlEtCl_2$, $ZnEt_2$, $BeEt_2$, and $CdEt_2$, and all of these have been available commercially in large volume or in experimental amounts.

For the polymerization of propylene with $TiCl_3$– or $TiCl_4$–AlR_3 catalysts, polypropylene stereoregularity decreases with increasing size of R (Table 4–4) (30, 31).

TABLE 4–4

Importance of Size of Hydrocarbon Substituent in AlR_3–$TiCl_x$ Catalyst (30, 31)

R	% Isotactic propylene	
	$TiCl_3$	$TiCl_4$
C_2H_5	85	48
$n\text{-}C_3H_7$	78	51
$i\text{-}C_4H_9$	—	30
C_6H_{13}	64	26
$C_{16}H_{33}$	59	16

$Al\text{-}i\text{-}Br_3$ has gained popularity because it is less pyrophoric than $AlEt_3$ and can be handled safely via syringing methods in higher concentrations.

A significant increase in yield of polymer was found when the size of R in RNa–$TiCl_4$ catalyst was increased (13) (Table 4–5).

TABLE 4–5

Importance of Size of Hydrocarbon Substituent in NaR–$TiCl_4$ Catalyst (13)

R in NaR	Polypropylene obtained (g)	$MW \times 10^4$	% not extractable with boiling heptane
n-Pentyl	16.5	65	52.1
n-Hexyl	21.5	79	50.8
n-Octyl	34.0	84	48.3
n-Dodecyl	44.5	62	32.4

b. AlR_2X and $AlRX_2$ Alkyls.   Replacement of one or two alkyl groups by halogens or radicals bearing heteroatoms usually dramatically alters the activity of the catalyst. Table 4–6 (32) compares the activity and stereoregulating ability of the AlR_2X–$TiCl_3$ catalyst for propylene polymerization where X = halogen or heteroatom-containing alkyl. Relative to the $AlEt_3$–$TiCl_3$ catalyst, all of these are less active but more stereoregulating.

TABLE 4–6

Varying X in $AlEt_2X$–$TiCl_3$ Catalysts when Used to Polymerize Propylene (32)

X	Rate of polymerization (relative to X = C_2H_5)	Stereoregularity (%)
C_2H_5	100	83
F	30	83
Cl	33	93
Br	33	95
I	9	98
OC_6H_5	0	—
SC_6H_5	0.25	95
SeC_6H_5	1.2	94
NC_5H_{10}	0	—

TABLE 4–7

Varying X and Donor in $AlEtX_2$–$TiCl_3$ Donor Catalyst when Used to Polymerize Propylene (32)

Organometallic compound	Yield (%)	Stereoregularity (%)
$Al_2(C_2H_5)_4Cl_2$	90	93
$Al_2(C_2H_5)_2Cl_4 + N(C_6H_5)_3$	20	92
$Al_2(C_2H_5)_2Cl_4 + (CH_3)_2NCHO$	15	92
$Al_2(C_2H_5)_2Cl_4[(CH_3)_2N]PO$	22	93
$Al_2(C_2H_5)_2Cl_4(C_2H_5)_3N$	8	93
$Al_2(C_2H_5)_2Cl_4(CH_3)_4NCl$	35	94
$Al_2(C_2H_5)_2Cl_4 + KCl$	40	93
$Al_2(C_2H_5)_2Br_4 + (n\text{-}C_4H_9)_4NBr$	18	95.5
$Al_2(C_2H_5)_2I_4 + (n\text{-}C_4H_9)_4NI$	20	98

Substituting $AlEtX_2$ for $AlEt_2X$ decreases considerably the activity of the catalyst, with an accompanying moderate decrease in stereoregulating ability. By the addition of electron donors, however, high activities can be restored and very highly isotactic polypropylene is formed (Table 4–7).

Other examples of both types of catalysts are described in Section III.

Other workers found the $AlEtCl_2$–$TiCl_3$ catalyst to be less stereoregulating than shown in Table 4–7 (32).

c. Exotic Metal Alkyls.   Table 4–8 (33–40a) collects some examples of exotic metal alkyls when combined with different transition metals to form active catalysts for specific olefins. In most cases, each of these variations presents a novel twist to Ziegler–Natta catalysts, but there appear to be no far reaching consequences.

TABLE 4–8

Catalysts from Exotic Metal Alkyls

Metal alkyl	Transition metal salt	Monomer	Ref.
$R_2AlO(CH_2)_3$–AlR_2 and R_4AlSO_4	$TiCl_4$	Propylene	33
Pyrolyzed AlR_3 or AlR_2H	$TiCl_4$ or $\beta TiCl_3$	Ethylene or 4-methyl-1-pentene	34
Et_2Al-X (X = α-picolyl)	$TiCl_4$ or $TiCl_3$	Ethylene or propylene	35
$EtAlN$–$[Si(CH_3)_3]_2$	$TiCl_4$ or VCl_4	Ethylene	36
Et_2Al–OCH_2CH_2–NRR'	$TiCl_3AA$ or $TiCl_4$	Propylene	37
$EtAlClOEt$	$TiCl_3AA$	Propylene	38
$H_2AlN(CH_3)_2 + AlCl_3$	$TiCl_4$	Ethylene	39
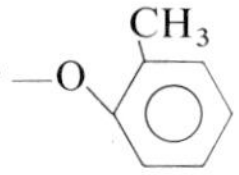 Et_2AlX	$TiCl_3$	Propylene	40

X = NEt$_2$ or —O—⟨CH$_3$ ring⟩

$(C_5H_5)_2Ti[CH_2Al(CH_3)Cl_2]_2$	$\delta TiCl_3$	Propylene	40a

B. TRANSITION METAL SALTS

Relative to the metal alkyl, this component of the catalyst has been more extensively examined. In fact, in the case of titanium trichlorides, their synthesis and use in polymerization has become so specialized that they are described separately (see Section II,B,5). While individual factors such as choice of metal, ligand, valence, mode of reaction, etc., are very important, it is ultimately the combined effect of these factors that decides the contribution of the transition metal salt. This was already recognized by 1957 in Natta's school, as shown in Table 4–9 (41).

TABLE 4–9

Effect of Transition Metal Compound (with AlEt$_3$) on Yield of Crystalline Polypropylene (41)

Compound	% Crystalline polymer	Compound	% Crystalline polymer
$TiCl_2$	80–90	$TiCl_4$	48
$\alpha TiCl_3$	85	$TiBr_4$	42
$\beta TiCl_3$	40–50	TiI_4	46
$TiBr_3$	44	$Ti(OR)_4$, $Ti(OH)_4$	trace yield
TiI_3	10	$ZrCl_4$	51.5
$ZrCl_2$	55	VCl_4	48
VCl_3	73	$VoCl_3$	32
$CrCl_3$	36		

Some of the individual factors that have been identified are now discussed. The importance of the oxidation state is discussed in Chapter 11.

1. Choice of Metal

Nearly all of the transition metals have formed active metal salts for one or more monomers. Yet most of the more important results have been obtained with a relatively small number of metals. The number of active metals is greatest for ethylene and diminishes for propylene and higher α-olefins.

The availability of $TiCl_4$ as an inexpensive material in the 1950's undoubtedly promoted considerable research aimed at industrial applications. It is not surprising that many papers and patents have appeared that use $TiCl_4$ or a derivative, especially $TiCl_3$'s and $TiCl_2$'s of varying compositions and crystal structures. As a result, many interesting and important findings developed with these $TiCl_3$'s, and this led to more research on elucidation of mechanism, etc. The same was true for aluminum alkyls.

Other transition metals, such as Sc, V, Cr, Nb, Zr, and W have been examined. Vanadium salts attracted much attention because they predominantly led to random alternating copolymers, in contrast to blocky polymers obtained with titanium salts (see Tables 20–1 and 20–2).

2. Ligands

Active catalysts for ethylene, propylene, and higher α-olefins have been made from transition metal salts bearing ligands of varied structures [see Boor *et al.* (1–8) for specific examples of catalysts]. These ligands include the groupings shown below.

Ligands
—Cl, Br, I or F
—OR(R = alkyl such as Bu, Me)
—SR(R = alkyl such as Bu, Me)
—NR$_2$(R = alkyl such as Bu, Me)
acetylacetonate
nitroso
phosphate
chromate
π-C$_5$H$_5$-Cp
indenyl
arenes
O—C—R (R = CH$_3$) $\parallel$ O
oxide
sulfide (disulfide)
sulfate
carbon monoxide

From a practical point of view, not all of these are always attractive. One must select each transition metal salt on the basis of the monomer polymerized, as well as such other constraints as yields, stereoregularity, copolymer composition, morphology, or a combination of these.

Thus, the following ground rules are suggested for the selection of a proper transition metal salt.

(a) Because it is the most active olefin and is not hampered by stereochemical limitations, ethylene can be polymerized to highly linear products with catalysts containing the largest array of transition metal salts, that is, those having different ligand structures. Economics and the balance of properties shown by the polyethylene probably dictate the choice of the salt. The variation in ligand structure is great, including sulfides, oxides, oxychlorides, dialkylamine, alkoxy, acetylacetonate, arene, cyclopentadienyl, halide ($=$ Cl, Br, I, or F), phosphate, sulfate, and so on. Because in many of these catalysts the transition metal salt and the metal alkyl undergo exchange of ligands in varying degrees, the activity of the catalyst is sensitive to the molecular ratio of the two components. In addition, the ligand structure of the active center may be significantly different from that of the starting salt, and only little insight is gained by consideration of the structure of the parent salt. One should be wary about comparisons of activities of the transition metal salts bearing different ligands if the polymerizations are not done under similar conditions.

TABLE 4–10

Polymerization with $Ti(NMe_2)_nCl_{4-n}$–$AlEt_3$ Systems (42)

n	Catalyst system	Weight of polymer formed (g)	$[\eta]^a$
4	$Ti(NMe_2)_4$–$AlEt_3$	6.4	3.7
3	$Ti(NMe_2)_3Cl$–$AlEt_3$	9.7	
2	$Ti(NMe_2)_2Cl_2$–$AlEt_3$	12.1	2.4
1	$Ti(NMe_2)Cl_3$–$AlEt_3$	18.3	
0	$TiCl_4$–$AlEt_3$	24.0	1.8

^{a} Inherent viscosity at 135°C in tetralin solution.

Tajima and co-workers (42) have done such a comparison for ethylene and found the order of catalytic activity of $Ti(NMe_2)_nCl_{4-x}$- and $Ti(NMe_2)_n(O\text{-}i\text{-}Pr)_{4-n}$-based catalysts increased as the content of chloride increased but decreased when the content of O-i-Pr increased (Tables 4–10 and 4–11).

TABLE 4–11

Polymerization with Ti(NMe$_2$)$_n$(O-i-Pr)$_{4-n}$–AlEt$_3$ System (42)

n	Catalyst system	Weight of polymer formed (g)	$[\eta]^a$
4	Ti(NMe$_2$)$_4$–AlEt$_3$	6.4	3.7
3	Ti(NMe$_2$)$_3$(O-i-Pr)–AlEt$_3$	5.0	
2	Ti(NMe$_2$)$_2$(O-i-Pr)$_2$–AlEt$_3$	4.3	3.8
1	Ti(NMe$_2$)(O-i-Pr)$_3$–AlEt$_3$	3.1	
0	Ti(O-i-Pr)$_4$–AlEt$_3$	2.4	4.1

^{a} Inherent viscosity at 135°C in tetralin solution.

Schreyer (43) examined mixtures of LiAl(decyl)$_4$ and TiCl(O-i-Pr)$_3$, TiCl$_2$(O-i-Pr)$_2$ or TiCl$_3$(O-i-Pr) as polymerization catalysts for ethylene. These salts were prepared by reacting TiCl$_4$ and Ti(O-i-Pr)$_4$ in the required amounts, either in or outside of the reactor. Polyethylenes having narrower molecular weight distributions than obtained with TiCl$_4$ were claimed.

(b) Propylene is less active by an order of magnitude and higher, and it can be polymerized to either isotactic or syndiotactic structures. For synthesis of isotactic structures, the preferred ligand is chlorine, especially in combination with Ti, V, Cr, or Nb as in TiCl$_3$, VCl$_3$, CrCl$_3$, and NbCl$_5$. Aluminum alkyls that have the AlR$_2$X or AlR$_3$ structures are preferred.

Syndiotactic polypropylene of high crystallinity has been prepared only with vanadium salts and here the ligands can be different so long as the salt remains soluble; for example, VCl$_4$, VOCl$_3$, VO(OEt)$_2$Cl, and VO(OEt)$_3$. Only aluminum alkyls having a structure AlR$_2$Cl can be used as cocatalysts. The catalyst remains soluble during the polymerization.

Higher α-olefins, such as 1-butene, are less active than propylene. In general, catalysts that are highly isotactic-regulating for propylene are also highly isotactic-regulating for higher α-olefins (compare Tables 4–13 and 4–14.) However, 1-butene and higher α-olefins do not polymerize with the syndiotactic-specific catalysts (Chapter 15, Section IV).

The choice of ligand in the copolymerization of ethylene and propylene closely follows the choice for polymerization of propylene to isotactic polypropylene if blocky copolymers are sought; the choice of ligand also follows the choice for syndiotactic polypropylene if random or prevalently alternating copolymers are sought. In the latter case, AlR$_3$ and AlRCl$_2$ alkyls have also been used in addition to AlR$_2$Cl. While the syndiotactic polymerization is done below -50°C, copolymerizations are carried out above 0°C. An important role of the ligand in the copolymerization is to

help form heterogeneous catalysts for blocky copolymerizations or homogeneous (colloidal) catalysts if random alternating copolymerizations are
sought.

3. Elemental Composition

The transition metal salt can have a neat composition if its synthesis
did not use other metals or metal compounds.

$$TiCl_4 + H_2 \longrightarrow \alpha TiCl_3$$

$$TiCl_4 + radiation \longrightarrow \beta TiCl_3 \qquad\qquad (4\text{--}1)$$

$$VCl_4 + heat \longrightarrow VCl_3 + Cl_2\uparrow \text{ (pumped off)}$$

However, a number of important titanium and vanadium transition metal
salts have been prepared by reduction of $TiCl_4$ or VCl_4 with metal alkyl
($AlEt_3$, Al-i-Bu_3, $AlEt_2Cl$, or $AlEtCl_2$ and related compounds) or with
metals and mixtures of metals and metal chlorides (Al or $AlCl_3$–Al mixtures).
Sometimes these catalysts were prepared in the polymerization reactor,
often in the presence of the olefin being polymerized. Much work also
has been devoted to isolating these salts and identifying their compositions.
Indeed, it was found that they were contaminated with aluminum compounds, especially $AlCl_3$. In many cases, chemisorbed aluminum alkyls
were probably also present when the reduction was done with an aluminum
alkyl.

These compositions are indicated in this book as $TiCl_3 \cdot xAlCl_3$, where
value of x will vary according to conditions of reduction and is usually
about 0.2 to 0.4. The presence of $AlCl_3$ and chemisorbed aluminum alkyls
can favorably or adversely influence the performance of the catalyst, as
will be described in this chapter in the section on specialized preformed
titanium trichlorides.

4. Crystalline Modification

Four crystalline modifications have been identified for $TiCl_3$ and
$TiCl_3 \cdot xAlCl_3$ salts, namely the α, β, γ, and δ forms. Activity and sometimes
stereochemical control can be influenced (see Section II,B,5 of this chapter
and Chapters 13 through 17).

5. Specialized Preformed Titanium Trichlorides
   (α, β, γ, and δ Crystalline Modifications)

Because so much of the Ziegler–Natta literature is based on $TiCl_3$'s of
different crystalline modifications, a special section is appropriate. Furthermore, if the patent literature is an indication of the types of catalysts that

are used in the commercial products of polyolefins, it is likely that $TiCl_3$'s are widely used. In fact, they may be used exclusively for production of isotactic polypropylene. It is understandable then why so much effort has been made to develop $TiCl_3$'s of higher activity and stereoregulating ability and of a particular morphology. The latter becomes important because a greater throughput in the reactor can be achieved if the formed polymer particles have a particular morphology (see Chapter 8). It is important to recognize that the actual performance of a $TiCl_3$ modification can vary significantly according to the particular conditions of synthesis. For example, the initial and constant rate of polymerization will depend on the average size of the $TiCl_3$ crystallites because the surface of the latter largely determines the number of active centers. The polymer particle shapes and sizes (hence the powder bulk density and ability of the powder to flow) will depend on the sizes and shapes of the catalyst particles. The exact concentrations of reactants, mode of mixing, temperature of reaction, etc., will determine the characteristics of the particular catalyst.

The reader may find it useful to read some earlier review papers on this subject (1–6, 8), especially the early basic publications of Natta and co-workers (44, 45).

a. Synthesis, Structural Features, and Use of $TiCl_3$'s. As stated earlier, $TiCl_3$ exists in four crystalline modifications, the α, β, γ, and δ forms. These $TiCl_3$'s can have the simple composition (one Ti per three Cl) or a more complex structure whereby a second metal is cocrystallized as an alloy in the $TiCl_3$ crystal. The particular method of reduction determines both the composition and crystalline modification. Table 4–12 (46–57) summarizes the synthesis of the most important $TiCl_3$'s that have been reported in Ziegler–Natta literature. Much credit must go to the Natta school, which did so much to elucidate the structures of the various modifications, their catalytic activity, and stereoregulating ability.

The early investigations of propylene polymerizations of the Natta school were done with the pure $\alpha TiCl_3$ modification combined with $AlEt_3$. The $\alpha TiCl_3$ was synthesized by reduction of $TiCl_4$ with H_2 at elevated temperatures (500° to 800°C) (46). $TiCl_4$ can be reduced with aluminum powder at lower temperatures (about 250°C); the formed $TiCl_3$ also has the α-crystal structure but, in addition, contains Al cations in the crystal lattice. In later polymerization studies, these $\alpha TiCl_3$ were replaced by more active modifications, namely the γ and δ forms. The γ modification is formed by heating the β modification (at 100° to 200°C); the higher the temperature, the shorter is the time required for the transformation. For a β modification containing one Al per three Ti, the transformation takes only several hours at about 140°C.

TABLE 4–12

Preparation of the α, β, γ, and δ Crystalline Modifications of Titanium Trichlorides (TiCl₃ and TiCl₃ · xAlCl₃ Types)

Expt. no.	Preparation	Crystalline form	Composition	Ref.
1a.	TiCl$_4$ + H$_2$ at high temp. (800°C)	α	TiCl$_3$	46
b.	TiCl$_4$ + Al at 100°–200°C in aromatic solvent	α	TiCl$_3$ · 0.33 AlCl$_3$	47, 48
c.	βTiCl$_3$ heated to 300°–400°C	α	TiCl$_3$	49
d.	3 TiCl$_4$ + Al + 0.5 AlCl$_3$ > 190°C (no solvent)	α	TiCl$_3$ · 0.5 AlCl$_3$	49a
e.	2 TiCl$_4$ + Mg + 2 AlCl$_3$ > 200°C (no solvent)	α	2 TiCl$_3$ · 0.5 MgCl$_2$ · AlCl$_3$	49a
f.	9 TiCl$_4$ + Ti + 2 Al > 200°C (no solvent)	α	10 TiCl$_3$ · 0.2 AlCl$_3$	49a
2a.	CH$_3$TiCl$_3$ heated > 25°C (AlMe$_2$Cl or AlMe$_3$ + TiCl$_4$ → CH$_3$TiCl$_3$)	β	TiCl$_3$	50
b.	TiCl$_4$ + H$_2$ exposed to silent electric discharge at low temp. or to γ radiation	β	TiCl$_3$	51
c.	TiCl$_4$ + AlR$_3$ or AlR$_2$Cl or AlEt$_2$Cl < 25°C	β	TiCl$_3$ · xAlCl$_3$	52
d.	TiCl$_4$ + activated Al below 100°C (e.g. in benzene at 80°C)	β	TiCl$_3$ · 0.33 AlCl$_3$	48
e.	TiCl$_4$ + Al, AlCl$_3$: aromatic solvent and ether treatment	β	TiCl$_3$ (98%)	53
3a.	βTiCl$_3$, Δ120°–200°C, up to several hours	γ	TiCl$_3$	54
b.	TiCl$_3$ · xAlCl$_3$, Δ120°–200°C, up to several hours	γ	TiCl$_3$ · AlCl$_3$	54
c.	TiCl$_4$ + Al + AlCl$_3$, inert solvent	γ	TiCl$_3$ · xAlCl$_3$	55
d.	TiCl$_4$ + Al at 160°C in 9:1 C$_{10}$H$_{22}$/xylene, solvent	γ	TiCl$_3$ · 0.33 AlCl$_3$	48
4a.	αTiCl$_3$ mechanical grinding, 25°C	δ	TiCl$_3$	56, 56a
b.	TiCl$_3$ · xAlCl$_3$, mechanical grinding, 25°C	δ	TiCl$_3$ · xAlCl$_3$	56, 56a
c.	αTiCl$_3$ + AlCl$_3$, mechanical grinding	δ	TiCl$_3$ · xAlCl$_3$	57, 56a

The β modification (free of Al) is synthesized by decomposition of CH$_3$TiCl$_3$, but this form does not form a highly active and highly stereoregulating catalyst for polymerization of propylene to isotactic polypropylene. The γ modification derived from it also has a low activity. The preferred compositions contains Al cations, and they are synthesized by reducing one TiCl$_4$ with about $\frac{1}{3}$ AlEt$_3$, one AlEt$_2$Cl, or 1–2 Al$_2$Et$_3$Cl$_3$.

The δ crystalline modification is synthesized by prolonged grinding of the α and γ forms. Because the γTiCl$_3$ (containing AlCl$_3$) is already highly active, it is not practical to transform it to the δ form. On the other hand, the activity of the αTiCl$_3$ form (containing AlCl$_3$) is much improved by transforming it to the δ form.

Figure 4–3 summarizes the observed transformations of the various crystalline modifications. Tables 4–13 to 4–16 compare the activities of these modifications for polymerization of propylene and 1-butene to isotactic polymers.

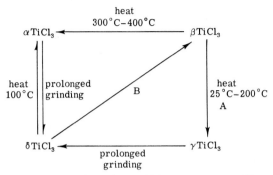

Fig. 4–3. Crystal–crystal transformations which occur most readily with the alloyed compositions TiCl$_3 \cdot x$AlCl$_3$. (A) The rate of transformation increases with increasing temperatures, (B) the possibility of converting δTiCl$_3$ to βTiCl$_3$ and αTiCl$_3$ was suggested by Wilchinsky, Looney, and Törnqvist (58).

TABLE 4–13

Polymerization of Propylene^a (9)

Organometallic compound	Mmoles	TiCl$_3$	Mmoles	Time (hr)	Wt.% converted	Isotacticity index^b
Et$_3$Al	5.4	γ	1.1	4	100	77
Et$_2$AlCl	5.4	α	1.8	20	51	90
	5.4	β	1.5	20	97	87
	5.4	γ	1.1	4	97	92
EtAlCl$_2$	5.0	γ	1.1	4	17	88
Et$_2$Cd	6.0	β	1.5	20	33	89
	4.0	γ	1.5	20	ca. 30	91
Et$_2$Zn	ca. 1.0^c	α	1.5	20	—	89
	6.3	α	1.4	20	45	79
	ca. 1.0^c	β	1.5	20	20	68
	5.0	β	1.5	20	80	68
	ca. 1.0^c	γ	1.5	20	20	82
	5.0	γ	1.5	20	80	63
EtZnCl	4.7	β	1.5	20	13	83
	ca. 5.0	γ	1.5	20	ca. 15^c	91

^{a} Conditions: 8 oz bottle, 100 ml heptane.
^{b} Ratio of infrared absorption at 10.02 μ and 10.28 μ × 10^2.
^{c} Estimate only.

TABLE 4–14

Polymerization of 1-Butene[a] (9)

Organometallic compound	Mmoles	TiCl$_3$	Mmoles	Wt.% converted	Isotacticity index[b]
Et$_2$Cd	6.0	γ	1.5	25	92
EtZnCl	4.7	γ	1.5	5	87
Et$_2$Zn	6.0	γ	1.5	48	57
Et$_2$Zn	2.5	γ	1.3	13	77
Et$_2$Zn	1.5	γ	1.5	8	92
Et$_3$Al	4.0	γ	1.1	100	60
Et$_2$AlCl	5.0	γ	1.1	90	95
Et$_2$Cd	6.0	β	1.5	10	89
EtZnCl	4.7	β	1.5	4	83
Et$_2$Zn	6.0	β	1.5	59	57
Et$_2$Zn	2.5	β	1.3	12	89

[a] Conditions: 8 oz bottle, 100 ml heptane, 50°C, 20 hr.
[b] Percent insoluble in boiling ether.

TABLE 4–15

Stereospecificity Indices of the Catalytic Systems (α, γ, and δ) Crystalline Violet TiCl$_3$[a] (66a)

	TiCl$_3$				
	α			δ	
	Free of AlCl$_3$	Containing AlCl$_3$[b]	γ[c]	Free of AlCl$_3$	Containing AlCl$_3$[b]
Al(C$_2$H$_5$)$_2$Cl	92	91	93	93	94
Al(C$_2$H$_5$)$_2$Br	95	95	96	96	96
Al(C$_2$H$_5$)$_2$I	96	96	98	97	97
Al(C$_2$H$_5$)$_3$	84	82	80	83	81

[a] With or without AlCl$_3$ in solid solution-Al(C$_2$H$_5$)$_2$X-toluene, in polymerization of propylene at 70°C and $p(C_3H_6) = 2000$ mm Hg as percent of isotactic polymer, nonextractable with boiling n-heptane. The reported data are average values obtained from runs performed twice.

[b] Containing 4.5% Al as AlCl$_3$ in solid solution.
[c] Containing 1% Al as AlCl$_3$ in solid solution.

TABLE 4–16

Indices of Stereospecificity[a] of Catalytic System Prepared from (α, γ or δ) Crystalline Violet $TiCl_3$ in Polymerization of Propylene[b] (62)

	Index of stereospecificity	
Organometallic compound	Polym. runs at 15°C	Polym. runs at 70°C
$Al(C_2H_5)_2I$	99–100	96–98
$Al(C_2H_5)_2Br$	97–98	94–96
$Al(C_2H_5)_2Cl$	96–98	91–94
$Al(C_2H_5)_3$	80–85	80–85
$Be(C_2H_5)_2$	94–96	93–95

[a] As percent of isotactic polypropylene, nonextractable with boiling *n*-heptane.

[b] Tests carried out at different pressures (1 to 10 atm) with different concentrations of reactant ($TiCl_3$, 5–30 mmole/liter; organometallic compound, 10–30 mmole/liter).

Wilchinsky, Looney, and Törnqvist (58) showed that a large increase in polymerization activity was produced if the $TiCl_3$ samples are ball-milled under dry conditions (Figs. 4–4 and 4–5 and Tables 4–17 and 4–18). Crystallographic changes took place as a result of strong shear forces causing sliding at the metal-free interfaces of the Cl–M–Cl double layer. Good linear correlations were established between catalyst activity and measured surface area and crystallite size. They presented direct evidence that the activating effect resulting from extensive dry ball-milling of the

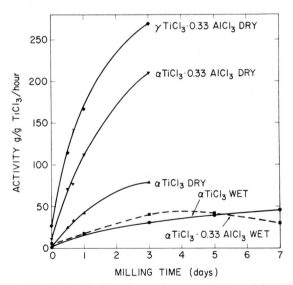

Fig. 4–4. Effect of milling time and type on catalyst activity (58).

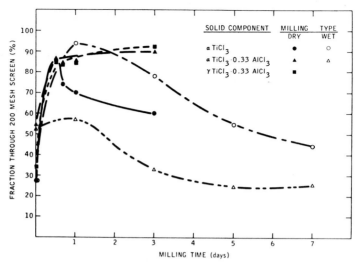

Fig. 4–5. Effect of milling time and type on particle size (58).

TABLE 4–17

Activity vs. Particle Size for Milled and Unmilled Catalyst Components (58)

	$\alpha TiCl_3$		$\alpha TiCl_3 \cdot 0.33\ AlCl_3$	
Solid catalyst component[a] milling time (days)	0	6[b]	0	4[c]
Particle fraction	Through 200 mesh	On 200 mesh	Through 200 mesh	On 200 mesh
Catalyst activity (g/g $TiCl_3$/hr)	4.7	122.0	23.6	251.3

[a] Structure before milling.
[b] 500 g charge in 7.7 liter jar.
[c] 340 g charge in 2.5 liter jar.

TABLE 4–18

Effect of Ball-Milling on Measured Surface Area (58)

	Surface area (m²/g)				
Milling time (days)	$\alpha TiCl_3$[a]		$\alpha TiCl_3 \cdot 0.33AlCl_3$[a]		$\gamma TiCl_3 \cdot 0.33AlCl_3$[a]
	Dry	Wet	Dry	Wet	Dry
0	1.28	1.28	1.13	1.07	5.0 ± 3.9 (3)
$\frac{1}{2}$	7.6	—	10.4	—	1.78 ± 0.20 (2)
$\frac{2}{3}$	12.7	—	11.7	—	4.2 ± 3.4 (2)
1	24.3	5.3	15.5	8.1	2.23 ± 0.99 (3)
3	32.2	13.3 ± 0.5 (2)[b]	30.6	12.2	7.1 ± 2.8 (4)
5	—	15.9		14.5	
7	—	25.5 ± 6.0 (2)		19.4 ± 1.4 (2)	

[a] Original component; dry or wet milling method.
[b] Numbers in parentheses refer to number of determinations involved in calculating the averages and the standard deviations.

layer-lattice $TiCl_3$ and $TiCl_3 \cdot \frac{1}{3}AlCl_3$ was due to capability of these salts to undergo extensive crystallite size reduction far beyond what would normally be expected for such a treatment. This large reduction in crystallite size was very effective because of the layer nature of these salts and the porous nature of the catalyst particle that is made up of these crystallites. For given catalyst preparations, the activity increased with decreasing crystallite size until a minimum average size was reached, corresponding to a value of about 50 Å for the D_{300} and D_{003} dimensions. They suggested that further reduction in crystallite size resulted in a less active catalyst, probably because of the tendency of very small crystallites to change their crystal structure to a less active β form.

Titanium dichloride has also been prepared with cocrystallized $AlCl_3$ (59). Four different synthetic methods were used: (1) reduction of liquid $AlCl_3$ by titanium metal; (2) reduction of $TiCl_3$ or $TiCl_4$ in presence of liquid $AlCl_3$ by titanium metal; (3) reduction of $TiCl_3$ or $TiCl_4$ by aluminum metal with or without the presence of $AlCl_3$; (4) reaction of $TiCl_2$ and liquid $AlCl_3$. All reactions were carried out at temperatures in the range of 200° to 300°C. Crystal structure determinations showed $TiCl_2 \cdot 2AlCl_3$ [for $Ti(AlCl_4)_2$] to have the titanium coordinated to six chlorides in a distorted octahedral arrangement.

The patent literature contains a great many examples of synthesis of $TiCl_3 \cdot xAlCl_3$-type salts based on reduction of $TiCl_4$ with a metal alkyl or a metal, usually aluminum alkyls and aluminum powder (48, 52, 60). These syntheses claim advantage in higher catalyst activities, higher isotactic specificities, improved polymer powder morphology, etc. The conditions described are declared to be critical and, in many cases, the total synthesis is said to require many steps at different temperatures and time durations. Some examples will be cited here for illustrative purposes.

Stauffer's commercially available $TiCl_3A$ and $TiCl_3AA$ cocatalysts are prepared by reduction of $TiCl_4$ with Al metal, but the details of synthesis are not known. This synthetic approach has apparently been investigated in other industrial laboratories (48, 49a, 56a).

The simplest procedure requires heating in a hydrocarbon solvent $TiCl_4$ and aluminum powder (61, 62). However, more active $TiCl_3 \cdot xAlCl_3$ compositions were claimed if aromatic solvents were used (63), the product was subsequently ball-milled (61–63), the aluminum powder was first activated (48, 63), or $AlCl_3$ was also present (55, 61). Aluminum powder was activated by ball-milling it alone (55) or in the presence of $AlCl_3$ and $AlEt_3$ (48). The $TiCl_3 \cdot xAlCl_3$ was also activated by grinding it in the presence of $AlEt_3$, $AlEt_2Cl$, or $ZnEt_2$ (64).

A $TiCl_3 \cdot xAlCl_3$ was prepared by adding a $TiCl_4$ solution to a $AlEt_3$ solution (3:1 molar ratio) at $-30°C$ and below and gradually warming the mixture to ambient and higher temperatures (155°C) to convert the β-product to a γ crystalline modification (60). Other examples are found in Chapter 8.

$TiCl_3 \cdot xAlCl_3$ compositions can be synthesized by mechanically alloying $TiCl_3$ and $AlCl_3$. These products are also more active than the parent $TiCl_3$ prepared by reduction of $TiCl_4$ with H_2. Kanetaka, Takagi, and Keii found activity was optimized for their preparations when the $TiCl_3 : AlCl_3$ ratio was about 10:1 (propylene polymerization in gas phase) (65).

Recently, similar compositions have been prepared with Grignard reagents and $TiCl_4$ and used for polymerization of ethylene with Al-i-Bu$_3$ as co-catalyst (66). They were reported to be more active by two orders of magnitude.

b. Structural Differences.   The α, γ, and δ modifications are purple and have a layer structure, while the β modification is brown and has a linear (chainlike) structure. The layer structure characterizing the α and γ forms is shown in Fig. 4–6, and the chainlike structure characterizing the β form is shown in Fig. 4–7. The α and the γ forms differ in that the layers of the $\alpha TiCl_3$ are stacked to produce a hexagonal close packing arrangement of the chloride anions, while in the γ form the layers are stacked to produce cubic close packing of the chloride anions. When the adjacent layers of the α and γ forms are shifted by prolonged grinding, a structurally disordered α modification is formed. This δ form is intermediate between the α and γ forms. Prolonged grinding of the β modification as another path to a δ form has not yet been reported.

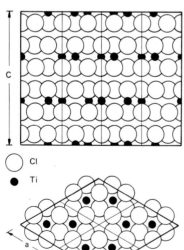

Cl
Ti

Fig. 4–6.   Model of the structural layer which characterizes the layer modification (α) of $TiCl_3$. It is interesting to point out that both in the β and layer or violet modifications, every titanium atom is placed in a position that can coordinate six chlorine atoms octahedrally. Natta, Corradini, and Allegra report the following mode of packing: $\alpha TiCl_3$, close hexagonal; γ, close cubic; and δ, a random alternation of α and γ packing. (Figure reproduced from *Atti Accad. Naz. Lincei, Cl. Sci. Fis., Mat. Nat., Rend.* through the courtesy of the Editor.)

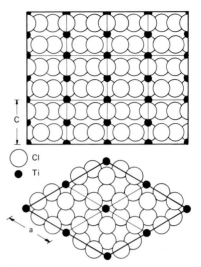

Fig. 4–7.   Model of the structure of the β form of TiCl$_3$. According to Natta, Corradini, and Allegra, the β modification of TiCl$_3$ shows a fiber-shaped structure consisting of structural units of the type:

$$-\underset{/}{\overset{\backslash}{Ti}}\overset{Cl}{\underset{Cl}{\diagdown\diagup}}Ti-Cl-Ti\overset{Cl}{\underset{Cl}{\diagup\diagdown}}Ti-$$

(Figure reproduced from *Atti Accad. Naz. Lincei, Cl. Sci. Fis., Mat. Nat., Rend.* through the courtesy of the Editor.)

The AlCl$_3$ is present as a solid solution in the crystal lattice. An iso-morphous substitution for titanium atoms of aluminum occurs during the formation of the crystal. The aluminum cation is not essential for the titanium trichloride to be active, but when present the activity of titanium trichloride is significantly increased. For example, the TiCl$_3 \cdot x$AlCl$_3$ com-positions ($x = 0.33$) are about two to seven times more active than any of the aluminum-free TiCl$_3$ compounds (67, 68). Highest activities have been reported when x is about 0.33 in these alloy compositions.

The presence of AlCl$_3$ was not found to influence the stereoregulating ability of the α, γ, and δ types of TiCl$_3$ in the polymerization of 1-butene (Table 4–14) (9) or propylene (Table 4–15) (66a). In contrast, AlCl$_3$-free βTiCl$_3$ when combined with AlEt$_2$Cl was less active and less isotactic-specific for propylene polymerization than was the AlCl$_3$-containing com-position (9, 68).

As will be shown later, the composition, structures, and morphology of the TiCl modifications will play an important role in the mechanism of isotactic propagation (Chapter 15) and in the gross morphology of the polymer particles obtained (Chapter 8). Other aspects of TiCl$_3$ chemistry will be considered in the appropriate sections.

C. THEORETICAL ASPECTS

Both the metal alkyl and the transition metal salt components play important roles in the generation of active centers. Naturally, work was directed to correlate some of the fundamental properties of each component of the mixture to polymerization activity or catalyst stereoregulating ability. This section describes some of the factors that were suggested by different workers to be relevant. For general information about the chemistry of metal alkyls and transition metal salt, the reader has available a number of excellent books and review papers (28a–28c, 69, 70).

1. Diameter of Metal in Metal Alkyl

Natta (71) suggested that the most stereoregulating catalysts were those in which the metal of the metal alkyl component had a very large diameter. For example, Table 4–19 compares the ionic radii of several metals with the isotactic content of the formed polymer. Contrary evidence was reported by Firsov and co-workers (22, 72a) and Boor (9), who showed that highly isotactic polypropylene could be made with zinc alkyls.

TABLE 4–19

Relationship between Ionic Radius of M in MR_3 and Isotacticity[a]

Metal alkyl	Ionic radius (Å)	Isotactic polypropylene[b] (%)
$BeEt_2$	0.35	94–97
$AlEt_3$	0.51	80–90
$MgEt_2$	0.66	78–85
$ZnEt_2$	0.74	30–40

[a] When used to polymerize propylene at 75°C ($\alpha TiCl_3$ = co catalyst).
[b] Measured as percent polypropylene insoluble in boiling heptane.

The apparent discrepancy arose because the polymers prepared with zinc alkyls had much lower molecular weights, for example $[\eta] \simeq 0.1$ to 0.4 dl/g vs. 2 to 12 for polymers obtained with aluminum and zinc alkyls, respectively. The zinc alkyl-derived polymers, even though they had the same isotactic crystallinity, were soluble in boiling heptane. This was clearly shown by Firsov in Table 4–20, which compares the crystallinity and molecular weight (as intrinsic viscosity $[\eta]$) for the three fractions isolated by extraction procedures. [The low molecular weight polypropylenes were formed when $ZnEt_2$ concentration was high; to prepare polymers with intrinsic viscosities (IV) $\simeq 2$ dl/g, lower $ZnEt_2$ concentrations were required.]

TABLE 4–20

Properties of Polypropylene Obtained over the Catalyst System $\alpha TiCl_3$–$Me(C_2H_5)_n$ (22, 72a)

$Me(C_2H_5)_n$	Ionic radius of metal (Å)	Content of fraction (%)	Crystallinity, infrared spectrum (%)	Crystallinity X-ray spectrum (%)	$[\eta]^a$
	Fraction I (polymer insoluble in boiling n-heptane)				
$Be(C_2H_5)_2$	0.35	91–98	60.5–69.5		2–11.8
$Al(C_2H_5)_3$	0.51	77–80	57–62	54–57	3–4
$Zn(C_2H_5)_2$	0.74	47–66	b	58–66.5	0.1–0.4
	Fraction II (polymer soluble in boiling and insoluble in cold n-heptane)				
$Al(C_2H_5)_3$		8–12	55	55	
$Zn(C_2H_5)_2$		17–29	55–58	61–64	
	Fraction III (polymer soluble in cold n-heptane)				
$Al(C_2H_5)_3$		8–13	18–32		
$Zn(C_2H_5)_2$		13–34	29–31		

^{a} Intrinsic viscosity of a representative sample of the polymer.
^{b} Because of the extremely high optical density of a film of the polymer, no infrared spectrum could be obtained.

Boor (9) also used infrared spectroscopy on thin films of the whole polymer product to evaluate the isotactic content ($I_i = A_{10.02\mu}/A_{10.28\mu}$; see Chapter 3, Section III for a description of methods). The isotactic values were not sensitive to differences in polymer molecular weight in the range examined. When low $ZnEt_2$–$TiCl_3$ ratios were used, the isotactic content of polypropylene was comparable to that obtained with the $AlEt_3$–$TiCl_3$ catalyst (Table 4–21). Also, the $ZnEtCl$–$TiCl_3$ catalyst was comparable to the $AlEt_2Cl$–$TiCl_3$ catalyst in stereoregulating ability.

TABLE 4–21

Comparisons of Isotactic Indices^a for Polypropylene Obtained with Zinc and Aluminum Alkyls (9)

Transition metal salt	Metal alkyl			
	$ZnEtCl(0.04\ M)$	$ZnEt_2(0.01\ M)$	$AlEt_2Cl(0.06\ M)$	$AlEt_3(0.05\ M)$
$\alpha TiCl_3$	—	0.88	0.90	—
$Al_xTi_yCl_z$–A	—	0.90	0.92	—
$Al_xTi_yCl_z$–70/160	—	0.68	0.73	—
$Al_xTi_yCl_z$–160	0.90	0.83	0.92	0.77
$\beta Al_xTi_yCl_z$–25	0.83	0.68	0.87	—

^{a} Indices of 0.65 and 0.90 correspond approximately to percent crystallinities of 35% and 70%, respectively.

Natta, Pasquon, Zambelli, and Gatti (72) later showed that the use of insolubility values as indices of isotacticity became invalid only when the polymers had low molecular weights. If the "as prepared" polymer had a sufficiently high steric regularity and a molecular weight higher than about 50,000 to 100,000, the index of isotacticity could be considered independent of the molecular weight. The insolubility index method failed for polymers of low stereoregularity. Similarly, it failed if the crude polymer contained a significant amount of highly isotactic polypropylene with molecular weights lower than 10,000 to 20,000. In both cases, the polymers were soluble in boiling heptane. These workers prepared the test samples by use of transfer agents, different temperatures of polymerization, and by thermally degrading high molecular weight highly isotactic polymer.

Zinc and cadmium alkyls are whimsical because isotacticity depends on the concentration of each metal alkyl present and on the particular titanium chloride used.

2. Complexing Ability of Metal Alkyl

The data in Ziegler (28a) and to a lesser degree, Coates (28b), Mole and Jeffery (28c), and Schindler (69), were largely used to collect the following background information.

Aluminum alkyls have a strong tendency to associate into dimers and sometimes to trimers and higher aggregates. This strong tendency to associate is attributed to the electron deficiency of the aluminum atom in compounds with coordination number 3, as shown below,

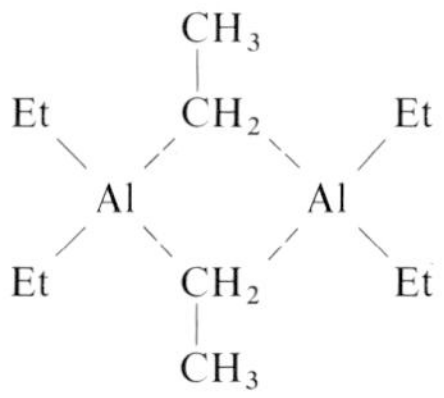

$$
\begin{array}{ccc}
& CH_3 & \\
& | & \\
& CH_2 & \\
Et & \diagdown \quad \diagup & Et \\
\diagdown & Al \qquad Al & \diagup \\
\diagup & \diagup \quad \diagdown & \diagdown \\
Et & CH_2 & Et \\
& | & \\
& CH_3 & \\
\end{array}
$$

I

where——is an "electron deficient" ("half") bond. The bridged complex or poly complex exists in equilibrium with the monomeric metal alkyl shown in Eq. 4–2. The position of this equilibrium varies according to the ligands

$$(AlR_3)_2 \rightleftharpoons 2\,AlR_3 \tag{4–2}$$

attached to the aluminum atoms: (1) If R is n-alkyl, the degree of dissociation is low but increases as the size of R increases. (2) If R is branched (for examples, isobutyl, isopropyl, or neopentyl) the degree of dissociation is very high, and the corresponding metal alkyls exist in the monomeric state (cryscopic measurements). Apparently, steric compression hinders the formation of electron-deficient bonds. (3) If one of the R's is a hydride, there is a tendency to form trimeric complexes with H-bridges. The adverse effect of branched alkyl substituents is absent. (4) Dimerization and trimerization also occur if one or two alkyls are replaced by halogens or radical bearing O, N, or P atoms as shown in structure II, III and IV.

mation of electron-deficient bonds. (3) If one of the R's is a hydride, there is

Not all metal alkyls show this strong tendency to associate to bridged dimers nor to complex with electron-donating compounds. For example, dialkylzinc and dialkylcadmium are normally monomeric and do not form molecular complexes with most amines. However, exceptions have been found: $ZnMe_2$ forms complexes with pyridine (73) ($ZnEt_2$ does not), and $ZnEt_2$ complexes with chelating donors. EtZnBr and EtZnCl form tetramers in benzene, but in ethyl ether the latter exist in monomeric form (74).

One consequence of the equilibrium nature of these complexes is the rapid exchange of ligands between the complexed metal alkyl molecules. This is most readily demonstrated when different aluminum alkyls are mixed and cryscopic measurements indicate a change in molecular weight.

$$2\,\text{Al-}i\text{-Bu}_3 + \tfrac{1}{2}\,\text{Al}_2(\text{CH}_3)_6 \rightleftharpoons 3\,\text{Al-}i\text{-Bu}_2\text{CH}_3 \qquad (4\text{--}3)$$

$$\underset{\text{monomeric}}{}\qquad \underset{\text{dimeric}}{}$$

The reaction is exothermic and because $\text{Al-}i\text{-Bu}_2\text{CH}_3$ can associate through CH_3 bridges, two and a half particles are converted to one and a half particles. Proof that the components exist in equilibrium is the ability to recover the starting materials by distillation (28a).

Because the bridging atoms can be atoms other than carbon, an alkyl interchange can take place with halogens, or radicals bearing oxygen, nitrogen, or phosphorus atoms, such as $-\text{OEt}$, $-\text{NEt}_2$, or $-\text{PEt}_2$.

These associated structures are stable, often to fairly high temperatures. They can, however, be separated by addition of electron donors such as ethers, amines, etc., as shown in Eq. 4–4. Several types of donors have been

$$(\text{AlEt}_3)_2 + 2\,\text{Et}_2\text{O} \rightleftharpoons 2\,\text{AlEt}_3\cdot\text{Et}_2\text{O} \qquad (4\text{--}4)$$

effective, including inorganic halides such as NaCl or KF. Many of these aluminum alkyl donor complexes are so strongly held that they can be distilled without decomposition.

Conductivity measurements suggest that metal alkyl is present in part in an ionic form (75), as shown in Eq. 4–5.

$$2 \text{ AlEt}_3 \; \rightleftharpoons \; (\text{AlEt}_3)_2 \; \rightleftharpoons \; \text{AlEt}_2{}^+\text{AlEt}_4{}^-$$

$$\text{ion pair} \qquad\qquad (4\text{–}5)$$

$$2 \text{ ZnEt}_2 \; \rightleftharpoons \; \text{ZnEt}^+\text{ZnEt}_3{}^-$$

Aluminum alkyls can exist with a coordination number of 4. Compounds of this type are easily made, for example, by addition of group I metal alkyl to AlR_3, as shown in Eq. 4–6.

$$\text{NaEt} + \text{AlEt}_3 \longrightarrow \text{NaAlEt}_4 \qquad\qquad (4\text{–}6)$$

Only when the alkyl ligand is large do these become appreciably soluble in hydrocarbon solvents.

It will be seen later that the chemistry of the aluminum alkyl has been used to explain not only catalyst formation but also its use in polymerization. But caution must be used in reaching dogmatic conclusions. For example, we might be tempted to conclude that facile ligand exchange reactions between the metal alkyl molecules and their ability to form electron-deficient bridge complexes are related. But zinc and cadmium, unlike aluminum alkyls, undergo fast exchange reactions even though they do not form strong bridged dimers. All of these metal alkyls form active catalysts. Thus, site formation cannot be directly associated with the ability of metal alkyl to form electron-deficient structures. Hoeg's suggestion that four-center transition states rather than distinct intermediates may actually be involved in ligand exchange is a reasonable explanation (3).

3. Ionic Nature of Catalyst

Bushick and Stearns (76) found that the rate of polymerization correlated well with measured electrical conductivities of catalyst components. Their data supported the view that Ziegler–Natta catalysts are highly ionic in nature and that the active centers that are responsible for polymerization have ionic character.

Figure 4–8 shows the dependence of polymerization of propylene on equivalent conductivity $\Lambda_0{}'$ for catalysts formed by combining TiCl_3 and AlEtCl_2, $\text{AlEt}_{1.5}\text{Cl}_{1.5}$, AlEt_2Cl, or AlEt_3. A dependence on conductance was also found for catalysts formed by combining Al-i-Bu$_2$Cl and VO(OEt)$_3$, VO(OEt)Cl$_2$, or VO(OEt)$_2$Cl when these were used to copolymerize ethylene and propylene. Similarly, when ethylene was polymerized with $\text{Al}_2\text{Et}_3\text{Cl}_3$–$\text{TiCl}_4$ catalyst solvents containing varying amounts of benzene and heptane, rate increased about fourfold on going from pure heptane to pure benzene.

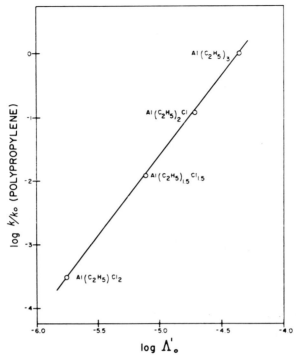

Fig. 4–8. Relative rate of propylene polymerization as a function of equivalent conductivity at Λ'_0 (76).

4. Reducing Ability of Metal Alkyl

It is generally recognized that metal alkyls have different abilities to reduce transition metal salts in a higher valence state. It is believed that reduction occurs via alkylated transition metal species, and since the latter can be active centers, higher catalyst activities should perhaps result with metal alkyls that have greater reducing power. Zambelli and co-workers (77) examined the polymerization of propylene with mixtures of VCl_3 and a variety of different metal alkyls, but they found no direct correlation of catalyst activity with metal alkyl reducing power. The Zambelli data will be elaborated in Chapter 13 in a mechanistic context.

5. First Ionization Potential of Transition Metals

Natta and co-workers (78) characterized the most active salts as those having a metal whose first ionization potential lies below 7 eV and whose work function was less than 4 eV (< 1.7 values in Pauling's electronegativity scale), for example, Ti, V, Cr, and Zr (Table 4–22).

TABLE 4–22

Dependence on Work Function and Ionization Potential

Property	Ti[a]	V[a]	Cr[a]	Mn	Fe	Ni	Zr	Mo	W
Work function (in eV)	3.9	3.8	3.7	3.8	4.7	5.0	3.7	4.1	4.5
First ionization potential	6.8	6.7	6.7	7.4	7.8	7.6	6.9	7.1	7.9

[a] Metals which form the most active salts.

In general, this proposal has been substantiated for olefins, but it is not applicable to diene polymerizations. Nickel and cobalt catalysts are highly active for butadiene.

6. Number of d Electrons and Energies of Metal and Olefin Orbitals

In his molecular orbital (MO) treatment of the propagation step, Cossee suggested that transition metal salts containing zero to three d electrons should be most active. This has generally been found true for olefin polymerizations. Catalysts made from transition metal salts that have four or more d electrons were about 10^{-3} to 10^{-5} as active (79). In combination with $AlEt_2Cl$, the following order of activity was found: $CrCl_2 > FeX_2$ (X = Cl, Br, or acac) $> MnX_2$ (X = Cl or acac). These divalent salts contain four, six, and five d electrons, respectively.

Cossee concluded from his MO calculations that a metal center was active if the energies of the metal and olefin orbitals were approximately similar. The influence of ligands in transition metal salts was seen to alter the relative energy of the metal orbitals and, in this way, affect the activity of the metal center. The Cossee model will be discussed in detail in Chapter 13.

7. Comments

While helpful in understanding some of the features of the catalyst, none of these theoretical factors are so revealing that the structure of the active center or the mechanistic path is unveiled. We must look for additional evidence, and this is done in the subsequent chapters.

III. Some Important Classes of Catalysts for Olefins

The worker has on hand a very large number of different combinations of metal alkyls and transition metal salts to make Ziegler–Natta catalysts that are active for polymerization of ethylene, propylene, and higher α-

TABLE 4–23

Examples of Catalysts for Olefins

Catalyst	Olefin	Structure
1. $TiCl_4$ + $AlEt_3$, Al-i-Bu_3 or $AlEt_2Cl$	Ethylene	Linear PE
2. $TiCl_3$ + $AlEt_3$, Al-i-Bu_3 or $AlEt_2Cl$	Ethylene	Linear PE
	Propylene, α-olefins	Isotactic PP
	Ethylene + olefins	Blocky copolymers
3. $TiCl_3$ + $AlEtCl_2$ + donor	Propylene	Isotactic PP
4. $(C_5H_5)_2TiCl_2$ + $AlEt_3$, $AlEt_2Cl$, or $AlEtCl_2$	Ethylene	Linear PE
	Ethylene + propylene or 1-butene	Copolymers
5. VCl_4 + $VO(OR)_xCl_{3-x}$ + $AlEt_2Cl$ (below $-45°C$)	Propylene only	Syndiotactic PP
	Propylene + ethylene or 1-butene	Copolymers
6. VCl_4 or $VO(OR)_xCl_{3-x}$ + $AlEt_3$, $AlEt_2Cl$, or $AlEtCl_2$	Ethylene and propylene	Random copolymers

olefins. Indeed, the literature contains what appears to be an endless number of examples.

This section discusses several classes of such catalysts. The chosen examples have been widely investigated as potential commercial catalysts or to gain mechanistic insight. Specific examples are shown in Table 4–23. Related catalysts also have been made by variation of ligands in both components, and these catalysts are also described. Supported catalysts also could have been included in this chapter, but they are more appropriately described in Chapter 6.

A. CATALYSTS FROM $TiCl_4$ AND $AlEt_3$, Al-i-Bu_3, OR $AlEt_2Cl$

These catalysts were among the first to be investigated by Ziegler and co-workers for polymerization of ethylene and continue to be examined widely in both academic and industrial laboratories. Emphasis has been placed on ethylene polymerizations, but examples using propylene, higher linear, and branched α-olefins have also been disclosed. Many kinetic studies were reported.

Significant information has come from only a few workers. However, these ($TiCl_4$ and $AlEt_3$, Al-i-Bu_3, or $AlEt_2Cl$ systems) do not appear today to be attractive catalysts for commercial production of ethylene if recent patent examples are an indication. In the laboratory, they have found use in the polymerization of monomers such as styrene, which are otherwise sluggishly polymerizable.

The literature is too extensive to review here. For the reader who is interested, Schindler has reviewed much of this data (69, pp. 153–163; 70, selected pages in Chapters III and IV). Mostert (80) also reviewed some of the literature, along with some original investigations.

A few salient features of these catalyst types are presented here: (1) They are very unstable, and this has produced many conflicting experimental results. (2) Their behavior is very sensitive to a number of experimental factors, such as ratio of Al/Ti, temperature and time of mixing of all components, absolute and relative concentrations of reactants, the time elapsed between mixing and use in polymerization, aging temperature, duration of polymerization, etc. (3) Because $AlEt_3$, $Al-i-Bu_3$, and $AlEt_2Cl$ differ in alkylating and reducing power, different conditions are necessary to produce active catalysts.

As already described in Section II,B,5, specialized $TiCl_3 \cdot xAlCl_3$ composition is formed from $TiCl_4$ by a very specific reduction process involving the aluminum alkyl or aluminum metal. This reduction involves the alkylation of $TiCl_4$ with aluminum alkyl molecules followed by a dealkylation reduction to a trivalent state.

Alkylation reactions

$$TiCl_4 + AlEt_3 \longrightarrow EtTiCl_3 + AlEt_2Cl$$

$$TiCl_4 + AlEt_2Cl \longrightarrow EtTiCl_3 + AlEtCl_2$$

$$TiCl_4 + AlEtCl_2 \longrightarrow EtTiCl_3 + AlCl_3$$

Reduction reactions

$$EtTiCl_3 \longrightarrow TiCl_3 + Et\cdot$$

$$2\ Et\cdot \longrightarrow ethane + ethylene$$

(4–7)

Under drastic conditions, $TiCl_3$ can be similarly reduced to $TiCl_2$. The above equations only suggest the basic steps which occur; the actual mechanistic process is very complex and not well understood.

The actual $TiCl_3$ product formed is a composition containing $AlCl_3$ as alloy and probably some chemisorbed $AlEt_2Cl$ and $AlEtCl_2$. If the reduction is incomplete, the unreacted $AlEt_3$ and $TiCl_4$ are also present. Some $EtTiCl_3$ species may also be adsorbed on the $TiCl_3$.

It is understandable that different catalytic mixtures will be formed when the components $AlEt_3$ (or $AlEt_2Cl$) and $TiCl_4$ are mixed and used under different conditions. The nature of the catalyst changes with polymerization time. Some workers stabilized the preparations by a prepolymerization aging at ambient or higher temperatures, but this produced only a partial improvement. The AlR_3–$TiCl_4$ systems are not catalysts that the author would recommend for mechanistic studies of olefins to one who is beginning Ziegler–Natta polymerizations. Undoubtedly, the shift to preformed titanium chlorides (see specialized $TiCl_3$'s) was provoked by the many disadvantages of these unstable catalysts.

B. CATALYSTS FROM PREFORMED $TiCl_3$'S
AND ALUMINUM ALKYLS

Polypropylenes made with the $TiCl_4$-based catalysts discussed above had a low isotactic crystallinity; for example, about 40 to 60% of the whole polymer was extracted by a hydrocarbon solvent. Moreover, these catalysts had low activity for propylene.

The Natta school first recognized that much higher stereoregulating catalysts were made if a preformed $TiCl_3$ was used. The particular $TiCl_3$ first used by them was the α crystalline modification made by reduction of $TiCl_4$ with H_2 at high temperatures.

Later the Natta group, as well as other investigators worldwide, developed elaborate synthesis for other crystalline modifications of $TiCl_3$, the β, γ, and δ forms, which usually also contained cocrystallized $AlCl_3$ (see Section II,B,5). Moreover, in addition to $AlEt_3$ and $Al\text{-}i\text{-}Bu_3$, AlR_2X metal alkyls (especially $AlEt_2Cl$) were shown to produce very highly stereoregulating catalysts for propylene and higher α-olefins.

Catalyst	% Polymer not extractable in boiling heptane
$TiCl_4$–$AlEt_3$	∼50
$TiCl_3$–$AlEt_3$	∼85
$TiCl_3$–$AlEt_2Cl$	∼95

The $AlEt_3$–$TiCl_3$ catalyst was, however, about three times more active relative to the one containing $AlEt_2Cl$ (67). It appears from the patent and journal literature that the catalysts containing preformed catalysts were favorably used for polymerization of ethylene (81), 4-methyl-1-pentene (82), 1-butene (83), and other α-olefins.

Catalysts based on preformed $TiCl_3$ have been investigated widely in mechanistic and kinetic studies, and the corresponding findings of these investigations are described in the appropriate chapters of this book. Without doubt, these represent the most important class of catalysts for olefin polymerizations, especially for propylene and 1-butene.

A word of caution, however: not all transition metal salts that are in their highest valence state lead to poorly stereoregulating catalysts. For example, the stereoregulating ability of the heterogeneous catalyst $AlEt_3$ + $NbCl_5$ is comparable to that obtained with $AlEt_3$ + $TiCl_3$ in a propylene polymerization (84).

C. CATALYSTS FROM PREFORMED $TiCl_3$, $AlEtCl_2$, AND AN ELECTRON DONOR

A close family of catalysts has also been made from the same preformed $TiCl_3$ salts except that a mixture of an electron donor and $AlEtCl_2$ was used in place of $AlEt_3$ or $AlEt_2Cl$. Interest in this catalyst developed about 1960 when Coover and Joyner (6, 85, 86) reported that a polypropylene was made with it that had a high melting point ($T_m = 183°C$), a high crystallinity, and was said to be distinctly different from isotactic polypropylene made by other Ziegler–Natta catalysts such as $AlEt_2Cl + TiCl_3$. Coover and Joyner designated it as stereosymmetric. Later characterization showed it to be isotactic, although some differences in tacticity characteristics were demonstrated (87, 88). Its melting point was also similar to that obtained for polypropylenes prepared by Natta and co-workers with their most specific catalysts.

Binary catalysts made from only $TiCl_3$ and $AlEtCl_2$ are not very attractive. When $TiCl_3$ (that made from $TiCl_4$ by H_2 reduction) was used, the catalyst produced only oils from propylene or had very low activity (Table 4–24). In presence of other $TiCl_3$ modifications (such as γ-, and $\delta TiCl_3 \cdot xAlCl_3$ types), propylene is polymerized at a low rate, but the polypropylene has a reasonable isotactic content (comparable to $AlEt_3$–$\alpha TiCl_3$ product in isotactic content).

Remarkable increases in activity occur (from five to tenfold and greater) if controlled amounts of electron donors are added, such as amines, ethers, and phosphines. The isotactic content also increases such that the products are comparable to those prepared with the $AlEt_2Cl$–$TiCl_3$ catalysts.

Coover and Joyner (87) found that an optimum rate of polymerization was obtained only if the donor/$AlEtCl_2$ molar ratio was near 0.7 (Fig. 4–9). Stereospecificity is also maximized in this system at about 0.6 HPT (hexamethylphosphoric triamide)/$AlEtCl_2$ and intrinsic viscosity begins a distinct departure from a plateau to higher values at about the same molar ratio of donor to Al (Figs. 4–10 and 4–11).

TABLE 4–24

Effect of Catalyst Composition on Stereospecificity of $TiCl_3/(C_2H_5)_2$ AlX/Third Component Catalyst System[a] (92)

X	Third component	Rate (mmoles/liter-sec)	Crystallinity index	$[\eta]$
F	None	1.89	36	1.34
F	HPT (0.4 mole ratio)	0.11	93	5.58
Cl	None	1.51	90	2.49
Br	None	0.98	93	2.80
I	None	0.73	96	2.60

[a] Monomer pressure was 40 psig and temperature was 70 C.

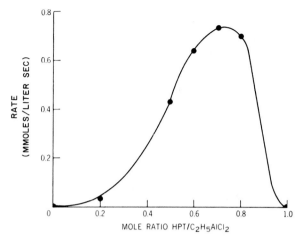

Fig. 4–9. Effect of varying HPT/$C_2H_5AlCl_2$ mole ratio on rate of propylene polymerization at 40 psig and 70°C with the use of equimolar quantities (3.33 mmole) of violet $TiCl_3$ and $C_2H_5AlCl_2$ (87).

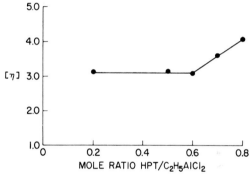

Fig. 4–10. Effect of varying HPT (hexamethyl phosphoric triamide)/$C_2H_5AlCl_2$ mole ratio on the inherent viscosity of the polypropylene produced at 40 psig propylene pressure and 70°C with the use of equimolar quantities (3.33 mmole) of violet $TiCl_3$ and $C_2H_5AlCl_2$ (87).

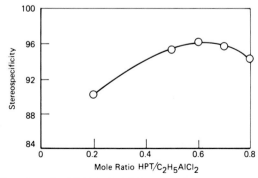

Fig. 4–11. Effect of varying HPT/$C_2H_5AlCl_2$ mole ratio on the stereo-specificity index of the three-component catalyst $C_2H_5AlCl_2$–HPT–$TiCl_3$ in the polymerization of propylene at 40 psig and 70°C with the use of equimolar quantities (3.33 mmole) of violet $TiCl_3$ and $C_2H_5AlCl_2$ (87).

At a ratio of unity, no polymerization was observed for this donor. Other workers confirmed that ratios in the range 0.5 to 0.7 produced optimum catalyst activities (89–91).

Donors that had a wide range of structures have been found effective. Coover and Joyner (92) disclosed amines, phosphines, KNH_2, phosphoramides, ethers, esters, acetals, and ketones (see Table 4–25 for examples). Other workers disclosed many other effective donors: hexamethyl disilazane (93), mesitaldehyde (94), tetrahydrofuran (60), $Si[N(CH_3)_2]_4$ (95), and Na_2SO_4 (94). This is only an arbitrarily chosen partial list.

TABLE 4–25

Effect of Catalyst Composition on Stereospecificity of $TiCl_3/RAlX_2/$Third Component Catalyst System[a] (92)

| Crystalline form of $TiCl_3$ | X | R | Third component | | Rate (mmoles/ liter-sec) | Crystallinity index | $[\eta]$ |
			Mole ratio	Cpd.			
α	Cl	C_2H_5	—	None	0	—	—
α	Cl	C_2H_5	0.6	HPT	0.68	96.4	3.14
β	Cl	C_2H_5	0.6	HPT	<0.09	75	3.72
α	Br	C_2H_5	0.6	HPT	0.53	98.3	3.14
α	I	C_2H_5	0.6	HPT	0.57	99.2	2.50
α	Cl	C_2H_5	0.7	$(C_4H_9)_3N$	0.93	95	3.06
α	Cl	C_2H_5	0.7	HPT	0.74	95	3.62
α	Cl	C_2H_5	0.7	$(C_4H_9)_3P$	0.73	97	3.11
α	Cl	C_2H_5	0.7	$(C_4H_9)_2O$	0.39	94	2.96
α	Cl	C_2H_5	0.7	$(C_4H_9)_2S$	0.15	97	3.16
α	Cl	CH_3	0.6	HPT	0.106	97	3.97
α	Cl	$i\text{-}C_4H_9$	0.6	HPT	0.085	95	3.85
α	Cl	C_8H_{17}	0.6	HPT	0.106	96	3.71

[a] Monomer pressure was 40 psig and temperature was 70°C.

In addition to $AlEtCl_2$, other $AlRX_2$ metal alkyls have been used, including $BuAlBr_2$ and $PrAlCl_2$ (93). A host of related catalysts where alkylaluminum sesquihalides were substituted for $AlEtCl_2$ have been reported (86). (These are 50:50 mixtures of $RAlX_2$ and R_2AlX.) Other particular examples include $Al_2Et_3Br_3$ (96) and $Al_2Et_3Cl_3$ (94).

Coover reported that polypropylene prepared with the $AlEtCl_2$–$TiCl_3$–hexamethyl phosphoramide catalyst had a higher stereoregularity than prepared with $AlEt_2Cl$–$TiCl_3$ or with $AlEt_3$–$TiCl_3$ (Fig. 4–12). The polypropylenes prepared with different donors had high molecular weights, and these varied according to the donor used.

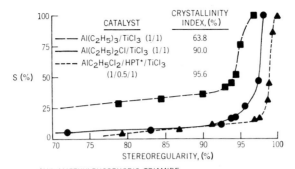

*HEXAMETHYLPHOSPHORIC TRIAMIDE.

Fig. 4–12. Stereoregularity distribution of polypropylene obtained with various catalysts at 70°C and 40 psig propylene pressure with the use of equimolar quantities (3.33 mmoles) of alkylaluminum and $TiCl_3$ (88).

Two roles have been assigned to the donor. Coover and co-workers believe that a donor $AlEtCl_2$ complex functions in a bimetallic mechanism involving both Al and Ti atoms. In contrast, Zambelli and co-workers report that the $AlEtCl_2$ is converted to $AlEt_2Cl$ by the reaction in Eq. 4–8. Other

$$2\ AlEtCl_2 + D \longrightarrow AlEt_2Cl + AlCl_3 \cdot D \qquad (4\text{–}8)$$

workers have taken sides with supporting evidence. The corresponding experimental work relating to the true mechanism of the donor is discussed in Chapter 9, which is devoted entirely to the effect of donors on Ziegler–Natta polymerization.

D. CATALYSTS BASED ON Cp_2TiCl_2
AND ALUMINUM ALKYLS

These soluble catalysts, first disclosed by Natta and co-workers (97) and Breslow and Newburg (98) for polymerization of ethylene, have been used extensively for mechanistic studies. The mechanistic interpretations reached with the use of these catalysts will be discussed later in Chapters 13 and 14, and only the scope of the catalyst is presented here. Typical catalysts are made by combining Cp_2TiCl_2 and $AlEt_3$ or $AlEt_2Cl$, but many related catalysts have also been examined (99–124). Aspects of this catalyst have previously been described in reviews by Schindler (125) and Zavorokhin (126).

Efforts to homopolymerize olefins other than ethylene have not been generally successful. Reichert and co-workers (110) concluded from kinetic studies using ^{14}C-labeled $AlEt_2Cl$ that styrene polymerized by insertion in metal–carbon bonds. In contradiction, Amass, Hay, and Robb (114) in an extensive kinetic investigation concluded that a radical mechanism prevailed and produced only atactic polymer.

Successful copolymerizations were made with ethylene and propylene (119) and 1-butene (105, 109, 117). Two types of centers were suggested (105) and, in general, the α-olefin acted as a transfer agent (108) and the copolymer contained it only in small amounts [for example, 7 mole% 1-butene (109)]. Marti and Reichert (106) reported that styrene was also polymerized in presence of the same catalyst ($Cp_2TiCl_2 + AlEt_2Cl$) and concluded that two kinds of copolymers with different sequence distributions were formed. Natta and co-workers (97) found that vinyl ethers were homopolymerized with combinations of $Cp_2TiCl_2 + AlEt_3$, $AlEt_2Cl$, and $AlCl_3$ but the mechanism was cationic.

In addition to $AlEt_3$ and $AlEt_2Cl$, other metal alkyls have been examined: $AlEtCl_2$ (92, 99, 100, 102, 112), Al-i-Bu_2Cl (101, 108), $AlMe_3$ (113, 118), $AlMeCl_2$ (100, 102), $Al(C_6H_5)_3$ (as etherate) (104), and LiEt (99).

While Cp_2TiCl_2 has been mostly used, other derivatives have been examined: $Cp_2Ti(R)Cl$ (where R = Me, Et and higher alkyl) (100, 102), $Cp_2Ti(C_6H_5)_2$ (104) and $(R'Cp)_2TiCl_2$ where R' = Me or Et (103). Fushman and co-workers (123) found the trivalent derivative was active in combination with $AlEt_2Cl$ or $AlEtCl_2$ in 1,2-dichloroethane. Cyclopentadienyl (Cp) derivatives of other transition metals have also been found active, such as Cp_2VCl_2 by De Liefde Meijer et al. (99) and Cp_2Cr and Cp_2ZrCl_2 by Breslow and Newburg (98).

Waters and Mortimer (100, 102) extensively studied the $Cp_2(R)Cl$–$RAlCl_2$ catalysts to elucidate the variation in activity of the catalyst due to differences in metal alkyl structure. They concluded that the first insertion of ethylene into a Ti–Me or Ti–C_6H_5 bond center was more difficult than if the active center was a Ti–Et bond. Differences in bond stabilities were cited as the reason: Ti(C_6H_5) and $TiCH_3 > $ TiEt (least stable). These results supported the earlier findings of Karapinka and Carrick (120) that insertion into a Ti–C_6H_5 bond was observed only if there was no other type of metal–carbon bond present. Waters and Mortimer (100) cite other literature findings as support: (1) a low number of polymer chains per Ti atom which, however, increases gradually with time; (2) the gradual rather than instantaneous disappearance of $AlMe_3$ or $Al(C_6H_5)_3$ during polymerization; and (3) the low initial polymerization rates which increase with time (the reverse was found when $AlEt_3$ or $AlEt_2Cl$ were used). In agreement is the finding by Natta and co-workers (78) that $Cp_2Ti(C_6H_5)_2$ alone is not active for polymerizing ethylene but becomes active upon addition of $AlEt_3$. The formed polyethylene, however, does not contain phenyl end groups.

Hocker and Saiki (103) found that cyclopentadienyltitanium compounds substituted in the ring with Me or Et led to lower catalyst activities. Inductive effects producing higher electron densities on the Ti atom were said to destabilize the Ti–C bond.

Aliphatic, aromatic, and halogenated solvents have been used for these polymerizations, with toluene and benzene being preferred. Apparently, the kinetic behavior of the catalyst and molecular weight distribution of the formed polyethylene is affected by the type of solvent. Aromatic solvents have been favored because the components of the catalyst and the catalyst itself remain in solution. Waters and Mortimer, however, state that toluene is a reactant and, therefore, obscures the observed kinetics. Chloroethane also has been widely used, and it is found that the rapid deactivation found in aromatic solvents can be avoided by use of halogenated solvents (115, 120, 127). Breslow and Newburg found traces of oxygen were necessary to maintain high activity. The presence of other olefins was shown to increase the consumption of ethylene in the presence of the Cp_2TiCl_2–$AlEt_2Cl$–chloroethane system (98).

It is generally concluded that the Ti in the most active Cp_2TiCl_2–aluminum alkyl catalyst is in the tetravalent state (128). Natta and co-workers identified the structure shown in V for the product of Cp_2TiCl_2 and $AlEt_3$ (97). It was

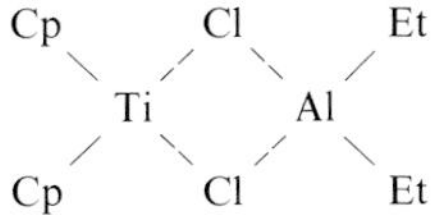

V

relatively less active for ethylene in comparison with freshly mixed mixtures. The Ti in this complex was trivalent, as Cp_2TiCl could be recovered as a donor complex by addition of $(CH_3)_3N$ (115). In agreement is the finding of Bartlett and Seidel that the trivalent derivative Cp_2TiCl in combination with Al-i-Bu_2Cl had only low activity for ethylene (101).

Drucker and Daniel found that if the titanium bore only one cyclopentadienyl group (for example, $CpTiCl_3$), it did not form an active catalyst in combination with $AlEt_3$ until some HCl was added (129).

E. CATALYSTS FROM VCl_4 AND $AlEt_2Cl$ ($-78°C$)

This catalyst and its modifications have been used exclusively for the polymerization of propylene to highly syndiotactic polypropylene at temperatures below about $-50°C$. Because this catalyst is uniquely different from the isotactic-specific ones, the author found it more useful to describe the experimental background in this section and the stereochemical mechanisms in Chapters 15, 16, and 17.

The potential synthesis and characterization of syndiotactic polypropylene was discussed by Natta long before its discovery. On the basis of theoretical arguments, it was anticipated that the melting point of polypropylene having the syndiotactic structure would be higher than found for the isotactic structure. Such a polypropylene, if it could be made, would have an important commercial value, because it could penetrate many high temperature applications then closed to isotactic polypropylene. Melting points above 190°C were suggested. Furthermore, if highly syndiotactic polypropylene could be made, physical chemists would have a polymer available for fundamental studies that could be made in any degree of isotacticity and syndiotacticity. But how does one alter a catalyst that has a very specific propensity to form long sequences of the isotactic configuration to one in which the configuration alternates perfectly adjacent propylene units? This section describes special Ziegler–Natta catalysts that have this ability.

Several reviews describing the synthesis and characterization, as well as the mechanistic features of a syndiotactic polypropylene polymerization, were published in 1967 and 1971 (5, 7, 8, 130).

1. Catalysts

Syndiotactic polypropylene has been synthesized with both V-based homogeneous and Ti-, V-, of Cr-based heterogeneous catalysts (131–142). Highly crystalline syndiotactic polymers that are completely free of the isotactic structure, however, can be made only with the soluble vanadium catalysts and only at very low temperatures.

a. Heterogeneous Catalysts. Syndiotactic polypropylene, albeit only partially crystalline, was first synthesized about 1959–1960 with heterogeneous Ziegler–Natta catalysts that ordinarily produce polymer containing an amorphous fraction, such as $AlEt_2F + \gamma TiCl_3 \cdot xAlCl_3$ (131–135). Several percent of partially syndiotactic polypropylene were isolated by extracting the crude polymer with boiling solvents such as ethyl ether, pentane, and hexane. These fractions had unique X-ray and infrared (131–135) spectra by which syndiotactic polypropylene was identified for the first time (this new polypropylene could be easily identified in the infrared spectrum by the presence of the 11.53 μ band).

Decreasing the temperature of polymerization from 0° to −78°C was reported to produce greater fractions of the syndiotactic polymer as an impurity in the isotactic product (132–134). However, one catalyst, $AlEt_2F + TiCl_3 \cdot AlCl_3$, gave a fairly high fraction of low crystalline syndiotactic polymer even at +70°C (132, 133). Addition of Et_3N to this catalyst prevented the syndiotactic polymer from forming (136), even though highly isotactic polymer was still formed. This suggests that syndiotactic polymer was formed at the more exposed Ti centers that lacked isotactic-regulating ability.

It should be mentioned that Natta and co-workers succeeded about this time in preparing a syndiotactic poly-1-butene by hydrogenating a syndiotactic polybutadiene polymer having a 1,2 structure (137). However, it was obtained only as an amorphous solid.

b. Soluble Catalysts.   Natta and co-workers (134) (examples 9 and 10) report that when $V(acac)_3$ is combined with $AlEt_2F$ or $AlEt_2Cl$, the formed polypropylene shows X-ray reflections only corresponding to lattice distances 7.25, 5.3, and 4.3 Å, characteristic of syndiotactic polymer. This appears to be the first report that soluble vanadium catalysts can produce the syndiotactic polymer free of the isotactic structure.

A more effective catalyst, based on combinations of AlR_2Cl, VCl_4, and optionally, anisole, were later (1962) reported by Natta, Pasquon, and Zambelli (Table 4–26) to produce very highly syndiotactic polypropylene that also was free of the isotactic structure (138). These catalysts were effective only at low polymerization temperatures, about $-78°C$.

TABLE 4–26

Polymerization of Propylene to a Syndiotactic Polymer at -78 C^a (138)

Vanadium compound^b	Organometallic compound	Al/V ratio	Solvent	Crystallinity index of syndiotactic polymer^c	Polymer obtained (g)
VA_3	$Al(C_2H_5)_2F$	5	Toluene	0.80	3
VA_3	$Al(C_2H_5)_2Cl$	5	Toluene	0.65	1.2
VA_3	$Al(C_2H_5)_2Cl$	10	Toluene	0	10.0
$VCl_4 \cdot AN$	$Al(C_2H_5)_2Cl$	5	Toluene	1	7.5
$VCl_4 \cdot AN$	$Al(C_2H_5)_2Cl$	10	Toluene	0.90	10.0
$VCl_4 \cdot AN$	$Al(C_2H_5)_2Cl$	2	Toluene	0	0.1
$VCl_4 \cdot AN$	$Al(i\text{-}C_4H_9)_2Cl$	5	Toluene	1.30	10.0
$VCl_4 \cdot AN$	$Al(neo\text{-}C_5H_{11})_2Cl$	5	Toluene	1.85	2.3
$VCl_4 \cdot AN$	$Al(i\text{-}C_4H_9)_2Cl$	5	n-Heptane	2.05	2.5
$VCl_4 \cdot AN$	$Al(neo\text{-}C_5H_{11})_2Cl$	5	n-Heptane	2.05	2.4

^{a} Length of time of runs, 20 hr; temperature of preparation of catalyst and of polymerization, $-78°C$; starting monomer, 90 g.

^{b} A = acetylacetonic residue; AN = anisole.

^{c} Crystallinity index for the syndiotactic polymer is referred to the polymer obtained by the system $VCl_4 \cdot AN$; $Al(C_2H_5)_2Cl$; Al/V = 5.

Infrared spectroscopy (on thin films) was routinely used to evaluate syndiotacticity and to detect if any isotactic polymer was also present. The syndio index, IS, was defined as

$$IS = A_{11.53\mu}/\tfrac{1}{2}(A_{2.32\mu} + A_{2.35\mu})$$

(Boor used the notation γ_N for IS; see Fig. 3–10.)

Subsequent work by the Natta group and others demonstrated that the syndiospecific catalyst was extremely delicate and restricted to very few combinations of vanadium compounds and aluminum alkyls.

Only vanadium compounds have been found effective (136, 138, 139). VCl_4 makes the most active catalyst but $V(acac)_3$ and various vanadates $VO(OR)_xCl_{3-x}$ ($x = 1, 2,$ and 3) were used but under more limited conditions. In combination with VCl_4, only AlR_2X alkyls were effective; other metal alkyls such as $AlEt_3$ or $ZnEt_2$ reduced the soluble vanadium chloride to a heterogeneous catalyst (131, 133, 134,). R in AlR_2X could be methyl, ethyl, isobutyl, neopentyl, phenyl, or methyl styryl (136, 138, 139). The size of R may influence syndiotacticity only under certain conditions of catalyst preparation. For example, $AlMe_2Cl + VCl_4$ forms a highly syndiospecific catalyst only when the concentration of the components are kept low (136). The syndiospecificity of the $AlR_2Cl–VCl_4$ catalyst was reported to increase as R was increased; that is, IS values of 1.0, 1.4, and 1.8 were found when R was ethyl, isobutyl, and neopentyl, respectively (138, 139). Other workers, however, found small differences for R = methyl, ethyl, or phenyl in the $AlR_2Cl–VCl_4$ catalyst (136). $AlEt_2Br–VCl_4$ produced a mixture of syndiotactic and isotactic polypropylenes (136).

Most effective catalysts were made when the AlR_2Cl/V compound ratios were between 3 and 10 (5, 138). In general, only when X in AlR_2X was Cl were syndiospecific catalysts formed. An exception is the combination of $V(acac)_3$ and $AlEt_2F$ (132, 133); however, $AlEt_2F + VCl_4$ produces a heterogeneous isotactic-specific catalyst (136).

Certain electron donors were added in controlled amounts to increase the syndiospecificity (IS) of $AlR_2Cl–VCl_4$ catalyst, that is, for the Al-i-$Bu_2Cl–VCl_4$ catalyst (134).

Donor	IS
Anisole	1.65
Furan	1.25
Et_2O	0.95
Cyclohexanone	0.90
Ethyl acetate	0.65
Thiophene	0.50
Pyridine	Inactivates catalyst
None	0.40

Anisole was most effective at an anisole/VCl_4 ratio of 1 (138, 139a). Only at low conversions of propylene was it effective in heptane solvent, but this was not true in toluene (136).

Cyclohexene, oxygen, and *tert*-butyl perbenzoate were also reported effective but only if used according to a specific order of addition (136).

2. Polymerization Conditions and Kinetics

Because it is very delicate, the syndiospecific catalyst is very sensitive to changes in solvent, temperature of synthesis and of polymerization, concentration of catalyst components, impurities, time of polymerization, etc. It requires stringent control over most variables and meticulous handling, as the following examples show.

a. Temperature and Time of Aging of Catalyst.

Only if the AlR_2Cl–VCl_4 catalyst is prepared and used at very low temperatures are the highest syndiotacticities obtained. This catalyst begins to precipitate quickly above $-40°C$, but even below $-40°C$ some deterioration occurs as the time of aging and polymerization increases (136). For example, the syndio index *IS* decreased from 1.8 to 1.2 in a 140 hour polymerization at $-78°C$ with the $AlEt_2Cl$–VCl_4 catalyst, while at $-45°C$ the index decreased from 0.9 to 0.5 in 3 hours.

If the preparation and aging are carried out at higher temperatures, the resulting catalysts are less syndiospecific. For example, the catalyst from $AlEt_2Cl$–VCl_4 was less syndiospecific if prepared and aged at $-45°C$ but used at $-78°C$ than if it was prepared, aged, and used entirely at $-78°C$ (28). In comparison, catalysts prepared, aged, and used at $-45°C$ were least syndiospecific. No isotactic polypropylene was found in the less syndiospecific samples.

It is apparent that low temperatures favor both the formation of the syndiospecific centers and the syndiotactic propagation. Figure 4–13 shows

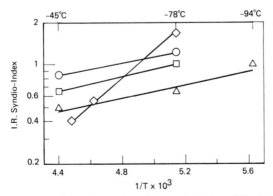

Fig. 4–13. Temperature dependence of syndiotacticity index (136). Catalyst for $-77°C$ and $-94°C$ polymerizations aged at $-45°C$; catalyst for $-45°C$ polymerization aged at $-78°C$. Aging time/polymerization time shown in brackets: $\Diamond$, Al-*i*-Bu$_2$Cl/VCl$_4$/anisole [0/24]; $\bigcirc$, AlEt$_2$Cl/VCl$_4$ [0.5/0.5]; $\square$, AlEt$_2$Cl/VCl$_4$ [1.5/1.5]; $\triangle$, AlEt$_2$Cl/VCl$_4$ [3.0/3.0].

a plot of *IS* vs. temperature of polymerization (136). The temperature dependence found by Zambelli, Natta, and Pasquon ($\diamond$ points) is greater than found by Boor and Youngman ($\bigcirc$, $\square$, $\triangle$ points) because it reflects combined effects while the latter attempts to separate the two effects.

b. Kinetic Features.   Zambelli and co-workers have carried out extensive investigations to understand the kinetics of the syndio catalyst (139, 141). Their work clearly shows that the kinetic behavior of this catalyst is very dependent on the precise conditions by which the catalyst is prepared and used for polymerization. The following salient features characterize this catalyst.

(1) Reaction time increased.   After an induction period, the amount of polymer formed is proportional to reaction time, polymer molecular weight increases, and syndiotacticity decreases.

(2) Concentration of VCl_4 increased.   When Al/V is kept high, the amount of polymer obtained is a linear function of the amount of VCl_4 used. Polymer molecular weight and syndiotacticity are not affected appreciably.

(3) Concentration of $AlEt_2Cl$ increased.   Above a certain limit and keeping other conditions the same, polymer yield, molecular weight, and syndiotacticity are decreased.

(4) Concentration of $AlEtCl_2$ and $AlEt_3$ contaminants increased. Upon addition of $AlEtCl_2$, polymer molecular weight and yield increases, maximizes, and then decreases while syndiotacticity decreases during the entire addition. Addition of $AlEt_3$ causes a rapid decrease in syndiotacticity but a steep increase in yield and molecular weight (presumably a heterogeneous isotactic catalyst is formed).

(5) Concentration of monomer increased.   Polymer yield and molecular weights maximize after a certain concentration of propylene is added, and then they decline. Syndiotacticity decreases during the initial addition but levels off. Monomer apparently has an effect on the catalyst formation or decomposition reactions.

(6) Solvent varied.   Polymer yields were higher in toluene than in heptane; however, syndiotactic indices were higher in heptane. Zambelli and co-workers suggested that catalytic complexes were more dissociated in aromatic solvent, resulting in higher propagation constants and lower syndiospecificity. Small amounts of impurities in the solvent markedly affect tacticity; for example, 0.6 mole % of impurities in toluene lowered the syndio index from 1.8 to 1.3 when the Al-*i*-Bu_2Cl– VCl_4–anisole catalyst was used (139).

(7) Al/V ratio increased.   Optimum syndiotacticity was obtained at a ratio equal to unity. At high ratios, the polymer yield decreased but molecular weight was mostly unchanged.

3. Olefins Polymerized

The syndiotactic-specific catalyst is active for polymerization of ethylene, propylene, higher α-olefins, dienes, and other monomers. Only with propylene, however, were syndiotactic polymers obtained (138, 139). Branched α-olefins such as 3-methyl-1-butene and 3,3-dimethyl-1-butene have been polymerized, but infrared and pyrolysis data showed them to have structures very similar to polymers prepared by cationic initiators, that is, mixtures of 1,2- and 1,3-addition units (136).

Copolymers have been synthesized from the comonomer pairs ethylene-propylene, ethylene-1-butene, and propylene-1-butene (139, 142). When added simultaneously, ethylene and propylene formed random copolymers but, when added sequentially, they formed block polymers. Evidence for block polymer formation was based on extraction data. The heptane-soluble extract contained long blocks of syndiotactic polypropylene and linear polyethylene; polyethylene formed under similar conditions does not dissolve in heptane.

The binary copolymers of ethylene, propylene, and 1-butene showed a tendency to an alternating sequence distribution when syndiospecific catalysts were used (Table 4–27) (142). In contrast, isotactic-specific catalysts under similar conditions produced copolymers having a random distribution.

TABLE 4–27

r_1r_2 Products for Ethylene–Propylene Copolymerization with Syndiotactic and Isotactic-Specific Catalysts (142)

Catalyst	$r_1r_2{}^a$
Syndiotactic	
$AlEt_2Cl$–VCl_4	0.25 ± 0.1
$AlEt_2Cl$–VCl_4–anisole	0.28 ± 0.02
Isotactic	
$AlEt_3$–VCl_4	1.0 ± 0.04
$GaEt_3$–VCl_4	1.14 ± 0.03

[a] $r_1r_2 \gg 1$ suggests a mixture of homopolymers; $r_1r_2 = 0$ suggests an alternation distribution; $r_1r_2 = 1$ suggests a random distribution.

The r_1r_2 products for ethylene-1-butene and propylene-1-butene copolymers indicated that alternation was favored.

	r_1	r_2	r_1r_2
Ethylene-1-butene	35.5	0.006	0.21
Propylene-1-butene	0.7	0.7	0.49

Syndiotactic poly-1-butene was formed at a very low rate. It resembled a product synthesized indirectly by hydrogenating syndiotactic poly-1,2-butadiene. Both were amorphous and are characterized by a 10.63μ band in the infrared spectrum.

Copolymers of propylene and 1-butene show syndiotactic crystallinity when the 1-butene content is as high as 30 mole%. It appears that CH_3 can be replaced by C_2H_5 in these syndiotactic copolymers without completely preventing crystallization. Similar findings were observed earlier for isotactic copolymers made from certain branched α-olefins (see Chapter 20, Section IV,C).

F. VCl_4 OR $VO(OR)_xCl_{3-x} = AlR_3$, $AlEt_2Cl$, OR $AlEtCl_2$

Binary mixtures from the above components and related structures have been examined as the preferred catalysts for copolymerizations of ethylene and propylene to products that had a random to alternating structure. Catalysts based on preformed $TiCl_3$ or $TiCl_4$ produced copolymers that were random to blocky. These catalysts are discussed in detail in Chapter 20.

IV. Conclusion

As a result of the extensive effort to find the most effective catalysts, a rather large number of metal alkyls and transition salts have been examined in olefin polymerizations. These investigations revealed that: (1) The activity and stereochemical behavior of the Ziegler–Natta catalyst is affected by the structure of both the metal alkyl and the transition metal salt. (2) Neither the metal alkyl nor the transition metal salt are inherently most stereoregulating or most active. (3) Aluminum alkyls followed distantly by zinc alkyls are preferred as the alkyl component. (4) Titanium and vanadium salts are preferred as the other component. By judicious choice of the binary mixture, the worker can formulate catalysts that are best suited for synthesis of highly linear polyethylene, highly isotactic polymer from propylene and higher α-olefins, highly syndiotactic polypropylene and either random, alternating, or blocky copolymers of ethylene and propylene. In fact, about a half dozen important classes of catalysts have emerged (see Chapter 5), whose polymerization chemistry has been extensively studied. Because titanium chlorides have been so widely used, much information about their synthesis and structure has become known. Vanadium-based catalysts, however, are necessary for the synthesis of syndiotactic polypropylene and random copolymers. Theoretical studies have been fruitful in both guiding catalyst research programs and in the understanding of the catalyst structures.

References

1. N. G. Gaylord and H. F. Mark, "Linear and Stereoregular Addition Polymers," pp. 90–106. Wiley (Interscience), New York, 1959.
2. D. O. Jordan, *in* "The Stereochemistry of Macromolecules" (A. D. Ketley, ed.), Chapter 1. Dekker, New York, 1967.
3. D. F. Hoeg, *in* "The Stereochemistry of Macromolecules" (A. D. Ketley, ed.), Chapter 2. Dekker, New York, 1967.
4. F. W. Breuer, L. E. Geipel, and A. B. Loebel, *in* "Crystalline Olefin Polymers" (R. A. Raff and K. W. Doak, eds.), Chapter 3. Wiley (Interscience), New York, 1965.
5. A. Zambelli, *Macromol. Prepr., Int. Congr. Pure Appl. Chem., 23rd, 1971* Vol. 1, pp. 124–128.
6. H. W. Coover, Jr., R. L. McConnell, and F. B. Joyner, *Macromol. Rev.* **1**, 91–118 (1966).
7. E. A. Youngman and J. Boor, *Macromol. Rev.* **2**, 33–69 (1967).
8. J. Boor, *Macromol. Rev.* **2**, 115–268 (1967).
9. J. Boor, *J. Polym. Sci., Part C* **1**, 237 (1963).
10. G. Natta, G. Mazzanti, D. DeLuca, U. Giannini, and F. Bondini, *Makromol. Chem.* **76**, 54 (1964).
11. M. Frankel, J. Rabani, and A. Zilkha, *J. Polym. Sci.* **28**, 387 (1958).
12. A. Zilkha, N. Calderon, A. Ottolenghi, and M. Frankel, *J. Polym. Sci.* **40**, 149 (1959).
13. P. Longi and A. Roggero, *Ann. Chim. (Rome)* **51**, 1013 (1961).
14. A. Zilkha, A. Ottolenghi, and M. Frankel, *J. Polym. Sci.* **39**, 347 (1959).
15. A. Zilkha, N. Calderon, and M. Frankel, *J. Polym. Sci.* **33**, 141 (1958).
16. A. L. J. Raum, Great Britain Patent 841,527, July 20, 1960, The Distillers Co., Ltd.
17. G. D. Brindell and A. F. Lundeen, Great Britain Patent 855,757, December 7, 1960, Continental Oil Company (equivalent to U.S. Patent 3,025,245).
18. K. A. Kocheshkov, V. A. Kargin, N. I. Sheverdina, T. I. Sogolova, I. Ye. Paleyeva, and O. A. Paleyev, *Polym. Sci. USSR (Engl. Transl.)* **4**, 1564 (1963).
19. O. A. Paleyev, N. I. Sheverdina, T. I. Sogolova, I. Ye. Paleyeva, V. A. Kargin, and K. A. Kocheshkov, *Polym. Sci. USSR (Engl. Transl.)* **8**, 6 (1966).
20. A. Verheyden and P. Ochsner, U.S. Patent 2,940,962, June 14, 1960, Union Chimique Belge, SA.
21. G. Natta, G. Mazzanti, and P. Longi, Great Britain Patent 882,560, November 15, 1961, Montecatini (equivalent to U.S. Patent 3,259,613).
22. A. P. Firsov, B. G. Kashporov, Yu. V. Kissin, and N. M. Chirkov, *Polym. Sci. USSR (Engl. Transl.)* **4**, 325 (1963).
22a. Great Britain Patent 856,645, December 21, 1960, California Research Corporation.
23. G. Natta, I. Pasquon, A. Zambelli, and G. Gatti, *J. Polym. Sci.* **51**, 387 (1961).
24. J. Boor, unpublished data.
25. T. Tanaka, T. Matsuda, K. Kimijima, I. Okuzumi, and K. Akutsu, *Kogyo Kagaku Zasshi* **73**, 1061 (1970).
26. Great Britain Patent 865,322, April 12, 1961, Farbenfabriken Bayer Aktiengesellschaft.
27. D. A. Fraser and A. L. J. Raum, Great Britain Patent 844,769, August 17, 1960, The Distillers Co., Ltd.
28. Ziegler *et al.* (28a–28c) contain excellent background material for these metal alkyls.
28a. K. Ziegler, *in* "Organometallic Chemistry" (H. Zeiss, ed.), American Chemical Society Monogr. Ser, No. 147, p. 194–269. Van Nostrand: Reinhold, Princeton, New Jersey, 1960.
28b. G. E. Coates, "Organo-Metallic Compounds," 2nd ed., pp. 126–143. Wiley, New York, 1960.

28c. T. Mole and E. A. Jeffery, "Organoaluminum Compounds." Elsevier Amsterdam, 1972.
29. H. N. Friedlander and K. Oita, *Ind. Eng. Chem.* **49**, No. 11, 1885 (1957).
30. N. Gaylord and H. F. Mark, "Linear and Stereoregular Addition Polymers," p. 129. Wiley (Interscience), New York, 1959.
31. F. Dawans and Ph. Teyssie, *Bull. Soc. Chim. Fr.* No. 10, p. 2376 (1963).
32. F. Danusso, *J. Polym. Sci., Part C* **4**, No. 4, Part 3, 1497 (1964).
33. Toray, Japanese Patent 36.848/71, October 21, 1971.
34. A. E. Borchert, A. B. Mekler, and R. W. Sauer, U.S. Patent 3,485,770, December 23, 1969, Atlantic Richfield Company.
35. M. Taniguchi, K. Ozaki, and N. Kawabata, *Polym. Rep. (Jpn.)* **24**, No. 263 (1967).
36. G. A. Baranova, N. N. Korneyev, B. A. Krentsel, and L. L. Stotskaya, *Polym. Sci. USSR (Engl. Transl.)* **9**, 1409 (1967).
37. H. Higashi and K. Watabe, *J. Polym. Sci., Part B* **5**, 681 (1967).
38. G. Natta, U. Giannini, E. Pellino, and D. De Luca, *Eur. Polym. J.* **3**, 391 (1967).
39. S. Cesca, W. Marconi, and L. Zerilli, *Chim. Ind. (Milan)* **51**, 976 (1969); W. Marconi, A. Mazzei, and L. Gatto, *Eur. Polym. J.* **5**, 747 (1969).
40. A. Langer, U.S. Patent 3,418,304, December 24, 1968, Esso Research and Engineering Company.
40a. D. C. Feay, private communication, see Chapter 12, Feay (92).
41. G. Natta, P. Pino, and G. Mazzanti, *Gazz. Chim. Ital.* **87**, 528 (1957); G. Natta, P. Pino, G. Mazzanti, P. Longi, *ibid.* pp. 549 and 570.
42. Y. Tajima and E. Kunioka, *J. Polym. Sci., Part A-1* **6**, 241 (1968).
43. R. C. Schreyer, U.S. Patent 2,986,531, May 30, 1961, E. I. du Pont de Nemours & Company.
44. G. Natta, P. Corradini, I. W., Bassi, and L. Porri, *Atti Accad. Naz. Lincei, Cl. Sci. Fis., Mat. Nat., Rend.* [8] **24**, 121 (1958).
45. G. Natta, P. Corradini, and G. Allegra, *J. Polym. Sci.* **51**, 399 (1961).
46. W. Klemm and E. Krose, *Z. Anorg. Chem.* **253**, 209 (1947).
47. "Gmelin's Handbuch der anorganischen Chemie," p. 841, Verlag Chemie, Weinheim, 1951.
48. E. Törnqvist, U.S. Patent 3,424,774, January 28, 1969, Esso Research and Engineering Company; also E. G. M. Törnqvist and A. W. Langer, U.S. Patent 3,032,513. May 1, 1962.
49. A. A. Korotkov and L. Tszun-Chan, *Polym. Sci. USSR (Engl. Transl.)* **3**, 621 (1962).
49a. E. Törnqvist, C. W. Seelbach, and A. W. Langer, U.S. Patents 3,128,252, April 7, 1964, and 3,001,951, Sept. 25, 1961, Exxon Chemical Company.
50. C. Beerman and H. Bestian, *Angew. Chem.* **71**, 618 (1959).
51. F. Bock and L. Moser, *Monatsh. Chem.* **34**, 1825 (1913); **33**, 1407 (1912).
52. Great Britain Patent 960,232, June 10, 1964, Hoechst; E. J. Vandenberg, U.S. Patent 3,108,973, October 29, 1963, Hercules; E. Törnqvist and A. W. Langer, U.S. Patent 3,032,510, May 1, 1962, Esso Research and Engineering Company.
53. G. H. Smith and D. C. Perry, *J. Polym. Sci., Part A-1* **7**, 707 (1969).
54. G. Natta, P. Corradini, and G. Allegra, *Atti Accad. Naz. Lincei, Cl. Sci. Fis., Mat. Nat., Rend.* [8] **26**, 159 (1959).
55. E. Törnqvist and A. W. Langer, Great Britain Patents 847,661, September 14, 1960, and 850,910, October 12, 1960, Esso Research and Engineering Company (equivalent to U.S. Patent 3,032,510).
56. G. Natta, I. Pasquon, and E. Giachetti, *Angew. Chem.* **69**, 213 (1957); *Makromol. Chem.* **24**, 258 (1957).
56a. E. Törnqvist and A. W. Langer, U.S. Patents 3,032,510, May 1, 1962 and 3,130,003, April 21, 1964. Exxon Chemical Company.

57. G. Natta, P. Corradini, and G. Allegra, *J. Polym. Sci.* **51**, 399 (1961).
58. Z. W. Wilchinsky, R. W. Looney, and E. G. M. Törnqvist, *J. Catal.* **28**, 351 (1973).
59. J. Brynestad, S. Von Winbush, H. L. Yakel, and G. P. Smith, *Inorg. Nucl. Chem. Lett.* **6**, 889 (1970).
60. M. H. de Jong and P. Van Prooijen, U.S. Patent 3,562,239, February 9, 1971, Shell Oil Company.
61. E. Törnqvist and A. W. Langer, Great Britain Patent 878,373, September 27, 1961, Esso Research and Engineering Company (equivalent to U.S. Patent 3,032,513).
62. L. Luciani and G. Corsi, U.S. Patent 3,560,146, February 2, 1971, Montecatini, Edison.
63. E. Törnqvist, U.S. Patent 3,531,420, September 29, 1970, Esso Research and Engineering Company.
64. L. Luciani and G. Corsi, U.S. Patent 3,461,083, August 12, 1969, Montecatini Edison.
65. S. Kanetaka, T. Takagi, and T. Keii, *Kogyo Kagaku Zasshi* **67**, No. 9, 1436 (1964).
66. E. W. Duck, D. Grant, A. V. Butcher, and D. G. Timms, *Eur. Polym. J.* **10**, 77 (1974).
66a. G. Natta, I. Pasquon, A. Zambelli, and G. Gatti, *J. Polym. Sci.* **51**, 388 (1961).
67. Stauffer Chemical Company, "Titanium Trichlorides," Tech. Bull. Stauffer Chem. Co., Speciality Chem. Co., Westport, Conn. 1962.
68. E. G. M. Törnqvist, *Ann. N.Y. Acad. Sci.* **155**, 447 (1969).
69. A. Schindler, *in* "Crystalline Olefin Polymers" (R. A. Raff and K. W. Doak, eds.), Chapter 5. Wiley (Interscience), New York, 1965.
70. L. Reich and A. Schindler, "Polymerization by Organometallic Compounds," Chapters III-1 and IV. Wiley (Interscience), New York, 1966.
71. G. Natta, *J. Polym. Sci.* **34**, 21 (1959).
72. G. Natta, I. Pasquon, A. Zambelli, and G. Gatti, *Makromol. Chem.* **70**, 191 (1964).
72a. A. P. Firsov, B. N. Kashprov, Y. V. Kissin, and N. M. Chirkov, *J. Polym. Sci.* **62**, S104 (1962).
73. J. G. Noltes and J. W. G. van Den Hurk, *J. Organomet. Chem.* **1**, 377 (1964).
74. J. Boersma and J. G. Noltes, *Tetrahedron Lett.* **14**, 1521 (1966).
75. E. Bonitz, *Chem. Ber.* **88**, 742 (1955).
76. R. D. Bushick and R. S. Stearns, *J. Polym. Sci., Part A-1* **4**, 215 (1966).
77. A. Zambelli, I. Pasquon, A. Marinangeli, G. Lanzi, and E. R. Mognaschi, *Chim. Ind. (Milan)* **46**, No. 12, 1464 (1964).
78. G. Natta, F. Danusso, and I. Pasquon, *Collect. Czech. Chem. Commun.* **22**, Spec. Issue, 191 (1957).
79. J. Boor, Note in *J. Polym. Sci., Part A-1* **9**, 3075 (1971).
80. S. Mostert, Doctorate Thesis presented to the Technische Hogeschool Twente Enschede, March 14, 1968, The Netherlands-Druco Krukkerijbedrijven N.V., Leiden.
81. B. M. Grieveson, *Makromol. Chem.* **84**, 93 (1965).
82. A. S. Hoffman, B. A. Fries, and P. C. Condit, *J. Polym. Sci., Part C* **4**, Part 1, 109 (1963).
83. I. D. Rubin, "Poly(1-butene)- Its Preparation and Properties," Gordon & Breach, New York, 1968.
84. See ref. 31 in J. Boor, Macromol. *Rev. 2*, p. 115–268, 1967.
85. H. W. Coover, Jr., F. B. Joyner, and N. H. Shearer, Jr., Belgian Patent 577,214, July 16, 1959, Eastman Kodak Company; also see U.S. Patent 3,549,608, December 22, 1970, issued to H. W. Coover and F. B. Joyner.
86. H. W. Coover, Jr., *J. Polym. Sci., Part C* **4**, 1511 (1964).
87. H. W. Coover, Jr. and F. B. Joyner, *J. Polym. Sci., Part A* **3**, 2407 (1965).
88. R. C. Combs, D. F. Slonaker, F. B. Joyner, and H. W. Coover, Jr., *J. Polym. Sci., Part A-1* **5**, 215 (1967).
89. C. W. Moberly and G. R. Kahle, U.S. Patent 3,554,993, January 12, 1971, Phillips Petroleum Company.

90. J. L. Jezl, H. Khelghatian, and L. D. Hague, U.S. Patent 3,441,551, April 29, 1969, AviSun.
91. W. R. Watt and C. D. Fisher, *J. Polym. Sci., Part B* **6**, 109 (1968); *Part A-1* **7**, 2815 (1969).
92. H. W. Coover, Jr. and F. B. Joyner, U.S. Patent 2,969,345, January 24, 1961; U.S. Patent 3,230,208, January 18, 1966, both assigned to Eastman Kodak Company.
93. H. W. Coover, Jr. and F. B. Joyner, U.S. Patent 3,629,222, December 21, 1971, Eastman Kodak Company.
94. K. Matsumura, Y. Atarashi, and O. Fukumoto, *J. Polym. Sci., Part. A-1* **9**, 485 (1971).
95. K. Matsumura, Y. Atarashi, and O. Fukumoto, *J. Polym. Sci., Part A-1* **7**, 311 (1969).
96. H. W. Coover Jr., and N. H. Shearer, Jr., U.S. Patents 2,951,066, August 30, 1960, and 2,956,999, October 18, 1960, Eastman Kodak Company.
97. G. Natta, P. Pino, G. Mazzanti, and R. Lanzo, *Chim. Ind. (Milan)* **39**, 1032 (1957); G. Natta, P. Pino, G. Mazzanti, and U. Giannini, *J. Inorg. Nucl. Chem.* **8**, 612 (1958).
98. D. S. Breslow and N. R. Newburg, *J. Am. Chem. Soc.* **79**, 5073 (1957); D. S. Breslow and N. R. Newburg, *ibid.* **81**, 81 (1959).
99. H. S. De Liefde Meijer, M. J. Jansen, and G. J. Van der Kerk, Thesis by H. J. De Liefde Meijer at University of Utrecht, November 1963.
100. J. A. Waters and G. A. Mortimer, *J. Polym. Sci., Part A-1* **10**, 1827 (1972).
101. P. J. Bartlett and B. Seidel, *J. Am. Chem. Soc.* **83**, 551 (1961).
102. J. A. Waters and G. A. Mortimer, *J. Polym. Sci., Part A-1* **10**, 895 (1972).
103. H. Höcker and K. Saiki, *Makromol. Chem.* **148**, 107 (1971).
104. G. Natta, *Chim. Ind. (Milan)* **39**, 19 (1957).
105. R. E. Wiman and I. D. Rubin, *Makromol. Chem.* **94**, 160 (1966).
106. M. G. Marti and K. H. Reichert, *Makromol. Chem.* **144**, 17 (1971).
107. P. Ye. Matkovskii, G. P. Belov, A. P. Lisitskaya, L. N. Russiyan, Kh. M. A. Brikenshtein, M. P. Gerasina, and N. M. Chirkov, *Polym. Sci. USSR (Engl. Transl.)* **12**, 1890 (1970).
108. I. D. Rubin, *J. Polym. Sci., Part A-1* **5**, 1119 (1967).
109. P. Ye. Matkovskii, G. P. Belov, L. N. Russiyan, A. P. Lisitskaya, Y. V. Kissin, T. I. Solov'eva, A. A. Brikenshtein, and N. M. Chirkov, *Polym. Sci. USSR (Engl. Transl.)* **12**, 2590 (1971).
110. K. H. Reichert, J. Berthhold, and V. Dornow, *Makromol. Chem.* **121**, 258 (1969).
111. G. P. Belov, A. P. Lisitskaya, N. M. Chirkov, and V. I. Tsvetkova, *Polym. Sci. USSR (Engl. Transl.)* **9**, 1417 (1967).
112. K. Meyer and K. H. Reichert, *Angew. Makromol. Chem.* **12**, 175 (1970).
113. A. Kader, M. Hilal, E. Naga, and H. Abou, *J. Chem. UAR* **12**, 447 (1969).
114. A. J. Amass, J. N. Hay, J. C. Robb, *Brt. Polym. J.* **1**, 273 (1969) (see also pp. 277 and 282).
115. G. P. Belov, N. B. Bogomolova, V. I. Tsvetkova, and N. M. Chirkov, *Kinet. Catal.* **8**, 229 (1967).
116. G. P. Belov, A. P. Lisitskaya, T. I. Solovyeva, and N. M. Chirkov, *Eur. Polym. J.* **6**, 29 (1970).
117. Great Britain Patent 875,078, August 16, 1961, Montecatini.
118. M. D. C. Feay, French Patent 1,378,890, November 20, 1964, Dow Chemical Company.
119. F. J. Karol and W. L. Carrick, *J. Am. Chem. Soc.* **83**, 2654 (1961); G. W. Phillips and W. L. Carrick, *ibid.* **84**, 920 (1962); *J. Polym. Sci.* **59**, 401 (1962).
120. G. L. Karapinka and W. L. Carrick, *J. Polym. Sci.* **55**, 145 (1961).
121. D. S. Breslow and W. P. Long, U.S. Patent 3,013,002, December 12, 1961, Hercules.
122. D. S. Breslow, U.S. Patent 2,924,593, February 9, 1960, Hercules.
123. E. A. Fushman, V. I. Tsvetkova, and N. M. Chirkov, *Izv. Akad. Nauk SSSR, Khim. Ser.* p. 2075 (1965).
124. P. Pino, G. Mazzanti, U. Giannini, and S. Cesca, *Atti Accad. Naz. Lincei, Cl. Sci. Fis., Mat. Nat. Rend.* [8] **27**, 392 (1959).

125. A. Schindler, *in* "Crystalline Olefin Polymers" (R. A. Raff and K. W. Doak, eds.), Chapter 5, pp. 163–167. Wiley (Interscience), New York, 1965.

126. N. D. Zavorokhin, *Tr. Inst. Khim. Nauk, Akad. Nauk Kaz. SSR* **23**, 3–35 (1969).

127. C. E. H. Bawn and R. Symcox, *J. Polym. Sci.* **34**, 139 (1959).

128. Evidence for this conclusion is presented in Chapter 11.

129. A. Drucker and J. Daniel, *J. Polym. Sci.* **37**, 553 (1959).

130. J. Boor, *Ind. Eng. Chem., Prod. Res. Dev.* **9**, No. 4, 437–456 (1970).

131. G. Natta, P. Corradini, I. Pasquon, U. Pegoraro, and M. Peraldo, U.S. Patent 3,258,455, June 28, 1966, Montecatini; see also G. Natta, P. Longi, G. Mazzanti, and A. Roggero, U.S. Patent 3,257,370, June 21, 1966, Montecatini.

132. G. Natta, I. Pasquon, P. Corradini, M. Peraldo, M. Pegoraro, and A. Zambelli, *Atti Accad. Naz. Lincei, Cl. Sci. Fis., Mat. Nat., Rend.* [8] **28**, 539 (1960).

133. G. Natta, P. Corradini, I. Pasquon, M. Pegoraro, and M. Peraldo, Australian Patent 253,856, September 9, 1964, Montecatini.

134. Irish Patent Appl. 430/60, June 7, 1960, Montecatini and K. Ziegler (data from Natta and co-workers).

135. E. J. Addink and J. Beintema, *Polymer* **2**, 185 (1961).

136. J. Boor and E. A. Youngman, *J. Polym. Sci., Part A-1* **4**, 1861 (1966).

137. G. Natta, L. Porri, P. Corradini, D. Moreno, and I. Borghi, *Atti Accad. Naz. Lincei, Cl. Sci. Fis., Mat. Nat., Rend.* [8] **28**, 452 (1960).

138. G. Natta, I. Pasquon, and A. Zambelli, *J. Am. Chem. Soc.* **84**, 1488 (1962).

139. A. Zambelli, G. Natta, and I. Pasquon, *J. Polym. Sci., Part C* **4**, 411 (1963).

139a. A. Zambelli, I. Pasquon, R. Signorini, and G. Natta, *Makromol. Chem.* **112**, 160 (1968).

140. G. Natta, A. Zambelli, G. Lanzi, I. Pasquon, E. R. Mognaschi, A. L. Segré, and P. Centola, *Makromol. Chem.* **81**, 161 (1965).

141. A. Zambelli, G. Natta, I. Pasquon, and R. Signorini, *J. Polym. Sci., Part C* **16**, 2485 (1967).

142. A. Zambelli, A. Lety, C. Tosi, and I. Pasquon, *Makromol. Chem.* **115**, 73 (1968).

5

The Chemical Description
of Catalysts for Conjugated Dienes

I. Introduction

The uniqueness, versatility, and delicateness of Ziegler–Natta catalysts is most dramatically exhibited in the polymerization of dienes. Butadiene, isoprene, and 1,3-pentadiene have been most extensively studied, and these have been polymerized to products exclusively, or in very high concentration, containing one of the many possible stereochemical structures such as *cis*-1,4, *trans*-1,4, 1,2-isotactic or 1,2-syndiotactic, and 3,4 structures. Polybutadiene has been made exclusively in all four of the possible structures; not all of the possible structures were synthesized in isoprene and 1,3-pentadiene polymers.

The preferred catalyst usually consists of an aluminum alkyl and a transition metal salt based on Ti, V, Cr, Mo, Ni, or Co metals. The stereochemical and kinetic characteristics of a catalyst for a specific diene depend on the particular combination of the above components and on the way they are reacted and used in a polymerization.

In general, the catalysts and conditions of polymerization that are most attractive for dienes are not correspondingly attractive for olefins, and for this reason these two monomer types are discussed in separate chapters. The literature is voluminous, especially for butadiene, which was investigated extensively in both academic and industrial laboratories. A number of excellent reviews have been published: in 1967 by Cooper and Vaughan (1) and by Marconi (2), in 1969 by Natta and Porri (3), in 1970 by I. Pasquon (3a) and Cooper (4), and in 1971 by Dolgoplosk (5). Reference 1 collects numerous examples in tabular form according to the transition metal and diene polymerized. References 1 to 5 contain other important aspects of diene polymerizations that are outside the scope of this book.

Metal alkyl-free catalysts that are active and stereoregulating for dienes are discussed later, in Chapter 12.

This chapter focuses mainly on the activity and specificity features of the catalysts. Once an attractive catalyst is found, other important problems must be solved before the process is commercialized, such as controlling gel contents, molecular weights, and their distributions. Often these pose the most difficult barriers to commercialization, and their solutions are just as important, if not more, as those that preceded them. Reference 5a illustrates this for *cis*-1,4-polyisoprene.

II. Factors Determining Isomerism

Section III,B in Chapter 3 describes the different isomeric microstructural units that are possible in polymers of butadiene, isoprene, and 1,3-pentadiene (such as *cis*-1,4, *trans*-1,4, 1,2, etc.). Understandably, one asks: What rules or generalizations have been found in selecting a specific catalyst that will predominantly give only one of these structures? Unfortunately, a unique list of generalizations is not at hand. True, some catalysts exclusively produce a particular structure and are virtually insensitive to reaction conditions as do vanadium-based catalysts, for example. Yet, other catalysts produce the desired structure only under very narrow conditions, and the same catalyst will produce still another structure in high concentration if one deviates from these conditions. The product may contain more than one structure. In this case, a fraction that has one of these structures in a very high concentration can usually be isolated by solvent extraction methods. This suggests that different centers are present in the catalyst.

Three important factors are presented now that, for specific catalysts, have been found to influence the type of isomerism that the diene molecules developed for specific systems during the growth step. These factors include: (1) choice of metal alkyl—group number, nature of R or Ar, other ligands; (2) choice of transition metal salt—group number, valence, ligands, crystalline forms; (3) the way the catalyst components are reacted—ratio of components, presence of monomer, preformed polymer, donor and certain solvents, order of mixing, and time and temperature of aging.

For comparison, it is useful to note the microstructures of polybutadienes and polyisoprenes synthesized with the help of radical, anionic, and cationic catalysts, as shown in Table 5–1 (1).

TABLE 5–1

Microstructure of Polydienes Prepared with Radical, Anionic, and Cationic Initiators[a] (1)

Initiator	Polybutadiene	Polyisoprene
Radical[b]		
cis-1,4	19	18
trans-1,4	60	75
1,2	21	—
3,4	—	5
Anionic[c] (LiBu)		
cis-1,4	35	93
trans-1,4	52	0
1,2	13	—
3,4	—	7
Cationic[d,e]		
cis-1,4	—	—
trans-1,4	28	88.8
1,2	7.6	6.1
3,4	—	5.1

[a] Variations in conditions of polymerization and choice of initiators may result in other distributions of microstructures.

[b] Carried out at 50°C; addition of ethers causes an increase in 1,2 content in polybutadiene.

[c] Carried out at 25° to 50°C; sodium alkyls would give higher contents of 1,2-microstructure in polybutadiene.

[d] Carried out at 30° to −78°C.

[e] Some cyclized insoluble polymer is also formed with both dienes.

A. METAL ALKYL

The nature of the metal alkyl influences the stereochemistry of the polydiene and may be varied by choice of base metal and by the groups or ligands attached to the metal.

1. Group of the Metal

Aluminum alkyls have been preferred overwhelmingly, and, in fact, the major steorechemical structures have been synthesized with an aluminum alkyl as the metal alkyl component of the Ziegler–Natta catalyst.

Metal alkyls from groups I and II have also been used with interesting results. For example, whereas $AlEt_3$–VCl_4 produces a high trans-1,4-polybutadiene (6), the catalyst NaR–VCl_4 produces polybutadiene with only 21% trans-1,4 structures, the rest being 69% 1,2 and 10% cis-1,4 (7, 8). There appears to be a tendency for group I metal alkyls to favor 1,2-addition. Exceptions exist; in combination with $TiCl_4$, metal alkyls from Li (8), Zn (9), Mg (8, 9), and Cd (8, 10) metals produce mainly *trans*-1,4-polybutadiene. Because group I metal alkyls can polymerize dienes in the absence of transition metal salts, the results obtained with the corresponding Ziegler–Natta catalysts should be interpreted with caution, especially when the metal alkyl is used in excess.

2. Nature of the Ligands

Important findings were made with aluminum alkyls that had the structures AlR_3, AlR_2Cl, $AlRCl_2$, and $LiAlR_4$, where R was usually ethyl. The catalyst activity was affected by the structure of the metal alkyl as well as the stereochemical nature of the obtained polymer. New metal alkyls often arise from exchange reactions. Some examples are given below.

Bresler and co-workers (11) showed that $\beta TiCl_3 \cdot xAlCl_3$ in combination with Al-*i*-Bu$_3$ was twelve times more active for isoprene polymerization than was the same salt with Al-*i*-Bu$_2$Cl. The more active catalyst, however, underwent a faster decline in activity.

The reducing and alkylating power for the following aluminum alkyls decreases in the order: $AlEt_3 > AlEt_2Cl > AlEtCl_2$, and $AlEt_3 > AlEt_2I$. One consequence for a butadiene polymerization is a low activity for the $AlEt_2I$–$TiCl_4$ catalyst, which otherwise would be an attractive catalyst because it gives a high *cis*-1,4 content (12). By the addition of a more powerful reducing metal alkyl such as $AlEt_3$ ($AlEt_3/TiCl_4 \geq 3$ and $AlEt_2I/TiCl_4 \geq 2$), a highly active ternary catalyst is formed that produces a polybutadiene with about 90% *cis*-1,4 structure content if aromatic solvents are used. Only 70 to 80% of the *cis*-1,4-polymer is obtained in aliphatic solvents. Evidence was presented which demonstrated that an exchange reaction occurred between $AlEt_2I$ and $TiCl_4$ in addition to the normal alkylation–reduction steps, and a TiI_3 salt (probably $TiI_3 \cdot xAlX_3$) was formed that favored a *cis*-1,4 addition of butadiene (12). When $AlEt_3$ was used, the formed Ti salt was $TiCl_3$ ($TiCl_3 \cdot xAlCl_3$). As a point of reference, mixtures of $TiCl_4$ and $AlEt_3$ produce a 91% *trans*-1,4 content in a butadiene polymerization. If $AlEt_2I$ is used in place of $AlEt_3$, the polymer contains 90% or greater *cis*-1,4 structures (12).

Dolgoplosk and co-workers (13) showed for butadiene and isoprene that substituting Cl for Et in the aluminum alkyl (for example, $AlEt_2Cl$ vs. $AlEt_3$) resulted in only modestly different contents of the possible isomeric structures when $CoCl_2$ was used as the transition metal salt (see Table 5–2). Takahashi and co-workers (14) demonstrated that addition of alkyl halides to the $AlEt_3$–cobalt salt catalysts altered their ability to form the *cis*-1,4 units more favorably at the expense of 1,2 units in a butadiene polymerization.

TABLE 5-2

Dependence of Microstructure on Ligands in Metal Alkyl

Heterogeneous catalyst	Polybutadiene			Polyisoprene			
	cis-1,4	trans-1,4	1,2	cis-1,4	trans-1,4	1,2	3,4
$AlEt_2Cl + CoCl_2$	87	7	6	70	15	2	13
$AlEt_3 + CoCl_2$	74	10	16	64	11	0	25

Since $AlEt_2Cl$–cobalt salt catalysts give cis-1,4 structures and since $AlEt_3$ and alkyl halides react to form $AlEt_2X$ derivatives, the explanation offered was that a new $AlEt_2X$–cobalt salt catalyst was formed (compare data in Table 5–2).

Attractive catalysts are made with the use of aluminum hydride derivatives in place of aluminum alkyls. A few examples illustrate their usefulness. $LiAlH_4$ in combination with $TiCl_4$ in ether favors 1,2-addition in a butadiene polymerization, while trans-1,4 addition occurs predominantly with TiI_4 (15). If ether was not used, cis-1,4 addition took place with mixtures of $LiAlH_4$ and TiI_4, TiI_2Cl_2, or $TiCl_4$ (16).

Rather exotic aluminum alkyls have also been reported to be effective. For example, the polyiminoalane

$$-[\overset{\displaystyle H}{\underset{\displaystyle |}{Al\text{-}NR}}]_n-$$

I

was combined with AlI_3 and $TiCl_4$, $TiCl_2I_2$, or $Co(acac)_3$–$AlBr_3$ to form an active catalyst for butadiene that had a prevailingly cis-1,4 structure (17).

The nature of the ligand may determine if the catalyst will become soluble or heterogeneous (18). For example, halogen ligands favor formation of heterogeneous catalysts, while ligands bearing oxygen and nitrogen atoms favor homogeneous catalysts (see Section II,B,2). Aloxanes, formed by partial hydrolysis by $AlEt_2Cl$, have been combined with $Co(acac)_3$ to form a highly specific catalyst for butadiene (cis-1,4 structure) (19).

Porri and Carbonaro (20) concluded that the first stage of the reaction between cobalt acetylacetonate $[Co(acac)_2]$ and $AlEt_2Cl$ consisted of an exchange between the Cl bonded to Al and the acac ligands bonded to Co, as shown in reaction 5–1.

$$\underset{Et}{\overset{Et}{>}}Al\overset{Cl}{<} + \underset{acac}{\overset{acac}{>}}Co \longrightarrow \underset{Et}{\overset{Et}{>}}Al\text{---}acac + \underset{Cl}{\overset{acac}{\underset{|}{Co<}}} \qquad (5\text{-}1)$$

They proposed that the active complexes can contain the same Co–Cl species, independent of the structure of the starting Co compound. Usually, high Al/Co ratios are used.

Cucinella and co-workers (20a) report the synthesis of high cis-1,4-polyisoprene (96%) and high trans-1,4-polybutadiene (94%) in the presence of $AlRCl_2$–$Ti(OR)_4$ catalyst. At an Al/Ti ratio of 4, $\beta TiCl_3$ is formed, and the maximum activity and stereospecificity for isoprene is found. At lower ratios, chloroalkoxide derivatives of Ti with different compositions separate, but these were inactive for both dienes. High trans-1,4-polybutadiene (94%) was found at a ratio of Al/Ti $\simeq$ 10. These authors proposed that the exchange of ligands (OR and Cl) occurred in varying degrees. In fact, they could also produce 96% cis-1,4-polyisoprene by combining preformed $\beta TiCl_3$ and Al(OEt)ClEt. Kollar, Schnecko, and Kern (20b), who studied the interaction of $Ti(OPr)_4$ and $AlEt_2Cl$, calculated an overall activation energy of catalyst formation of about 17 kcal/mole.

B. TRANSITION METAL SALT STRUCTURE

The structure of the transition metal salt is extremely important, and thus a variety of transition metal salts differing in metal group number, valence, ligands, and crystalline forms have been studied.

1. The Group of the Metal

Vanadium halides in combination with aluminum alkyls almost exclusively produce high trans-1,4 structures from butadiene and isoprene, regardless of polymerization conditions (6). For example, 95 to 99% trans-1,4-polybutadienes have been produced with catalyst consisting of $VOCl_3$ or VCl_4 and $AlEt_3$ or $AlEt_2Cl$. Similarly, $AlEt_3$–$VOCl_3$ and $AlEt_3$–VCl_3 produced polyisoprenes that had 91 to 93% and 99% trans-1,4 contents, respectively. A few exceptions are found. When salts containing no halides are used in combination with $AlEt_3$ [for example, $V(acac)_3$], the prevailing structures are 1,2 in polybutadiene and 3,4 in polyisoprene (21, 22).

Cobalt salts in combination with AlR_2X or $AlRX_2$ (X = halogen = Cl) alkyls produce high cis-1,4-polybutadiene (for example, see Table 5–6). Polymer structure is independent of the anion attached to the Co atom and is largely insensitive to the Al/Co ratio. A variety of ligands have been examined, including chloride, bromide, iodide, alcoholate, acetate, naphthenate, aryl sulfonate, phosphate, carbonate, etc. When the salt is insoluble, it can be solubilized with donors, such as $CoCl_2$ plus ethanol, pyridine, or tributyl phosphate. Polyisoprene that is formed with cobalt-based catalysts contains a lower content of cis-1,4 structure, about 50 to 75% (23).

Chromium and molybdenum salts containing nonhalogen ligands favor the formation of 1,2-polybutadiene, such as $AlEt_3$ + $Cr(acac)_3$ (23a) or

$AlR_3 + MoO_2(OR)$ (23b). Nickel and iron salts also produce high *cis*-1,4-polybutadienes (Ni > Fe) (24).

2. Ligands Attached to the Transition Metal

In some catalysts, the ligand that is attached to the transition metal plays an important role in determining to what degree a particular stereochemical structure is formed.

An example is the polymerization of butadiene with the AlR_3–TiX_4 catalyst where X = I, Br, or Cl (24a, 25). The *cis*-1,4 content decreases in the order I > Br > Cl (see Table 5–3).

TABLE 5–3

Dependence of Microstructure on Ligand of Transition Metal Salt

	X in TiX_4		
Structure	I	Br	Cl
cis-1,4	95	88	6
trans-1,4	2	3	91
1,2	3	9	3

The catalyst $LiAlH_4$–TiI_4 polymerized isoprene to a high *trans*-1,4 content (26). The binary mixture $AlEt_3$–TiI_4, however, was inactive for the polymerization of isoprene. Substituting Cl for I or Br altered the ability of the catalyst to produce prevailingly *trans*-1,4 instead of *cis*-1,4 structures (27, 28). Other ligands in these catalysts, such as X = OR (28) or X = NEt_2 (29), produced high contents of the 1,2 structure (90 to 100% and 85%, respectively). There is a tendency for halide containing catalysts (X = I, Br) to produce 1,4 structures and polar ligands such as alcoholates (27, 28), acetylacetonates, etc., to produce 1,2 structures in polybutadiene and 3,4 structures in polyisoprene (27, 28).

The nature of the ligand may decide if the catalyst will become soluble or heterogeneous (18). For example, halogen ligands favor formation of heterogeneous catalysts, while ligands bearing oxygen and nitrogen atoms favor homogeneous catalysts, as shown in the following tabulation.

Heterogeneous catalysts	Homogeneous catalysts
$AlEt_3$ or $AlEt_2Cl$ + $TiCl_4$, $TiCl_3$, VCl_3, $VOCl_3$, or VCl_4	$AlEt_3 + TiCl_4$, $VO(OR)_3$, $V(acac)_3$, $Cr(acac)_3$, or $Co(acac)_2$ $AlEt_2Cl + CoCl_2 \cdot 2$ pyridine, $Co(acac)_3$, or $VCl_3 \cdot THF$

Complexes of the transition metal salt with electron donors can alter the solubility of the catalyst, such as complexes with carbon monoxide, amines, isonitriles, nitriles, and other heteroatom-containing organic molecules (30).

While the homogeneous $(\pi\text{-}C_5H_5)_2TiCl_2$–$AlEt_2Cl$-type catalysts are inactive for polymerization of dienes, indenyl–titanium dihalides are active and stereospecific in combination with R_2AlX (31), as shown in structure II.

II

When $X = I$, high *cis*-1,4-polybutadiene is formed, while with $X = Br$, mainly *trans*-1,4-polybutadiene is obtained. By substituting $RMgX$ for AlR_2X in the latter catalyst, a 1,2-addition is induced.

3. Crystalline Form of the Transition Metal Salt

The crystalline form of $TiCl_3 \cdot xAlCl_3$ can influence the stereochemical polymerization (32, 33). The layer structures (α, γ, and δ) give prevailingly *trans*-1,4-polybutadiene, while the β chain structure gives *trans*-1,4 polymer along with some of the *cis*-1,4 structure. When $AlEt_2I$ was used in place of $AlEt_2Cl$ in combination with $\beta TiCl_3 \cdot xAlCl_3$, the fraction of *cis*-1,4 structure was considerably greater (about 90%) (12). Similarly, in combination with $AlEt_3$, α- and $\beta TiCl_3$ salts gave 91% *trans*-1,4 and 85% *cis*-1,4-polyisoprenes, respectively (32).

4. Initial Valence of the Transition Metal

A small decrease (from 99 to 95%) in the *trans*-1,4 content of polybutadiene was noted on going from VCl_3 to VCl_4 to $VOCl_3$ (6, 22); $AlEt_3$ was used as cocatalyst. When tetravalent titanium compounds are reduced with aluminum alkyls ($Al/Ti = 1$), insoluble solids are formed which contain the titanium in a trivalent state. Examples are $TiCl_4$, TiI_4, $TiCl_2I_2$, $Ti(OR)_4$, and $TiCl_n(OR)_{4-n}$ [see Cooper and Vaughan (1, p. 127)]. The active valence state has not been generally established (34).

C. THE INTERACTION AND USE OF CATALYST COMPONENTS

The catalyst activity and stereochemistry is frequently dependent on such factors as the ratio of the components, the presence of the donor, monomer, or polymer during preparation, as well as the solvent, order of mixing, and other conditions of polymerization.

1. Ratio of Components

For some catalysts, the ratio of metal alkyl to transition metal salt determines the prevailing isomeric structures. For example, at low Al/Ti ratios for both butadiene and isoprene, the AlR_3-TiCl_4 catalyst produces polymers of high trans-1,4 structure (35–37). Because the polymers have low molecular weight and contain cyclized material, the participation of a cationic mechanism has been suggested. Yet at Al/Ti ratios $\simeq 1$, isoprene has been polymerized to a product containing greater than 96% cis-1,4 structure.

In comparison, when Al/Ti ≥ 2 was used for polymerization of butadiene, the "as obtained product" contained both cis- and trans-1,4 structures. But a fraction containing 95% cis-1,4 structure could be isolated from this crude mechanical mixture by crystallization from ethyl ether.

The importance of selecting a proper ratio is also demonstrated for the $AlEt_3-TiX_4$ catalyst (butadiene polymerization). The ratio needed to obtain maximum activity increased as the electronegativity of X in TiX_4 decreased, that is $Cl < Br < I < OR$ (38). Gibbs and co-workers (39) showed that for isoprene polymerization with the $AlEt_3-TiCl_4$ catalyst, monomer conversion increased with increasing Ti/Al ratios (ratios between 0.6 and 1 were examined).

With the $AlEt_3-Cr(CNC_6H_5)_6$ catalyst at an Al/Cr ratio of 2, a 1,2-syndiotactic product free of isotactic structure was obtained (30). If the ratio was increased up to 8, a mixture of isotactic and syndiotactic structures was obtained. At ratios above 10, only isotactic 1,2-polybutadiene was obtained (see also Section 5).

In contrast, the ratio of $AlEt_2Cl$ to Co salt in most of the solubilized versions of this catalyst has no effect over a rather wide range. High cis-1,4-polybutadiene is formed independent of the ratio (40).

Susa (41) found a maximum syndiotacticity and conversion when the $AlEt_2Cl/AlEt_3$ ratio was equal to 0.66 for the polymerization of butadiene with the quaternary system: $AlEt_3-AlEt_2Cl-CoCl_2$-pyridine.

2. Presence of a Donor

When butadiene was polymerized in benzene with $AlEt_2Cl$–cobalt salt catalysts, a high cis-1,4-polybutadiene was obtained (42). However, if Et_3N or Me_3N was also present, but only in high concentrations, a trans-1,4-polybutadiene was exclusively obtained.

Addition of ethers to the $AlH_3 \cdot TiCl_4$ catalyst (ether/Ti > 1) causes activity and cis-1,4 contents to decrease (43). The final limiting cis-1,4 contents that are obtained varied with the structure of the ether, for example, 30% for Me_2O, 50% for MeOEt, and 70% for Et_2O.

Other examples are cited in Chapter 9 and in Section III,A,1.

3. Presence of a Diene Monomer or a Diene Polymer

By preparing the catalyst $AlR_3-Cr(acac)_3$ in the presence of butadiene monomer, a 1,2-polymer was obtained that was free of the isotactic structure. If butadiene was not present, the 1,2-polymer contained equal amounts of the isotactic and syndiotactic structures (30). This effect is most unique when working with dienes (44).

The presence of butadiene in the $AlEt_2Cl$ (or $AlEt_3$)–Co(acac)$_2$ catalyst (THF as solvent) prevented reduction of the cobalt to metallic cobalt. The resulting catalyst was able to produce in low yields a *trans*-1,4-polybutadiene (45).

The presence in small amounts of high *cis*-1,4-polybutadiene or polyisoprene in the $TiCl_4$ when $AlEt_2Cl$ was added in benzene solvent produced a highly active catalyst for isoprene; for example, a 97% *cis*-1,4 content was obtained (46).

Cobalt catalysts prepared in the presence of butadiene or an aromatic solvent were found to be more stable. The following explanation was offered by Matsuzaki and Yasukawa (47). Aryl or alkenyl derivatives of cobalt are more stable than aliphatic-substituted cobalt, as shown below.

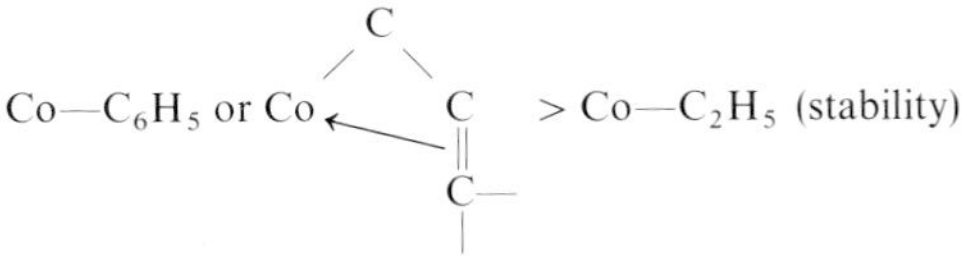

$$Co-C_6H_5 \text{ or } Co\underset{\underset{C-}{\overset{C}{\underset{\|}{C}}}{\overset{C}{\diagup}}}{\longleftarrow} \quad > Co-C_2H_5 \text{ (stability)} \qquad (5\text{-}2)$$

Because π-electrons from the aromatic ring or from the alkenyl group can be donated into the vacant d orbital of the Co atom, the Co–C bond is stabilized against decomposition. Alternately, when the ligand is the alkenyl group, a stable π-allyl bond can form. In practice, when $AlEt_2Cl$ and a cobalt salt are mixed in an aliphatic solvent, reduction occurs readily and metallic cobalt is produced. Apparently, π electron-donating molecules are able to stabilize the Co–C bond by a similar mechanism.

4. Type of Solvent

Horne and Carman (48) showed that the $(RO)_4Ti$–$AlEtCl_2$ catalyst can give either *cis*- or *trans*-1,4-polyisoprene, depending on the structure of the titanate and solvent, as shown in Table 5–4.

When $AlEt_2Cl$–Co(acac)$_3$ catalyst is used for polymerization of 1,3-pentadiene, a 1,2-syndiotactic polymer is obtained in aliphatic solvents, while in aromatic solvents, a syndiotactic-*cis*-1,4-polypentadiene is made (49).

TABLE 5–4

Dependence of Microstructure on Solvent

Titanate	Solvent	Structure
Primary R	Aliphatic or aromatic	*cis*-1,4
Secondary R	Aliphatic	*cis*-1,4
Secondary R	Aromatic	*trans*-1,4
Tertiary R	Aliphatic or aromatic	*trans*-1,4

TABLE 5-5

Dependence of Microstructure on Aging

	Structure	
Catalysts	Unaged	Aged
AlR_3–$Cr(acac)_3$	15% syndiotactic	15% isotactic
AlR_3–$CrCNC_6H_5$	86% amorphous	54% amorphous
	7% syndiotactic	46% isotactic
	7% isotactic	

5. Aging of Catalyst

Certain chromium-based catalysts are unique in that they produce syndiotactic 1,2-polybutadiene when unaged and isotactic polybutadiene when aged (23a), as shown in Table 5–5.

6. Conditions of Polymerization

Saltman and Kuzma (50, 51) report that the *trans*-1,4 content of a polymer chain was higher during the initial period of growth (Fig. 5-1). Butadiene was polymerized with the $Ni(octanoate)_2$–$BF_3 \cdot Et_2O$–$AlEt_3$ catalyst. Varying amounts of butadiene per Ni were used. With increasing BD/Ni ratios, molecular weight and *cis*-1,4 content increased. At the lowest molecular weight, only a 50 to 60% *cis*-1,4 content was obtained (about 10 to 15% vinyl, 25 to 30% *trans*-1,4). At high molecular weights ($\bar{M}_n > 10^4$), over 90% *cis*-1,4 structure was obtained, reaching values of 96% when $\bar{M}_n > 10^5$.

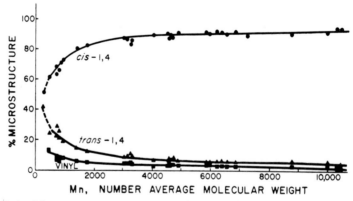

Fig. 5–1. Microstructure of poly(butadiene)s (Bd) obtained with $AlEt_3/NiOct_2/BF_3 \cdot Et_2O$ catalyst at 17/1/15 mole ratio at varying Bd/Ni ratios from 10 to 2000/1. Polymer microstructure as a function of number average molecular weight: ○, *cis*-1,4 content; △, *trans*-1,4 content; and □, vinyl content (51).

7. Order of Mixing

For many catalysts, the order of mixing does not affect the behavior of a catalyst. An effect, however, was observed for the $AlEt_3$–$TiCl_4$ catalyst when used to polymerize isoprene. Addition of $AlEt_3$ to $TiCl_4$ is the preferred mixing since the reverse addition yields a less active catalyst, and the polymer product contains higher contents of gel (52).

III. Important Catalysts Identified for Specific Conjugated Dienes

Sections A to C separately discuss the most important catalysts for butadiene, isoprene, and 1,3-pentadiene. Other dienes will be described in Chapter 19.

A. BUTADIENE

Some of these catalysts are specific only under narrow conditions, while others are essentially insensitive and produce the desired structure under variable conditions. The literature sometimes contains discrepancies as to the exact experimental variables necessary to prepare the desired product with some of these sensitive catalysts.

1. cis-1,4-Polybutadiene

High cis-1,4-polybutadienes (95 to 100%) have been synthesized with Co-, Ni-, and Ti-based Ziegler–Natta catalysts (Table 5–6) (19, 53–61). These form the most important practical systems.

TABLE 5–6

Cobalt Catalysts for High cis-1,4-Polybutadiene (53, 54)

Catalyst	Structure			Ref.
	cis-1,4	trans-1,4	1,2	
$AlEt_2Cl$ + bis(salicylaldehyde) Co(II)	99.3	0.4	0.3	53
AlR_2Cl + $CoCl_2 \cdot EtOH$	97	2	1.0	54
$LiBH_4$ + $CoCl_2 \cdot AlCl_3$	96.8	1.6	1.6	55
$AlEt_2Cl \cdot CoCl_2 \cdot (RO)_3PO$	98.1	0.9	1.0	56
$AlEt_2Cl \cdot$ Cobalt salts	88–98	—	—	57
$AlEt_2Cl$ + $\beta TiCl_3 \cdot xAlCl_3$	mainly	—	—	58
Aloxane + $Co(acac)_3$	98	—	—	19
$LiAlH_4$ + Co octanoate–$AlCl_3$ complex	94.9	3.7	1.5	59
$AlEt_2Cl$ + $Co(Co)_4$	86	—	—	60
$AlEt_2Cl$ + cobalt dinonyl naphthalene sulfonate	98	—	—	61

Cobalt-based systems have been most widely investigated, probably because of their attractiveness as potential commercial catalysts. The most important catalysts contained solubilized cobalt salts plus $AlEt_2Cl$ (24, 40, 53–67) such as $CoCl_2 \cdot 2$ pyridine and $CoCl_2 \cdot (RO)_3PO$ (see Table 5–1 for other examples). Chelates of divalent cobalt were also effective, such as bis(salicylaldehyde)Co: six kinds of chelates containing electron donor atom pairs such as (O,O), (O,N), (N,N), (O,S), (N,S), and (S,S) were found equally effective (53). Cobalt(II) dinonylnaphthalene sulfonate when combined with $Al_2Et_3Cl_3$ was also effective in producing highly cis-1,4-polybutadiene (95%) (61). Polybutadiene possessing 98% cis-1,4 structure has been easily prepared. Molecular weights were easily controlled by variation of the polymerization temperature (57) and by use of H_2 (35), olefins (64), nonconjugated dienes (65), or $ZnEt_2$ (66) (see Chapter 10).

If the reaction system is completely anhydrous and free of impurities, these catalysts have low or no activity, according to Gippin (40, 67) and Balas and Porter (56). Addition of water, oxygen, $AlCl_3$, alcohols, alkyl halides, and other reactants activates these catalysts or increases their existing activities. Donor influence is discussed in Chapter 9.

The preferred solvents are benzene or toluene, but other solvents such as heptane and cyclohexane have been used. One advantage of the aromatic solvent is to keep the polymer and catalyst in soluble form.

Nickel-based catalysts, for example, $AlEt_3$ + Ni salt, also produce high cis-1,4-polymers (95 to 98%) but at a lower rate. The addition of BF_3 and other metal fluorides was found by Ueda and co-workers to significantly increase the activity of this catalyst, and this system has apparently gained importance at the Bridgestone Laboratories as a potential commercial catalyst (68–75). The activating effect of the fluoride component BF_3 is to react with the metal alkyl and generate some $AlEt_2F$ (69–71). In addition to BF_3, other fluoride compounds used were HF (75), SnF_4, PF_3, ZrF_4 (71), and HSO_3F (73). In addition to $AlEt_3$, LiBu was used as metal alkyl in combination with nickel isopropylsalicylate (72). Also, nickel salts containing ligands other than halogen were used, including naphthenate (69), octanoate (71), and hexafluoroacetonate (74). The Bridgestone catalyst gives only 56 to 60% cis-1,4-polyisoprene (76).

An apparently related system consisting of $Ni(II)(acac)_2$, BF_3 etherate, and $AlEt_3$ was studied by IR and ESR in great detail by Tkac, Stasko, and Adamcik (77). A trimetallic complex $(acac)BF_2 \cdot Ni(I)C_2H_5 \cdot FAlEt_2$ fixed to a colloidal Ni(0) surface was proposed as the active center for the stereospecific polymerization of butadiene.

High cis-1,4-polybutadienes have also been synthesized with Ti-based catalysts if they contain compounds that can serve as a source of reactive iodine, such as $AlEt_3$, $TiCl_4 + I_2$, AlI_3, or $AlEt_2I$. Considerable evidence has been put forth to show that TiI_3 is formed through exchange reactions and that this is the active component (38). Comparable activities and stereochemical control have been found if one starts with preformed Ti salts containing iodine ligands, such as TiI_4, TiI_3Cl, or TiI_2Cl_2 (78). To maximize the effect of iodine exchange, it is important that specific ratios of the components be reacted under the chosen conditions. It has not been established unequivocally whether the actual active species are soluble or part of the heterogeneous phase. Natta and Porri (3, p. 616) believe that they are homogeneous on the basis of two pieces of experimental data: (1) the cis content of the polymer increases with decreasing Ti concentration, and (2) the molecular weight increases greatly with decreasing Ti concentration.

TABLE 5-7

Titanium Catalysts Containing Iodide[a]

Catalyst	Ref.
Al-i-Bu$_3$–I$_2$–TiCl$_4$	81
AlEt$_3$–AlEt$_2$I–TiCl$_4$	82
Al hydrides–AlI$_3$–TiCl$_4$	82, 83
[e.g. AlHCl$_2$ · Et$_2$O, AlH$_2$N(CH$_3$)$_3$,	
AlH$_2$Cl · N(CH$_3$)$_3$ and AlH$_3$ · N(CH$_3$)$_3$]	
LiAlH$_4$–TiCl$_2$I$_2$	16
LiAlH$_4$–I$_2$–TiCl$_4$	16
AlEt$_2$I–TiCl$_3$ · xAlCl$_3$	82
Al-i-Bu$_2$I–TiI$_3$	77

[a] AlEt$_2$I in combination with the layer-structured titanium trichlorides, such as αTiCl$_3$ or γTiCl$_3 \cdot \frac{1}{3}$AlCl$_3$, formed some of the most highly isotactic polypropylenes in Natta's hands (80). Whether iodine and chlorine ligands exchange between the two components has not been established for the olefin catalysts.

These characteristics are similar to those found for soluble Co catalysts and dissimilar to those found for heterogeneous Ti catalysts. In addition, they point out that the bimolecular termination found for the Al-i-Bu$_3$–TiI$_4$–(i-Pr)$_2$O system by Henderson (79) is more acceptable for a homogeneous system. In agreement is the finding that the solubilized Ti species formed by reacting AlEt$_2$I$_3$ and TiCl$_3 \cdot \frac{1}{3}$AlCl$_3$ forms a polybutadiene in high yield that has up to 99% cis-1,4 units (12). Some examples of these catalysts are collected in Table 5-7 (16, 77, 80–83).

2. $trans$-1,4-Polybutadiene

Highly crystalline $trans$-1,4-polybutadienes have been prepared with V, Ti, and Co catalysts.

Both homogeneous and heterogeneous catalysts have been successfully used. Homogeneous catalysts such as $AlEt_2Cl + V(acac)_3$ or $VCl_3 \cdot 3THF$ complex produce trans-1,4 contents above 98% (84). Very high trans-1,4 contents are also obtained with the heterogeneous catalyst $AlEt_3 + VCl_3$ (85). In contrast, catalysts using $VOCl_3$ or VCl_4 are more active, but they produce a higher content of amorphous product. If these catalysts ($AlEt_2Cl + VOCl_3$ or VCl_4, and $AlEt_3 + VOCl_3$ or VCl_4) are first washed and only the washed precipitate used, then only high trans-1,4 contents (about 98%) are obtained.

Catalysts prepared from $TiCl_4$ and $TiCl_3$ are similar in that a product is obtained which contains a considerable amount of amorphous polymer. If the polymer is extracted with ethyl ether, the insoluble residue (60 to 70% of the total product) is found to have 95% trans-1,4-structure (32). Cozewith and Törnqvist prepared 91% trans-1,4 product with the $AlEt_3$–$TiCl_3$–AlI_3 mixture in presence of tetrahydrothiophene modifier (86). High contents of trans-1,4 product were reported recently when butadiene was polymerized at 5°C with the Al-i-Bu$_3$–TiCl$_4$ catalyst (87). Furukawa and co-workers reported that $CdEt_2$, in combination with $TiCl_4$ (Cd/Ti = 5), gives 98% trans-1,4 structure (10).

Trans-1,4 polymers have also been synthesized with $AlEt_2Cl$–$Co(acac)_2$ and $AlEt_3$–$Co(acac)_2$ catalysts in THF solvent if the two components are mixed in the presence of the monomer (88).

3. 1,2-Syndiotactic Polybutadiene

Polybutadiene containing this structure in various degrees was synthesized with Ti-, V-, Cr-, Co-, and Mo-based catalysts, usually under special conditions. In many of these syntheses, the highly crystalline fraction must be isolated by extraction with ethyl ether.

The first synthesis of 1,2-syndiotactic polybutadiene was reported by Natta and Porri (89) using homogeneous catalysts, that is, $AlEt_3$ and acetylacetonate or alcoholate of Ti, V, or Cr reacted under specific conditions. Other halogen-free transition metal compounds, such as $Cr(acac)_3$, $Cr(CNR)_6$, $Cr(CO)_6$, in combination with AlR_3 (30), were found effective in later investigations. One chlorinated catalyst is effective ($MoCl_5 + AlEt_3$); other molybdenum compounds were found similarly active and stereospecific [for example, $MoO_2(OR)_2$, $MoO_2(acac)_2$, and $Mo(acac)_3$ (88)].

Iwamoto and Yuguchi (90) used the mixture ($AlEtCl)_2SO_4 + Co(SCN)_2$ to polymerize butadiene to 95% 1,2 product. Infrared spectroscopy suggested the structure to be syndiotactic. Mazzei and co-workers (29) obtained highly crystalline syndiotactic polybutadiene ($T_m = 135°C$) with mixtures of $Ti(NEt_2)_4$ and $AlHCl_2 \cdot N(CH_3)_3$, $AlHCl_2 \cdot O(Et)_2$, or $AlH_2N(CH_3)_2$.

4. Isotactic 1,2-Polybutadiene

A number of Cr-based catalysts produce either syndiotactic or isotactic polymer, depending on conditions of preparation (30, 91). Two variables are important: the Al/Cr ratio and the time of aging. Formation of syndiotactic polymer is favored by short aging times and low Al/Cr ratios or both. As Al/Cr ratio and aging time increase, isotactic polymer is also formed and eventually only isotactic is obtained. The effective catalysts were combinations of $AlEt_3$ and $Cr(CNC_6H_5)_6$, $Cr(acac)_3$, $Cr(CO)_6$, and derivatives.

B. ISOPRENE

Table 5–8 collects some examples of catalysts that have been reported to polymerize isoprene to highly stereospecific cis-1,4 polymer (89, 93–102).

TABLE 5–8

Stereospecific Catalysts for Isoprene

Catalyst	cis-1,4	trans-1,4	3,4	Ref.	Comment
$ZnEt + TiCl$ $(Zn/T = 1.1)$	94	—	5.2	93	also $(CaH_2)_x \cdot (CaEt_2) \cdot ZnEt_2$
$HAl(NR_2)_2 + TiCl_4$	91–97	—	3–4	94	gel free
$AlEt_2Cl + TiCl_4 + RSiOR$	97	—	—	95	
$AlEt_2F + TiCl_4$	95.9	—	—	96	
$Al\text{-}i\text{-}Bu_3 + TiCl_4 + \phi_2O$	98	—	—	97	$Al/Ti = 0.9$, yield ∼ 1500 g/gTi
$AlR_3 \cdot NR_3 + TiCl_4$	96	—	—	98	$Al/Ti = 1.1$
$AlEt_2OEt + CoI_2$	44	—	56	99	
$Al\text{-}i\text{-}Bu_2Cl + CoO$	65	18	17	100	
$AlEt_3 + Fe(acac)_3$	50	—	46	101	Modifier = 2 cyanopyridine
$AlR_3 + Ti(OR)_4$	—	—	∼99	89	
$C_4H_9MgI + (C_4H_9)_2Mg + TiCl_4$	0	98	2	102	

1. cis-1,4-Polyisoprene

Only Ti-based catalysts have been found to give this structure a high concentration. In fact, the $AlEt_3$–$TiCl_4$ catalyst, first reported by Horne, Gibbs, and Carlson in 1954, still appears to be one of the most active and stereospecific (103). Conversion, microstructure, and molecular weight are intimately dependent on the Al/Ti ratio; the optimum ratio is about 1. Polymer formed at a lower ratio is mostly cross-linked (probably formed by cationic mechanisms), while the polymer formed at a high ratio (about 2.5) is oligomeric. Figure 5–2 (52) shows the effect of varying the Al/Ti ratio. The oligomerization reaction could be controlled by addition of carbon disulfide [see Chapter 9, Perry et al. (70)].

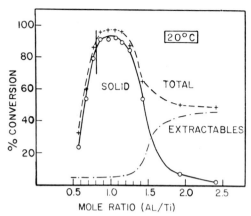

Fig. 5–2. Dependence of conversion on Al/Ti ratio in isoprene polymerization with AlEt$_3$–TiCl$_4$ catalyst (52).

It is believed that the active TiCl$_3$ formed in the specific catalyst for *cis*-1,4 addition has the β structure (104). When preformed TiCl$_3$ is used, only the β structure leads to high *cis*-1,4 structures. In 1969, Smith and Perry (105) prepared a nearly pure βTiCl$_3$ by reduction of TiCl$_4$ with Al in the presence of AlCl$_3$ and aromatic compound and subsequent treatment with ethers. They combined it with AlEt$_3$, Al-*i*-Bu$_3$, and Al-*n*-Bu$_3$ to produce highly active and stereospecific catalysts.

A combination of TiCl$_4$ and a polyiminoalane also polymerized isoprene to a high *cis*-1,4 content (about 97%) (94). The same catalyst also produced a high *cis*-1,4 polybutadiene (17).

2. *trans*-1,4-Polyisoprene

Ti- and V-based catalysts have been found effective.

Highly crystalline products (99 to 100% *trans*-1,4 units) were synthesized with AlR$_3$–VCl$_3$ (106) or AlR$_3$–VCl$_4$ catalysts (107). More active catalysts were made by supporting VCl$_3$ on inert substrates such as silica or alumina (108).

3. 3,4-Polyisoprene

The most active and stereospecific catalyst is made by mixing AlEt$_3$ and Ti(OR)$_4$ in aliphatic or aromatic solvents (88, 109, 110). The structure of the titanium alcoholate does not seem to affect 3,4 content. Products containing 94, 97, and 99% 3,4 units are prepared at 25°, 0°, and −10°C, respectively, with the AlEt$_3$–Ti(O-*n*-Pr)$_4$ catalyst (Al/Ti = 6) (109). Poly-isoprenes containing high contents of 1,2 units are not known.

TABLE 5–9

Stereochemical Structures Obtained in Poly(1,3-pentadiene)(III)

	Properties		
Structure	Melting point (°C)	Density (g/cm^3)	Identity period (units)
Isotactic *trans*-1,4	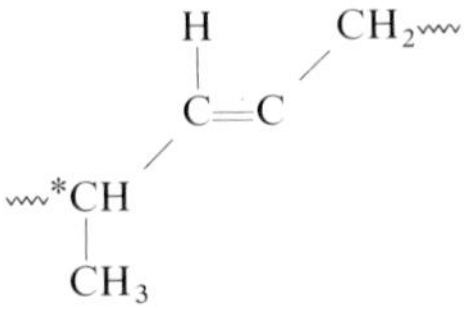 96	0.98	4.85
Isotactic *cis*-1,4	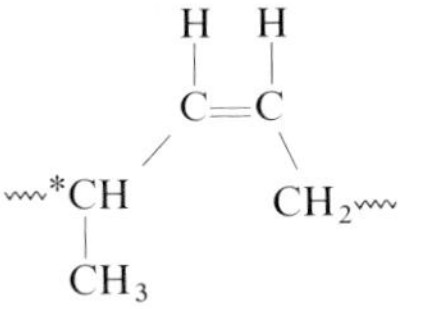 44	0.97	8.1
Syndiotactic *cis*-1,4 (same as structure above)	53	1.01	8.5
Syndiotactic 1,2	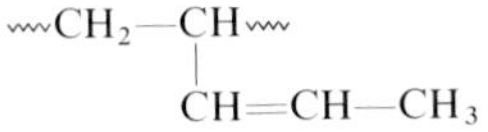 amorphous		

Isotactic *trans*-1,4 structure:
$$\underset{\underset{\displaystyle CH_3}{|}}{\overset{\displaystyle H}{\underset{\sim\sim^*CH}{}}}\diagup \overset{\displaystyle C=C}{}\diagup \overset{\displaystyle CH_2\sim\sim\sim}{}$$

Isotactic *cis*-1,4 structure:
$$\overset{H\ \ \ H}{C=C} \diagup \underset{\underset{\displaystyle CH_3}{|}}{\sim\sim^*CH}\ \ \diagdown CH_2\sim\sim\sim$$

Syndiotactic 1,2 structure:
$$\sim\sim CH_2-CH\sim\sim\sim \ \ |\ \ CH=CH-CH_3$$

C. 1,3-PENTADIENE (PIPERYLENE)

1. Scope

This diene is commercially available as a mixture of the cis and trans isomers, which often show different polymerization abilities. Of the large number of isomeric products possible, four have been isolated and characterized, primarily by the Natta school (Table 5–9) (111). Specific catalysts that have been found for the different stereochemical structures are discussed next. Infrared and NMR spectra are reported in Ciampelli *et al.* (112).

2. Isotactic *trans*-1,4-Polypentadiene

The most specific and active catalyst found is a mixture of AlEt$_3$ and VCl$_3$ (113, 114). Although the content of *trans*-1,4 units was 98 to 99%, the isotacticity was not very high. Less stereospecific catalysts included AlEt$_3$ + TiCl$_3$ (113, 114). In addition to VCl$_3$, VCl$_4$ and VOCl$_3$ were effective (115). Soluble catalysts such as AlEt$_2$Cl + V(acac)$_3$ or VCl$_3 \cdot$ THF were less stereospecific (116), but this does not preclude other soluble catalysts being potentially effective. It is well established that isotactic *trans*-1,4-polypentadiene has been made with soluble lithium alkyls (117).

3. Isotactic *cis*-1,4-Polypentadiene

This isomeric form is obtained with the aid of the $AlR_3-Ti(OR)_4$ catalyst; it must be separated from the crude product by solvent fractionation (118– 120). Both isomers of 1,3-pentadiene polymerize with titanium-based catalysts.

4. Syndiotactic *cis*-1,4-Polypentadiene

The same catalysts that were found to polymerize butadiene to *cis*-1,4 structures in aromatic solvents also polymerized 1,3-pentadiene (only the trans isomer) to *cis*-1,4 structures. Catalysts such as $AlEt_2Cl$ plus cobalt salts were used (121). The cis isomer did not polymerize.

5. 1,2-Syndiotactic Polypentadiene

The catalysts used to prepare syndiotactic *cis*-1,4-polypentadiene in aromatic solvents give 1,2 structures containing syndiotactic configuration if the polymerization of the trans isomer is done in aliphatic solvents (49). If water is present ($Al/H_2O = 2$), the same catalyst gives *cis*-1,4 structures in aliphatic solvents from *trans*-1,3-pentadiene (122). The proposal was made that the cobalt species in the 1,2 syndiospecific catalyst has a tetrahedral or planar coordination, whereas in the *cis*-1,4 and *trans*-1,4 specific catalysts, an octahedral configuration exists (122). By altering the structure of the metal alkyl ($AlEt_2Cl$, $Al_2Et_3Cl_3$, and alkyls containing Al–O–Al bonds), specificity of the catalyst was altered (123).

IV. Conclusion

Considering the large number of stereochemical structures that have been synthesized in high purity from butadiene (4), isoprene (3), and 1,3-pentadiene (4), one has to be very much impressed with the versatility of the Ziegler–Natta catalyst. By and large, these results came from systematic experimental studies where the trial and error approach was used. These investigations have resulted in a general knowledge that now guides us in the selection of catalysts. Aluminum alkyls are preferred as one of the components, for example, AlR_3, AlR_nCl_{3-n}, or alkoxyaluminum compounds. There is a much larger choice concerning the selection of the transition metal salt. Here, not only is the primary choice of the transition metal important, but one must also consider the ligands attached to it, as well as the chemistry the salt can undergo when contacted with the metal alkyl and the diene under specific conditions. Several striking observations have been made. Some of these systems are almost (but not completely) insensitive to changes in structure and conditions of polymerization, and they seem to produce a particular stereochemical structure in spite of all obstacles; for example, vanadium salts prevail in *trans*-1,4 structures while cobalt salts lead to *cis*-1,4 structures. Yet other catalysts are so sensitive that small changes induce completely different stereochemical behavior. Much credit goes to the various investigators cited in this chapter for their careful and diligent work, which revealed this very significant feature of the Ziegler–Natta catalyst. No attempt was made here to identify the nature of the active center and how it is able to produce a particular stereochemical structure. This is done in Chapters 13 through 17.

References

1. W. Cooper and G. Vaughan, *Prog. Polym. Sci.* **1**, 93–160 (1967).
2. W. Marconi, Polymerization of dienes by Ziegler-Natta catalysts *in* "Stereochemistry of Macromolecules," pt. 1, (A. D. Ketley, ed.), p. 239–307, Marcel Dekker, New York, 1967.
3. G. Natta and L. Porri, "Diene Elastomers" *in* Polymer Chemistry of Synthetic Elastomers, (J. P. Kennedy and E. G. M. Törnqvist, eds.) pt. 2, p. 597, Interscience, 1969.
3a. I. Pasquon, *Encycl. Polym. Sci. Technol.* **13**, 13–86 (1970).
4. W. Cooper, *Ind. Eng. Chem., Prod. Res. Dev.* **9**, No. 4, 457–466 (1970).
5. B. A. Dolgoplosk, *Polym. Sci. USSR (Engl. Transl.)* **13**, 367–393 (1971); *Vysokomol. Soedin., Ser. A* **13**, 325–347 (1971).
5a. W. M. Saltman, F. S. Farson, and E. Schoenberg, *Rubber Plast. Age*, **46**, 502 (1965).
6. G. Natta, L. Porri, and A. Mazzei, *Chim. Ind. (Milan)* **41**, 116 and 398 (1959).
7. See also Table 19, Part B of Cooper and Vaughan (1).
8. W. Cooper, *Rubber Plast. Age* **44**, No. 1, 44 (1963).
9. J. Furukawa and T. Tsuruta, *J. Polym. Sci.* **36**, 275 (1959).
10. J. Furukawa, T. Tsuruta, T. Saegusa, A. Onishi, A. Kawasaki, and T. Fueno, *J. Polym. Sci.* **28**, 450 (1958).
11. S. E. Bresler, M. I. Mosevitskii, I. Ya. Poddubnyi, and Kuan-I Shi, *Vysokomol. Soedin.* **3**, 1591 (1961).
12. W. Marconi, M. Araldi, A. Beranger, and M. De Malde, *Chim. Ind. (Milan)* **45**, 522 (1963); W. Marconi, *in* "The Stereochemistry of Macromolecules" (A. D. Ketley, ed.), p. 246. Dekker, New York, 1967.
13. B. A. Dolgoplosk, B. L. Erusalimskii, E. N. Kropasheva, and E. I. Tinyakova, *J. Polym. Sci.* **38**, 1333 (1962).
14. A. Takahashi, K. Takahashi, and T. Hirose, *J. Polym. Sci., Part B* **5**, 415 (1967).
15. R. C. Farrar, Great Britain Patent 906,266, September 19, 1962, Phillips Petroleum Company (equivalent to U.S. Patent 3,066,129).
16. W. Marconi, A. Mazzei, M. Araldi, and M. De Malde, *Chim. Ind. (Milan)* **46**, 245 (1964).
17. Italian Patent 2355–65, 1965, SNAM, Laboratory Ruineti Studi e Ricerche.
18. This conclusion is drawn from statements made by various workers about the apparent physical state of the catalyst being investigated. Some of these catalysts may be colloidal.

19. P. Racanelli and L. Porri, *Eur. Polym. J.* **6**, 751 (1970); see also C. Longiave and R. Castelli, *IUPAC Int. Symp. Macromol. Chem., 1963, J. Poly. Sci., Poly. Symp.*, **C4**, p. 387–398. (1963).

20. L. Porri and A. Carbonaro, *Makromol. Chem.* **60**, 236 (1963).

20a. S. Cucinella, A. Mazzei, W. Marconi, and C. Besetto, *J. Macromol. Sci., Chem.* **A4**, no. 7, 1549 (1970).

20b. L. Kollar, H. Schnecko, and W. Kern, *Makromol. Chem.* **142**, 21 (1971).

21. Table 19 of Cooper and Vaughan (1) collects many examples of vanadium catalysts and should be consluted for comparisons of original experimental results. See also Natta *et al.* (22) for example using V(acac)$_3$.

22. G. Natta, L. Porri, P. Corradini, and D. Moreno, *Chim. Ind. (Milan)* **40**, 362 (1958).

23. H. Noguchi and S. Kambara, *J. Polym. Sci., Part B* **2**, 593 (1964).

23a. G. Natta, L. Porri, G. Zanini, and A. Palvarini, *Chim. Ind. (Milan)* **41**, 12 (1959).

23b. G. Natta, *Nucleus* **3**, 211 (1963).

24. G. Longiave, R. Castelli, and C. F. Groce, *Chim. Ind. (Milan)* **43**, 625 (1961).

24a. G. Natta, L. Porri, A. Mazzei, and D. Moreno, *Chim. Ind. (Milan)* **41**, 398 (1959); W. M. Saltman and T. H. Link, *Ind. Eng. Chem., Prod. Res. Dev.* **3**, 199 (1964); Great Britain Patent 824,201, November 25, 1959, Chemische Werke Huels.

25. R. P. Zelinski, D. R. Smith, G. Nowlin, and H. D. Lyons, Belgian Patent 551,851, April 17, 1957, Phillips Petroleum Company.

26. R. C. Farrar, Great Britain Patent 906,266, September 19, 1962, Phillips Petroleum Company (equivalent to U.S. 3,066,129).

27. G. Natta, *J. Polym. Sci.* **48**, 219 (1960).

28. G. Natta, L. Porri, and A. Carbonaro, *Makromol. Chem.* **77**, 126 (1964).

29. A. Mazzei, D. Cucinella, W. Marconi, and M. De Malde, *Chim. Ind. (Milan)* **45**, 528 (1963).

30. G. Natta, L. Porri, G. Zanini, and A. Palvarini, *Chim. Ind. (Milan)* **41**, 1163 (1959).

31. W. Marconi, M. L. Santostasi, and M. De Malde, *Chim. Ind. (Milan)* **44**, 235 (1962).

32. G. Natta. L. Porri, and L. Fiore, *Gazz. Chim. Ital.* **89**, 761 (1959).

33. A stereochemical mechanism suggested by Arlman is discussed in Chapter 16.

34. This aspect is discussed in Chapter 11.

35. G. Natta, L. Porri, A. Mazzei, and D. Moreno, *Chim. Ind. (Milan)* **41**, 398 (1959).

36. E. Schoenberg, D. L. Chalfant, and T. L. Hanlow, *Adv. Chem. Ser.* **52**, 6 (1966).

37. E. Schoenberg, D. L. Chalfant, J. B. Pyke, and R. H. Mayor, presented at *Am. Chem. Soc. Div. of Ind. Eng. Chem. Meeting*, New York, 1963.

38. P. H. Moyer and M. H. Lehr, *J. Polym. Sci., Part A* **3**, 217 (1965).

39. G. F. Gibbs, S. E. Horne, J. J. Macey, and H. Tucker, *Kautschuk Gummi* **13**, WT336 (1960).

40. M. Gippin, *Ind. Eng. Chem., Prod. Res. Dev.* **1**, 32 (1962).

41. E. Susa, *J. Polym. Sci., Part C* **4**, 399 (1963).

42. W. Cooper, G. Degler, D. E. Eaves, R. Hank, and G. Vaughan. *Adv. Chem. Ser.* **52**, 46 (1966).

43. B. S. Turov, P. A. Vinagradov, B. A. Dolgoplosk, E. N. Khranina, and S. I. Kostina, *Dokl. Akad. Nauk SSSR* **146**, 1141 (1962).

44. This is unique to catalysts that are active for dienes. Very little advantage has been observed when olefin-active catalysts were used.

45. See reference 61 (unpublished work) in Natta and Porri (3).

46. A. Bonfardeci, U.S. Patent 3,652,528, March 28, 1972, Montecatini-Edison SpA.

47. K. Matsuzaki and T. Yasukawa, *J. Polym. Sci., Part B* **3**, 393 (1965).

48. S. E. Horne, Jr. and C. J. Carman, *J. Polym. Sci., Polym. Chem. Ed.* **9**, 3039 (1971).

49. G. Natta, L. Porri, and G. Sovarzi, *Eur. Polym. J.* **1**, 81 (1965).
50. W. M. Saltman and L. J. Kuzma, *Pap. 103rd Meet. Rubber Div., Am. Chem. Soc., 1973*
51. W. M. Saltman and L. J. Kuzma, *Rubber Chem. Technol.* **46**, 1055 (1973).
52. E. Schoenburg, D. L. Chalfant, and R. H. Mayor, *Rubber Chem. Technol.* **37**, 103 (1964).
53. A. Takahashi and S. Kambara, *J. Polym. Sci., Part B* **3**, 279 (1965).
54. B. A. Dolgoplosk, E. N. Kropacheva, E. N. Kennikova, E. I. Kuznetsova, and K. G. Goldova, *Dokl. Akad. Nauk SSSR* **135**, 847 (1960).
55. J. G. Balas, U.S. Patent 3,067,189, December 4, 1962, Shell Oil Company.
56. J. Balas and L. M. Porter, U.S. Patent 3,040,016, June 19, 1962, Shell Oil Company.
57. Table 19 in Cooper and Vaughan (1) collects other cobalt-based catalysts, including heterogeneous examples.
58. V. A. Khodzhemirov, E. V. Kristal'nyi, Ye. V. Zabolotskaya, and S. S. Medvedev, *Polym. Sci. USSR (Engl. Transl.)* **9**, 603 (1967).
59. M. Gippin, U.S. Patent 3,159,612, July 7, 1970, Firestone Tire and Rubber Company.
60. R. N. Kovalevskaya, Ye. I. Tinyakova, and B. A. Dolgoplosk, *Polym. Sci. USSR (Engl. Transl.)* **4**, 414 (1963).
61. J. Boor and F. Fowkes, U.S. Patent 3,234,198, February 8, 1966, Shell Oil Company.
62. G. Natta, L. Porri, and L. Fiore, Italian Patent 587,976, July 5, 1959, Montecatini.
63. G. Marullo, A. Baroni, U. Maffezzoni, C. Longiave, and E. Susa, Italian Patent 587,968, July 27, 1959, Montecatini.
64. C. Longiave, R. Castelli, and M. Ferraris, *Chim. Ind. (Milan)* **44**, 725 (1962).
65. E. Giachetti and W. Bortolini, Italian Patent 687,758, February 12, 1963, Montecatini.
66. E. A. Youngman, K. Nozaki, and J. Boor, U.S. Patent 3,084,148, April 2, 1963, Shell Oil Company.
67. M. Gippin. *Ind. Eng. Chem., Prod. Res. Dev.* **4**, 160 (1965).
68. K. Ueda, A. Onishi, T. Yoshimoto, J. Honoso, and K. Maeda, U.S. Patents 3,170,904–3,170,907, February 23, 1965, Bridgestone Tire Company, Ltd., Japan.; K. Ueda, A. Onishi, T. Yoshimoto, J. Honoso, and T. Matsumoto, U.S. Patent 3,178,403, April 13, 1965, Bridgestone Tire Company, Ltd., Japan.
69. K. Ueda, A. Onishi, T. Yoshimoto, J. Honoso, K. Maeda, and T. Matsumoto, *Kogyo Kagaku Zasshi* **66**, 1103 (1963).
70. T. Matsumoto and A. Onishi, *Kogyo Kagaku Zasshi* **71**, 2059 (1958).
71. M. C. Throckmorton and W. M. Saltman, U.S. Patent 3,446,788, May 27, 1969, The Goodyear Tire and Rubber Company; see also U.S. Patent 3,446,787.
72. C. Dixon, E. W. Duck, D. P. Grieve, D. K. Jenkins, and M. N. Thornber, *Eur. Polym. J.* **6**, 1359 (1970); **7**, 55 (1971).
73. M. C. Throckmorton, U.S. Patent 3,542,751, November 24, 1970, Goodyear Tire and Rubber Company.
74. M. C. Throckmorton, U.S. Patent 3,640,989, February 8, 1972, Goodyear Tire and Rubber Company.
75. M. C. Throckmorton and F. S. Farson, *Rubber Chem. Technol.* **45**, 268 (1972).
76. S. Anzai and A. Onishi, *Kogyo Kagaku Zasshi* **72**, 2068 (1969); also see pp. 2058, 2076, 2081, 2090, 2101, and 2113.
77. A. Tkac and A. Stasko, *Collect. Czech. Chem. Commun.* **37**, 1006 (1972); A. Tkac and V. Adamcik, *ibid.* **38**, 1346 (1973).
78. W. Marconi [*Prog. Polym. Sci.* **1**, 247–249 (1967)] and G. Natta and L. Porri [*High Polym.* **23**, 615–617 (1969)] collect earlier references which illustrate this finding.
79. J. F. Henderson, *J. Polym. Sci., Part C* **4**, 233 (1964).
80. See Tables 4–6 and 4–25.
81. J. Darcy, Belgian Patent 610,400, May 16, 1962, Polymer Corporation, Ltd.

82. R. P. Zelinski and D. R. Smith, U.S. Patent 3,050,513, August 21, 1962, Phillips Petroleum Company; see also Belgian Patent 551,851, April 17, 1957, by R. P. Zelinski, D. R. Smith, G. Novlin, and H. D. Lyons.

83. A. Mazzei, M. Araldi, W. Marconi, and M. De Malde, *J. Polym. Sci., Part A* **3**, 735 (1965).

84. G. Natta, L. Porri, and A. Carbonaro, *Atti Accad. Naz. Lincei, Cl. Sci. Fis., Mat. Nat., Rend.* [8] **31**, 189 (1961).

85. G. Natta, L. Porri, P. Corradini, and D. Moreno, *Chim. Ind. (Milan)* **40**, 362 (1958); see also Natta *et al.* (6).

86. C. Cozewith and E. G. M. Törnqvist, U.S. Patent 3,642,758, February 15, 1972, Esso Research and Engineering Company.

87. E. J. Carlson and S. E. Horne, U.S. Patent 3,657,209, April 18, 1972, Goodrich-Gulf Chemicals, Inc.

88. See reference 61 of Natta and Porri (3) and relevant discussion in text.

89. G. Natta and L. Porri, Belgian Patent 549,554, January 14, 1957, Montecatini.

90. M. Iwamoto and S. Yuguchi, *J. Polym. Sci., Part B* **5**, 1007 (1967).

91. G. Natta, L. Porri, and A. Palvarini, Italian Patent 563,507, April 14, 1956, Montecatini.

92. Table 19 of Cooper and Vaughan (1) collects other examples.

93. W. Marconi, A. Mazzei, S. Cucinella, and M. De Malde, *Chim. Ind. (Milan)* **44**, 121 (1962).

94. A. Mazzei, S. Cucinella, and W. Marconi, *Makromol. Chem.* **122**, 168 (1969).

95. A. Bonfardeci and L. Porri, U.S. Patent 3,557,076, January 19, 1971, Montecatini-Edison SpA.

96. H. Antropiusova, K. Mach, B. Matsyska, J. Trneny, and C. Vyroubal, U.S. Patent 3,607,854, September 21, 1971, Checkoslovak Academy.

97. E. Schoenberg and D. L. Chalfant, Belgian Patent 622, 492, December 31, 1962, Goodyear Tire and Rubber Company.

98. H. Kahn and S. E. Horne, Jr., U.S. Patent 3,165,503, January 12, 1965, Goodrich-Gulf Chemical Company.

99. F. Bougival, U.S. Patent 3,491,078, January 20, 1970, Institute Francais du Petrole.

100. R. N. Kovalevskaya, E. I. Tinyakova, and B. A. Dolgoplosk, *Vysokomol. Soedin.* **4**, 1338 (1962).

101. J. Bozik, H. Swift, and C.-Y. Wu, U.S. Patent 3,565,875, February 23, 1971, Goodrich-Gulf Chemical Company.

102. B. L. Erusalimskii, F. S. Wang, and A. P. Kavunenko, *J. Polym. Sci.* **53**, 27 (1961).

103. S. E. Horne, Jr., F. Gibbs, and E. J. Carlson, Great Britain Patent 827,365, February 3, 1960, Goodrich-Gulf Chemical Company; W. M. Saltman, W. E. Gibbs, and J. Lal, *J. Am. Chem. Soc.* **80**, 5615 (1958).

104. S. E. Bresler, M. I. Mosevitskii, I. Ya. Poddubnyi, and S. Kuan-I Shi, *Vysokomol. Soedin.* **3**, 1591 (1961).

105. G. H. Smith and D. C. Perry, *J. Polym. Sci., Polym. Chem. Ed.* **7**, 707 (1969).

106. G. Natta, L. Porri, P. Corradini, and D. Moreno, *Chim. Ind. (Milan)* **40**, 362 (1958); Italian Patent 553,904, December 22, 1955, Montecatini; *Atti Accad. Naz. Lincei, Cl. Sci. Fis., Mat. Nat., Rend.* [8] **20**, 728 (1956); G. Natta, L. Porri, and L. Fiore, *Gazz. Chim. Ital.* **89**, 761 (1959).

107. J. W. C. Crawford and M. F. Vincent, Belgian Patent 547,699, April 18, 1956, Imperial Chemical Ind., November 9, 1956.

108. J. S. Lasky, H. K. Garner, and R. H. Ewart, *Ind. Eng. Chem., Prod. Res. Dev.* **1**, 82 (1962).

109. G. Natta, L. Porri, and A. Carbonaro, *Makromol. Chem.* **77**, 126 (1964).

110. G. Wilke, *Angew. Chem.* **68**, 306 (1956); H. Breil, P. Heimbach, M. Kroner, H. Muller, and G. Wilke, *Makromol. Chem.* **69**, 18 (1963).
111. G. Natta, *Chim. Ind. (Milan)* **46**, 397 (1964).
112. F. Ciampelli, M. P. Lachi, M. Venturi, and L. Porri, *Eur. Polym. J.* **3**, 353 (1967).
113. G. Natta, L. Porri, P. Corradini, G. Zanini, and F. Ciampelli, *Atti Accad. Naz. Lincei, Cl. Sci. Fis., Mat. Nat., Rend.* [8] **29**, 257 (1960); see also Natta *et al.* (85).
114. G. Natta, L. Porri, P. Corradini, G. Zanini, and F. Ciampelli, *J. Polym. Sci.* **51**, 463 (1961).
115. G. Natta, L. Porri, and M. C. Gallazi, *Chim. Ind. (Milan)* **46**, No. 10, 1158 (1964).
116. G. Natta, L. Porri, A. Carbonaro, *Atti Accad. Naz. Lincei, Cl. Sci. Fis., Mat. Nat., Rend.* [8] **29**, 189 (1960).
117. L. Porri, M. C. Gallazi, and G. Natta, *Atti Accad. Naz. Lincei, Cl. Sci. Fis., Mat. Nat., Rend.* [8] **36**, 752 (1964).
118. G. Natta, L. Porri, G. Stoppa, G. Allegra, and F. Ciampelli, *J. Polym. Sci., Part B* **1**, 67 (1963).
119. G. Natta, L. Porri, A. Carbonaro, and G. Stoppa, *Makromol. Chem.* **77**, 114 (1964).
120. K. Bujadoux, J. Jozefonvicz, and J. Neel, *Eur. Polym. J.* **6**, 1233 (1970).
121. G. Natta, L. Porri, A. Carbonaro, F. Ciampelli, and G. Allegra, *Makromol. Chem.* **51**, 229 (1962); Italian Patent 659,704.
122. L. Porri, A. di Corato, and G. Natta, *J. Polym. Sci.* **5**, 321 (1967).
123. L. Porri, A. di Corato, and G. Natta, *Eur. Polym. J.* **5**, 1 (1969).

6

Initial Physical State
of the Catalyst

I. Introduction

Olefins and dienes have been polymerized with soluble, colloidal, and heterogeneous Ziegler–Natta catalysts. Each offers a particular advantage for the monomer being polymerized, such as higher activity and type of stereoregularity or microstructure. Some supported catalysts have very high activities for polymerization of ethylene, a feature of much commercial importance. In addition, the physical state of the catalyst is an important factor in a polymerization.

One of the most fascinating aspects of the polymerization with heterogeneous Ziegler–Natta catalysts involves the growth of the polymer particle from the catalyst particle. Chapter 8 is entirely devoted to describing the practical and mechanistic features of this subject.

II. Importance of the Physical State of the Catalyst

The physical state of the catalyst can be an important factor in determining stereoregularity, microstructure, copolymer composition, and the morphology and efficiency of the catalyst to polymerize.

A. STEREOCHEMICAL CONTROL

Highly isotactic polymers of α-olefins have been made only with heterogeneous Ziegler–Natta catalysts, such as $AlEt_3 + VCl_3$ (1). In this catalyst, the VCl_3 component is insoluble, and the $AlEt_3$ alkyl is soluble in the hydrocarbon solvent. The components interact on the surface of the VCl_3.

Very little effort has been made to develop inverse catalysts in which the metal alkyl remains insoluble. One example comes from Bandermann and co-workers (2), who polymerized ethylene with $TiCl_4$ and an insoluble polymeric $-(Al(CH_3)CH_2-)_n$ alkyl. Apparently, there was no advantage, since other workers have not sought to exploit this approach.

Syndiotactic polypropylene can be synthesized in high steric purity only with a soluble catalyst, such as $AlEt_2Cl + VCl_4$ at $-78°C$ (3). A small yield of low syndiotactic crystalline polypropylene can be made at the nonisotactic regulating centers of a heterogeneous catalyst, for example, $AlEt_2F + \gamma TiCl_3$. Highly syndiotactic poly-1,2-butadienes have been synthesized with a soluble catalyst, such as $AlEt_3 + Cr(acac)_3$ (4). While highly cis-1,4-polyisoprene has been made only with a heterogeneous catalyst ($AlEt_3 + TiCl_4$), highly cis-1,4-polybutadiene has been made with both heterogeneous and soluble catalysts (5). Highly trans-1,4-polydienes (isoprene and butadiene) seem to be made equally well with vanadium-based catalysts, regardless of the catalyst physical state.

B. COMONOMER DISTRIBUTION IN COPOLYMERS

When ethylene and propylene are copolymerized, random copolymers are formed with soluble vanadium-based catalysts, but the copolymers contain long blocks of each olefin when heterogeneous titanium-based catalysts are used. Detailed descriptions are found in Chapter 20.

C. MORPHOLOGY OF THE FORMED POLYMER PARTICLES

Polymer particles are faithful replicas of the parent catalyst particles when heterogeneous catalysts are used to polymerize propylene, 1-butene, and branched α-olefins, provided certain polymerization conditions are met. Soluble catalysts produce polyethylene that has a gelatinous physical state (see Chapter 8).

D. EFFICIENCY OF THE CATALYST

The efficiency of a catalyst is often (but not always) dependent on its physical state. In principle, if every available transition metal atom acted as a center, soluble catalysts would always be more efficient than heterogeneous catalysts.

The following catalysts are cited as examples of highly active systems. Balas and Porter reported that butadiene was readily polymerized with the $AlEt_2Cl–CoCl_2$–tributylphosphate catalyst when the cobalt concentration was only 2×10^{-5} moles/liter in the reaction solvent (6). The $(C_6H_5)_4$ $Sn–VCl_4–AlBr_3$ system ($AlEtBr_2$ forms) was shown by Carrick to be very efficient for polymerization of ethylene when the vanadium concentration was only 5×10^{-5} moles/liter (7). To obtain similar activities by heterogeneous catalysts, it would be necessary to increase the concentration of the transition metal by a factor of 500 or more.

The addition of benzene sulfonyl chloride to the soluble catalyst $AlEtCl_2$–VCl_4 produced a very highly active catalyst for ethylene polymerization, that is, about 390,000 g polymer formed per hour per atmosphere ethylene per g VCl_4 (8). These are fantastically high rates relative to those normally obtained with heterogeneous catalysts. It is not clear, however, how long they are sustained (one hour syntheses were done above). Very often, after a high initial rate, the catalyst undergoes a "die out."

While less active than catalysts cited above, the Union Carbide catalyst made by reacting chromium(III) 2-ethyl hexanoate with a mixture of Al-i-Bu_3 and a preformed aloxane (Al-i-$Bu_3 + H_2O \rightarrow (Al$-$i$-$BuO)_n$ also had a higher activity relative to heterogeneous catalysts (9). This catalyst may, however, be at least partially colloidal. Ethylene was polymerized at a rate of 3,280 g per hour per atmosphere ethylene per g Cr compound (60°C and 1 atm ethylene).

In the heterogeneous catalyst, $AlEt_3$–$TiCl_3$, only a small fraction of the titanium atoms are active at any one time, about 1% or less. In effect, the remaining $TiCl_3$ acts as a support for these few active centers. This is not a very efficient use of the $TiCl_3$ component. This prompted some workers to fix a small number of transition metal salt molecules on a high surface support to make a more efficient catalyst, such as $TiCl_4$ on $Mg(OH)Cl$ or SiO_2. Several very highly active systems have been reported (Section V).

These three types of Ziegler–Natta catalysts are now separately discussed.

III. Soluble Catalysts

Several soluble catalysts have achieved importance either for mechanistic studies or for the manufacture of a polymer, such as $(\pi$-$C_5H_5)_2TiCl_2 +$ aluminum alkyls (Chapter 4), $VCl_4 + AlEt_2Cl$ at $-78°C$ (Chapter 4), vanadyl esters + aluminum chloroalkyls (Chapter 20), and solubilized cobalt salts + aluminum chloroalkyls (Chapter 5). Since these were already described in this and other chapters, no further elaboration is necessary here.

IV. Colloidal Catalysts

When certain aluminum alkyls are combined, transition metal salts bearing alcoholate, acetylacetonate, or similar organic ligands catalysts are formed that have been described as "soluble" or "apparently soluble". For the polymerization of butadiene, these catalysts have been quite versatile. By the proper choice of a metal and ligand, the ratio of the two components and temperature of polymerization, the stereoregularity, or microstructure can be controlled. For example, butadiene was polymerized at 18°C to predominantly 1,2 polymer, which contained mostly syndiotactic units when Ti(O-i-Bu)$_4$ was mixed with AlEt$_3$ at an Al/Ti ratio below 6 (10, 11). Under similar conditions, isoprene was predominantly polymerized by 3,4-addition, and cis- or $trans$-1,3-pentadiene was converted at 0°C to predominantly cis-1,4-isotactic polypentadiene. Polypropylene made with catalysts of this type have a low degree of isotacticity.

The fact that these catalysts are actually colloidal was suggested by Natta and co-workers. Furthermore, once the polymerization begins, it is difficult to make meaningful visual observations about the physical state of the catalyst.

Sometimes, if the components of the catalysts are allowed to interact for an ample time, precipitation occurs. An example was reported by Berger and Fletcher (12), who found that the filtrate from a mixture of Al-i-Bu$_2$OEt + TiCl$_4$ was active for ethylene polymerization after an induction period of 60 minutes. In a separate experiment in which ethylene was excluded, Berger and Fletcher observed precipitation after about 60 minutes. It appears that the "soluble" catalyst was converted to a heterogeneous one.

Hirai and co-workers (13) correlated ESR spectra of catalysts derived from (n-BuO)$_4$Ti and AlEt$_3$ at Al/Ti ratios between 1 and 10 in toluene from $-78°$ to $+25°$C with their corresponding ability to polymerize butadiene and styrene. A larger number of signals was found at higher temperatures. The catalytic activity of styrene was at its maximum at Al/Ti = 1.4 and negligible at Al/Ti = 3. In contrast, the catalyst was most active at Al/Ti = 2. They suggested that active species for polymerization of styrene were different from species that polymerized butadiene. They proposed a titanium(III) species containing one alkyl group and two or more alkoxy groups polymerized styrene.

These results are in agreement with the earlier findings of Takeda and co-workers (14), who detected two types of trivalent titanium species by ESR spectroscopy in the (n-BuO)$_4$Ti –AlEt$_3$ catalyst. At low Al/Ti ratios (about 0.5), (n-BuO)$_2$TiEt was produced (green solution). When the Al/Ti ratio was increased to 1.2, (n-BuO)$_2$TiEt complexed with 2 n-BuOAlEt$_2$ to form a reddish-brown solution. Both Ti species polymerize styrene.

Farina and Ragazzini (15) showed for the AlEt$_3$–(i-PrO)$_4$Ti system that both 1-butene and high polymer can be obtained in variable amounts when the Al/Ti ratio is changed from 10 to 50.

A colloidal catalyst has been made from a heterogeneous one. When 1-octene was polymerized with AlEt$_2$Cl–βTiCl$_3$ · xAlCl$_3$, the reaction mixture had the appearance of a ruby-red solution. Supplementary work showed the catalyst to be colloidal form. A proposal was made that the product of Al(Octyl)$_3$ and VCl$_4$ produces a colloidal catalyst (16).

TABLE 6–1

Selected Examples of Supported Catalysts–Polymerization of Ethylene

Authors	No.	Company patent no. example no.	Catalyst			Polym. cond.	Catalyst act. $2/hr \cdot atm \cdot g\ TXn$	g Polymer	g Polymer
			Metal alkyl	TXn	Support	$hr/°C/atm$		g TXn	g Support
R. N. Johnson F. J. Karol L. A. Pilato	1	Union Carbide Corp. U.S. 3,642,749 2/15/72 (Ex. 4)	$AlEt_2OEt$	$RO—Cr—OR$ with $O=$ and $=O$; $R = 2$ Methyl borneal	0.4 g SiO_2 (350 m²/g)	1 90 20	750	15,000	225
W. L. Carrick	2	Union Carbide Corp. U.S. 3,324,095 6/6/67 (Ex. 1)	$AlEt_3 +$ $AlEt_2OEt$	$RO—Cr—OR$ with $O=$ and $=O$; $R = \phi_3Si$	0.4 g SiO_2	2 95 27	363	19,600	690
	3	Union Carbide Corp. Br. 1,253,872 11/17/71 (Ex. 1)	$AlEt_3 +$ $AlEt_2OEt$	$Cl—Cr—Cl$ with $O=$ and $=O$	0.4 g SiO_2	1.5 92 20	1,066	32,000	600

No.	Author	Reference	Cocatalyst	Catalyst	Support						
4	A. Delbouille J. L. Derroitte	Solvay & Cie Fr 3,658,722 4/25/72 (Ex. 9)	Al-i-Bu$_3$	TiCl$_4$	MgO (Special Treatment)	2	80	2	2,750	22,000	110
5		Solvay & Cie Fr 1,448,320 (Ex. 4)	Al-i-Bu$_3$	TiCl$_4$	Mg(OH)Cl	2	90	10	1,550	31,000	1,520
6	N. Kashiwa T. Tokuzumi H. Fujimura	Mitsui Ind. U.S. 3,642,746 2/15/72 (Ex. 1)	Al-i-Bu$_3$	TiCl$_4$	MgCl$_2$ + MeOH	2	90	3.5	5,000	35,000	1,400
7	B. Diedrich K. D. Keil	Hoechst U.S. 3,644,318 2/22/72 (Ex. 1)	AlEt$_3$	TiCl$_2$(OPr)$_2$	Mg(OEt)$_2$	7	85	7	179	8,600	11,700
8	V. D. Aftandilian	Cabot U.S. 3,222,296 12/7/65	Al-i-Bu$_3$	TiCl$_4$	SiO$_2$	1	80	14	189	2,650	500
9	W. J. Craven	Grace U.S. 3,322,691 5/30/67	None	CrO$_2$Cl$_2$	SiO $\cdot$ Al$_2$O$_3$ + (NH$_4$)$_2$SiF$_6$	1	100	30	1,250	37,800	760
10	W. C. Lanning	Phillips U.S. 2,963,470 12/6/60	None	CrO$_3$	SiO$_2 \cdot$ Al$_2$O$_3$	1	104	30	1,200	36,000	1,800
11	(Typical Range)	Conventional Heterogeneous Cat.	AlEt$_3$	TiCl$_3$	—	5	80	10	50–200	1,000–3,000	—

V. Heterogeneous Catalysts Including Supported Types

Two types should be differentiated. In type 1, which is the typical catalyst usually reported in the literature, the transition metal salt is a pure solid compound, such as $\alpha TiCl_3$ or VCl_3, or an alloy with a metal halide, for example, $TiCl_3 \cdot xAlCl_3$. These catalysts are described in detail in Chapter 4 and in Chapter 8 and therefore do not require further elaboration here. This section is devoted to supported catalysts. Karol (16a) has recently compared the Phillips, Union Carbide, and Solvay supported catalysts in a review.

In recent years, supported catalysts have received considerable attention because several were shown to be highly efficient for polymerization of ethylene. The transition metal salt can be physically and/or chemically fixed to the surface of a high surface carrier. If done properly, the metal atoms remain isolated, and all have the potential to become active centers.

The reader should be warned that, while many examples have been claimed in the patent literature, only a few have practical or scientific value. Unfortunately, many of the claimed supported catalysts are mixtures of an unknown structure. It is understandable from the historical developments why this should be. Many workers, motivated by the continued success of the Phillips $Cr_3O-SiO_2-Al_2O_3$ supported catalyst, hurriedly attempted to show that supported Ziegler–Natta catalysts would also be highly efficient. There was an important practical objective in accomplishing this. If the yield of polymer per g catalyst (excluding the carrier) was sufficiently high, then the expensive catalyst removal step could be eliminated. The innocuous support would be left in the polymer.

Many cursory experiments were made, but only a few gave meaningful information.

Table 6–1 shows selected examples of supported catalysts that have been reported for polymerization of ethylene to high density polyethylene. References 17 to 20 contain examples of the more important supported catalysts, and these will be elaborated according to the company that developed them. But first some general comments are made.

A. GENERAL DESCRIPTION OF SUPPORTED CATALYSTS

In general, these polymerizations were done under conditions similar to those used for regular Ziegler–Natta catalysts, such as low pressures (2 to 27 atm olefin pressure), intermediate temperatures (80° to 100°C), short residence times (1 to 8 hours), and usually with molecular hydrogen present to control polymer molecular weight. The catalyst activities are given as grams of polyethylene formed per hour (h), per atmosphere (atm) olefin, and per gram (g) transition metal salt. A conventional nonsupported catalyst

($AlEt_2Cl + TiCl_3$) has an activity value in the range 50 to 200 for ethylene, depending on the particular Ziegler–Natta catalyst used.

In comparison, the supported catalysts in Table 6–1 have activity values usually ranging from 250 to 7,051, but a few higher values have been reported. This corresponds to approximately 3,200 to 85,000 g polyethylene per g transition metal compound compared to 100 to 2,000 g/g for regular non-supported catalysts (comparable residence times). It should be noted, however, that the yields of polymer per g support are much lower (usually about 500 to 1,000), but since the support is innocuous it need not be removed. The Ziegler–Natta supported catalysts compare favorably with the Phillips-type supported catalysts.

While a variety of supports have been reported, those based on Mg and Si have produced the most active catalyst. In most of these catalysts, the transition metal salt is said to become bonded via oxygen bridges to the surface of the support. This is illustrated for the Solvay and Cabot catalysts in Eqs. 6–1 and 6–2.

Solvay Catalyst

$$—Mg(Cl)OH + TiCl_4 \longrightarrow —Mg(Cl)O–TiCl_3 + HCl \qquad (6–1)$$

Cabot Catalyst

$$—\overset{|}{\underset{|}{Si}}OH + TiCl_4 \longrightarrow —\overset{|}{\underset{|}{Si}}–O–TiCl_3 + HCl \qquad (6–2)$$

When metal alkyl is added (such as $AlEt_3$), alkylation and reduction occur and an active catalyst is formed. The specific nature of the active center has not been elucidated.

B. SELECTED EXAMPLES OF SUPPORTED CATALYSTS

Selected examples of supported catalysts include the Solvay, Cabot, and Union Carbide catalysts, as well as a few other examples.

1. Solvay Catalysts

Solvay workers (17, 18, 21–29) have extensively investigated the supported $Mg(OH)Cl$ and related catalysts including $MgSO_4$, $MgOSiO_2$, and MgO (18a). It is not clear whether all of the Ti centers in their supported catalysts are isolated. The high activities obtained for polymerization of ethylene suggest a high degree of isolated centers. Yet, isotactic polypropylenes of low (21) and fairly high (26) crystallinity were prepared with some of these catalysts, especially $Mg(OH)Cl + TiCl_4 + AlEt_3$. This suggests that small $TiCl_3$ crystallites were present or, alternately, one would have to postulate that the surface of $Mg(OH)Cl$ at the center has the required geometry to induce an isotactic placement.

TABLE 6–2

Efficiency of Ti in the Polymerization of Ethylene at Different Loadings on a Mg(OH)Cl Support[a] (22)

Loading (mg Ti per g support)	g Polyethylene formed	
	(per g supported[b] catalyst)	(per mg Ti[c])
12	1,400	117
12	1,500	125
17	835	49
22	550	25
40	248	6

[a] Conditions: Al-i-Bu$_3$ metal alkyl, at 90°C for 2 hr using 10 atm each of H$_2$ and ethylene.

[b] Includes weight of Mg(OH)Cl and Ti salt.

[c] Calculated by author from data contained in patent.

There appears to be at least a tendency for centers to become isolated as the amount of Ti on the support decreases. Delbouille and Speltinckx (22) increased the rate of polymer formation from 6 to about 120 g per mg Ti used on decreasing the loading of Ti from 40 to 12 mg Ti per g support (Table 6–2).

Solvay workers were able to obtain very high rates with specific supports if these were first pretreated and if specific metal alkyls were used. Some examples are cited here.

If MgO was reacted with gaseous HCl, HCl diluted with N$_2$, Cl$_2$, HBr, HI, and TiCl$_4$ at temperatures between 50° and 200°C, activities between 6,710 and 23,800 g polyethylene per hr per atm ethylene per g Ti were obtained (29). MgSO$_4$, treated with HCl, TiCl$_4$, or a mixture of the two, was comparably effective, but microspheroidal magnesium silicates, calcium oxide, cobalt oxide, and nickel oxide were less effective.

If the reaction product of Al-i-Bu$_2$H, Al-i-Bu$_3$, and butadiene was used as an activator for a TiCl$_4$–MgO catalyst, very high activities were obtained, for example, up to 13,600 g PE/hr/atm C$_2$/g Ti (23). The MgO was prepared by calcining MgCO$_3$ at 540°C for 16 hours.

The activity of the supported catalyst depended on the temperature of calcination; for example, for MgO prepared by calcining 3MgCO$_3$ Mg(OH)$_2$ 3H$_2$O, an optimum activity was obtained if the catalyst was calcined at about 500°C (28).

Supported catalysts were used to copolymerize ethylene and propylene (21) and homopolymerize butadiene (24). High *cis*-1,4 or high *trans*-1,4 polymers were obtained when Co and Ti transition metal salts were fixed on the Mg(OH)Cl, MgO, or Mg(OEt)$_2$ supports.

When polymeric material such as 80% *trans*-1,4-polybutadiene was used as support in place of the inorganic compound, lower rates were obtained (27); resins of formaldehyde and melamine were also used (27a), but these did not appear advantageous.

Supports bearing an oxygenated organic nitrogen compound of divalent metal (Mg, Ca, Zn) were used; the organic radical was an oxime hydroxamic acid or *N*-substituted hydroxylamine (18).

According to trade journal accounts (18a), Solvay & Cie is using one of these catalysts in the commercial production of high density polyethylene. These polymers were characterized by narrow molecular weight distributions.

2. Cabot Catalysts

Cabot workers investigated silica as the support for both Ziegler–Natta (19, 20, 30–35) and metal alkyl-free catalysts (see Chapter 12). In a series of patents issued in 1965, Orzechowski, MacKenzie, and co-workers described novel chemical modifications aimed at improving catalyst activities.

The pyrogenic silica used had an oversize particle diameter of about 20 μm and a hydroxyl group content on surface between 2.3 and 2.7 mg per g (19).

The preferred transition metal salts were $TiCl_4$ (20) and $MnCl_6$ (30); $TiCl_4$ was significantly better than $(n\text{-PrO})_4Ti$ (32).

Improved catalysts were claimed if the silica surface was first reacted with LiBu (31), NaH (33), Na metal (33), or Ga_2H_6 (34).

$$-\overset{\mid}{\underset{\mid}{Si}}-OH + LiBu \longrightarrow -\overset{\mid}{\underset{\mid}{Si}}-OLi$$

surface hydroxyls   A

$$A + Mn\,Cl_6 \longrightarrow -\overset{\mid}{\underset{\mid}{Si}}-OMnCl_5 \qquad\qquad (6\text{--}3)$$

B

$$B + AlEt_3 \longrightarrow catalyst$$

Addition of amine, presumably to neutralize HCl, resulted in improved catalysts (35).

Chien (36) found that in the Cabot catalyst all of the bonded titaniums became reduced when metal alkyl was added if the initial Cab–O–Sil SiO_2 support was annealed, but only 45% reduction took place if rehydrated Cab–O–Sil was used. The annealed SiO_2 had 1.4 Si–OH per 100 Å^2 surface, while the unannealed SiO_2 had 4.6 SiOH per 100 Å^2. He explained the different degree of reduction in terms of disposition of the silanol groups on the surface of these two silicas.

The reactions of silica surfaces with $TiCl_4$ and $Al(CH_3)_3$ vapor, respectively, have been studied by Murray and co-workers (36a) using infrared and ESR spectroscopy. Various reaction complexes were identified.

3. Union Carbide Catalysts

Union Carbide workers examined a number of chromium compounds on SiO_2 as catalysts (37–40). Such chromium compounds as CrO_2Cl_2 and $CrO_2(OR)_2$ where $-OR$ = 2-methylborneol and ϕ_3SiO were effective. $AlEt_3$ or $AlEt_2OEt$ or mixtures of both were used as activators (37–39).

Carrick and co-workers recently elaborated further on the bis(triphenyl-silyl)chromate–SiO_2–Al alkyl catalyst (40). The component bis(triphenyl-silyl)chromate is mildly active even if SiO_2 and Al–alkyls are absent; for example, see Table 6–3, example 1. Upon addition of SiO_2 and an aluminum alkyl, substantial increases in activity were obtained for each component. Carrick and co-workers showed that porous supports with pore diameters 47 Å made poor catalysts, while those with pore diameters 67 Å made excellent catalysts. They suggested that pores must be above a certain size (67 Å) for the catalyst components to be able to penetrate and react effectively. They found that the adsorption of the silyl chromate on the support was a slow process requiring several hours for completion.

TABLE 6–3

Bis(triphenylsilyl)chromate-based Catalyst

Example	$[\phi_3SiO]_2 CrO_2$ (g)	SiO_2 (g)	Al–Rn	Ethylene (psig)	Temp. (°C)	Time (hr)	Polymer (g)
1	1.0	None	None	20,000	150	6	55
2	0.02	1.1	None	600	135	6	65
3	0.01	0.4	0.18 mm $AlEt_3OEt$	300	89	1.5	195

After addition of the organoalumina compound, the Cr component was reduced in valence below the hexavalent state. The rate of polymerization was linear with ethylene pressure.

4. Other Selected Examples

A few other supported catalysts have been reported by Kashiwa and co-workers (41, 42), who reacted $MgCl_2$ with methanol to produce $MgCl_2 \cdot \frac{1}{10}MeOH$, followed by reaction with $TiCl_4$ at 125°C for 1.5 hours. $Al-i-Bu_3$ was used as activator (see example 6, Table 6–1). In another example, $MgCO_3$ was used as support (43). High conversions of ethylene (up to 616,000 g per g Ti) were reported by Susa and Mayr (44) for a catalyst prepared by reacting $TiCl_4$ and $Al(isohexyl)_3$ in the presence of $Mg(OH)Cl$ at $-78°C$ and then letting the reaction temperature rise to ambient temperatures. The supported catalyst contained 0.27 wt% Ti. Long (43) described a catalyst made by reacting $(C_6H_5CH_2)_4Ti$ and SiO_2 and using $AlEt_2Cl$ as activator. Peters (45) prepared a catalyst by calcining calcium manganese–SiO_2 (4.7 wt% Mn) and adding Al-i-Bu$_3$ as activator. Horvath (46) described a particle forming catalyst by supporting 8% $(Et_2N)_4Ti$ on SiO_2, calcining the product, and combining it with $AlEt_2Cl$ as activator. An anion exchange resin (Amberlite IRA 904) was mixed with $Al_2Et_3Cl_3$ and $TiCl_3$ to form a catalyst that made high isotactic polypropylene (47). The exchange resin may have functioned as a donor (see Chapter 9).

Diedrich and co-workers (48, 49) described supported catalysts based on $Mg(OEt)_2$ and $Mg(OH)_2$ as supports and $TiCl_4$ or $TiCl_2$ as the transition metal component. These catalysts, however, did not have the high activity reported earlier by Solvay workers (compare examples 7 and 4 in Table 6–1).

Long (50) obtained a yield of 1610 g polyethylene in 22 hours at 52°C and 2.1 atm ethylene using as catalyst $AlEt_2Cl$, tetrabenzyltitanium on SiO_2 (0.072 mmole per g SiO_2). $Mg(OH)_2$ was used to support $TiCl_4$, and $AlEt_3$ was added to the reaction product to form a highly active catalyst for ethylene (51).

The use of supported catalyst has apparently not been exploited in the polymerization of dienes. A few examples are cited here. Lasky and co-workers (52) reported an increase of 10 times in the polymerization activity with the use of a supported VCl_3 catalyst, such as VCl_3 on TiO_2 or kaolin mixed with $Ti(OR)_4$ and $AlEt_3$ for isoprene. A nickel catalyst supported on SiO_2 Al_2O_3 or active carbon diatomaceous earth was active for polymerizing butadiene to high cis-1,4 polymers (53) in presence and absence of activators.

In 1972, Montecatini (54) announced a low-pressure polyethylene process in a conventional stirred reactor at 50° to 100°C and 1 to 15 atm pressure. The catalyst was an unspecified supported Ti system. Yields of 200 to 400 kg per g Ti were claimed; if no residue was removed, this would leave only 3 to 5 ppm Ti in the polymer.

VI. Conclusion

The finding that Ziegler–Natta catalysts are active in a soluble, colloidal, and heterogeneous form is commercially and mechanistically very important. Each type allows a certain type of a commercial process to be developed, giving much flexibility to the manufacture of olefin and diene polymer products. From a mechanistic point of view, it seems remarkable that the active center is able to function so well, often exceptionally well, in such diverse environments. To have the active center in such apparently different molecular environments is in itself instructive, and this offers the investigator unique opportunities to probe its capability in more detail. Consider, for example, the supported type catalysts. Here the active centers have the isolated feature of soluble systems (centers are not linked by atom bridges) and the fixed feature of heterogeneous catalysts. Because both features offer a practical benefit, it is like having the best of two worlds.

References

1. However, Giannini and Zuccini have reported the synthesis of isotactic polymers with soluble metal alkyl-free catalyst based on benzyl derivatives of transition metals. See Chapter 12, Section III,B,2,b.
2. F. Bandermann, P. Vesper, P. Raulinat, and H. Sinn, *Chem.-Ing.-Tech.* **40**, No. 23, 1168 (1968).
3. See Chapter 4, Section III,E.
4. See Chapter 5, Section III,A.
5. See Chapter 5, Section III,A,B.
6. J. Balas and L. Porter, U.S. Patent 3,040,016, June 19, 1962, Shell Oil Company.
7. W. L. Carrick, *J. Am. Chem. Soc.* **80**, 6455 (1958); see also Great Britain Patent 873,498, July 26, 1961, Union Carbide Corporation.
8. K. Nakaguchi, M. Hirooka, and T. Fujita, U.S. Patent 3,328,366, June 27, 1967; Great Britain Patent 1,022,931, March 16, 1966, Sumitomo Chemical Company.
9. R. M. Manyik, W. E. Walker, T. P. Wilson, and G. F. Hurley, U.S. Patent 3,242,099, March 22, 1966, Union Carbide Corporation.
10. G. Natta, L. Porri, A. Carbonaro, and G. Stoppa, *Makromol. Chem.* **77**, 114 (1964).
11. G. Natta, L. Porri, and A. Carbonaro, *Makromol. Chem.* **77**, 126 (1964).
12. M. N. Berger and F. Fletcher, *Polymer* **2**, 441 (1961).
13. H. Hirai, K. Hiraki, I. Noguchi, and S. Makishima, *J. Polym. Sci., Part A-1* **8**, 147 (1970).
14. M. Takeda, K. Iimura, Y. Nozawa, M. Hisatome, and N. Koide, *J. Polym. Sci., Part C* **23**, 741 (1968).
15. M. Farina and M. Ragazzini, *Chim. Ind. (Milan)* **40**, 816 (1958).
16. For more detailed descriptions, see Section IV,A in Chapter 8.
16a. F. J. Karol, *Encycl. Polym. Sci. Technol.* to be published.
17. A. Delbouille and H. Toussaint, U.S. Patent 3,526,616, September 1, 1970, Solvay & Cie.
18. OLS Patent 2,205,102, August 31, 1972, Solvay & Cie.
18a. *Mod. Plast.* **47**, no. 7, 26 (1970); *Chem. Week* **100**, no. 22, 61 (1967); see also *Plast. Week* **27**, No. 24, (1967).
19. A. Orzechowski and J. C. MacKenzie, U.S. Patent 3,166,542, January 19, 1965, Cabot Corporation.
20. A. Orzechowski, U.S. Patent 3,220,959, November 30, 1965, Cabot Corporation.
21. Great Britain Patent 1,024,336, March 30, 1966, assigned to Solvay & Cie; see also French Patents 1,516,800, March 15, 1968, and 1,529,845, June 21, 1968.
22. A. Delbouille and R. Speltinckx, U.S. Patent 3,454,547, July 8, 1969, assigned to Solvay & Cie; see also French Patent 1,448,320, August 5, 1966.
23. Belgian Patent 757,847, April 22, 1976, Solvay & Cie.
24. P. Baekelmans and E. Leblon, U.S. Patent 3,642,760, February 15, 1972, Solvay & Cie.

25. F. Bloyaert, A. Delbouille, and J. Stevens, U.S. Patent 3,624,059, November 30, 1971, Solvay & Cie.
26. A. Delbouille and H. Toussaint, U.S. Patent 3,594,330, July 20, 1971, Solvay & Cie.
27. A. Delbouille and J. L. Derroitte, U.S. Patent 3,600,367, August 17, 1971, Solvay & Cie.
27a. H. Lefebvre and R. Dechenne, U.S. Patent 3,488,333, January 6, 1970, Solvay & Cie.
28. A. Delbouille, Y. Gobillon, and J. Stevens, U.S. Patent 3,663,660, May 16, 1972, Solvay & Cie.
29. A. Delbouille and J. L. Derroitte, U.S. Patent 3,658,722, April 25, 1972, Solvay & Cie.
30. A. Orzechowski and J. C. MacKenzie, U.S. Patent 3,205,178, September 7, 1965, Cabot Corporation.
31. A. Orzechowski and J. C. MacKenzie, U.S. Patent 3,205,177, September 7, 1965, Cabot Corporation.
32. V. D. Aftandilian, U.S. Patent 3,285,891, November 15, 1966, Cabot Corporation.
33. A. Orzechowski and J. C. MacKenzie, U.S. Patent 3,221,002, November 30, 1965, Cabot Corporation.
34. J. A. Yancey, U.S. Patent 3,202,645, August 24, 1965, Cabot Corporation.
35. V. D. Aftandilian, U.S. Patent 3,222,296, December 7, 1965, Cabot Corporation.
36. J. C. W. Chien, *J. Catal.* **23**, 71 (1971).
36a. J. Murray, M. J. Sharp, and J. A. Hockey, *J. Catal.* **18**, 52 (1970).
37. W. L. Carrick, G. L. Karapinka, and R. J. Turbett, U.S. Patent 3,324,095, June 6, 1967, Union Carbide Corporation.
38. Great Britain Patent 1,253,872, November 17, 1971, Union Carbide Corporation.
39. R. N. Johnson, F. J. Karol, and L. A. Pilato, U.S. Patent 3,642,749, February 15, 1972, Union Carbide Corporation.
40. W. L. Carrick, R. J. Turbett, F. J. Karol, G. L. Karapinka, A. S. Fox, and R. N. Johnson, *J. Polym. Sci., Part A-1* **10**, No. 9, 2609 (1972).
41. N. Kashiwa, T. Tokuzumi, and H. Fujimura, U.S. Patent 3,642,746, February 15, 1972, Mitsui Petrochemical Industries.
42. N. Kashiwa, U.S. Patent 3,647,772, March 7, 1972, Mitsui Petrochemical Industries.
43. W. Long, U.S. Patent 3,635,935, January 18, 1972, Hercules, Inc.
44. E. Susa and A. Mayr, U.S. Patent 3,634,384, January 11, 1972, B. F. Goodrich Company.
45. E. F. Peters, U.S. Patent 3,607,856, September 21, 1971, Standard Oil Company.
46. B. Horvath, U.S. Patent 3,646,000, February 29, 1972, Phillips Petroleum Company.
47. S. Nakano, M. Murayama, T. Watanabe, and T. Okawa, U.S. Patent 3,595,849, July 27, 1971, Mitsubishi Chemical Industries.
48. B. Diedrich and K. D. Keil, U.S. Patent 3,644,318, February 22, 1972, Farbwerke Hoechst.
49. B. Diedrich and W. Dummer, U.S. Patent 3,654,249, April 4, 1972, Farbwerke Hoechst.
50. W. P. Long, OLS Patent 2,049,477, October 8, 1970, Hercules.
51. Great Britain Patent 1,288,962, September 13, 1972, Farbwerke Hoechst.
52. J. S. Lasky, H. K. Garner, and R. H. Ewart, *Ind. Eng. Chem., Prod. Res. Dev.* **1**, No. 2, 82 (1962).
53. K. Ueda, A. Onishi, T. Yoshimoto, J. Honoso, K. Maeda, and T. Matsumoto, *Kogyo Kagaku Zasshi* **66**, 1103 (1963).
54. E. Susa, *Hydrocarbon Process.* **51**, no. 7, 115 (1972).

7

Physical State of the Polymer during Polymerization

I. Introduction

The choice of a process is one of the most important decisions a worker makes at the beginning of a polymerization. Depending on a number of factors, such as the structure of the polymer, nature of the solvent, and temperature of polymerization, the polymer may be formed in one of several physical states: (1) in solution (the solution process); (2) in the form of solid particles suspended in a liquid solvent (the slurry process); (3) in the form of solid particles suspended in a fluidized bed (vapor phase process). These three conditions form the basis of the three main types of processes that can be used for commercial production of olefin and diolefin polymers. Each process has led to a commercial application with its own advantage for a particular monomer–polymer catalyst system.

These processes can be done in a batch or a continuous operation. In the batch operation, polymerization takes place in the reactor for a specific time and then the entire polymer is removed and isolated. In a continuous process, on the other hand, some polymer is continuously removed while fresh catalyst and olefin are simultaneously added. Processes possessing features of both are possible. Because of its simplicity, the batch operation is mostly used in the small laboratory synthesis.

This chapter describes typical examples of each process that have been practiced on a laboratory scale. It is beyond the scope of this chapter to elaborate potential commercial processes. The interested reader might consult assessments of patent literature by Sittig (1), reviews by Doak and Schrage (2), Heggs (3), Albright (4), and Vandenberg and Repka (5).

II. The Solution Process

A. GENERAL COMMENTS

In a solution process, the polymer is dissolved throughout the polymerization. The catalyst, however, may be soluble or insoluble in the reaction medium.

The conditions for maintaining the polymer in solution depend on the structure of the polymer, the reaction solvent, and the temperature of polymerization. For example, poly-*cis*-1,4-butadiene and poly-*cis*-1,4-polyisoprene remain in solution at room temperature when the solvent is benzene. Poly-α-olefins containing long pendant groups, such as 1-octene, are soluble in aromatic and aliphatic solvents at ambient temperatures. Highly isotactic poly-1-butene becomes soluble in hydrocarbon solvents at about 70° to 80°C. At this temperature, isotactic polypropylene is still in particle form, and the reaction must be heated above 120°C to maintain the formed polymer in solution. High density polyethylene requires similar conditions. Lower temperatures (about 10° to 25°C lower) often suffice if better solvents are used, for example, cyclohexane. As a point of interest, when molecular weights are determined by viscometric methods, decalin or tetralin at 135° to 150°C are used as solvents for polyethylene and polypropylene. The heat of polymerization in a solution process is easily transferred by the solvent to the walls of the reactor, which are cooled by circulating water.

Solution processes for production of high density polyethylene have several advantages. One is better control over molecular weight and molecular weight distribution. Products that have low molecular weights and narrow molecular weight distributions have been easily made. These products are best suited for use as injection molding grade polyethylenes. Slurry processes are better suited for production of high molecular weight materials for use as extrusion and blow molding grades.

B. SELECTED EXAMPLES

Hoeg and Liebman (6) characterized the solution polymerization of propylene in cyclohexane at temperatures between 120° and 175°C using the AlR_3–$TiCl_3$ (or $TiCl_2$) catalysts (Figs. 7–1 and 7–2). They found that: (1) the isotacticity of polypropylene formed at 120°C was comparable to samples prepared similarly at 75°C but that isotactic content decreased slightly on raising the temperature to 175°C; (2) the molecular weight of the insoluble fraction (the part of the polymer not soluble in boiling *n*-heptane) decreased from 11×10^5 to 1×10^5 on raising the temperature from 75° to 175°C; (3) spontaneous dissociation of the polymer chains from the active centers was proposed to account for the major portion of the transfer processes.

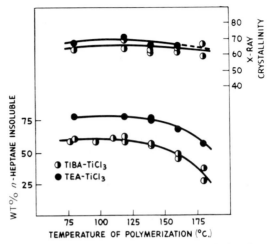

Fig. 7–1. Effect of reaction temperature on stereospecificity of catalyst (6).

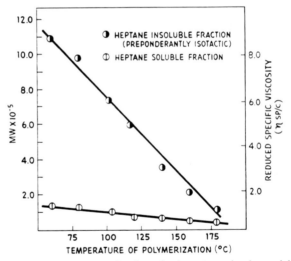

Fig. 7–2. Effect of temperature of polymerization on molecular weight of polymer (Al-i-Bu$_3$–TiCl$_3$ catalyst) (6); $[\eta] = 1.0 \times 10^{-4} \, [M_w]^{0.8}$; Chiang's equation was used to relate M_w and $[\eta]$.

Hagemeyer, Hull, and Park (7) took advantage of a solution polymerization to prepare isotactic polypropylene with a low catalyst residue content (about 0.001 wt% ash). Using a four-component catalyst (LiAlH$_4$, TiCl$_3$, NaF, and MgO, 0.8 : 1.0 : 0.5 : 0.1 molar ratio), they polymerized the propylene at 150°C in mineral spirits solvent and a pressure of 1,000 psig to a 20 wt% solution (Fig. 7–3). After the propylene was flashed off, the

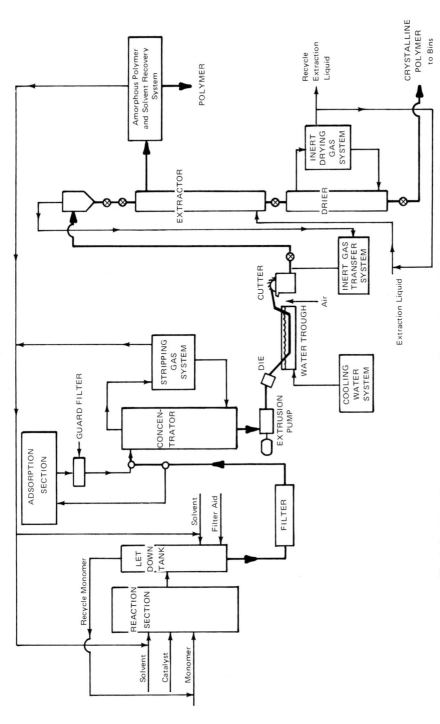

Fig. 7–3. Flow diagram of reactor system used for solution synthesis of polypropylene (7).

solution was diluted, heated to 230°C, and finally filtered to remove insoluble catalyst residues. After the solvent was removed, the polymer was extruded into pellets. A blocky propylene–ethylene copolymer (designated as a polyallomer by the Eastman workers) was prepared similarly; this copolymer is described in Chapter 3, Section V.

Muller (8) has described the polymerization of ethylene by a continuous solution process at 120°C and 20 atm pressure that uses a four-component catalyst [AlEt$_3$Cl, (C$_6$H$_5$)$_2$Mg, TiCl$_4$, and decanol].

Two very highly active solution processes have been reported for production of polyethylenes that have intermediate to high densities. One of these is the Du Pont process (9), which is operated above 200°C and at reactor pressures exceeding 1,000 psi. The catalysts are believed to be mixtures of TiCl$_4$, VCl$_4$, or VOCl$_3$ + AlR$_3$, and the residence times are only several minutes duration. The other is the Dutch State Mines (Stamicarbon) process (9, 10), which is believed to contain a mixture of a transition metal compound and a Grignard reagent as the metal alkyl component. Since production of polymer exceeds 100 kg/g transition metal, deashing is not necessary.

The solution process has been used to copolymerize ethylene and propylene. Baldwin and Ver Strate (11) have compared the advantages and disadvantages of the solution and slurry processes for this copolymerization.

High cis-1,4-polybutadiene and high cis-1,4-polyisoprene have been preferentially prepared by solution processes using aromatic solvents (benzene) of cyclohexane (see Chapter 5). In the case of AlEt$_2$Cl–cobalt salt donor catalysts, aromatic solvents were preferred since the catalyst was soluble in them.

III. The Slurry Process

A. GENERAL COMMENTS

In a slurry process, almost all of the polymer separates as a separate phase from the reaction solvent, which can be an aliphatic or aromatic solvent or even the monomer itself. The slurry process has found wide use for the synthesis of polyolefins but not polydienes. The polymer particles can vary in size, shape, and density. Soluble and colloidal catalysts produce a polyethylene that has a flocculant, gelatinous appearance in the reaction solvent. Only a limited amount of polymer, about 3 to 6 wt%, is present in the reaction mixture. In comparison, if heterogeneous catalysts are used, slurries containing 10 to 40 wt% polymer in the solvent can be achieved. The throughput of polymer (for example, polyethylene, polypropylene, or poly-1-butene) in a reactor is increased as the slurry concentration is increased.

The slurry process has most often been used for laboratory investigations of Ziegler–Natta catalysts and for synthesis of polyolefins. Usually temperatures in the range of 25° to 100°C are used, but the upper temperature depends on the polymer and solvent; for example, for isotactic poly-1-butene, it is about 50°C in heptane. Usually aliphatic solvents have been used, but more recently processes have been reported in which the olefin being polymerized also acts as the reaction medium, such as for propylene (12) and 1-butene (13). The heat of polymerization is easily removed via solvent to the reactor walls, which are cooled by circulating water.

B. SELECTED EXAMPLES

Since many examples of slurry-type processes are given throughout this book, it is not necessary to further elaborate here. A typical slurry process would be the polymerization of propylene at 50°C in heptane solvent and in the presence of $AlEt_3$–$TiCl_3$ catalyst. Other examples are cited in Tables 4–13, 4–14, 4–15, 4–16, and 4–25. Chapter 8 is devoted entirely to this process. Examples of reactors in which slurry polymerization can be done are given in Chapter 3, Section IV,A. Most of the catalysts described in Chapter 18 are of this type.

In recent years, very highly active slurry processes have been developed that use supported Ti catalyst, especially by Solvay, Montecatini–Edison, and Mitsubishi (see Chapter 6). Typically, these polymerizations are done between 30° and 90°C and at 100 to 500 psig ethylene pressure. Between 100 and 500 kg polymer is obtained per gram of titanium used. The residual Ti ash is so low that its removal is not required.

IV. Vapor Phase Process

A. GENERAL COMMENTS

At the present time, only a few commercial polymers are known to be polymerized by a vapor phase process. It may, however, be the process of the future.

Unlike the slurry process, a liquid phase is absent in the vapor phase process, and the polymer particles must, therefore, be fluidized by alternate means. Fluidization is usually achieved by one of several possible ways: (1) a fast recycle of the olefin being polymerized or a mixture of the olefin and an inert gas; (2) a combination of the first method and mechanical agitation; (3) entirely by mechanical agitation. By fluidization, we mean that the polymer particles are maintained in a suspension in the reactor to such a degree that there is a constant collision between polymer particles, as well as between polymer particles and the wall of the reactor. The fluidized bed is continuously charged with catalyst while polymer is simultaneously removed. The catalyst becomes uniformly dispersed by the fluidization. Sometimes, the vapor phase process is called a solventless process, but this does not mean the reactor is completely free of non-olefinic hydrocarbons. The catalyst components are frequently introduced into the reactor in a small amount of an alkane solvent, such as heptane, which becomes vaporized.

In the solution and slurry processes, the heat of polymerization is transferred from the catalyst–polymer particle via solvent to the water-jacketed reactor walls, where heat exchange then takes place with cooling water. In the vapor phase process, the heat of polymerization can be removed, in part, by the same way and in part by other paths; the preferred way depends on the olefin being polymerized. For ethylene polymerization, heat is largely removed by the recycled gas. This recycled gas is cooled before it is readmitted into the reactor. If the olefin is propylene, it is admitted as a liquid, and the heat of polymerization is used to vaporize the liquid propylene. Part of this gaseous propylene is vented to a condenser and readmitted as a liquid. Several such cycles may be required to remove the total heat of polymerization.

Temperature control has been reported to be more difficult than in a slurry or a solution process. For this reason, the vapor phase process has not been used in the laboratory to investigate Ziegler–Natta catalysts. Most of the papers and patents have come from industrial laboratories whose prime interest was to develop a more economical way to synthesize polyolefins.

B. SELECTED EXAMPLES

Only one commercial source of isotactic polypropylene produced by a vapor phase process is known, which is Badische's "Novolen" (14). In comparison with other commercial polypropylenes, Novolen contains a higher content of polymer soluble in heptane (about 10%), indicating a lower crystallinity. Several superior properties relative to conventional polypropylenes are claimed, including toughness, resistance to impact, and special uses such as fabrication of flat and tubular films, uniaxial and biaxial stretched films, and pressed plates. The increased impact strength was attributed to differences in molecular weight distributions of heptane and toluene-soluble fractions.

The Badische patents and publications (15–19) describe reactors utilizing a combination of stirring and gas recycle (15) or only stirring (16–18) as means of fluidization. Figure 7–4 shows a drawing of the reactor used by

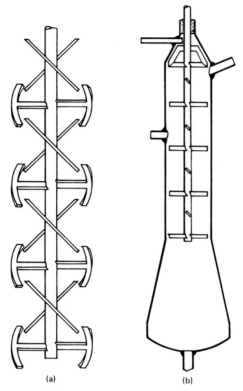

Fig. 7–4.   Reactor used for vapor phase polymerization (15).

Schmid and co-workers (15). In a typical experiment, ethylene was polymerized at 6 atm and 50°C in the presence of Al-i-Bu$_3$–TiCl$_3$ catalyst; 20 parts of ethylene and 0.019 parts of catalyst were supplied to the apparatus per hour. The flow velocity of the ethylene in the cylindrical portion of the reactor was 6 to 7 cm/sec, and the stirring speed was 123 rpm. The polyethylene product had a molecular weight of 100,000 and an ash content of 0.12%. To facilitate addition of the catalyst, 20 parts of TiCl$_3$ was mixed with 2,000 parts of powdered polyethylene, and the mixture was combined with 77 parts of Al-i-Bu$_3$ in 400 ml pentane. After the pentane was evaporated off, the catalyst–polyethylene powder was added to the reactor. Polypropylene was similarly polymerized with an AlEt$_3$–TiCl$_3$ donor catalyst.

More recently, Trieschmann and co-workers reported vapor phase polymerizations of propylene whereby liquid propylene was introduced into a stirred reactor, vaporized, and then polymerized with a catalyst having a

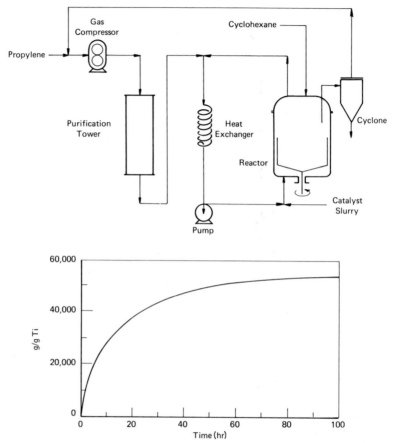

Fig. 7–5.   Top: Design of reactor system by which propylene was polymerized via vapor phase process (17). Bottom: Dependence of conversion on hours of polymerization.

high activity (16–18). Figure 7–5 shows a reactor scheme used in ref. 17. In a typical experiment, propylene was first purified over aluminum oxide, liquefied, and injected into the polymerizing powder layer, which was mechanically agitated by a stirrer. Complete evaporation and heating up to the reactor temperature took place with a concomitant absorption of heat of polymerization. Excess gaseous propylene was recondensed and recycled into the reactor. A 4.5-fold excess of propylene is required at a reaction pressure of 35 atm and a reaction temperature of 90°C to absorb the entire heat of polymerization. The $TiCl_3 \cdot \frac{1}{3}AlCl_3$ and $AlEt_3$ components were added in cyclohexane to the propylene and to the top of the reactor, respectively. The polypropylene formed contained 28% polymer, which was soluble in boiling heptane, indicating the whole product to have only an intermediate crystallinity. Polymer conversions up to about 20,000 g per g $TiCl_3 \cdot \frac{1}{3}AlCl_3$ per hour were reported. At these high yields of polymer per g of catalyst deashing of the product may not be necessary.

Union Carbide workers have recently reported the successful vapor phase polymerization of ethylene. According to Rasmussen (20), gas phase polymerization of ethylene to high density product saves 15% on investment and 10% in operating costs compared to processes that operate in the liquid phase. The savings occur because solvent recovery and drying equipment are not required. At least two plants were on stream using the vapor phase technology. The catalyst used by Union Carbide workers is not a Ziegler–Natta type, but it was said to be a specific chromium compound deposited on dehydrated silica or related supports. Productivity of about 600,000 lb of polymer per lb of metallic chromium was reported, thus eliminating the need for catalyst removal. They reported that it was necessary to keep impurities below 0.1 ppm. Formation of low molecular weight polymer was slight. The isolated polymer contains 1 to 3 ppm of entrained catalyst. The specific chromium compound was recently stated to be chromacene, $(C_5H_5)_2Cr$, and Union Carbides' publications on this metal alkyl-free catalyst are described in Chapter 12, Section III,B,9.

In a typical polymerization of gaseous ethylene, with the catalyst in the form of dry powder, comonomers such as propylene and 1-butene and hydrogen were fed into the reactor. The reaction pressure was kept at about 20 atm and temperature was controlled between 85° and 100°C. The circulating gas served to fluidize the bed of growing polymer powder, supplying the olefin and removing the heat of polymerization.

Vapor phase polymerizations have been studied in the laboratory (usually in the absence of agitation) by several workers.

For polymerization of ethylene in gas phase using CH_3TiCl_3–$TiCl_3$ catalyst, Edgecombe (21) found only the initial rate to be first-order. Activation energy for overall polymerization was 10 kcal/mole.

Kanetaka, Takagi, and Keii (22) examined the kinetics of polymerization of propylene with four kinds of $TiCl_3$ salts, with and without $AlCl_3$. A conventional constant absorption apparatus (120 cm³) and a constant pressure absorption apparatus with mercury–gas burette were used. Different activation energies were found for temperature ranges of 0° to 45°C vs. 45° to 70°C, that is, about 0.9 kcal/mole vs. 5.6 kcal/mole, respectively.

Rosen and Mason (23) synthesized 1-butene by a vapor phase polymerization with the $TiCl_3$–$AlEt_2Cl$ in a rotating 4' × 10' reactor (3 rpm). During the formation of the first 10% polymer, the reactor was kept at 24°C, but later it was cascaded up to 40°C. The product was a fine powder, of which 68% had a particle size of 44 to 300 μm.

V. Conclusion

In recent years, producers of high density polyethylene have reported on the successful development of superactive processes and/or catalysts (20, 24–28). The main motivation appears to be lower costs related to polymer isolation and workup rather than to savings on cost of the catalyst itself. The detailed nature of these processes is not disclosed; at best, hints are dropped relevant to some feature of the process. One can guess from the description of the different types of catalysts and processes, described in Chapters 4, 5, 6, and 7, what some of these innovations may be. It is certain that different routes have been taken.

Solventless processes would save on capital expenditure of equipment necessary to recover and recycle the reaction solvent. Yet, solution processes at higher temperatures of polymerization might have the advantage of higher rates and shorter residence times. Smaller reactors might suffice. Supported catalyst would produce very high yields of polymer per unit catalyst, especially if they could be used at higher temperatures. Removal of catalyst residues might be eliminated; this is an operation that can lower polymer costs substantially. Slurry processes can produce polymer powders that have very high bulk densities (see Chapter 8). For some applications, these powders might be used directly. Normally, polymer powders are fused and extruded into small nibs; this is a step requiring extra capital expenditure and operating costs.

Another important factor has to be considered in choosing a process, namely, the character of the polymer that is formed and its advantageous use for certain end-use applications. Such is the case for polyethylene, where the particular process determines a characteristic molecular weight and molecular weight distribution for the polymer.

Thus, it is not possible to conclude that one type of process is inherently better than another. The successful marriage of two or more operating features, coupled with product performance, may decide the choice of process for a particular producer. As novel process designs are innovated, changes in the practical processes will naturally occur.

The successful marriage of two or more of the above obviously would lead to superior processes; problems such as this will continue to challenge new investigators.

References

1. M. Sittig; "Polyolefin Processes" *Chem. Process. Rev.* **2**, (1967); "Stereorubber and Other Elastomer Processes" **3**, (1967); "Catalysts and Catalytic Processes 1967." **7** Noyes Dev. Corp., Park Ridge, New Jersey, 1967.

2. K. W. Doak and A. Schrage, *in* "Crystalline Olefin Polymers" (R. A. Raff and K. W. Doak, eds.), Chapter 8. Wiley (Interscience), New York, 1965.

3. T. G. Heggs, *in* "Block Copolymers" (D. C. Allport and W. H. Janes, eds.), Chapter 4. Wiley, New York, 1973.

4. M. A. Albright, "Processes for Major Addition-Type Plastics." McGraw-Hill, New York, 1974.

5. E. J. Vandenberg and B. C. Repka, *in* "Polymerization Processes" *High Polymer Ser.*, **29**, (C. Schildknecht, ed.), p. 337. Wiley (Interscience), New York, 1975.

6. D. F. Hoeg and S. Liebman, *Ind. Eng. Chem., Process Des. Dev.* **1**, No. 2, 120 (1962).

7. H. J. Hagemeyer, Jr., D. C. Hull, and S. J. Park, U.S. Patent 3,600,463, August 17, 1971, Eastman Kodak Company.

8. F. S. Muller, Netherlands Patent 680 6890, November 18, 1969, Stamicarbon N.V.

9. J. P. Forsman, *Hydrocarbon Process.* **51**, 13 (1972).

10. S. D. deBree, *Chem. Eng. (N.Y.)* **79**, No. 28, 72 (1972).

11. F. P. Baldwin and G. Ver Strate, *Rubber Chem. Technol.* **45**, No. 3, 709–881 (1972).

12. Great Britain Patent 1,040,669, September 1, 1966, Rexall Drug and Chemical Company.

13. Mobil process.

14. H. Mueller-Tamm, *Soc. Plast. Eng., Tech. Pap.* **27**, No. 15 (1969).

15. K. Schmid, J. Stedefeder, G. John, M. Haeberle, H. Lautenschlager, and H. G. Trieschmann, U.S. Patent 3,300,457, January 24, 1967, Badische Anilin- & Soda-Fabrik Aktiengesellschaft (BASF).

16. H. G. Trieschmann, W. Rau, H. Mueller-Tamm, and H. Pfannmueller, U.S. Patent 3,634,382, January 11, 1972, BASF.

17. H. G. Trieschmann, K. Wisseroth, R. Scholl, and R. Herbeck, U.S. Patent 3,639,377, February 1, 1972, BASF.

18. H. G. Trieschmann, W. Rau, T. Jacobesen, and H. Pfannmueller, U.S. Patent 3,652,527, March 28, 1972, BASF.

19. K. Wisseroth, *Angew. Makromol. Chem.* **8**, 41 (1969).

20. D. M. Rasmussen, *Chem. Eng. (N.Y.)* **79**, No. 21, 104 (1972); *Eur. Chem. News* (1972).

21. F. C. H. Edgecombe, *Can. J. Chem.* **41**, 1265 (1963).

22. S. Kanetaka, T. Takagi, and T. Keii, *Kogyo Kagaku Zasshi* **67**, 1433 and 1436 (1964).

23. M. K. Rosen and C. D. Mason, U.S. Patent 3,580,898, May 25, 1971, Allied Chemical Corporation.

24. Solvay, Hoechst and Stamicarbon, *Chem. & Eng. News* **49**, No. 19, 38 (1971).

25. Stamicarbon, *Chem. & Eng. News* **48**, No. 17, 36 (1970).

26. Montedison, *Chem. & Eng. News* **49**, No. 29, 34 (1971).

27. Mitsui, Phillips, *Mod. Plast.* **47**, No. 7, 22 (1970).

28. Amoco, *Mod. Plast.* **48**, No. 1, 20 (1971).

8

Growth of the Polymer Particle

I. Introduction

The growth of the polymer particle has aroused much interest in both academic and industrial laboratories. The reasons are twofold: on one hand, it was felt that if we understood the mechanistic path by which olefin is converted into polymer particles, then our knowledge of Ziegler–Natta catalysts will be more complete. We might, for example, learn how active centers are distributed on the surface of the catalyst.

The second reason was more practical and of much importance for the commercial producers of polyolefins. The industrial motivation stemmed from the observation that heterogeneous Ziegler–Natta catalysts are uniquely capable of replicating their morphology into the morphology of the progeny polymer particles. This discovery was very important because it enabled workers to easily control the polymer particle morphology by controlling the morphology of the transition metal salt, such as $TiCl_3$. By morphology, we mean the size, shape, density, and texture which the particle has.

During the formation of the polymer, the $TiCl_3$ particle becomes dispersed throughout the polymer particle. The architecture of the $TiCl_3$ particle is largely determined by the method used for its synthesis. Secondary $TiCl_3$ particles formed by reduction of $TiCl_4$ with aluminum alkyls can be very large (40 μm) but they are in fact comprised of many much smaller primary particles, about 0.01 to 0.1 μm diameters. Growth of the polymer takes place on these primary particles, and they remain within the same polymer particle because the formed polymer acts to cement them. Several substructures have been identified. How the polymer chains are assembled to form the substructures and eventually the optically visible polymer particle is not yet completely understood. A fiber with a diameter of 0.5 μm may be a basic building unit.

A logical beginning of this chapter, then, is a description of the replication phenomena, since this aspect of the Ziegler–Natta polymerizations provoked more detailed studies of the architecture of the $TiCl_3$ crystal and the way in which the polymer particle was formed. Later, when workers began to study the detailed physical characteristics of this polymer particle, research was directed toward understanding how polymer chains are assembled into the larger structures that comprise the polymer particle. All of these subjects are subsequently discussed. Some industrial applications are mentioned at the end of the chapter.

Many aspects of this subject have been critically discussed recently in an excellent review paper by Chanzy, Fisa, and Marchessault (1). General reviews on morphology are also available (2).

II. Replication

Replication can occur when ethylene, propylene, and many higher α-olefins are polymerized by a slurry or a vapor phase process with heterogeneous catalysts. Figures 8-1 and 8-2 show selected examples of micrographs

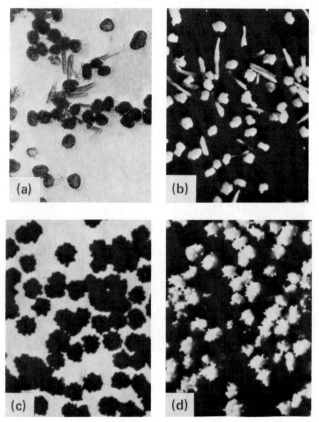

Fig. 8-1. Replication in an ethylene polymerization (5). (a) β-TiCl$_3$ ($\times 800$); (b) high-density polyethylene made from catalyst shown in (a); (c) β-TiCl$_3$ ($\times 400$); (d) high-density polyethylene made from catalyst shown in (c) ($\times 25$).

Fig. 8–2. Replication in a propylene polymerization using aluminum (a, b) and hydrogen (c, d) reduced $TiCl_3$ (4). (a) AA 215–55 $TiCl_3$; magnification: × 120. Translucent compactions with flat surfaces and angular edges. (b) Polymer from AA 215; magnification: × 50. On 80 mesh screen. Dense particles with flat sides and sharp edges. (c) H 104–1740 $TiCl_3$; magnification: × 230. Crystals stacked like mica, uneven shape, few hexagonal sides. (d) Polymer from H 104–1740 $TiCl_3$ on 80 mesh screen; magnification: × 50. Polymer is a rough, porous, agglomeration of strings.

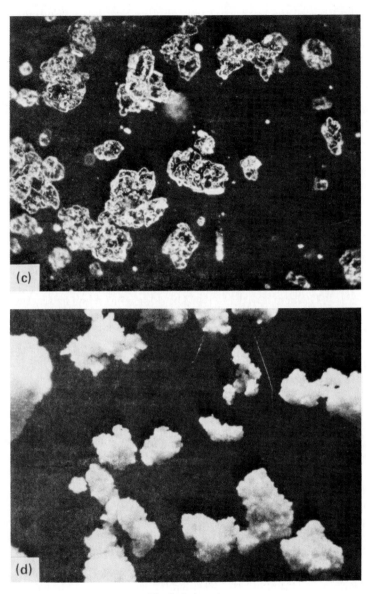

Fig. 8–2. (*cont.*)

obtained by light microscopy of polyethylene and polypropylene powders, respectively, and the corresponding catalyst powders from which they were made.

The light micrographs reveal that regardless of the shape of the parent catalyst particles, whether they are irregular, stringy, blocklike, or nearly spherical, the progeny polymer particles appear to be faithful replicates of them (3–6). Also, the relative sizes of the polymer particles depend on the relative sizes of the catalyst particles; that is, smaller catalyst particles produce smaller polymer particles relative to larger catalyst particles. Furthermore, if the catalyst particle is dense or porous, the corresponding polymer particle will be dense or porous, respectively (3).

III. Architecture of TiCl₃ Particles

The synthetic methods which have been used to prepare different titanium trichlorides are described in Chapter 4, Section II,B,5. Of special interest are αTiCl_3 and the β, γ, and δ modifications that have the composition $\text{TiCl}_3 \cdot x\text{AlCl}_3$ where $x = 0.2$ to 0.4. The β form has a linear structure, while the α, γ, and δ forms have a layer structure.

The morphology of these titanium chloride particles can vary considerably, the variation being controlled by choice of the reducing metal alkyl or metal and the conditions of reduction during their synthesis.

The αTiCl_3 modification has been prepared in various sizes, shapes, and degrees of perfection. Nearly perfect crystals were prepared by Rodriguez and co-workers by reduction of TiCl_4 with H_2 (7, 8), by Guttman and Guillet (9), and also by Carradine and Rase (10) by sublimation techniques. The crystals resembled hexagonal shaped platelets whose lateral faces were oblique to the basal face (7, 8), at an angle of 62° (9). Crystals as large as 10 mm in width were prepared (10).

The large crystals were especially useful for observation of polymer growth by light microscopy. But the αTiCl_3 samples that were prepared in the laboratory or purchased from commercial sources (4) consisted of smaller and considerably less perfect crystals; for example, Stauffer's TiCl₃H grade has particles 50 to 300 μm in diameter. These had the appearance of being agglomerates of small crystallites which appeared to have diameters of greater than several microns. Considerable variation was shown in the size of the large particles described in Ref. 4.

When subjected to a mild shear force, such as agitation, the αTiCl_3 crystals defoliate easily, producing smaller flat leaflets (1). This occurs easily because adjacent chlorine layers are held together by weak cohesive forces. It is probable that the smaller crystallites are agglomerated into a larger particle via these weak cohesive forces between chlorine layers.

The β-, γ-, and δTiCl₃ modifications that are prepared by reduction of TiCl₄ with an aluminum alkyl have a different architecture. They have not yet been prepared as large single crystals. Hock (3) was the first to propose and support experimentally the idea that the large secondary particles in the δ modification, which are visible under an optical microscope, are actually made up of much smaller primary particles (Fig. 8–3). The particular

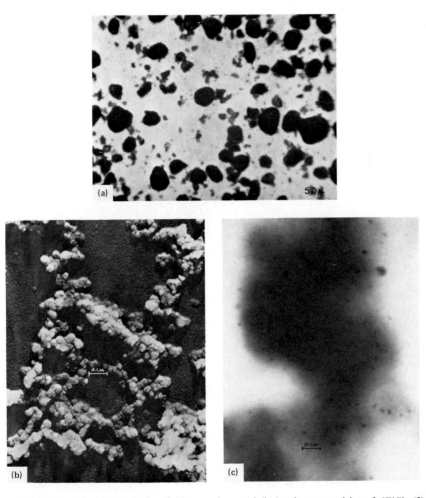

Fig. 8–3. Microphotographs of (a) secondary and (b,c) primary particles of δTiCl₃ (3). (a) Secondary ($\sim 30\ \mu$m) δTiCl₃ particles. Microphotograph was taken by light microscopy before polymerization. (b) Primary δTiCl₃ particles which are bound together into the larger ($\sim 30\ \mu$m) secondary particles. Microphotograph was taken by electron microscopy (metal shadowed) before polymerization. (c) Primary δTiCl₃ particles in an ultrathin section of an unpurified polymer flake obtained from a polymerization of propylene with this TiCl₃ and an aluminum alkyl. Micrograph was taken by electron microscopy at × 144,000 magnification.

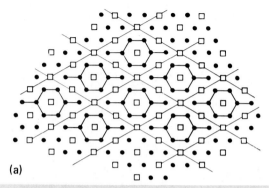

Fig. 8–4. Proposed models showing primary particles are Ti_8Cl_{24} units (13). (a) Titanium layer in $\gamma TiCl_3$ showing cleavage to Ti_8Cl_{24} units. (b) Oblique view of two Ti_8Cl_{24} units separated by Al_2Cl_6.

$\delta TiCl_3$ catalyst that he examined consisted of secondary particles that were 20 to 40 μm in diameter and had a nearly spherical shape. The primary particles were described as flat polygons whose diameters varied from 300 to 1,000 Å. Berger and co-worker later described a similar morphology for a $\gamma TiCl_3$ modification (11). Buls and Higgins (12, 13) concluded from kinetic studies and electron microscopy that the primary particles in $TiCl_3$–AA and $\gamma TiCl_3$ (both of which contain $AlCl_3$) consisted of a unit cell equivalent Ti_8Cl_{24} units, which were agglomerated into the larger secondary particle by Al_2Cl_6 linkages (Fig. 8–4a and b). Two active centers were said to be present per primary particle. Graff, Kortleve, and Vonk (14) derived a diameter of 1,000 Å and a thickness of 175 Å for primary particles of a catalyst prepared by reduction of $TiCl_4$ with $Al_2Et_3Cl_3$.

As far as the architecture is concerned, the $\alpha TiCl_3$ modification appears to lack the very small ($< 1,000$ Å) primary particles present in the β, γ, and δ

forms that are synthesized by reduction of $TiCl_4$ with Al or Al alkyls. When the $\alpha TiCl_3$ consists of smaller particles, the latter appear to range up to several microns in diameter (4).

Hock related porosity and denseness of catalyst and polymer particles by means of light micrographs, as shown in Fig. 8–5a and b. Another level of

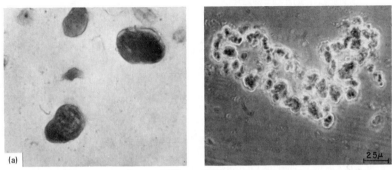

Fig. 8–5a. Sections through two different secondary catalyst particles showing differences in texture: (a) dense catalyst; (b) porous catalyst (5).

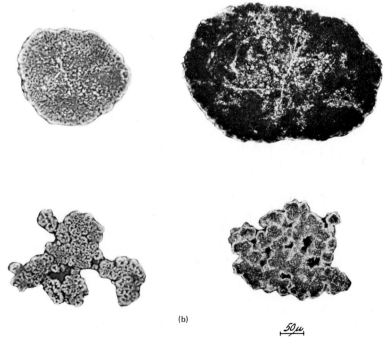

Fig. 8–5b. Sections through flake polymer showing differences in texture: (a) dense flake; (b) porous flake. Photographed by transmitted light and by dark field illumination (3).

Fig. 8–6.   Electron scanning photomicrograph of $\beta TiCl_3 \cdot 0.4AlCl_3$; magnification: × 2050 (15).

sophistication has been applied to the study of the architecture of $TiCl_3$ particles with the use of electron scanning microscopes (ESM). These have a resolving power of 0.015 μm compared to 1 to 2 μm for light microscopes. Wristers (15) took advantage of this method and his ability to manipulate the sensitive $TiCl_3$ sample in the apparatus to produce several elegant and informative micrographs of several $TiCl_3$ samples. Figure 8–6 shows that this particular $\beta TiCl_3 \cdot 0.4AlCl_3$ particle is actually composed of subparticles. When examined at higher magnification, the surface of this $TiCl_3$ particle appeared very irregular and consisted of small globules from 0.1 to 0.7 μm in diameter. Wristers stated that this was typical of catalyst surfaces produced by reduction of $TiCl_4$ with aluminum alkyls.

Wristers also showed that the microstructure of the $TiCl_3$ particle was dependent on the environment in which $TiCl_4$ was reduced by the aluminum alkyl. Figure 8–6 shows $\beta TiCl_3 \cdot 0.4AlCl_3$ particles that have a close structure; an $\alpha TiCl_3 \cdot 0.4AlCl_3$ particle that has an open structure is shown in Fig. 8–7. Polymer particles that were obtained from the latter had a low particle surface area, as measured by mercury porosimetry. In contrast, catalysts that had a closed structure formed polymer particles that had higher particle densities, lower porosities, and lower surface areas (see Table 8–1).

Wristers also concluded that the catalyst's final macrostructure was determined early in its synthesis, while the microstructure was susceptible to modification by changes in the reaction environment during the intermediate and later stages of particle growth.

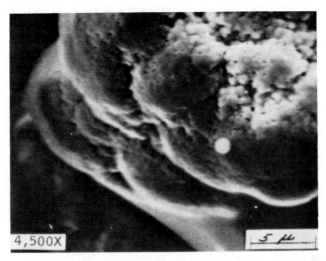

Fig. 8–7.   Electron scanning photomicrograph of αTiCl$_3$ · 0.4AlCl$_3$; magnification: × 4500 (15).

TABLE 8–1

Comparison of Polypropylene Particle Properties (15)

Structure of αTiCl$_3$ · 0.4AlCl$_3$ particle	Density (g/cm^3)	Porosity (cm^3/cm^3)	Surface area (m^2/g)
Closed	0.83	10.3	5.3
Open	0.74	16.8	8.2

The literature states clearly that the synthesis of TiCl$_3$ particles of the desired morphology requires control of the environment in which TiCl$_4$ is reduced. Many reaction parameters appear to be important, including the structure of the reducing metal alkyl, relative concentration of the reactants, stirring speed, etc. Two patent examples are cited.

A Hoechst patent (16) illustrates this nicely for a TiCl$_3$ prepared by reduction of TiCl$_4$ with AlEt$_2$Cl. As a diagnostic tool, they measured bulk density of the formed polyethylene powder. Section V elaborates on this measurement. The Hoechst workers found that the bulk density increased from 0.35 to 0.48 g/cm^3 as the TiCl$_4$/AlEt$_2$Cl ratio was decreased from 1.66 to 1.0, was increased from 0.26 to 0.49 g/cm^3 as the stirring speed was increased from 30 to 152 rpm, was increased from 0.31 to 0.49 g/cm^3 as the weight percent of TiCl$_4$ in solution was increased from 15 to 43%, and was increased from 0.36 to 0.46 g/cm^3 as the temperature of reduction was decreased from $30°$ to $0°C$.

A Union Carbide patent (17) reports that catalysts of highly geometrically uniform and essentially spherical form were prepared when $TiCl_4$ was reduced at high stirring speeds with various metal alkyls.

IV. How Does the Polymer Particle Grow?

With the above knowledge of the architecture of the $TiCl_3$ particle in mind, two views on how the polymer particle grows will be examined. First, the gross changes occurring as the small catalyst particle grows into the large polymer particle, which still contains the residue catalyst in some form, will be observed. Then, one must attempt to define the sequence that occurs as polymer chains are assembled into larger morphological structures and which comprise the large discrete polymer particle that one can see under an optical microscope.

A. GROSS CHANGES

In a polymerization with each of these $TiCl_3$ modifications, the optically visible catalyst particles disintegrate into smaller particles. If the cited architecture of the $\alpha TiCl_3$ is correct, then disintegration would be expected to be largely produced by defoliation but also, to some extent, by fragmentation of the larger crystals. In contrast, for the β-, γ-, or $\delta TiCl_3$ particles, disintegration would be expected to be largely the separation of the primary particles by the formed polymer. Different catalyst activities might be expected. The growth of the polymer particle in these two situations is next discussed.

On the basis of their kinetic studies, Natta suggested that, under the mechanical action of the growing polymer chains, the loosely held aggregates comprising the large $\alpha TiCl_3$ particles were broken up into the smaller crystalline particles (18). The same forces were said to continue to cleave the smaller crystallite into still smaller particles. As mentioned, the disintegration of these smaller crystallites is presumed to take place by cleavage along the loosely held chlorine–chlorine layers. The kinetic features of this change are described in Chapter 18.

Mechanical grinding can result in catalyst cleavage and a concomitant increase in catalyst activity; this was also demonstrated by Vecchi and co-workers (19). They reported that mechanical grinding of a $TiCl_3$ increased its surface area from 10 to 40 m^2/g and that the activity of the catalyst was simultaneously increased. In accordance with Natta's finding that the steady-state rate was not changed by grinding the $TiCl_3$ sample, the above represents only the changes in the initial activities of the catalysts.

Stauffer workers (4) recognized first the significance of the relationship between disintegration of catalyst particles and how progeny particles are formed. They reported that polymer growth occurs on the individual smaller crystals in such a way that the polymer growth seems to be uniform throughout the mass of the particle. They calculated that for reasonably high conversions, the volume of the polymer particle was about seven times greater than the volume of the parent catalyst particle, but otherwise the shape and density of the progeny polymer particle was a faithful replica of the parent catalyst.

According to Hock (3), the active centers of the $\delta TiCl_3$ are found in the primary particles. When polymerization takes place, then primary particles become separated and the voids become filled with polymer. This polymer acts as a cement keeping the primary particles and the formed polymer intact. What used to be an optically visible secondary particle now becomes a larger catalyst–polymer particle. Hock showed that at sufficiently high conversions, the primary particles become uniformly dispersed throughout the catalyst–polymer particle. Berger and Grieveson (20), from their study of ethylene polymerizations catalyzed by $AlEt_2Cl-\delta TiCl_3$, observed under an optical microscope that the outer surfaces of the catalyst–polymer particle still had a strong purple color of $TiCl_3$ after the polymerization commenced (Fig. 8–8). They concluded that the catalyst particles were not completely occluded in the layer of pure polymer but, rather, as the polymer grew it entrained the catalyst. This picture is consistant with Hock's view.

Buls and Higgins (12, 13) confirmed Hock's finding that secondary catalyst particles break up during polymerization and become dispersed in the polymer matrix. Figure 8–9 shows the polypropylene catalyst particle after 2, 6, and 32 minutes. The small dark portions are the catalyst fragments.

When ethylene and propylene are polymerized with heterogeneous catalysts below 90° to 100°C, the polymer chains which form at the reaction centers crystallize into solid particles. If higher temperatures are used, e.g. 130° to 200°C, both highly linear polyethylene and isotactic polypropylene remain in solution. In the case of isotactic poly-1-butene, a temperature as low as 50° to 80°C suffices to keep the polymer in solution.

If higher linear α-olefins such as 1-octene are polymerized, the polymer remains in solution even at room temperature. The catalyst also appears to be in solution. But is it? For 1-octene monomer and $AlEt_2Cl-\beta TiCl_3$ catalyst, the catalyst has a colloidal physical state according to centrifuging experiments at high g values (21) and according to electron microscopy (22). From the g value required to bring down a solid, an average particle size of 1,100 Å was calculated (21). Electron microscopy showed particles ranging in size from about 400 to several thousand Å (22). These findings are consistent

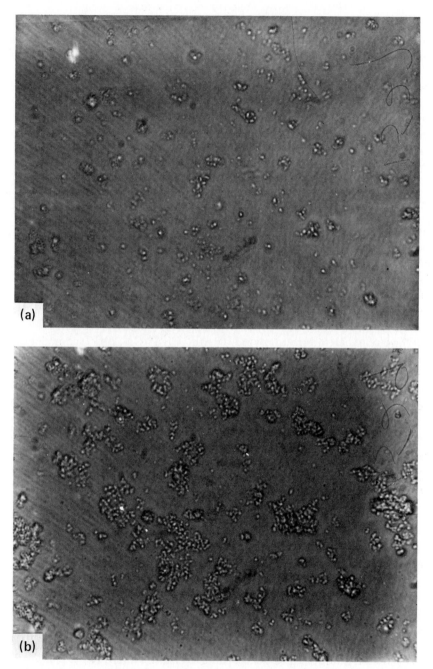

Fig. 8–8. Microphotographs of a Ziegler polymerization catalyst system: $\gamma TiCl_3/AlEt_2Cl$, room temperature. Magnification: $\times$ 700. (a) Before ethylene addition; (b) 2 hr reaction time; (c) 5 hr reaction time; (d) 70 hr reaction time (20).

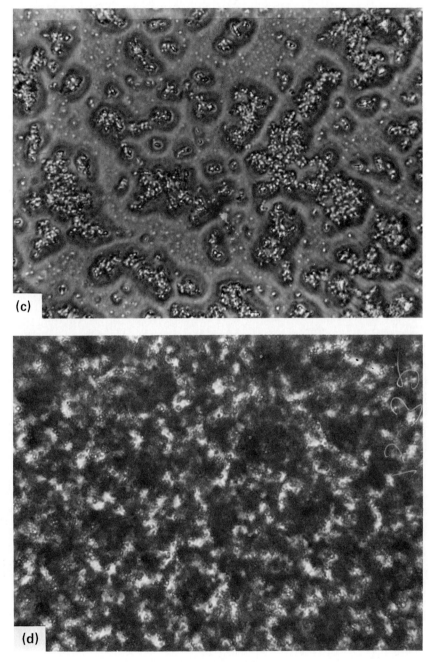

Fig. 8–8. (*cont.*)

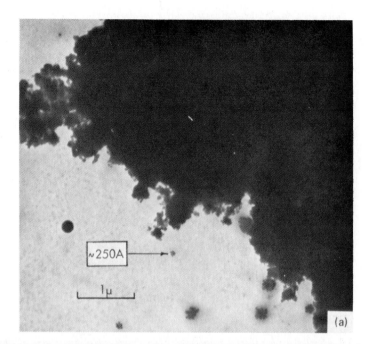

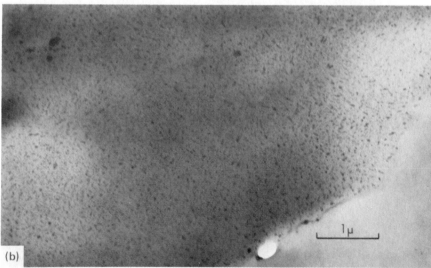

Fig. 8–9. Buls–Higgins observations on disintegration of secondary particles (12). (a) Alkyl-reduced catalyst; (b) polypropylene particle at 2 min; (c) polypropylene particle at 6 min.

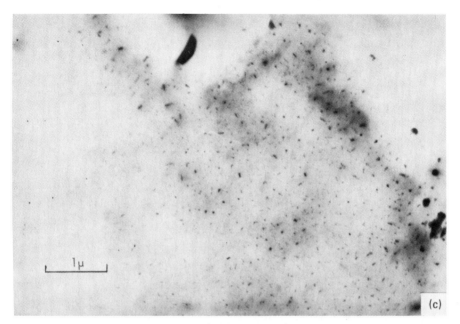

Fig. 8–9. (*cont.*)

with the conclusions of Stauffer workers, Hock, and later investigators based on slurry and vapor phase polymerizations.

B. HOW POLYMER CHAINS BECOME ASSEMBLED

The previous section described only the gross changes that took place when the olefin monomer was converted to the polymer–catalyst particle; for example, olefin + catalyst particle → polymer–catalyst particle. Various investigators have been concerned about the intermediate steps that occur, particularly how the polymer chains assemble into higher structures which then combine into still higher structures until the observed polymer particle is formed.

The picture that has emerged is not complete, and this account represents only an interim report. The particular experimental work required is difficult to do, and the findings are difficult to interpret unequivocally. Anomolies occur frequently. Each worker has examined some aspect of the problem and has given his view of what is happening. The findings are similar pieces of a puzzle which are coupled to form the whole picture; only here, some critical pieces are still missing. Yet, considerable progress has been made which deserves comment.

Fig. 8-10. Observation of initial polymerization of propylene on hexagonal leaflets of TiCl$_3$ as seen by electron microscopy investigations. Rodriguez and Gabant propose that the dots are traces of polymer that are localized along a spiral which they believe to be a crystal growth spiral (26).

In order to develop a possible scheme, one must start with the formation of polymer chains at the reaction centers. It will be helpful to first comment on the helical nature of polymer chains and on the loci of active centers. Stereoregular polymer chains, in order to minimize steric interactions between lateral substituents, form helices [see Natta and Corradini (23) for critical survey]. Polyethylene chains assume a zigzag conformation. Chain models of isotactic polypropylene, syndiotactic polypropylene, syndiotactic 1,2-polybutadiene, and syndiotactic cis-1,4-polypentadiene in the helical forms are shown in Chapter 3, Section II.

It is generally, but not universally, believed that the active centers are located on the lateral rather than on the basal planes of the $TiCl_3$ crystal (24), or on natural and induced defects of the crystal. The actual observation of the initial growth of polymer has been made by a number of workers using light and electron microscopy. The earliest experiments were done by Hargitay, Rodriguez, and Miotta, who used optical microscopy to study the initial polymerization of ethylene (25). Later, Rodriguez and Gabant (26), using electron microscopy, observed the formation of hemspherical particles of polypropylene along the spiral dislocations and on surface defects of the $\alpha TiCl_3$ crystal face (Fig. 8–10). Arlman's theoretical studies are described elsewhere (27). Kollar and co-workers (27a) observed widely different surfaces on two $TiCl_3$ samples, even though their surface areas (measured by adsorption) and their catalytic activities were identical. They explained this result by suggesting that the surfaces of the free lateral planes and the number of free edges were the same.

Guttman and Guillet (9) extended this polymerization of propylene and observed that the hemispherical particles elongated into hairlike protuberances. When less crystalline $\alpha TiCl_3$ was used, the polypropylene growths were randomly placed on the surface of the $TiCl_3$.

Carradine and Rase (10) examined by optical and electron microscopy the polymerization of ethylene on large sublimed leaflets of $\alpha TiCl_3$. After treatment with $AlEt_3$, a network of cracks developed on which polymer grew upon introduction of ethylene. When the $\alpha TiCl_3$ sample was mechanically deformed, polymer grew preferentially on the crushed areas and cleaved edges in preference to grown edges.

A rather novel experiment was devised by Baker and co-workers (28) who continuously examined the polymerization of ethylene in the presence of $AlEt_2Cl/TiCl_3$ at $25°C$ in the electron microscope. The catalyst was supported on graphite, and the reaction was carried out in gas phase at ethylene pressures between 10 and 20 torr. The transmission image was recorded on a videotape from a television camera located below the plate chamber of the microscope (Fig. 8–11). They were able to observe polymer growth at existing cracks and voids within the large catalyst particles, and thus they

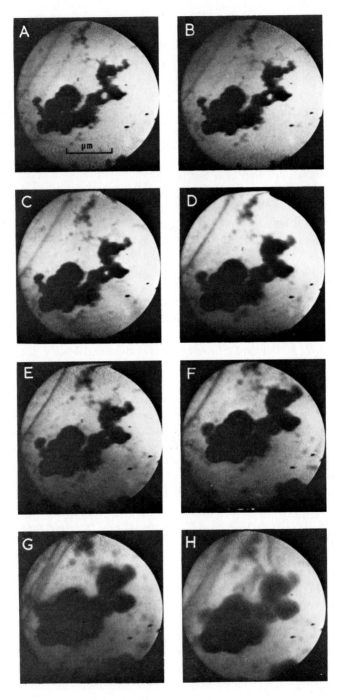

Fig. 8–11. (A–H) Baker's transmission images from videotape observations (28).

confirmed earlier findings that faults and defects in the particles were preferential sites for polymer growth.

The different proposals that have been made to explain how polymer chains become assembled to form the larger morphological structures will now be discussed. From physical measurements on the polymer particle, two different models have been suggested: the extended chain fibril model and the folded chain fibrillar crystal model. According to the extended chain model as initially proposed by Ingram and Schindler (29), Chanzy, Day, and Marchessault (30), the polymer chains are formed and crystallized nearly simultaneously. Thus, the fibrils that form from the catalyst surface consist of aggregates of extended chains. Chain folding is absent and thus the lamellar crystallite does not form. As evidence for this morphology, the higher melting point of the particle polyethylene was cited in comparison to a polyethylene that has a lamellar (chain folding) structure (138° to 140°C vs. 130°C). Also cited were a narrower melting region in the differential thermal analysis curve and the absence of small angle X-ray maxima, which is characteristic of a lamella. Thermal treatment of the particle polymer below its melting point converts it at least in part to the lamellar structure. But stretching polyethylene that has lamellar structure leads to an extended chain structure (29). Recent calorimetric studies by Chanzy, Bonjour, and Marchessault (30a) give support to a model that involves successive polymerization and crystallization when ethylene was polymerized on various substrates, such as glass, mica, and starch granules. The melting point of the polyethylene increased with time of polymerization.

According to the folded chain model, as proposed by Blais and St. John Manley, polymer chains undergo folding to form a lamellar structure (31, 32, and later papers 33–36). In their early papers, they proposed that fibrils were formed directly on the catalyst surface through crystallization of polymer chains growing from the active centers. As evidence, Blais and Manley cited electron micrographs that showed fibrils of 200 to 1,000 Å in width and an undetermined length (Fig. 8–12). From low angle X-ray, scattering measurements were taken that determined that polymer chains were folded and that the molecular chain axis was oriented perpendicular to the plane of the fibril lamella. Basic to this model is the presence of lamellar crystallites consisting of folded chains. The lamellae combine to form the larger structures.

These two early proposals are important because they focused on the extended and folded chains as basic building units for the various higher structures which form the polymer particle. Several important questions are logically raised. What are these higher structures? What is the sequence of their formation? How do they combine to form the next larger aggregate? What is the role played by the catalyst surface in determining if chain folding

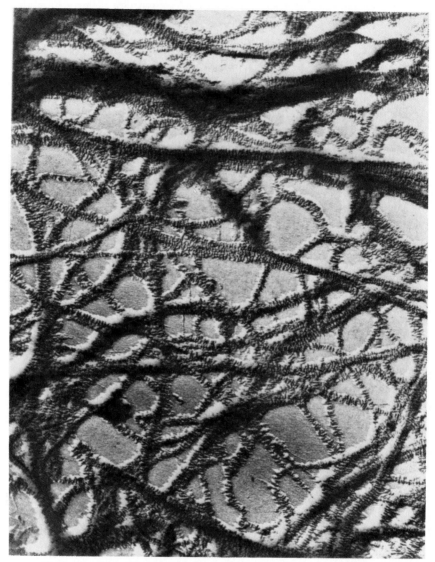

Fig. 8–12. Shish-kebob-type fibrils observed by Blais and St. John Manley (32).

or extended chain takes place preferentially and in the formation of the larger structures? Does the aggregation process itself create forces which alter the morphology of the polymer chain? How about factors such as temperature of polymerization and the solvent that is used? It is questions such as these that recent publications have tried to answer.

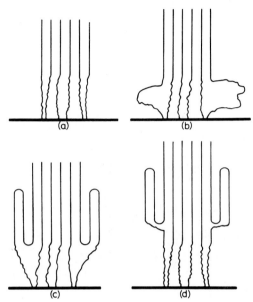

Fig. 8–13. (a–d) Different growth patterns suggested for polymer chains by Keller and Willmouth (37).

Keller and Willmouth (37) compared fibrils present in the polyethylene particle with those found in polyethylene crystallized from stirred solutions. They concluded that both extended chain and chain folded structures were present in both materials. Two possible mechanisms were offered for growth during polymerization. According to one proposal, the fibrils in the formed particle are flow-induced. In the alternative mechanism, chain folding is explained in terms of unequal growth rates of molecules from a cluster of active sites (Fig. 8–13). In accordance with the latter scheme, the proximity of growing chains will determine the structure of the fibrils. If the chains are distant, crystallization will involve chain folding. If the chains are in close proximity, crystallization will favor extended chains. This means that, even with soluble catalysts, crystallization of extended chains can occur if the concentration of chains is high enough. Because active centers in Ziegler–Natta catalysts appear clustered on lateral faces and defects, the concentration of chains is favored.

It is important to point out that when polyethylene or propylene is crystallized from solution under quiescent conditions, the crystalline polymer has a lamellar structure in which chain folding is present. Polyethylene synthesized with soluble catalysts also has a lamellar structure. Stirring can affect the rate of crystallization. For example, isotactic polystyrene crystallizes only slowly out of solution or from a melt under quiescent conditions, but it crystallizes quickly with rapid stirring. A sample of freshly prepared isotactic polystyrene has a very high crystallinity even though a short polymerization time was used. Yet the same sample after dissolution or fusion requires days to achieve the same crystallinity.

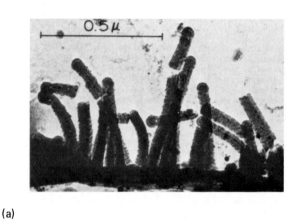

(a)

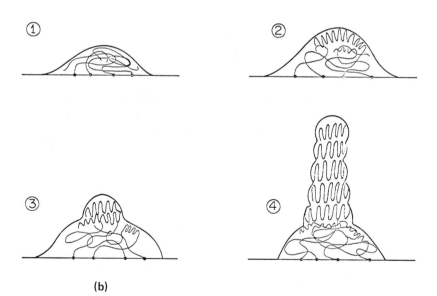

(b)

Fig. 8–14. Formation of fibrils during Ziegler–Natta polymerization as proposed by Guttman and Guillet (38). (a) Fractured polypropylene fibrils. Carbon replica; electron micrograph. (b, 1–4) Proposed mechanism of fibril growth.

The interior morphology of polyethylene particles was examined by Davidson (37a), who fractured the polymer particle in liquid nitrogen. Numerous complex structures, including thin filaments, platelets, nodules, worms, polyps, and helices were identified. The synthesis of the experimental polymer was, however, carried out at 90° to 110°C and the author feels that these structures may have resulted from secondary phenomena, such as solution and recrystallization.

Guttman and Guillet (38) proposed a model on the basis of observations made when propylene was subjected to a vapor phase polymerization with the $AlMe_3$–$\alpha TiCl_3$ catalyst (Fig. 8–14). According to their view, the rate of polymerization is initially more rapid than the rate of crystallization, resulting in an initial structure having the shape of a hemispherical mass of amorphous polymer. As the size of this globule increases, the rate of polymerization decreases, and, at this time, crystallization occurs at the apex of the globule in the form of folded chain lamellae. This model was based on electron micrographs of the polymerization product in the initial stages. A large number of polymer fibrils (up to 15,000 Å in length and 400 to 600 Å in diameter) were seen in the lateral edges and on dislocation lines of the catalyst crystal.

Marchessault, Lamande, and Chanzy (39) studied the spherulitic morphology of thick (10 to 100 μm) polyethylene membranes formed with the $AlEt_3$–VCl_3 catalyst and found that it was strongly dependent on the polymerization conditions and the manner of catalyst deposition. The observed texture was explained in terms of two schemes: each VCl_3 crystal is encapsulated by the growing polyethylene chains (the crystal surface induces a tangential orientation of polyethylene about itself); and the "warty" surface originates from the continuous cracking and movement of VCl_3 particles.

Graff, Kortleve, and Vonk (14) observed by electron scanning microscopy that the surface structures of polypropylene and polyethylene particles differed (Fig. 8–15). The polypropylene surface showed spheres of average diameter 0.4 μm, in agreement with the earlier findings by Hock. In contrast, the polyethylene surface showed a cobweb structure. The proposed growth process for the cobweb structure is shown in Fig. 8–16. According to these authors, the rate of polymerization at the surface of a catalyst particle is initially much higher than in the interior, whereas in a later stage the reverse is true. Because a greater part of the polymer is produced at or near the surface, a crust is formed around the conglomerate. But as the rate of growth later increases as one goes toward the center of the particle the crust breaks, but formerly adjacent parts of the crust remain interconnected by cold-drawn threads of polymer. Average volumes for the primary particles were calculated to be about 0.13×10^9 to 0.18×10^9 Å, corresponding to a diameter of 1,000 Å.

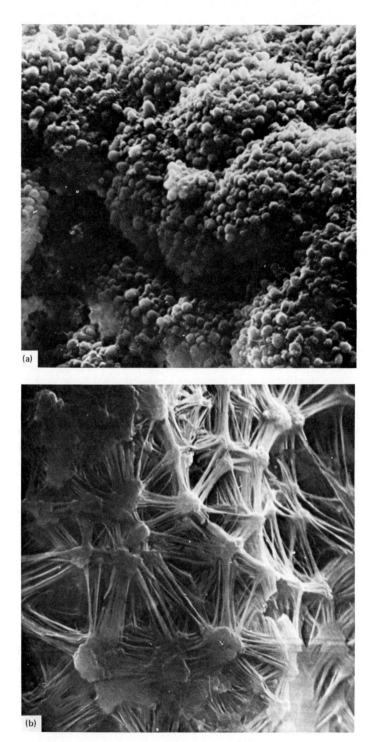

Fig. 8–15. Scanning electron micrograph of the external surface of a polypropylene (a) and a polyethylene (b) particle (14).

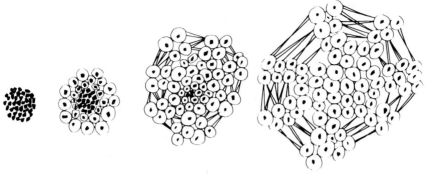

Fig. 8–16. Formation of "cobweb" morphology in a polyethylene particle (14).

TABLE 8–2

Structure of Polyolefins (40)

Monomer	Catalyst	Polymerization		Polymer structure		Microscope
		Reaction medium	Temp. (°C)	Size (μm)	Shape	
C_3-	$\alpha TiCl_3$, $Al(CH_3)_3$	Gas phase	R.T.[a]	0.03	Sphere	Electron transmission
C_3-	$\alpha TiCl_3$, $Al(CH_3)_3$	Gas phase	R.T.	0.03–0.05	Fibril	Electron transmission
C_3-	$TiCl_4$, $AlEt_3$	Toluene	84	0.04	Fibril	Electron transmission
C_2-	$TiCl_3$, $AlEt_3$	C_7	40–80	0.02–0.03	Fibril	Electron transmission
C_2-	$TiCl_4$, $AlEt_3$	Toluene	100	0.05–0.1	Fibril	Electron transmission
C_2-	Vanadium ester, $AlEt_2Cl$	Octanes	R.T.	0.05–0.1	Fiber	Electron transmission
C_2-	VCl_4, $AlEt_3$	Toluene	90	0.4–0.5	Fibril	Electron transmission
C_2-	$\alpha TiCl_3$, $AlEt_3$	C_6–C_{10} alkanes	R.T.	0.1–0.5	Ribbon	Optical and electron transmission
C_3-	$TiCl_4$, $Al_2Et_3Cl_3$	Diluent		0.4	Sphere	Electron scanning
C_2-	$TiCl_4$, $Al_2Et_3Cl_3$	Diluent		1.0	Globule	Electron scanning
C_2-	CrO_3/SiO_2	C_4	100	0.5–0.85	Worm and helix	Electron scanning
C_2-	VCl_3, $AlEt_3$	C_7	R.T.	23–70	Polyps	Optical
C_2-	$TiCl_3$, $AlEt_2Cl$	Hydrocarbon	50–80	160, 40	Sphere and needle	Optical
C_3-	$TiCl_4$, $AlEt_3$	Toluene	84	500	Globule	Optical

[a] R.T., room temperature.

In Table 8–2, Wristers has collected and identified the different polymer structures that have been proposed by various workers for polyolefins prepared with Ziegler–Natta catalysts. Admittedly, a clear picture has not emerged.

A reasonable unifying scheme has recently been proposed by Wristers (40) to show the many degrees of order in a polymer particle (Fig. 8–17). The

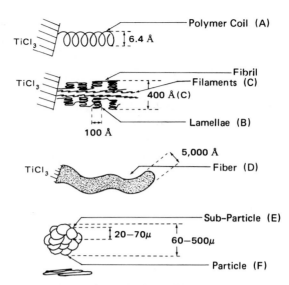

Fig. 8–17.   Consecutive levels of order in a polyolefin particle (40).

consecutive structures according to his order are: polymer coil ($d = 6.4$ Å), fibril (400 Å) which consists of filaments and lamellae (100 Å thick), fibers ($d = 5,000$ Å), subparticles 20 to 70 μm, and particles (60 to 500 μm) [1 μm = 10,000 Å].

Wristers suggested that the fiber is the basic morphological structure in all polyolefin powders made with the heterogeneous Ziegler–Natta catalyst and that the primary particle was responsible for the fiber structure.

TABLE 8–3

Size of Polypropylene Fibers with Different Crystallite Size (40)

Catalyst	Polymerization phase	Catalyst efficiency (g $HC_3^=$/g Ti catalyst)	Polymer structure	Fiber size (μm)	Catalyst crystal breadth (Å)
$TiCl_3$(H_2-reduced)	Vapor	85	Irregular	2.0	533
$TiCl_3$(H_2-reduced and ball-milled)	Vapor	300	Globular and fibrous	0.37	50
$TiCl_3$(Al-reduced)	Vapor	150	Fibrous	1.2	365
$TiCl_3$(Al-reduced and ball-milled)	Heptane	600	Fibrous	0.55	70
$TiCl_3$($AlEt_2$Cl-reduced)	Vapor	500	Fibrous	0.46	105

When the $TiCl_3$ catalyst was ball-milled, the size of the fiber decreased (Table 8–3). In earlier work, it was shown that ball-milling lowers the size of the primary particle (see Ref. 58 in Chapter 4).

Wristers showed that polymer prepared at low catalyst efficiencies had a globular shape, while that prepared at high catalyst efficiencies had a fiber shape. Intermediate catalyst efficiencies resulted in intermediate shapes (the three structures are compared in Figs. 8–18a, b, and c).

According to Wristers, more than one primary particle of $TiCl_3$ was responsible for the formation of a fiber. As evidence, he pointed out that fibers had branches. Not all primary particles had equivalent activities. Polymer growth from the surface of the primary particle was said to occur only in a limited number of quadrants, and thereby the primary particle was not densely encapsulated. Wristers explained the easy removal of $TiCl_3$ residues from the polymer particle by this model. For example, 90% of $TiCl_3$ was removed in 30 minutes by contacting the polymer with alcohol. Also, removal of the $TiCl_3$ from the polymer was not affected by increased efficiencies.

Wristers' findings strongly suggest that the basic morphological fiber structure is present in the polyolefin products that he prepared. Indeed, this work also pointed to the importance of primary particles in the formation of the fiber.

In the earlier investigations, some workers speculated that the catalyst surface (the concert action of adjacent centers) exerted a unique driving force that results in the formation of "nascent polymers" [polymer substructures consisting of chains that were simultaneously (or very nearly simultaneously) synthesized and crystallized without first being fused or dissolved].

Other workers held that, in addition to any affect which the catalyst surface may exert, factors such as heat of polymerization, stress and compression forces, solvent, temperature of polymerization, and stirring also contribute their affect. The observed morphology was, accordingly, due to the composite action of all of the factors, and it is understandable that variations in observed morphology would occur because of differences in procedures used for the polymerization.

The author agrees with Chanzy and co-workers (1) in that the forces of polymerization leave polymer chain segments that are close to reaction center in a partially swollen or fused state. Due to the heat of polymerization, local hot spots near the center may exist. Only the further removed (and cooler) segments can crystallize. The crystallization can induce stretching on the segments closer to reaction centers and cause them to crystallize in an extended form. If the polymerization is done at higher temperatures

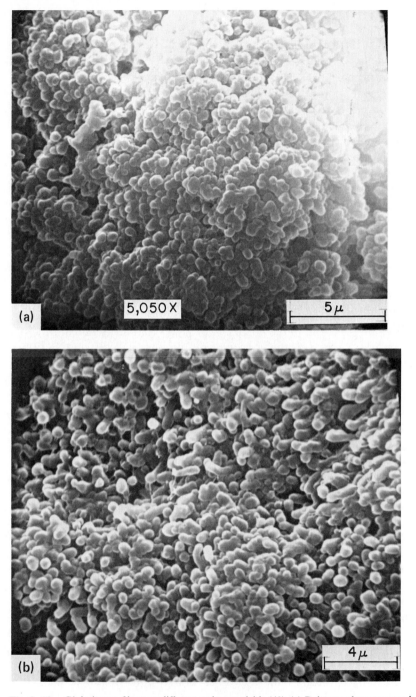

Fig. 8–18. Globules on fibers at different polymer yields (40). (a) Polypropylene prepared at low catalyst efficiency (hemisphere shapes). Magnification: × 5050. (b) Polypropylene prepared at intermediate catalyst efficiency in gas phase (elongated sphere shapes). Magnification: × 5000. (c) Polypropylene prepared at high catalyst efficiency (fiber shapes). Magnification: × 5500.

Fig. 8–18. (*cont.*)

(100° vs. 25°C) and in good solvents (toluene vs. heptane), then there is an opportunity for the polymer to dissolve and then recrystallize. Depending on the degree of stirring, extended chain formation and/or chain folding will occur. Also, chain transfer by metal alkyl molecules will release the polymer chain from the catalyst surface and leave it in a more independent position, not necessarily aligned with the chains fixed to the surface by Ti–C bonds. These chains can crystallize from their vantage position.

Such possibilities cause us to hold back before accepting or rejecting any of these models or to speak of the formed polymer by the term "nascent," as defined above.

V. Practical Importance

Control over catalyst (hence polymer) particle morphology can be very important in an industrial process. This section briefly describes the different ways.

A. EFFECT ON POLYMERIZATION KINETICS

A number of workers speculated that once the polymer forms and encapsulates the active center, the passage of olefin to these centers will be diffusion-limited (see Chapter 18, Section IV). If, however, the catalyst particle is porous, then the polymer particle is porous and passage of olefin to the center will be enhanced. Porosity would appear to be an especially important property as the size of the catalyst and polymer particles get larger.

B. RATE OF PRODUCTION OF THE POLYMER
IN THE REACTOR

Economics dictates that industrial reactors have a limiting size and require a certain residence time that the catalyst particle remains in the reactor. It is clear that the production of polymer would increase as the concentration of polymer particles in the reactor increases. Thus, the greater the bulk density of the polymer powder, the greater the throughput of polymer through the reactor. It does not matter whether the process is batch or continuous. Polymer powders that are highly dense usually have a regular shape and tend to have higher bulk densities. For example, irregularly shaped polypropylene powders that consist of small particles have bulk density values near 0.10 to 0.20 g/cm^3, while relatively larger, regularly shaped and denser polymer particles can have bulk density values 0.20 to 0.40 g/cm^3 and higher. Normally, polyolefins are sold as nibs (short fused cylinders of polymer). For some applications, powders that have a specific bulk density and other morphological characteristics might be preferred.

C. ENCAPSULATION

Herman and co-workers (41) encapsulated cellulose fibers by first reacting these fibers with $TiCl_4$ and metal alkyl and then polymerizing ethylene. Chanzy and co-workers (42) encapsulated wood pulp with polyethylene using VCl_4–aluminum alkyl catalyst deposited on the pulp surface. Studies using cellulose (43) and glass (44) surfaces were made by other workers.

D. CATALYST REMOVAL

Wristers (40) found that rate of removal of $TiCl_3$ was independent of the amount of polymer associated with the catalyst. He concluded that the primary catalyst particle is responsible for the fiber structure and accounts for the ready removal of catalyst residues from the finished polymer particle.

VI. Conclusion

In conclusion, a number of factors determine the morphology of the polymer particle synthesized in the presence of heterogeneous Ziegler–Natta catalysts. Replication follows from the architecture of $TiCl_3$ particles and from the fact that the $TiCl_3$ breaks down into smaller components during the polymerization while retaining the intact shape of the catalyst particle. This happens because the polymer acts as a cement for the smaller catalyst particles that bear the reaction centers. The whole particle consists of substructures that are formed from polymer chains by one or more paths. Very probably, the paths by which polymer chains are assembled are affected by the catalyst surface, the mechanics of growth and crystallization, and external conditions such as stirring. Considerable understanding of the mechanistic paths by which this occurs has been achieved through the careful and sophisticated work of a small numbers of workers. Continued interest in this important area is expected.

References

1. H. D. Chanzy, B. Fisa, and R. H. Marchessault, *Crit. Rev. Macromol. Science*, **1** (3), 315 (1973).
2. B. Wunderlich, *Adv. Polym. Sci.* **5**, 568 (1968); *Angew. Chem., Int. Ed. Engl.* **7**, 912 (1968).
3. C. W. Hock, *J. Polym. Sci., Part A-1* **4**, 3055 (1966).
4. Stauffer Chemical Company, Anderson Chemical Division, Preliminary Titanium Trichloride Technical Data (Collection of Technical Bulletins), 1962.
5. P. Mackie, M. N. Berger, B. M. Grieveson, and D. Lawson, *J. Polym. Sci., Part B* **5**, 493 (1967).
6. J. Boor, unpublished data.
7. L. A. M. Rodriguez and J. A. Gabant, *J. Polym. Sci., Part C* **4**, 125 (1964).
8. L. A. M. Rodriguez and H. M. Van Looy, *J. Polym. Sci., Part A-1* **4**, 1971 (1966).
9. J. Y. Guttman and J. E. Guillet, *Macromolecules* **1**, No. 5, 461 (1968).
10. W. R. Carradine and H. F. Rase, *J. Appl. Polym. Sci.* **15**, 889 (1971).
11. M. Berger, G. Boocock, and R. N. Haward, *Adv. Catal. Relat. Subj.* **19**, 211 (1969).
12. V. W. Buls and T. L. Higgins, *J. Polym. Sci., Part A-1* **8**, 1037 (1970).
13. V. W. Buls and T. L. Higgins, *J. Polym. Sci., Part A-1* **8**, 1025 (1970).
14. R. J. L. Graff, G. Kortleve, and C. G. Vonk, *J. Polym. Sci., Part B* **8**, 735 (1970).
15. J. Wristers, *J. Polym. Sci., Polym. Phys. Ed.* **11**, 1619 (1973).
16. Great Britain Patent 960,232, June 10, 1964, Hoechst.
17. A. K. Ingberman, U.S. Patent 3,179,604, April 20, 1965, Union Carbide Corporation.
18. G. Natta and I. Pasquon, *Adv. Chem. Ser.* **11**, 1 (1959).
19. E. Vecchi, R. Zannetti, L. Luciani, and P. Barbe, *Chim. Ind.* (*Milan*) **43**, 741 (1961).
20. M. N. Berger and B. M. Grieveson, *Makromol. Chem.* **83**, 80 (1965).
21. J. Boor, J. N. Wilson, and W. S. Anderson, unpublished data.
22. D. Winkler and R. Meisenheimer, unpublished data.
23. G. Natta and P. Corradini, *Nuovo Cimento* **15**, Suppl. 1, p. 9 (1960).
24. J. Boor, *J. Polym. Sci., Part C* **1**, 237 (1963); E. J. Arlman and P. Cossee, *J. Catal.* **3**, 99 (1964); G. Allegra, *Makromol. Chem.* **145**, 235 (1971).

25. B. Hargitay, L. A. M. Rodriguez, and M. Miotta, *J. Polym. Sci.* **35**, 359 (1959).
26. L. A. M. Rodriguez and J. A. Gabant, *J. Polym. Sci., Part C* **4**, 125 (1963); *ibid., Part A-1* **4**, 1971 (1966).
27. Chapter 15 describes models reported by E. J. Arlman.
27a. L. Kollar, A. Simon, and A. Kallo, *J. Polym. Sci., Part A-1* **6**, 937 (1968).
28. R. T. K. Baker, P. S. Harris, R. J. Waite, and A. N. Roper, *J. Polym. Sci.* **11**, 45 (1973).
29. P. Ingram and A. Schindler, *Makromol. Chem.* **111**, 267 (1968).
30. H. D. Chanzy, A. Day, and R. H. Marchessault, *Polymer* **8**, 567 (1967).
30a. H. D. Chanzy, E. Bonjour, and R. H. Marchessault, *Colloid Polym. Sci.* **252**, 8 (1974).
31. P. Blais and R. St. John Manley, *Science* **153**, 539 (1966).
32. P. Blais and R. St. John Manley, *J. Polym. Sci., Part A-1* **6**, 291 (1968).
33. A. G. Wikjord and R. St. John Manley, *J. Macromol. Sci., Phys.* **2**, 501 (1968).
34. A. G. Wikjord and R. St. John Manley, *J. Macromol. Sci., Phys.* **4**, 397 (1970).
35. A. G. Wikjord, D. H. Page, and R. St. John Manley, *J. Macromol. Sci., Phys.* **4**, 413 (1970).
36. R. St. John Manley, *Am. Chem. Soc., Div. Org. Coat. Plast. Chem., Pap.* **30**, No. 1, 176 (1970).
37. A. Keller and F. M. Willmouth, *Makromol. Chem.* **121**, 42 (1969).
37a. T. Davidson, *J. Polym. Sci., Part B* **8**, 855 (1970).
38. J. Y. Guttman and J. E. Guillet, *Am. Chem. Soc., Div. Org. Coat. Plast. Chem., Pap.* **30**, No. 1, 177 (1970).
39. R. H. Marchessault, A. Lamande, and H. D. Chanzy, *Am. Chem. Soc., Div. Org. Coat. Plast. Chem., Pap.* **30**, No. 1, 183 (1970); see also *IUPAC Meet., 1968, Prepr. Pap.* A 6.12 (1968).
40. J. Wristers, *J. Polym. Sci., Polym. Phys. Ed.* **11**, 1601 (1973); see also *Polym. Prepr., Am. Chem. Soc., Div. Polym. Chem.* **13**, No. 2, 1007 (1972).
41. D. F. Herman, U. Kruse, and J. J. Brancato, *J. Polym. Sci., Part C* **11**, 75 (1965).
42. H. D. Chanzy, W. A. Cote, Jr., and R. H. Marchessault, *Text. Res. J.* **38**, 583 (1968).
43. A. Dankovics, J. Erdelyi, and M. Koltai, *J. Appl. Polym. Sci.* **13**, 1809 (1969).
44. T. E. Lipatova and I. S. Skorynina, *J. Polym. Sci., Part C* **16**, 2341 (1967).

9

Modification of Ziegler–Natta Catalysts by Third Components

I. Introduction

The number of journal and patent examples of Ziegler–Natta catalysts that are modified by third components is too large to evaluate individually. This chapter contains selected examples to illustrate the various ways that the Ziegler–Natta catalyst can be modified and the proposed mechanistic paths by which these occur. Several reviews were written earlier on this subject (1).

Why does such a wide-scale interest exist? One reason is that the use of third components served as a powerful diagnostic tool to probe the mechanistic features of Ziegler–Natta catalysis. The other reason is practical and probably of significant commercial importance. The patent literature shows many examples of catalysts that attained their high stereoregulating ability only after the third component was added. It is understandable why so much effort was spent in screening literally all types of molecules as possible third components.

These third components can be complexing molecules, such as amines, and ethers, or molecules that have the potential of complexing and reacting with the components of the catalyst or the reaction centers, such as alcohols and water. It was realized early in Ziegler–Natta history that the active centers consisted of metal–carbon bond centers and that the metal atoms had electron-deficient character. It was inviting to add molecules such as amines or ethers and expect a change in the behavior of the catalyst.

The reader may have read what appears to be contradictory effects caused by the same third component (oxygen and water are good examples). The explanation may simply be that these third components work the stated way only under very narrow conditions and for specific catalysts only. For this reason, the broad claims of patents should be ignored when interpreting data, unless they are confirmed by experimental examples.

The positive effect of third components is usually an increase in activity and/or stereoregulating ability of the catalyst and/or an increase in polymer molecular weight. Other effects have also been reported. Many mechanistic explanations have been offered to account for these effects. Considering the wide scope of catalysts, monomer structures, and conditions of polymerizations, it is not likely that a common explanation exists.

Only a few catalysts have been investigated where the third component reacts or complexes preferentially with only one of the catalyst components. These are relatively the most easy to interpret. The nature of most Ziegler–Natta catalysts makes a cogent interpretation of the polymerization data difficult. For reasons stated in Chapter 4, aluminum alkyls have been most frequently used in Ziegler–Natta polymerizations. Most of the third components cited in the literature can complex or react with aluminum alkyls (2,3). When they also complex or react with the transition metal salt or with the reaction by-product formed when the aluminum alkyl and the transition metal salt react, several equilibria or reaction paths are possible. Consider, for example, the following complexes formed when Et_3N is added to a mixture of $AlEt_2Cl$ and $TiCl_3$.

$$Et_3N + TiCl_3 \rightleftharpoons Et_3N \cdot TiCl_3 \text{ (as a surface complex)}$$

$$Et_3N + AlEt_2Cl \rightleftharpoons Et_3N \cdot AlEt_2Cl \text{ (in solution)} \qquad (9\text{–}1)$$

$$Et_3N + AlEtCl_2 \rightleftharpoons Et_3N \cdot AlEtCl_2 \longrightarrow AlEt_2Cl + Et_3N \cdot AlCl_3 \text{ (in solution)}$$

($AlEtCl_2$ can be formed during site formation, that is, alkylation by $AlEt_2Cl$ of the $TiCl_3$ species located on surface of the $TiCl_3$ crystallite.)

If the third component reacts with both the metal alkyl and transition metal salt, then the actual catalyst is a mixture of several reactants, as shown in Eq. 9–2.

$$EtOH + TiCl_4 \longrightarrow TiCl_n(OEt)_{4-n}$$

$$EtOH + AlEt_3 \longrightarrow AlEt_2OEt \longrightarrow AlEt(OEt)_2$$

$$EtOH + AlEt_2Cl \longrightarrow AlEt(OEt)Cl \qquad (9\text{–}2)$$

$$EtOH + TiCl_3 \longrightarrow TiCl_2(OEt) \text{ (as a surface Ti species)}$$

($AlEt_2Cl$ and/or $AlEtCl_2$ may be formed during the reduction of $TiCl_4$ by $AlEt_3$.)

Depending on the concentration of reactants, mode of mixing, temperature and time of reaction and use in polymerization, and other factors, the working catalyst may have a variable composition. It is most difficult, if not impossible, in many of these cases to explain the role of the third component precisely.

II. Discussion of Donors: Types of Third Components Added

The apparently simpler cases, in which the third component reacts or complexes preferentially with only the metal alkyl or the transition metal salt, are first discussed as types 1 and 2, respectively. The more complicated case, in which the donor can react favorably with both catalyst components, is discussed as type 3. Some examples are borderline, and it is not certain in which type they should be classified.

A. TYPE 1. THE COMPONENT REACTS OR COMPLEXES PREFERENTIALLY WITH A METAL ALKYL COMPONENT

Not many examples are known of this type. Perhaps the most clearcut case involves the activation of the $AlEtCl_2$–$TiCl_3$ catalyst by addition of specific inorganic halides, such as KCl and NaF (4). Increases in polymer conversions between 10 to 20 times have been obtained when these inorganic salts are present. The unmodified catalyst usually has a very low activity or not at all, depending on the origin and composition of the $TiCl_3$. Compositions such as β-, γ-, and $\delta TiCl_3 \cdot xAlCl_3$ (aluminum alkyl- or aluminum-reduced $TiCl_4$) produced more active base and activated catalysts than did $\alpha TiCl_3$.

Since there is no apparent reaction of the solid inorganic halide and the solid $TiCl_3$, the increase in activity must be attributed to its reaction with $AlEtCl_2$. Zambelli and co-workers (4–6) suggested that upon mixing $AlEtCl_2$ and KCl, the $AlEtCl_2$ is converted to the more active $AlEt_2Cl$ alkyl plus $AlCl_3 \cdot KCl$ complex. In contrast, Coover, McConnell, and co-workers (7, 8) proposed that a complex is formed, $AlEtCl_2 \cdot KCl$, which acts as a better activator. Complexes of this type have been well established [see Table 5–1 in Ziegler (8a)]. A similar debate developed when an organic donor was used, and it was for the latter case that the most evidence was reported. It is not certain if the latter belongs here or in the type 3 classification. It is included here because the main function of the donor appears to be the modification of the $AlEtCl_2$ component. Some of the donors may complex with a few of the Ti centers and also affect catalyst behavior in this way (see type 2 donors).

The organic compound, like the inorganic salts NaF or KCl, increased the activity of the $AlEtCl_2$–$TiCl_3$ catalyst dramatically. A wide range of compounds were found effective, including hexamethylphosphortriamide (HMT), triethyl amine, tributyl phosphine (see Tables 4–7, 4–24, and 4–25 for other examples and Chapter 4, Section III,C for background). The ternary catalyst was also reported to be more isotactic-specific. Coover, McConnell, Joyner, and co-workers (7, 8) also attributed the higher activity to the formation of a stable complex of $AlEtCl_2$ and the donor, hexamethylphosphortriamide (HMT), $AlEtCl_2 \cdot HMT$, which they said formed a more active and highly stereoregulating catalyst than obtained with $AlEtCl_2$. The tacticity was said to be different, and they designated these polymers as "stereosymmetric."

Zambelli, Natta, and co-workers (4–6) took a different view and suggested that the donor converted $AlEtCl_2$ into $AlEt_2Cl$ even at room temperature, as shown in Eq. 9–3.

$$2\ AlEtCl_2 + HMT \longrightarrow AlEt_2Cl + AlCl_3 \cdot HMT \tag{9-3}$$

Furthermore, they examined the polypropylene formed with this catalyst and found that it had the isotactic structure. Others confirmed the latter finding.

Experimental evidence was presented for each view of the aforementioned two groups (4–8); later by other investigators (9–11).

The products from the reaction of 2 moles $AlEtCl_2$ and 1 mole donor are soluble in benzene. When heptane was added to this solution, two phases were formed (4–6). One phase contained the solvent and a solute which formed a highly isotactic-specific catalyst for propylene when combined with $TiCl_3$. The other phase formed an inactive system. The Cl/Al ratios in a number of $AlEtCl_2$–donor systems ranged from 1.06 to 1.6 for the hydrocarbon-soluble phase but were always greater than 2 for the oily semisolid phase. It was concluded that the hydrocarbon phase was rich in $AlEt_2Cl$, while the oily semisolid mostly contained the $AlCl_3 \cdot$donor complex. In other experiments, $AlEt_2Cl$ was isolated by distillation, and cryoscopic measurements failed to show complex compounds of aluminum dihalides.

McConnell, Coover, and co-workers (7, 8) ruled out the presence of $AlEt_2Cl$ in their products on the basis of far-infrared spectroscopic data. None of the binary $AlEtCl_2 \cdot$donor mixtures showed the weak 18.2 to 1.84 μ absorbtion band (about 549 cm^{-1}), which is characteristic of $AlEt_2Cl$. Zambelli criticized their experimental technique, as $AlEtCl_2$ may react with KBr optics and $AlCl_3 \cdot$ donor also absorbs strongly in the same region.

Watt (9) modified the infrared technique to overcome these problems and examined the systems $AlEtCl_2 \cdot Et_3N$ and $AlEtCl_2 \cdot HMT$. In his hands, the heptane-soluble phases in both catalysts showed the 538 cm^{-1} band, and highly isotactic-specific catalysts were formed when each was combined with $TiCl_3$ (the 538 cm^{-1} band was equivalent to 549 cm^{-1} band in McConnell's spectra). These results support the views of Zambelli and co-workers. Watt, however, also observed a band at 480 cm^{-1}, which he attributed to the complex $AlEtCl_2 \cdot Et_3N$ on the basis of earlier assignments by Zambelli.

Matsumura and co-workers (10, 11) reported that only a 2:1 complex was formed when tetrakis(dimethylamino)silane (TDSI) was mixed with $AlEtCl_2$. When this 2:1 complex was reacted with $AlCl_3$, $AlEtCl_2$ but not $AlEt_2Cl$ was released, supporting their conclusion that the complex was not the dismuted form $AlEt_2Cl \cdot AlCl_3 \cdot TDSI$.

It appears that both sides may be right. Considering the reaction path, shown in Eq. 9–4,

$$2 \text{ AlEtCl}_2 + \text{donor} \rightleftharpoons 2 \text{ AlEtCl}_2 \cdot \text{donor complex} \rightarrow \text{AlEt}_2\text{Cl} + \text{AlCl}_3 \cdot \text{donor} \quad (9\text{–}4)$$

the presence of AlEt_2Cl and $2\text{AlEtCl}_2 \cdot$ donor may be determined not only by the nature of donor but also on the conditions under which the mixing occurs, including temperature, time, and solvent. Probably, in some cases both components are present.

Another system which in principle qualifies in this category consists of water, a trivalent titanium chloride, and aluminum alkyl. It is known that water reacts with divalent but not with trivalent Ti. Thus, if the conditions of catalyst preparation and use in polymerization are such that the reduction Ti(III) to Ti(II) does not occur, then the interaction of water will exclusively be with the metal alkyl component. When strong alkylating and reducing metal alkyls are used, such as AlEt_3 or Al-i-Bu_3, some reduction undoubtedly occurs (see Chapter 11). Unfortunately, the reported cases use these metal alkyls, and the author is skeptical about disregarding the interaction of water with the divalent Ti in the TiCl_3 and thereby influencing the polymerization results. These results, therefore, are included in the type 3 group.

B. TYPE 2. THE COMPONENT COMPLEXES PREFERENTIALLY WITH A TRANSITION METAL SALT

Two Ziegler–Natta catalysts ($\text{ZnEt}_2 + \text{TiCl}_3$ and AlEt_3 or $\text{AlEt}_2\text{Cl} + \text{TiCl}_3$) and one metal alkyl-free catalyst (TiCl_3 + donor) were extensively investigated by Boor and co-workers for the polymerization of propylene (12–17). Third components were chosen that did not complex with the metal alkyl (such as trialkylamines with ZnEt_2) or only weakly, if at all (azulene with AlEt_2Cl). By avoiding the complexing of the donor with the metal alkyl, Boor was able to interpret the effect of the amine donor on the polymerization directly in terms of interactions between the donor and the TiCl_3. Complementary experiments with similar donors in the metal alkyl-free catalysts (no added metal alkyl) served as a check on the conclusions drawn from the Ziegler–Natta catalysts.

1. The ZnEt_2–TiCl_3–Donor System (12–14)

Table 9–1 shows the activating and depressing effect of Et_3N on conversion of propylene for several titanium chloride–ZnEt_2 catalysts (12). At low $\text{Et}_3\text{N}/\text{TiCl}_3$ ratios (<0.2), Et_3N significantly lowered the conversion of propylene to isotactic polymer. At higher ratios (>1.6 to 4.5), Et_3N increased the activity of the catalyst. The effective ratios and corresponding changes in activity depended on the particular catalyst.

TABLE 9–1

The Activating and Depressing Effect of Et$_3$N on Conversion for Several Titanium Chloride–ZnEt$_2$ Systems^a (12)

Titanium chloride^b	Low Et$_3$N			High Et$_3$N		
	Expt. no.	Et$_3$N/Ti	Conversion (wt %)	Expt. no.	Et$_3$N/Ti	Conversion (wt %)
βAl$_x$Ti$_y$Cl$_z$–25	8	0	80	18	0	39
	9	0.06	48	19	4.5	91
γAl$_x$Ti$_y$Cl$_z$–160	10	0	83	20	0	26
	11	0.04	35	21	4.5	70
Al$_x$Ti$_y$Cl$_z$–70/160	12	0	57	22	0	32
	13	0.2	8	23	1.6	73
Al$_x$Ti$_y$Cl$_z$–A	14	0	30	24	0	30
	15	0.003	~15	25	4.5	89
αTiCl$_3$ (Stauffer)	16	0	50–70	26	0	50–70
	17	0.004	~30	27	2.9	95

^{a} Materials: 8 oz bottles, 100 ml heptane, 20 hr at 50°C, 27 g propylene (catalyst amount selected to give desired conversions for control runs).

^{b} x, y, and z have approximate relative values of 0.3, 1.0, and 4, respectively.

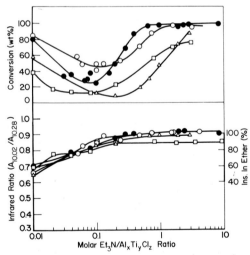

Fig. 9–1. Dependence of conversion and infrared ratios on Et$_3$N/Al$_{0.3}$TiCl$_4$ ratio (12). Polypropylene: (○) 1.2 mmole βAl$_x$Ti$_y$Cl$_z$-25-4.8 mmole ZnEt$_2$; (●) 1.5 mmole γAl$_x$Ti$_y$Cl$_z$-160-4.8 mmole ZnEt$_2$; (△) 2.0 mmole Al$_x$Ti$_y$Cl$_z$-70/160-6.5 mmole ZnEt$_2$. Poly-1-butene: (□) 1.5 mmole γAl$_x$Ti$_y$Cl$_z$-160-5.5 mmole ZnEt$_2$. Materials: 8 oz bottle, 100 ml heptane, 20 hr at 50°C, and 27 g propylene or 25 g 1-butene. Each point represents a separate experiment.

TABLE 9–2

The Influence of Et$_3$N at High Concentration in Propylene Polymerization^a (12)

Expt. no.	Et$_3$N/Ti	Conversion (wt %)	Infrared ratio^b	IV (dl/g)
1	0	72	0.65^c	1.5
2	2.5	98	0.92	2.7
3	8.4	98	0.90	3.0
4	20.6	96	0.89	3.9
5	60.0	82	0.91	4.9
6	150.0	45	0.92^d	6.3
7	330.0	<1	—	Insol.

^{a} Conditions: 8 oz bottles, 100 ml heptane, 4.2 mmole ZnEt$_2$, 1.3 mmole γAl$_x$Ti$_y$Cl$_z$-160 (composition is approximately Al$_{0.3}$TiCl$_4$), 20 hr at 50°C, 27 g propylene.

^{b} $A_{10.02\mu}/A_{10.28\mu}$, an index of crystallinity.

^{c} About 35% crystalline.

^{d} About 70% crystalline.

In Table 9–1, the catalyst concentrations and ratios were chosen so that the rate increase or decrease could be most effectively shown. In Fig. 9–1, changes are shown that take place when the concentration of Et$_3$N was increased steadily from Et$_3$N/TiCl$_3$ ratio = 0.01 to 10 in three propylene and 1-butene runs. In all cases, the initial small addition of Et$_3$N depressed catalyst activity (lower polymer conversions) and, above certain threshold amounts of Et$_3$N, the additional Et$_3$N resulted in increased activities. Polymer stereospecificity (measured by the infrared $A_{10.02\mu}/A_{10.28\mu}$ ratios for polypropylene) increased as the amount of Et$_3$N was increased, especially during the initial additions. The increases in the isotactic crystallinity of polypropylene samples were substantial, about 20 to 30%. This infrared method is discussed in Chapter 3, Section III,A. Moreover, the molecular weight of the polypropylene increased significantly; that is, intrinsic viscosity values increased from 1.5 dl/g (no amine) to 6.3 dl/g when Et$_3$N/TiCl$_3$ = 150 (Table 9–2). While polymer isotacticity remained at its optimum value in the entire Et$_3$N/TiCl$_3$ range used, the activity of the catalyst decreased at ratios near 60.

As the steric encumbrance of the amine increased, its ability at low amine/TiCl$_3$ ratios to depress catalyst activity and increase the stereoregulating ability of the catalyst decreased. Thus, while Et$_3$N at Et$_3$N/TiCl$_3$ = 0.1 lowered the conversion from 80 to 40% and increased polypropylene isotactic crystallinity from about 35 to 70%, the more bulky n-octyl$_3$N produced negligible changes. Intermediate changes were observed with n-Pr$_3$N, n-Bu$_3$N, and n-pentyl$_3$N.

In contrast, the apparent activating affect of these amines was in the opposite order: n-octyl$_3$N > n-Bu$_3$N > Et$_3$N. By "apparent activating abilities" of the amine was meant the absolute increase in conversion for an experiment containing no amine to one containing the high amine/TiCl$_3$

TABLE 9–3

Site Removal and Site Activation by Donors at Low and High Donor Concentration, Respectively[a]

Strong	Moderate–Weak	Ineffective
	Relative abilities of donors to remove sites	
Azulene	n-Octyl$_3$N	Benzene
Et$_3$N	Acenaphthylene	Naphthalene
n-Pr$_3$N	Mesitylene	Acenaphthene
Pyridine and quinoline compounds	Durene	Anthracene
(C$_6$H$_6$)$_2$Cr	Pentamethylbenzene	Pyrene
Cyclopentadienyl compounds of V, Cr, Ni,	Hexamethylbenzene	Fluorene
and Co	Tetramethylethylene	Hexaethylbenzene
Butadiene		1,3-Cyclooctadiene
Isoprene		(C$_6$H$_5$)$_3$N
1,4-Pentadiene		
2-Butyne		
Cyclooctatetraene		
Dimethylfulvene		
Diphenylfulvene		
2,5-Norbornadiene		
trans-Cyclooctene		
PF$_3$		
1,5-Cyclooctadiene		
n-Bu$_3$N		
	Relative abilities of donors to activate catalyst	
R$_3$N (R = ethyl to n-octyl)	Azulene	
	Diphenylphosphine	Benzene
(C$_6$H$_5$)$_3$P	2-Ethylpyridine	Naphthalene
n-Bu$_3$P	2-Propylpyridine	Acenaphthene
	2,4,6-Trimethylpyridine	Anthracene
		Pyrene
		Fluorene
		Hexaethylbenzene
		1,3-Cyclooctadiene
		(C$_6$H$_5$)$_3$N

[a] Under the conditions examined in this investigation. At much higher donor concentrations, different results might be obtained.

ratio (about 3). The real activating abilities of amines, however, were nearly independent of their steric encumbrance. By "real activating abilities" was meant the absolute increase in conversion for an experiment containing a small amount of amine (amine/$TiCl_3$ ratio = 0.1) to one containing the higher amine/$TiCl_3$ ratio (about 3). In other words, when one corrects for the depressing effect of the less bulky amines, the activating abilities become nearly similar.

Other electron donors were examined for depressing and activating abilities (Table 9–3).

The most effective activators were trialkylamines (R = ethyl to n-octyl were examined), $(C_6H_5)_3P$, and n-Bu_3P; weak to moderate activators were azulenes, diphenylphosphine, 2-ethylpyridine, 2-propylpyridine, and 2,4,6-trimethylpyridine; triphenylamine, aromatic, and diene compounds were ineffective. As Table 9–3 shows, some compounds that are good activators have poor depressing abilities and vice versa.

The interaction of amines with $TiCl_3$ involves chemisorption (Table 9–4); that is, heats of mixing (kcal/mole amine) of 26 and 32 were determined for

TABLE 9–4

Heats of Interaction of Catalyst Components (14)

	Heat liberated (gcal/mmole catalyst)	
Compound[a,b,c]	$\gamma Al_x Ti_y Cl_z$-160[d]	$ZnEt_2$
Azulene	0.59	0
TCNE	0.29	ca. 95
Trioctylamine	3.06	—
2,8-Dimethylquinoline	1.68	0
Triphenylphosphine	2.18	0
Triethylamine	2.96	0
Pyridine	2.96	0
$AlEt_2Cl$	0.88	—
$ZnEt_2$	2.40	—
$(C_6H_5)_2Ni$	—	0
Acenaphthylene	—	0
Dimethylfulvene	—	0

[a] The compounds 1,4-pentadiene, naphthalene, cyclooctatetraene, hexamethylbenzene, and triphenylamine did not react with $\gamma Al_x Ti_y Cl_z$-160 to give measurable heats.

[b] Excess of these components added to $ZnEt_2$ or $\gamma Al_x Ti_y Cl_z$-160.

[c] The thermometric titrations were performed at ambient temperatures; the polymerizations in which the effect of these compounds was studied were performed at 50°C.

[d] Because the reaction of the donor takes place on a surface, stoichiometry is not possible.

Et_3N and n-octyl$_3N$, respectively. The chemisorption, however, involved only a small amount of amine, about 0.1 mole amine per mole $TiCl_3$ at ambient temperatures (polymerizations were usually done at 50°C).

A correlation was not established between the activating ability of a donor and fundamental parameters describing the donor, for example, ionization potential or m, the calculated energy of the highest filled molecular orbital. Because steric compression affects the donor's ability to complex with the Ti centers, this observation was not unexpected.

2. Al–Alkyl–TiCl₃–Azulene (14, 17)

Azulene and dimethylfulvene were also examined in several Al–alkyl-based catalysts for polymerization of propylene.

When the catalyst already produced very highly isotactic polypropylene ($AlEt_2Cl + TiCl_3 \cdot xAlCl_3$), azulene lowered the catalyst activity but did not increase its isotactic-regulating ability (Table 9–5).

TABLE 9–5

Azulene in Highly Specific Catalysts ($AlEt_2Cl + \gamma TiCl_3$ at 60°C)[a]

Expt. no.[b]	Azulene (mmole)	$\dfrac{\text{Azulene}}{\gamma TiCl_3}$	Conversion (wt %)	$\dfrac{A_{10.02\mu}}{A_{10.28\mu}}$[b]
1	0	0	89	0.90
2	0.1	0.15	83	0.90
3	0.5	0.77	79	0.91
4	1.5	2.3	74	0.91

[a] Experimental conditions: 8 oz bottles as reaction vessels, 100 ml heptane, 3.0 mmole $AlEt_2Cl$, 0.65 mmole $\gamma TiCl_3$, 30 g propylene rotated in thermostated baths for 50 min at 60°C.

[b] Indices which are related to polymer isotacticity. A value of 0.90 corresponds to a crystallinity of about 65%. See ref. 12 for methods used to determine these values.

When moderately stereoregulating aluminum alkyl-based catalysts were used, azulene was effective in increasing their isotactic-regulating abilities (Table 9–6). In all cases, azulene also lowered catalyst activities.

If the stereoregulating ability of the $AlEt_2Cl–TiCl_3 \cdot xAlCl_3$ catalyst was also first decreased by use at high polymerization temperatures (Table 9–7) or by addition of $ZnEt_2$ (Table 9–8), then azulene became effective in increasing its stereoregulating ability to a very high level.

TABLE 9–6

Azulene in Partially Specific Catalysts Based on AlEt$_2$Cl, AlEt$_2$F, AlEt$_3$, VCl$_3$ or βTiCl$_3$ and γTiCl$_3$[a]

Expt. no.	Metal alkyl	Transition metal compound	Azulene (mmole)	Conversion (wt %)	$\dfrac{A_{10.02\mu}}{A_{10.28\mu}}$
1	AlEt$_2$Cl	βTiCl$_3$	0	41	0.84 (IV = 2.3)
2	AlEt$_2$Cl	βTiCl$_3$	0.5	22	0.90 (IV = 3.9)
3	AlEt$_2$F	γTiCl$_3$	0	96	0.66
4	AlEt$_2$F	γTiCl$_3$	4.0	58	0.87
5	AlEt$_3$	VCl$_3$	0	83	0.60
6	AlEt$_3$	VCl$_3$	4.0	95	0.70
7	AlEt$_3$	βTiCl$_3$	0	92	0.69
8	AlEt$_3$	βTiCl$_3$	4.0	21	0.84
9	AlEt$_3$	γTiCl$_3$	0	95	0.74
10	AlEt$_3$	γTiCl$_3$	4.0	74	0.86

[a] Experimental conditions: 8 oz bottles as reaction vessels, 100 ml heptane, 3.0 mmole AlR$_3$ or AlR$_2$X, 0.65 mmole transition metal salt (except 1.3 mmole VCl$_3$), 30 g propylene; rotated in thermostated baths for 50 min at 60°C.

TABLE 9–7

Azulene in Partially Specific Catalysts (Polymerization at 115°C)[a,b]

Expt. no.[a]	Azulene (mmole)	Polymer (g)	$\dfrac{A_{10.02\mu}}{A_{10.28\mu}}$	IV (dl/g)
1	0	20.7	0.68	1.1
2	0.05	6.1	0.90	1.4
3	0.25	8.1	0.90	1.5
4	0.50	5.2	0.92	1.9
5	1.00	5.6	0.94	2.1
6	0.25	6.0	0.93	—

[a] These polymerizations were done in a Magne–Dash stainless steel autoclave at a propylene pressure of 140 psig. Heptane was used in experiments 1 to 5 and cyclohexane in experiment 6.

[b] Catalyst used: AlEt$_2$Cl + γTiCl$_3$ · xAlCl$_3$.

TABLE 9–8

Azulene in Partially Specific Catalysts (AlEt$_2$Cl + ZnEt$_2$ + γTiCl$_3$ at 80°C)^{a,b}

Expt. no.	ZnEt$_2$ (mmole)	Azulene (mmole)	Conversion (wt %)	$\dfrac{A_{10.02\mu}}{A_{10.28\mu}}$	IV (dl/g)
1	0	0	80	0.90	2.0
2	0.5	0	58	0.85	0.60
3	0.5	0.5	33	0.98	0.89
4	1.0	0	52	0.78	0.57
5	1.0	0.5	32	0.98	1.08
6	2.0	0	48	0.66	0.69
7	2.0	0.5	29	0.96	1.30
		Dimethylfulvene (mmole)			
8	1.5	0	47	0.61	0.71
9	1.5	0.85	4	0.96	1.80

^{a} Experimental conditions: 8 oz bottles as reaction vessels, 100 ml heptane, 3.0 mmole AlEt$_2$Cl, 0.65 mmole γTiCl$_3$, 20 g propylene; rotated in a thermostated bath for 30 min at 80°C.

^{b} γTiCl$_3$ contained alloyed AlCl$_3$.

3. Metal Alkyl-Free Catalysts (13, 16, 17)

In the absence of a metal alkyl, TiCl$_3$ salts (such as used in combination with ZnEt$_2$ and AlEt$_2$Cl) produced only low molecular oils when contacted with propylene in heptane solvent. Upon addition of an electron donor in specific concentrations, the TiCl$_3$ became active and high molecular weight isotactic polypropylene was formed. For the same concentration of trialkyl amine, the following dependence of activity and stereospecificity was observed.

For activity: n-octyl$_3$N > n-Bu$_3$N > Et$_3$N

For isotacticity: Et$_3$N > n-Bu$_3$N > n-octyl$_3$N

For the TiCl$_3$–n-Bu$_3$N (also containing H$_2$) system, maximum activity was obtained at a ratio of 1 mmole n-Bu$_3$N per 1 g TiCl$_3$AA (Fig. 9–2). When H$_2$ was absent, the catalyst was less active, but the maximum rate still occurred at the same ratio of amine to TiCl$_3$AA. Thermometric titrations showed that amine became strongly adsorbed in amounts near 0.4 to 0.6 mmole per g TiCl$_3$AA.

In addition to trialkyl amines, other donors activated the γTiCl$_3$ and TiCl$_3$AA compositions (both contain AlCl$_3$). Table 9–9 compares these

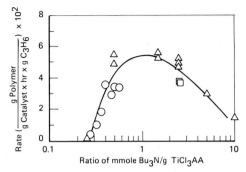

Fig. 9–2. Dependence of polymerization rate on $n\text{-Bu}_3\text{N}/\text{TiCl}_3\text{AA}$ ratio when either $n\text{-Bu}_3\text{N}$ or TiCl_3AA is kept separately constant: ($\bigcirc$) $n\text{-Bu}_3\text{N}$ constant (0.8 mmole), TiCl_3AA varied; ($\triangle$, $\square$) TiCl_3AA kept constant (2.0 g), concentration of $n\text{-Bu}_3\text{N}$ varied. H_2 present in all experiments.

TABLE 9–9

Relative Ability of Donors to Activate TiCl_3AA and γTiCl_3 Metal Alkyl-Free Catalysts[a]

Good activators[b]	Weak activators[c]
R_3N (R $\geq$ Et)	$(\text{C}_6\text{H}_5)_2\text{PH}$
R_3P (R $\geq$ Pr)	2,6-Dimethylpyridine
$(\text{C}_6\text{H}_5)_3\text{P}$	Azulene
Et_4NBH_4	α-R-pyridine (R = Me, Et)
$(\text{C}_{10})_3\text{N} + n\text{-BuI}$ (1:1)	α-Me-quinoline
	Pyridine (very poor)

[a] These donors were examined with both the TiCl_3AA and the γTiCl_3 compositions for the polymerization of propylene (100 ml heptane, 30 g propylene, 2.0 g TiCl_3, 10 mmole donor for 144 hr at 25°C in 8 oz bottles).

[b] 1 to 10 g polymer was formed, the exact amount varying with the donor.

[c] 0.03 to 0.3 g polymer was formed, the exact amount varying with the donor.

donors. The donors that were the best activators in the Ziegler–Natta systems were also the best activators in the metal alkyl-free systems, such as R_3N, R_3P, and $(\text{C}_6\text{H}_5)_3\text{P}$ compounds. Those compounds that showed weak activating ability in Ziegler–Natta catalysts formed only sluggishly active metal alkyl-free catalysts.

Boor interpreted the donor affects by proposing that the active centers were exposed Ti's of variable complexing ability and stereoregulating ability.

At low donor/TiCl$_3$ concentrations in the Ziegler–Natta ZnEt$_2$–TiCl$_3$ system, the donor preferentially complexed and inactivated the more exposed Ti centers. This was experimentally observed as a decrease in conversion of propylene to polymer. Because the more exposed and less isotactic-regulating Ti centers (which are better acceptors) were preferentially complexed and inactivated, the remaining active centers, which were less exposed and more highly isotactic-regulating, produced polymer that was more highly isotactic and had a higher molecular weight.

As more donor was added, it became chemically adsorbed on the chlorine surface, and this resulted in the activation process. The activation was attributed to two potential mechanistic paths: (1) the electron donor, when chemisorbed on the surface near the center, enhanced its activity; and (2) the chemisorbed electron donor promoted the disruption of larger catalyst aggregates (and secondary particles) of TiCl$_3$ to smaller crystallites or primary particles. In the latter case, the smaller particles were stabilized by the chemisorbed amine, and more centers became available as the number of the small particles increased. While Boor favored the latter, no evidence was presented. Some support for this view came from electron microscopic studies by Marchessault and Chanzy (17a), who observed directly that electron donors promoted the disintegration of VCl$_3$ during the polymerization of ethylene with the AlR$_3$–VCl$_3$ system.

In the metal alkyl-free catalysts, the donor prevented the fast cationic polymerization of propylene by also complexing with cationic centers and allowed the existing isotactic-specific centers to polymerize propylene over an extended period. But similarly to the ZnEt$_2$–TiCl$_3$ system, the less isotactic-specific catalysts were preferentially deactivated by less bulky amines (Et$_3$N gave higher isotactic polymer than did n-Bu$_3$N). But also similarly to the ZnEt$_2$–TiCl$_3$ system, fewer sites were deactivated by less bulky amines, and so TiCl$_3$ formed a more active catalyst with n-Bu$_3$N than with Et$_3$N.

The behavior of azulene in the AlEt$_2$Cl–TiCl$_3$ system was also explained similarly. Boor assumed that azulene did not complex with AlEt$_2$Cl. In contrast, Langer (17b), who also reported an increase in isotactic crystallinity for polypropylene when azulene was added to the AlEt$_3$–TiCl$_3$ catalyst, proposed complex formation but did not elaborate further. The effectiveness of azulene at 115°C argued against complexing with the metal alkyl; complexing with the transition metal centers was more likely.

The above mechanistic explanation is based on the view that isotactic-regulating ability is derived from the asymmetric character of the active center on the TiCl$_3$ surface. This idea is further elaborated upon in Chapter 15, Section III,A.

Tanabe and Watanabe (17c) found the acidity of four commercial titanium trichlorides $(TiCl_3HR, TiCl_3AR, TiCl_3ARA, TiCl_3HRA)$ to be anomalously large. For example, the acidities of ARA and HR-type $TiCl_3$ samples having an acid strength equal to or lower than $H_o = 4.0$ were about 10 and 3.8 mmole/g, respectively. These values were ten to one hundred times as large as those found for silica–alumina and nickel sulfate; they were even greater when compared on the basis of surface area. From these values, these authors calculated that the n-butylamine molecule that was used as the titrant would occupy only 0.38 $Å^2$ for $TiCl_3ARA$ and 0.66 $Å^2$ for $TiCl_3HR$. These values were too small, since the cross-sectional area of n-butylamine was about 20 $Å^2$ per molecule. The suggestion was made without elaboration that the increased activity resulting from added basic organic compounds was due to covering up of acid sites having strong acid strengths.

C. TYPE 3. THE COMPONENT COMPLEXES OR REACTS WITH A METAL ALKYL AND A TRANSITION METAL SALT

This is the most complicated type to interpret. It is also the most likely case to be encountered in the journal and patent literature, because the preponderance of Ziegler–Natta catalysts contains aluminum alkyls and transition metal salts that are both complexed by most donors. While some proposals are largely speculative, others are based on careful experimental work and, in the opinion of the author, are very probably correct or at least closely reflect what is happening.

This material is presented in two ways: (1) selected findings will be presented to show the different ways a donor can affect a Ziegler–Natta polymerization and (2) selected mechanistic proposals will be described to show the various interpretations that were given by individual authors.

1. The Different Ways That a Donor (or Reactant) Can Affect a Ziegler–Natta Polymerization

a. Isotactic Crystallinity in Polyolefins. Table 9–10 (18–28) collects examples of different donors or reactants and shows their affect on isotactic crystallinity in a propylene and 1-butene polymerization. For polypropylene and higher α-olefins, crystallinity or some property related to crystallinity was measured. The more crystalline a polymer, the higher was its tensile strength, the higher were the insoluble fractions in boiling heptane, the lower was the xylene soluble fraction, and higher was the flexural modulus. Electron donors usually increased the isotactic-regulating ability of a catalyst, but not always.

TABLE 9–10

The Effect of Donors and Reactants on Isotactic-Regulating Ability in the Polymerization of Propylene and 1-Butene

Third component	Catalyst	Improvement	Ref.
Methyl methacrylate	$AlEt_2Cl + TiCl_3$	% Insol. in Heptane increased from 88.7 to 95%	Fukuda et al. (18)
Copper ethyl mercaptide	$AlEt_2Cl + TiCl_3$	Higher crystallinity	Sugiura et al. (19)
1,1-Dimethyl-2-formyl hydrazine	$AlEt_2Cl + TiCl_3$	Higher insoluble fraction in boiling heptane	Nicco (20)
Tetrakis (haloalkyl) phosphonium halide, CS_2, $AlCl_3 \cdot \phi_3P$ complex, $\phi N_2 + PF_6^-$, PBr_3, I_2	$AlEt_2Cl + TiCl_3$	Higher flexural modulus, higher tensile strength, and lower xylene soluble fraction	Fodor (21)
$\phi_3P \rightarrow 0$, $[(CH_3)_2N]_3PO$ or γ-picoline	$AlEt_2Cl$ or $AlEt_3 + TiCl_3$	Higher insolubles in boiling heptane	Staiger and Stedefeder (22)
Acetic acid	$AlEt_3 + TiCl_4$	Higher insolubles in boiling heptane	McClaflin (23)
Diallylaminophosphorus dichloride	$AlEt_2Cl$ or $AlEt_3 + TiCl_3$	Lower xylene solubility, higher density, higher tensile strength	Moberly (24)
Phenetole[b]	$AlEt_3$ or $AlEt_2Cl^{[a]} + TiCl_3$	Increased crystallinity	Mardykin and Antipova (25)
H_2O or Mixture of $H_2O + Et_3N$	$AlEt_2Cl + \gamma TiCl_3$	Lower content of atactic polymer	Cheney (26)
$I_2^{[c]}$	$AlEt_2Cl + TiCl_3$	Higher isotacticity	Eichenbaum and Murray (27)
H_2O	$AlEt_2Cl + \gamma TiCl_3$	Increased isotacticity	Boiko and Ivanyukov (28)

[a] $TiCl_3$ usually contained alloyed $AlCl_3$ unless specified.
[b] Styrene and propylene were both investigated.
[c] 1-Butene instead of propylene was used.

b. Activity. Most frequently a claim is made that the activity of the catalyst is increased. A variety of reactants or donors have been reported. Organic nitro compounds ($C_6H_5NO_2$) increased the activity of the catalysts $Al_2Et_3Cl_3$–$VOCl_3$, $VO(OBu)_3$, or $V(acac)_3$ for the polymerization of ethylene or copolymerization of ethylene and propylene (29). Silicon hydrides added to the $AlEt_3/TiCl_4$ catalyst increased its activity for the polymerization of ethylene (30). The activity of the CH_3TiCl_3–$TiCl_3$ catalyst for polymerizing propylene was increased by n-Bu_3N (17). The presence of ethylorthosilicate in a catalyst prepared by reduction of $TiCl_4$ with $AlEtCl_2$ and activated with $AlEt_2Cl$ was reported to increase its activity, and a highly crystalline polypropylene having a high bulk density was formed (31). More active catalysts for ethylene were formed when transition metal salts were complexed with amines and then combined with Al-i-Bu_2Cl (32). Similarly, complexes of $CrCl_3$ with amines, alcohols, and ketones were said to lead to a more active catalyst for ethylene when combined with $AlEt_2Cl$ (33). A salt made by reacting vanadium oxides and phenyl phosphonic acid at 80°C was combined with $AlEt_2Cl$ to make an active catalyst for polymerization of ethylene (34). Dienes were very effective in lowering the activity of the $AlEt_3$–$\alpha TiCl_3$ catalyst for the polymerization of ethylene; for example, butadiene > isoprene > styrene > isobutylene > methyl styrene (35). Higher rates were claimed for polymerization of propylene when peroxides were added to $AlEt_3$–$TiCl_4$ (36). Zgonnik, Dolgoplosk, and co-workers (37) established that the rate of polymerization of butadiene and the molecular weight of the formed polymer maximized at a critical concentration of water in the Al-i-Bu_2Cl–$CoCl_2$–pyridine system. A simultaneous affect on stereoregularity was reported by others (38).

c. **Copolymerization Behavior.** The copolymerization behavior of a catalyst can be altered. The activity of the $AlEt_2Cl$–$TiCl_4$ catalyst was increased by addition of isopropanol, according to Erofeev (39). Also, the copolymer molecular weight was higher when isopropanol was first mixed with $TiCl_3$ and was lower if first mixed with $AlEt_2Cl$. An improved copolymerization catalyst was reported if VCl_3 was reacted with R_4NCl, as shown in Eq. 9–5, and the product was combined with Al-i-Bu_2Cl (40). The

$$VCl_3 + R_4NCl \xrightarrow{C_2H_5OH} R_4\overset{+}{N}(VCl_4 \cdot HOC_2H_5)^- \qquad (9\text{–}5)$$

molecular weight of the ethylene–propylene copolymer was regulated by use of pyridine N-oxide when the catalyst $Al_2Et_3Cl_3$–$VOCl_3$ was used (41).

Cornelis and co-workers (42) found alcohols increased the rate of copolymerization of ethylene, propylene, and dicyclopentadiene with the $Al_2Et_3Cl_3$–$VOCl_3$ catalyst. Controlled addition of oxygen to the Al-i-Bu_2Cl–$V(acac)_3$ catalyst increased the copolymerization of ethylene and propylene, according to Seidov (43).

Duck and co-workers (44) found improved elastomeric properties when isoamylether was present in the polymerization of ethylene and propylene with the LiBu–$VOCl_3$–$AlCl_3$ catalyst.

d. **Molecular Weight Control.** As early as 1956, Dost and co-workers (45) recognized that electron donors can increase or lower molecular weights of formed polyethylene when added to Ziegler–Natta catalysts; for example, aldehydes, esters, and amines were added to the AlR_3–$TiCl_4$ catalyst. Since then, many other examples have been reported for olefins and diolefins. Typically, the electron donor increases the molecular weight of the formed polymer (see examples in previous section). Claims have also been made that the molecular weight distribution can be altered by presence of donors, and some examples are given here.

A narrowing was reported by Erofeev (39) in the "width" of the molecular weight distribution by the presence of anisole in the $AlEt_2Cl–TiCl_4$ catalyst. This was attributed to diminution of low molecular weight fractions. $\bar{M}_w/\bar{M}_n$ values 4.3 to 9.4 were observed, compared to values 13 to 16 for unmodified catalyst. Mottus (46) also reported a narrower molecular weight distribution when water was added to the $Al-i-Bu_3–TiCl_4$ catalyst and used to polymerize propylene. Kircheva and co-workers (47) noted that the polyethylene product had a narrower molecular weight distribution when $Ti(OR)_4$ (where R = Et, i-Pr, or Bu) or alcohols were added to the $AlEt_2Cl–TiCl_4$ catalyst. Hirooka and co-workers (48), however, did not significantly affect the molecular weight distribution of polypropylene by addition donors (pyridine, hexamethylphosphoramide, CS_2, etc.) to $AlEt_3–TiCl_3H$, $AlEt_2Cl–TiCl_3AA$, and $AlEtCl_2–TiCl_3AA$ catalysts.

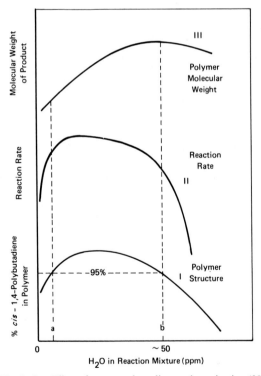

Fig. 9–3. Effect of water on butadiene polymerization (38).

e. **Microstructure of Diene Polymer.**   Turov (49) reported that as Bu_3N or Et_3N was added to the $Al\text{-}i\text{-}Bu_3\text{-}TiCl_4$ catalyst in a butadiene polymerization, the *trans*-1,4 content increased at the expense of the *cis*-1,4 content. Cooper did not observe this for isoprene polymerization when $AlEt_3\text{-}VCl_3$ catalysts were used (50). Parfenova and co-workers (51) did not alter the microstructure of polybutadiene by the addition of amines to the $Al\text{-}i\text{-}Bu_2Cl\text{-}CoCl_2\text{-}BuOH$ system. However, the polymers had a lower molecular weight. In contrast, Carlson (38) showed that *cis*-1,4 content and reaction rate in polybutadiene maximized at a certain concentration of water with the aluminum alkyl-soluble cobalt system ($CoCl_2 \cdot AlCl_3 + AlEt_3$). Polymer molecular weight maximized when the concentration of water was greater than needed to achieve optimum *cis*-1,4 content and reaction rates (Fig. 9–3). Figure 16–6 shows the effect of bases on *cis*-1,4 content in polybutadiene synthesized with cobalt-based catalysts.

f. **Other Changes.**   Better fabrication of bottles by blowing techniques was reported by Roberts (52) if NH_3 was present in the $Al\text{-}i\text{-}Bu_3\text{-}\beta TiCl_3$ catalyst when used to polymerize ethylene. By complexing $Al\text{-}i\text{-}Bu_3\text{-}TiCl_4$ with carbon black (catalyst was chemisorbed on its surface), Donnet obtained a polyethylene that contained a highly dispersed carbon black (53).

2. Selected Mechanistic Proposals

The change in the behavior of a catalyst has been explained by individual authors in a number of ways. This section collects some of the more cogent suggestions.

a. **Inactivation of the Poison by a Donor.**   A poison is formed during site formation, and the donor removes this poison by complexing it. This argument was put forth by Caunt to explain the activating affect of inorganic halides (KCl) and electron donors (amines) when added to the $AlEt_2Cl\text{-}TiCl_3$ catalyst (54). Specifically, he proposed that the third components complex and inactivate the poison $AlEtCl_2$, which was formed when $AlEt_2Cl$ and $TiCl_3$ were mixed. It has been well established by many groups that intentionally adding $AlEtCl_2$ to such a catalyst lowers its activity, and the deactivating effect is negated by the addition of donors. The exact path may involve formation of a complex or dismutation, as already discussed in Section II,A.

Burfield and Tait (54a) examined in great detail the effect of Et_3N on the polymerization of 4-methyl-1-pentene in benzene solvent with the aid of $VCl_3 + Al\text{-}i\text{-}Bu_3$. The activating–deactivating effect of Et_3N was explained in the following way: activation at low amine concentration was due to the removal of all adsorbed chloroaluminum species from the catalyst surface, where they blocked potential active centers. The chloroaluminum alkyls

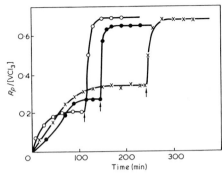

Fig. 9–4. Variation of rate activation with $[VCl_3] \cdot [4-MP-1] = 2.00$ mole/liter; $[Al-i-Bu_3] = 37.0$ mmole/liter; $[NEt_3] = 27.8$ mmole/liter; temp. = 30°C. $[VCl_3]$: x, 3.63 mmole/liter; ●, 18.5 mmole/liter; ○, 44.9 mmole/liter; ↓denotes addition of donor (54a).

were by-products of the site formation step. The above is shown by the reactions in sequence 9–6.

$$VCl_3 + Al-i-Bu_3 \longrightarrow VCl_2-i-Bu + Al-i-Bu_2Cl \text{ (solution)}$$

$$VCl_3 + Al-i-Bu_2Cl \longrightarrow VCl_3 \cdot Al-i-Bu_2Cl \text{ (at surface centers) (surface blocked)} \quad (9-6)$$

$$VCl_3 \cdot Al-i-Bu_2Cl + Et_3N \longrightarrow VCl_3 + Al-i-Bu_2Cl \cdot Et_3N \text{ (solution) (surface free)}$$

The first equation shows the formation of the active center $i\text{-}BuVCl_2$. Site formation is prevented when the surface of VCl_3 is complexed with the chloroaluminum alkyl. A similar role was suggested for excess $Al-i-Bu_3$, as shown in Eq. 9–7.

$$VCl_3 \cdot Al-i-Bu_2Cl + Al-i-Bu_3 \longrightarrow VCl_3 + Al-i-Bu_3 \cdot Al-i-Bu_2Cl \quad (9-7)$$

In agreement with this scheme were their findings that: (1) Rp/VCl_3 decreased as the concentration of VCl_3 increased and that Rp/VCl_3 values with Et_3N present were the same, regardless of the concentration of VCl_3 (Fig. 9–4); and (2) site counting showed no increase in the number of centers. On the basis of kinetic arguments, they attributed the deactivating effect of Et_3N at higher concentrations to competitive adsorption at the active centers by Et_3N and 4-methyl-1-pentene molecules.

Burfield and Tait showed that more active catalysts were formed if VCl_3 was first contacted with 4-methyl-1-pentene before Et_3N was added. This agrees with the results of Vesely (55), Mezhikovskii (55a), Keii (55b), Overberger (55c), and Hoeg (55d), and their co-workers, who said that having monomer present during site formation significantly affects the behavior of the catalyst. While Burfield and Tait could interchange the order of olefin and $Al-i-Bu_3$ without affecting the rate, Keii found that first mixing propylene and $TiCl_3$ before addition gave higher activities. Burfield and Tait attributed this apparent discrepancy to the stronger adsorbing ability of $AlEt_3$ to $Al-i-Bu_3$.

b. Lowered Concentration of the Metal Alkyl. An electron donor complexes the metal alkyl and decreases its effective concentration. Processes such as site formation and chain transfer reactions are affected. This explanation was invoked by Vesely (55) and Milovskaya (56) to explain the lower catalyst activity and higher molecular weight of the polymer formed when electron donors were added to the $AlEt_3$–$\alpha TiCl_3$ and $AlEt_2Cl$–$\beta TiCl_3$ catalysts, respectively. Pirogov and Chirkov (57) observed that the $AlEt_3$–$TiCl_3$ was inactive when a sufficient amount of pyridine was added to complex these components.

Vinogradov and co-workers (58) studied the effect of electron donors and acceptors on the polymerization of butadiene in the presence of the Al-i-Bu_2Cl–$CoCl_2$ catalyst. They suggested that the effect of the additives was to shift the reaction as shown in Eq. 9–8.

$$2\, R_2AlCl \rightleftharpoons RAlCl_2 + R_3Al \qquad (9\text{–}8)$$

Those donors which shifted the equilibrium to the right (RSR, ROR, R_3N, KCl, and NaCl) enhanced the reduction efficiency of the system. Consequently, the catalyst was less active, lower molecular weight polymers were formed, and 1,2-microstructure was favored. But if donors were added that shifted the equilibrium to the left (such as $AlCl_3$, $RAlCl_2$, Br_2, H_2O, and aloxane of $AlEt_2Cl$), an increase in catalyst activity, an increase in cis-1,4 structure, and higher molecular weight polymers were obtained.

c. Diminished Reactivity of the Metal Alkyl. When the metal alkyl is complexed with an electron donor, its coordination capacity is diminished and, consequently, the reactivity of the metal–carbon bond is decreased. Bacskai (59) proposed this idea to explain the activity of the $AlEt_3$–$TiCl_3$–pyridine system for the copolymerization of propylene and 6-chlorohexene-1. The chlorine in 6-chlorohexene-1 reacts readily with uncomplexed $AlEt_3$ but not with the $AlEt_3 \cdot$ pyridine complex.

d. Formation of the New Metal Alkyl. The third component reacts with the metal alkyl to generate a new metal alkyl; for example

$$AlEt_3 + EtOH \rightarrow AlEt_2OEt + ethane$$

The catalyst may exclusively contain the new metal alkyl or a mixture of $AlEt_3$ and $AlEt_2OEt$. Any compound that can react with the active metal–carbon bonds of the metal alkyl may act similarly, such as EtSH, COS, CS_2, O_2, RCHO, R_2CO, alkyl halides [see Mole and Jeffery (2) and Coates (60) for reactions of aluminum alkyls]. If the reaction of the third component and the metal alkyl is incomplete, then unreacted molecules may also react with the active centers (the transition metal–carbon bonds). Thus, a decline in rate due to the reactant may also occur in this way. Inorganic salts such as KCl and NaF form complexes with aluminum alkyls (8a), and it has been proposed that the complex is inactive (2) or the complex is more active (see discussion of type 1 donor). Again, the choice of donor and conditions of use are critical (61, 62).

Gippin found that in the absence of water the soluble $AlEt_2Cl$–$CoCl_2$–pyridine catalyst was inactive for polymerization of butadiene (63, 64). Zgonnik, Dolgoplosk, and co-workers reported that the polymerization rate and molecular weight of polybutadiene reached a maximum and then declined as the concentration of water was increased (37). Balas (65) found that the catalyst activity and *cis*-1,4 content of polybutadiene increased when a critical amount of HCl was present in the $AlEt_2Cl$–$CoCl_3 \cdot AlCl_3$ catalyst (HCl/Al = 1.1/2.2) (see Fig. 9–5).

Gippin showed that the water effect could be achieved with other additives, such as organic hydroperoxides, halogens, tertiary alcohols, organic halides, and aluminum. He suggested that these reacted with $AlEt_2Cl$ to form coactivating aluminum alkyls as shown in Eq. 9–9.

$$AlEt_2Cl + HCl \longrightarrow AlEt(OH)Cl$$

$$AlEt_2Cl + H_2O \longrightarrow (AlClEt)_2O$$

(9–9)

The metal alkyl component of the catalyst consists of $AlEt_2Cl$ and the new metal alkyls. Consistent with this explanation, Gippin reported that water was not necessary when an $AlEt_2Cl$–$AlEtCl_2 \cdot Co$–octoate system was used.

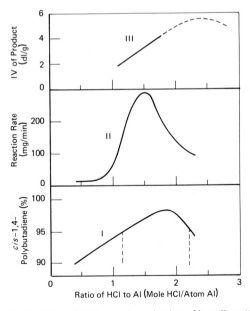

Fig. 9–5.   Effect of HCl on polymerization of butadiene (65).

Another view was expressed by Van de Kamp (66), who proposed that the water stabilized lower valent Co(I) species, which are the active centers.

Timofeyeva and co-workers (67) also showed that the catalyst $AlEt_2Cl$–$CoCl_2 \cdot 2$ pyridine was activated by the addition of $AlCl_3$, $SnCl_4$, or $AlEtCl_2$.

e. Formation of the New Catalyst.   Gippin found that, for the polymerization of isoprene with the $Al\text{-}i\text{-}Bu_3$–$TiCl_4$ catalyst ($Al/Ti \geq 1$), water and diphenyl ether acted synergistically in raising the catalyst efficiency (68). Table 9–11 shows the relationship between the Al/Ti ratio, diphenyl ether,

TABLE 9–11

Interaction Study. The Relationship between Al/Ti Ratio, Diphenyl Ether, and Water toward Percent Polymer Conversion (68)

	Polymer conversion (%)				
ϕ_2O/Al	$H_2O/Al = 0^a$	0.2	0.5	0.8	1.0
		$Al/Ti = 1.1$			
0	55.0	42.3	42.9	50.1	52.2
1	54.9	64.5	78.6	76.8	80.6
2	56.5	97.2	96.7	84.7	74.8
4	55.4	98.8	98.0	79.2	29.3
6	52.8	97.1	95.9	98.3	89.2
		$Al/Ti = 1.0$			
0	98.3	79.9	69.5	57.8	52.7
1	89.5	97.3	97.8	89.5	88.8
2	95.5	95.6	95.6	54.0	39.7
4	98.2	92.7	97.9	44.5	37.5
6	99.0	96.9	96.9	98.2	66.3
		$Al/Ti = 0.9$			
0	55.2	92.4	67.1	38.2	42.1
1	97.7	97.0	72.5	68.1	38.6
2	98.9	91.4	76.6	33.4	8.8
4	97.9	97.2	94.1	25.3	16.4
6	97.2	97.5	94.6	62.1	25.7
		$Al/Ti = 0.8$			
0	22.1	49.8	43.8	30.6	28.9
1	64.9	56.0	49.9	40.5	38.4
2	93.7	45.0	41.9	18.0	14.9
4	96.0	79.2	49.5	13.2	0
6	97.3	90.7	59.6	35.4	25.7

^{a} Ratio actually was 0.046 due to residual H_2O in monomer–solvent blend.

and water toward the percent of polymer conversion. According to Gippin, these catalyst components react to form the real catalyst, as shown in Eq. 9–10.

$$\text{Al-}i\text{-Bu}_3 + \text{TiCl}_4 \longrightarrow \beta\text{TiCl}_3 + \text{Al-}i\text{-Bu}_2\text{Cl} \tag{9–10}$$

Examination of Table 9–11 shows that when the Al-i-Bu$_3$/TiCl$_4$ ratio is either less or greater than 1, lower efficiencies are obtained than if this ratio was 1. The maximum efficiency was obtained at lower ratios by the addition of diphenyl ether; for example $(C_6H_5)_2O/Al \geq 2$ was needed. When the ratio was 1.1, however, both diphenyl ether and water were needed to restore the maximum efficiency, that is, $(C_6H_5)_2O/Al = 2$ and $H_2O/Al = 0.2$. Other combinations of ether and water concentrations are also effective. Gippen stated that water acts to hydrolyze and deactivate the poison i-BuTiCl$_2$, which would form from the excess Al-i-Bu$_3$ when the Al-i-Bu$_3$/TiCl$_3$ ratio was 1.1. (According to him, i-BuTiCl$_2$ causes isoprene to form dimers, trimers, and low molecular weight oils.) Diphenyl ether acts to complex Al-i-Bu$_2$Cl and thereby prevents the formation of higher Lewis acids by hydrolysis; for example, i-BuAl(OH)Cl or i-Bu Al(Cl)–O–Al(Cl)-i-Bu. These other components were believed to be responsible for unwanted gel formation and perhaps cyclization by a cationic mechanism. In developing a practical commercial process, the action of those components must be eliminated.

f. Deactivation of the Active Centers.   This explanation was discussed in detail in the type 2 case.

Mezhikovskii, Kissin, and Chirkov (69) examined the effect of water and alcohols on the polymerization of propylene when the AlEt$_3$–αTiCl$_3$ catalyst was used. Rather complex relationships were obtained when the rate of polymerization and isotacticity were measured as the concentration of water was increased. The observed induction period was attributed to adsorption of water on the surface of the αTiCl$_3$ (sites were blocked). But as this adsorbed water was reacted with AlEt$_3$, and thereby removed from the surface of TiCl$_3$, the polymerization commenced. Because the adsorption was fast, the induction period was tantamount to the time required for the adsorbed water to be reacted off the surface by AlEt$_3$ molecules. The reaction product, AlEt$_2$OH, however, is also a poison and can be removed by reaction with additional AlEt$_3$ molecules. These workers speculated from their kinetic data that higher concentrations of water promoted the polymerization by stabilizing reaction centers, which they assumed to be trivalent alkylated titanium alkyl–aluminum complexes.

Perry, Farson, and Schoenberg (70) found that carbon–disulfide prevents the formation of oligomers when present in small amounts in the polymerization of isoprene with Al-i-Bu$_3$–TiCl$_4$ catalysts to 97% cis-1,4 polymer. They suggested that CS$_2$ inactivates by preferentially complexing those sites that produce oligomers, that is, o-sites. The sites that produce cis-1,4-polyisoprene (p-sites) were not affected significantly. The number of o-sites remained effectively constant within the Al/Ti range of 1 to 2, which they examined. The molecular weight and microstructure of the polyisoprene product was not affected by the addition of CS$_2$. Oligomerization increased as the Al/Ti ratio was increased above 1. In absence of competition by o-sites, the remaining p-sites produced a greater yield of polymer. The author agrees with their conclusion. Because CS$_2$ was so specific, one wonders about the influence of COS and CO$_2$ on this system.

g. Activation of Dead Centers.   Under the condition of the polymerization, the active centers undergo reduction to a lower valence state metal species that are no longer active for polymerization of the olefin. The third component oxidizes this inactive species to a higher valence state compound, which is easily realkylated to form an active center. The effect of two different third components is often explained in this way.

Oxygen in controlled amounts increases the activity of a number of catalysts when used to polymerize ethylene: CH$_3$TiCl$_3$–TiCl$_3$ (71); AlBr$_3$–VXn–Sn(C$_6$H$_5$)$_4$ (72, 73); AlR$_3$–TiCl$_4$ (74); and Cp$_2$TiCl$_2$ · AlR$_3$ (75). Many of these catalysts undergo a fast die-out after the polymerization begins, and the addition of oxygen reactivates the catalyst.

Halogenated organic molecules have also been found to be very efficient activators (promoters) in vanadium-based catalysts. Sumitomo workers found extremely high polymerization rates for ethylene when benzene sulfonyl chloride (C$_6$H$_5$SO$_2$Cl) was added to the AlEt$_2$Cl–VCl$_4$ catalyst in an ethylene polymerization (76) (see Chapter 6). Other promoters disclosed for vanadium compounds include carbon tetrachloride (77–79), hexa-chlorocyclopentadiene (80), phenyl trichloromethane, hexachloroacetone, and trichloroacetic acid [see Christman (81) for compilation of earlier references].

Christman (81) examined the effect of a series of polyhaloorganic molecules on the activity of several soluble vanadium-based catalysts for the polymerization of ethylene at 120°C in some detail. In the absence of a promotor, yields were low. When promotor was added continuously throughout the polymerization, high yields of polymer per mmole vanadium were obtained. For example, in a 2 hour run using (t-BuO)$_3$VO, Al$_2$Et$_3$Cl$_3$, and ethyl trichloroacetate as promotor, 31,200 g polymer were obtained per mmole vanadium. Up to 266 polymer chains were formed per vanadium atom. The product polyethylene had a very narrow molecular weight distribution, $Q = \overline{M}_w/\overline{M}_n \simeq 2.1$ to 2.6. As a result of the addition of metal trichloroacetate to the $AlBr_3$–$(C_6H_5)_4Sn$–$VOCl_3$, 6.6 kg polymer per mmole vanadium (30 to 60 chains per vanadium atom) were obtained (also $Q = 2.4$ by GPC). To support this view that the halogen compound acted as an oxidizing agent to reactivate dead divalent vanadium centers, Christman showed that CCl_3–$COOR$ compounds do oxidize di- and trivalent vanadium. Similar results were obtained with other vanadium compounds [VCl_4 and $VO(OEt)_2Cl$] and metal alkyls ($AlEt_3$, $AlEt_2Cl$, and $AlEtCl_2$).

The following activating mechanism was proposed, as shown in sequence 9–11.

$$VCl_3 + AlEt_2Cl \longrightarrow EtVCl_2 + AlEtCl_2$$

$$EtVCl_2 + nCH_2{=}CH_2 \longrightarrow \underset{A}{Et{-}(CH_2{-}CH_2)_n{-}VCl_2} \qquad (9-11)$$

$$2\,A \longrightarrow VCl_2 + Et(CH_2{-}CH_2)_{n-1}CH{=}CH_2 + Et(CH_2{-}CH_2)_{n-1}CH_2{-}CH_3$$

$$VCl_2 + Cl_3CCO_2R \longrightarrow VCl_3 + \cdot Cl_2\overset{\cdot}{C}{-}CO_2R$$

Activation of titanium-based catalysts by chloroalkanes was observed by other workers. Fushman and co-workers (81a) found the activities of the $(C_5H_5)_2TiCl$–$AlEt_2Cl$ and $(C_5H_5)_2TiCl$–$AlEtCl_2$ catalysts were greater in the presence of chlorinated solvents and showed that the oxidation Ti(III) to Ti(IV) had occurred. Matkovskii and co-workers (81b) showed by ESR that the inactive blue complex, $(C_5H_5)_2TiCl \cdot AlEtCl_2$, was oxidized with triphenylchloromethane. These authors also concluded that some unsaturated compounds, such as 1-butene, caused an increase in the rate of recovery of active centers in the $(C_5H_5)_2TiCl_2$–$AlEt_2Cl$–$EtCl$ system with a simultaneous accumulation of $AlEtCl_2$ (81c). As the concentration of 1-butene in ethylene increased, the stationary rate of polymerization and the number of macromolecules increased.

More recently, Duck and co-workers (81d) used butylperchlorocrotonate as activator for the polymerization of ethylene in the presence of $VOCl_3$ and $Al_2Et_3Cl_3$ (a homogeneous catalyst) between 15° and 85°C. Catalyst efficiencies (g polymer/g vanadium hr atm ethylene in cap gas) of $10^{6.1}$ were obtained. This activator was more effective (1.3) in their hands than hexachlorocyclopentane (0.2), methyltrichloroacetate (0.5), and ethyltrichloroacetate (0.7). They proposed that the activator acted as a chlorinating agent, that is, chlorinating inactive V(II) species to catalytically active V(III) ones.

h. Change in the Physical State of the Transition Metal Salt.   Several distinct schemes were suggested.

Minsker and Bykhovskii (82) suggested that amine molecules that complexed on the surface of $TiCl_3$ facilitate the breakdown of the crystals along the plane of cleavage. By virtue of a greater surface area, more sites can form for polymerization of propylene. Similarly, Cooper and co-workers (83, 84) proposed that isopropyl ether in the $AlEt_3$–VCl_3 catalyst (isoprene polymerization) assisted in the breakup of the VCl_3 aggregates by penetration of the crystals lattice or by a solvation effect, which also prevented reaggregation of the smaller particles. Boor also proposed a similar explanation for amine donors in $ZnEt_2$–$TiCl_3$ systems (see type 2 case).

Razuvaev, Minsker, and Chernovskaya (85, 86) proposed that aromatic-type electron donors in the $AlEt_3$–$TiCl_3$ catalyst (propylene) changed the energetic conditions at the surface of $TiCl_3$ and the strength of the $AlEt_3$–$TiCl_3$ complex. Compounds at concentrations of 0.2 to 3 moles per liter solvent increased activity but lowered polymer isotacticity and molecular weight. But in other experiments, they observed that ethers increased catalyst activities and molecular weights and proposed that the donor aids in the complexing of propylene at the center (87). Korotkov and co-workers (88) proposed that chain-breaking reactions were affected by donors during the polymerization of 1-pentene with the $AlEt_2Cl$–$\beta TiCl_3$ catalyst.

Ambroz and Hamrik (89) suggested that donors increased the number of surface defects at which polymerization takes place. They investigated the polymerization of propylene with the $AlEt_3$–$TiCl_3$ catalyst, with amine donors being used.

III. Conclusion

This chapter shows that a third component can have a pronounced effect on the polymerization behavior of a Ziegler–Natta catalyst. A variety of third components have been examined for the polymerization of olefins and dienes, and many different mechanistic explanations have been offered.

If one takes into consideration the different types of third components studied, the structure of the monomer used, the chemical behavior of the catalyst components, and the conditions of catalyst preparation and use in polymerization, it is extremely unlikely that a general mechanistic explanation exists. It is more probable that these third components affect the Ziegler–Natta catalyst in different ways, and in some cases the third component may perform two or more functions, such as removing poisons and inactivating less specific centers.

For some Ziegler–Natta catalyst–monomer systems, the third component is an integral part of the polymerization. It is a powerful handle for controlling the stereochemical and kinetic behavior of the catalyst. In other catalyst systems, the third component has served as a powerful diagnostic probe to study the nature of the active transition metal center in both soluble and heterogeneous catalysts.

References

1. J. Boor, *Macromol. Rev.* **2**, 115–268 (1967); E. Pajda, M. Uhniat, and A. Pajda, *Chemik* **22**, 249–251 (1969).
2. The chemistry of aluminum metal alkyls has been reviewed recently by T. Mole and E. A. Jeffery, "Organoaluminum Compounds." Elsevier, Amsterdam, 1972.
3. Chapter 4, Section IIC1 should be consulted for background information.
4. A. Zambelli, J. DiPietro, and G. Gatti, *J. Polym. Sci., Part A* **1**, 403 (1963).
5. A. Zambelli, J. DiPietro, and G. Gatti, *Chim. Ind. (Milan)* **44**, 529 (1962).
6. G. Natta, A. Zambelli, I. Pasquon, G. Gatti, and D. DeLuca, *Makromol. Chem.* **70**, 206 (1964).
7. H. W. Coover, Jr. and F. B. Joyner, *J. Polym. Sci., Part A* **3**, 2407 (1965).
8. R. L. McConnell, M. A. McCall, G. O. Cash, Jr., F. B. Joyner, and H. W. Coover, Jr., *J. Polym. Sci., Part A* **3**, 2135 (1965).
8a. K. Ziegler, *in* "Organometallic Chemistry," (H. Zeiss, ed.) *Monogr. Ser. No.* 147, p. 194–269, Van Nostrand-Rheinhold, Princeton, New Jersey (1960).
9. W. R. Watt, *J. Polym. Sci., Part A-1* **7**, 787 (1969).
10. K. Matsumura, Y. Atarashi, and O. Fukumoto, *J. Polym. Sci., Part A-1* **7**, 311 (1969).
11. K. Matsumura, Y. Atarashi, and O. Fukumoto, *J. Polym. Sci., Part A-1* **9**, 485 (1971).
12. J. Boor, *J. Polym. Sci.* **62S**, 45 (1962); *Part C* **1**, 237 and 257 (1963).
13. J. Boor and E. A. Youngman, *J. Polym. Sci., Part B* **2**, 265 (1964).
14. J. Boor, *J. Polym. Sci., Part B* **3**, 7 (1965); *Part A* **3**, 995 (1965).
15. J. Boor, Pap., *150th Meet., Am. Chem. Soc., 1965*; *Polym. Prep., Am. Chem. Soc., Div. Polym. Chem.* **6**, No. 2, 890 (1965).
16. E. A. Youngman and J. Boor, *J. Polym. Sci., Part B* **3**, 577 (1965).
17. J. Boor, *J. Polym. Sci., Part A-1* **9**, 617 (1971).
17a. R. H. Marchessault and H. D. Chanzy, *J. Polym. Sci., Part C* **30**, 311 (1970).
17b. A. W. Langer, U.S. Patent 3,278,511, October 11, 1966, Esso Research and Engineering Company.
17c. K. Tanabe and Y. Watanabe, *J. Res. Inst. Catal., Hokkaido Univ.* **11**, No. 2, 65 (1963).
18. J. Fukuda, K. Tashiro, M. Koga, T. Hori, H. Ogawa, E. Takeshita, and K. Akaiwa, U.S. Patent 3,622,552, November 23, 1971, Mitsubishi Petrochemical Company, Ltd.
19. S. Sugiura, H. Ueno, H. Ishikawa, and T. Yamo, U.S. Patent 3,574,179, April 6, 1971, Ube Industries, Ltd.
20. A. Nicco, U.S. Patent 3,514,433, May 26, 1970, Soc. Normande De Matieres Plastiques.
21. L. M. Fodor, U.S. Patent 3,429,862, February 25, 1969; L. M. Fodor and G. R. Kahle, U.S. Patent 3,513,147, May 19, 1970; L. M. Fodor, U.S. Patent 3,514,434, May 26, 1970.
22. G. Staiger and J. Stedefeder, U.S. Patent 3,639,375, February 1, 1972, Badische Anilin & Soda-Fabrik.
23. G. G. McClaflin, U.S. Patent 3,405,113, October 8, 1968, Continental Oil Company.
24. C. W. Moberly, U.S. Patent 3,484,424, December 16, 1969, Phillips Petroleum Company.
25. V. P. Mardykin and A. M. Antipova, *Vysokomol. Soedin., Ser. A* **11**, 1600 (1969).
26. H. Cheney, U.S. Patent 3,081,289, March 12, 1963.
27. R. Eichenbaum and J. G. Murray, U.S. Patent 3,635,839, January 18, 1972, Mobil Oil Corporation.
28. V. V. Boiko and D. V. Ivanyukov, *Plast. Massy* No. 7, p. 8 (1970).
29. H. K. Garner and D. N. Mathews, U.S. Patent 3,441,546, April 29, 1969, Uniroyal, Inc.
30. G. V. Sorokin, N. S. Nametkin, and V. N. Perchenko, *Polym. Sci. USSR (Engl. Transl.)* **8**, 520 (1960).
31. L. Hague, H. M. Khelghatian, J. L. Jezl, and J. A. Price, U.S. Patent 3,506,591, April 14, 1970, Avisun Corporation.

32. P. E. Matkovskii, A. D. Promogailo, G. A. Beikhol'd, and N. M. Chirkov, *Vysokomol. Soedin.*, *Ser B* 11, 610 (1969).

33. T. Arakawa and M. Hashimoto, *Kogyo Kagaku Zasshi* 72, 1724 (1969).

34. J. W. Bayer, S. W. Gagnon, and W. C. Grinonneau, U.S. Patent 3,574,174, April 6, 1971; J. W. Bayer and E. Santiago, U.S. Patent 3,440,179, April 22, 1969, Owens-Illinois, Inc.

35. L. M. Lanovskaya, A. R. Gantmakher, and S. S. Medvedev, *Polym. Sci. USSR (Engl. Transl.)* 3, 335 (1962).

36. Great Britain Patent 876,093, August 30, 1961, Ethyl Corporation.

37. V. N. Zgonnik, B. A. Dolgoplosk, N. I. Nikolayev, and V. A. Kropachev, *Polym. Sci. USSR (Engl. Transl.)* 7, 338 (1965).

38. G. J. Carlson, W. Dong, T. L. Higgins, and C. H. Wilcoxen, Jr., U.S. Patent 3,066,127, November 27, 1962, Shell Oil Company.

39. B. V. Erofeev, V. M. Zapletnyak, A. D. Pechenkin, V. A. Klyushnikov, V. A. Sererova, V. D. Kozochkina, Z. V. Arkhipova, and I. N. Andreeva, *Dokl. Akad. Nauk Beloruss. SSR* 13, 621 (1969).

40. A. L. Barney and R. L. Morgan, U.S. Patent 3,418,303, December 24, 1968, E. I. duPont de Nemours & Company.

41. D. N. Mathews and R. J. Kelly, U.S. Patent 3,405,107, October 8, 1968, Uniroyal, Inc.

42. E. P. V. Cornelis, J. L. R. Graff, and W. J. Hendriks, U.S. Patent 3,506,629, April 14, 1970.

43. N. M. Seidov, M. A. Dalin, A. I. Absov, and R. M. Osman-Zade, *Dokl. Akad. Nauk Az. SSR* 24, 46 (1968).

44. E. W. Duck, J. H. Farmer, and B. J. Ridgewell, U.S. Patent 3,644,309, February 22, 1972, International Synthetic Rubber Company.

45. N. Dost, D. Y. Waddan, and H. E. Strauss, Great Britain Patent 851,113, October 12, 1960, Petrochemicals, Ltd.

46. E. H. Mottus, U.S. Patent 3,440,237, April 22, 1969, Monsanto Company.

47. R. S. Kircheva, F. D. Radenkov, L. I. Petkov, M. Kh. Mikhailov, S. S. Karaenev, P. K. Prodanov, and N. P. Shestak, *Plast. Massy* No. 7, p. 3 (1971).

48. M. Hirooka, H. Kanda, and K. Nakaguchi, *J. Polym. Sci., Part B* 1, 701 (1963).

49. B. S. Turov, P. A. Vinogradov, B. A. Dolgoplosk, and S. I. Kostina, *Dokl. Akad. Nauk USSR* 151, No. 5, 1118 (1963).

50. W. Cooper, D. E. Eaves, G. D. T. Owen, and G. Vaughan, *J. Polym. Sci., Part C* 4, 211 (1963).

51. G. A. Parfenova, V. A. Krol, and R. N. Karelina, *Kauch. Rezina* 28, 1 (1969).

52. R. F. Roberts, U.S. Patent 3,644,325, February 22, 1972, Dow Chemical Company.

53. J. B. Donnet, J. P. Wetzel, and G. Riess, *J. Polym. Sci., Part A-1* 6, 2359 (1968).

54. A. D. Caunt, *J. Polym. Sci., Part C* 4, 49 (1963).

54a. D. R. Burfield and P. J. T. Tait, *Polymer* 15, 87 (1974).

55. K. Vesely, J. Ambroz, R. Vilim, and O. Hamrik, *J. Polym. Sci.* 55, 25 (1961).

55a. S. M. Mezhikovskii, Yu. V. Kissin, and N. M. Chirkov, *Vysokomol. Soedin.*, *Ser. A* 9, 2006 (1967).

55b. T. Keii, K. Soga, and N. Saiki, *J. Polym. Sci., Part C* 16, 1507 (1967).

55c. C. G. Overberger and P. A. Jarovitsky, *J. Polym. Sci., Part C* 4, 37 (1963).

55d. D. F. Hoeg and S. Liebman, *Ind. Eng. Chem., Process Des. Dev.* 1, No. 2, 120 (1962).

56. E. B. Milovskaya and P. I. Dolgopolskaya, *Vysokomol. Soedin.* 4, 1049 (1962).

57. O. N. Pirogov and N. M. Chirkov, *Vysokomol. Soedin.* 7, 491, 1965; *Polym. Sci. USSR (Engl. Transl.)* 7, 544 (1965).

58. P. A. Vinogradov, B. A. Dolgoplosk, V. N. Zgonnik, O. P. Parengo, E. I. Tinyakova, and B. S. Turov, *Dokl. Akad. Nauk USSR* 163, No. 5, 1147 (1965).

59. R. Bacskai, *J. Polym. Sci., Part A* 3, 2491 (1965).

60. See also G. E. Coates, "Organo-Metallic Compounds, 2nd ed., pp. 126–143. Wiley, New York, 1960.
61. Unfortunately, many authors do not give adequate descriptions of the way the modifier and catalyst were reacted, making it difficult to reproduce their results.
62. Unless a highly purified modifier is used, the observed results may be partially or wholly due to a contaminant.
63. M. Gippin, *Ind. Eng. Chem., Prod. Res. Dev.* **1**, 32 (1962).
64. M. Gippin, *Rubber Chem. Technol.* **39**, 508 (1966).
65. J. Balas, U.S. Patent, 3,111,510, November 19, 1963, Shell Oil Company.
66. P. F. Van de Kamp, *Makromol. Chem.* **93**, 202 (1966).
67. G. V. Timofeyeva, N. A. Kokorina, and S. S. Medvedev. *Polym. Sci. USSR (Engl. Transl.)* **11**, 677 (1970).
68. M. Gippin, *J. Appl. Polym. Sci.* **14**, 1807 (1970).
69. S. M. Mezhikovskii, Y. V. Kissin, and N. M. Chirkov, *Polym. Sci. USSR (Engl. Transl.)* **10**, 2592 (1968); **9**, 1383 and 2267 (1967).
70. D. C. Perry, F. S. Farson, and E. Schoenberg, *Polym. Prepr., Am. Chem. Soc., Div. Polym. Chem.* **13**, No. 2, 1097 (1972); *Pap., 164th Meet., Am. Chem. Soc., 1972*; *J. Polym. Sci., Polym. Chem. Ed.* **13** (5), 1071–81 (1975).
71. C. Beerman and H. Bestian, *Angew. Chem.* **71**, 618 (1959).
72. H. J. de Liefde Meijer, J. W. G. van der Hurk, and G. J. M. van der Kerk, *Recl. Trav. Chim. Pays-Bas* **85**, 1018 and 1025 (1966).
73. G. W. Phillips and W. L. Carrick, *J. Am. Chem. Soc.* **84**, 920 (1962); *J. Polym. Sci.* **59**, 401 (1962).
74. Z. S. Smolyan, A. I. Graevskii, O. I. Demin, V. K. Fukin, and G. N. Matveeva, *Vysokomol. Soedin.* **3**, 81 (1961).
75. D. S. Breslow and N. R. Newburg, *J. Am. Chem. Soc.* **79**, 5073 (1957); **81**, 81 (1959).
76. K. Nakaguchi, M. Hirooka, and T. Fujita, U.S. Patent 3,328,366, June 27, 1967; Great Britain Patents 1,022,931, March 16, 1966, Sumitomo Chemical Company; 1,077,931, August 2, 1967.
77. W. P. Long, U.S. Patent 3,203,940, August 31, 1965, Hercules.
78. H. Weber and P. Schneider, French Patent 1,415,238, October 22, 1965, Badische Anilin & Soda Fabrik.
79. G. Boocock, Great Britain Patent 1,059,865, February 22, 1967.
80. A. Gumboldt, E. Junghanns, G. Schleitzer, and H. D. Stemmer, French Patent 1,370,358, August 21, 1964, Hoechst (equivalent to U.S. Patent 3,349,064).
81. D. L. Christman, *J. Polym. Sci., Part A-1* **10**, 471 (1972).
81a. E. A. Fushman, V. N. Tsvetkova, and N. M. Chirkov, *Dokl. Akad. Nauk USSR* **164**, 1085 (1965).
81b. P. Ye. Matkovskii, T. S. Dzhabiev, F. S. D'yachkovskii, G. A. Beikhol'd, Kh. M. A. Birkenshtein, and N. M. Chirkov, *Polym. Sci. USSR (Engl. Transl.)* **13**, 1981 (1971).
81c. P. Ye. Matkovskii, G. P. Belov, A. P. Lisitskaya, L. N. Russiyan, Kh. M. A. Birkenshtein, M. P. Gerasina, and N. M. Chirkov, *Polym. Sci. USSR (Engl. Transl.)* **12** 1890 (1970).
81d. E. W. Duck, D. Grant, J. R. Horder, D. K. Jenkins, A. E. Marlow, S. R. Wallis, A. G. Doughty, J. M. Maradon, and G. A. Skinner, *Eur. Polym. J.* **10**, 481 (1974).
82. K. S. Minsker and V. K. Bykhovskii, *Vysokomol. Soedin.* **2**, 535 (1960).
83. W. Cooper, D. E. Eaves, G. D. T. Owen, and G. Vaughan, *J. Polym. Sci., Part C* **4**, 211 (1963).
84. W. Cooper, R. K. Smith, and A. Stokes, *J. Polym. Sci., Part B* **4**, 309 (1966).
85. G. A. Razuvaev, K. S. Minsker, and R. P. Chernovskaya, *Dokl. Akad. Nauk USSR* **147**, 636 (1962); also with G. I. Burlakova, *Polym. Sci. USSR (Engl. Transl.)* **7**, 39 (1965).

86. R. P. Chernovskaya, K. S. Minsker, and G. A. Razuvaev, *Vysokomol. Soedin.* **6**, 1656 (1964).
87. G. A. Razuvaev, K. S. Minsker, G. T. Fedoseeva, and V. K. Bykhovskii, *Polym. Sci. USSR (Engl. Transl.)* **2**, 299 (1961).
88. A. A. Korotkov, I. S. Lishanskii, and L. S. Semenova, *Polym. Sci. USSR (Engl. Transl.)* **2**, 175 (1961).
89. J. Ambroz and O. Hamrik, *Collect. Czech. Chem. Commun.* **28**, 2550 (1963).

10

Termination
of Polymer Chain Growth

I. Introduction

Termination of chain growth in a Ziegler–Natta polymerization can be caused by specific reactants and complexing agents or by thermal cleavage. Only a small number of these reactants are known that behave as transfer agents; that is, they do not simultaneously destroy the transition metal bond center. Examples of transfer agents are H_2 and $ZnEt_2$, and these have become the practical agents for control of polymer molecular weight in most olefin polymerizations. In contrast, simple olefins function as the most effective transfer agents in a butadiene polymerization with cobalt-based catalysts.

This chapter will describe some of the reactions that take place at the transition metal–carbon bond centers. These reactions lead to termination of growth of the polymer chain, with the practical benefit of control over the molecular weight of the polymer product.

Using Pn to represent the growing polymer chain, these reactions can be expressed in two ways: (1) those in which the active center is simultaneously killed, as shown in Eq. 10–1, where the active Ti–C bond is converted to an

$$\text{Ti—Pn} + \text{BuCl} \longrightarrow \text{Ti–Cl} + \text{Bu–Pn}$$
$$\underset{\text{center}}{\text{active}} \qquad\qquad \underset{\text{center}}{\text{dead}} \qquad\qquad\qquad (10\text{–}1)$$

inactive Ti–Cl bond and Pn is exchanged for Cl and (2) those in which the active center continues to polymerize, as shown in Eq. 10–2. The polymer chain, Pn, becomes detached from the Ti center by exchange of Pn for Et.

$$\text{Ti—Pn} + \text{ZnEt}_2 \longrightarrow \text{Ti—Et} + \text{PnZnEt}$$

active center   active center   (10–2)

It is important to distinguish between primary reactions involving the active centers, as in Eqs. 10–1 and 10–2, and secondary reactions involving the metal alkyl component, as in Eq. 10–3.

$$\text{AlEt}_2\text{Cl} + \text{BuCl} \longrightarrow \text{AlEtCl}_2 + \text{hexane} \qquad (10\text{–}3)$$

The formation of a second alkyl can also modify the catalyst in a way that changes in molecular weight of polymer may occur, but the mechanistic path is different. The activity of the catalyst may also become different if the site-forming ability of the new alkyl is different.

In practice, viscosity methods, because they were routinely most easily done in the laboratory, were used to observe changes in molecular weight. For example, the Natta group (1) determined the molecular weight (MW) of polypropylene in a solution of tetralin at 135°C by means of a Desreux–Bischoff viscometer, applying the formula

$$[\eta] = 1.75 \times 10^{-4} \times \bar{M}_w^{0.74} \qquad (10\text{–}4)$$

where $[\eta]$ is the intrinsic viscosity (units = dl/g).

II. Chain Termination by Organic and Inorganic Compounds

Many workers attempted to control molecular weight by terminating the growth of the polymer chain at the metal centers by the addition of reactive organic and inorganic compounds, by reactions of the type shown in Eq. 10–1. Typical compounds included alcohols, alkyl halides, alkane thiols, H_2S, $SnCl_4$, $SiCl_4$, CCl_4, NaX (X = Br, Cl) and BCl_3 (1a).

In practice, none of these resulted in attractive molecular weight control agents, for two reasons. Firstly, in all cases the reaction center was simultaneously destroyed, with a concomitant decrease in catalyst activity. Only if the activity of the catalyst was very high could the lowering of molecular weight be justified by the loss in polymer yield. Secondly, a problem arose because the compound reacted so fast with the metal alkyl that little was available for termination of chain growth. This reaction often resulted in the generation of the second metal alkyl that could function as a site former, and in the loss of the first metal alkyl. It is easy to see that any number of effects can be obtained, depending on the rates of these reactions and the nature of the formed metal alkyl.

Some of the experimental findings are now reported.

Pozamantir (2) showed that the more reactive the alkyl halide, the greater was the lowering of molecular weight when ethylene was polymerized by the $AlEt_3-TiCl_4$ (1:1) catalyst. Alkyl halides such as *tert*-BuCl, benzyl chloride, and *i*-PrCl were more effective than *n*-BuBr or EtCl. Pozamantir suggested that the growing macromolecular chain, which was attached to the aluminum atom, was displaced by the halide atom of the alkyl halide. As evidence, he cited that $SnCl_4$ and CCl_4 also acted as effective terminating agents, while $SiCl_4$ did not. In separate experiments, he showed that $SnCl_4$ and CCl_4, but not $SiCl_4$, reacted with $AlEt_3$ and $AlEt_2Cl$. The author supports Pozamantir's view that these compounds react with the aluminum alkyl component but suggests that the reaction is with $AlEt_3$ and not with a polymer-bearing aluminum alkyl growth center. The later finding of Pozamantir and co-workers (3) that the molecular weight of polyethylene decreased in the order: $AlEt_3/TiCl_4 > AlEt_2Cl/TiCl_4 > AlEtCl_2/TiCl_4$ is cited as partial evidence. Clearly, the greater the reaction of the alkyl halide to produce the more chlorinated metal alkyl, the lower the expected polymer molecular weight. In support of this conclusion, Schnecko and co-workers (4) found that catalyst activity for polymerization of propylene became lower as significant concentrations of chlorine-containing alkyls were added to the $AlEt_3-TiCl_3$ catalyst. Catalyst activity decreased for $AlEt_xCl_{3-x}-TiCl_3$ system in the order: $AlEt_3 > AlEt_2Cl > Al_2Et_3Cl_3 > AlEtCl_2$.

In contrast, Schindler reported that addition of CCl_4 to the Al-*i*-Bu_2H/$TiCl_4$ catalyst resulted in a lower polyethylene molecular weight without a loss in rate [see Schindler (5, p. 186)]. No chlorine was found in the polymer.

Weber and Kiepert (6) examined the effect of several organic chlorides on the $Al_2Et_3Cl_3-(n-PrO)_2TiCl_2$ catalyst when used for polymerization of ethylene. While allyl chloride, cinnamyl chloride, crotyl chloride, and *tert*-butyl chloride were effective terminators, methallyl chloride was ineffective and 2-phenyl-3-bromo-1-propene actually increased polymer molecular weight. The latter finding may be due to the formation of bromine-containing aluminum alkyls. In light of Pozamantir's findings, it is not clear whether the lower molecular weights are due to actual termination or to formation of $AlEtCl_2$ from $AlEt_2Cl$. The inertness of methallyl chloride is anomalous since it is known to react with $AlEt_2Cl$.

In all of the previously mentioned catalyst systems, the alkyl halide could react with the aluminum alkyl component. Boor (7) reported when the $ZnEt_2-TiCl_3$ catalyst was used, this problem could be avoided or minimized if certain alkyl halides were used, namely *n*-BuCl, *sec*-BuCl, allyl chloride, and methallyl chloride. In the time of polymerization (about 20 hours), these alkyl halides reacted negligibly ($<5\%$) with $ZnEt_2$. The decreases obtained in polypropylene molecular weight were attributed directly to the reaction of these alkyl chlorides with the reaction centers, that is, Ti–Pn bonds where Pn is a polymer chain. The order of terminating efficiency was: allyl chloride $\simeq$ methallyl chloride > *sec*-butyl chloride > *n*-butyl chloride. If the highly reactive *tert*-BuCl or benzyl chloride was used, these reacted with $ZnEt_2$ to form some $ZnEtCl$; the polymerization results were in agreement with this secondary reaction.

In support of his conclusions obtained with the $ZnEt_2$–$TiCl_3$ catalyst, Boor also reported similar terminating efficiencies for the above chloro compounds when propylene was polymerized with the metal alkyl-free catalyst $TiCl_3$ + amine. In both the $ZnEt_2$–$TiCl_3$ and the $TiCl_3$–amine systems, the alkyl halide preferentially terminated the nonisotactic-specific sites, leading to more isotactic polymer. Apparently, polymer chains attached to the more exposed Ti centers are more susceptible to reaction by these alkyl halides.

III. Chain Termination by Metal Alkyls

Metal alkyls can act as terminators of chain growth and, under suitable conditions, they function as true transfer agents. Natta and Pasquon (8) first recognized that in a polymerization of propylene with the $AlEt_3$–$\alpha TiCl_3$ catalyst the molecular weight decreased at higher concentrations of $AlEt_3$. Figure 10–1 shows the changes in molecular weight (as $1/[\eta]^{1.35}$ values) as a function of the square root of the $AlEt_3$ concentration. To support their view that variations in molecular weight with $AlEt_3$ concentrations were due to chain transfer with participation of the Et groups of $AlEt_3$, polymerizations were done with ^{14}C-labeled $AlEt_3$ and the polymer was analyzed for radioactivity. Figure 10–2 shows that the counts per minute for individual polymers increased with increasing $AlEt_3$ concentration.

Fig. 10–1. Dependency of the reciprocal of the degree of polymerization (proportional to $1/[\eta]^{1.35}$) of the nonatactic polypropylene fraction on the square root of the aluminum alkyl concentration [temp., 70°C, $p_{C_3H_6}$, 950 mm Hg, ground $\alpha TiCl_3$: sample (a) (8)]. (Taking $[\eta]^{1/0.74} = 1/[\eta]^{1.35}$, this factor is assumed to be proportional to reciprocal of degree of polymerization.)

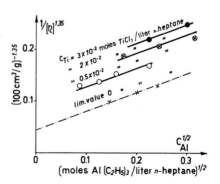

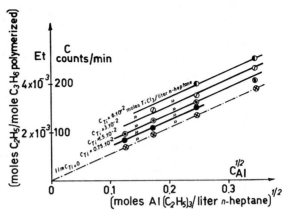

Fig. 10–2. Specific radioactivity (and corresponding values of $-C_2H_5$ mole per mole of polymerized C_3H_6) of the nonatactic polypropylene fraction, as function of the square root of the alkylaluminum concentration. [Tests performed with ^{14}C-labeled $Al(C_2H_5)_3$ at temp., 70°C, $p_{C_3H_6}$, 450 mm Hg, ground $\alpha TiCl_3$: sample (a) (8).]

Natta and co-workers (9, 10) later demonstrated that $ZnEt_2$ was a considerably more effective transfer agent when added to the $AlEt_3$–$TiCl_3$ catalyst and used for polymerization of propylene. They could easily lower molecular weights from about 6×10^5 to 2×10^5 without lowering overall polymerization rate or the isotacticity of the polymer (Table 10–1). At higher $ZnEt_2$ concentrations, a decrease in rate of polymerization and an increase in percent solubilities accompanied the decrease in polymer molecular

TABLE 10–1

Polymerization of Propylene in the Presence of Zinc Diethyl[a] (9, 10)

$Zn(C_2H_5)_2$ (ml)	Polymer obtained (g)	Ether extract (g)	Ether extract (%)	Heptane extract (g)	Heptane extract (%)	Residue (g)	Residue (%)	Polymeric residue from ether extract $\bar{M}_v$	Polymeric residue from ether extract $[\eta]$
0	14.50	1.67	11.5	0.94	6.5	11.89	82.0	570,000	3.18
0.10	14.21	1.21	8.5	1.35	9.5	11.76	82.0	222,000	1.58
0.35	13.00	1.36	10.5	2.60	20.0	9.04	69.5	91,000	0.82
0.70	11.57	1.12	9.7	3.43	29.6	7.02	60.7	60,500	0.60
1.40	9.14	1.14	12.5	3.62	39.6	4.38	47.9	29,400	0.35
2.00	6.94	0.94	13.5	3.50	50.5	2.50	36.0	21,100	0.29
2.80	5.05	1.01	20.0	2.74	54.5	1.30	25.5	14,800	0.23

[a] Experiments with varying concentrations of $Zn(C_2H_5)_2$. 0.31 g $\alpha TiCl_3$; 0.45 ml $Al(C_2H_5)_3$; 250 ml n-heptane; temp., 70°C; $P_{C_3H_6}$, 950 mm Hg; duration, 1 hr.

weight. (Because very low molecular weight polypropylene was produced, the high percent heptane solubilities may reflect its increased ability to dissolve, not the lowering of isotacticity as Natta's paper implied. This has already been discussed in Chapter 3, Section II.)

The process by which aluminum and zinc alkyls terminate chain growth was shown by Natta and co-workers to involve alkyl exchange between ethyl groups of the Et_2Zn and the growing polymer chains on the active centers. Using Natta's notation, [cat] = the catalyst complex on which the polymer chain Pn is growing, the reaction may be represented as follows.

$$[cat]-Pn + ZnEt_2 \longrightarrow [cat]-Et + PnZnEt \qquad (10-5)$$

Its use left the question of the metal open in the active metal–carbon bond, that is, Al–C or Ti–C in the $AlEt_3$–$TiCl_3$ catalyst. The rate of this exchange reaction was first-order with respect to the $ZnEt_2$ concentration and was independent of the partial pressure of propylene. The formed Pn–$ZnEt$ alkyl does not further participate in the polymerization. This view was supported by their observation that $ZnEt_2$ was consumed during the polymerization. To obtain a polymer having a constant average molecular weight throughout the polymerization, additional $ZnEt_2$ must be added to replace the consumed portion.

Supplementary experiments by the Natta group showed the number of zinc atoms and ethyl groups coming from $ZnEt_2$ were nearly the same (10). The ethyl groups were determined by radiolabeling methods, and the analyzed polymer was isolated from the catalyst (see Table 10–2).

TABLE 10–2

End Groups in Polypropylene Obtained with the Catalyst System[b] (10)

$\bar{M}_v{}^a$	Zn end groups/mole C_3H_6 polymerized	C_2H_5 end groups/mole C_3H_6 polymerized
30,000	10.0×10^{-3}	—
31,000	—	9.0×10^{-3}
39,000	8.6×10^{-3}	—
50,000	7.0×10^{-3}	—
55,000	—	4.0×10^{-3}
69,000	2.7×10^{-3}	—
126,000	1.0×10^{-3}	—
140,000	—	1.0×10^{-3}

^{a} Determined by viscometry and calculated by use of the equation $[\eta] = 1.75 \times 10^{-4} \bar{M}_v{}^{0.74}$

^{b} Conditions: 0.31 g $\alpha TiCl_3$/0.45 cm^3 $Al(C_2H_5)_3$/0.1 to 0.7 cm^3 $Zn(C_2H_5)_2$ in 250 cm^3 n-heptane at a C_3H_6 pressure of 450 mm. Hg at 70°C.

Natta attributed the lower transfer ability of aluminum alkyls to their being closely associated as dimers, whereas $ZnEt_2$ exists in the undimerized form. The effectiveness of $ZnEt_2$ as a transfer agent was also demonstrated by Boor (11) in the catalyst system $ZnEt_2–TiCl_3$, where the α-, γ-, and $\delta TiCl_3$ modifications were used (but no aluminum alkyl). By increasing the $ZnEt_2$ concentrations from 0.5 to 15 moles/liter, the polypropylene intrinsic viscosity decreased from 3.0 to 0.8 dl/g. A similar dependence was shown for ZnEtCl. In addition, $ZnEt_2$ and ZnEtCl were found to be effective transfer agents for the $AlEt_2Cl–TiCl_3$ catalyst when used for the polymerization of propylene.

Vilim (12) showed that, in addition to $ZnEt_2$, $CdEt_2$ and $HgEt_2$ acted as transfer agents for the polymerization of propylene with the $AlEt_3–\alpha TiCl_3$ catalyst.

Firsov and co-workers (13), who also studied the kinetic effect of $ZnEt_2$ in the $AlEt_3–TiCl_3$ catalyst for propylene polymerization, reported that molecular weight could be decreased 20 times in its presence. The reciprocal value of the average degree of polymerization $\bar{v}$ was linearly dependent on the concentration of $ZnEt_2$ (see Table 10–3).

TABLE 10–3

Dependence of Molecular Weight on $ZnEt_2$[a] Concentration (13)

Et_2Zn(moles/liter)	0.0	0.035	0.344	0.563	1.05
$[\eta]$(dl/g)	3.1	2.8	2.8	0.24	0.155
$\bar{v}$	2,950	2,660	2,660	2,288	147

[a] Conditions: 36 moles/liter $AlEt_3$; 0.04 g $\alpha TiCl_3$; 50°C; $P_{propylene}$, 8.6 atm.

Boor (13a) showed that $CrCl_2$ was less active than $CrCl_3$ for the polymerization of ethylene when combined with $AlEt_3$. When combined with $AlEt_2Cl$ it was slightly more active.

$ZnEt_2$ was shown by Youngman and co-workers (14) to be an active transfer agent for butadiene polymerization with a solubilized cobalt chloride–aluminum alkyl chloride catalyst. The use of organometallic compounds as transfer agents for copolymerization of ethylene and propylene was disclosed by Giachetti and Scalari (15).

A dual role for $AlEtCl_2$ as a terminator and transfer agent was suggested by Borisova and co-workers (15a), who concluded this from kinetic studies of the polymerization of ethylene with the catalyst $(C_5H_5)_2TiEtCl + AlEtCl_2$. The number of polymer chains formed per $(C_5H_5)_2TiClEt$ added initially exceeded one. They disagreed with the earlier conclusion of Henrici–Olive and Olive (15b, p. 22) that deactivation was due to formation of the complex $(C_5H_5)_2TiEtCl \cdot (AlEtCl_2)_2$.

IV. Chain Termination by Molecular Hydrogen

Since the use of molecular hydrogen as a chain transfer agent was first described by Vandenberg (16), Ettore and Luciano (17), and Natta (18), it has been widely used to lower molecular weight of polymers synthesized with Ziegler–Natta catalysts.

As a terminating agent, molecular hydrogen is most attractive because it does not simultaneously kill reaction centers. It is a true transfer agent. Side reactions, however, can occur that will affect catalyst activity; these are described later in this chapter.

Natta (18) proposed that chain termination involves hydrogenolysis of the live metal–carbon bond centers, as shown in Eq. 10–6. By infrared

$$[cat]\text{–}Pn + H_2 \longrightarrow [cat]\text{–}H + PnH \qquad (10\text{–}6)$$

spectroscopy, they showed the presence of isopropyl end groups in the polymer when propylene polymerized. The active center (M–C bond) is regenerated (Eq. 10–7) when an olefin molecule adds to the M–H bond.

$$[cat]\text{–}H + CH_3CH=CH_2 \longrightarrow [cat]\text{—}CH_2CH_2CH_3 \qquad (10\text{–}7)$$

The patent literature strongly suggests that hydrogen is the preferred transfer agent for decreasing molecular weights of polyolefins. Without hydrogen, most catalyst processes would produce polyolefins that have too high a molecular weight for most practical applications (a few applications, however, do use very high molecular weight polyolefins). The hydrogen method is inexpensive, easy to use, and does not leave a residue in the polymer. It does not appear, however, to be the choice method when cobalt-based catalysts are used to polymerize dienes (see Section V).

Vandenberg (16) showed that hydrogen was effective for lowering molecular weights of polyethylene, polypropylene, and styrene using $TiCl_4$–containing catalysts (Al-i-Bu$_3$, AlEt$_2$Cl, and AlEt$_3$ were used as co-catalysts). Typically, to lower the reduced specific viscosity values in polyethylene from 9.5 to 2.7, about 17 psig H_2 (13 mole%) hydrogen was needed per 56 to 67 psig ethylene. The melting point of the formed polyethylene remained the same, 137° to 138°C. When propylene was polymerized, only several mole percent hydrogen was needed to lower the reduced specific viscosities from 5.1 to 1.2 dl/g.

Natta (18) investigated the influence of hydrogen on the molecular weight and on the rate of polymerization of ethylene and propylene with catalysts prepared from $AlEt_3$ and $\alpha TiCl_3$. He found that, under his experimental conditions, the decreased polymer molecular weight $\overline{PM}$ depended on the square root of the partial pressure of hydrogen and was independent of the ratio of partial pressure of monomer to hydrogen. The experimental curves were described by Eq. 10–8,

$$\overline{PM} = 1/(K_1 + K_2 \sqrt{P_{H_2}}) \qquad (10\text{–}8)$$

where $K_1 = 2.15 \times 10^{-6}$ and $K_2 = 15.35 \times 10^{-6}$ for the propylene polymerization and $K_1 = 1.538 \times 10^{-6}$ and $K_2 = 3.105 \times 10^{-6}$ for the ethylene polymerization.

TABLE 10–4

Variation of the Index of Isotacticity of Polypropylene as a Function of the Partial Pressure of the Hydrogen Present during the Polymerization[a] (18)

$P_{C_3H_6} = 4$ atm		$P_{C_3H_6} = 6$ atm	
P_{H_2} (atm)	Fraction insoluble in boiling heptane (%)	P_{H_2} (atm)	Fraction insoluble in boiling heptane (%)
0	81.3	0	80.2
0.2	79.7	0.1	79.4
0.4	78.8	0.4	78.6
0.7	76.5	0.7	75.2
1.0	76.5	1.0	75.0

[a] Conditions: Temp., 75°C; solvent: 250 cm^3 of n-heptane; 2.63 mmoles $AlEt_3$; 3.24 mmoles $TiCl_3$.

TABLE 10–5

Variation of the Standard Rate of Polymerization of Propylene as a Function of the Hydrogen Partial Pressure[a,b] (18)

P_{H_2}(atm)	0	0.2	0.4	0.5	0.7	0.8	1
g C_3H_6 polymerized/ g $TiCl_3$-hr ($= V$)	15	11	9.5	9	7.5	6.7	6

[a] The curve describing this data corresponds to the expression $V_H = V_0 - \sqrt{P_{H_2}}/0.11$.

[b] Conditions of polymerization: the hydrogen was introduced in every experiment after the standard rate of polymerization had been reached in the absence of hydrogen. Constant propylene pressure, 4 atm; temp., 75°C; solvent: 250 cm^3 of n-heptane; 5.26 mmoles $AlEt_3$; 1.94 mmoles $TiCl_3$.

Tables 10–4 and 10–5 show the variation in isotacticity and rate in a propylene polymerization as the partial pressure of hydrogen was increased from 0 to 1 atmosphere. Isotacticity decreased from 80.2 to 75%, and rate decreased from 15 to 6 (grams polymer per gram TiCl₃ per hour). Natta and co-workers indicated that, if the propylene–hydrogen mixture is removed and replaced by propylene at the same partial pressure, the rate of polymerization rapidly returns to the standard rate previously observed in the absence of hydrogen (Fig. 10–3). Hydrogen apparently does not irreversibly alter the catalyst.

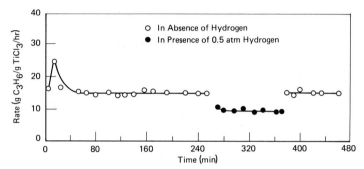

Fig. 10–3. Propylene polymerization velocity in the absence and in the presence of H₂ (18). Polymerization conditions: propylene pressure, 4 atm.; temp., 75°C; catalyst prepared from 5.26 millimoles of Al(C₂H₅)₃ and 1.94 millimoles of TiCl₃; 250 cm³ n-heptane.

According to Natta, the addition of olefin to the cat–H bond is slow and accounts for the decrease in the rate of polymerization.

Following publication of these reports, other investigators generally confirmed these findings, but new data also revealed some surprising results.

The mechanistic scheme proposed by Natta was supported by the findings of Hoffman, Fries, and Condit (19), who incorporated tritium as tracer for the hydrogen. Polymerizations of propylene and 4-methyl-1-pentene were made with the AlEt₂X–TiCl₃ catalyst (X = Cl or I). The polymers were fractionated and, for each fraction, a tritium count and intrinsic viscosity were determined. For 95% of the whole product, the data indicated two hydrogens (tritiums) per polymer molecule. This would be the case if only the cat–H centers continued to polymerize the olefin.

The tritium tracer method was used to estimate the relationship between intrinsic viscosity and the number average molecular weight via Eq. 10–9.

$$[\eta] = 1.94 \times 10^{-4} (\overline{M_n})^{0.81} \qquad (10\text{–}9)$$

Figure 10–4 shows this relationship.

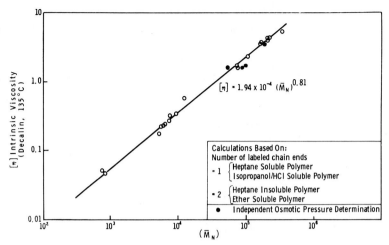

Fig. 10-4. Poly-4-methyl-1-pentene, $[\eta]$ vs. $\bar{M}_n$ (19).

Bourat and co-workers (20), on the other hand, found only one hydrogen (tritium) per polymer molecule when propylene was polymerized with the $AlEtCl_2$–$TiCl_3$ donor-type catalyst and a mixture of tritium–hydrogen gas as transfer agent. The explanation offered for this observation was that the cat–H center did not continue to polymerize propylene.

In contrast to Natta's earlier findings that hydrogen lowered the polymerization rate (when $\alpha TiCl_3$ was used), other workers, using γ- and $\delta TiCl_3 \cdot xAlCl_3$ crystalline modifications, found that hydrogen actually increased catalyst activity (21–24). The first suggestion was made in 1963 by Rayner (21), who commented that, if $TiCl_3$'s other than the α-type were used, higher rates were found. Okura and co-workers (22) found that hydrogen produced an increase in the rate of polymerization if toluene but not heptane was used. However, an increase took place only if a $TiCl_3$ containing $AlCl_3$ was used, such as $TiCl_3 \cdot xAlCl_3$ (Stauffer Company $TiCl_3AA$ was used). They suggested that new centers were formed on the surface of the $TiCl_3AA$ as a result of washing out $AlCl_3$ by toluene and of reducing fresh surface by $AlEt_2Cl$ with the aid of hydrogen.

The rate activating effect was also observed when 1-butene and 4-methyl-1-pentene monomers were used with the same catalysts.

Mason and Schaffhauser (23), also using an aluminum reduced $TiCl_3$ ($TiCl_3AA$), found a rate increase for 1-butene polymerization due to hydrogen, as shown in Table 10–6. An explanation was not offered.

Pijpers and Roest (24) reported that 4-methyl-1-pentene polymerized in the presence of the $AlEt_2Cl$–$\gamma TiCl_3$ catalyst (latter contains $AlCl_3$) at a

TABLE 10-6

Effect of Hydrogen on the Polymerization Rate of 1-Butene Using an $\alpha TiCl_3AA$ Catalyst[a] (23)

Hydrogen partial pressure (psi)	Reaction temp. (°C)	Cocatalyst	Al/Ti ratio	Rate, (g polymer/g catalyst/hr)
0	70	TIBAL[b]	2.0	111
5	70	TIBAL[b]	2.0	170
10	70	TIBAL[b]	2.0	200
15	70	TIBAL[b]	2.0	226
0	90	TIBAL[b]	2.0	94
2	90	TIBAL[b]	2.0	119
5	90	TIBAL[b]	2.0	155
10	90	TIBAL[b]	2.0	169
0	100	DEAC[c]	3.3	122
5	100	DEAC[c]	3.3	156
0	70	TEA[d]	3.5	66
5	70	TEA[d]	3.5	92

[a] All preparations were carried out for 5 hr in Esso Isopar (mixture of isooctanes) as the solvent. Reaction pressure was held constant at 40 psig.
[b] TIBAL: triisobutylaluminum.
[c] DEAC: diethylaluminum chloride.
[d] TEA: triethylaluminum.

higher rate if hydrogen was present. Also, when hydrogen was present, the catalyst lost its activity via a second-order reaction, while in the presence of nitrogen, a first-order reaction was observed. They explained the increase in the following way: in the absence of hydrogen, the polymer chain cannot migrate from the active center because it contains a double bond at its end that is capable of forming a π complex with the Ti center and thereby preventing the next growth step. It was assumed that the major termination in absence of hydrogen involved the reaction shown in Eq. 10–10.

$$Ti-CH_2-CH-Pn \longrightarrow TiH + CH_2=C-Pn$$

$$\begin{array}{cc} CH_2 & CH_2 \\ CHCH_3 & CHCH_3 \\ CH_3 & CH_3 \end{array} \qquad (10-10)$$

When termination by H_2 occurred, the polymer chain end was saturated and complexing could not occur. Migration of the polymer chain away from the center was stimulated by the exothermic reaction of H_2, which afforded translational energy. They concluded from their data that 2 and 15 polymer chains were formed per 5 Ti atoms when hydrogen was absent and present, respectively.

The introduction of hydrogen into the Ziegler–Natta system may cause other reactions to occur. For example, the aluminum alkyl can be partially hydrogenated to AlR_2H (18, 25); propylene has been reported to be hydrogenated, but if pyridine was present this side reaction was prevented (25a).

Schindler and Strong reported other reactions using $AlEt_2Cl–TiCl_4$ to polymerize ethylene in the presence of hydrogen and deuterium. They observed three kinds of deuterated groups in the polymer: CH_2D groups resulting from transfer reactions, as well as CHD and CD_2 groups arising from exchange reactions involving the β-hydrogen of the growing chain. These reactions are discussed in Chapter 11, Section II.

Hydrogen was also an effective transfer agent for the polymerization of butadiene with the $AlEt_2Cl–CoCl_2 \cdot$ pyridine catalyst without loss in conversion of cis-1,4 content (26). The amount of hydrogen required was fairly high, and was comparable to ethylene polymerizations, as shown in Table 10–7.

TABLE 10–7

Hydrogen as a Chain Transfer Agent[a] (26)

P_{H_2} (atm)	Conversion (%)	$[\eta]$	Content (cis-1,4) (%)
0	99	5.50	97.6
0.33	100	4.86	97.4
1.13	100	3.80	97.4
5	98	2.95	97.5
15	100	1.94	97.7
40	95	1.15	97

[a] Reaction conditions: 0.036 moles $AlEt_2Cl$; 0.0925 moles $CoCl_2$; 1,000 cm^3 benzene; 100 g butadiene; reaction temp., 15°C; time, 1 hr.

V. Chain Termination by Unsaturated Hydrocarbons

Unsaturated compounds have been found effective as chain terminators in a butadiene polymerization when cobalt-based catalysts were used.

Longiave and co-workers (26) reported that ethylene, propylene, 1-butene, allene, and 1,2-butadiene lowered molecular weight with the $AlEt_2Cl–CoCl_2 \cdot AlCl_3$ catalyst without a noticeable loss in cis-1,4 structure or catalyst activity. Higgins and Wilcoxen showed dimethylacetylene to be an effective transfer agent (27). Sometimes a crude feed containing some of these components can be advantageously used, such as butadiene containing 1-butene and *cis-* and *trans*-2-butene (28).

The mechanism by which ethylene acts as a transfer agent involves its insertion into the Co–π-allyl bond, as shown in Eq. 10–11.

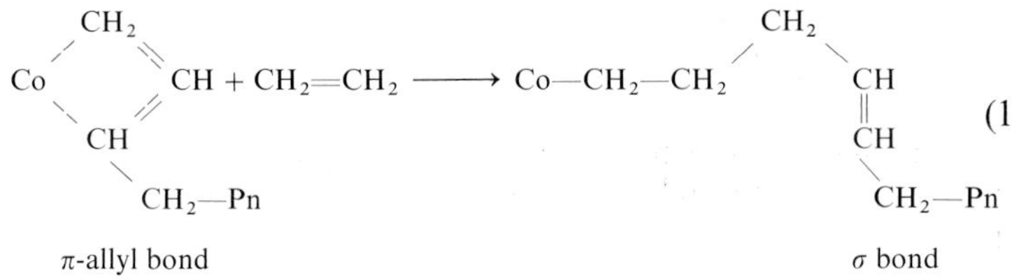

$$
\begin{array}{ccc}
& CH_2 & \\
& \diagup \parallel & \\
Co & \quad CH + CH_2{=}CH_2 \longrightarrow Co{-}CH_2{-}CH_2 \\
& \diagdown \diagup & \\
& CH & \\
& \diagdown & \\
& CH_2{-}Pn &
\end{array}
\qquad
\begin{array}{c}
CH_2 \\
\diagup \diagdown \\
\quad CH \\
\parallel \\
CH \\
\diagdown \\
CH_2{-}Pn
\end{array}
\qquad (10\text{--}11)
$$

π-allyl bond   σ bond

This was experimentally shown by radiolabeled ethylene (29). It is believed that the Co–C σ bond is unstable and disproportionates to Co–H and a detached polymer chain. When the last added group was butadiene, the polymer remains attached because π-allyl bond stabilization is possible. It is well established that ethylene does not polymerize to high molecular weight polymers with this particular catalyst.

Sakata and co-workers (30) examined the effect of a number of aliphatic hydrocarbons on the polymerization of butadiene to *cis*-1,4 polymer with a nickel naphthenate–borontrifluoride etherate–AlEt$_3$ catalyst. In agreement with the cobalt-based polymerizations, they found that acetylene, allene, and vinylcyclohexene acted as terminators but at the expense of the polymerization rate.

There is strong evidence that an olefin complexed at the transition metal center can cause termination of chain growth. Henrici–Olive and Olive showed that the decomposition of $(C_5H_5)_2TiEtCl \cdot AlEtCl_2$ [Ti(IV) to Ti(III)] reduction was enhanced if nonpolymerizable olefins were present (15b, p. 22). This finding is elaborated in Chapter 13 (see Fig. 13–23).

This type of mechanism may amount for the inability of the above catalyst to homopolymerize propylene and other α-olefins, as well as a tendency of α-olefins to act as terminators of chain growth when added to a polymerization of ethylene (13a). Henrici–Olive and Olive found that when the alkyl attached to the Ti was an octyl, the catalyst was more easily reduced. They attributed this to the lower dissociation energy of the C–H bond in CH$_2$ (R = octyl) compared to the C–H bond in CH$_3$ (R = Et).

The β-hydrogen would be a tertiary hydrogen in the polymerization of propylene (R = polymer chain). Since this would be more easily transferred, reduction would occur more readily and the polymer chain would not grow. Copolymerization would occur, because more reactive ethylene molecules would decrease the time an α-olefin would be bonded to the Ti as the last added unit.

One would expect, however, that termination of chain growth would increase with an increase in the concentration of the α-olefin. This was found by Wiman and Rubin (31), who reported intrinsic viscosities of 1.9 to 2.2 dl/g for polyethylene, whereas copolymers of ethylene and 1-butene had intrinsic viscosities of 0.3 to 0.5 dl/g.

VI. Chain Termination by Thermal Cleavage

Two chain termination reactions have been postulated to explain the formation of vinylidene and vinyl end groups (32), as shown in Eqs. 10–12 and 10–13.

$$M-CH_2-\underset{\underset{CH_3}{|}}{CH}-Pn \xrightarrow[\text{temp.}]{\text{low}} M-H + CH_2=\underset{\underset{CH_3}{|}}{C}-Pn \tag{10-12}$$

$$M-CH_2-\underset{\underset{CH_3}{|}}{CH}-Pn + CH_3CH=CH_2 \xrightarrow[\text{temp.}]{\text{high}} M-CH_2-CH=CH_2 + Pn-\underset{\underset{CH_3}{|}}{CH}-CH_3$$

$$M-CH_2-CH=CH_2 \xrightarrow[nC_3H_6]{\text{propylene}} M-CH_2-\underset{\underset{CH_3}{|}}{CH}\left(CH_2-\underset{\underset{CH_3}{|}}{CH}\right)_n CH_2-CH=CH_2 \tag{10-13}$$

Longi and co-workers (32) showed that the ratio between vinyl and vinylidene end groups increased from 1.0 at 100° to 2 at 200°C.

Spontaneous dissociation type of chain cleavage (Eq. 10–12) apparently is insignificant for Ti-based catalysts at low polymerization temperatures, that is, below 60° to 80°C. In fact, Natta, Bier, and their respective co-workers have shown that nearly all of the polymer chains remain attached to metal atoms at lower temperatures (8, 33). They concluded from their polymerization studies of propylene at 120°C with Al-i-Bu$_3$–TiCl$_3$ catalysts that the spontaneous dissociation reaction was the major transfer process and that the propylene played no part (see Chapter 4, Section II).

Polypropylene products having a high degree of unsaturation were claimed to have outstanding film forming and flow properties. These were prepared at 150°C with LiAlH$_4$–TiCl$_3$ catalysts (34).

The mechanistic path by which reaction 10–12 occurs involves a β-hydrogen abstraction, according to the proposals of Natta and co-workers, and this has generally been accepted by other workers for polymerizations with conventional Ziegler–Natta catalysts. This is shown in Eq. 10–14.

$$\underset{\underset{H}{\diagup}\,\underset{Pn}{\diagdown}}{M}\overset{\overset{CH_2}{\diagup}\,\overset{CH_3}{\diagdown}}{C} \longrightarrow M-H + \underset{\underset{C-CH_3}{\|}}{CH_2} \xrightarrow{CH_3CH=CH_2} M-CH_2CH_2CH_3 \tag{10-14}$$

Recently, a six-center bicyclic, highly polar transition state involving ethylene was suggested by Henrici–Olive and Olive (35) to account for the unusual activation parameters and rate law they found in their kinetic studies when ethylene was polymerized in the presence of $(EtO)_3Ti/AlEtCl_2$ catalyst. This is presented in Eq. 10–15.

$$
\begin{array}{ccc}
CH_2{=}CH_2 & \overset{-}{C}H_2\text{----}\overset{+}{C}H_2 & \overset{CH_2-CH_3}{\underset{|}{Ti}} \\
| & | \qquad | & \\
Ti{-}CH_2{-}CH_2{-}R \;\longrightarrow\; & Ti^+\text{----} H^- \;\longrightarrow\; & +\qquad (10\text{–}15)\\
& | \qquad | & CH_2{=}CH{-}R \\
& \underset{+}{C}H_2\text{----}\underset{}{C}H{-}R &
\end{array}
$$

VII. Chain Termination by Mechanical Forces

Kissin (36) presented evidence for his proposal that mechanical termination takes place during the stereospecific polymerization of propylene. Because mechanical termination is proportional to polymerization rate, it is not easily distinguishable from chain transfer through the monomer. Kissin suggested that the cleavage of the metal–carbon bond occurred during a temporary discontinuation of monomer addition at the center. This bond broke because of stretching of segments of the polymer chains in the crystallite bundles.

References

1. G. Moraglio, *Chim. Ind.* (*Milan*) **41**, 879 (1959); G. Natta, I. Pasquon, A. Zambelli, and G. Gatti, *J. Polym. Sci.* **51**, 387 (1961).
1a. The reactions of aluminum alkyls with organic and inorganic compounds are reviewed comprehensively by T. Mole and E. A. Jeffery, "Organo-aluminum Compounds," Chapters 12 and 13, Elsevier, Amsterdam, 1972.
2. A. G. Pozamantir, *Vysokomol. Soedin.* **2**, 1026 (1960); *Polym. Sci. USSR* (*Engl. Transl.*) **3**, 217 (1962).
3. A. G. Pozamantir, A. A. Korotkov, and I. S. Lishanskii, *Vysokomol. Soedin.* **3**, 1769 (1961).
4. H. Schnecko, K. Weirauch, M. Reinmoller, and W. Kern, *Makromol. Chem.* **77**, 159 (1964).
5. A. Schindler, *in* "Crystalline Olefin Polymers" (R. A. V. Raff and K. W. Doak, eds.), p. 147, Wiley (Interscience), New York, 1965.
6. H. Weber and K. Kiepert, *Makromol. Chem.* **70**, 54 (1965).
7. J. Boor, *Polym. Prepr., Am. Chem. Soc., Div. Polym. Chem.* **6**, No. 2, 690 (1965).
8. G. Natta and I. Pasquon, *Adv. Catal.* **11**, 1–65 (1959).
9. G. Natta, E. Giachetti, I. Pasquon, and G. Pajaro, *Chim. Ind.* (*Milan*) **42**, No. 10, 1091 (1960).
10. G. Natta, I. Pasquon, and L. Giuffre, *Chim. Ind.* (*Milan*) **43**, No. 8, 871 (1961).

11. J. Boor, *J. Polym. Sci., Part C* **1**, 237 (1963).
12. R. Vilim, *Chem. Prum.* **12**, 102 (1962).
13. A. P. Firsov, A. D. Ter-Gazaryan, N. M. Chirkov, *Vysokomol. Soedin.* **6**, 417 (1964); *Polym. Sci. USSR (Engl. Transl.)* **6**, 462 (1964).
13a. J. Boor, *Ind. Eng. Chem., Prod. Res. Dev.* **9**, 437–456 (1970).
14. E. A. Youngman, K. Nozaki, and J. Boor, U.S. Patent 3,084,148, April 2, 1963, Shell Oil Company.
15. E. Giachetti and F. Scalari, U.S. Patent 3,242,149, March 22, 1966, Montecatini.
15a. L. F. Borisova, E. A. Fushman, E. I. Vizen, and N. M. Chirkov, *Eur. Polym. J.* **9**, 953 (1973).
15b. G. Henrici-Olive and S. Olive, *Adv. Polym. Sci.* **6**, 421–472 (1969).
16. E. J. Vandenberg, U.S. Patent 3,051,690, August 28, 1962, Hercules Powder Company (appl. July 29, 1955); see also Belgian Patent 549,910.
17. B. Ettore and L. Luciano, Italian Patent 554,013, January 5, 1957, Montecatini (see also Great Britain Patent 584,794, Italian Patent 557,013, and Great Britain Patent 850,585).
18. G. Natta, *Chim. Ind. (Milan)* **41**, No. 6, 519 (1959).
19. A. S. Hoffman, B. A. Fries, and P. C. Condit, *J. Polym. Sci., Part C* **4**, 109 (1963).
20. G. Bourat, J. Ferrier, and A. Perez, *J. Polym. Sci., Part C* **4**, 103 (1963).
21. L. S. Rayner, see comment made on page 125 of Hoffman *et al.* (19) discussion period following lecture by Hoffman and co-workers.
22. I. Okura, K. Soga, A. Kojima, and T. Keii, *J. Polym. Sci., Part A-1* **8**, 2717 (1970).
23. C. D. Mason and R. J. Schaffhauser, *J. Polym. Sci., Part B* **9**, 661 (1971).
24. E. M. J. Pijpers and B. C. Roest, *Eur. Polym. J.* **8**, 1151 (1972).
25. K. S. Minsker, V. I. Biryukov, A. I. Graevskii, and G. A. Razuvaev, *Izv. Akad. Nauk SSSR, Otd. Khim. Nauk* p. 637 (1963).
25a. O. N. Pirigov and N. M. Chirkov, *Polym. Sci. USSR (Engl. Transl.)* **8**, 1985 (1966).
26. C. Longiave, R. Castelli, and M. Ferraris, *Chim. Ind. (Milan)* **44**, No. 7, 725 (1962).
27. T. L. Higgins and C. H. Wilcoxen, Jr., U.S. Patent 3,068,217, December 11, 1962, Shell Oil Company.
28. E. A. Youngman, U.S. Patent 3,066,128, November 27, 1962, Shell Oil Company.
29. M. Dubini, C. Longiave, and R. Castelli, *Chim. Ind. (Milan)* **45**, 923 (1963).
30. R. Sakata, J. Honoso, A. Onishi, and K. Ueda, *Makromol. Chem.* **139**, 73 (1970).
31. R. E. Wiman and I. D. Rubin, *Makromol. Chem.* **94**, 160 (1966).
32. P. Longi, G. Mazzanti, A. Roggero, and A. M. Lachi, *Makromol. Chem.* **61**, 63 (1963).
33. G. Bier, *Makromol. Chem.* **70**, 44 (1964).
34. H. J. Hagenmayer and M. B. Edwards, U.S. Patent 3,412,078, November 19, 1968, Eastman Kodak Company.
35. G. Henrici-Olive and S. Olive, *Polym. Prepr., Am. Chem. Soc., Div. Polym. Chem.* **15**, 368 (1974); *J. Polym. Sci., Polym. Lett. Ed.* **12**, 39 (1974).
36. Y. Kissin, *J. Polym. Sci. USSR (Engl. Transl.)* **11**, 1779 (1970).

11

Oxidation State of Catalysts and Active Centers

I. Introduction

While the measured oxidation state probably often describes the oxidation state of individual active centers, conclusive proof is lacking. Indirect evidence suggests that, for many of the important transition metals, for example Ti, V, and Cr, more than one oxidation state leads to an active catalyst. The ligand environment of the active center plays a dominant role in deciding which oxidative state is active for a particular monomer.

Because the oxidation state of the transition metal of the active center significantly affects the structure of that center, much work has been done to establish its value for different Ziegler–Natta catalysts. It is, however, important to distinguish between the measured average oxidation state of the whole catalyst and the oxidation state or states of the individual centers. The first can usually be determined easily, but a direct assignment of oxidation state of the active centers is more difficult. First of all, the fraction of the total transition metal atoms that are active centers is small, about 1% or less. Suggestions that traces of the transition metal in an unmeasured oxidation state actually form the active centers in some catalysts cannot be lightly dismissed.

Yet, evidence is accumulating that the measured oxidation state reflects the oxidation state of active centers in many of the investigated Ziegler–Natta catalysts. There is justification, then, to devoting a chapter on the various experimental efforts made by different workers to establish the oxidation state of the active centers and to describe the consequences of this data.

Selected examples are presented to show the range of oxidation states that have been found for some of the more important or more widely investigated catalysts. The findings are given according to the transition metal. An attempt is made in the last section to reach some conclusion about the meaning of the experimental findings.

II. Catalysts Containing Selected Period 2 Transition Metal Salts

This section discusses many of the transition metals of period 2 catalysts whose oxidation state has been investigated.

A. SCANDIUM

Only the Sc(III) oxidation state has been reported to be active. Arlman (1) suggested that Sc in the $AlEt_3$–$ScCl_3$ catalyst, when used for propylene polymerization, was exclusively trivalent. He based his conclusion on the observation that no discoloration of white $ScCl_3$ crystals occurred during the polymerization and that no Sc(II) ions were detected by ESR spectroscopy.

B. TITANIUM

Active oxidation states for Ti(II), Ti(III), and Ti(IV) have been reported for Ziegler–Natta catalysts. No evidence has been presented that lower oxidation states are active.

1. Ti(IV) Centers

The first example of an active Ti(IV) center was suggested by Breslow and Newburg (2) for the catalyst $(C_5H_5)_2TiCl_2$–$AlEt_2Cl$ in ethylene polymerizations. An active bimetallic complex having the structure shown in Eq. 11–1 was visualized.

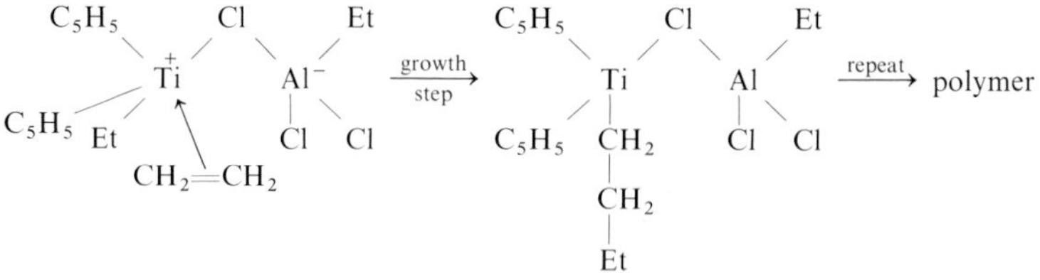

$$\begin{array}{c}
C_5H_5\quad\quad Cl\quad\quad Et \\
\diagdown\diagup\diagdown\diagup \\
\overset{+}{Ti}\quad\quad Al^- \\
C_5H_5\ Et\ \diagup\quad\diagdown\ \\
\quad\quad\uparrow\quad\quad Cl\ \ Cl \\
CH_2{=}CH_2
\end{array}
\xrightarrow[\text{step}]{\text{growth}}
\begin{array}{c}
C_5H_5\quad\quad Cl\quad\quad Et \\
\diagdown\diagup\diagdown\diagup \\
Ti\quad\quad Al \\
\diagup\ \diagdown\quad\diagdown \\
C_5H_5\ \ CH_2\quad Cl\ \ Cl \\
\quad\quad|\quad \\
\quad\quad CH_2 \\
\quad\quad|\quad \\
\quad\quad Et
\end{array}
\xrightarrow{\text{repeat}} \text{polymer}
\qquad (11\text{–}1)$$

Tetravalent Ti site at
which ethylene was complexed

Their evidence for Ti(IV) centers was twofold: (1) the polymerization of ethylene decreased as the fraction of Ti(IV) decreased, and (2) molecular oxygen activated the catalyst, presumably by oxidizing inactive Ti(III) centers to active Ti(IV) centers.

More recently, Henrici–Olive and Olive (3) combined a kinetic study of the polymerization of ethylene with ESR and magnetic susceptibility measurements to relate catalyst activity, the configuration of the active complex, and the oxidation state of Ti in the soluble catalysts $(C_5H_5)_2TiCl_2$–$AlEtCl_2$ and $(C_5H_5)_2TiEtCl$–$AlEtCl_2$. From this extensive study, they concluded active centers were Ti(IV) because the polymerization rate decreased to the same extent as Ti(IV) was reduced to Ti(III) (Fig. 11–1). The proposed structure of their catalyst is shown below, where Cp = cyclopentadienyl = C_5H_5.

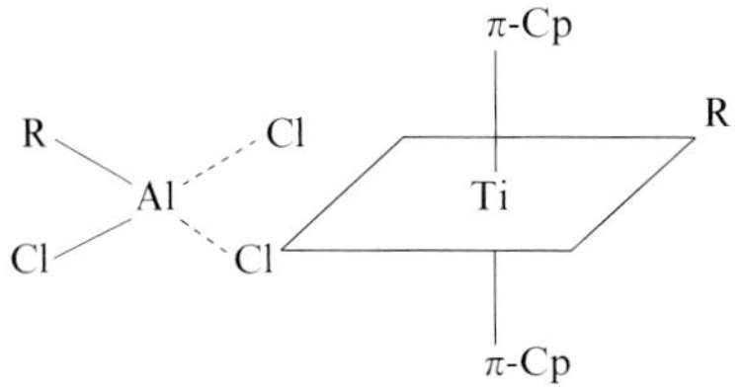

I

In contrast to the above model, a Ti(IV) bimetallic complex with a bridging ethyl group instead of a chlorine atom was suggested by Reichert and Schubert (4).

The most extensive work, aimed at showing that both trivalent and tetravalent titanium centers are active in the $AlEt_3$–$TiCl_4$ and $AlEt_2Cl$–$TiCl_4$ catalysts when used for polymerizing ethylene, was reported by Schindler (5–11). Schindler used deuterium gas (D_2) as a probe to differentiate types of centers in these catalysts. The polyethylene that was formed under these conditions had CH_2D, CHD, and CD_2 groups present according to infrared spectroscopic measurements. The CH_2D group was readily explained by a transfer reaction, as shown in Eq. 11–2. The CHD and CD_2

$$\text{cat–}CH_2CH_2CH_2R + D_2 \longrightarrow \text{cat–}D + DCH_2\text{–}CH_2CH_2R \qquad (11\text{–}2)$$

groups were explained by an exchange reaction involving D_2 and the labile β-hydrogen, as shown in Eq. 11–3. Because he did not observe this exchange

$$\text{cat–}CH_2CH_2CH_2R + D_2 \longrightarrow \text{cat–}CH_2CHDCH_2R + HD \qquad (11\text{–}3)$$

reaction with catalysts that were free of Ti(IV), such as $AlEt_2Cl + TiCl_3$, Schindler concluded that the active center that produced these results contained Ti(IV). This assessment was based on the earlier findings of Bestian and Clauss that the β-hydrogen of a polyethylene chain which is attached to a Ti(IV) center is very labile (12). A double exchange would give CD_2.

Using this diagnostic probe, Schindler examined in detail several catalysts prepared at different Al/Ti ratios for the relative contents of Ti(III) and Ti(IV) centers (Table 11–1).

Also, he described a method by which the M_n and M_w values of two polyethylene polymers formed on Ti(III) and Ti(IV) centers were determined (Table 11–2). This method involved the evaluation of CH_3 or CHD group distributions in fractionated polymers prepared in the absence or presence of deuterium, respectively. In addition to Ti(III) and Ti(IV) centers, inactive Ti–H centers were proposed.

TABLE 11–1

The Effect of Catalyst Ratio on the Distribution of Titanium Valence States in Different Catalyst Systems[a] (10)

	Catalyst system			
Al/Ti ratio	$Et_3Al/TiCl_4$	$Et_3Al/TiCl_3$	$Et_2AlCl/TiCl_4$	$Et_2AlCl/TiCl_3$
Low	IV + III	III	IV + III	III
Medium	III	—	—	—
High	III + II	III + II	III	III

[a] Roman numerals refer to the particular oxidation state of Ti.

TABLE 11–2

Molecular Weights[a] Calculated from the Evaluation of Fractionation Data from Polymer Growth on Ti(III) and Ti(IV) Sites (11)

Polymer number	Wt% polymer derived from Ti(IV) sites	Total polymer		Polymer from Ti(III) sites		Polymer from Ti(IV) sites	
		M_w	M_n	M_w	M_n	M_w	M_n
P–I	40.1	137.3	14.1	210.7	41.8	29.4	6.9
P–II	38.9	144.6	13.3	207.2	39.3	45.6	6.5
P–III	45.4	112.8	9.8	173.9	32.3	39.1	5.3
P–IV	39.9	79.6	10.8	90.5	17.7	63.4	6.7
P–V	38.4	54.3	9.0	64.2	17.3	39.1	5.1
P–VI	40.9	33.5	5.9	37.2	8.7	27.7	4.0
P–VII	16.0	239.9	18.2	281.1	45.3	22.2	4.4
P–VIII	60.1	82.1	8.8	130.3	30.7	50.6	6.0
P–IX	81.8	182.8	18.6	323.7	96.6	150.1	15.6
P–X	79.0	127.2	11.2	259.0	57.3	92.0	9.2

[a] Molecular weights in units of 1,000.

Beerman and Bestian (13) proposed that an alkylated tetravalent titanium in the presence of some solid $TiCl_3$ was the active catalyst when ethylene was polymerized with partially decomposed CH_3TiCl_3. The latter was not active unless some of the CH_3TiCl_3 decomposed to form solid $TiCl_3$. The activity of the catalyst was enhanced by the addition of controlled amounts of molecular oxygen.

Some workers have argued that the oxygen oxidizes inactive lower valent state sites to active tetravalent sites. Others have proposed that the activating effect of oxygen is actually due to the greater ability of tetravalent titaniums to be realkylated in contrast to lower valent sites (14). According to the latter view, inactive tetravalent alkylated centers are subsequently reduced to active lower valent alkylated sites; for example, $R_2TiCl_2 \rightarrow RTiCl_2 + R \cdot$ radical.

A soluble quadrivalent alkylated titanium species, Ti(IV)–Rn which could be adsorbed and/or chemisorbed on the surface of the $TiCl_3$ particle, was also reported by Overberger and co-workers (15, 16) who used deuterated styrene and propylene monomers in their investigation. They proposed that highly isotactic polymers of high molecular weight were formed at the Ti(III) sites, while amorphous polymer of lower molecular weight was formed by the adsorbed Ti(IV) centers. The heterogeneous $AlEt_3$–$TiCl_4$ catalyst was used.

A tetravalent titanium was suggested by Roha and co-workers (17) for the soluble catalyst formed by mixing Al-i-Bu_2O-i-Bu, $TiCl_4$, and a Lewis acid ($AlCl_3$). This catalyst was used for polymerizing ethylene. Uelzmann (18) was able to polymerize ethylene but not propylene at $-78°C$ with a dark red soluble catalyst formed by mixing $TiCl_4$ and Al-i-Bu_3. It contained 91% of the Ti in the tetravalent oxidation state.

Smolyan and co-workers (19) reported a different finding. When $TiCl_4$ was mixed with $AlEt_2Br$, $AlEt_2Cl$, or $AlEt_2OEt$, a yellow solution formed that did not contain tetravalent titanium. This solution was inactive for the polymerization of ethylene even if an electrophilic agent such as $AlEtCl_2$ was also added. However, the catalyst became active as soon as it partially precipitated, with formation of about 2 to 5% trivalent titanium. The activity of this catalyst was also increased when oxygen was added.

Further support for tetravalent centers has come more recently from the extensive magnetic (susceptibility and ESR) measurements, as well as ethylene polymerization kinetics by Henrici–Olive and Olive (3). The plot R_p vs. Ti(IV) concentration, shown in Fig. 11–1, clearly shows that a close dependence exists. They also concluded that the two Ti(III) complexes that they observed by ESR spectroscopy originated from two or more Ti(IV)–Al alkyl complexes, which first formed when the catalyst components were mixed, for example, $(C_5H_5)_2TiCl_2 + AlEtCl_2$.

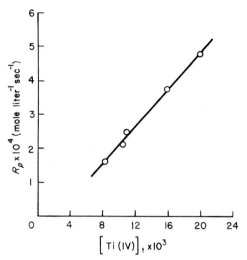

Fig. 11–1. Initial rate of polymerization of ethylene, R_p, as a function of actual concentration of Ti(IV) (3).

2. Ti(III) Centers

The journal and patent literature suggests that titanium trichlorides have been most widely used in the polymerizations of olefins. Typical salts have been described in Chapter 4, Section II,B. They can be simple structures such as αTiCl_3 or βTiCl_3 or they can be contaminated with AlCl_3 and aluminum alkyls. Preferred metal alkyls have been AlEt_3 (or $\text{Al-}i\text{-Bu}_3$), AlEt_2Cl, and AlEtCl_2.

If the titanium trichloride and the aluminum alkyl are mixed and used for polymerization under mild conditions, the oxidation state of the Ti remains III. Favoring such conditions are low mixing and polymerization temperatures, low concentrations of metal alkyl, and use of a metal alkyl that has a low reducing capacity, such as AlEt_2Cl or AlEtCl_2. Partial reduction to divalent state occurs when the catalyst components are mixed and used for polymerization at higher temperatures, as well as when the metal alkyl is used at a high concentration or if a strongly reducing alkyl is used, including AlEt_3 or ZnEt_2.

An investigation of the $\text{ZnEt}_2\text{–}\gamma\text{TiCl}_3 \cdot x\text{AlCl}_3$ catalyst showed it can contain only trivalent or a mixture of divalent and trivalent titaniums (20). The Ti(II) content varied between 0 and 47%, depending on experimental conditions. No direct relationship, however, was found between the Ti(II) content and the activity of the catalyst or molecular weight or isotacticity for this system (Table 11–3).

TABLE 11–3

Ti(II) Content of $ZnEt_2$–$\gamma TiCl_3$ after the Heating and Polymerization of Propylene (20)

Components[a]				Conversion, g polymer	Ti(II) (mole %)[b]
$\gamma TiCl_3$	$ZnEt_2$	Propylene	Modifier		
+[c]	–	–	–	–	0
+[c]	–	–	Et_3N +	–	0
+[c]	+	–	–	–	4.9 ± 0.5
+[d]	+	+	–	2.4 ± 0.3	5.5 ± 1.0
+[c]	+	–	Et_3N +	–	27.5 ± 2.5
+[d]	+	+	Et_3N +	9.8 ± 0.8	14 ± 1.2
+[d]	+	+	Azulene +	2.5	0

[a] + denotes presence and – denotes absence of component during reaction or polymerization.

[b] Analysis for Ti(II) and polymerizations carried out in high vacuum line. Ti(II) content established from reaction products of $2\,Ti^{2+} + D_2O \rightarrow 2\,Ti^{3+} + 2\,(OD)^- + D_2$. D_2 and other gaseous products identified by mass spectroscopy.

[c] Components heated for 4 hours at 50°C in absence of propylene.

[d] Propylene polymerized at 1 atm for 4 hours at 50°C.

In the above catalyst, the oxidation state of the starting $TiCl_3$ was III. A number of workers concluded from their investigations that the active Ti centers were also trivalent if the oxidation state of the starting Ti salt was IV. These are described next.

Adema (21) and Bartelink and co-workers (22) showed by ESR spectroscopy that for $TiCl_4$–$AlEtCl_2$, $TiCl_3CH_3$–$AlEtCl_2$, and $TiCl_4$–$AlEt_2Cl$, the concentration of unpaired electrons was equal to the concentration of Ti(III) in solution. When ethylene was polymerized with the $TiCl_4$–$AlEt_2Cl$ (or $AlEtCl_2$) catalyst, the rate of polymerization was directly related to the concentration of the unpaired electrons. A soluble complex was suggested that had the structure, as shown in II.

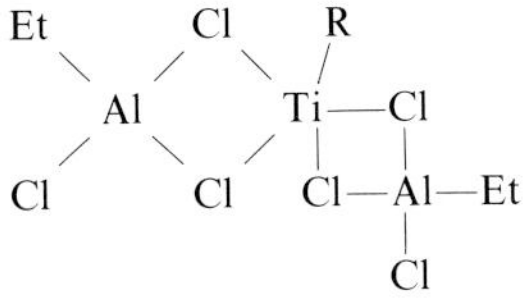

II

Adema suggested that a similar complex was formed on the surface of the $TiCl_3$ crystal that also was active for polymerization of ethylene.

Moyer (23) showed that replacement of two chlorines in $TiCl_4$ by O-i-Bu or O-i-Pr groups resulted in an increase in rate of reduction to Ti(III) by $AlEt_2Cl$. Ti(II) was not detected. Some Ti(III) species remained in colloidal solution even after reduction was complete and the filtrates were deeply covered. After 18 hours, all of the Ti(III) precipitated, and the filtrates became colorless. Reduction was accompanied by exchange of alkoxy groups from Ti to Al.

Moyer considered the reduction of the mixed titanates to occur by the path similar to that of $TiCl_4$, as shown in Eq. 11–4. Substitution of

$$(R_2AlCl)_2 + 2\ TiCl_4 \rightleftharpoons (RAlCl_2)_2 + 2\ RTiCl_3$$

$$2\ RTiCl_3 \longrightarrow 2\ TiCl_3 + 2\ R \cdot$$

$$(11–4)$$

Cl or OR would make $RTi(OR)_3$ more stable and reduction would be hindered. Acceleration of reduction was favored because Cl in $Ti(OR)Cl_3$ or $Ti(OR)_2Cl_2$ has a higher partial negative charge than Cl in $TiCl_4$, and the above equilibrium is shifted to the right.

No divalent Ti was detected when $TiCl_4$ was combined with $AlEt_2Br$ (24), $AlEtCl_2$ (25, 26), or $AlEt_2Cl$ (27). These catalysts were active for polymerization of ethylene and propylene. Watt, Fry, and Pobiner (28) concluded that the same catalyst intermediate was formed when Ti(III) acetylacetonate and oxobisacetylacetotitanate (IV) were reacted with $AlEt_2Cl$. Relatively high ratios (Al/Ti > 20) were required to produce active catalysts for polymerizing ethylene. Precipitates formed in the catalyst mixture, but dilute catalyst solutions in which no precipitate was visible showed activity. Kollar and co-workers (29) concluded that maximum activity for the $AlEt_3$–$TiCl_4$ catalyst, when used to polymerize ethylene, occurred when titanium was trivalent.

Takeda and co-workers (30) reported a maximum activity for polymerization of styrene was obtained at an Al/Ti ratio of 1.5 for the catalyst $AlEt_3$–$Ti(O$-n-$Bu)_4$. Active Ti(III) centers were proposed. Further reduction of Ti to a valency of less than III produced inactive catalysts. From their IR and ESR data, they concluded two types of active centers.

$Ti(O$-n-$Bu)_2Et$	$Ti(O$-n-$Bu)_2Et \cdot AlEt_2O$-n-Bu
Al/Ti < 0.5	Al/Ti > 1.2

The measured oxidation state may be affected by other parameters. Henrici–Olive and Olive (31) increased the reduction of Ti(IV) to Ti(III) by the addition of an olefin to the soluble $(C_5H_5)_2TiClEt$–$AlEtCl_2$ catalyst. In contrast, the presence of propylene in the heterogeneous $ZnEt_2$–$\gamma TiCl_3 \cdot xAlCl_3$ catalyst lowered the reduction of Ti(III) to Ti(II) from 27.5 to 14 mole% (see Table 11–3).

The method of analysis may be important. A very interesting observation was reported by Czajlik and co-workers (32), who found that the amount of Ti(III) present when $TiCl_4$ was reduced with $AlEt_2Cl$ was dependent on the temperature at which the catalyst was decomposed (hydrolyzed or killed with alcohol). They suggested that one or more intermediate products were decomposed differently at high and low temperatures.

3. Ti(II) Centers

Ludlum and co-workers (33) reported that maximum catalytic activity was obtained for ethylene polymerization with the $Al-i-Bu_3–TiCl_4$ catalyst when the average valence of the catalyst was II. A divalent active center, RTiCl, was proposed.

C. VANADIUM

Active oxidation states for V(II), V(III), but not V(IV) have been reported for vanadium-based catalysts.

1. V(III) Centers

Junghanns and co-workers (34) found that V(III) was the active species in the catalyst $Al_2Et_3Cl_3–VOCl_3$ when used to copolymerize ethylene and propylene at 20°C. Reduction to V(II) species resulted in lower copolymerization rates.

Gumboldt and co-workers (35) observed that addition of hexachlorocyclopentadiene to the $Al_2Et_3Cl_3–VOCl_3$ catalyst increased its activity for copolymerization of ethylene and propylene. They suggested that the active vanadium centers were trivalent and that the hexachlorocyclopentadiene oxidized inactive bivalent vanadium centers to the active V(III) centers. The latter became active only after realkylation with the metal alkyl.

Taking advantage of ESR to detect VCl_2, Zambelli and co-workers (36) made a comparative study of the reductive powers of different metal alkyls toward VCl_3 and of the polymerization activity of the corresponding isotactic-specific catalysts. Different orders were obtained for reduction and for polymerization.

For reduction: $LiBu > CdEt_2 > ZnEt_2 > BeEt_2 > AlEt_3 >$
$AlEt_2Cl > AlEt_2I$
For polymerization: $AlEt_3 > BeEt_2 > GaEt_3 > AlEt_2I > AlEt_2Cl >$
$CdEt_2 > LiBu > ZnEt_2$

Cooper, Smith, and Stokes (36a) found that a direct relationship between activity and degree of reduction does not exist for the polymerization of isoprene with the $AlEt_3$–VCl_3 catalyst in the presence of electron donors. They observed that the addition of isopropyl ether and other donors greatly increased the reduction of VCl_3 to VCl_2. Assuming that the active centers were $RVCl_2$ species, they qualitatively associated the extent of reduction with the number of centers. Not all donors that facilitated reduction, however, resulted in higher catalyst activities. This failure was attributed to the preferential complexing of the active center with the donor instead of with the isoprene being polymerized.

No divalent vanadium was detected in the catalyst formed by combining $AlEt_2I$ and VCl_3 (37). But mixtures of $AlEt_2Cl$ and $VOCl_3$, when used to copolymerize ethylene and propylene, contained an average oxidation state between II and III (38).

Natta and co-workers (39) did not find divalent vanadium in the soluble $AlEt_2Cl$–VCl_4 catalyst by ESR spectroscopy under conditions when it was used to synthesize syndiotactic polypropylene. Since a syndiotactic-specific catalyst could also be formed by starting with a trivalent vanadium salt such as $V(acac)_3$, they concluded the vanadium in the active site was trivalent. When both types of catalysts were heated above $-50°C$, reduction to divalent vanadium occurred, and activity for producing highly syndiotactic polypropylene was lowered and eventually lost (39a).

Lehr and Carman (40, 41) studied the active oxidation state of vanadium in the AlR_2Cl–VCl_4 catalyst under conditions which were used to polymerize ethylene and propylene to syndiotactic polymer (Tables 11–4 and 11–6). VCl_4 was rapidly reduced with a variety of AlR_2Cl alkyls at $-78°C$ (Table 11–4). After 5 minutes, less than 10% of the tetravalent vanadium was left, and V(II) was formed in small amounts ($\leq 15\%$) only after 16 hours.

TABLE 11–4

Reduction of VCl_4 by R_2AlCl at $-78°C^a$ (40, 41)

R	Al/V	Time	V(IV) (%)	V(III) (%)	V(II) (%)
Et	5	5 min	5.1	94.9	—
Et	5	5 min	1.3	98.7	—
i-Bu	5	2.5 min	7.6	92.4	—
i-Bu	5	2.5 min	8.5	91.5	—
i-Bu	5	2.5 min	3.1	96.9	—
Et	4	17.4 hr	—	90.0	10.0
Et	4	16.0 hr	—	85.5	14.5
i-Bu	4	17.4 hr	—	98.4	1.6
i-Bu	4	16.0 hr	—	99.8	0.2

^{a} In toluene. $[VCl_4] = 0.019\ M$.

TABLE 11–5

Reduction of V(III) by R_3Al in Toluene at $-78°C^a$ (40, 41)

R	R_3Al/V	Time (min)	V(III) (%)	V(II) (%)
Et	0.5	2.5	98.9	1.1
Et	0.5	120	96.9	3.1
i-Bu	0.5	2.6	95.5	4.5
i-Bu	0.5	120	93.0	7.0
i-Bu	1.0	2.5	84.7	15.3
i-Bu	1.0	240	72.7	27.3
i-Bu	2.0	2.6	31.6	68.4
i-Bu	2.0	120	19.4	80.6

^{a} V(III) prepared by reaction of VCl_4 with i-Bu_2AlCl (Al/V = 4) at $-78°C$. $[VCl_4] = 0.019$ M.

TABLE 11–6

Relative Rates of Polyethylene Formation at $-78°C$ with the V(III) and V(II) Catalyst Systems^a (40, 41)

Amount (g-atom %)			Reaction time (min)	Amount	
V(IV)	V(III)	V(II)		g/min/mole of V/l	g/min/mole of V(III)/l
3.9	96.1	—	6.8	4.3	4.5
2.1	97.9	—	6.8	2.5	2.6
4.1	95.9	—	9.0	2.8	3.0
4.4	95.6	—	9.0	2.1	2.2
				2.9^c	3.1^c
4.5	95.5	—	2.6	7.2	7.5^b
4.4	95.6	—	2.6	7.2	7.5^b
—	10.2	89.8	60	0.053	0.53
—	9.5	90.5	44	0.039	0.40
—	8.0	92.0	60	0.065	0.81
—	8.4	91.6	60	0.065	0.78
				0.055^c	0.63^c
—	5.2	94.8	12.5	0.017	0.32^b
—	3.1	96.9	60	0.022	0.70^b

^{a} $[VCl_4] = 0.011$ M in toluene.
^{b} Ethylene introduced approximately seven times faster.
^{c} Average.

In contrast, when AlR_3 was used instead of AlR_2Cl, reduction was more rapid and a large amount of V(II) was formed, for example, 80.6% in 2 hours when $AlR_3 = Al\text{-}i\text{-}Bu_3$ and $Al/V = 2.0$ (Table 11–5).

From data based on visible and ESR spectroscopy, they concluded that the active oxidation state of vanadium, a V(III) species, came from an inactive V(III) species, rather than from a V(IV) precursor. The following scheme was suggested for the formation of active centers.

$$R_2AlCl + VCl_4 \xrightarrow[\text{fast}]{k_1} RVCl_3 + RAlCl_2$$

$$4\,R_2AlCl + 2\,RVCl_3 \xrightarrow[\text{fast}]{k_2} 2\,VCl_3 \cdot (R_2AlCl)_2 + 2\,R$$

$$VCl_3 \cdot (R_2AlCl)_2 \xrightarrow[\text{slow}]{k_3} RVCl_2 \cdot (R_{1.5}AlCl_{1.5})_2 \qquad (11\text{--}5)$$

$$RVCl_2 \cdot (R_{1.5}AlCl_{1.5}) + nM \xrightarrow[\text{slow}]{k_p} RM_nVCl_2 \cdot (R_{1.5}AlCl_{1.5})_2$$

$$2\,RM_nVCl_2 \cdot (R_{1.5}AlCl_{1.5})_2 \xrightarrow[\text{slow}]{k_t} 2\,[RM_n] + 2\,VCl_2 \cdot (R_{1.5}AlCl_{1.5})_2$$

Because V(IV) in a solution of VCl_4 is difficult to detect by ESR at temperatures above $9°K$, Lehr and Carman (41) quenched the test sample with ethanol in order to convert V(IV) species to VO^{2+}. The latter shows a characteristic 8-line spectrum.

2. V(II) Centers

VCl_2 and VBr_2 can be made to dissolve in cyclohexane by complex formation with $AlBr_3$. These complexes become active for polymerization of ethylene if an aluminum alkyl is added, thus demonstrating that V(II) centers can be active. In other experiments, Carrick (42–44) showed that a soluble vanadium catalyst containing the components VCl_4, $AlBr_3$, and $AlR_3[\text{or }(C_6H_5)_4Sn]$ was active for the polymerization of ethylene. By means of polarographic analysis, they showed the vanadium of this catalyst to be exclusively divalent. The reduction of VCl_4 to the V(II) species occurred in less than a minute. Only when some V(II) centers were formed did these workers observe activity in these catalysts.

These workers proposed a V(II) aluminum alkyl complex to be the active center, as shown in structure III, where $R' = $ alkyl or Pn (the growing

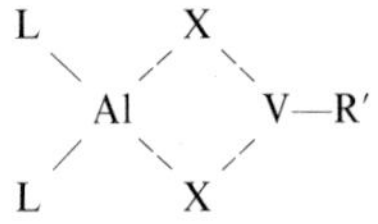

III

polymer chain); $L = $ halogen or alkyl. When $(C_6H_5)_4Sn$ was used in place of AlR_3, alkylation of $AlBr_3$ occurred, producing $C_6H_5AlBr_2$, as demonstrated directly by van der Kerk and confirmed by showing that VCl_4–$AlEtCl_2$–$AlBr_3$ produced a similar catalyst.

There is no published evidence that V(0) or V(I) centers are active.

TABLE 11–7

Fractions of V(III) and V(II) after the Reaction of VO(acac)$_2$ with (C$_2$H$_5$)$_2$AlCl (31)

Al/V	Unpaired electrons	$\chi_{mol} \times 10^6$,[a] e.m.u.	V(III) (%)	V(II) (%)
20	2.3	4110	75	25
50	2.7	5360	33	67
100	2.95	6150	6	94
300	2.95	6150	6	94

[a] Paramagnetic susceptibility/mole in electromagnetic units.

Henrici–Olive and Olive (31) found that the soluble system VO(acac)$_2$–AlEt$_2$Cl–triethylphosphite contained both V(III) and V(II) centers and that the relative fraction of the V(II) center increased as the Al/V ratio increased (see Table 11–7).

The polymerization rate had a maximum at an Al/V ratio of 50 (Fig. 11–2). These workers concluded that the V(III) center was the more active species. To explain the low rate at Al/Ti = 20, they proposed the active V(III) species had to be complexed with aluminum alkyl molecules and that at low Al/Ti ratios, the concentration of AlEt$_2$Cl was too low to displace the equilibrium in favor of the complex. From the ESR data, they concluded that acac ligands did not form part of the reduced V(II) complex. The following structure was proposed.

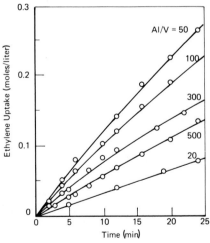

Fig. 11–2. Polymerization of ethylene with VO(acac)$_2$/(C$_2$H$_5$)$_2$AlCl in the presence of (C$_6$H$_5$)$_3$P, at different ratios Al/V (numbers next to curves). [V] = 1 × 10^{-4} moles/liter; P/V = 6; toluene, 20°C, ethylene pressure = 70 atm (31).

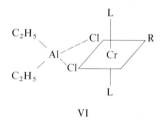

IV   V

In contrast, if $AlEt_3$ had been used, a V(0) complex would be formed.

D. CHROMIUM

Active catalysts have been reported in which the oxidation state of the starting Cr salt was III and II, but only very little work was done to establish the oxidation state of the active catalyst.

Combining magnetic susceptibility, ESR, and kinetic measurements, Henrici–Olive and Olive (31) concluded that the active Cr centers in the $AlEt_2Cl–Cr(acac)_3$ catalyst were divalent and had the structure shown in VI.

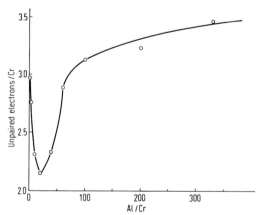

VI

Fig. 11–3. Average number of unpaired electrons per chromium center in the system $Cr(acac)_3/(C_2H_5)_2AlCl/(C_2H_5O)_3P$; $P/Cr = 6$; temp. $= 20°C$ (31).

At Al/Ti ratios <200, triethylphosphite had to be added to prevent precipitation, but above 200, the binary catalyst remained homogeneous. Figure 11–3 shows that as the Al/Cr increases from 0 to 20, the number of unpaired electrons decreases from 3 to 2; but at higher ratios (about 70), it again reaches a value of 3 and tends asymptotically to 4 at an Al/Cr ratio of 400.

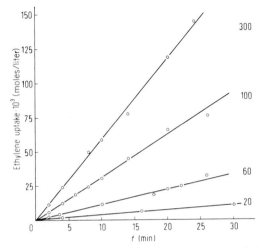

Fig. 11–4. Polymerization of ethylene with $Cr(acac)_3/(C_2H_5)_2AlCl$ in the presence of $(C_2H_5O)_3P$ at different ratios Al/Cr (numbers next to curves) (31). $[Cr] = 2 \times 10^{-4}$ moles/liter; $P/Cr = 6$; toluene, 20°C; ethylene pressure = 700 atm (31).

This corresponds to the formation of low spin complexes when Al/Cr ≤ 20 and high spin complexes at higher ratios (Fig. 11–3). They proposed that at Al/Cr < 20, reduction of Cr(III) to Cr(II) occurred (low spin complex), but as Al/Cr is increased, acac ligands were replaced by weaker ligands, presumably chlorine. Because chlorine ligands stabilize the Cr(II) complex, no further reduction occurs at very high Al/Cr ratios. Only the high spin Cr(II) complex was found highly active, whereas the low spin complex (when Al/Cr $\simeq 20$) had relatively a lower activity (Fig. 11–4). This strongly suggests that acac must be replaced by Cl and C_2H_5 ligands in order for Cr(II) to be highly active. The $CrCl_2$–$AlEt_2Cl$ system was also active for polymerization of propylene to isotactic polymer.

Boor (45) showed that $CrCl_2$ was less active than $CrCl_3$ for polymerization of ethylene when combined with $AlEt_3$ but, when combined with $AlEt_2Cl$, it was slightly more active (Table 11–8). Polarographic measurements of $CrCl_2$ samples before and after ball-milling showed some Cr(III) was present.

TABLE 11–8

Polymerization of Ethylene with $CrCl_2-$ and $CrCl_3$–Al Alkyl Catalysts (45)

Chromium chloride	Al alkyl	Time (hr)	Polyethylene formed (g)
$CrCl_3$	$AlEt_3$	2	>28
$CrCl_3$	$AlEt_2Cl$	4	1.2
$CrCl_2$	$AlEt_3$	2	4.2
$CrCl_2$	$AlEt_2Cl$	4	2.6

The obtained rate enhancement, however, was in line with the observed increase in surface area of the sample rather than its Cr(III) content. The yield of polymer increased about 17 times while surface area increased about 20 times and, concomitantly, the Cr(III) content decreased from 0.73 to 0.36 wt%.

E. COBALT

As described in Chapter 14, Natta, Porri, and Carbonaro (46) concluded that the role of $AlEt_2Cl$ in the cobalt-based catalyst was to reduce Co(II) to Co(I). The active center was a Co(II)–π-allyl bond formed by an oxidative addition of butadiene to the Co(I) species.

III. Conclusion

The experimental findings cited in this chapter suggest the following: (1) active centers in a Ziegler–Natta catalyst can be exclusively in one oxidation state. More frequently, the catalyst can have the active centers in two or three oxidation states. In the case of titanium-based catalysts, active centers in the di-, tri-, and tetravalent oxidation state were reported; (2) the relative activity of a transition metal center in a specific oxidation state is influenced strongly by the ligands attached to it.

Carrick and co-workers related catalyst activity and oxidation state with the suggestion that two opposing forces come into play when the oxidation state of the transition metal was decreased (43, 44). On one hand, as its oxidation state is lowered, the electronegativity of the transition metal center is decreased. This makes the metal–carbon bond more polarized and thereby more able to insert the complexed olefin. But with the progressive filling of the transition metal orbitals with electrons as its oxidation state is lowered, the metal loses its acceptor characteristics, and its ability to complex an olefin molecule simultaneously decreases.

One of the most important factors is stabilization of the active center by ligands. In heterogeneous systems, the ligands are the anions in the crystal structure, such as Cl in $TiCl_3$. In soluble catalysts, ligand stabilization may arise not only from the groups attached to the transition metal but also from the metal alkyl that is complexed to the transition metal center (see structure VI). Some workers hold the latter to be true for heterogeneous catalysts as well. Because exchange reactions between the metal alkyl and the transition metal salt can occur, the ligands that are finally attached to the transition metal center can be a mixture of both.

The ligand causes a change in the electronic structure of the center that is coordinated to the olefin being polymerized. This effect is exerted through the σ- and π-electron systems. The ligand is able to alter the activation of the coordinated olefin and the activity of the transition metal–carbon bond center to insert the olefin. Thus, by varying the steric and electronic nature of the ligands, it has been possible to alter the activity of the center. While tailoring catalysts at will from elementary beginnings is not yet possible, one can begin to understand the experimental observations related to ligand changes in Ziegler–Natta catalysts (see also Chapter 13, Section IV,B).

References

1. E. J. Arlman, *J. Polym. Sci.* **62**, S30 (1962).
2. D. S. Breslow and N. R. Newburg, *J. Am. Chem. Soc.* **79**, 5073 (1957); **81**, 81 (1959).
3. G. Henrici-Olive and S. Olive, *J. Polym. Sci., Part C* **22**, 965 (1967).
4. K. H. Reichert and E. Schubert, *Makromol. Chem.* **123**, 58 (1969).
5. A. Schindler, *J. Polym. Sci., Part B* **3**, 147 and 793 (1965).
6. A. Schindler, *Makromol. Chem.* **90**, 284 (1966).
7. A. Schindler, *J. Polym. Sci., Part B* **4**, 193 (1966).
8. A. Schindler and R. B. Strong, *Makromol. Chem.* **93**, 145 (1966).
9. A. Schindler, *Makromol. Chem.* **102**, 263 (1967).
10. A. Schindler, *Makromol. Chem.* **118**, 1 (1968).
11. A. Schindler, *Makromol. Chem.* **114**, 77 (1968).
12. H. Bestian and K. Clauss, *Angew. Chem.* **75**, 1068 (1963).
13. C. Beerman and H. Bestian, *Angew. Chem.* **71**, 618 (1959).
14. G. W. Phillips and W. L. Carrick, *J. Am. Chem. Soc.* **84**, 920 (1962); *J. Polym. Sci.* **59**, 401 (1962).
15. C. G. Overberger, P. A. Jarivistsky, and H. Mukamal, *J. Polym. Sci., Part A-1* **5**, 2487 (1967).
16. A. Simon, P. A. Jarivitsky, and C. G. Overberger, *J. Polym. Sci., Part A-1* **4**, 2513 (1966).
17. M. Roha, L. C. Kreider, M. R. Frederick, and W. L. Beears, *J. Polym. Sci.* **38**, 51 (1959).
18. H. Uelzmann, *J. Polym. Sci.* **37**, 561 (1959).
19. Z. S. Smolyan, A. I. Graevskii, O. I. Demin, V. K. Fukin, and G. N. Matveeva, *Vysokomol. Soedin.* **3**, 81 (1961); *Polym. Sci. USSR (Engl. Transl.)* **3**, 18 (1962).
20. J. Boor and G. A. Short, *J. Polym. Sci., Part A-1* **9**, 235 (1971).
21. E. H. Adema, *J. Polym. Sci., Part C* **16**, 3643 (1968).

22. H. J. M. Bartelink, H. Bos, J. Smidt, C. H. Vrinssen, and E. H. Adema, *Recl. Trav. Chim. Pays-Bas* **81**, 225 (1962).
23. P. H. Moyer, *J. Polym. Sci., Part A* **3**, 199 (1965).
24. R. Havinga and Y. Y. Tan, *Recl. Trav. Chim. Pays-Bas* **79**, 56 (1960).
25. A. Malatesta, *Can. J. Chem.* **37**, 1176 (1959).
26. J. Stedefeder, Dissertation, Technische Hochschule, Aachen (1957).
27. A. Schindler, *in* "Crystalline Olefin Polymers" (R. A. V. Raff and K. W. Doak, eds.), Part 1, p. 147. Wiley (Interscience), New York, 1964.
28. W. R. Watt, F. H. Fry, and H. Pobiner, *J. Polym. Sci., Part A-1* **6**, 2703 (1968).
29. L. Kollar, A. Simon, and J. Osvath, *J. Polym. Sci., Part A-1* **6**, 919 (1968).
30. M. Takeda, K. Iimura, Y. Nozawa, M. Hisatome, and N. Koide, *J. Polym. Sci., Part C* **23**, 741 (1968).
31. G. Henrici-Olive and S. Olive, *Angew. Chem., Int. Ed. Engl.* **10**, 776 (1971).
32. I. Czajlik, J. Varadi, and A. Baan, *J. Polym. Sci., Part B* **4**, 661 (1966).
33. D. B. Ludlum, A. W. Anderson, and C. E. Ashby, *J. Am. Chem. Soc.* **80**, 1380 (1958).
34. E. Junghanns, A. Gumboldt, and G. Bier, *Makromol. Chem.* **58**, 18 (1962).
35. A. Gumboldt, J. Helberg, and G. Schleitzer, *Makromol. Chem.* **101**, 229 (1967).
36. A. Zambelli, I. Pasquon, A. Marinangeli, G. Lanzi, and E. R. Mognaschi, *Chim. Ind. (Milan)* **46**, 1464 (1964).
36a. W. Cooper, R. K. Smith, and A. Stokes, *J. Polym. Sci., Part B* **4**, 309 (1966).
37. G. Natta, G. Mazzanti, D. De Luca, U. Giannini, and F. Bandini, *Makromol. Chem.* **76**, 54 (1964).
38. J. Obloj, M. Uhniat, and M. Nowakowshka, *Vysokomol. Soedin.* **7**, 939 (1965); *Polym. Sci. USSR (Engl. Transl.)* **7**, 1040 (1965).
39. G. Natta, A. Zambelli, G. Lanzi, I. Pasquon, E. R. Mognaschi, A. L. Segré, and P. Centola, *Makromol. Chem.* **81**, 161 (1965).
39a. I. Collamati and G. Sartori, *Chim. Ind. (Milan)* **47**, No. 4, 368 (1965).
40. M. H. Lehr, *Macromolecules* **1**, 178 (1968).
41. M. H. Lehr and C. J. Carman, *Macromolecules* **2**, 217 (1969).
42. W. L. Carrick, *J. Am. Chem. Soc.* **80**, 6455 (1958).
43. W. L. Carrick, A. G. Chasar, and J. J. Smith, *J. Am. Chem. Soc.* **82**, 5319 (1960).
44. G. L. Karapinka, J. J. Smith, and W. L. Carrick, *J. Polym. Sci.* **50**, 143 (1961).
45. J. Boor, *J. Polym. Sci., Part A-1* **9**, 3075 (1971).
46. G. Natta, L. Porri, and A. Carbonaro, *Atti Accad. Naz. Lincei, Cl. Sci. Fis., Mat. Natl., Rend.* [8] **29**, 491 (1960).

12

Metal Alkyl-Free Catalysts

I. Introduction

Olefins and dienes have been polymerized with catalysts that contain a transition metal salt but not a metal alkyl of group I, II, or III metals. While many transition metal salts are active without modification, others require the presence of an organic or inorganic component or, in some cases, a specific physical modification. To differentiate these catalysts from Ziegler–Natta ones and to stress that they do not contain added base metal alkyl, they have been designated as metal alkyl-free catalysts (MAF).

Many of the MAF catalysts are isotactic-specific for the polymerization of α-olefins and can polymerize dienes to polymers that contain prevailingly one of the possible microstructures: *cis*-1,4, *trans*-1,4, or 1,2 units. While MAF and Ziegler–Natta catalysts share many common features, they are not equivalent in all respects, and this points to the importance of the metal alkyl as an alkylating agent, a site former, a reducing agent, a site stabilizer, a poison scavenger, and a transfer agent. It is generally believed that in both catalysts, the active center is a transition metal–carbon bond, and similar stereochemical mechanisms take place.

This entire chapter is devoted to metal alkyl-free (MAF) catalysts, which are catalysts that consist of a transition metal salt, alone or in combination with a second component.

Why should such a detailed account be made of these catalysts in a book about Ziegler–Natta catalysts? The author considers it most appropriate since he shares the view of many workers that the very existence of metal alkyl-free catalysts for polymerization of olefins and dienes is the most cogent evidence we have for the view that active centers in Ziegler–Natta catalysts are transition metal–carbon bonds. A study of common and dissimilar features of MAF and Ziegler–Natta catalysts offers an opportunity to clarify the role which the two components of the latter catalyst play in the stereospecific polymerization.

The chapter begins with a description of the Phillips and Standard of Indiana catalysts for synthesis of high density polyethylene (Section II). These are treated separately from the other MAF catalysts to emphasize their commercial importance and the fact that they were contemporary to the Ziegler–Natta discoveries. Section III describes the many metal alkyl-free catalysts that were discovered after 1955.

II. Phillips and Standard of Indiana Catalysts

A. THE PHILLIPS CATALYST

This catalyst has been investigated extensively world-wide, especially in the Phillips' laboratory where it was discovered by Hogan and Bank in the early 1950's (1).

It is ironical that the discovery of this catalyst followed a course that was very similar to that which preceded Ziegler's discovery (1a). The Phillips workers, while trying to make liquid fuels from ethylene in the presence of NiO supported on $SiO_2 \cdot Al_2O_3$, were plagued with excess formation of 1-butene. By replacing NiO with CrO_3, however, all of the ethylene was consumed in the synthesis of high density polyethylene.

The Phillips Company has licensed many other manufacturers to produce high density polyethylene with this catalyst, and today more than 60% of the world's production of high density polyethylene is made with this catalyst (2, 2a). The remainder is made by Ziegler–Natta and Standard of Indiana catalysts. The catalyst is used exclusively to make polyethylene. Several extensive review papers have been published (2b–7). Rather than compile a long list of references, these review papers are cited for leading references. The following comprise only a glimpse of this very important catalyst.

The most widely investigated Phillips catalyst was prepared by impregnating a silica–alumina (87:13 composition) or a silica support with an aqueous solution of CrO_3 (8). High surface supports were used, about 400 to 600 m^2/g (5, 8). Other supports such as alumina, zirconia, and thoria were also effective but showed no advantages. After the water was removed, the powdery catalyst was fluidized and activated by a stream of dry air at temperatures of 400° to 800°C to remove the bound water. The impregnated catalyst contained 1 to 5 wt% chromium oxides. When this catalyst was heated in the presence of carbon monoxide, a more active catalyst was obtained (5). It was stored under dry air or an inert gas because it is poisoned by polar compounds.

The Phillips catalyst is specific for polymerization of ethylene to high density polyethylene. Higher α-olefins, notably propylene, 1-butene, 1-pentene, and 1-hexene, are polymerized to branched high polymers ranging from semisolids to viscous liquids (8). When branching is closer than the 4-position (3-methyl-1-butene), dimers and trimers are formed.

Usually, for many applications, users of polyethylene want a less crystalline material than can be made with the Phillips catalyst. This problem was easily solved by the synthesis of copolymers of controlled amounts of an α-olefin, usually several percent, such as 1-butene.

The polymerization of ethylene can be done in a solution, slurry, or vapor phase process (Chapter 7 describes comparable Ziegler–Natta processes). The early patent literature suggested that a solution process was most favored, such as cyclohexane solvent at 125° to 175°C and 20 to 30 atm ethylene pressure. The low concentration of polymer that could be tolerated was offset by the high polymerization rates, and thus commercial production of polyethylene was feasible. Catalyst support residues were removed by filtering or centrifuging the reaction solution.

When the slurry process was used, a solvent such as 1-pentene was preferred, and the reaction was held below 110°C to prevent solution of the polymer. An advantage of the slurry process is that a high concentration of polyethylene ($\geq 30\%$) can be formed in the reactor. The catalyst support particle becomes scattered throughout the polymer particle during polymerization (5, 8). Because the polymerization can be carried to high yields of polymer per gram of catalyst, the amount of catalyst remaining in polymer is small and need not be removed (5, 8). Claims have been made that from 5,000 to 50,000 pounds of polymer are made with a pound of catalyst (8a).

Both types of processes are operable in laboratory and commercial plant scales. Some attention has been given to vapor phase processes, but its commercial application is not known to the author (2b).

In addition to developing highly active catalysts, considerable work was done to understand how this catalyst works. A few salient features are described in the following paragraphs.

1. Stabilization of the Chromium Oxide by the High Surface Support (3, 8)

The support is not merely an inert diluent. Chromium oxide by itself is stable only as Cr_2O_3 at the activating temperature, which is about 500°C. A mixture of CrO_3 and activated silica–alumina is inactive for polymerizing ethylene but becomes increasingly active as the temperature is increased from 196° to 400°C (CrO_3 melts at 196°C). Interaction of chromium oxide with both SiO_2 and Al_2O_3 was proposed on the basis of experimental data, and, in the presence of each support of the mixture, the chromium is stabilized at the Cr(VI). A chromium oxide–silica alumina catalyst (containing 2.49 wt% Cr) contains 96% Cr(VI) when activated at 540°C with dry air (8). Hogan calculated that for a silica support having 600 m^2/g and about 5% Cr as Cr(VI), the average distance between adjacent Cr(VI) atoms was 10 Å. This corresponded to the accepted population of silanol groups on this silica after calcination. On the basis of this, he proposed CrO_3 could be stabilized, as shown in reactions 12–1 and 12–2. He concluded that the

$$
\begin{array}{l}
\overset{\displaystyle O\quad\ \ O}{\underset{\displaystyle O\quad\ \ O}{\diagdown\ \diagup}}\\
\text{Cr}\\
\end{array}
$$

$$
\underset{|}{\overset{|}{\underset{\text{Si}}{\overset{\text{OH}}{|}}}}\text{—O—}\underset{|}{\overset{|}{\underset{\text{Si}}{\overset{\text{OH}}{|}}}} + \text{CrO}_3 \longrightarrow \underset{|}{\overset{|}{\underset{\text{Si}}{\overset{O}{|}}}}\text{—O—}\underset{|}{\overset{|}{\underset{\text{Si}}{\overset{O}{|}}}} + \text{H}_2\text{O} \qquad (12-1)
$$

$$
\text{O=Cr—O—Cr=O}
$$

$$
\underset{|}{\overset{|}{\underset{\text{Si}}{\overset{\text{OH}}{|}}}}\text{—O—}\underset{|}{\overset{|}{\underset{\text{Si}}{\overset{\text{OH}}{|}}}} + 2\text{CrO}_3 \longrightarrow \underset{|}{\overset{|}{\underset{\text{Si}}{\overset{O}{|}}}}\text{—O—}\underset{|}{\overset{|}{\underset{\text{Si}}{\overset{O}{|}}}} + \text{H}_2\text{O} \qquad (12-2)
$$

chromate structure, rather than the dichromate, was preferentially formed on the basis of the water formed per available CrO_3.

2. Only a Small Number of Total Chromium Atoms Were Active Centers

It had been calculated that between 0.1 and 0.4 wt% of the total chromium form active centers (5). Using various independent methods, values ranging from 1.7×10^{-5} to 3.0×10^{-5} moles per gram of catalyst were established for the number of active centers (about 1 wt% Cr catalyst) (8–10). Hogan calculated that 2.8 polymer molecules were produced per second per active site (8). Several thousand polymer chains were produced in a typical run at the same site. Ayscough, Eden, and Steiner (9a), however, established a value of 2.3×10^{-7} mole sites per gram of catalyst (2.3 wt% Cr and 0.1 wt% of total Cr are active centers).

3. The Catalyst Contains Cr in Several Valence States

Valences of II, III, IV, V, and VI have been established. But because of the small number of total chromium that are active centers, it has not been possible to unequivocally assign the active valence or valences (12, 13). Some workers (11) suggested that the catalyst must have Cr in two valence states to be active, and it must contain at least some Cr(VI). Clark and Hogan describe the problem of valence state and list leading references (6). Recently,

Krauss (9b) concluded that the reduction of hexavalent chromium centers (which were linked to support) produced coordinately unsaturated Cr(II) surface compounds. Infrared spectra of 1:1 CO complexes indicated no back donation of electrons, thus supporting the oxidation state of II. Mass spectra of the low polymer–Cr products suggested the presence of a σ Cr–C bond; stable $[Cr(CH_2)_n^+]$ fragments were found.

4. The Cr Exists in Tetrahedral and Octahedral Coordination

The tetrahedral configuration was favored in high silica supports that are usually used for Phillips catalysts. Both configurations occur in silica–alumina mixtures. To explain the aromatization of acetylene to benzene, Hogan suggested that the Cr center has three open coordination positions (8).

5. Cr–C Bonds Are Active Centers

The presence of metal–carbon bonds was established by radiolabeling. Also in agreement is the observation that hydrogen acts as a transfer agent for the Phillips catalyst. Polymer growth involves insertion of olefin molecules into Cr–C bonds to form long chains.

Some of the CrO_3 on the surface existed as bulk CrO_3. Hogan found that, as the loadings of CrO_3 on the support increased, there simultaneously was an increase in the formation of bulk CrO_3 and a decrease in Cr(VI) centers. For example, at 10.1 wt% Cr he measured 55% Cr(VI), while at 2.49 wt% Cr he obtained 90% Cr(VI). The polymerization efficiency of the Cr decreased with increased loadings of CrO_3 on the silica support (8).

6. The Origin of Cr–C Bond Centers Has Not Been Established

Hogan speculated that the hydride end group might arise from the "refuse" of the oxidation–reduction step, which occurs when ethylene is first admitted and reacts at the Cr center. The oxygenated products of the reaction are displaced by excess ethylene molecules and polymerization begins. But Hogan appears to doubt this, on the basis of observing that the CO treated catalyst is even faster in initiation (presumably less H^- would be generated in the latter case upon addition of ethylene). Pecherskaya and co-workers (14) suggested that initiation consisted of a reaction between Cr and ethylene whereby hydrogen from ethylene is transferred to an oxygen attached to Cr and the ethylene becomes σ-bonded to the Cr atom (14), according to structure I.

$$
\begin{array}{c}
\text{OH} \quad CH=CH_2 \\
\diagdown \; | \; \diagup \\
\text{Cr} \\
\diagup \; | \; \diagdown \\
O \quad O \quad O \\
\end{array}
$$

I

7. Termination Involves Cleavage of Cr–Pn Bond (8)

An active center produces several thousand polymer chains. This means that when chain growth is terminated, the center remains active to initiate the growth of a new chain. The predominant form of termination is believed to involve a transfer of hydrogen to the Cr center with the simultaneous formation of an unsaturated end group, as shown in Eqs. 12–3 and 12–4,

$$Cr—CH_2—CH_2—Pn \longrightarrow Cr—H + CH_2{=}CH—Pn \qquad (12\text{--}3)$$

$$Cr—CH_2—\underset{\underset{H}{|}}{\overset{\overset{CH_3}{|}}{C}}—Pn \longrightarrow CrH + CH_2{=}C\overset{\diagup CH_3}{\diagdown Pn} \qquad (12\text{--}4)$$

when the last added olefin unit was ethylene (Eq. 12–3) and propylene (Eq. 12–4), respectively (in the latter comonomer is present). Pn again represents the growing polymer chain. Termination is increased with increasing polymerization temperatures. Molecular hydrogen and α-olefins act as chain terminators and can be used to lower the molecular weights of the polyethylene product.

B. STANDARD OF INDIANA CATALYST

Unlike the Phillips catalyst, which uses chromium as the transition metal, the Standard of Indiana catalysts can have any wide range of compositions (7). The early catalysts consisted of Ni–charcoal and Co–charcoal mixtures (15). Later, reduced molybdenum oxide on alumina was disclosed, and when promotors were added, higher activities were obtained for polymerization of ethylene. Alkali or alkaline earth metals or their compounds were used as promotors, including NaH, BaH_2, CaH_2, $LiAlH_4$, and $LiBH_4$. Group V and VI transition metal oxides on alumina were preferred as catalysts. The molybdenum oxide–alumina catalyst was calcined at about 500°C and then reduced in the presence of CO or H_2 at about 450°C. When promotors were used, they were added next. An active catalyst was obtained without reduction by H_2 or CO if promotors were used. Reduction converted molybdenum partly to valences below VI.

The Ni–charcoal catalyst cited above was made by heating charcoal that was impregnated with $NiNO_3$ ($2\ NiNO_3 \rightarrow Ni_2O_3$) at 260°C (15). The activation was done in the presence of hydrogen. The polyethylene made by this catalyst had a low molecular weight (specific viscosity 0.1 to 0.2 dl/g), but the crystallinity was high (d = 0.95 g/cm³ and crystallinity = 80%).

More recently, Tadokoro (16) modified the procedure to produce higher molecular weight products (viscosity value was about 1 dl/g).

III. Metal Alkyl-Free Catalysts Discovered after 1955

Metal alkyl-free (MAF) catalysts that were discovered after 1955 are now described. It is important to recognize that not every transition metal salt is active. If one leaves out the base metal alkyl from a typical Ziegler–Natta catalyst, only rarely is that transition metal salt active. Only specific transition metal salts have been found active and often only subsequent to a special chemical or physical modification. Nevertheless, the absolute number of different types of MAF catalysts is large and increasing each year.

The general features of MAF catalysts are first described, and then specific examples of catalysts according to a common characteristic are presented.

A. GENERAL FEATURES AND SCOPE

This section focuses on the general features and scope of metal alkyl-free catalysts, especially in relation to: (1) the active transition metal, (2) effective ligands, (3) active oxidation states, (4) physical state of catalyst, (5) the importance of chemical or physical modification of transition metal salt, and (6) the different monomers that have been polymerized.

1. Transition Metals that Form Catalysts

Catalysts that are active for monoolefins or diolefins have been made from many transition metals, especially Ti, V, Cr, Co, Ni, Zr, Nb, Mo, W, Pd, Rh, and Ru. In general, monoolefins are more favorably polymerized with transition metal salts on the left side of the periodical table, while dienes prefer transition metal salts on the right side. There are exceptions. For example, ethylene has been polymerized with a bis(1,3-cyclooctadiene)nickel(0) catalyst to a highly linear polyethylene, while butadiene has been polymerized to a high 1,2-polybutadiene in the presence of tris(π-allyl)chromium. A similar relationship was shown between these monomers and the Ziegler–Natta catalysts (see Chapters 4 and 5).

2. Types of Ligands

Active catalysts have been formed where the ligand was a halogen or a group linked to the metal by a heteroatom or a hydrocarbon group, which was attached to the metal by σ- or π-allyl type bonds. Sometimes the active

TABLE 12–1

Selected Examples of Active Salts with Different Ligands[a]

Ligand	Salt	Monomers
Cl	$TiCl_2$	Ethylene, propylene
Br	$(\pi\text{-allyl})_3ZrBr$	Butadiene
I	$(\pi\text{-allyl})_2CoI$	Butadiene
ϕ_3SiO	$(\phi_3SiO)_2CrO_2$	Ethylene
CN	$Pd(CN)_2$	Propylene
Oxide	NiO–charcoal	Ethylene
π-Allyl	$(\pi\text{-allyl})\,Cr$	Ethylene, butadiene
π-Crotyl	π-Crotyl–Ni–Cl	Butadiene
π-Methallyl	π-Methallyl–Ni–Cl	Butadiene
2,6,10-Dodecatriene	π-2,6,10-Dodecatriene NiCl	Butadiene
π-Cyclopentadienyl(Cp)	Cp_2Cr	Ethylene
Benzene (C_6H_6)	$(C_6H_6)_2Cr$	Ethylene
Benzyl (ϕCH_2-)	$(\phi CH_2)_4Ti$	Ethylene, propylene
1,5-Cyclooctadiene (COD)	$(COD)_2Ni$	Ethylene

[a] The microstructure of the polybutadienes was shown to be especially sensitive to the structure of the ligands attached to the transition metal center.

transition metal bore both ligands. Table 12–1 collects selected examples of these catalysts, which are described in more detail in Section III,B.

3. Oxidation State

Active catalysts have been prepared in which the initial oxidation state of the transition metal was solely 0, I, II, III, IV, V, or VI, or a mixture of two oxidation states. Because in most of these catalysts (more so than in Ziegler–Natta catalysts) the number of active centers is small, it has not been possible to directly identify the oxidation state of the active center. Obviously, when coreactants and monomer are added, the possibility of a change in the oxidation state is great, so more direct measurements are necessary.

The following examples (which will be elaborated upon in Section III,B) illustrate an example of each oxidation state. A 0-valent nickel salt was used to polymerize butadiene, that is, bis(1,3-cyclooctadiene) nickel plus an organic acid, CF_3CO_2H. $NiCl_2$ containing a small amount of monovalent nickel was reported by Anderson (17) to be active for the polymerization of butadiene to a high cis-1,4 product. Divalent titanium chlorides were reported by Werber, Benning, and co-workers (18) to polymerize ethylene and propylene, the latter to a partially isotactic polymer. Trivalent titanium chlorides, combined with amines, polymerize propylene to highly isotactic polypropylene, $TiCl_3HA$ or $TiCl_3AA$ plus $n\text{-Bu}_3N$. Tetravalent benzyl and π-allyl derivatives of Ti and Zr were found active for polymerization of ethylene, propylene, and butadiene. The Cr in the bis(triphenylsilyl) chromate-supported catalyst is hexavalent.

4. Physical State of Catalyst

Active soluble and heterogeneous MAF catalysts have been disclosed for olefin and diolefin monomers. Several examples of effective soluble catalysts have already been cited, namely, the π-allyl and benzyl derivatives of Ti, Zr, Cr, and Ni. Two types of heterogeneous catalysts have been identified: (1) a compound salt such as $TiCl_2$ or $TiCl_3$, sometimes alloyed with $AlCl_3$, and (2) a supported catalyst such as chromocene on SiO_2.

5. Chemical and Physical Modification

While many transition metal salts were active in their unmodified form, others required the presence of a second component to make them active. Often the second component increased the activity of the transition metal salt or modified its stereochemical ability. An example of the first is $CoCl_2$, which requires the presence of an $AlCl_3$ to make it active for the polymerization of butadiene to high cis-1,4 polymer in benzene solvent. An example of the second is the addition of Friedel–Crafts acids to π-allyl nickel compounds.

Sometimes, a physical modification was essential to affect or increase activity, such as ball-milling of transition metals in absence and presence of hydrogen.

6. Monomers Polymerized

Most workers selected ethylene, propylene, or butadiene as the monomer in their studies of metal alkyl-free catalysts, and, consequently, most of our information about these catalysts is based on these monomers. Sometimes the reported metal alkyl-free catalysts were shown active only for one of these or another particular monomer, and this remains one of the mysteries of these catalysts (such mysteries also prevail with Ziegler–Natta catalysts). Yet, a few examples of polymerizations of other monomers have been reported, including acetylene, cyclobutene, cyclopentene, 1,3-cyclohexadiene, 1-butene, 4-methyl-1-pentene, and isoprene.

B. SPECIFIC EXAMPLES OF MAF CATALYSTS

Specific examples of MAF catalysts are now presented. They are classified according to a dominating structural feature or the mode of preparation. For many of these catalysts, proposals have been made for the origin of the active metal–carbon bond and the structure of the active center. These will be presented at the same time that each catalyst is described.

1. Catalysts Containing Elemental Transition Metal or Metal Hydride with Chemical and Mechanical Modification

Matlack and Breslow (19) reported the polymerization of ethylene in a vibratory ball-mill at low pressures and temperatures with the following catalysts: (1) a transition metal alone, such as V, Nb, and Ta (but they also found active La and Ce of the rare earth series and Th and U from actinide series), (2) a mixture of metal plus alkyl halide, such as Ti + ethyl bromide, and (3) certain divalent transition metal hydrides, such as TiH_2, ZrH_4, NbH_5, TaH, reaction products of Ti + H_2 or V + H_2. High melting (melting point about 130°C) and high molecular weight (reduced specific viscosities $\simeq 1$ dl/g) polyethylenes were formed.

Propylene was polymerized with some of these catalysts to low yields of partially isotactic polymer (melting point $\simeq 160°$ to 163°C for the heptane-insoluble residues). These products had high molecular weights. Some examples of catalysts that were found active included: (1) Ti or V + ethyl bromide, (2) Ti + $TiCl_3$ and V + VCl_3, and (3) Ti + HCl or I_2.

Matlack and Breslow proposed that alkylated species were formed *in situ*, and these were the active centers. For example, using the known formation of anion radical when styrene is polymerized with alkali metal (such as Na^+ $\cdot CH_2 - \bar{C}H - C_6H_5$), they proposed that V + $CH_2 = CH_2 \rightarrow V - CH_2CH_2 \cdot$. The radical might dimerize, add to another molecule of olefin, or acquire an electron from V, as shown in Eq. 12–5. When metals were reacted with H_2

$$V-CH_2-CH_2 \cdot \begin{array}{l} \nearrow V-CH_2CH_2-CH_2CH_2-V \\ \longrightarrow V-CH_2CH_2-CH_2CH_2 \cdot \\ \searrow V-CH_2-CH_2-V \end{array} \qquad (12-5)$$

or alkyl halide, the formation of M–H and M–alkyl bonds was proposed. The function of the vibratory ball-mill was to provide a continuous supply of fresh catalyst surface during polymerization.

A number of other similar catalysts were disclosed by other workers. D'Alelio reported that catalysts could be made active for the polymerization of ethylene by reacting the transition metal (Ti and Zr) with a second component, such as I_2 (21) and $ZrCl_4$, $ZrCl_2$, or $TiCl_3$ (22). Mulley and Small (23) formed active catalysts by reacting transition metals and EtBr–$HgCl_2$, I_2–$HgCl_2$, or allyl chloride–EtBr mixtures (23). Mixtures of $TiH_{1.75}$ and $TiCl_4$ were active for ethylene polymerization if sufficiently heated (17, 29a). Coover and Joyner polymerized propylene to partially crystalline polymer with a four-component catalyst consisting of transition metal (Ti, Zr, V, Cr, or Mo), base metal (M, Zn, or Mg), halogen (Cl_2, Br_2, I_2, or ICl), and a donor (a carboxylic amide) (24).

These were some of the earliest found systems. They have not been exploited commercially or for mechanistic studies.

2. Catalysts Containing an Alkylated Transition Metal Compound of Known Structure

Beerman and Bestian synthesized and isolated CH_3TiCl_3 (melting point $= 28°C$) by reaction of $(CH_3)_2AlCl$ and $TiCl_4$ (25). Ethylene, however, could be polymerized (80°C) only after partial decomposition occurred and a visible precipitate appeared in the reaction vessel, $CH_3TiCl_3 \rightarrow TiCl_3 +$ alkane. The active catalyst apparently was a mixture of $\beta TiCl_3 + CH_3TiCl_3$. The polyethylene had a low molecular weight, $M_n \simeq 9,400$.

Karapinka, Smith, and Carrick (26) made mixtures of CH_3TiCl_3 and VCl_4 or $VOCl_3$ and polymerized ethylene to high molecular weight products (melt index $= 1.5$) at 80°C. Boor (27) polymerized propylene to highly isotactic polymer using CH_3TiCl_3 and $TiCl_3AA$ or $\gamma TiCl_3 \cdot xAlCl_3$ as catalysts. More active catalysts were obtained when amine (3 mmole $n\text{-}Bu_3N$ per 1 mmole Ti) was added. The activity was increased 10 times for $TiCl_3AA$ and 50 times for $\gamma TiCl_3 \cdot xAlCl_3$ catalysts. Several other binary mixtures were reported active, including CH_3TiCl_3 plus a chromium halide (28) and $EtTiCl_3$ plus lead alkyls (29).

Natta and co-workers (30, 31) and Butts (34), and later Hoeg and co-workers (32) reported the polymerization of ethylene and (or) propylene with binary mixtures of $(C_5H_5)_2Ti(CH_3)_2$ or $(C_5H_5)_2Ti(C_6H_5)_2 + TiCl_4$ or $TiCl_3$. The polymerizations were done at 60°C and at 8 atm olefin pressure for durations up to 18 hours. These findings by Natta and co-workers in 1957 appear to be the first reported examples of a metal alkyl-free catalyst.

Lanovskaya and co-workers (33) polymerized ethylene to polymer, $[\eta] = 0.92$ dl/g, with binary mixtures of CH_3TiCl_3 and $TiCl_3$ (α- or β-modification). Different activities were found for the different preparations of these titanium trichlorides.

The above polymerizations were done above room temperature. At $-78°C$, CH_3TiCl_3 affected a slow oligomerization of ethylene. Apparently, reduction had not occurred to any great extent since the Ti(IV) compounds were recovered. But the trivalent salt CH_3TiCl_2, under similar conditions, was found by Kuhlein and Clauss (33a) to polymerize ethylene to a high molecular weight crystalline polyethylene. The CH_3TiCl_2 was synthesized by the route shown in Eq. 12–6. Higher olefins did not polymerize under

$$2 CH_3TiCl_3 + Hg[Si(CH_3)_3]_2 \longrightarrow Hg + 2 CH_3TiCl_2 + 2 (CH_3)_3SiCl \quad (12\text{–}6)$$

similar conditions.

Relative to CH_3TiCl_3, higher alkyl derivatives are less stable; for example, $R = C_2H_5$, etc. (33b).

Two views have been expressed to explain the activity of these catalysts. The first is the obvious suggestion that the alkylated transition metal compound functions as the base metal alkyl does in a Ziegler–Natta catalyst: it alkylates the transition metal compound and these Ti–C bonds act as active centers. The other view, expressed recently by Henrici–Olive and Olive (33c), attributes activity to a change in the configuration of the alkylated titanium salt when the other transition metal becomes complexed to it, as shown in Eq. 12–7. This author suggests that the latter may undergo

$$
\begin{array}{c}
\pi\text{-}C_5H_5 \quad CH_3 \\
\diagdown \diagup \\
Ti \\
\diagup \diagdown \\
\pi\text{-}C_5H_5 \quad CH_3
\end{array}
+ TiCl_4 \longrightarrow Cl_3Ti
\qquad (12\text{-}7)
$$

Inactive compound
(tetrahedral configuration)

Active complex
(octahedral configuration)

an exchange of CH_3 and Cl and both Ti's may act as centers, as shown in

II

structure II where ☐ represents the open octahedral position through which the olefin can complex to the Ti center.

a. Benzyl Derivatives (Bz $= CH_2$–C_6H_5). Giannini, Zucchini, Albizzati, and D'Angelo synthesized a number of soluble benzyl compounds of Ti and Zr, which they found active for polymerization of ethylene, and some for higher α-olefins (37–40). These compounds showed unusual thermal stability in comparison with other transition metal alkyls. For example, $(CH_3)_4Ti$ decomposes above $-80°C$ while $Ti(CH_2C_6H_5)_4$ is stable at $25°C$. Typical compounds examined included $TiBz_4$, $TiBz_3Cl$, $ZrBz_4$, $ZrBz_3Cl$, $ZrBz_2Cl_2$, $TiBz_3Br$, $TiBz_3F$, and $TiBz_3CH_3$ (37, 39, 40). With $TiBz_3Cl$, 15 grams polyethylene per mmole Ti was formed in toluene solvent at $20°C$ during 4 hours and 10 atm pressure. These catalysts were less active for polymerization of propylene and 4-methyl-1-pentene but, nevertheless, the formed polymers were isotactic-crystalline. Propylene polymerizations required some halogen to be present, such as $TiBz_3Cl$, while 4-methyl-1-

pentene was polymerized with solely $TiBz_4$ or $ZrBz_4$. If the claims of these authors are valid, then these represent the first known soluble catalysts for isotactic polymerizations.

Ballard and co-workers (40a–d) have extensively studied the polymerizations of ethylene, propylene, and vinyl monomers with benzyl and π-allyl derivatives of transition metals under different experimental conditions. Ballard (40e) discusses the earlier work in an excellent comprehensive review that covers the polymerizations catalyzed by π and σ transition metal–carbon compounds in great detail. Table 12–2 collects some of the benzyl Zr, Hf, and Ti derivatives that were found active for polymerization of ethylene; the corresponding work with π-allyl derivatives is given in the next section.

TABLE 12–2

Polymerization of Ethylene in the Dark by Transition Metal Benzyl Compounds[a] (40e)

Catalyst	Relative activity
Zirconium benzyls	
Zr(benzyl)$_4$	1.0
Zr(benzyl)$_4$(C$_5$H$_5$N)$_3$	0.0
Zr(4-Me-benzyl)$_4$	1.2
Zr(4-MeO-benzyl)$_4$	0.4
Zr(4-F-benzyl)$_4$	2.5
Zr(3-Cl-benzyl)$_4$	0.08
Zr(1-methylene-1-naphthyl)$_4$	2.25
Titanium benzyls	
Ti(benzyl)$_4$	1.0
Ti(4-Me-benzyl)$_4$	4.0
Ti(2-Cl-benzyl)$_4$	2.1

[a] In toluene at 80°C; ethylene partial pressure = 10 atm.

Ballard writes the active centers of the π-allyl and benzyl derivates as shown in structures III and IV. On the basis of the low measured bond angles

III   IV

(Table 12–3), he speculated that some interaction occurred between the aromatic nucleus and the metal, this most probably being the donation of π-electrons of the aromatic nucleus to the vacant d orbitals of the metal atom,

TABLE 12–3

Relevant Bond Angles and Bond Lengths (at $-40°C$) of Tetrabenzyl Compounds (40e)

	Bond angle, metal–C–C (deg)	Bond length, metal–C (Å)
Sn(benzyl)$_4$	111	2.18
Ti(benzyl)$_4$	103	2.13
Zr(benzyl)$_4$	91	2.28
Hf(benzyl)$_4$	93	2.25
Ni(2-1 Me-allyl)$_2$ (8)	72	2.00

as shown in structure V (40b). This idea was used to explain the stability of

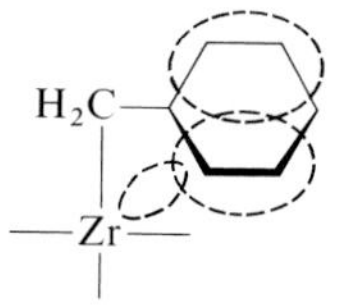

V

benzyl derivatives of Zr, Ti, and Hf metals in contrast to the methyl and $C_6H_5CH_2CH_2$ derivatives that decompose above $-80°C$.

From kinetic studies, Ballard and co-workers (40b) concluded that the polymerization of styrene was of the coordinated-anionic type, in which only one of the four zirconium benzyl bonds was active. Only one polymer chain formed at each active center, and site termination involving β-hydrogen abstraction or a bimolecular reaction involving monomer occurred.

To explain why styrene was polymerized by the benzyl but not π-allyl derivatives, Ballard suggested that styrene cannot compete with the allyl group for the coordination site on the metal atom, in Eq. 12–8, whereas it

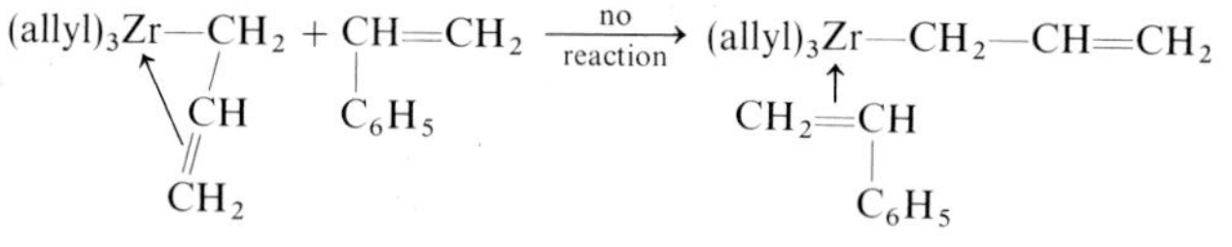

$$(12\text{–}8)$$

can in the corresponding benzyl derivative, shown in Eq. 12–9. Apparently,

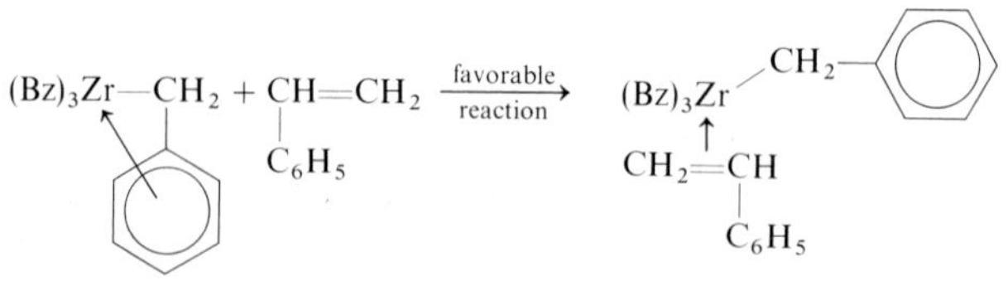

$$(12\text{–}9)$$

ethylene coordination is facile in both derivatives since it is polymerized almost equally by both.

Substitution in the benzyl derivatives affects the activity of the resulting derivative differently for ethylene and styrene. In the case of ethylene, substituents that lower electron density at the aliphatic carbon favor polymerization; that is, those substitutions that give a more stable benzyl anion such as $Zr(4\text{-F-benzyl})_4$ and $Zr(1\text{-methylene-1-naphthyl})_4$. The opposite effect was observed for styrene.

Polymerizations of styrene with labeled Zr derivatives, such as $Zr(^{14}CH_2C_6H_5)_4$, showed that approximately 1.5 benzyls were present per polymer chain. This showed that chain transfer did not occur and that monomer insertion took place in the $Zr\text{–}CH_2C_6H_5$ bond. The extra 0.5 benzyl per chain was accounted for by a side alkylation reaction (this was confirmed by showing 0.35 benzyl per chain had incorporated into a polystyrene chain when preformed polystyrene was mixed with tetrabenzyl (^{14}C) zirconium).

Ballard found that transition metal alkyl compounds such as $Zr(CH_2C_6H_5)_4$, $Zr(CH_2CH{=}CH_2)_4$, and $Ti[CH_2Si(CH_3)_3]_4$ react with OH group on surface of silica in alumina to give highly active polymerization catalysts of long life (40b). Active centers having the following structures were suggested:

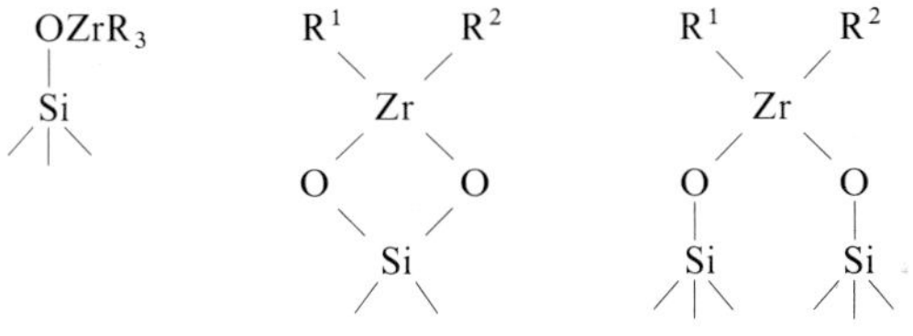

VI VII VIII

where R^1 can be $CH_2C_6H_5$, allyl, or $CH_2Si(CH_3)_3$; R^2 can be R^1 or halogen; and Zr can be replaced by Ti(IV), Cr(IV), or Cr(III). In later work (40c), compounds were prepared with the silanols

$$(C_6H_5)_3SiOH,\quad (C_6H_5)_2Si(OH)_2,\quad (C_6H_5)_2\text{—}\underset{\displaystyle OH}{\underset{|}{Si}}\text{—}O\text{—}\underset{\displaystyle OH}{\underset{|}{Si}}\text{—}(C_6H_5)_2,\quad (CH_3)_3SiO\text{—}\underset{\displaystyle OH}{\underset{|}{\overset{\displaystyle CH_3}{\overset{|}{Si}}}}\text{—}O\text{—}Si(CH_3)_3$$

The combination of tetrabenzyl zirconium and 1,1,3,3-tetraphenyldisiloxane 1,3-diol was found particularly active for the polymerization of ethylene; the other derivatives were inactive. The silicon compound behaved as a monodentate ligand (the Si–O–Si bond was 180°). To explain the activity of the Zr derivative obtained with $(C_6H_5)_2\text{–}\underset{\displaystyle OH}{\underset{|}{Si}}\text{–}O\text{–}\underset{\displaystyle OH}{\underset{|}{Si}}\text{–}(C_6H_5)_2$ but

TABLE 12-4

Polymerization of Propylene by Transition Metal Alkyl Compounds[a] (40e)

Compound	Activity $(gm \cdot mM^{-1} \cdot atm^{-1} \cdot hr^{-1})$	Polymers soluble in toluene			Polymers insoluble in toluene		
		Amount (%)	Isotactic (%)	Syndiotactic (%)	Amount (%)	Isotactic (%)	Syndiotactic (%)
Zr(allyl)$_4$	0.0014	36	50	50	64	81	19
Zr(allyl)$_3$Cl	0.03	60	37	63	40	82	18
Zr(allyl)$_3$Br	0.013	44	39	61	56	77	23
Zr(benzyl)$_4$	0.0007	100	68	21	0	—	—
Ti(benzyl)$_4$	0.004	98	75	23	2	—	—
SiO$_2$/Zr(allyl)$_4$	0.16	40	31	69	60	74	26
Al$_2$O$_5$/Ti(benzyl)$_4$	5.5	30	40	60	70	80	20
SiO$_2$/Zr(allyl$_3$Br)	2.7	55	45	55	45	84	16
SiO$_2$/Zr(benzyl)$_3$I	1.03	52	47	53	48	71	29
SiO$_2$/Zr(benzyl)$_4$	5.8	30	40	60	70	80	20
Zr[CH$_2$ · Si(CH$_3$)$_3$]$_3$/Al$_2$O$_3$	6.0	30	40	60	70	80	20
TiCl$_3$/AlEt$_2$Cl	4.0	5	—	—	95	—	—

[a] Conditions: toluene as solvent; temp. = 65°C; ethylene pressure = 10 atm.

not the others, Ballard speculated that ligands derived from stronger acids would provide more electronegative oxygen atoms that would, in turn, make the ZrO bond more ionic, and intensify the electropositive charge on the transition metal. This would not only enhance the coordination of the olefin to the Zr but would make the Zr–C bond more polarized and more active for insertion of the coordinated olefin.

This was nicely supported later (40d) by showing that Zr derivatives of α-pinacol (a weak acid) were inactive, whereas those of perfluoro-α-pinacol (a strong acid) formed active catalysts, as shown in structures IX and X.

$$
\begin{array}{ll}
(CH_3)_2\!-\!C\!-\!O \quad\quad CH_2C_6H_5 & (CF_3)_2\!-\!C\!-\!O \quad\quad CH_2C_6H_5 \\
\qquad\qquad\;\; \diagdown \;\diagup & \qquad\qquad\;\; \diagdown \;\diagup \\
\qquad\qquad\;\;\; Zr & \qquad\qquad\;\;\; Zr \\
\qquad\qquad\; \diagup \;\diagdown & \qquad\qquad\; \diagup \;\diagdown \\
(CH_3)_2\!-\!C\!-\!O \quad\quad CH_2C_6H_5 & (CF_3)_2\!-\!C\!-\!O \quad\quad CH_2C_6H_5 \\
\qquad\qquad Inactive & \qquad\qquad Active \\
\qquad\qquad\quad IX & \qquad\qquad\quad X
\end{array}
$$

[The activity of $(C_6H_5CH_2)_4Zr$ was also increased by substitution of some of the benzyl group with halogens.]

Ballard (40e) showed that partially isotactic polypropylenes were formed in very low yields with apparently soluble derivatives of Zr. More active catalysts were formed when these Ti and Zr derivatives were supported in SiO_2 or Al_2O_3 (see Table 12–4).

Whereas $Ti(CH_2C_6H_5)_4$ gave 60% 1,2-addition in a butadiene polymerization, $Ti(CH_2C_6H_5)_3I$ gave 73.5% cis-1,4-addition (40f), and $Zr(CH_2C_6H_5)_4$ on Al_2O_3 gave a 90% trans-1,4-addition with butadiene but only a 50% with isoprene.

Long found $TiBz_4$–SiO_2-supported catalysts were active for ethylene polymerization (41). Guzman and co-workers (42) have polymerized butadiene predominantly to 1,2-polymer with $TiBz_4$ and to cis-1,4-polymer with $TiBz_3I$.

b. Allyl Compounds of Transition Metals. Ethylene, α-olefins, and butadiene have been polymerized with specific π-allyl compounds of transition metals.

Wilke and co-workers reported in 1966 what is probably the first example of polymerization of ethylene with a π-allyl–transition metal compound, namely $(\pi\text{-allyl})_3Cr$ (43). Specific binary mixtures of allyl derivatives and transition metal salts were also found active by Herwig, Gumboldt, and Weissermel (44) for polymerization of not only ethylene to high molecular weight polymer but also propylene and 4-methyl-1-pentene. The polypropylene was crystalline and had a high molecular weight (reduced viscosity was 3.3 dl/g). Typical catalysts consisted of π-allyl derivatives of Cr, W, Mo, and Zr and halides, oxyhalides, esters halides, or esters of Ti, Cr, and Fe.

TABLE 12–5

Transition Metal π Complexes.
Polymerization of Ethylene in Toluene in the Dark (40e)

Initiator	Amounts (mM/liter)	Temperature (°C)	Partial pressure (atm)		Activity (gm·mM^{-1}· atm^{-1}·hr^{-1})
			Ethylene	Hydrogen	
Zr(allyl)$_4$	3.0	80	10	10	2.00
Cr(allyl)$_3$	3.0	80	10	10	0.3
Ti(2-Me-allyl)$_4$	8.0	50	53	0	0.52
Hf(allyl)$_4$	3.0	160	27	10	1.00
Nb(allyl)$_4$	3.0	160	27	10	0.11
V(allyl)$_3$	2.0	−80 to 65	10	0	1.4
αTiCl$_3$/AlEt$_2$Cl	2.0	80	10	10	20.0

One-component catalysts using metals other than Cr, such as Ti, Zr, Hf, Nb, and V, were disclosed by Ballard (45) and Job (46) for the polymerization of ethylene to high molecular weight polymers (Table 12–5). Some typical soluble catalysts include (π-allyl)$_3$ZrBr (46), (π-allyl)$_3$HfBr (46), (2-methallyl)$_4$Ti (45), (π-allyl)$_4$Nb (45), and (π-allyl)$_3$V (45). Ballard found (π-allyl)$_2$Ni, (π-allyl)$_2$Cr, (π-allyl)$_2$Pd, and (π-allyl)$_3$Mn inactive. Halogen derivatives, made by reaction with the allyl halides (X = F, Cl, Br, or I) were more active. Ballard proposed that the π-allyl bond is not the active center, but when ethylene becomes complexed to the metal center it does so by converting the π-allyl bonding to σ bonding, as shown in Eq. 12–10, where Lp = ligands attached to the transition metal M. Often

$$\text{Lp M} \underset{\text{CH}_2}{\overset{\text{CH}_2}{\diagdown}} \text{CH} + \text{CH}_2\text{=CH}_2 \rightleftarrows \text{Lp M} \underset{\text{CH}_2}{\overset{\text{CH}_2}{\diagdown}} \text{CH=CH}_2 \quad (12\text{–}10)$$

the π-allyl bond is written as shown in structure XI. According to his view,

$$\text{Lp} \overset{\text{CH}_2}{\underset{\text{CH}_2}{\diagup\diagdown}} \text{CH}$$

XI

π-allyl bonds are inactive because they are too stable as a result of delocalization of electrons. Once two or more ethylenes add, the double bond becomes too distant to revert back to the π-allyl structure. Substitution of alkyl groups (such as methyl) at the π-allyl groups results in high catalytic activity, while substitution of electron-withdrawing groups (such as phenyl) lowers activity. For the latter case, more stable but less active compounds are formed.

The use of π-allyl–transition metal compounds as catalysts for polymerizing dienes apparently had its start in 1963 when Wilke reported the trimerization of butadiene to cyclooctadecatriene by bis-π-allyl–Ni (47). The following year, Porri, Natta, and Gallazzi reported the polymerization of butadiene to crystalline *trans*-1,4-polymers by π-allyl–NiBr (48). Also, in 1964 and 1965, Babitskii and co-workers described polymerizations of butadiene by π-allyl-Ni compounds (49). Concomitant with their investigation of π-allyl-transition metal compounds as oligomerization catalysts, Wilke and co-workers discovered that π-allylchromium polymerized butadiene predominantly to 1,2-polybutadiene and to *cis*-1,4-polybutadiene with $(\pi$-allyl$)_2$CoI. [The polymerization of ethylene was mentioned earlier (43).]

The versatility (and delicateness) of polymerizations with allyl–transition metal compounds is exemplified in the results of Porri, Natta, and Gallazzi (50, 51), as shown in the polymerization of butadiene to low molecular weight polymers ($M_w < 1500$) (see Table 12–6).

TABLE 12–6

Polymerization of Butadiene with π-Allylnickel Salts (50, 51)

Catalyst	Solvent	Predominant structure
π-allyl–NiX(X = I, Br)	Benzene	*trans*-1,4
π-allyl–NiCl	Benzene	*cis*-1,4
π-allyl–NiBr + AlBr$_3$, Al(i-PrO)$_3$, BF$_3$, or TiCl$_2$(i-PrO)$_2$	Benzene	*cis*-1,4
π-allyl–NiCl	HF	*trans*-1,4

When allylnickel bromide was reacted with AlBr$_3$, the ionic structure shown in XII was proposed as the active center on the basis of analysis on the product.

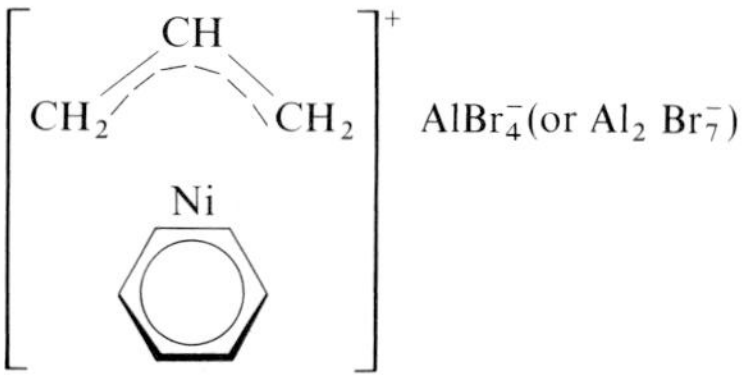

XII

Porri and co-workers also showed variations in polybutadiene microstructure when the halide in (cycloocta-1,5-diene)NiX was changed from Br(*cis*-1,4 units) to I(*trans*-1,4 units) (50).

Lazutkina and co-workers (52) more recently reported that the polymerization of butadiene in the presence of allylpalladium chloride, allylpalladium bromide, or allylpalladium iodide in benzene gave 98, 98, and 30% 1,2-polybutadiene, respectively. As evidence for their view that the active center was a Ni–C bond, Porri and co-workers (51) cited the following: (1) the presence of vinyl groups in polybutadiene obtained by allyl Ni–X (X = I, Br, Cl), (2) the presence of ester groups in polybutadiene obtained with the *in situ* catalyst $Ni(CO)_4$ and allyl halide, and (3) catalysts do not exhibit radical or cationic behavior toward butadiene and other monomers.

An interesting aspect of these polymerizations is the use of hydroxylated solvents such as ethanol. This was already demonstrated for other catalysts, such as the polymerization of butadiene in alcohols and water by salts of Rh, Ru, and Os (53, 54).

Polymers of higher molecular weight and in greater conversions were reported by Dawans and Teyssie in 1965 when bis(cyclooctadienyl)nickel(0), $Ni(COD)_2$, was reacted with specific metal halides, including $MoCl_5$, $TiCl_4$, $SbCl_5$, $FeCl_3$, and even $NiCl_2$ (55, 56). Moreover, the microstructure obtained with the binary catalyst was different from that obtained with only the $Ni(COD)_2$ complex. Optimum conversions and highest intrinsic viscosities were obtained with $Ni/MX_n = 2$ ($[\eta] = 0.68$ dl/g). Organic and inorganic acids as well as halogen compounds were also effective (57).

These investigators (including their co-workers Durand and Marechal) described the considerable versatility of π-allyl–transition metal and related compounds in a series of papers (58). Some of the salient features of their findings include the following points.

In a homogeneous series of π-allylnickel halides, the overall rate constants for the polymerization of 1,3-butadiene appeared to vary in the order: 2,6,10-dodecatriene-2-yl > allyl ≥ methallyl > crotyl (59).

Substitution of the halogen in π-allyl derivatives of transition metals by anions of halogen-substituted carboxylic acids leads to increase in activity (55). The catalytic activities increased sharply in the following sequence: $CH_3CO_2^- < CH_2ClCO_2^- < CHCl_2CO_2^- < CCl_3CO_2^- < CF_3CO_2^-$.

Addition of electron-donating materials led to decreased catalyst activities and to an increase in the amount of *trans*-1,4 (or 1,2) units.

The transition metal in the salt plays an important role in deciding the type and degree of stereoregularity (58) (Table 12–7).

New polymers designated as equibinary polydienes from butadiene or isoprene were synthesized that had equimolar amounts of two of the possible

TABLE 12–7

Influence of Transition Metal on Microstructure of Polybutadienes Produced by π-Allylic Catalysts (58)

	Microstructure of polybutadiene		
Catalytic systems	cis-1,4 (%)	trans-1,4 (%)	1,2 (%)
$(\pi\text{-}C_4H_7)_2Ni + NiCl_2$	95	3	2
$(\pi\text{-}C_3H_5)_3Co + I_2$	90	2	8
$(\pi\text{-}C_4H_7)_3Fe + FeCl_3$	92	6	2
$(\pi\text{-}C_4H_7)_3Rh$	0	94	6
$[(\pi\text{-}C_4H_7)_2Mo]_2 + MoCl_5$	15	4	81
$(\pi\text{-}C_4H_7)_3Cr$	0	10	90
$(\pi\text{-}C_4H_7)_3Nb$	1	2	97
$(\pi\text{-}C_4H_7)_3Cr + HCl$	90	5	5
$(\pi\text{-}C_4H_7)_3Nb + HCl$	91	5	4

geometric isomers (60). For example, using the $CF_3CO_2H/Ni(COD)_2$ catalyst at a 1:1 ratio produces predominantly cis-1,4 polymer. But increasing the ratio above 5 (up to 50 was examined) causes the microstructure to change to 50% cis-1,4 and 50% trans-1,4. The polymer is not a mixture of the two types of polybutadienes. Each polymer chain has both geometric isomer units. Similarly, equibinary polyisoprenes were made by modifying cobalt catalysts. Recent ^{13}C-NMR studies in different laboratories led to the conclusions that the equibinary polybutadienes were block copolymers containing long sequences of the cis- and trans-1,4 units (60a,b). However, Julemont, Teyssie, and co-workers (60c) concluded via ^{13}C-NMR spectroscopy that the units were distributed statistically along the chain or showed a tendency toward an alternating structure depending on the solvent used.

We might interpose here that an equibinary (cis-1,4–1,2)polybutadiene was reported by Grant and Paul using the Ziegler–Natta catalyst, $MoCl_3(OR)_2$–$AlEt_3$ (60d). Furukawa and co-workers, using ^{13}C-NMR spectroscopy concluded, however, that the sequence distribution of cis-1,4 and 1,2 units was random (60e).

Dawans and Teyssie (58) have collected much of the π-allyl-type polymerization literature in an excellent comprehensive review, in which various mechanistic schemes are critically discussed.

Some of the possible reaction paths for butadiene are summarized in Eqs. 12–11 and 12–12.

$$(allyl)_3Cr \underset{\overset{0.5\,O_2}{\longrightarrow}}{\overset{+CCl_3-COOH}{\longrightarrow}} \begin{array}{l} \textit{cis}\text{-1,4-PBD (93\%)} \\ \text{1,2-PBD (90\%)} \\ \textit{trans}\text{-1,4-PBD (92\%)} \end{array} \qquad (12\text{--}11)$$

$$(allyl\ NiOOCCF_3)_2 \underset{\overset{C_6H_6}{\longrightarrow}}{\overset{(C_6H_5O)_3P}{\longrightarrow}} \begin{array}{l} \textit{trans}\text{-1,4-PBD (99\%)} \\ \textit{cis}\text{-1,4-PBD (96\%)} \\ 50\%\ \textit{cis}\text{-1,4-} \\ 50\%\ \textit{trans}\text{-1,4-PBD} \end{array} \qquad (12\text{--}12)$$

Dolgoplosk (60f), who also has examined the mechanistic aspects of diene polymerizations with metal alkyl-free and Ziegler–Natta catalysts in a comprehensive review, made a number of interesting observations. On the basis of the formation of propylene and 1-butene (10% yield based on original concentration of allyl groups), he suggested that chain transfer by butadiene (BD) had taken place, as shown in Eq. 12–13. Spontaneous

TABLE 12–8

Oxidation State of the Transition Metal and Stereospecificity of Action in Butadiene Polymerization (60f)

Catalyst	Number of electrons in outer orbitals	Content of bond type (%)		
		cis-1,4	*trans*-1,4	1,2
$(\pi\text{-}C_4H_7)_3Cr$	15	0	19	81
$(\pi\text{-}C_4H_7)_3Cr + 2HCl$	11	90	5.5	4.5
$(\pi\text{-}C_4H_7)_3Cr + CCl_3COOH$	—	93	4	3
$(\pi\text{-}C_4H_7)_3Nb$	14	0	0	100
$(\pi\text{-}C_4H_7)_3Nb + HCl$	12	91	5.5	4.5
$(\pi\text{-}C_4H_7)_2Ni$	16	cyclododecatriene		
$(\pi\text{-}C_4H_7)NiCl$	14	91	5	4
$\pi\text{-}C_4H_7NiOCOCCl_3$	14	95	4	1
$(\pi\text{-}C_4H_7)_3Co$	18	oligomers		
$(\pi\text{-}C_4H_7)_3Co + 2HCl$	13–14	91.5	1.5	7
$(\alpha\text{-}C_4H_7)_{3-4}Ti$	—	0	17	83
$(\pi\text{-}C_4H_7)_{3-4}Ti + TiI_4$	—	88.5	7.5	4
$(\pi\text{-}C_4H_7)_3Rh$	—	0	94	6
$(\pi\text{-}C_4H_7)_4Mo + CCl_3C\overset{\diagup Cl}{\underset{\diagdown O}{}}$	—	0	0.5	99.5
$(\pi\text{-}C_4H_7)_4Mo + chloranil$	—	6.5	16.5	77
$(\pi\text{-}C_3H_5)_4W + HCl$	—	—	8	92
$(\pi\text{-}C_3H_5)_4W + CH_2 = CHCH_2Cl$	—	—	5	95
$(\pi\text{-}C_3H_5)_4W + CCl_3COOH$	—	—	13.5	86.5

$$
\begin{array}{c}
\text{CH}_2 \\
\parallel \\
\text{CH} \qquad \text{Cr} \\
\diagdown \quad \diagup \\
\text{CH} \\
\mid \\
\text{Pn}
\end{array}
+ \text{BD} \longrightarrow \text{Pn—CH=CH—CH=CH}_2 +
\begin{array}{c}
\text{CH}_2 \\
\parallel \\
\text{CH} \qquad \text{Cr} \\
\diagdown \quad \diagup \\
\text{CH} \\
\mid \\
\text{CH}_3
\end{array}
\xrightarrow{\text{H}_2\text{O}} \text{1-butene}
$$

A

$$(12\text{–}13)$$

decomposition of A, followed by addition of butadiene to the $>$CrH product, was also considered possible. In fact, active catalysts for the polymerization of butadiene with transition metal hydride derivatives were described as shown in sequence 12–14. Dolgoplosk has collected selected

$$
\begin{aligned}
&\text{NiH}_2 + \text{BD} \longrightarrow \pi\text{-allyl}_2\text{Ni} \xrightarrow[\text{butadiene}]{\text{NiCl}_2} \textit{cis}\text{-1,4-polybutadiene} \\
&\text{CoH}_2 + \text{BD} \longrightarrow \pi\text{-allyl}_2\text{Co} \xrightarrow[\text{butadiene}]{\text{CoCl}_2} \textit{cis}\text{-1,4-polybutadiene} \\
&\text{CrH}_3 + \text{BD} \longrightarrow \pi\text{-allyl}_3\text{Cr} \xrightarrow[\text{butadiene}]{} \text{1,2-polybutadiene} \\
&\text{HNiX} + \text{BD} \longrightarrow \pi\text{-allylNiX} \xrightarrow{\text{NiX}_2} \textit{cis}\text{-1,4-polybutadiene}
\end{aligned}
$$

$$(12\text{–}14)$$

examples of catalysts (Table 12–8). More recently, Dolgoplosk and co-workers (60g) synthesized the major structures from butadiene and isoprene by suitable modification of tris (π-allyl)chromium. Supported systems (silica–alumina or silica gel) produced high *trans*-1,4 structures.

The effect of specific reactants was investigated by a number of workers. High *cis*-1,4 ($> 90\%$) polybutadiene ($[\eta] = 0.89$ to 2.14 dl/g) were synthesized with $\text{CF}_3\text{CO}_2\text{NiCl}$ prepared by the reaction shown in Eq. 12–15. The

$$(\text{CF}_3\text{CO}_2)_2\text{Ni} + \text{SO}_2\text{Cl} \xrightarrow{\text{ether}} \text{CF}_3\text{CO}_2\text{NiCl} \cdot \text{ether} + \text{SO}_2 + \text{CF}_3\text{CO}_2\text{H} \quad (12\text{–}15)$$

etherate complex liberates ether molecules upon heating to 75°C, and the uncomplexed form was used for the polymerization (61).

When benzoyl peroxide or oxygen was added to π-allylnickel halide, the microstructure of the polybutadiene changed from predominantly *trans*-1,4 units to a mixture of *cis*- and *trans*-1,4 units (62). Wallace and Harrod (63) concluded from their experiments that the benzoyl peroxide simply destroyed the catalyst to produce largely inactive products. They suggested that radicals generated during the destruction of the catalyst initiate some polymerization and give an apparent change in stereospecificity. Of relevance may be the report of Komatu and co-workers that Ni(OH)_2 or Ni peroxide, sintered at 300°C in the presence of air and then mixed with AlCl_3, was active for polymerization of butadiene to high *cis*-1,4 polymer (64). Other workers, however, reported that binary mixtures of π-allylnickel chloride or bromide plus benzoyl peroxide yield an active catalyst bearing a benzoyloxy group of a conjugated structure (64a). Predominantly *cis*-1,4-polybutadiene was obtained when the halogen was Cl or Br, but X = I produced *trans*-1,4 polymer. Studies with oxygen in place of benzoyl peroxide were also reported.

Reactions of $(allyl)_4Zr$ with $TiCl_4$ involved the migration of allyl ligands to the titanium chloride with subsequent reduction of the latter compound (64b). Complexes of $(allyl)_4Zr$ or $(allyl)_3ZrCl$ and $TiCl_3$ polymerized isoprene to a *cis*-1,4 product, but $(allyl)_2ZrCl_2$ led to cationic polymerizations.

Very high *cis*-1,4-polybutadienes (about 98 to 99%) have been reported recently by Lugli, Mazzei, and Poggio (64c), using tris(π-allyl)uranium halides. Increased activities were also obtained if Lewis acids were added, such as $TiCl_4$, $AlEtCl_2$, etc.

c. Cyclopentadienyl Compounds of Transition Metals.
Chromacene, $(C_5H_5)_2Cr$, was found active by Karol and co-workers only if it was chemisorbed on a high surface support. This system, however, is described in Section III,B,9 along with other supported metal alkyl-free catalysts. The author is not aware of any other acenes of transition metals that have been shown active for polymerization of olefins or dienes in absence of base metal alkyls.

d. Arene Compounds.
Tsutsui and Koyano (65) synthesized highly linear polyethylene (d = 0.966 g/cm^3, 88% crystalline) with an intrinsic viscosity of 0.99 dl/g with dibenzene chromium. Drastic polymerization conditions of 250°C and 2,800 psi ethylene pressure, however, were necessary. It was proposed that the dibenzenechromium decomposes to metallic chromium, which abstracts two hydrogens from two coordinated ethylene molecules to form a CrH_2 species. The mechanism proposed, however, was a radical type, and it did not use the Cr–H bonds as reaction centers.

Another arene catalyst was disclosed by Karol (66), who polymerized ethylene in the presence of chromium dicumene deposited on SiO_2–Al_2O_3 supports.

Isotactic polypropylene was reported to be formed when tetraaryl dichromium was reacted with titanium chlorides in the presence of organic molecules containing N, O, P, or S (67).

3. Transition Metal Salts Modified by Mechanical Means

Ball-milled transition metal halides, oxides, and nitrides formed active catalysts for ethylene and sometimes also for propylene (18, 19, 68–70). The most extensive investigation was made by Werber, Benning, Wszolek, and Ashby, who activated $TiCl_2$ by controlled ball-milling (18, 69, 70). In the absence of mechanical activation, $TiCl_2$ was an extremely sluggish catalyst (71). $TiCl_2$ was synthesized by reducing $TiCl_3$ with Ti at $>450°C$, disproportionation of $TiCl_3$ at 700°C to $TiCl_2$ and $TiCl_4$, and by reduction of $TiCl_4$ with Na. Figure 12–1 shows the affect of ball-milling time on catalyst

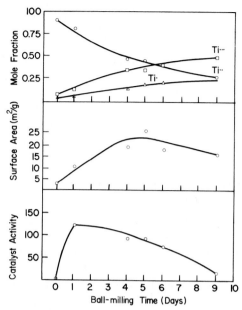

Fig. 12–1. Dependence of polymerization activity and properties of $TiCl_2$ on duration of ball-milling.

activity, surface area, and oxidation state of Ti. Maximum activity occurred after 1 day ball-milling, even though highest surface area was achieved after 5 days. The Ti was largely in a divalent state. Ball-milling caused Ti(II) to disproportionate partly to Ti(0) and Ti(III) species. Apparently, divalent Ti is the active catalyst; they found $TiCl_3$, formed by reducing $TiCl_4$ with H_2, was inactive when ball-milled similarly. The polyethylene had properties comparable to Grace's commercially produced polymer via a Phillips-type catalyst. Propylene was polymerized at 120°C in cyclohexane solvent to a low crystalline isotactic polymer (about 30% insolubles in heptane); 1-butene would not polymerize. Copolymers of ethylene with propylene and 1-butene were easily made.

When $TiCl_2$ was ball-milled in the presence of ethylene (200 psig), an optimum activity for ethylene polymerization was obtained when the $TiCl_2$ contained 3.5% ethylene (72), as shown in the tabulation below.

Wt% ethylene on $TiCl_2$	Relative rate
0 ($TiCl_2$ not ball-milled)	1.4
0 ($TiCl_2$ was ball-milled)	33.6
3.5 ($TiCl_2$ was ball-milled)	160
5.1 ($TiCl_2$ was ball-milled)	0

When $TiCl_3$ was ball-milled in the presence of 110 psig propylene, about 8% propylene was adsorbed, but the $TiCl_3$–propylene (while active for ethylene) was negligibly active for polymerization of propylene (73). Minsker and co-workers made only low molecular weight oils when ball-milled $TiCl_3$, VCl_3, and $CrCl_3$ salts were used to polymerize propylene (74).

More recently, Khodzhemirov and co-workers polymerized ethylene, butadiene, and isoprene with $TiCl_2$. The polybutadiene had a 70% trans-1,4 and 27% cis-1,4 structure (75). The $TiCl_2$ (surface area = 17.3 m^2/g) was prepared by reduction of $TiCl_4$ with Ti, followed by pulverization in a vibratory mill.

4. Transition Metal Salts Modified by Light, Electrolysis, Heat, or Irradiation

An active catalyst for the polymerization of ethylene was made by Schutze and co-workers (76) by irradiating $TiCl_4$ with 2 MeV electrons. Oita and Nevitt used gamma radiation on $TiCl_4$ to make an active catalyst (77). A polyethylene with a density of 0.96 g/cm^3 and a melting point of 128°C was made at 300 psig pressure and ambient temperatures. Anderson (77a) reported that butadiene was polymerized to 98% cis-1,4 polymer with photolyzed Co(II) and Ni(II) halides. He attributed activity to the presence of less than 0.1 mole% Ni or Co monovalent halides in the catalyst.

TABLE 12–9

Activation by Thermal Treatment[a] (78)

	Microstructure		
Metal halide heated	cis-1,4	trans-1,4	1,2
$NiCl_2$	93	5	2
NiI_2	17	75	8
$CoCl_2$	90	6	4
$\beta TiCl_3$	82	15	3
$NbCl_3$	88	8	4
NiF_2	97	2	1

[a] Under vacuum at 200° to 300°C.

Certain transition metal salts became active and stereospecific for polymerization of butadiene after subjected to thermal treatment under vacuum at 200° to 300°C (78) (see Table 12–9).

Dolgoplosk and co-workers proposed that the active center was a π-allyl metal bond formed when the diene molecule reacted with the metal subhalide, as shown in Eq. 12–16.

$$C_4H_6 + M\!-\!\!\!-\!Cl \longrightarrow C \overset{\overset{\displaystyle C}{|}}{\underset{C}{\diamond}} M\!-\!\!\!-\!Cl \qquad (12\text{--}16)$$

Guainazzi and co-workers (78a) initiated the polymerization of ethylene to oils or to high molecular weight polyethylene by coupling aluminum and transition metal (chromium) anodes in an anhydrous electrolytic solution. The latter consisted of 1,2-dichloroethane and a tetraalkylammonium salt. The polymerization was done in a pressure resistant cell, which allowed ethylene pressures over 30 atm. Yields of 16.4 grams polyethylene contaminated with 0.3 grams of oil were prepared.

An active catalyst was produced by electrolysis of the NiI_2 in the presence of maleic anhydride in THF or alcoholic solutions (78b). $Ni(II)I_2$ was electrochemically reduced to zero-valent nickel, which then reacted with $Ni(II)I_2$ to yield a π-maleic anhydride $Ni(I)I$. Maleic anhydride was then displaced by excess butadiene, and one molecule of butadiene on the nickel atom derived a hydrogen atom from the solvent, forming π-crotyl $Ni(I)I$. An insertion of butadiene into the π-crotylnickel bond initiates the polymerization, and a *trans*-1,4 polymer was found.

5. Transition Metal Salts Modified by Organic and Inorganic Compounds

An active catalyst for ethylene, consisting of $TiCl_2$ and 4,4-diazoheptane, was disclosed by Arlman (79), who proposed that an alkylated trivalent chloride was formed by oxidation, as shown in Eq. 12–17.

$$2\,TiCl_2 + RN{=}NR \longrightarrow 2\,TiCl_2\cdot R + N_2\!\uparrow \qquad (12\text{--}17)$$

Coover and co-workers reported active catalysts for ethylene and propylene by reacting $TiCl_3$ and hexaphenyl ethane or 1,4-dihydronaphthalene (80, 81).

Ethylene and propylene were copolymerized to a rubbery polymer in the presence of ball-milled VCl_3 and diazomethane (82). Ethylene was polymerized in the presence of the reaction product of $TiCl_2$ (or Ti) and ethyl bromide (20). Presumably, an alkylated Ti was formed by oxidative addition, as shown in Eq. 12–18.

$$TiCl_2 + EtBr \longrightarrow EtTiCl_2Br \qquad (12\text{--}18)$$

Otsu and Yamaguchi reported an active catalyst for the polymerization of butadiene to 80% *cis*-1,4 polymer by reacting reduced Ni with CH_3SiCl_3 and other organic halides (83).

Kormer and co-workers (83a) polymerized cyclobutene and cyclopentene with π-allyl complexes of Mo, W, and Ni alone or in the presence of Lewis acids; the polybutadiene contained over 90% *cis*-1,4 units. Chernenko and co-workers (83b) polymerized 1,3-cyclohexadiene with a π-crotylnickel chloride–chloranil complex.

Highly isotactic-specific catalysts consisting of $TiCl_3$ and electron donors were extensively investigated by Boor and co-workers (84) as a diagnostic tool to elucidate Ziegler–Natta mechanisms (see Chapter 9).

Dolgoplosk and co-workers (84a) formed crystalline polymers from 1,3-cyclohexadiene and 2,3-dimethylbutadiene in the presence of π-allyl-nickel derivatives, such as π-allylnickel bromide + chloranil.

The hydridonickel coordination compounds prepared by reacting tetrakis (triethyl phosphate) nickel and CF_3CO_2H in ethanol produced a mixture consisting of 2-methylene vinylcyclopentane, 4-vinylcyclohexene, 1,5-cyclooctadiene, and octadiene (84b). In contrast, when an inorganic acid such as H_2SO_4 or $HClO_4$ was used, only *trans*-1,4-polybutadiene was formed. The hydridonickel(I) coordination compound $[HNiLn]^+$ was proposed as the active species in the linear dimerization of butadiene.

Lewis acids change the stereoregulating ability of π-crotyl nickel bromide and iodide, favoring *cis*-1,4 addition. Increased catalyst activities have also been obtained in their presence (84c). Binary mixtures of anhydrous $NiCl_2$ or $CoCl_2$ and Lewis acids such as $AlCl_3$, $TiCl_4$, and $SnCl_4$ polymerized ethylene to a linear polymer with a melting point in the range 127° to 134°C and intrinsic viscosities of 1.56 dl/g. Low yields were obtained in spite of the high polymerization temperatures (160°C) and high ethylene pressures (150 to 280 atm) that were used (85).

Novel halogen-free nickel catalysts were reported by Bauer, Chung, Keim, and co-workers (85a), which consisted of bis(cyclooctadiene)nickel(0) and various bidendate ligands. For example, a rate of 2,530 grams of polyethylene per gram of Ni per hour was obtained at 40°C and 800 psi ethylene when *o*-diphenylphosphinobenzoic acid was used. The polyethylene was highly linear and had a density of 0.96 g/cm³. Control of molecular weight by hydrogen was not reported, nor did α-olefins copolymerize effectively. Unlike Ziegler–Natta catalysts, these catalysts could be used in water, as well as in organic solvents. Other effective ligands included trisubstituted phosphorus-containing acids, heteroacids, substituted benzoic acids, heterocyclic acids, and phosphorus containing ylids (see Table 12–10). Specific ligand/Ni ratios were required for optimum rate. Maximum activities were obtained at about 60°C. Not all ligands produced high molecular

TABLE 12–10

Polymerization of Ethylene with Biscyclooctadiene-1,5-nickel and a Ylid Compound[a] (85a)

Example no.	Ylid structure (ϕ is phenyl)	Molar ratio (nickel:ylid)	Polymer yield (g)	Inherent viscosity (dl/g)	Density (g/ml)
II	$\phi_3-P=C-\overset{\displaystyle O}{\overset{\|}{C}}-O-C_2H_5$, with CH_3 on the $=C$	1:1	18.7	0.17	
III	Same as above	1:4	32.1	0.29	
IV	$\phi_3-P=C-\overset{\displaystyle O}{\overset{\|}{C}}-OC_2H_5$, with ϕ on the $=C$	1:4	12.2	0.24	0.969
V	Same as above	1:1	15.1	0.33	
VI	$\phi_3-P=C-\overset{\displaystyle O}{\overset{\|}{C}}-CH_3$, with H on the $=C$	1:2	77.5	0.06	
VII	$\phi_3-P=C-\overset{\displaystyle O}{\overset{\|}{C}}-CH_3$, with CH_3 on the $=C$	1:2	5.3	0.14	
VIII	$\phi_3-P=C-\overset{\displaystyle O}{\overset{\|}{C}}-\phi$, with ϕ on the $=C$	1.4	9.6	4.1	0.962
IX	$\phi_3-P=C$ branching to $\overset{\displaystyle O}{\overset{\|}{C}}-CH_3$ (upper) and $\overset{\displaystyle O}{\overset{\|}{C}}-CH_3$ (lower)	1.2	1.40	0.34	0.963
X	$\phi_3-P=$ (cyclic ring with $\overset{\displaystyle O}{\overset{\|}{C}}-O$)	1.4	0.80	0.10	

[a] Solvent: 30 ml toluene/120 ml hexane; temp = 60°C; pressure = 900–1,000 psi; time = 1 hr.

weight polymers. Highest molecular weights were obtained with *o*-diphenyl-phosphinobenzoic acid.

Hiraki and Hirai (85b) reported soluble catalysts for butadiene by addition of triphenylphosphine or tributylphosphine to σ-allyl complexes of dichloro-2,6,10-dodecatriene-1,12-diylruthenium.

6. Transition Metal Salts Containing Metals or Metal Compounds from Group I to III Metals

A number of catalysts have been reported in which aluminum metal or an aluminum compound is converted into a metal alkyl.

Van Helden and co-workers (86) reacted K, $AlCl_3$, and $TiCl_4$ to form an active catalyst for ethylene. A metal alkyl having the structure Cl_2Al-CH_2CH_2-$AlCl_2$ was identified. Its origin was attributed to subhalides of aluminum which were formed when K reacted with ethylene and $AlCl_3$.

Active ethylene catalysts have been reported when Al, $AlCl_3$, and $TiCl_4$ were reacted. The first example actually preceded the Ziegler–Natta discoveries, namely Fischer's observation in 1953 (87) that, in the presence of this ternary mixture at 150°C and 50 atm ethylene, some solid polyethylene was made in addition to low molecular weight oils (88).

Upon reacting Al, $AlCl_3$, and $TiCl_3$, Al, HCl, and $TiCl_3$, or Al, Ti, and Cl_2 in benzene solvent, a product was obtained that polymerized ethylene at 60° to 90°C and 10 atm pressure (89). When benzene was substituted by an aliphatic solvent, the catalyst was inactive.

Al–C bonds may arise in these catalysts by reaction of Al metal with the partially chlorinated solvents. These preparations are usually done at 100° to 200°C, and it is feasible that chlorination of the solvent by $TiCl_4$ or $FeCl_3$ may have occurred. $FeCl_3$ was sometimes found in $TiCl_4$ in low concentrations.

Natta and co-workers showed that when Al, $AlCl_3$, and $TiCl_4$ are reacted in benzene, a composition having the structure shown in XIII was formed.

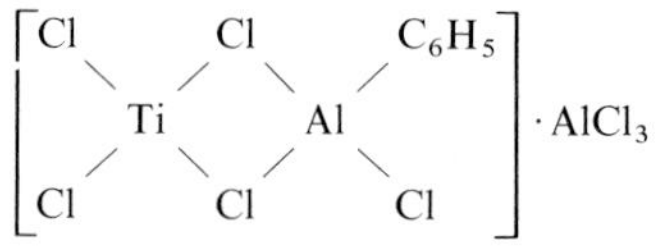

XIII

It was active for polymerization of ethylene (90). The $AlCl_3$ was liberated from the complex by addition of tetrahydrofuran (to form $AlCl_3 \cdot THF$), and the remaining Ti–Al complex was used to polymerize propylene to a partially isotactic crystalline polymer, about 10% insoluble in heptane.

The compound $Al[N(C_6H_5)_2]_3$ in combination with $TiCl_3$ has been shown by Natta and co-workers to be an isotactic-specific catalyst for polymerization of propylene (91). They showed, however, that an aluminum alkyl was formed by the reaction shown in Eq. 12–19.

$$-\overset{|}{Al}-N(C_6H_5)_2 + CH_2{=}CH{-}CH_3 \xrightarrow{\geq 100^\circ C} -\overset{|}{Al}-CH_2{-}CH{=}CH_2 \quad (12{-}19)$$

Feay (92) reported that a compound having the structure

$$(C_5H_5)_2Ti[CH_2Al(CH_3)Cl]_2$$

was formed when $(C_5H_5)_2TiCl_2$ and $Al(CH_3)_3$ were reacted. The isolated compound was mixed with $TiCl_3$ to form a highly isotactic-specific catalyst for propylene, about 90 to 98% insoluble.

The soluble catalyst $VCl_4{-}(C_6H_5)_4Sn{-}AlBr_3$ was found active for polymerization of ethylene and copolymerization of ethylene and propylene (93). It was inactive for polymerization of propylene. Subsequent work showed the presence of $C_6H_5AlBr_2$. A similar catalyst for ethylene was made by reacting $Sn(C_4H_9)_4$, $AlCl_3$, and $(C_5H_5)_2TiCl_2$ (94).

Other examples have been reported:

Balas, DeLaMare, and Schissler (95) have obtained 98% cis-1,4-poly-butadiene with a catalyst obtained by reacting $CoCl_2$ and $AlCl_3$ in benzene solvent, as shown in structure XIV.

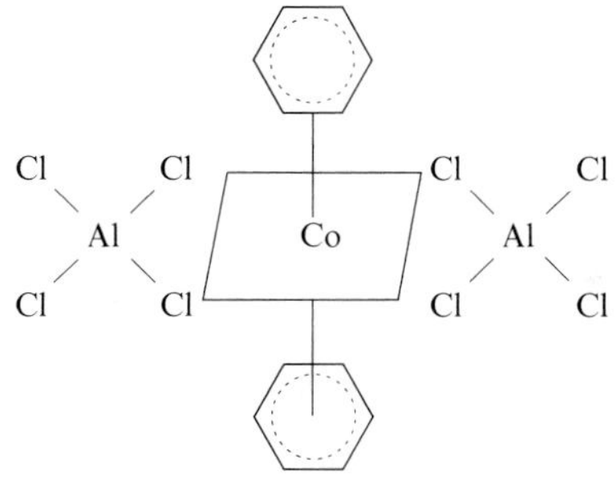

XIV

Jenkins and Timms prepared a high cis-1,4 (up to 92%) polybutadiene with a ternary catalyst composed of $Ni(PCl_2)_4$, $TiCl_4$, and $AlCl_3$ (96).

Boron hydrides and alkyls in combination with $TiCl_3$ have been shown to be active for ethylene and propylene, such as $Ti(BH_4)_3$, $Ti[B(C_6H_5)_4]_3$, and $TiCl_3 + Et_3N \cdot BH_3$ (97, 98). Isotactic polypropylene was obtained with the first two compounds.

Isotactic polypropylene was also reported to be made with the following: (1) Ti, Al, ICl plus a carboxylic amide (99); (2) Mg or Al + $TiCl_3$ or $TiCl_4$ (100, 101); (3) Al, $TiCl_3$ ball-milled in presence of H_2 (102); (4) diazomethane, $AlCl_3 + VCl_3$, $TiCl_3$ or $VCl_3{-}TiCl_3$ mixture (103); and (5) $TiCl_4$, Al + hexamethyl phosphoramide (104).

A polybutadiene polymer having predominantly *trans*-1,4 structure was reported by Marshall and Ridgewell to be formed when cyclopentene was polymerized with a mixture of WCl_6 and $AlBr_3$ (105).

A binary mixture of $TiCl_4$ (or $TiBr_4$, TiI_4) and a zirconium or hafnium borohydride with or without $AlCl_3$ was reported as a catalyst for ethylene (106).

A ternary mixture $TiCl_3$–Na–$(C_5H_5)_2TiCl_2$ (in absence and presence of H_2) was found active and stereospecific for the polymerization of propylene (107). Weickman found the ternary mixture of $TiCl_4$, Al, and an alcohol (ethanol) active for polymerizing ethylene (108). Coover and Joyner found the product of Na, $TiCl_4$, and hexamethylphosphoramide active (109). Apotheker and co-workers (109a) copolymerized ethylene and propylene in presence of the mixture $Zr(BH_4)_4$, VCl_4, and $AlBr_3$ to a product that had 57% propylene, an inherent viscosity of 3.49 dl/g, and was completely soluble in cyclohexane. High *cis*-1,4-polybutadiene was synthesized with a nickel diisopropylsalicylate–BF_3 Et_2O–$LiBBu_4$ catalyst (109b). Green and co-workers have synthesized polyacetylene by treating bisphosphine complexes of nickel and cobalt chloride with borohydride (109c).

An extensive investigation was made at the Sumitomo Laboratories and Kyoto University by Fukui and co-workers (110–112) to develop practical catalyst by reactions of Al, Zn, or Mg metals with transition metal salts such as $TiCl_4$, $TiCl_3$, $TiCl_2$, and $FeCl_3$. Temperatures near 150°C and pressures up to 150 atm were used for durations of at least several hours. Polyethylenes with densities near 0.94 to 0.95 g/cm³ and crystallinities of about 70 to 76% were made, comparing favorably with commercial products. The rates of polymerizations, however, were considerably lower than obtained with Ziegler–Natta and Phillips catalysts. Propylene was polymerized to solids that had melting points near 140° to 150°C and molecular weights of about 3×10^4 with $[\eta] = 0.90$ dl/g. The $VOCl_3$–Al catalysts were examined by Hirooka (113, 114) who concluded that two kinds of catalyst species were present:

$$C_2H_5(C_2H_4)_n-C_2H_4^+ \quad ^-VOCl_4 \cdot Al$$
$$C_2H_5(C_2H_4)_n-C_2H_4^+ \quad ^-VOCl_4-C_2H_4 \cdot Al$$

Joyner and co-workers (115) also examined the Al–$TiCl_4$ catalyst in the synthesis and use for ethylene polymerization. A high density polyethylene (d = 0.96 g/cm³) with an inherent viscosity of 1.03 dl/g was obtained in a yield of 1,873 grams per gram of catalyst. The reaction conditions were 18 atm pressure, 55° to 75°C, and about 20 hours reaction time. The $TiCl_4$ was added to Al (1 : 1 molar) powder in mineral oil either at ambient temperature or at 200°C, but a final heating at 200°C was done in both cases.

Liquid aliphatic hydrocarbons were found more effective than aromatic hydrocarbons as the media for the polymerization.

7. Transition Metal Salts Modified by Metal Alkyls from Group IV, V, and VI Metals

Many examples of catalysts for polymerization of ethylene have been reported; some of these binary mixtures must be heated to high temperatures to make them active. The following examples are cited here: (1) $TiCl_4$–$SnEt_4$ (116); (2) $TiCl_4$–Et_3SiH (117); (3) WCl_5 or $MoCl_5$ + $Sn(n\text{-}Bu)_4$ (118, 118a); and (4) VCl_4–$PbEt_4$ (119). When propylene was polymerized with $SnEt_4$–$TiCl_4$ (ratio 0.54 to 0.91), a partially crystalline isotactic polymer was formed at 30°C (120). A polyethylene with a molecular weight greater than 39,000 was made with mixtures of $TiCl_4$ and trialkyl(aryl)tin hydrides or trialkyl(aryl)-tinphenoxide (30°C and 1 atm ethylene for $\frac{1}{2}$ hour) (121). A patent, issued to Farbenfabriken Bayer AG, describes a catalyst for ethylene polymerization consisting of $TiCl_2$ and one of several reactants, such as $PbCl_2$, $BiCl_3$, $CrCl_3$, $SnCl_2$, $FeCl_3$, or $ZrCl_2$ (122). Also, a catalyst from $ZrCl_2$ and CuCl was described. Crystalline polyethylene and polypropylene in the presence of mixtures of triethylsilane and $TiCl_4$ or $TiCl_3$ were reported to be made if the polymerizations were carried out at elevated temperatures and pressures (122a).

8. Transition Metal Compounds Prepared without Modification

$TiCl_2$ prepared by reduction of $TiCl_4$ with H_2 formed 1.5 gram solid polyethylene per 2.7 gram $TiCl_2$ during 1 hour at 197°C and 33 atm pressure (123). Natta also reported the activity of $TiCl_2$ for ethylene homopolymerization and ethylene–propylene copolymerization, but not for propylene polymerization (124). $TiCl_3$ produces only low molecular weight oil from propylene (125). Titanium nitride was active for the polymerization of ethylene at 140°C and 500 to 700 psig (126). A copolymer (7% propylene and 93% ethylene units) was synthesized by Ketley and Braatz from propylene with $Pd(CN)_2$ at 60°C, 50 psig pressure, and 114 hours (127). An isomerization-polymerization mechanism was proposed. Ermakov, Bukatov, and Zakharov (128) prepared a $TiCl_2$ by decomposing $TiCl_3$ in vacuum at 500°C or in the presence of H_2 at 700°C; it had a specific surface area of about 25 m^2/g (by BET method). They found that only about 0.1% or less of the total number of titaniums adjacent to the surface was active; both ethylene and propylene were polymerized, the latter to a partially isotactic product. Addition of $AlEt_2Cl$ was found to lower the activity of the $TiCl_2$, and the isotacticity of the formed polymer was increased. They suggested that the aluminum alkyl complexed and preferentially inactivated the less isotactic-specific centers. This author finds the decreased activity due to $AlEt_2Cl$ surprising.

9. The Transition Metal on a Support

Orzechowski, MacKenzie, and Aftandilian (128a–134) at the Cabot Laboratories prepared a number of supported metal alkyl-free catalysts using Cab–O–Sil pyrogenic silica, $TiCl_4$, and organometallic compounds of Si or Sn, such as SiO_2–$TiCl_4$–$(C_6H_5)_3SnH$ (129, 130). While the ternary mixture polymerized ethylene, neither binary pair [$SiO_2 + TiCl_4$ nor $(C_6H_5)_3SnH + TiCl_4$] was active. Other reducing agents included Cl_2SiH_2, $SiCl_2H$ (131), $ClSiH_3$ (133), and $(C_6H_5)_2SbH$ (134). A solid polymer was also obtained when Alon pyrogenic alumina, $VOCl_3$, and $(C_6H_5)_2SbH$ was used to form the catalyst (134). The yields of polyethylene were low relative to Ziegler–Natta and Phillips catalysts.

Long (135) modified the Phillips catalyst by reducing the activated catalyst with CO at 135° to 260°C and by adding metal alkyls from Sn, Pb, and Ge. While the latter increased activity for polymerization of propylene, the isotacticity remained nearly the same (about 24 to 46% insoluble). Compared to the unmodified Phillips catalyst, this catalyst was considerably more isotactic-specific. When similar catalysts were modified by addition of AlR_3 and mixtures of H_2O (or alcohol), higher rates of polymerization and higher isotactic polymers were claimed (135a).

Propylenes with higher isotactic contents were made by Bulatnikova and co-workers (136) with special chromia-support catalysts synthesized by them; for example, see Table 12–11. Earlier, Topchiev and co-workers showed that with chromium oxide-based catalysts, largely atactic polymer was formed [see references 3, 4, and 9, Clark *et al.* (9)]. Shmonina and co-workers (136a) found that triallylchromium on an aluminosilicate salt

TABLE 12–11

Stereospecific-Supported Catalysts^a (137)

Catalyst	% Polypropylene insoluble in boiling heptane	$[\eta]$ (dl/g)	Melting temp. (°C) (residue)
Chromium silicate	13	0.76	155–160
Chromoaluminum silicate	—	—	—
Chromomagnesium silicate	24	0.8	160–165
Chromoaluminomagnesium silicate	68	1.36	165–170
Chromoaluminomagnesium silicate prepared by ion exchange	76	0.9	165–170

^{a} Synthesis at 100° to 105°C and 32 to 35 atm propylene.

support gave predominantly *cis*-1,4 units in a butadiene polymerization, in contrast to 1,2 units reported earlier by others for unsupported systems.

Two interesting catalysts have been discovered by Carrick, Karol, and their co-workers for polymerization of ethylene (137–142).

The first to be disclosed consisted of bis(triphenylsilyl) chromate alone in a homogeneous solution or supported on high surface silicas (see Table 12–12) (137, 139–142). Alone it was only sluggishly active; 55 grams polyethylene formed during 6 hours at 150°C from 1 gram catalyst and at 20,000 psig ethylene. About 65 grams polymer ($[\eta] = 1.5$ dl/g) could be formed if 0.02 gram of the chromate was deposited on 1.1 gram SiO_2 and ethylene was polymerized for hours at 135°C and at only 600 psig pressure. If aluminum alkyls were also present, even higher activities were obtained (see Table 6–3, Chapter 6). However, when aluminum alkyls were added only to the chromate compound in absence of silica, only a feeble low pressure catalyst for ethylene polymerization was found. Hydrogen acted as a transfer agent in these catalysts.

TABLE 12–12

Ethylene Polymerization with Bis(triphenylsilyl) Chromate (138)

$(\phi_3SiO)_2CrO_2$ (g)	Pressure (psig)	Silica-alumina (g)[a]	Temp. (°C)	Time (hours)	Yield (g)	Melt index (deg/min)	HLMI[b] (deg/min)	HLMI/ MI	$[\eta]$ (dl/g)
1.00	20,000[c]	None	150	6	55	0.04	4.8	120	—
5.00	21,000[c]	None	130	16	170	0.002	0.8	400	—
0.20	600	2.0	160	5	140	4.5	221	49	1.50
0.05	600	2.0	136	3	41	3.7	424	114	—
0.02	600	1.1	135	6	65	1.0	88	88	—
0.10	600	0.5	158	2	30	0.7	54	77	2.02
0.10	600	0.3	160	6	90	0.7	60	86	2.15
0.20	600	1.5	153	2	110	0.4	44	110	2.41
0.05	600	1.0	135	6	110	0.3	41	137	—

[a] Supports were dehydrated at 500°C.
[b] High load melt index.
[c] See reference 2 in Karol *et al.* (138).

Carrick proposed that in the absence of the metal alkyl, there occurred an oxidation–reduction reaction between the chromate and ethylene to form a low valent Cr, as shown in Eq. 12–20.

$$\begin{array}{c} R\!-\!O \diagdown \quad \diagup\!\!\diagup O \\ Cr \\ R'\!-\!O \diagup \quad \diagdown\!\!\diagdown O \end{array} + CH_2\!=\!CH_2 \longrightarrow \begin{array}{c} RO \diagdown \\ Cr + 2\,CH_2O \\ R'O \diagup \end{array} \qquad (12\text{--}20)$$

$$R = R' = \phi_3Si \text{ or } R' = \text{silica gel surface}$$

A chemical transesterification reaction path was suggested when the chromate was fixed to SiO_2, as shown in Eq. 12–21. Alternately, a physical

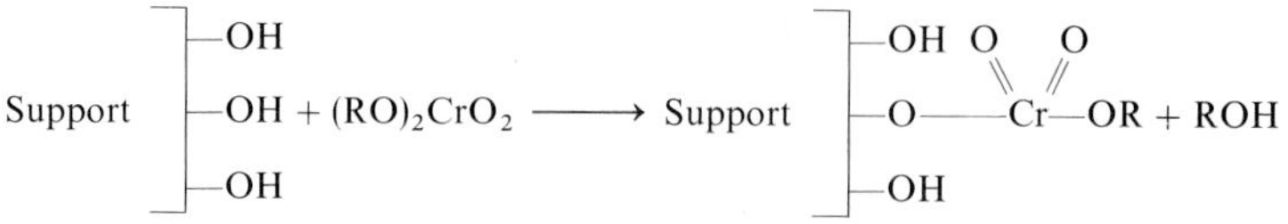

$$\text{Support} \begin{bmatrix} -OH \\ -OH \\ -OH \end{bmatrix} + (RO)_2CrO_2 \longrightarrow \text{Support} \begin{bmatrix} -OH \quad O \quad O \\ -O \quad\quad Cr-OR + ROH \\ -OH \end{bmatrix} \quad (12\text{--}21)$$

adsorption path was also considered. Nevertheless, upon the addition of ethylene, oxidation–reduction occurred forming the structure shown in XV.

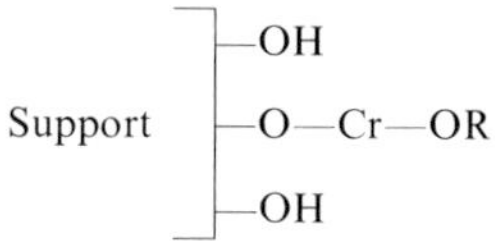

$$\text{Support} \begin{bmatrix} -OH \\ -O-Cr-OR \\ -OH \end{bmatrix}$$

XV

Carrick then proposed that when no metal alkyl was added, a possible initiation involved insertion of the monomer into a Cr–O bond to form $Cr-CH_2-CH_2-$ centers (137). This is reasonable in the light of the high temperature and pressure used. When ethylene was used below 300 psi and no SiO_2 present, little or no polyethylene was formed. Apparently, this insertion is more facile for a Cr–O bond where the oxygen is attached to the SiO_2 surface rather than to $Si(C_6H_5)_3$. Continued insertion of ethylene into Cr–O bonds constitutes the polymerization step. Such a mechanism may occur in the polymerization with the Phillips catalyst.

The other catalyst disclosed by Karol and co-workers consisted of chromocene, $(C_5H_5)_2Cr$, supported on high surface silicas (138). Recent announcements by Union Carbide suggest that this catalyst has been commercialized for production of high density polyethylene by a vapor phase process. Like the chromate catalyst (as well as the Ziegler–Natta and Phillips catalyst), the chromocene catalyst has a high response to hydrogen; for example, k_H/k_M at 90°C was 3.6×10^3 (H = H_2 and M = monomer).

They could not polymerize ethylene with homogeneous solutions of chromocene at pressures and temperatures up to 500 psi and 140°C, respectively. However, upon deposition of chromocene from hydrocarbon solutions on amorphous silicas of high surface area, a highly active catalyst was obtained. Polymer properties and kinetic parameters could be varied by using silicas of different surface areas and/or pore size, by controlling the loading of chromocene on the silica, and by the choice of temperatures at which the silicas were dehydrated (see Table 12–13). When SiO_2 and chromocene were reacted, cyclopentadiene was evolved, but some remained bound to the catalyst. Polymerization temperatures ranged from 30° to 170°C; both solution (>120°C) and slurry processes (<100°C) could be used. A maximum in activity was observed at about 60°C (Fig. 12–2).

TABLE 12–13

Influence of Silica Dehydration Temperature on Chromocene Catalysis[a] (139)

Dehydration temp. (°C)	$(C_5H_5)_2Cr$ (mole $\times 10^3$)	CrH_4 (psi)	Yield (g)	Normalized activity (g/mmole $(C_5H_5)_2Cr$/ 100 psi C_2H_4)	Relative activity
200	0.26	460	156	130	0.08
300	0.13	460	102	171	0.10
300	0.26	460	278	232	0.14
400	0.066	460	258	849	0.51
670	0.066	185	204	1671	1.00

[a] Grade 56 silica; 15 psi H_2; polymerization at 60°C for 1 hr.

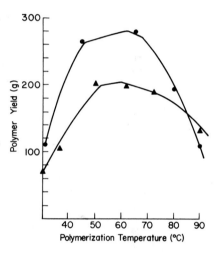

Fig. 12–2. Effect of polymerization temperature on activity: ● 0.26 mmole $(C_5H_5)_2Cr$, 0.4 g grade 56 (300°C), 460 psi C_2H_4, 1 hour reaction: ▲ 0.066 mmole $(C_5H_5)_2Cr$, 0.4 g grade 56 (670°C), 185 psi C_2H_4, 1 hour reaction.

Karol and co-workers proposed that the chromocene reacts with the hydroxyl group on the silica surface via the reactions shown in Eqs. 12–22 and 12–23. Several possibilities were suggested for the formation of Cr–C

$$
\begin{array}{c}
\text{—Si—OH} + Cp_2Cr \longrightarrow \text{—Si—O—Cr} + CpH
\end{array}
\qquad (12\text{–}22)
$$

$$
\begin{array}{c}
\text{—Si—OH} \\
\text{—Si—OH}
\end{array}
+ Cp_2Cr \longrightarrow
\begin{array}{c}
\text{—Si—O} \\
\quad\quad\;\; Cr + 2CpH \\
\text{—Si—O}
\end{array}
\qquad (12\text{–}23)
$$

or Cr–H bonds: (1) insertion of ethylene into Cr–O bond, and (2) addition of –H, –OH, hydrocarbon fragment or hydrogen from ethylene. Once the center is formed, propagation consists of continued addition of ethylene into the Cr–C bond, as shown in Eq. 12–24. A reactivity ratio $r = 72$ was

$$
\left[\begin{array}{c} \quad \\ \mathrm{Si-O-Cr-R} \\ \quad \end{array}\right] \begin{array}{c} O \\ | \\ | \\ O \end{array} \begin{array}{c} Cp \\ | \\ \quad \end{array} + CH_2{=}CH_2 \longrightarrow \left[\begin{array}{c} \quad \\ \mathrm{Si-O-Cr} \\ \quad \end{array}\right] \begin{array}{c} O \\ | \\ | \\ O \end{array} \begin{array}{c} Cp \\ | \\ \quad \end{array} CH_2-CH_2-R \longrightarrow etc.
$$

polymer

$$(12-24)$$

estimated for an ethylene–propylene copolymerization, showing that the catalyst does not homopolymerize α-olefins. The chromocene–SiO_2 catalyst produces polyethylenes with a relatively narrow molecular weight distribution. Above 100°C, unsaturated branched polymers or oligomers are formed by a simultaneous polymerization–isomerization process.

Karol attributed considerable importance to the presence of cyclopentadienyl ligand attached to the active center. When these catalysts were heated, some cyclopentadiene was liberated, and this had a pronounced affect on the overall polymerization behavior of the chromocene-supported catalyst (138a).

The overall polymerization process between 30° and 56°C had an apparent activation energy of 10.1 kcal/mole, which is in the range reported for Ziegler–Natta catalysts (139). At 90°C, $k_H/k_M = 3.6 \times 10^{-3}$, showing that the catalysts had a high response to hydrogen as a chain transfer agent relative to the monomer. The ratio k_H/k_P was 4.65×10^{-1}.

One other catalyst is worth mentioning. Bower and Long (143) polymerized ethylene is the presence of tetrakis(1-bicycloheptyl)chromium upon exposure to light. The initiator could be used in solution or on a SiO_2 support. Also found active on SiO_2 support was an allylchromium(II) (144).

IV. Relationship to Ziegler–Natta Catalysts

One can only be impressed by the large number and variety of metal alkyl-free catalysts that have been disclosed for the polymerization of olefins and diolefins. In the absence of knowledge about the mechanisms by which each operates, it is not at all obvious how such diverse compositions of transition metal salts can polymerize these monomers. Furthermore, evidence is strong that not all of the transition metal atoms in a catalyst act as active center. In fact, only a very small fraction become active centers. For this reason, the relationship between the structure of the active center and the starting transition metal salt is not easy to elucidate, and much is left to speculation.

One common thread that can link the MAF catalysts together and to Ziegler–Natta catalysts is the view that the active center is a transition metal–carbon bond. The existence of such a bond center can in fact be shown in most cases by radiolabeling and other diagnostic tests.

In a Ziegler–Natta catalyst, this transition metal–carbon bond can be formed by alkylation of the transition metal salt by the metal alkyl. Furthermore, the metal alkyl can function to reduce the transition metal salt to a lower valent compound, and this may be necessary for some salts. Another function of the metal alkyl is a chain transfer role (see Chapter 10). Finally, the metal alkyl acts as a scavenger of impurities that may poison the active centers. The stereochemical behavior of the catalyst is apparently dependent on the metal alkyl. All of these have been recognized as important functions of the metal alkyl in the Ziegler–Natta catalyst.

When then is the origin of the transition metal–carbon bonds that act as active centers in metal alkyl-free catalysts? For alkyl, allyl, and benzyl derivatives of the transition metal, this appears easy, for one can state safely that some of these bonds can become the active centers. Radiolabeling experiments in a few cases confirmed this view. But whether the starting material or some species derived from it is the active species is not certain. Not all transition metal salt molecules in the soluble catalysts, such as $(C_6H_5CH_2)_4Zr$, are simultaneously active. Perhaps Ballard's suggestion that the first addition of the monomer is slow accounts for the low activity of this catalyst.

The origin of active transition metal–carbon bond centers on the other MAF catalysts (catalysts described in Sections III,B,1,3,4, and 9) is more difficult to ascertain, and the suggestions made by different authors are very speculative.

But the experimental observations are indisputable in that the stereochemical characteristics of Ziegler–Natta catalysts are also present in the metal alkyl-free catalysts. True, the latter are not usually so versatile and manageable as are Ziegler–Natta catalysts, but as their number has increased and as their chemistry has become better understood, the differences have narrowed considerably.

The author considers the existence of these MAF catalysts and their polymerization capability as the most cogent and strongest evidence for the view that the activity and stereoregulation ability of the active center in the Ziegler–Natta catalysts are due to the transition metal component. This view is properly discussed in Chapter 13, Section II,A.

References

1. J. P. Hogan and R. L. Bank, Belgian Patent 530,617, January 24, 1955; U.S. Patent 2,825,721, March 4, 1958, Phillips Petroleum Company.
1a. A. Clark, *Ind. Eng. Chem.* **59**, 29 (1967).
2. See Chapter 3, Section V.
2a. See also *Mod. Plast.* **47** (7), 26 (1970).
2b. A. Clark, *Adv. Chem. Ser.* **91**, 387–398 (1969).
3. J. P. Hogan, Chap. III *in* "Copolymerization," George Ham, Ed., **18**, 89–111, Wiley (Interscience), New York, 1964.
4. M. Sittig, Catalysts and Catalytic Processes. Noyes Data Corp, New Jersey, 1967. pp. 138–175.
5. A. Clark, *Catal. Rev.* **3**, 145–174 (1969).
6. A. Clark and J. P. Hogan, *in* "Polyethylene" (A. Renfrew and P. Morgan, eds.), 2nd ed., p. 29. Wiley (Interscience), New York, 1960.
7. O. O. Juveland, E. F. Peters, and J. W. Shepherd, *Polym. Prepr., Am. Chem. Soc., Div. Polym. Chem.* **10**, 263–270 (1969).
8. J. P. Hogan, *J. Polym. Sci., Part A-1* **8**, 2637 (1970).
8a. *Chem. Week* **110** (19), 41 (1972).
9. A. Clark, J. N. Finch, and B. H. Ashe, *Proc. Int. Congr. Catal., 3rd, 1964* Vol. 11, p. 1010 (1965).
9a. P. B. Ayscough, C. Eden, and H. Steiner, *J. Catal.* **4**, 278 (1965).
9b. H. L. Krauss, *Proc. Int. Congr. Catal., 5th, 1972* Vol. 1, p. 207 (1973).
10. C. Eden, H. Feilchenfeld, and Y. Haas, *J. Catal.* **11**, 263 (1968).
11. K. G. Miesserov, *J. Polym. Sci., Part A-1* **4**, 3047 (1966).
12. Yu. I. Yermakov and V. A. Zakharov, *Proc. Int. Congr. Catal., 4th, 1968* Preprint 16 (1971).
13. P. Cossee and L. L. Van Reijen, *Actes Congr. Int. Catal., 2nd, 1960* Paper 82 (1961).
14. Yu. I. Pecherskaya, V. B. Kazanskii, and V. V. Voevodskii, *Actes Congr. Int. Catal., 2nd, 1961* Paper 108 (1961).
15. E. F. Peters, A. Zletz, and B. Evering, *Ind. Eng. Chem.* **49**, 1879 (1957).
16. Y. T. Tadokoro, K. Kasano, and T. Hosoi, *Kogyo Kagaku Zasshi* **70**, No 2, 144 (1967); Y. T. Tadokoro, K. Kasano, T. Hosoi, M. Katsurayama, T. Hashimoto, and H. Honishi, *ibid.* No. 10, p. 1615 (translated for Shell Development by E. E. Daub).
17. W. S. Anderson, Shell Development Company, private communication.
18. F. X. Werber, C. J. Benning, W. R. Wszolek, and G. E. Ashby, *Polym. Prepr., Am. Chem. Soc., Div. Polym. Chem.* **3**, No. 1, 122 (1962); *J. Polym. Sci., Part A-1* **6**, 743 (1968).
19. A. S. Matlack, U.S. Patents 2,891,041–4, June 16, 1959; 2,913,442, November 17, 1959; 2,938,020, May 24, 1960; 3,004,962, October 17, 1961; Great Britain Patent 848,285, September 14, 1960; D. S. Breslow and A. S. Matlack, U.S. Patent 2,961,435, November 22, 1960; see also Matlack and Breslow (20) for scientific paper.
20. A. S. Matlack and D. S. Breslow, *J. Polym. Sci., Part A* **3**, 2853 (1965).
21. G. F. D'Alelio, U.S. Patent 3,095,383, June 25, 1963, W. R. Grace and Company.
22. G. F. D'Alelio, U.S. Patents 3,092,589, June 3, 1963; 3,038,864, June 12, 1962; Great Britain Patent 891,575, Dal Mon Research Company.

23. R. D. Mulley and P. A. Small, Great Britain Patent 838,723, June 22, 1960, Imperial Chemical Industries, Ltd.
24. H. W. Coover and F. B. Joyner, U.S. Patent 3,140,277, July 7, 1964, Eastman Kodak Company.
25. C. Beerman and H. Bestian, *Angew. Chem.* **71**, 618 (1959).
26. G. L. Karapinka, J. J. Smith, and W. L. Carrick, *J. Polym. Sci.* **50**, 143 (1961).
27. J. Boor, *J. Polym. Sci., Part A-1* **9**, 617 (1971).
28. W. L. Carrick, F. J. Karol, and G. L. Karapinka, U.S. Patent 3,173,902, March 16, 1965, Union Carbide Corporation.
29. C. E. H. Bawn and J. Gladstone, *Proc. Chem. Soc., London* p. 227 (1959).
29a. Great Britain Patent 872,861, July 12, 1961, W. R. Grace and Company.
30. G. Natta, P. Pino, G. Mazzanti, and R. Lanzo, *Chim. Ind. (Milan)* **39**, 1032 (1957); P. Pino and G. Mazzanti, U.S. Patent 3,000,870, September 19, 1961, Montecatini.
31. P. Pino and G. Mazzanti, U.S. Patents 3,000,870, September 19, 1961, Montecatini; 2,992,212, July 11, 1961, E. H. DeButts, Hercules.
32. D. F. Hoeg, F. X. Werber, and W. R. Wszolek, U.S. Patent 3,150,107, September 22, 1964, W. R. Grace and Company.
33. L. M. Lanovskaya, N. A. Pravikova, A. R. Gantmakher, and S. S. Medvedev, *Polym. Sci. USSR (Engl. Transl.)* **11**, 1315 (1969).
33a. See reference 36 (of review) by A. Gumboldt, *Fortschr. Chem. Forsch.* **16**, No. 3–4, 299–328 (1971).
33b. H. de Vries, *Recl. Trav. Chim. Pays-Bas* **80**, 866 (1961).
33c. G. Henrici-Olive and S. Olive, *Fortschr. Hochpolym.-Forsch.* **6**, 421–472 (1969).
34. E. H. DeButts, U.S. Patent 2,992,212, July 11, 1961, Hercules, Incorporated.
35. K. Shikata, K. Nishino, and K. Azuma, *Kogyo Kagaku Zasshi* **68**, 352 (1965).
36. K. Azuma, K. Shikata, S. Oba, K. Nishimo, and T. Matsumura, *Kogyo Kagaku Zasshi* **68**, 347 (1965).
37. U. Giannini, U. Zucchini, and E. Albizzati, *J. Polym. Sci., Part B* **8**, 405 (1970).
38. U. Giannini, E. Albizzati, and U. Giannini, *J. Organomet. Chem.* **26**, 357 (1971).
39. U. Giannini and U. Zucchini, *Chem. Commun.* p. 940 (1968).
40. U. Zucchini, U. Giannini, E. Albizzati, and R. D'Angelo, *J. Chem. Soc.* p. 117 (1969).
40a. D. G. H. Ballard and P. W. Van Lienden, *Makromol. Chem.* **154**, 177 (1972).
40b. D. G. H. Ballard, J. V. Dawkins, G. M. Key, and P. W. Van Lienden, *Makromol. Chem.* **165**, 173 (1973).
40c. D. G. H. Ballard, N. Heap, B. T. Kilbourn, and R. J. Wyatt, *Makromol. Chem.* **170**, 1 (1973).
40d. D. G. H. Ballard, *Polym. Prepr., Am. Chem. Soc., Div. Polym. Chem.* **15**, No. 1, 364 (1974).
40e. D. G. H. Ballard, *Adv. Catal.* **23**, 263–325 (1973).
40f. D. G. H. Ballard, *Int. Congr. Pure Appl. Chem., 23rd, 1971, Spec. Lect.* Vol. 6, p. 219 (1971).
41. W. P. Long, OLS Patent 2,049,477, April 29, 1971, Hercules, Incorporated.
42. I. Sh. Guzman, O. K. Sharaev, E. I. Tinyakova, B. A. Dolgoplosk, *Izv. Akad. Nauk SSSR, Ser. Khim.* No. 3, p. 661 (1971).
43. G. Wilke, B. Bogdanovic, P. Hardt, P. Heimbach, W. Keim, M. Kroner, W. Oberkirch, V. Tanaka, E. Steinrucke, D. Walter, and H. Zimmermann, *Angew. Chem., Int. Ed. Engl.* **5**, 151 (1966); G. Wilke, U.S. Patent 3,379,706, April 22, 1968, Studiengesellschaft, Gm.b.H.
44. W. Herwig, A. G. M. Gumboldt, and K. Weissermel, U.S. Patent 3,501,415, March 17, 1970, Farbwerke Hoechst Aktiengesellschaft, Vormal Meister Lucius & Bruning.

45. D. G. H. Ballard, E. Jones, T. Modinger, and A. J. Pioli, *Makromol. Chem.* **148**, 175 (1971).
46. B. E. Job, A. J. P. Pioli, and T. Medinger, U.S. Patent 3,654,254, April 4, 1972, Imperial Chemical Industries, Ltd.
47. G. Wilke, *Angew. Chem., Int. Ed. Engl.* **2**, 105 (1963).
48. L. Porri, G. Natta, and M. C. Gallazzi, *Chim. Ind.* (*Milan*) **46**, 426 (1964).
49. B. D. Babitskii, B. A. Dolgoplosk, V. A. Kramer, M. I. Lobach, E. I. Tinyakova, N. N. Chesnokova, and V. A. Yakovlev, *Vysokomol. Soedin.* **6**, 2202 (1964); *Dokl. Akad. Nauk SSSR* **161**, 583 (1965).
50. L. Porri, M. C. Gallazzi, and G. Vitulli, *J. Polym. Sci., Part B* **5**, 629 (1967).
51. L. Porri, M. C. Gallazzi, and G. Vitulli, *J. Polym. Sci., Part C* **16**, 2525 (1967).
52. A. I. Lazutkina, L. Ya Alt, T. L. Matveeva, and A. M. Lazutkin, *Kinet. Katal.* **11**, 1591 (1970).
53. R. E. Rinehart, H. P. Smit, H. S. Witt, and H. Romeyn, *J. Am. Chem. Soc.* **83**, 4864 (1961).
54. A. J. Canale, W. A. Hewett, T. M. Shryne, and E. A. Youngman, *Chem. Ind.* (*London*) p. 1054 (1962).
55. F. Dawans and P. Teyssie, *J. Polym. Sci., Part B* **3**, 1045 (1965).
56. F. Dawans and P. Teyssie, *C. R. Hebd. Seances Acad. Sci.* **263**, 1512 (1966).
57. J. P. Durand, F. Dawans, and P. Teyssie, *J. Polym. Sci., Part B* **5**, 785 (1967).
58. F. Dawans and P. Teyssie, *Ind. Eng. Chem., Prod. Res. Dev.* **10**, No. 3, 261 (1971).
59. J. P. Durand, F. Dawans, and P. Teyssie, *J. Polym. Sci., Part A-1* **8**, 979 (1970).
60. J. P. Durand and P. Teyssie, *J. Polym. Sci., Part B* **6**, 299 (1968); J. C. Marechal, F. Dawans, and P. Teyssie, *ibid. Part A-1* **8**, 1993 (1970).
60a. J. Furukawa, E. Kobayashi, T. Kawagoe, N. Katsuki, and M. Imanari, *J. Polym. Sci., Part B* **11**, 239 (1973).
60b. J. M. Thomassin, E. Walckiers, R. Warin, and P. Teyssie, *J. Polym. Sci., Part B* **11**, 229 (1973).
60c. M. Julemont, P. Teyssie, J. M. Thomassin, E. Walckiers, and R. Warin, *Polymer Prepr., Am. Chem. Soc., Div. Polym. Chem.* **15**, No. 1 (1974); *Makromol. Chem.* **175**, 1673 (1974).
60d. D. M. Grant and E. G. Paul, *J. Am. Chem. Soc.* **86**, 2984 (1964).
60e. J. Furukawa, I. Kobayashi, N. Katsuki, and T. Kawagoe, *Makromol. Chem.* **175**, 237 (1974).
60f. B. A. Dolgoplosk, *Polym. Sci. USSR* (*Engl. Transl.*) **13**, 367–393 (1971); *Vysokomol. Soedin., Ser. A* **13**, 325–347 (1971).
60g. B. A. Dolgoplosk, E. I. Tinyakova, N. N. Stefanovskaya, I. A. Oreshkin, and V. L. Shmonina, *Eur. Polym. J.* **10**, 605 (1974).
61. F. Dawans, J. P. Durand, and P. Teyssie, *J. Polym. Sci., Part B* **10**, 493 (1972).
62. J. Furukawa and H. Morimura, *J. Polym. Sci., Part B* **7**, 541 (1969); **6**, 869 (1968).
63. L. R. Wallace and J. F. Harrod, *Macromolecules* **4**, No. 5, 656 (1971).
64. K. Komatu, J. Hirota, Y. Niromiya, and H. Yasunga, *Kogyo Kagaku Zasshi* **72**, 2630 (1969).
64a. T. Matsumoto, J. Furukawa, and H. Morimura, *J. Polym. Sci., Part A-1* **9**, 875 and 1971 (1971).
64b. V. A. Kormer, V. A. Vasilev, N. A. Kalinicheva, and O. I. Belgorodskaya, *J. Polym. Sci., Polym. Chem. Ed.* **11**, 2557 (1973).
64c. G. Lugli, A. Mazzei, and S. Poggio, *Makromol. Chem.* **175**, 2021 (1974).
65. M. Tsutsui and T. Koyano, *J. Polym. Sci., Part A-1* **5**, 683 (1967).
66. F. J. Karol, Belgian Patent 751,010, May 27, 1970, Union Carbide Corporation.
67. Idemitsu Kosan, Japan Publication 24, 153/70, 16.5.66, August 12, 1970.

68. Many of the earlier disclosures of this type of MAF catalyst were made in patents, where they unfortunately escaped scientific notice; it is most difficult to establish first disclosures.

69. C. J. Benning, W. R. Wszolek, and F. X. Werber, *Am. Chem. Soc. Meet., 1962, Polym. Prepr.* Vol. 3, No. 1, p. 138 (1962).

70. C. J. Benning, W. R. Wszolek, and F. X. Werber, *J. Polym. Sci., Part A-1* **6**, 755 (1968).

71. A. W. Anderson, J. M. Bruce, N. G. Merckling, W. L. Truett, W. F. Gresham, R. C. Schreyer, and W. N. Baxter, French Patent 1,134,740, E. I. du Pont de Nemours and Company.

72. C. J. Benning, U.S. Patent 2,956,050, October 11, 1960, W. R. Grace and Company.

73. C. J. Benning and W. R. Wszolek, U.S. Patent 3,046,266, July 24, 1962, W. R. Grace and Company; Great Britain Patent 901,720, July 25, 1962.

74. K. S. Minsker, G. T. Fedoseeva, and G. A. Razuvaev, *Vysokomol. Soedin.* **5**, 655 (1963).

75. V. A. Khodzhemirov, E. V. Zabolostaskaia, A. R. Gantmakher, and S. S. Medvedev, *Vysokomol. Soedin., Ser. B* **11**, 871 (1969).

76. H. G. Schutze, A. D. Suttle, Jr., and A. T. Watson, Belgian Patent 551,330, March 27, 1957, Esso Research and Engineering Company.

77. K. Oita and T. S. Nevitt, *J. Polym. Sci.* **43**, 585 (1960).

77a. W. S. Anderson, *J. Polym. Sci., Part A-1* **5**, 429 (1967).

78. I. Ya. Ostrovskaya, K. L. Makovetskii, B. A. Dolgoplosk, and Ye. I. Tinyakova, *Izv. Akad. Nauk SSSR, Ser. Khim.* p. 1632 (1967).

78a. M. Guainazzi, G. Filardo, G. Silvestri, S. Gambino, and R. Ercoli, *J. Chem. Soc., Chem. Commun.* 138 (1973).

78b. N. Yamazaki, M. Kase, and T. Ohta, *Polym. J.* **2**, No. 3, 364 (1971).

79. E. J. Arlman, Great Britain Patent 903,677, August 15, 1962, Shell International Research Maatschappij.

80. H. W. Coover, Jr., U.S. Patent 2,933,483, April 19, 1960, Eastman Kodak Company.

81. H. W. Coover, Jr., and W. C. Wooten, Jr., U.S. Patent 2,976,272, March 21, 1961, Eastman Kodak Company.

82. E. W. Duck and B. J. Ridgewell, *Eur. Polym. J.* **2**, 37 (1966).

83. T. Otsu and M. Yamaguchi, *J. Polym. Sci., Part A-1* **7**, 387 (1969).

83a. V. A. Kormer, T. L. Yufa, I. A. Poletaeva, B. D. Babitskii, and Z. D. Stepanova, *Dokl. Akad. Nauk SSSR* **185**, 873 (1969).

83b. G. M. Chernenko, T. I. Bevza, Yu. V. Korshak, S. I. Beihn, M. P. Teterina, and B. A. Dolgoplosk, *Vysokomol. Soedin., Ser. B* **13**, 288 (1971).

84. J. Boor and E. A. Youngman, *J. Polym. Sci., Part B* **2**, 265 (1964); J. Boor, *ibid. Part A* **3**, 995 (1965); *Polym. Prepr., Am. Chem. Soc., Div. Polym. Chem.* **6**, No. 2, 690 (1965); *J. Polym. Sci., Part B* **4**, 913 (1966); *Part A-1* **9**, 617 (1971); *Pure Appl. Chem.* **8**, 57 (1971); *IUPAC Meet., 1971.*

84a. B. A. Dolgoplosk, S. I. Beilin, Yu. V. Korshak, G. M. Chernenko, L. M. Vardanyan, and M. P. Teterina, *Eur. Polym. J.* **9**, 895 (1973).

84b. J. Furukawa, J. Kiji, H. Konishi, and K. Yamamoto, *Makromol. Chem.* **174**, 65 (1973).

84c. V. A. Kormer, B. D. Babitskii, M. I. Lobach, and N. N. Chesnokova, *J. Polym. Sci., Part C* **16**, 4351 (1969).

85. A. Vandi, F. Valeretto, and M. Ragazzini, U.S. Patent 3,546,196, December 8, 1970, Montecatini-Edison, SpA.

85a. H. Van Swet, R. S. Bauer, and W. Keim, U.S. Patent 3,644,564, February 22, 1972, Shell Oil Company; R. S. Bauer, H. Chung, P. W. Glockner, W. Keim, and H. Van Swet, U.S. Patents 3,635,937, January 18, 1972; 3,644,563, February 22, 1972, Shell Oil Company; R. S. Bauer, H. Chung, L. G. Cannell, W. Keim, and H. Van Swet, U.S. Patent 3,637,636, January 25, 1972, Shell Oil Company; R. S. Bauer, P. W. Glockner, W. Keim, and R. S. Mason, U.S. Patent 3,647,915, March 7, 1972, Shell Oil Company;

R. S. Bauer, H. Chung, K. W. Barnett, P. W. Glockner, and W. Keim, U.S. Patent 3,686,159, August 22, 1972, Shell Oil Company; R. S. Bauer, H. Chung, W. Keim, and H. Van Swet, U.S. Patent 3,759,889, September 18, 1973, Shell Oil Company.

85b. K. Hiraki and H. Hirai, *J. Polym. Sci., Part B* **7**, 449 (1969).

86. R. Van Helden, H. P. Braendlin, A. F. Bickel, and E. C. Kooyman, *Tetrahedron Lett.* No. 12, pp. 18 and 24 (1959).

87. M. Fisher, German Patent 874,215, March 12, 1953, Badische Anilin-Soda Fabrik und I.G. Farbenindustrie.

88. E. Törnqvist, U.S. Patent 3,420,809, January 7, 1969, Esso Research and Engineering Company.

89. G. A. Razuvaev, K. S. Minsker, and G. T. Fedoseeva, *Vysokomol. Soedin.* **4**, 1495 (1962); K. S. Minsker, G. T. Fedoseeva, N. A. Vorobieva, and G. A. Razuvaev, *Dokl. Akad. Nauk SSSR* **149**, 1351 (1963); *Vysokomol. Soedin.* **4**, 1361 (1962).

90. G. Natta, G. Mazzanti, and G. Pregaglia, *Gazz. Chim. Ital.* **89**, 2065 (1959).

91. G. Natta, G. Mazzanti, P. Longi, and F. Bernardini, *Chim. Ind. (Milan)* **42**, 457 (1960).

92. D. C. Feay, Great Britain Patent 995,395, June 16, 1965, Dow Chemical Company; also the receipt of (May 3, 1966) a preprint of proposed publication entitled "A New Catalyst for the Isotactic Polymerization of Propylene," is acknowledged.

93. W. L. Carrick, *J. Am. Chem. Soc.* **80**, 6455 (1958); W. L. Carrick, R. W. Kluiber, E. F. Bonner, L. H. Wartman, F. M. Rugg, and J. J. Smith, *ibid.* **82**, 3883 (1960).

94. A. Delbouille and H. Toussaint, U.S. Patent 3,424,737, January 28, 1969, Solvay & Cie.

95. J. G. Balas, H. DeLaMare, and D. O. Schissler, *J. Polym. Sci., Part A* **3**, 2243 (1965); J. G. Balas, U.S. Patent 3,067,189, December 4, 1962, Shell Oil Company.

96. D. K. Jenkins and D. G. Timms, U.S. Patent 3,414,555, December 3, 1968, International Synthetic Rubber Company.

97. H. W. Coover, Jr., U.S. Patent 3,057,835, October 9, 1962, Eastman Kodak Company.

98. Great Britain Patents 865,322, Farbenfabriken Bayer; 882,600, November 15, 1961, Pechiney.

99. H. W. Coover, Jr., U.S. Patent 2,962,487, November 29, 1960, Eastman Kodak Company.

100. J. T. Kummer, U.S. Patent 3,179,601, April 20, 1965, Dow Chemical Company.

101. F. B. Joyner, N. H. Shearer, Jr., and H. W. Coover, Jr., *J. Polym. Sci.* **58**, 881 (1962).

102. D. F. Hoeg, U.S. Patent 3,096,316, July 2, 1963, W. R. Grace and Company.

103. E. W. Duck and B. J. Ridgewell, *Eur. Polym. J.* **2**, 37 (1966).

104. H. W. Coover, Jr., U.S. Patent 2,962,487, November 29, 1960, Eastman Kodak Company.

105. P. R. Marshall and B. J. Ridgewell, *Eur. Polym. J.* **5**, 29 (1969).

106. D. R. Witt, U.S. Patent 3,189,589, June 15, 1965, Phillips Petroleum Company.

107. K. Machida, A. Ogawa, and T. Kiyotoshi, *Kogyo Kagaku Zasshi,* **70** (10), pp. 1812–1818 (1967).

108. A. Weickman, *Angew. Chem.* **72**, 866 (1960).

109. H. Coover, Jr. and F. B. Joyner, U.S. Patent 3,184,443, May 18, 1965, Eastman Kodak Company.

109a. D. Apotheker, A. L. Barney, and N. Brodoway, U.S. Patent 3,597,367, August 3, 1971, E. I. du Pont de Nemours & Company.

109b. C. Dixon, E. W. Duck, and D. K. Jenkins, *Eur. Polym. J.* **8**, 13 (1972).

109c. M. L. H. Green, M. Nehmer, and G. Wilkinson, *Chem. Ind. (London)* p. 1136 (1960).

110. K. Fukui, T. Kagiya, T. Shimidzu, T. Yagi, S. Machi, S. Yuasa, M. Hirata, and S. Kodoma, *J. Polym. Sci.* **37**, 341 (1959).

111. K. Fukui, T. Kagiya, T. Yagi, T. Shimidzu, and S. Yuasa, *J. Polym. Sci.* **37**, 353 (1959).
112. K. Fukui, T. Shimidzu, T. Yagi, S. Fukumoto, T. Kagiya, and S. Yuasa, *J. Polym. Sci.* **55**, 321 (1961).
113. M. Hirooka, S. Yuasa, and K. Fukui, *Kogyo Kagaku Zasshi* **65**, 1865 (1962).
114. M. Hirooka and K. Fukui, *Kogyo Kagaku Zasshi* **65**, 1869 (1962).
115. F. B. Joyner, N. H. Shearer, and H. W. Coover, Jr., *J. Polym. Sci.* **58**, 881 (1962).
116. Y. Takami, *Kogyo Kagaku Zasshi* **64**, 2049 (1961).
117. R. K. Freidlina, E. Ts. Chukovskaya, I. Tsao, and A. N. Nesmeyanov, *Dokl. Akad. Nauk. SSSR* **137**, 885 (1961).
118. W. F. Gresham and N. G. Merckling, U.S. Patent 2,872,439, February 3, 1959, E. I. du Pont de Nemours & Company.
118a. P. R. Hein, *J. Polym. Sci., Polym. Chem. Sect.* **11**, 163 (1973).
119. H. Kroeper, H. M. Wetz, and R. Platz, German Patent 1,076,369, February 25, 1960, BASF.
120. N. Ashikari and M. Honda, *Bull. Chem. Soc. Jpn.* **34**, 767 (1961).
121. K. Itoi, U.S. Patent 3,501,450, March 17, 1970, Kurashiki Rayon Company, Ltd., Japan.
122. Great Britain Patent 825,958, December 23, 1959, Farbenfabriken Bayer AG.
122a. S. Cesca, M. L. Santostasi, and G. Bertolini, *Chim. Ind. (Milan)* **51**, 1093 (1969).
123. Great Britain Patent 778,639, July 10, 1957, E. I. du Pont de Nemours & Company.
124. G. Natta, *Makromol. Chem.* **61**, 46 (1963).
125. J. Boor and E. A. Youngman, *J. Polym. Sci., Part B* **2**, 265 (1964).
126. F. X. Werber, D. F. Hoeg, and W. R. Wszolek, U.S. Patent 2,973,349, February 28, 1961, W. R. Grace and Company.
127. A. D. Ketley, U.S. Patent 3,535,302, October 20, 1970, W. R. Grace and Company; A. D. Ketley and J. A. Braatz, *J. Polym. Sci., Part B* **6**, 341 (1968).
127a. S. Otsuka, K. Mori, and M. Kawakami, *Kogyo Kagaku Zasshi* **67**, 1652 (1964).
128. Yu. I. Ermakov, V. A. Zakharov, and G. D. Bukatov, *Proc. Int. Congr. Catal., 5th, 1972* p. 399 (1973); G. D. Bukatov, V. A. Zakharov, and Yu. I. Ermakov, *Kinet. Katal.* **12**, 505 (1971).
128a. A. Orzechowski and J. C. MacKenzie, U.S. Patent 3,216,982, November 9, 1965, Cabot Corporation.
129. A. Orzechowski and J. C. MacKenzie, U.S. Patent 3,166,544, January 19, 1965, Cabot Corporation.
130. A. Orzechowski, U.S. Patent 3,243,421, March 29, 1966, Cabot Corporation.
131. V. D. Aftandilian and J. C. MacKenzie, U.S. Patent 3,243,422, March 29, 1966, Cabot Corporation.
132. V. D. Aftandilian, U.S. Patent 3,274,120, September 20, 1966, Cabot Corporation.
133. J. C. MacKenzie, U.S. Patent 3,280,096, October 18, 1966, Cabot Corporation.
134. V. D. Aftandilian, U.S. Patent 3,285,890, November 15, 1966, Cabot Corporation.
135. W. P. Long, U.S. Patent 3,639,379, February 1, 1972, Hercules.
135a. W. P. Long, U.S. Patent 3,639,378, February 1, 1972, Hercules.
136. E. L. Bulatnikova, A. S. Semenova, A. A. Buniyat-Zade, and N. A. Danilova, *Plast. Massy* **7**, 6–7 (1968).
136a. V. L. Shmonina, N. N. Stefanovskaya, E. I. Tinyakova, and B. A. Dolgoplosk, *Vysokomol. Soedin., Ser. B* **12**, 566 (1970).
137. W. L. Carrick, R. J. Turbett, F. J. Karol, G. L. Karapinka, A. S. Fox, and R. N. Johnson, *J. Polym. Sci., Part A-1* **10**, 2609 (1972).
138. F. J. Karol, G. L. Karapinka, C. Wu, A. W. Dow, R. N. Johnson, and W. L. Carrick, *J. Polym. Sci., Part A-1* **10**, 2621 (1972).

138a. F. J. Karol and C. Wu, *J. Polym. Sci., Polym. Chem. Ed.* **12**, 1549 (1974).

139. F. J. Karol, G. L. Brown, and J. M. Davison, *J. Polym. Sci., Polym. Chem. Ed.* **11**, 413 (1973).

140. L. M. Baker and W. L. Carrick, *J. Org. Chem.* **35**, 744 (1970).

141. L. M. Baker and W. L. Carrick, U.S. Patent 3,324,101, June 6, 1967, Union Carbide.

142. W. L. Carrick, G. L. Karapinka, and R. J. Turbett, U.S. Patent 3,324,095, June 6, 1967, Union Carbide.

143. B. K. Bower and W. P. Long, U.S. Patent 3,666,743, May 30, 1972, Hercules.

144. Netherlands Patent Appl. 6,918,920, Union Carbide.

13

Mechanisms for Initiation and Propagation of Olefins

I. Introduction

Many mechanistic schemes have been proposed to explain the way the two components of the Ziegler–Natta catalyst interact to form the active centers. The proposed mechanisms are often characterized by considerable detail, which can be appreciated only by reading the original papers. This chapter describes only the main features of each mechanism for a general comparison. Some of these mechanisms differ only slightly, and the differences may be only semantic.

While olefin and diolefin polymerizations hold many common features, the differences are sufficient to warrant separate treatments. The diene mechanisms are described in Chapter 14. The experimental evidence is directed toward differentiating between various types of mechanisms, rather than confirming the validity of any one individual mechanistic proposal.

Of the four types of mechanisms that have been proposed for polymerization of olefins, the most attractive ones have been those in which propagation takes place at a transition metal–carbon bond or at a base metal–carbon bond. The preponderance of experimental evidence favors the transition metal–carbon bond as the growth center. While in heterogeneous catalysts there is no apparent need for the alkylated growth center to be complexed to a base metal alkyl (already the center in part of a poly-metal atom crystal), there is evidence to support the view that, in some but not necessarily all homogeneous catalysts, the base metal alkyl stabilizes the alkylated transition metal center against decomposition by complexing with it. Experimental evidence for these views, which is based on Ziegler–Natta and metal alkyl-free catalysts, is discussed.

II. Nomenclature

The proposed mechanisms are classified according to the structure of the growth center: (1) the center is a transition metal–carbon bond, (2) the center is a base metal–carbon bond, (3) the center is a bound radical, and (4) the center is a bound anion (1).

The literature uses the terminology "monometallic" and "bimetallic" in describing both mechanisms and catalysts. When talking about mechanisms, the notations monometallic and bimetallic have been used to show that one or two metal atoms are involved (and are essential) in the growth step, respectively. On the other hand, some workers refer to metal alkyl-free catalysts (Chapter 12) and Ziegler–Natta catalysts as monometallic- and bimetallic-coordination anionic catalysts, respectively.

Natta and co-workers judiciously avoided this problem by using the simple designation Me–C (or M–C or cat–E) for the active center without requiring M to be the base metal or the transition metal (2). Figure 13–1 shows this scheme for a heterogeneous catalyst with the M–C bond in opposite polarizations. Natta and co-workers accepted the bottom scheme

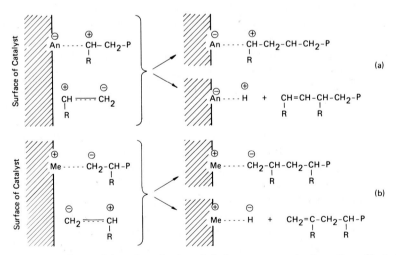

Fig. 13–1. Scheme of the polymerization of vinyl monomers under formation of isotactic polymers (2).

(b) as applicable to isotactic polymerizations and used it advantageously in kinetic treatments.

This mono- and bimetallic nomenclature can get cumbersome for certain mechanistic proposals. Consider, for example, three types of active centers that have been proposed for the active metal–carbon bond M–Pn (Pn is the growing polymer chain, M is Ti or Al, and Cl is chlorine ligands of $TiCl_3$ crystal surface or attached to Al) (see structures I, II, and III). A common

feature of growth at all three structures is the complexing of olefin at the Ti center. In structures I and II, however, growth takes place in the Ti–Pn bond, while in structure III, the Al–Pn is the growth bond. According to some authors, the presence of AlR_3 as part of the complex is essential for a number of reasons: (1) it stabilizes the transition metal–carbon bond and prevents its decomposition, (2) it makes the transition metal–carbon bond more active for polymerization, and/or (3) it is necessary to orient the complexed α-olefin in the isotactic configuration. Should the mechanism via structure II be designated as mono- or bimetallic, or is it better to classify the proposed mechanisms according to growth at the transition metal–carbon or at the base metal–carbon bond centers? The author has selected the latter approach in this book.

III. Proposed Mechanisms

The proposed mechanisms described in this section have served a worthwhile purpose in advancing Ziegler–Natta chemistry and deserve some attention even though most are now outdated. In many cases, the proposals provoked individual workers to perform experiments that resulted in valuable information, irrespective of mechanism.

The reader should keep in mind that many workers after publishing specific mechanistic proposals later modified their views, and some even reversed themselves completely in the light of new evidence (the author is an example). The proposed mechanistic schemes are presented in four classifications.

A. GROWTH OCCURS AT A TRANSITION METAL–CARBON BOND

Nenitzescu and co-workers (3) made the suggestion in 1956 that the transition metal–carbon bond was the active center, but they rejected this view in favor of a radical mechanism. According to Nenitzescu's rejected proposal, the olefin polymerized at a tetravalent alkylated titanium trichloride, according to the scheme in Eq. 13–1.

$$R\text{—}TiX_3 \longrightarrow R:]^-TiX_3]^+$$
$$R:]^-TiX_3]^+ + CH_2\text{=}CH_2 \longrightarrow RCH_2CH_2:]^-TiX_3]^+ \ldots$$

(13–1)

A mechanism employing a titanium in a lower valence state, such as RTiCl, was suggested in 1958 by Ludlum and co-workers (4) to be the active site. It was proposed that an alkyltitanium chloride with a valence of II complexed with an ethylene molecule, which then was inserted into the titanium–carbon bond, as shown in Eq. 13–2. These workers reported

$$ClTi\text{—}(C_2H_4)_nR + C_2H_4 \xrightarrow[\text{slow}]{k_p} \overset{\overset{\displaystyle CH_2\text{=}CH_2}{\displaystyle |}}{ClTi\text{—}(C_2H_4)_n\text{—}R} \xrightarrow{\text{fast}} Cl\text{—}Ti\text{—}(C_2H_4)_{n+1}\text{—}R$$

(13–2)

that the rate of polymerization of ethylene with $AlEt_3$–$TiCl_4$ (or Al-i-Bu_3–$TiCl_4$) catalyst was maximum when the average valence state of titanium was II.

Carrick (5) also proposed a mechanism in 1958 in which the alkylated species was in a divalent state, but which was also stabilized by complexing with a metal alkyl or an $AlCl_3$ molecule, as shown in Eq. 13–3, where

$$AlX_2R + VCl_4 \longrightarrow \begin{matrix} X & & X \\ & \diagdown \diagup & \diagdown \\ & Al & V\text{—}R \\ & \diagup \diagdown \diagup & \\ X & & X \end{matrix}$$

(13–3)

X = halogen, alkyl, or aryl, and R = alkyl or aryl (valence of V = II). According to this mechanism, coordination takes place at the vanadium atom followed by insertion of the ethylene into the V–R bond. Repetition of this two-step process constitutes the propagation step by which polymer chains are formed at the center.

Breslow and Newburg (5a) postulated growth at the Ti–C center for the soluble cyclopentadienyltitanium chloride-based catalyst, as shown in Eq. 13–4. Transfer of Et and insertion of ethylene into the Ti–C bond was

$$
\begin{array}{c}
\underset{\text{Et}}{\overset{\text{Cp}}{\underset{|}{\text{Ti}}}} \cdots
\end{array}
\quad (13\text{-}4)
$$

facilitated by weakening of Ti–Et bond when CH_2=CH_2 became complexed to the Ti center.

Cossee (6–9) developed these early proposals into a very sophisticated mechanism, which he supported by molecular orbital calculations. He proposed that the active site was a transition metal atom in an octahedral configuration (structure IV) and that one position was vacant, $\square$, due to a missing ligand, the remaining position being occupied by one alkyl and four ligands (6). In structure IV, M is the transition metal ion, R is the alkyl derived from metal alkyl or growing polymer chain, X is the ligands of crystal (Cl in TiCl$_3$), and $\square$ is the vacant octahedral position.

$$
\begin{array}{c}
\text{R} \\
| \quad X_2 \\
X_4\text{—M--}\square \\
X_1 \quad X_3
\end{array}
$$

IV

One growth step in Cossee's mechanism is shown in Eq. 13–5; here R is the growing polymer chain and $\square$ is the vacant octahedral position.

$$
\begin{array}{c}
\text{R} \\
| \quad X \\
X\text{—M--}\square \\
X \quad X
\end{array}
+ C_2H_4 \longrightarrow
\begin{array}{c}
\text{R} \\
| \quad X \quad CH_2 \\
X\text{—M} \quad \| \\
X \quad X \quad CH_2
\end{array}
\longrightarrow
\begin{array}{c}
\text{R} \quad CH_2 \\
| \quad X \\
X\text{—M---} \\
X \quad X \quad CH_2
\end{array}
\longrightarrow
$$

$$
\quad (13\text{-}5)
$$

$$
\begin{array}{c}
\square \quad R \\
| \quad X \quad CH_2 \\
X\text{—M——CH}_2 \\
X \quad X
\end{array}
\xrightarrow{\text{migration}}
\begin{array}{c}
R \\
| \\
CH_2 \\
| \\
CH_2 \quad X \\
X\text{—M---}\square \\
X
\end{array}
$$

Repetition of this growth step is responsible for the polymerization of the olefin to a high molecular weight polymer. In Cossee's scheme, the growing alkyl group and the vacancy exchange octahedral positions after each insertion of monomer, and the R group must migrate back to its former position. The stereochemical aspects of Cossee's mechanism are discussed in Chapter 15.

Cossee also suggested a driving force for the polymerization on the basis of molecular orbital calculations (6–9). The olefin is coordinated to the transition metal ion at the vacant octahedral position through π bonding.

Figure 13–2 shows a schematic picture of the spatial arrangement of the relevant orbitals in this π bond.

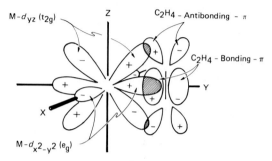

Fig. 13–2. Schematic picture showing spatial arrangement of the relevant orbitals in a "π bond" between a transition metal and C_2M_4 (6).

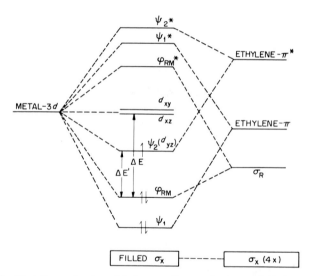

Fig. 13–3. Tentative molecular orbital energy diagram for the octahedral complex $RTiCl_4$. For reasons of simplicity, $4s$ and $4p$ orbitals are not considered and the Ti–Cl bond is taken to be 100% ionic (Cossee) (8).

Figure 13–3 describes the π complex in terms of a molecular orbital diagram, where M is titanium in the heterogeneous catalyst containing $TiCl_3$.

ϕ_{RM} is the energy level of the alkyl titanium bond. This bond is apparently stable when $RTiCl_4 \square$ is not complexed with an olefin because the catalyst can be stored for a long time. ΔE represents the energy that an electron in the Ti–R bond must be excited to in order to weaken the bond. ΔE is large enough an energy barrier to maintain the Ti–R bond intact in the absence of olefin.

When an olefin molecule is coordinated to the $RTiCl_4 \square$ octahedral complex, however, a new energy level $\Psi_2(d_{yz})$ is formed by mixing metal d orbitals and ethylene π^* antibonding orbitals. This new orbital is sufficiently below the energy level of the original metal $3d$ orbitals so that an electron from the metal–carbon bond Ψ_{RM} can be easily excited into it. If $\Delta E'$ is smaller than the critical energy gap in the Chatt and Shaw model, the alkyl group will be expelled as a radical. This radical attaches itself in a concerted process to the nearest carbon atom of the olefin while, at the same time, the other side of the olefin connects itself to the metal. Eq. 13–5 shows this reaction path.

This mixing of orbitals results in lowering the activation energy for the rearrangement in which the alkyl group moves from the transition metal toward the nearest carbon atom of the olefin. In order for the isolated transition metal–carbon bonds to be stable in the absence of coordinated olefin, the electronegativity of the metal ion must be sufficiently low. The metal–carbon bond, however, must be sufficiently destabilized when an olefin becomes coordinated in the vacant position.

Cossee's model predicts that only certain transition metals will be highly active as catalysts. According to his molecular orbital energy diagram, the catalyst activity of the transition metal compound does not depend on the presence of electrons in the d_{yz}, d_{xz}, and d_{zy} orbitals. No two electrons, however, are allowed to be present in the $\Psi_2(d_{yz})$ orbital; this means that transition metal ions having one to three unpaired electrons are most effective. The specific action of the transition metal is most favorable only when the energy of the metal $3d$ level is between the bonding and antibonding levels of the olefin. A modification was made in 1966 in which Cossee described nuclear motions (vibrations) and subsequently the electron configurations and the energy of the system as a function of the positions of the nuclei (8). However, similar conclusions were reached.

Cossee, Ros, and Schachtschneider (9) carried out semiempirical molecular orbital calculations on the catalytic site, the complex formed between the site and the olefin, and several intermediate situations along the reaction path (Table 13–1). The authors considered these calculations as "experiments of a new kind, which need to be refined by comparing the results with known experimental facts." Table 13–1 collects the charges and bond orders for the complex during one growth step.

TABLE 13–1

Charges and Bond Orders along the Reaction Path in Cossee's Model (9)

Reaction path $\longrightarrow$	$\left(\frac{Cl}{2}\right)_4$ Ti—CH₃ + C₂H₄	$\left(\frac{Cl}{2}\right)_4$ Ti⋯⋯CH₃	$\frac{1}{4}$	$\frac{2}{4}$	$\frac{3}{4}$	$\left(\frac{Cl}{2}\right)_4$ Ti—CH₂ / CH₂ / CH₃	$\left(\frac{Cl}{2}\right)_4$ Ti—CH₂⋯CH₂—CH₃
↓Charges							
Ti	+0.40	+0.42	+0.41	+0.38	+0.38	+0.38	+0.40
CH₃	−0.37	−0.35	−0.28	−0.11	+0.05	+0.08	−0.01
C(3)H₂	0	−0.01	−0.05	−0.14	−0.29	−0.31	−0.30
C(4)H₂	0	−0.01	−0.02	−0.01	−0.04	−0.06	−0.05
Total Cl	−2.03	−2.05	−2.06	−2.12	−2.10	−2.09	−2.04
Total propyl	−0.37	−0.37	−0.35	−0.26	−0.28	−0.29	−0.36
↓Bond orders							
Ti–C(2)	0.18	0.17	0.12	0.04	−0.03	−0.03	≈0
Ti–C(3)	—	≈0	0.02	0.05	0.13	0.16	0.18
Ti–C(4)	—	≈0	0.01	≈0	−0.04	−0.06	−0.04
C(3)–C(4)	0.64	0.58	0.54	0.49	0.38	0.33	0.34
C(2)–C(4)	—	≈0	≈0	0.08	0.23	0.35	0.34

Their calculations showed that the charge on the Ti and Cl's remained nearly constant, although both can store some negative charge temporarily. The CH_3 group, however, does not move as an anion, as other authors suggested earlier. Ti shows its largest negative charge in the beginning and loses the negative charge gradually once ethylene complexes with ethylene. When Ti, V, and Cr centers are compared by this treatment, the calculations suggest that the lability of the Ti–CH_3 bond increases on going from Ti to Cr. Simultaneously, there is in the same direction a decreasing interaction between metal and ethylene. While a great lability may be a favorable factor in propagation, it may become a limiting factor for the formation of active centers. The authors cite this as a reason for poorer overall performance of Cr in Ziegler–Natta catalysts.

The reviewer feels that the authors did not mean this to be generally true but only when the ligand used was a chloride. Cr-based catalysts are indeed very active in the Phillips and Union Carbide type catalysts where oxide ligands are present and in which Cr–C bonds are the active centers.

Begley and Pennella (10) calculated the electronic energy levels for the molecular orbitals of the octahedral center $RTiCl_4\square$ and the corresponding olefin complex, $RTiCl_4 \cdot$ olefin. They considered an electronic transition from the highest filled orbital to the half-filled orbital above to be the initial step in the olefin insertion reaction (the growth step). The energy of this transition was calculated to be about 10 to 14 kcal/mole, corresponding to the most widely found value for Ziegler–Natta polymerizations. They suggested, then, that this electronic transition is actually the controlling step in the reaction.

Barriers of 3.02, 1.07, and 0.7 eV were calculated for excitation of electrons from the highest filled orbital to the lowest empty or partly filled orbital when the Ti–C was present in the structures CH_3TiCl_3 (an isolated molecule), in $RTiCl_4\square$, and in $RTiCl_4 \cdot$ olefin complex, respectively. The latter two are Ti centers in the crystal lattice of the heterogeneous catalyst. Begley and Pennella agree with Cossee that insertion of the olefin molecule into the Ti–C bond can occur only if the Ti–C bond is weakened by complexing the $RTiCl_4\square$ center with the olefin. According to their calculations, the energy barrier was lowered from 1.07 to 0.7 eV.

In all of the above mechanisms, the center involves only one transition metal atom. Several mechanisms have been proposed in which propagation takes place at a Ti–C bond, but which involve an active participation of neighboring Ti or Al atoms.

A mechanism was proposed by De Bruijn (11) in which the olefin coordinates at one titanium but is interposed into a metal–carbon bond of another titanium atom, as shown in Eq. 13–6. The growing alkyl migrates

$$
\begin{array}{ccc}
\underset{\substack{| \\ R}}{\overset{\substack{Cl \\ |}}{Cl-Ti}}\uparrow \underset{\substack{| \\ H_2C\!\!=\!\!CH_2}}{\overset{\substack{Cl \\ |}}{Cl-Ti-Cl}} \longrightarrow
\underset{\substack{R \\ }}{\overset{\substack{Cl \\ |}}{Cl-Ti}} \longleftarrow\!\!-\!\!- \underset{\substack{| \\ CH_2 \\ \diagdown \\ CH_2}}{\overset{\substack{Cl \\ |}}{Cl-Ti-Cl}} \longrightarrow
\underset{\substack{| \\ CH_2\!\!=\!\!CH_2}}{\overset{\substack{Cl \\ |}}{Cl-Ti-Cl}}\uparrow \underset{\substack{| \\ CH_2 \\ | \\ CH_2 \\ | \\ R}}{\overset{\substack{Cl \\ |}}{Ti-Cl}}\cdots
\end{array}
\qquad (13-6)
$$

between two adjacent titaniums on the surface of the TiCl$_3$ crystal surface.

Rodriguez and van Looy (12) require a complexed aluminum alkyl as an integral part of the center, even though growth takes place at a transition metal–carbon bond, as shown in Eq. 13–7, where • denotes Ti and □ denotes a chlorine vacancy.

$$ (13-7) $$

This mechanism has features of the Cossee model, as well as of earlier bimetallic models proposed by Natta and co-workers, and Patat and Sinn (see next section). Steric interactions between the metal alkyl and the complexed α-olefin are said to be responsible for stereochemical control (see Chapter 15). Another example, which was proposed by Langer, is described in Chapter 15 (see Fig. 15–6a).

A very unusual mechanism was recently proposed by Buls and Higgins (12a) for the polymerization of propylene in the presence of AlEt$_2$Cl–TiCl$_3$ catalyst. According to this mechanism, the propylene is first inserted into the Al–H bond and then transferred into a Ti–C bond, the latter being the growth center. Two bimetallic complexes per unit cell Ti$_8$Cl$_{24}$ were suggested. Upon addition of propylene to the Al–H bond, Al–Pr is formed and the step is repeated.

B. GROWTH AT A BASE METAL–CARBON BOND

In the early days of Ziegler–Natta history, most workers believed the growth center to be the base metal–carbon bond, that is, the Al–C bond in the catalyst $AlEt_3$–$TiCl_3$. Many different mechanisms were proposed based on this view (13–20).

Of these mechanisms, those proposed in 1960 by Natta (Eq. 13–8) and in 1958 by Patat and Sinn (Eq. 13–9) contain the most cogent features.

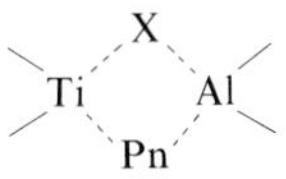

$$
\begin{array}{c}
\diagdown \quad\quad\overset{X}{\diagdown}\quad\quad\diagup \\
\diagup Ti \quad\quad Al \diagdown \\
\diagup \quad \underset{Pn}{\diagdown} \quad
\end{array}
$$

V

The structure shown in V (where Pn is the growing polymer chain and X is halogen) is common to both bimetallic mechanisms. The monomer is co-ordinated to the titanium while the Ti–polymer partial bond is simultaneously broken. In the Natta mechanism (13), the olefin becomes polarized and inserted into the aluminum–carbon bond.

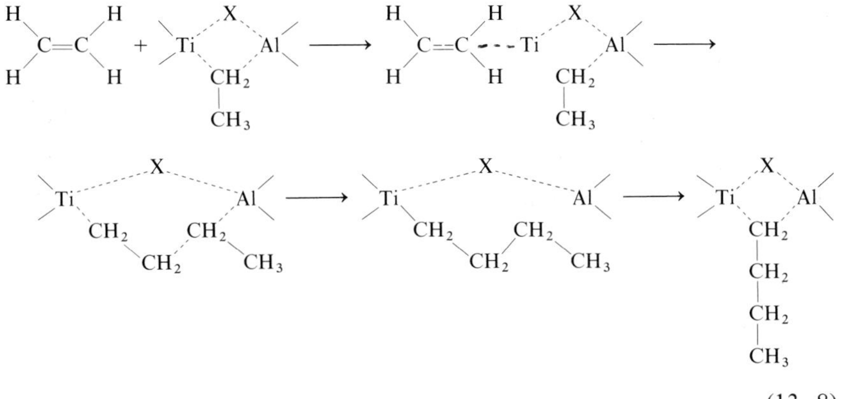

$$(13\text{–}8)$$

In the Patat–Sinn mechanism (14), the olefin becomes partially bonded between the methylene of the last added monomer unit and the titanium atom. The latter is partially bonded via an alkyl bridge to the aluminum. When a σ bond forms between the olefin and the methylene, the methylene unit becomes detached from the aluminum and now the new methylene unit becomes partially bonded to both metal atoms.

$$
\begin{array}{cccc}
& \overset{\displaystyle P}{\underset{\displaystyle\mid}{}} & \overset{\displaystyle P}{\underset{\displaystyle\mid}{}} & \overset{\displaystyle P}{\underset{\displaystyle\mid}{}} \\[4pt]
& CH_2 & CH_2 & CH_2 \\
& \mid & \mid & \mid \\
& CH_2 & CH_2 & {}^{-}CH_2 \\
CH_2{=}CH_2 + \;Ti \quad\quad Al & \longrightarrow \;Ti \quad\quad Al & \longleftrightarrow \overset{+}{Ti} \quad\quad Al & \longrightarrow \\
\qquad R & \uparrow \quad R & \uparrow \quad R \\
& CH_2{=}CH_2 & CH_2{=}CH_2
\end{array}
\qquad (13\text{–}9)
$$

$$
\begin{array}{ccccc}
& & CH_2{-}CH_2{-}P & & CH_2{-}CH_2{-}P \\
\overset{-}{C}H_2{-}\overset{+}{C}H_2 & & CH_2 & & CH_2 \\
& & \mid & & \mid \\
& & {}^{-}CH_2 & & CH_2 \\
\overset{+}{Ti} \qquad \overset{-}{C}H_2{-}CH_2{-}P & \longrightarrow \overset{+}{Ti} \qquad Al & \longrightarrow Ti \qquad Al \\
R{-}Al & R & R \\
\mid \text{\textbackslash}
\end{array}
$$

Other mechanistic schemes were proposed in which the Al–C was stated to be the growth center. These are mentioned briefly here, emphasizing only the salient feature of each proposal.

Eirich and Mark (15) also proposed a similar mechanism, which employed adsorbed layers of alkylaluminum chloride on the titanium chloride crystals.

Uelzmann (16) consolidated the essential features of the different proposed mechanisms and suggested the path in Eq. 13–10. $TiCl_3$ and $AlEt_2Cl$

$$
\begin{array}{ccc}
CH_2 & \overset{+}{C}H_2 \\
\parallel & \mid \\
CH_2 & CH_2 & CH_2{-}\overset{+}{C}H_2 \\
(TiCl_2)^{+}(AlR_3Cl)^{-} & \longrightarrow (TiCl_2)(AlR_3Cl)^{-} & \longrightarrow (TiCl_2)(AlR_3Cl)^{-} \longrightarrow \\
\text{Attraction} & \text{Activation} & \begin{array}{c}\text{Addition of R and}\\ \text{anionic migration}\end{array}
\end{array}
\qquad (13\text{–}10)
$$

$$
\begin{array}{c}
\qquad CH_2{-}R \\
\nearrow \\
CH_2 \\
\mid \\
(TiCl_2)^{+}(AlR_2Cl)^{-}
\end{array}
$$

Complex formation
(propagation)

combine to form an ion pair, and the cation $TiCl_2$ acts to polarize the olefin while the AlR_3Cl anion acts as the growth center.

Boor (17) proposed a concerted mechanism (Eq. 13–11), in which bond breaking and making occurred simultaneously, but he later rejected this view in favor of growth at a transition metal–carbon bond.

$$
\begin{array}{c}
\text{Et} \\
\text{Et}\diagdown \quad \diagup \text{Et} \\
\text{Al--Et} \\
\diagup \\
\text{Cl} \quad \overset{\text{H}}{\underset{\text{H}}{\diagdown}}\!\!C\!\!=\!\!C\!\!\overset{\text{H}}{\underset{\text{CH}_3}{\diagup}} \\
\uparrow \\
\text{Ti} \\
\text{Cl}\diagup \quad \diagdown\text{Cl}
\end{array}
\qquad\longrightarrow\qquad
\begin{array}{c}
\text{Et} \quad \text{CH}_3 \\
\text{Et}\diagdown \mid \quad \mid \\
\text{Al--CH}_2 \quad \text{H} \\
\text{Cl}\diagup \quad \overset{}{C}\!\!=\!\!C \\
\text{H}\diagup \uparrow \quad \diagdown\text{CH}_3 \\
\text{H} \quad \text{Ti} \\
\text{Cl}\diagup \quad \diagdown\text{Cl}
\end{array}
\qquad (13\text{--}11)
$$

$$
\begin{array}{c}
\text{Et} \\
\text{Et}\diagdown \mid \quad \text{Et} \\
\text{Al--CH}_2\text{--C---H} \\
\text{Cl}\diagup \quad \diagdown\text{CH}_3 \\
\\
\overset{\delta^+}{\text{Ti}} \\
\text{Cl}\diagup \quad \diagdown\text{Cl}
\end{array}
$$

Bier (18) and Gumboldt and Schmidt (19) proposed that aluminum alkyls are adsorbed on the $TiCl_3$ surface in two ways, as shown in Eq. 13–12.

$$
\begin{bmatrix}
\overset{\text{Cl}}{\underset{\text{Cl}}{\text{Cl}-\text{Ti}}}
\end{bmatrix}^{+}
\begin{bmatrix}
\overset{\text{R}}{\underset{\text{R}}{\text{Cl}-\text{Al}-\text{R}}}
\end{bmatrix}^{-}
\qquad \text{and} \qquad
\overset{\text{Cl}\ \ \text{R}}{\underset{\text{Cl}\ \ \text{R}}{\text{Cl}-\text{Ti}\cdot\text{Al}-\text{R}}}
\qquad (13\text{--}12)
$$

During the growth step, the growing polymer chain migrates from one aluminum atom and becomes attached to the olefin that is coordinated to an adjacent aluminum alkyl. A simultaneous electron transfer occurs between adjacent titanium atoms. The $TiCl_3$ surface acts to activate and stabilize the metal alkyl growth centers but does not coordinate the olefin itself. In a way, this mechanism can be regarded as an Aufbau reaction, which is enhanced by the $TiCl_3$ surface, as shown in Eq. 13–13.

$$
\begin{bmatrix}
\overset{\text{Cl}}{\underset{\text{Cl}}{\text{Cl}-\text{Ti}}} \\[4pt]
\overset{\text{Cl}}{\underset{\text{Cl}}{\text{Cl}-\text{Ti}}}
\end{bmatrix}^{+}
\begin{bmatrix}
\overset{\text{R}}{\underset{\text{R}}{\text{Cl}-\text{Al}-\text{P}}} \\[6pt]
\overset{\text{R}}{\underset{\text{R}}{\text{Cl}-\text{Al}^{-}-\text{CH}_2-\text{CH}_2{}^{+}}}
\end{bmatrix}^{-}
\longrightarrow
\begin{bmatrix}
\overset{\text{Cl}}{\underset{\text{Cl}}{\text{Cl}-\text{Ti}}} \quad \overset{\text{R}}{\underset{\text{R}}{\text{Cl}-\text{Al}}} \\[4pt]
\overset{\text{Cl}}{\underset{\text{Cl}}{\text{Cl}-\text{Ti}}}{}^{+} \quad \overset{\text{R}}{\underset{\text{R}}{\text{Cl}-\text{Al}-\text{CH}_2-\text{CH}_2-\text{P}}}
\end{bmatrix}^{-}
\qquad (13\text{--}13)
$$

Friedlander and Resnick (20) proposed a surface-coordinate mechanism, according to which the adsorbed olefin and the metal–carbon bond growth

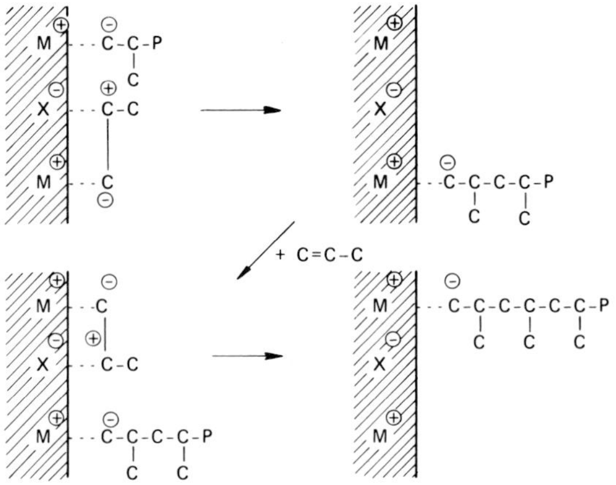

Fig. 13–4. Friedlander and Resnick's surface-coordinate mechanism (20).

centers shifted positions during each growth step (Fig. 13–4). Only one bond is made and broken at any one time.

A mechanism involving the breaking of only one bond and the formation of only one bond at each step was also proposed by Huggins (21), as shown in Eq. 13–14. This mechanism shares some features of a mechanism proposed earlier by Furukawa and Tsuruta (22).

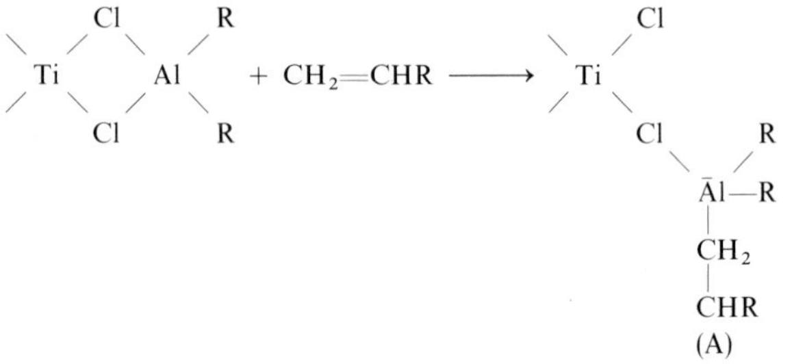

$$(13-14)$$

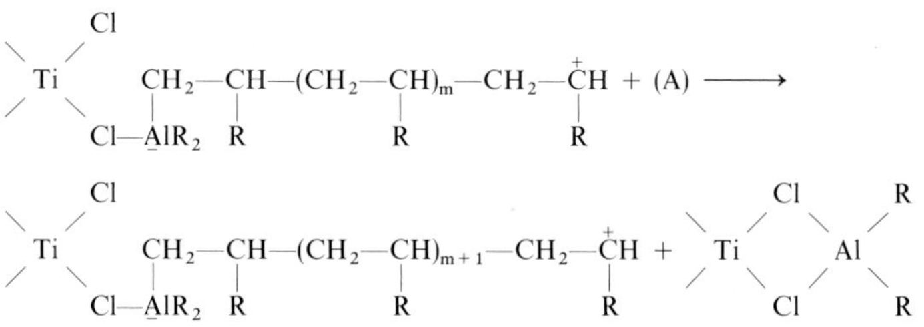

Vesely and co-workers (23) proposed a scheme in which the growth is isotactic if the active bond is an Al–C bond (A), but if it is a Ti–C bond (B) atactic polymer is formed. The corresponding catalyst reactants and centers are in the equilibrium, as shown in Eq. 13–15. The growth reaction

$$AlEt_3 + TiCl_3 \rightleftharpoons TiCl_2{}^+AlEt_3Cl^- \rightleftharpoons TiCl_2Et + AlEt_2Cl$$

$$\qquad\qquad\qquad\quad A \qquad\qquad\qquad\qquad B \qquad\qquad (13\text{–}15)$$

of an isotactic polymer at A was represented by the consecutive reaction shown in Eq. 13–16, where $(M)_n$ = growing polymer chain and M is the

$$(M)_n\text{—}\overset{\diagdown\;\diagup}{\underset{|}{Al}}\;\overset{-}{}\overset{+}{Ti}M \xrightarrow{M} (M)_n\text{—}\overset{\diagdown\;\diagup}{\underset{|}{Al}}\;\overset{-}{}\overset{+}{Ti}M_2 \longrightarrow (M)_{n+1}\text{—}\overset{\diagdown\;\diagup}{\underset{|}{Al}}\;\overset{-}{}\overset{+}{Ti}M \qquad (13\text{–}16)$$

monomer. They point out from earlier work that the dependence of reaction rate on (M) is not of the same order over a wide range of concentrations. In the range of low (M) this dependence is observed, but the olefin acts as a solvating agent.

The view that an active site is formed when a metal alkyl is adsorbed on the surface of the transition metal salt has been exploited in kinetic treatments with the application to Langmuir–Hinshelwood and Rideal rate laws. This is discussed in Chapter 18.

C. RADICAL TYPE

As already mentioned, Nenitzescu and co-workers (3) suggested the possibility of a radical mechanism, shown in Eq. 13–17.

$$nR\text{—}Al + TiCl_4 \longrightarrow R_nTiCl_{4-n} + nAlCl\ (n \leq 4)$$

$$RTiCl_3 \longrightarrow R\cdot + TiCl_3 \qquad\qquad\qquad (13\text{–}17)$$

$$R\cdot + CH_2{=}CH_2 \longrightarrow R\text{—}CH_2CH_2\cdot \longrightarrow polymer$$

About the same time (1956), Friedlander and Oita (24) proposed a free radical mechanism in which initiation occurred when an electron was transferred from the metal to the olefin (Fig. 13–5).

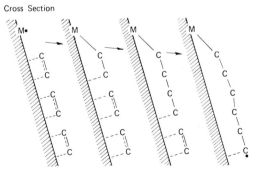

Fig. 13–5. Friendlander and Oita's radical mechanism where initiation is effected by the transfer of an electron from the transition metal to the olefin (24).

Topchiev, Krentsel, and co-workers (25) proposed that initiation occurred by a R· radical, which is bound to the surface of the $TiCl_3$, as shown in Eq. 13–18.

$$
\begin{array}{c}
R \\ | \\ R—Al—R \\ | \\ Cl \qquad Cl \\ \diagdown \underset{3+}{} \diagup \\ Ti \\ |
\end{array}
\longrightarrow
\begin{array}{c}
R \\ | \\ Cl—Al—R \\ | \\ R \qquad Cl \\ \diagdown \underset{3+}{} \diagup \\ Ti \\ |
\end{array}
\longrightarrow
\begin{array}{c}
R \quad CH_2—CH_2 \\ | \qquad | \qquad | \\ \qquad Cl \\ \underset{2+}{} \diagup \\ Ti \\ |
\end{array}
\longrightarrow
\begin{array}{c}
R—CH_2—CH_2 \\ | \\ Cl \\ \underset{2+}{} \diagup \\ Ti
\end{array}
\quad (13–18)
$$

The chemisorbed aluminum alkyl alkylates the $TiCl_3$ surface, and the formed Ti–R decomposes to yield an R· radical that adds to ethylene to form a new radical. The new radical also remains bound to the surface and adds another chemisorbed ethylene molecule. This process is repeated with the radical being transferred to the last added monomer each time. Repetition of this step constitutes the polymerization; both the radical and the reacting ethylene molecules are bound to the $TiCl_3$ surface.

The source of the radical is the aluminum in the mechanisms of Duck (26) and Van Helden, Kooyman, and co-workers (27). In the Duck mechanism, one aluminum center is involved (Eq. 13–19), while in the Van

$$
\begin{array}{c}
X \\ | \\ Cl \qquad Cl—Al—X \\ \diagdown \quad \swarrow \\ Al—CH_2—CH_2· + CH_2{=}CH_2 \\ \diagup \\ Cl \\ \underline{\qquad} \\ —TiCl_3—
\end{array}
\longrightarrow
\begin{array}{c}
X \\ | \\ Cl \qquad Cl—Al—X \\ \diagdown \quad \swarrow \\ Al—CH_2—CH_2—CH_2—CH_2· \\ \diagup \\ Cl \\ \underline{\qquad} \\ —TiCl_3—
\end{array}
\quad (13–19)
$$

Helden-Kooyman mechanism two aluminum atoms participate (Eq. 13–20). In both mechanisms, the $TiCl_3$ stabilizes the radical species.

$$
\begin{array}{c}
X \qquad\qquad Cl \qquad\qquad X \\
\diagdown \quad \swarrow \quad \diagdown \quad \diagup \\
Al \qquad\qquad Al \qquad + CH_2{=}CH_2 \longrightarrow \\
\diagup \qquad\qquad \diagdown \\
X \quad CH_2—CH_2· \qquad X \\
\underline{\qquad\qquad} \\
—TiCl_3—
\end{array}
$$

$$
\begin{array}{c}
X \qquad Cl \qquad X \\
\diagdown \quad \swarrow \quad \diagdown \quad \diagup \\
Al \qquad\qquad Al \\
\diagup \quad | \quad \diagdown \\
X \quad CH_2 \quad CH_2\,X \\
\qquad | \quad || \\
\qquad CH_2--CH_2 \\
\underline{\qquad\qquad} \\
—TiCl_3—
\end{array}
\longrightarrow
\begin{array}{c}
X \qquad Cl \qquad X \\
\diagdown \quad \swarrow \quad \diagdown \quad \diagup \\
Al \qquad\qquad Al \\
\diagup \quad | \quad \diagdown \\
X \quad CH_2 \quad ĊH_2\,X \\
\qquad | \quad \diagup \\
\qquad CH_2—CH_2 \\
\underline{\qquad\qquad} \\
—TiCl_3—
\end{array}
$$

$$(13–20)$$

D. ANIONIC TYPE

Gilchrist (28) suggested an anionic mechanism that differs from the coordinative–anionic types in that the anionic end of the polymer chain is not fixed to a metal center (Eq. 13–21). A transfer of R anion from the

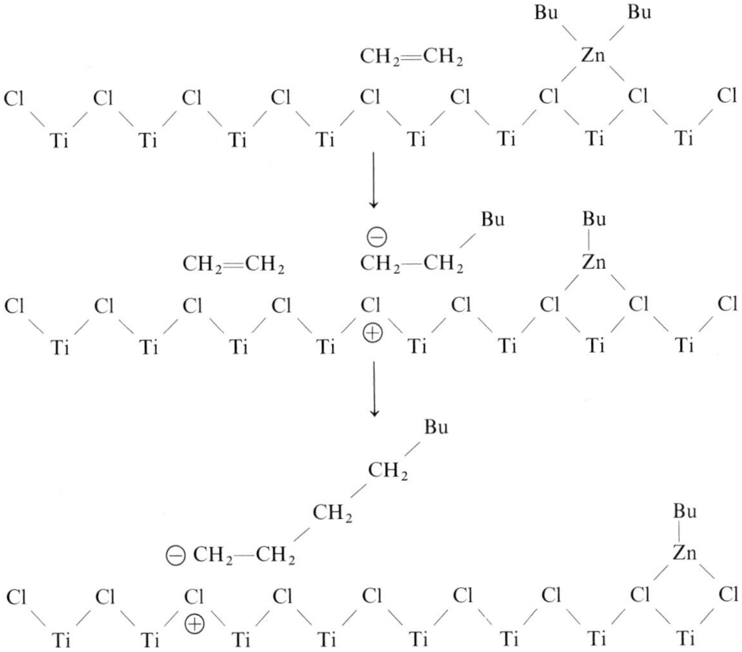

$$
\begin{array}{c}
\text{Bu} \quad \text{Bu} \\
\text{CH}_2{=}\text{CH}_2 \qquad \text{Zn} \\
\end{array}
$$

Cl   Cl   Cl   Cl   Cl   Cl   Cl   Cl   Cl
  Ti   Ti   Ti   Ti   Ti   Ti   Ti   Ti

$\downarrow$

$$
\text{CH}_2{=}\text{CH}_2 \qquad \overset{\ominus}{\text{CH}_2{-}\text{CH}_2} \qquad \text{Bu} \quad \text{Bu}
$$

Cl   Cl   Cl   Cl   Cl   Cl   Cl   Cl   Cl
  Ti   Ti   Ti   Ti   Ti   Ti   Ti   Ti   (13–21)

$\downarrow$

Bu
CH₂
CH₂
⊖ CH₂—CH₂   Bu
                 Zn

Cl   Cl   Cl   Cl   Cl   Cl   Cl   Cl   Cl
  Ti   Ti   Ti   Ti   Ti   Ti   Ti   Ti

adsorbed metal alkyl to the adsorbed olefin initiates the polymerization. Propagation occurs because this anionic end continues to add to adsorbed olefin molecules.

IV. Experimental Evidence

Evidence is first presented for the conclusion that the active center is a metal–carbon bond. Then experimental findings are described which various workers have cited as evidence to show that the active growth center is either a transition metal bond or a base metal–carbon bond. These data are presented according to the types of catalysts used in the diagnostic test. This evidence is assessed at the end of this section.

A. THE METAL–CARBON BOND IS THE ACTIVE CENTER

The radical and anionic mechanisms previously described do not resemble conventional radical and anionic systems. Diagnostic tests that are valid for the latter would not necessarily be applicable for the former, and so conventional tests might lead to ambiguous conclusions (29).

That the active centers were metal–carbon bonds was supported by the following diagnostic tests: (1) the isolated polymer contains bound tritium when the polymerization is terminated with a tritiated alcohol (such as a CH_3O^3H–CH_3OH mixture) (30); (2) hydrogen (as well as tritium and deuterium) act as transfer agents and lower the molecular weight of the formed polymer (31); (3) when tritium $^3H_2$ (32) and deuterium ($^2H_2$) (33) were used in the second test, bound deuterium and tritium were found in the polymer respectively; and (4) when labeled metal alkyls were used, such as $AlEt_3(^{14}C)$, ^{14}C was found in the polymer (34).

This data can be explained within the framework of reactions 13–22 to 13–24, where M–Et is the active center.

$$M\text{—}Et(^{14}C) + ethylene \longrightarrow M(CH_2\text{—}CH_2)_n\text{—}Et(^{14}C) \qquad (13\text{–}22)$$

$$A \begin{array}{l} \xrightarrow{RO^3H} MOR + {}^3H(CH_2\text{—}CH_2)_n\text{—}Et(^{14}C) \qquad (13\text{–}23) \\ \xrightarrow{Y={}^1H_2,\,{}^2H_2\,or\,{}^3H_2} M\text{—}Y + Y(CH_2\text{—}CH_2)_n\text{—}Et(^{14}C) \qquad (13\text{–}24) \end{array}$$

The mechanistic proposals of Nenitzescu and Topchiev would lead to results in tests 1 to 3, while the Friedlander, Duck, and Van Helden mechanisms would not produce the findings described in tests 2 to 4. While cationic mechanisms were not proposed for isotactic-specific catalysts, these can be ruled out on the basis of the above diagnostic tests.

A discussion of other features of the initiation and propagation steps is postponed to Chapter 15, namely, mode of complexing, primary vs. secondary addition, and cis- vs. trans-addition.

B. EXPERIMENTAL EVIDENCE FAVORING PROPAGATION AT A TRANSITION METAL–CARBON BOND CENTER

Considerable evidence has been published that supports the view that growth takes place at a transition metal–carbon bond. Evidence based on metal alkyl-free catalysts will be presented first, and then evidence based on the Ziegler–Natta catalyst will be discussed.

1. Evidence Based on Metal Alkyl-Free Catalysts

In the opinion of the author, the most cogent and strongest evidence for the view that the active center is a transition metal–carbon bond is based on polymerizations with metal alkyl-free catalysts. Because of their importance, Chapter 12 was devoted entirely to the polymerization chemistry of these catalysts for both olefins and diolefins. If it wasn't for the large amount of the material, the content of Chapter 12 might have properly been presented at this time. The salient features of metal alkyl-free catalysts will now be presented.

Olefins, especially ethylene and propylene, have been polymerized to linear polymers in the presence of transition metal compounds or salts which sometimes also contained other reactants, but base metal alkyls were not added. Some metal alkyl-free catalysts were active only for ethylene (this is sometimes true for Ziegler–Natta catalysts).

Only specific transition metal compounds or salts are active, and sometimes only by a special modification or addition of another component. Most Ziegler–Natta catalysts would be inactive if the base metal alkyl was left out.

The polymers that are formed with metal alkyl-free catalysts have molecular weights ranging from oligomers to greater than 1 million.

When the olefin is propylene and the metal alkyl-free catalyst is heterogeneous, the formed polymer has an isotactic structure. Partially to highly isotactic polypropylenes have been prepared.

Copolymers of ethylene and higher α-olefins have been synthesized.

Hydrogen acts as an efficient transfer agent and molecular weights are easily controlled for many of the metal alkyl-free systems.

Labeling experiments (termination with tritiated methanol) introduced bound tritium into the polymer.

Most metal alkyl-free catalysts are less active than Ziegler–Natta catalysts when compared on the basis of catalyst weight. A few, however, have comparable or even much higher activity. Labeling experiments with one catalyst showed the lower activity was due to fewer metal–carbon bonds being present. When activities were compared on the basis of the number of centers, this metal alkyl-free catalyst had an activity comparable to a typical Ziegler–Natta catalyst.

Metal alkyl-free catalysts based on the transition metals Ti, V, Cr, Ni, Co, Mo, W, Zr, and Hf have been disclosed for polymerization of olefins and dienes.

2. Experimental Evidence Based on Ziegler–Natta Catalysts

a. Proof for Alkylation. That the base metal alkyl compound alkylates the transition metal salt has been demonstrated in several cases.

Beerman and Bestian (35) found that $TiCl_4$ was alkylated easily at about $25°C$ with $Al(CH_3)_3$ to give CH_3TiCl_3 or CH_3TiCl_2, as shown in Eq. 13–25.

$$Al(CH_3)_3 + TiCl_4 \longrightarrow CH_3TiCl_3 + Al(CH_3)_2Cl \qquad (13–25)$$

By controlling the ratio of the metal alkyl to $TiCl_4$, they could get either product in greatest amount.

Gray and co-workers (36) studied the vapor phase reactions of $TiCl_4$ and $Al(CH_3)_3$ and $Zn(CH_3)_2$, respectively, by infrared spectroscopy to elucidate the nature of the active center and the reaction path by which it was formed. They found that the catalytic activity correlated with the methyl content of the solid phase, and they concluded that the modes of decomposition leading to catalytically active products were as shown in Eqs. 13–26 and 13–27.

$$(CH_3)_2TiCl_2 \longrightarrow CH_3TiCl_3 + CH_3 \cdot \qquad (13\text{–}26)$$

$$(CH_3)_3TiCl \longrightarrow CH_3TiCl_3 + 2\,CH_3 \cdot \qquad (13\text{–}27)$$

Arlman (37) investigated the interaction of $ScCl_3$ with $ZnEt_2$, which was labeled with ^{65}Zn and ^{14}C. The interaction product was washed to remove all soluble and uncomplexed (or unreacted) components, and the solid was analyzed for Et and Zn content by radiochemical methods. The Et/Zn ratio found was greater than two, supporting the premise that alkylation had occurred. If only complexing of $ZnEt_2$ had occurred on the surface, the ratio Et/Zn would have been two.

Koide and co-workers (37a) identified the two reaction products $Ti(NEt_2)_3CH_3$ and $Ti(NEt_2)_2(CH_3)_2$ by NMR spectroscopy when $Al(CH_3)_3$ and $Ti(NEt_2)_4$ were reacted; two kinds of paramagnetic species were detected by ESR. They proposed that the compound $Ti(NEt_2)_3CH_3$ played an important role as an active species for the polymerization of styrene.

Khodzhemirov and co-workers (37b) also identified alkylated surface structures from their calorimetric studies of reactions of $\alpha TiCl_3$ and $AlMe_3$ vapor.

Rodriguez, van Looy, and Gabant (38–43) carried out an extensive and thorough investigation of the reactions between solid α- and $\beta TiCl_3$ and the aluminum alkyls $Al(CH_3)_3$, $Al(CH_3)_2Cl$, and $AlEt_2Cl$. The experiments were done in the absence of solvents and at concentration levels higher than usual. From the stoichiometry data, CH_3/CD_3 exchange values, and infrared analysis combined with electron microscopy, they proposed the following reactions: (a) One chlorine atom for every two titanium atoms on the surface is replaced in a rapid exchange by an alkyl group via Eq. 13–28. (b) This

$$2\,TiCl_3 + AlR_3 \longrightarrow TiCl_3 \cdot TiCl_2R + AlR_2Cl \qquad (13\text{–}28)$$

alkylated titanium decomposes or complexes with another aluminum alkyl molecule, as shown in Eqs. 13–29 and 13–30. (c) Only a relatively small

$$TiCl_3 \cdot TiCl_2R \xrightarrow{\text{decomposition}} TiCl_3 + \text{hydrocarbon} \qquad (13\text{–}29)$$

$$TiCl_3 \cdot TiCl_2R \xrightarrow{\text{AlR}_3} TiCl_3 \cdot TiCl_2\text{–}R \cdot AlR_3 \qquad (13\text{–}30)$$

number of alkylated titaniums that are located on the lateral faces escape decomposition. These form active centers upon complexing with metal alkyl molecules (Fig. 15–7). (d) The complexed form of the alkylated titanium (A) is unstable and decomposes into a nonvolatile surface complex via Eq. 13–31.

$$TiCl_3 \cdot TiCl_2 - R \cdot AlR_3 \longrightarrow TiCl_3 \cdot TiCl_2 \cdot AlR_2 + \text{hydrocarbon products} \quad (13\text{–}31)$$

A   B

Rodriguez and co-workers concluded that this second complex (B) was inactive for olefin polymerization even though it was present in high concentration. Electron microscopy work suggested that polymerization occurred in the lateral rather than basal planes of the $TiCl_3$ surface. Their reaction and infrared spectroscopic data indicated strongly that the surface complexes (B) were present mostly on the inactive basal plane (95% of the total surface is due to the basal planes). Infrared spectra of the catalyst did not show a change in the type of bonds present on the surface. Their site counting experiments showed a number of centers on the surface of the dry active catalyst $[Al(CH_3)_3 + TiCl_3]$ were small in comparison to the Al–Ti complexes on the entire surface.

The rate of formation of centers was high even at low temperatures, and the presence of monomer is not required (43a). The efficiency of alkylation producing active centers is, however, low. The observed increase in the number of active centers with increasing amounts of aluminum alkyl at low Al/Ti ratios is probably due to the depletion of the aluminum alkyl by the alkylation process.

b. Heterogeneous Catalysts. Karol and Carrick (44) reported some copolymerization data which suggested to them that growth occurred on a transition metal–carbon bond. They found that the relative copolymerization reactivities of ethylene and propylene depended only on the transition metal salt used and were independent of the metal alkyl structure. The relative reactivity of propylene increased in the series $ZrCl_4 < TiCl_4 < VOCl_3 < VCl_4$ (see Fig. 13–6). The electronegativities of these salts increased in the same order. No significant changes in relative reactivities of the two olefins were noticed when the catalyst was made from VCl_4 and different metal alkyls, such as $Al\text{-}i\text{-}Bu_3$, $Zn(C_6H_5)$, $Zn\text{-}n\text{-}Bu_2$, CH_3TiCl_3. The metal alkyls differ considerably in structure (steric configuration, bond hybridization, electronegativity, valence, and size of metal). Karol and Carrick see this as strong evidence that the metal alkyl is not an integral part of the activity center.

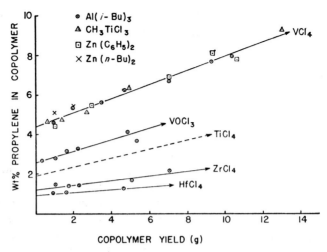

Fig. 13–6. Effect of catalyst structure on copolymer composition. Monomer contained 23.7 mole% propylene and 76.3 mole% ethylene.   Al-i-Bu$_3$; $\triangle$, CH$_3$TiCl$_3$; $\square$, Zn(C$_6$H$_5$)$_2$; $\times$, Zn(n-Bu)$_2$. (Figure reproduced from *J. Am. Chem. Soc.* through the courtesy of the Editor.)

The Karol–Carrick conclusions were challenged later by other workers. Schindler (45) correctly pointed out that copolymer composition would be controlled by the respective adsorption equilibria of each of the monomers at the reaction site if the complexing step was rate-determining. Thus the observed dependency on structure of the transition metal salt would be seen even if growth occurred at the Al–C bond. The author agrees with Cossee's conclusion that the observed activation energy is too high to account for putting a neutral molecule into a vacant position, and therefore, subsequent atom rearrangement process constitutes the rate-limiting step (see Chapter 18, Section V,A). For the catalysts examined, Carrick's conclusions appear valid. The interpretations are beclouded when one examines the meaning of reactivity ratios in the heterogeneous polymerizations (see Section III, Chapter 20 for a detailed discussion).

When other binary mixtures were examined, however, a similar dependency was not found. For example, Junghanns and co-workers (46) found a decrease in the relative reactivity of ethylene relative to propylene when AlEt$_3$ or AlEt$_2$Cl was replaced by AlEtCl$_2$ or Al$_2$Et$_3$Cl$_3$ (VCl$_4$ or VOCl$_3$ were coreactants).

In the early period, the Natta school supported a mechanism in which growth took place at a base metal–carbon bond, for example, Al–C in a bimetallic complex of AlR$_3$ and TiCl$_3$. In the period 1964 to 1967, the group (Natta, Zambelli, Pasquon, Marinangeli, and Giongo) modified its view so that it considered the transition metal–carbon bond to very likely be the growth bond, but it still (47–51) required it to be complexed with a metal alkyl. The metal alkyl had two roles in this complex: to stabilize the transition metal–carbon bond, and to influence the degree of isotatic growth. This view was based on the following evidence.

Starting with the expression:

$$\frac{1}{\bar{M}_v} = \frac{1}{k_p' m t} + A \tag{13–32}$$

where m is concentration of monomer, t, reaction time, M_n (number average molecular weight) $= x\bar{M}_v$ (viscosity average molecular weight), and the concentration of active centers and reagents was assumed to remain constant during time t, they plotted $1/\bar{M}_v$ vs. $1/t$ for different catalysts for the polymerization of butene. A typical curve is shown in Fig. 13–7 for VCl_3–$AlEt_3$, with a corresponding curve for $TiCl_3$–$AlEt_3$ in the inset. From these curves, a viscosity propagation rate constant, k_p', was evaluated for each of the catalyst systems examined. This viscosity propagation rate constant k_p' was found dependent only on the transition metal salt used and did not depend on the structure of the metal alkyl. Two explanations were considered: (1) the rate-determining step is the complexing of the olefin at the transition metal center (and so the above dependency results even if growth occurred at the Al–C bond), and (2) the growth center is an alkylated transition metal–carbon bond that is complexed to the metal alkyl, and the rate-determining step is the complexing of olefin at the transition metal center.

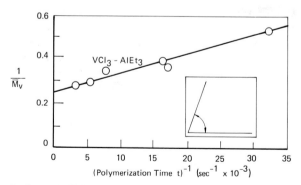

Fig. 13–7. Reciprocal of the viscosity average molecular weight vs. the reciprocal of the polymerization time for isotactic polybutenes (pentanic residues) obtained at 0°C by VCl_3–$Al(C_2H_5)_3$. Comparison with the systems based on violet $TiCl_3 \cdot C_4H_8$:333 g/liter toluene. Inset: average slope for systems based on violet $TiCl_3$ (50).

Pasquon favored the second explanation and further adds that the metal alkyl stabilizes the transition metal–carbon bond (50). Furthermore, on the basis of a detailed examination of the stereochemical structure of the corresponding polypropylenes prepared with different catalysts, he also concluded that the metal alkyl influences the stereochemical reaction. This important conclusion is discussed in Section III,B,1 of Chapter 15.

Other experimental findings have been suggested to support the view that the metal alkyl can stabilize a transition metal–carbon bond center or that the base metal in the bimetallic complex is the growth center. Zambelli and co-workers (47–50) found that different metal alkyls had different reducing and site-forming abilities. When propylene was polymerized with a mixture of VCl_3 plus the metal alkyl, activity decreased in the order: $AlEt_3 >$ $BeEt_2 > GaEt_3 > AlEt_2I > AlEt_2Cl > CdEt_2 > LiBu > ZnEt_2$. In contrast, the ability of these alkyls to reduce VCl_3 to VCl_2 decreased in the order: $LiBu > CdEt_2 > ZnEt_2 > BeEt_2 > AlEt_3 > AlEt_2Cl > AlEt_2I$.

The argument they presented assumed that reduction involved alkylated vanadium species, such as $RVCl_2 \rightarrow R\cdot + VCl_2$. If the active center was an alkylated vanadium, then it would follow that those alkyls that caused greatest reduction also generated the highest concentration of $RVCl_2$, and they should have produced the most active catalyst.

Ambroz and co-workers (51a) found the reaction rate at $50°C$ for propylene decreased in the series: $AlEt_3 > AlEt_2I > Al\text{-}i\text{-}Bu_3 > AlEt_2Cl >$ $AlMe_3 > AlEt_2Br \simeq AlEt_2I$ ($\alpha TiCl_3$ was used as a cocatalyst).

The above papers do not appear to exclude entirely some growth at a transition metal–carbon bond that is not complexed with a metal alkyl. But Pasquon, Zambelli, and co-workers made the point that at least some of the active complexes contain both metals. In the latter case, however, even if growth took place at an alkylated transition metal–carbon bond, the complexed metal alkyl influenced site stability and stereocontrol.

Alternate explanations were suggested by Boor (1). Not every alkylated vanadium has to function as an active center. For example, while alkylation and reduction could occur on the lateral ($10\overline{1}0$ face) and basal ($000\overline{1}$ face) planes (which form 5 and 95% of the total surface, respectively), polymerization is probable only on the lateral planes. There could be considerable variation with the degree of alkylation according to alkyl structure in these two types of surfaces. Also, reduction could have taken place so fast that polymerization could not occur. Combinations of LiBu and $\alpha TiCl_3$ do not polymerize propylene. Visual inspection of the $\alpha TiCl_3$ just after addition of LiBu shows the color changed from purple to yellow-green, suggesting an extensive surface reaction, probably overreduction. Under certain conditions, for example, at higher temperatures, $ZnEt_2$ decomposes rapidly when present in the $AlEt_3$–$TiCl_3$ catalyst (50a).

As illustrated earlier, Arlman showed alkylation of $ScCl_3$ with $ZnEt_2$ occurred. Yet, while this catalyst is active for ethylene, it is not active for propylene under comparable conditions. In contrast, $ScCl_3$ in combination with $AlEt_3$ is active for both olefins.

The absence of bimetallic complexes in one heterogeneous catalyst was concluded from NMR studies. Di Carlo and Swift examined the catalyst $AlEt_2Cl–TiCl_3$ by ^{27}Al nuclear magnetic spectroscopy (51b). Their analysis excluded structures in which Al was complexed to a paramagnetic titanium by a Cl bridge.

c. Soluble Cyclopentadienyl Titanium-Based Catalysts.

Very interesting experimental evidence was obtained with the use of soluble cyclopentadienyl-titanium chloride-based catalysts by a number of workers.

In 1958, Long and Breslow (52) had already demonstrated through spectroscopic studies on these catalysts that alkylation occurred by an exchange of alkyl and chloride ligands.

$$Cp_2TiCl_2 + 1/2\ (AlEt_2Cl)_2 \rightleftharpoons Cp_2TiCl_2 \cdot AlEt_2Cl$$

$$A\ (complex) \tag{13-33}$$

$$A \xrightleftharpoons{\text{exchange}} Cp_2TiClEt \cdot AlEtCl_2$$

$$B\ (complex) \tag{13-34}$$

$$B \rightarrow Cp_2TiCl \cdot AlEtCl_2 + Et \cdot \tag{13-35}$$

They concluded that the active site was B or some species derived from it or in equilibrium with it on the basis of a direct dependence between the initial rate of ethylene production and the concentration of B.

Zefirova and Shilov (53) speculated that the cationic species $[Cp_2TiEt]^+$ was responsible for the polymerization in Eq. 13–36. Later, Shilov, Shilova,

$$Cp_2TiEtCl \cdot AlEtCl_2 \rightleftharpoons [Cp_2TiEt]^+ [AlEtCl_3]^- \tag{13-36}$$

and Bobkov (54) found that the rate of polymerization of ethylene was proportional to the increase in electrical conductivity that occurred with the formation of B. (The behavior of other catalysts in this way was already discussed in Chapter 4.)

Ingenious experiments involving electrodialysis were done by Dyachkovskii and co-workers (55–61) to show the presence of the $[Cp_2TiEt]^+$ cation. In a typical experiment, a homogeneous solution of Cp_2TiCl_2 and $AlEt_2Cl$ catalyst in dichloroethane was placed in the left chamber and an $AlEt_2Cl$ solution was placed in the right chamber of the dialyzer (Fig. 13–8). A cellophane membrane separated the left and right chambers and ethylene was bubbled into the right chamber. Polymerization was observed in the

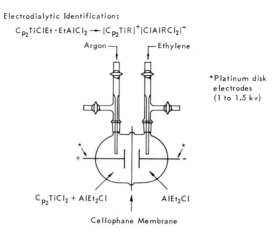

Fig. 13–8. Scheme of the dialyzer. Electrodialytic identification: $Cp_2TiClEt \cdot EtAlCl_2 \rightarrow$ $[CpTiEt]^+[ClAlEtCl_2]^-$ (55).

right chamber only after the disc platinum electrodes in the left and right chambers were connected with positive and negative poles, respectively. Also, $AlEt_2Cl$ had to be present in both chambers. Dyachkovskii proposed that the active catalyst was the Cp_2TiR^+ cation that formed in the left chamber (according to Eq. 13–36) and then passed, under the action of the electric field, through the membrane into the right chamber where it complexed with aluminum alkyl molecules already present.

In later work, Dyachkovskii and co-workers (55–61) supported this view with the following findings:

i. Activity is greater in more polar solvents, presumably because more ions can be present (56).

ii. To remove possible affects of platinum electrodes, similar experiments were done in a three-chamber dialyzer containing two cellophane membranes (56).

iii. The polyethylene that is formed had a weight average molecular weight about 40,000 and had a narrow molecular weight distribution. This catalyst apparently remains homogeneous under these conditions.

iv. When the complex $(C_5H_5)_2Ti(^{14}CH_3)Cl \cdot Al(^{14}CH_3)_2Cl$ was placed in the left (anode) chamber and electrodialysis begun, ^{14}C was found in the right chamber and also in the formed polymer. At an Al/Ti ratio of 2, the CH_3/Ti ratio approached 1 and one $^{14}CH_3$ per one or two polymer chains was found. Migration did not occur right to left if the Ti compound was placed in the right chamber. Apparently, rapid exchange of CH_3 in $[Cp_2TiCH_3]^+$ with CH_3's of $Al(CH_3)_2Cl$ did not occur in the right chamber.

v. The dialysis technique was used for two other Ziegler–Natta catalysts, $AlEt_2Cl + TiCl_4$ (51, 56) and $AlEt_3 + Ti(OBu)_4$, and for several anionic initiators (EtLi and naphthalene–Li) to show that the centers have ionic character (60, 61).

vi. $(C_5H_5)_2Ti \cdot THF$ oxidatively added CH_3I, to give $(C_5H_5)_2TiCH_3I$ and ethylene, is polymerized upon addition of $AlCH_3Cl_2$. The polymer, however, contains ^{14}C, supporting the view that the active center is the Ti–C bond (59).

Nuclear magnetic resonance spectroscopy was used by Dyachkovskii and co-workers (62) to show the presence of Ti–CH_3 in the catalyst $Cp_2TiCl_2 + Al(CH_3)_3$ or $Al(CH_3)_2Cl$. When phenylacetylene was added to the soluble catalyst, the CH_3 proton chemical shift disappeared. The reaction shown in Eq. 13–37 was suggested.

$$\geqslant Ti-CH_3 + C_6H_5C\equiv CH \longrightarrow Ti-CH=C\begin{array}{c} CH_3 \\ \diagup \\ \diagdown \\ C_6H_5 \end{array} \qquad (13-37)$$

Grigoryan, Dyachkovskii, and Shilov (63) found that the initial rates of polymerization of ethylene and deuteroethylene with the $Cp_2TiCl_2 \cdot Al(CH_3)_2Cl$ catalyst were identical for these catalysts. It was shown by spectrophotometric and ESR methods, however, that reduction of Ti(IV) was slowed down by a factor of 2.5 to 3.0 when deuteroethylene was used. Also, the molecular weight of the formed polyethylene was twice as great. This result argues against chain termination taking place as a bimolecular reaction between active centers, as suggested by a number of workers. An ionic mechanism involving the cationic species discussed above was proposed, as shown in Eqs. 13–38 and 13–39. If polymerization had occurred,

$$[Cp_2TiEt]^+[AlEt_2Cl_2]^- \longrightarrow Cp_2TiEt_2 + AlEtCl_2 \qquad (13-38)$$

$$Cp_2TiEt_2 \longrightarrow Cp_2Ti + C_2H_4 + C_2H_6 \qquad (13-39)$$

then one of the hydrocarbon products would have been the polymer chain.

A comprehensive study was made by Henrici–Olive and Olive (63–67) to relate catalyst activity, the configuration of the active complex, and the oxidation state of the Ti in the soluble catalysts (π-$C_5H_5)_2TiCl_2$–$AlEtCl_2$ and (π-$C_5H_5)_2TiEtCl$–$AlEtCl_2$. The investigation involved a kinetic study of the polymerization of ethylene coupled with ESR and magnetic susceptibility measurements. They concluded that the active centers were tetravalent because the polymerization rate decreased to the same extent that

Ti(IV) was reduced to Ti(III). The complex, $Cp_2TiClR-AlRCl_2$, shown in structure VI, was proposed for the active center, where Cp is C_5H_5, a cyclopentadienyl group.

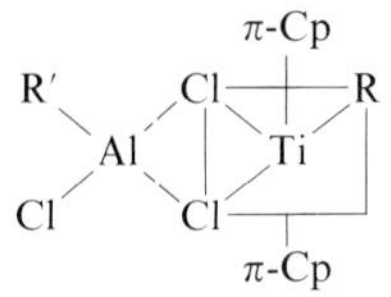

$$\pi\text{-Cp}$$

R'—Al(Cl)—Cl—Ti(R), π-Cp

VI

Using this model, these workers proposed a mechanism to account for the increase in reduction decomposition of $Cp_2TiClEt \cdot AlEtCl_2$ [Ti(IV) → Ti(III)] in the presence of nonpolymerizable olefins. The rate of reduction of Ti(IV) was increased in the series 1-octene < cis-2-octene < trans-2-octene. A reductive step involving two adjacent octahedral complexes (VII, VIII) was postulated.

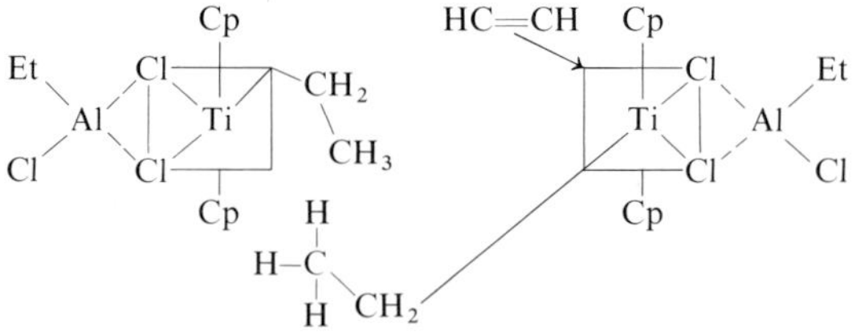

VII   VIII

The following mechanism was proposed. The β-hydrogen of the second unit occupies an empty octahedral position of the first unit. It is subsequently transferred to the ethyl group of the second unit to liberate ethane. The remaining radical splits off ethylene and forms a stable trivalent species. The role of the coordinated olefin is to weaken the Ti–Et bond and aid the reduction. The earlier findings of Shilov and co-workers (54), who reported that alkylation of α-olefins with the $Cp_2TiClMe-AlMeCl_2$ catalyst was always accompanied by reduction of Ti(IV) to Ti(III), were cited as supplementary evidence.

This argument can be further extended to show why this type of catalyst will not homopolymerize propylene or 1-butene even though it will homopolymerize ethylene and copolymerize the same α-olefins and ethylene. Henrici–Olive and Olive showed that when the alkyl attached to the Ti was octyl the catalyst was more easily reduced. This was attributed to the lower dissociation energy of the C–H bond in $CH_2(R = octyl)$ compared to the C–H bond in $CH_3(R = Et)$. For the homopolymerization of propylene or 1-butene, the β-hydrogen would be a tertiary hydrogen. Since this β-hydrogen would be even more easily transferred, reduction would be facile and this polymer chain should not grow. If the more active ethylene molecules are present, however, these insert faster and thus decrease the time that an α-olefin would be bonded to the Ti as the last added unit.

Henrici–Olive and Olive (68) carried out some oligomerization studies to establish the location of the active Ti–C bond and the rate of exchange of ligands between Ti and Al. Two catalysts were used: A, $Al(CH_3)_3$ + $Cp_2Ti(Bu)Cl$; and B, $Al\text{-}i\text{-}Bu_3$ + $Cp_2Ti(CH_3)Cl$. After contacting ethylene for 30 minutes at 0°C with each catalyst, they analyzed for the hydrocarbon products as shown in the following tabulation:

Catalyst A	Catalyst B
69.1% 1-alkenes	42.4% 1-alkenes
30.9% alkanes	57.6% alkanes

Because most of the hydrocarbon molecules had an even number of carbons in catalyst A, they ruled out a bridge position as the location of the active center. Also, exchange between CH_3 and Bu was not significant. Most of the alkene had less than eight carbons, and this suggested that chain termination by hydride transfer occurred, as shown in Eq. 13–40. The fact

$$Ti\text{-}CH_2CH_2R \longrightarrow TiH + CH_2{=}CH\text{-}R \qquad (13\text{-}40)$$

that even-chain hydrocarbon molecules formed most favorably in the catalyst B experiment suggests that a facile exchange between CH_3 and Bu must have occurred. Apparently CH_3 prefers a bridged position in these structures as it does in dimeric aluminum alkyls. In this case, termination by hydride transfer occurs less since a smaller fraction of alkenes forms.

In Chapter 11, it was already shown that the influence of ligands can profoundly affect the active center and, in fact, more than one oxidation state is active in a suitable ligand environment. The varying of ligands in the Ziegler–Natta catalyst has been recognized as a powerful tool that one has in making alterations in activity and stereochemical ability. Henrici–Olive and Olive have critically discussed this subject for soluble catalysts and their excellent review is highly recommended. Like Cossee and others, they have recognized the mechanistic similarities (such as hydroformylation) between Ziegler–Natta and other transition metal catalysts and have taken advantage of findings from both sides to evolve a more complete under- standing of the effect of the ligand on the activity and specificity of transition metal catalysts.

Using the Cossee–Chatt molecular orbital scheme (Fig. 13–3), they point out that the effect of the ligand is to cause a change in the electronic structure of the center, such as a decrease in the electronic density in certain regions of the complex, an increase in others, and a consequent change in the bonding. The literature of transition metal chemistry contains numerous other examples, which are presented under such labels as donor ability or basicity of the ligand, acceptor strength of the ligands, *trans*-effect, inductive and resonance effects, and the influence on the energies of the orbital. The different notation is inherent in the different theoretical treatments that have been used to describe the complexes.

The ligands can be ions such as halogen, hydride, alkoxy, cyanide and acetylacetonate, or neutral molecules such as CO, NO, amine, or phosphine. The metal alkyl molecule can act as a ligand when it complexes to the transition metal center by means of anion bridges, as shown in structure IX, where $Cp = \pi$-cyclopentadienyl, $R =$ alkyl, and R^1, $R^2 =$ alkyl or Cl.

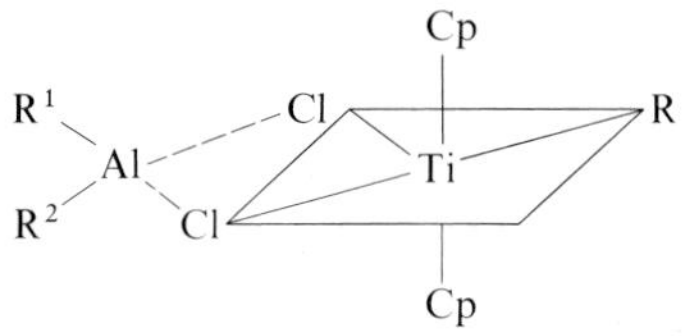

IX

The ligand can have the following effects by virtue of being able to alter the electron density at the center:

i. The stability of the transition metal–carbon is affected.
It has been recognized that certain ligands stabilized the transition metal–carbon bond so efficiently that these complexes can be isolated. Such excessive stability of the M–R bond is not desired, since it would not be active for polymerization. The job of the ligand, then, is to give the M–R bond the desired stability needed for high activity in the environment in which it is present. Henrici–Olive and Olive offer the following illustration: In structure IX, destabilization of the Ti–R bond increased in the order $R^1, R^2{=}Cl < R^1{=}Cl$, $R^2{-}CH_3 < R^1{=}Cl$, $R^2{=}C_2H_5 < R^1$, $R^2{=}C_2H_5$. The stronger the donor properties of the ligands R^1 and R^2, the stronger the destabilization of the Ti–C bond, that is, the complex was less stable when R^1, R^2 was C_2H_5 than Cl. Excessive stabilization and destabilization is not desired.

ii. The activity of the transition metal–carbon bond can be changed.
Altering the ligand in the Ziegler–Natta catalyst can induce this catalyst to form increasingly larger amounts of oligomers at the expense of high

TABLE 13–2

Dependence of the Molecular Weight of the Ethylene Polymer Obtained on the Ligands at Titanium[a]

System	Ti component	Oligomers[b] (%)	Polymers (%)
(16)	$(C_2H_5O)_4Ti$	31	69
(17)	$(C_2H_5O)_3TiCl$	56	44
(18)	$(C_2H_5O)TiCl_3$	77	23
(19)	$TiCl_4$	92	8

[a] The Al component is $C_2H_5AlCl_2$; $Al/Ti = 5$; the solvent is benzene; temp. $= 5°C$; $P_{C_2H_4} = 1$ atm.
[b] The oligomer is taken to be the part that is soluble in benzene at room temperature.

molecular weight polyethylene (Table 13–2). The relative frequency of the β-hydrogen transfer, shown in Eq. 13–41, was attributed to the

$$M-CH_2-CH_2-R \rightleftharpoons M\begin{matrix} H_2C \\ \diagdown \\ CH-R \\ \diagup \\ H \end{matrix} \longrightarrow CH_2=CH-R + M-H$$

$$(13–41)$$

electron affinity of the metal, that is, on the donor–acceptor character of the ligands in the complex. If the electron affinity of Ti was lowered by the donor ligand substitution of Cl by OEt, hydrogen transfer occurred less frequently, and the molecular weight of the polymer chain increased.

iii. The complexing ability of the center is altered.
When the olefin is coordinated to the transition metal center (structure IX), electron flow from the olefin to the Ti occurs through the σ bond and from the Ti to the antibonding orbitals of the olefin through the π bond. This activates the coordinated olefin. The extent of coordination can thus be influenced by the donor character of the ligands that are attached to the metal, since they will alter this electronic flow.

iv. The symmetry of the transition metal complex changes.
In order for the transition metal atom to be active, it must have an open coordination site to complex the olefin. One of the effects of the ligand is to change the coordination symmetry of the metal from one which lacks an open site to one which has one, for example, tetrahedral (Cp_2TiClR) to the octahedral compound (see Fig. 13–8).

C. Experimental Evidence Favoring Propagation at a Base Metal–Carbon Bond Center

Two factors probably encouraged workers in the early period of Ziegler–Natta chronology to propose that the propagation step takes place at the base metal–carbon bond. One was the obvious influence of work already done on the Aufbau reaction, by which ethylene was polymerized to low molecular weight polymers in presence of only a base metal alkyl (see Chapter 2, Section II). The other was the recognition that an olefin molecule can complex with the transition metal, and, because it becomes polarized, its ability to be inserted into the highly active base metal–carbon bond might be increased. By virtue of the bimetallic complex between the base metal alkyl and the transition metal compound, both metal centers are brought into a favorable position for the growth step to take place.

Some workers have suggested that the base metal–carbon bond is even activated by complexing. This view is not supported, however, by Ziegler's observation that the Aufbau reaction occurs only with monomeric species. By complexing the metal alkyl with electron donors, a lower rate of ethylene oligomerization was observed. As was already discussed in Chapter 4, base metal alkyls can exist in monomeric, dimeric, or higher aggregate form. But metal alkyls that do not dimerize form active Ziegler–Natta catalysts, such as $ZnEt_2 + TiCl_3$. Even Al-i-Bu$_3$, which exists predominantly in the monomeric state, is comparably efficient to $AlEt_3$, which exists predominantly in the dimeric state. We cannot, therefore, draw cogent conclusions about the active center from a relative comparison of abilities of metal-alkyl molecules to dimerize or form aggregates. The early workers recognized the deficiency of the above arguments and sought more convincing evidence.

Natta and co-workers (69–74) synthesized and identified bimetallic complexes having the structure shown in X by X-ray spectroscopy, where

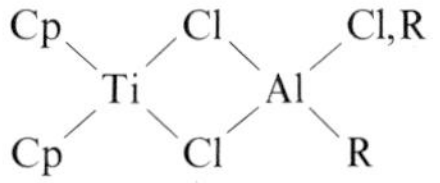

X

R = alkyl group (Et or Me) and Cp = π-cyclopentadienyl = π-C_5H_5. The starting materials were $Cp_2TiCl_2 + AlEt_3$ or $AlEt_2Cl$.

Because these bimetallic compounds were active for the polymerization of ethylene, it was concluded that growth took place at the Al–C bond. This structure was considered to be similar to that proposed earlier by Natta and co-workers for heterogeneous catalysts (Eq. 13–8 and Fig. 15–6). In both types of complexes, the Al–C was said to be stabilized and activated by virtue of its complexed state.

The Natta group (71) found that the substituent attached to the aluminum alkyl in these complexes became attached to the polymer chain.

Catalyst	End group
$Al(C_6H_5)_3 + Cp_2TiCl_2$	C_6H_5
$AlEt_3 + Cp_2Ti(C_6H_5)_2$	Et
$Al(C_6H_5)_3 + Cp_2Ti(C_6H_5)_2$	C_6H_5

$Cp_2Ti(C_6H_5)_2$ did not polymerize ethylene when used above. Natta and co-workers interpreted these findings to mean that the Al–C bond was the growth center.

Karapinka and Carrick (75) disagreed with this conclusion on the basis of their findings that (1) only Et end groups were found in the polyethylene when both $(C_6H_5)_3Al$ and Et_3Al were used in combination with Cp_2TiCl_2, and (2) that a facile exchange of C_6H_5 and Et groups takes place when $Cp_2Ti(C_6H_5)_2$ and $AlEt_2Cl$ are mixed. It is thus difficult to identify the growth center on the basis of end-group analysis. They proposed that, because $AlEt_3$ is a better alkylating agent than $Al(C_6H_5)_3$, no phenyl groups are present in the polymer when the mixture of Cp_2TiCl_2, $Al(C_6H_5)_3$, and $AlEt_3$ was used for the catalyst. Because the $Ti–C_6H_5$ bond only sluggishly polymerizes, they suggested that this allowed the exchange of Et and C_6H_5 ligands before much polymerization occurred at these centers.

It is easy to see why a worker might conclude that the metal alkyl is an integral part of the active center. The literature discloses a voluminous assortment of examples in which stereochemical control, kinetic features, and molecular weights of polymers are dependent on the structure of the metal alkyl. See the review by Coover and co-workers, for example (76). In some cases, the complexes of metal alkyl and transition metal salt were isolated and identified, such as $(C_6H_5)_3Al + Cr(acac)_3$ (76a) or $AlEt_2Cl + V(acac)_3$ (76b).

Some examples have already been described in detail in Chapter 4, namely (1) the dependence of isotacticity on diameter of the metal of the metal alkyl (however, not for metals); (2) the dependence of isotacticity on the nature of the ligands attached to the metal alkyl, such as hydrocarbon or heteroatom, size, branching, and number of each type; (3) the dependence of catalyst activity on complexing ability of metal alkyl; and (4) the dependence of catalyst activity on the ionic nature of the catalyst, for example, conductivity.

Other examples are now cited. Natta and co-workers found that when α-, γ-, and δ-crystalline modifications of $TiCl_3$ were mixed with a series of metal alkyls and used to polymerize propylene, the isotacticities obtained

TABLE 13-3

Stereospecificity Indices[a] of the Catalytic Systems (α, γ, and δ) Crystalline Violet TiCl$_3$ with or without AlCl$_3$ in Solid–Solution Al(C$_2$H$_5$)$_2$X in Polymerization of Propylene at 70°C and $P_{C_3H_6}$ = 2,000 mm Hg

| | TiCl$_3$ | | | | |
| | α | | | δ | |
Al(C$_2$H$_5$)$_2$X	Free of AlCl$_3$	Containing AlCl$_3$[b]	γ^c	Free of AlCl$_3$	Containing AlCl$_3$[b]
Al(C$_2$H$_5$)$_2$Cl	92	91	93	93	94
Al(C$_2$H$_5$)$_2$Br	95	95	96	96	96
Al(C$_2$H$_5$)$_2$I	96	96	98	97	97
Al(C$_2$H$_5$)$_3$	84	82	80	83	81

[a] As percent of isotactic polymer, nonextractable with boiling n-heptane. The reported data are average values obtained from runs performed twice.

[b] Containing 4.5% Al as AlCl$_3$ in solid solution.

[c] Containing 1% Al as AlCl$_3$ in solid solution.

TABLE 13-4

Indices of Stereospecificity[a] of Catalytic System Prepared from (α, γ, or δ) Crystalline Violet TiCl$_3$ in Polymerization of Propylene[b]

| | Index of stereospecificity | |
Organometallic compound	Polymerization runs at 15°C	Polymerization runs at 70°C
Al(C$_2$H$_5$)$_2$I	90–100	90–98
Al(C$_2$H$_5$)$_2$Br	97–98	94–96
Al(C$_2$H$_5$)$_2$Cl	96–98	91–94
Al(C$_2$H$_5$)$_3$	80–85	80–85
Be(C$_2$H$_5$)$_2$	94–96	93–95

[a] As % of isotactic polypropylene, nonextractable with boiling n-heptane.

[b] Tests carried out at different pressures (1 to 10 atm) with different concentrations of reactant (TiCl$_3$, 3 to 30 mmole/liter; organometallic compound, 10 to 30 mmole/liter).

were dependent only on the structure of the metal alkyl (77, 78). An identical order was always found: $AlEt_2I > AlEt_2Br > BeEt_2 \geq AlEt_2Cl > AlEt_3$. On the basis of these findings, they concluded that the individual active centers were substantially the same in the α-, γ-, or δ-forms of $TiCl_3$, and that stereochemical control depended only on the organometallic compound in the corresponding binary mixtures (Tables 13–3 and 13–4).

This was not, however, the case when VCl_3 was used. While $AlEt_2Cl + TiCl_3$ was more isotactic-specific than $AlEt_3 + TiCl_3$, the reverse was found when VCl_3 was used instead of $TiCl_3$ [see Chapter 4, Boor (9) and Natta et al. (10)].

At the time the Natta findings (Tables 13–3 and 13–4) were published, it was implied that the active center was a bimetallic complex that polymerized propylene according to the mechanism shown in Eq. 13–8. Later work by the Natta school led them to believe that the active center could be a bimetallic complex in which the titanium was alkylated, and while growth occurred in the transition metal–carbon bond, stereochemical control was influenced, at least in part, by the base metal–alkyl component. This conclusion was reached on the basis of careful and extensive examination of polypropylene fractions for tacticity. This work is described in detail in Chapter 15.

Coover (76) also argued that because stereochemical control is influenced by the structure of metal alkyl, growth was more probable at a base metal–carbon bond than at a transition metal–carbon bond. Once a few growth steps occurred, the influence of the alkyl ligand substituent would be lost if growth occurred at a transition metal–carbon bond.

Pozamantir (79) argued in favor of growth at the Al–C bond on the basis of his observation that the efficiency of alkyl chlorides as polymer chain terminators during polymerization followed the same order as their reactivity with $AlEt_3$, namely, $EtCl < n\text{-}PrCl < i\text{-}BrCl < t\text{-}BuCl <$ benzyl Cl. He suggested that the chloride displaced the growing polymer chain from the Al–polymer bond. As additional proof, he cited that $SiCl_4$, which does not react with $AlEt_3$, $AlEt_2Cl$, or $AlEtCl_2$, was not an effective terminator, while $SnCl_4$, which does react, was very effective.

Alkyl chlorides react readily with aluminum alkyls to form various derivatives, as shown in Eq. 13–42. Depending on the amount of the alkyl

$$AlEt_3 + RCl \longrightarrow AlEt_2Cl \longrightarrow AlEtCl_2 \longrightarrow AlCl_3 \qquad (13\text{--}42)$$

chloride added, the catalyst is probably a mixture of $AlEt_3$ and one or more of these derivatives. Since in a polyethylene polymerization, molecular weight is dependent on the chlorine content in the alkyl, the decrease in molecular weight observed by Pozamantir can be rationalized, at least in part, in this way (compare the Graf and Pozamantir data in Table 13–5) (80, 81).

TABLE 13–5

Dependence of Molecular Weight on Metal Alkyl Structure

	Polyethylene molecular weight	
Metal alkyl[a]	η_{sp}/C^b	$[\eta]^c$
$AlEt_3$	11.4	7.0
$AlEt_2Cl$	4.4	5.0
$AlEtCl_2$	2.0	1.5

[a] $TiCl_4$ used as cocatalyst.
[b] Data from Graf *et al.* (80).
[c] Data from Pozamantir *et al.* (81).

Boor examined chain termination by organic chlorides when propylene was polymerized with the $ZnEt_2$–$TiCl_3$ catalysts (82). These findings, already described in Chapter 10, Section II, support the above explanation. Boor interpreted his finding that an organic chloride functioned as a polymer chain terminating agent even though it did not react with $ZnEt_2$ to mean that the growth site was not a Zn–C bond but rather a Ti–C bond.

Schindler (83) reported that the relative amounts of CHD and CH_2D units in the polyethylene polymer formed when deuterium was present in the polymerization depended on the metal alkyl structure. He compared the different Ziegler catalysts based on mixtures of $AlEt_2Cl$ or $ZnEt_2$ with $TiCl_4$ or $VOCl_3$. The CH_2D unit was attributed to a transfer reaction shown in Eq. 13–43, while CHD unit was attributed to an exchange reaction shown in

$$(Cat)CH_2\!-\!CH_2\!-\!CH_2\!-\!R \;\longrightarrow\; (Cat)D + CH_2\!-\!CH_2\!-\!CH_2\!-\!R \quad (13\text{–}43)$$
$$+ \qquad\qquad\qquad\qquad\qquad\qquad\qquad |$$
$$D\text{–}D \qquad\qquad\qquad\qquad\qquad\qquad\qquad D$$

Eq. 13–44. Schindler suggested that either the metal atom of the reducing

$$(Cat)CH_2\!-\!CH_2\!-\!CH_2R \;\longrightarrow\; (Cat)CH_2\!-\!CH\!-\!CH_2\!-\!R \quad (13\text{–}44)$$
$$+ \qquad\qquad\qquad\qquad\qquad\qquad\qquad + \quad |$$
$$D\text{–}D \qquad\qquad\qquad\qquad\qquad\qquad D\text{–}H \quad D$$

agent represents the center of chain propagation, or the catalyst site is a bimetallic complex, strongly influenced in its reactivity by the participation of the reducing agent.

Lanovskaya and co-workers (84) studied the effect of the metal alkyl component structure on the polymerization kinetics of ethylene and polyethylene molecular weight when they were combined with α- and $\beta TiCl_3$ modifications. Various aluminum alkyls, including $MgEt_2$, $ZnEt_2$, LiEt, and CH_3TiCl_3 were used. For catalytic systems based on $\alpha TiCl_3$ and Al-i-Bu_3, $AlEt_2Cl$, or $ZnEt_2$, the stationary rate of polymerization increased and the induction period decreased as the concentration of the metal alkyl was increased. They interpreted this to support the composite (bimetallic) nature of the active centers.

Soga and Keii (85) proposed, on the basis of kinetic comparisons of the polymerization of propylene, that the active center is a Ti–C or Al–C bond, depending on the catalyst used, namely $AlEt_3$ + $TiCl_3$ (A) or $ZnEt_2$ + $TiCl_3$ (B). If catalyst A was used, the active center on the surface of $TiCl_3$ was bimetallic, as shown in structure XI. With catalyst B, however, the

$$R_2Al \underset{X}{\overset{R}{<}} \underset{X}{\overset{X}{>}} Ti$$

XI

active center was an alkylated transition metal. This conclusion was based on three findings: (1) the activation energy of the stationary rate of polymerization was 6.5 kcal/mole with $ZnEt_2$ and 13.8 kcal/mole with $AlEt_3$, (2) the growth rate of the polymer chains with $ZnEt_2$ was evaluated to be about three times slower at 43.5°C, and (3) the polymerization center formed with $ZnEt_2$ was considered less stable. On the basis of ESR spectroscopy, Ono and Keii (86) concluded that similar centers were obtained when $TiCl_3$ was mixed with $AlEt_3$ and $AlEt_2Cl$. In both cases, an absorption with a gram value 1.96 was observed in the solid phase.

Kinetic arguments in general are weak because rates are evaluated in terms of grams of catalyst rather than number of sites. Until definitive site-counting methods are employed, the rate data cannot be interpreted cogently to signify activity per site or stability per site.

V. Conclusion

The experimental evidence, taken collectively, argues strongly that the active center is a transition metal–carbon bond. The author, however, is not aware of one experimental finding that is unequivocally convincing in favor of this conclusion. The very existence of metal alkyl-free (MAF) catalysts is our strongest evidence that the growth center in olefin polymerization is an alkylated transition metal–carbon bond. One may add the many similarities between MAF and Ziegler–Natta catalysts as secondary evidence, especially when both systems were subjected to the same diagnostic probes, including site counting, termination by alkyl halides of different structures, effect of electron donors, termination with hydrogen, deuterium, or tritium, characterization of stereoregular polymer, etc. Kinetic studies on both systems help explain some of the observed differences. The role of the metal alkyl as a site-former, reducing agent, transfer agent, poison scavenger, etc., is recognized. Its role as a site-stabilizer has been debated widely.

The author, in assessing the Ziegler–Natta and MAF literature, draws the following conclusions about the stabilizing role of the metal alkyl. (1) In heterogeneous catalysts, the active center does not contain a complexed base metal alkyl, that is, sites with the structure as in IV are probable. Stabilization of the active transition metal–carbon bonds is affected by the polymetallic complex system of the crystal. Further complexing by base metal alkyl molecules might lead to excess stabilization as well as increased steric encumbrance with subsequent deactivation of the center. The $AlEt_3 +$ $TiCl_3$ is an example of such a catalyst. (2) In some soluble catalysts, because the stabilization of the crystal lattice is absent, the isolated alkylated transition metal species might indeed be too unstable for polymerization to occur. In this case, the transition metal–carbon bond might become sufficiently stable and hence active for polymerization if it becomes complexed with a base metal alkyl. The syndiotactic-specific catalyst, $AlEt_2Cl + VCl_4$, might be an example at $-78°C$. Complexing might also alter the symmetry of the transition metal complex so that it is changed from an inactive to an active configuration. Not all soluble catalysts require complexing by a base metal alkyl. Chapter 12 disclosed a number of metal alkyl-free catalysts in which the alkylated species or a derivative of it was active, such as tetrabenzylzirconium.

Some workers, while admitting that a small fraction of the polymer can be formed at alkylated transition metal sites that are not complexed to a base metal alkyl, argue that most of the polymer is formed at centers that contain both metals. The bimetallic complexes (Eq. 13–8) are said to be more active and more highly isotactic-specific.

The author is not aware of any direct experimental evidence that unequivocally rules out such a role for the metal alkyl. But the presence of such a complexed metal alkyl is not necessary to affect active and stereospecific catalysts for polymerizations of ethylene, propylene, and butadiene, as well as other olefin and diene monomers. Many examples have been cited in Chapter 12 in support of this claim.

References

1. J. Boor, *Macromol. Rev.* **2**, 115–268 (1967).
2. G. Natta, *Makromol. Chem.* **16**, 213 (1955).
3. C. D. Nenitzescu, C. Huch, and A. Huch, *Angew. Chem.* **68**, 438 (1956).
4. D. B. Ludlum, A. W. Anderson, and C. E. Ashby, *J. Am. Chem. Soc.* **80**, 1380 (1958).
5. W. L. Carrick, *J. Am. Chem. Soc.* **80**, 6455 (1958).
5a. D. S. Breslow and N. R. Newburg, *J. Am. Chem. Soc.* **81**, 81 (1959).
6. P. Cossee, *Tetrahedron Lett.* **17**, 12 (1960).

7. P. Cossee, *Proc. Int. Congr. Coord. Chem., 6th, 1961* p. 241 (1961).
8. P. Cossee, *J. Catal.* **3**, 80 (1964); *Rec. Trav. Chim. Pays-Bas* **85**, No. 9–10, 1152 (1966).
9. P. Cossee, P. Ros, and J. H. Schachtschneider, *Proc. Int. Congr. Catal., 4th, 1968* Paper I4 (1971).
10. J. W. Begley and F. Pennella, *J. Catal.* **8**, 203 (1967).
11. P. H. De Bruijn, *Chem. Weekbl.* **56**, 161 (1960).
12. L. A. M. Rodriguez and H. M. van Looy, *J. Polym. Sci., Part A-1* **4**, 1971 (1966).
12a. V. Buls and T. L. Higgins, *J. Polym. Sci., Polym. Chem. Ed.* **11**, 925 (1973).
13. G. Natta and G. Mazzanti, *Tetrahedron* **8**, 86 (1960).
14. P. Patat and Hj. Sinn, *Angew. Chem.* **70**, 496 (1958).
15. F. Eirich and H. Mark, *J. Colloid Sci.* **11**, 748 (1956).
16. H. Uelzmann, *J. Polym. Sci.* **32**, 457 (1958); *J. Org. Chem.* **25**, 671 (1960).
17. J. Boor, *J. Polym. Sci., Part C* **1**, 257 (1963).
18. G. Bier, *Kunststoffe* **48**, 354 (1958).
19. A. Gumboldt and H. Schmidt, *Chem.-Ztg.* **83**, 636 (1959).
20. H. N. Friedlander and W. Resnick, *Adv. Pet. Chem. Refin.* **1**, 526 (1958).
21. M. L. Huggins, *J. Polym. Sci.* **43**, 473 (1960).
22. J. Furukawa and T. Tsuruta, *J. Polym. Sci.* **36**, 275 (1959).
23. K. Vesely, J. Ambroz, J. Mejzlik, and E. Spousta, *J. Polym. Sci., Part C* **16**, 417 (1967).
24. H. N. Friedlander and K. Oita, *Abstr. Pap., 130th Meet., Am. Chem. Soc.* p. 138 (1956).
25. A. V. Topchiev, B. A. Krentsel, and L. G. Sidorova, *Dokl. Akad. Nauk SSSR* **128**, 732 (1959); A. V. Topchiev, B. A. Krentsel, and L. L. Stotskaya, *Russ. Chem. Rev. (Engl. Transl.)* **30**, 192 (1961).
26. E. W. Duck, *J. Polym. Sci.* **34**, 86 (1959).
27. R. Van Helden, H. P. Braendlin, A. F. Bickel, and E. C. Kooyman, *Tetrahedron Lett.* No. 12, p. 24 (1959).
28. A. Gilchrist, *J. Polym. Sci.* **34**, 49 (1959).
29. The reader will also find a review by A. Zambelli and C. Tosi instructive on this aspect [*Adv. Polym. Sci.* **15**, 32–60 (1974)].
30. C. F. Feldman and E. Perry, *J. Polym. Sci.* **46**, 217 (1960).
31. E. J. Vandenberg, U.S. Patent 3,051,690, August 28, 1962, Hercules.
32. A. S. Hoffman, B. A. Fries, and P. C. Condit, *J. Polym. Sci., Part C* **4**, 109 (1964).
33. A. Schindler, *J. Polym. Sci., Part B* **3**, 147 (1965).
34. G. Natta and I. Pasquon, *Adv. Catal.* **11**, 1–66 (1959).
35. C. Beerman and H. Bestian, *Angew. Chem.* **71**, 618 (1959).
36. A. P. Gray, A. B. Callear, and F. H. Edgecombe, *Can. J. Chem.* **41**, 1502 (1963).
37. E. J. Arlman, *Proc. Int. Congr. Catal., 3rd, 1964* p. 957 (1965).
37a. N. Koide, K. Iimura, and M. Takeda, *J. Polym. Sci., Polym. Chem. Ed.* **11**, 3161 (1973).
37b. V. A. Khodzhemirov, V. Ye. Ostrovskii, Ye. V. Zabolotskaya, and S. S. Medvedev, *Polym. Sci. USSR (Engl. Transl.)* **13**, 1871 (1971).
38. A. S. Hoffman, B. A. Fries, and P. C. Condit, *J. Polym. Sci., Part C* **4**, 125 (1963).
39. L. A. M. Rodriguez, H. M. van Looy, and J. A. Gabant, *J. Polym. Sci., Part A-1* **4**, 1905 (1966).
40. L. A. M. Rodriguez, H. M. van Looy, and J. A. Gabant, *J. Polym. Sci., Part A-1* **4**, 1917 (1966).
41. H. M. van Looy, L. A. M. Rodriguez, and J. A. Gabant, *J. Polym. Sci., Part A-1* **4**, 1927 (1966).
42. L. A. M. Rodriguez and H. M. van Looy, *J. Polym. Sci., Part A-1* **4**, 1951 (1966).
43. L. A. M. Rodriguez and H. M. van Looy, *J. Polym. Sci., Part A-1* **4**, 1971 (1966).
43a. J. Mejzlik, S. Petrik, and B. Kokta, *Collect. Czech. Chem. Commun.* **37**, 2920 (1972).

44. F. J. Karol and W. L. Carrick, *J. Am. Chem. Soc.* **83**, 2654 (1961); F. J. Karol, W. L. Carrick, G. L. Karapinka, and J. J. Smith, *ibid.* **82**, 1502 (1960).
45. A. Schindler, *J. Polym. Sci., Part B* **3**, 147 (1965).
46. E. Junghanns, A. Gumboldt, and G. Bier, *Makromol. Chem.* **58**, 18 (1962).
47. A. Zambelli, I. Pasquon, A. Marinangeli, G. Lenzi, and E. R. Mognaschi, *Chim. Ind. (Milan)* **46**, 1464 (1964).
48. G. Natta, A. Zambelli, I. Pasquon, and G. M. Giongo, *Chim. Ind. (Milan)* **48**, 1298 (1966).
49. G. Natta, A. Zambelli, I. Pasquon, and G. M. Giongo, *Chim. Ind. (Milan)* **48**, 1307 (1966).
50. I. Pasquon, *Pure Appl. Chem.* **15**, 465 (1967).
50a. A. Guyot, P. Sili, E. Agouri, and R. Laputte, *Makromol. Chem.* **171**, 28 (1973).
51. I. Pasquon, G. Natta, A. Zambelli, A. Marinangeli, and A. Surico, *J. Polym. Sci., Part C* **16**, Part 5, 2501 (1967).
51a. J. Ambroz, L. Ambroz, and D. Nahlikova, *Chem. Prum.* **14**, 648 (1964).
51b. E. N. DiCarlo and H. E. Swift, *J. Phys. Chem.* **68**, 551 (1964).
52. W. P. Long and D. S. Breslow, *J. Am. Chem. Soc.* **82**, 1953 (1960); W. P. Long, *ibid.* **81**, 5312 (1959).
53. A. K. Zefirova and A. E. Shilov, *Proc. Acad. Sci. USSR, Chem. Sect. Eng. Trans.* **136**, 77 (1961).
54. A. Y. Shilov, A. K. Shilova, and B. N. Bobkov, *Vysokomol. Soedin.* **4**, 1688 (1962); *Polym. Sci. USSR (Engl. Transl.)* **4**, 526 (1963).
55. F. S. D'yachkovskii, A. K. Shilova, and A. Y. Shilov, *Polym. Sci. USSR (Engl. Transl.)* **8**, 336 (1966).
56. F. S. D'yachkovskii, A. K. Shilova, and A. Y. Shilov, *J. Polym. Sci., Part C* **16**, 2333 (1967).
57. E. A. Grigoryan, F. S. D'yachkovskii, G. M. Khvostik, and A. Y. Shilov, *Polym. Sci. USSR (Engl. Transl.)* **9**, 1372 (1967).
58. F. S. D'yachkovskii, M. L. Yeritsyan, O. Ye. Kashireninov, B. Matiska, K. Makl, M. Shvestka, and A. Y. Shilov, *Polym. Sci. USSR (Engl. Transl.)* **11**, 617 (1969).
59. E. A. Nevel'skii and F. S. D'yachkovskii, *Vysokomol. Soedin.* **11**, 797 (1969).
60. U. N. Babkina and F. S. D'yachkovskii, *Vysokomol. Soedin., Ser. B* **12**, 301 (1970).
61. T. S. Dzhabiev, F. S. D'yachkovskii, and A. L. Khamrayeva, *Polym. Sci. USSR (Engl. Transl.)* **11**, 1881 (1970).
62. F. S. D'yachkovskii, P. A. Yarovitskii, and V. F. Bystrov, *Vysokomol. Soedin.* **6**, 659 (1964).
63. E. A. Grigoryan, F. S. D'yachkovskii, and A. Y. Shilov, *Polym. Sci. USSR (Engl. Transl.)* **1**, 158 (1965).
64. G. Henrici-Olive and S. Olive, *Adv. Polym. Sci.* **6**, 421 (1969).
65. G. Henrici-Olive and S. Olive, *J. Polym. Sci., Part C* **22**, Part 2, 965 (1969).
66. G. Henrici-Olive and S. Olive, *Makromol. Chem.* **121**, 70 (1969).
67. G. Henrici-Olive and S. Olive, *Angew. Chem.*, Int. Ed. *Engl.* **7**, 821 (1968).
68. G. Henrici-Olive and S. Olive, *J. Polym. Sci., Part B* **8**, 271 (1970).
69. G. Natta, P. Pino, G. Mazzanti, and R. Lanzo, *Chim. Ind. (Milan)* **39**, 1032 (1957).
70. G. Natta, P. Pino, G. Mazzanti, and U. Giannini, *J. Inorg. Nucl. Chem.* **8**, 612 (1958).
71. G. Natta, P. Pino, G. Mazzanti, U. Giannini, E. Mantica, and M. Peraldo, *J. Polym. Sci.* **26**, 120 (1957).
72. G. Natta, *Chim. Ind. (Milan)* **39**, 19 (1957).
73. P. Pino, G. Mazzanti, U. Giannini, and S. Cesca, *Atti Accad. Naz. Lincei, Cl. Sci. Fis., Mat. Nat., Rend.* [8] **27**, 392 (1959).
74. G. Natta and G. Mazzanti, *Tetrahedron* **8**, 86 (1960).
75. G. L. Karapinka and W. L. Carrick, *J. Polym. Sci.* **55**, 145 (1961).

76. H. W. Coover, Jr., *J. Polym. Sci., Part C* **4**, 1511 (1965); H. W. Coover, Jr., R. L. McConnell, and F. B. Joyner, *Macromol. Rev.* **1**, 91–118 (1966).

76a. G. Sartori and G. Costa, *Z. Elektrochem.* **63**, 105 (1959).

76b. I. Collamati and G. Sartori, *Chim. Ind. (Milan)* **47**, 368 (1965).

77. G. Natta, I. Pasquon, A. Zambelli, and G. Gatti, *J. Polym. Sci.* **51**, 387 (1961).

78. G. Natta, *Chim. Ind. (Milan)* **42**, No. 11, 1207 (1960).

79. A. G. Pozamantir, *Vysokomol. Soedin.* **2**, 1026 (1960).

80. R. Graf, H. Zimmerman, and H. Bestian, German Patent 1,019,466, November 14, 1957, Hoechst; Belgian Patent 545,087, August 9, 1956, Hoechst.

81. A. G. Pozamantir, A. A. Korotkov, and I. S. Lishanskii, *Vysokomol. Soedin.* **3**, 1769 (1961).

82. J. Boor, *Am. Chem. Soc., Polym. Prepr.* **6**, No. 2, 890 (1965).

83. A. Schindler, *J. Polym. Sci., Part B* **3**, 147 (1965).

84. L. M. Lanovskaya, N. V. Makletsova, A. R. Gantmakher, and S. S. Medvedev, *Polym. Sci. (Engl. Transl.)* **7**, 820 and 827 (1965); L. M. Lanovskaya, N. A. Pravikova, A. R. Gantmakher, and S. S. Medvedev, *ibid.* **11**, 1315 (1970).

85. K. Soga and T. Keii, *J. Polym. Sci., Part A-1* **4**, 2429 (1966); T. Keii, *Polym. Prepr., Am. Chem. Soc., Div. Polym. Chem.* **15**, 373 (1974).

86. Y. Ono and T. Keii, *J. Polym. Sci., Part A-1* **4**, 2441 (1966).

14

Mechanisms for Initiation
and Propagation of Dienes

I. Introduction

Chapter 13 described proposed mechanisms for olefin polymerizations, and it was concluded that the growth center was a transition metal–carbon bond. Often olefin workers implied that their mechanisms were just as valid for diene polymerizations, but without further elaboration. Section II of this chapter describes the mechanistic models that were specifically proposed for diene polymerizations.

Like the olefin workers, the diene workers also suggested potential mechanisms involving either a transition metal–carbon or a base metal–carbon bond as the active center. The growth center can have either a σ- or a π-allyl structure, depending on the choice of transition metal, as well as other factors. However, the mechanisms proposed for dienes were viewed differently and, for this reason, they are treated separately.

A number of excellent reviews that cover this subject from different vantage points are available (1–7).

II. Proposed Mechanisms

A. GROWTH AT A BASE METAL–CARBON BOND

According to these models, the transition metal is essential to coordinate the diene before it is inserted into the base metal–carbon bond, which means that the growth step requires two metal atoms, as shown in Eq. 14–1, where

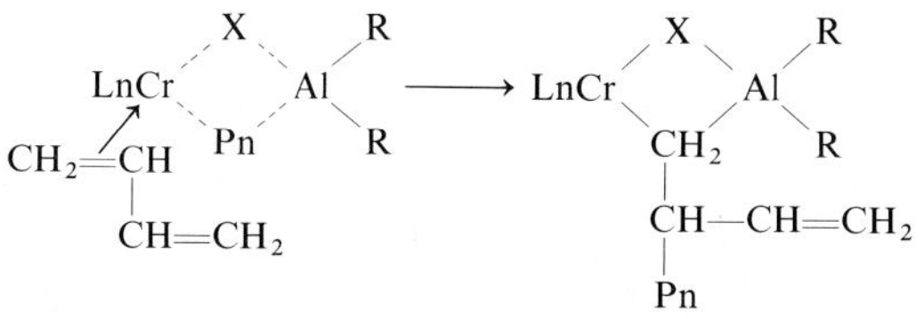

$$
\begin{array}{c}
\text{LnCr} \overset{X}{\underset{\underset{\text{CH}_2 = \text{CH}}{\nearrow}}{\diagdown}} \text{Al} \overset{R}{\underset{R}{\diagup}} \\
\overset{\text{Pn}}{\underset{\text{CH}=\text{CH}_2}{|}}
\end{array}
\longrightarrow
\begin{array}{c}
\text{LnCr} \overset{X}{\diagdown} \text{Al} \overset{R}{\diagup} \\
\overset{\text{CH}_2}{\underset{\text{CH}-\text{CH}=\text{CH}_2}{|}} \overset{R}{} \\
\underset{\text{Pn}}{|}
\end{array}
\qquad (14\text{--}1)
$$

Ln is a nonbridging ligand associated with Cr; Pn, a growing polymer chain; R, an alkyl attached to Al; and X, a halogen or bridging ligand. This bimetallic center can be part of a surface in a heterogeneous catalyst such as Saltman's model (8), or actually be the center in the homogeneous catalyst, as in Gippin's model (9). These specific models, however, vary in some respects, such as in the ionic character of the complex, as the following examples show.

Gippin (9) proposed a mechanism involving two metal alkyl units associated with a cobalt atom. According to him, the polymerization of butadiene in the soluble cobalt catalyst system to a high *cis*-1,4-polybutadiene required two alkylaluminum molecules differing in Lewis acid-type acidity, such as $AlEt_2Cl + AlEtCl_2$. The growth step shown in Eq. 14–2 was visu-

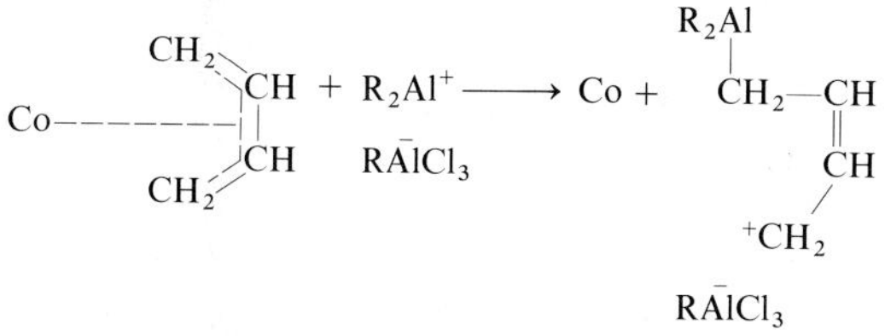

$$
\begin{array}{c}
\text{Co} \text{------} \begin{array}{c} \overset{\text{CH}_2}{\diagdown} \\ \underset{\text{CH}_2}{\diagup} \end{array} \begin{array}{c} \text{CH} \\ \| \\ \text{CH} \end{array} + \begin{array}{c} R_2Al^+ \\ \overline{R}\overline{Al}Cl_3 \end{array}
\longrightarrow
\text{Co} + \begin{array}{c} R_2Al \\ | \\ CH_2-CH \\ \| \\ CH \\ \diagup \\ {}^+CH_2 \\ \overline{R}\overline{Al}Cl_3 \end{array}
\end{array}
\qquad (14\text{--}2)
$$

alized. The second and subsequent growth steps involve a 1,4-addition of the coordinated butadiene to the butadienyl cobalt of the carbonium ion and its counter ion. The cobalt atom acts as carrier of the cis-coordinated butadiene and confers increased polarizability upon the butadiene molecule.

Saltman (8) proposed for polymerization of isoprene with the $AlEt_2Cl$–$\beta TiCl_3$ catalyst that $AlEt_2Cl$ becomes epitactically adsorbed on the $\beta TiCl_3$ surface. While the isoprene molecule coordinates to both Ti and Al centers, as shown in structure I, it becomes inserted in the Al–C bond (see also

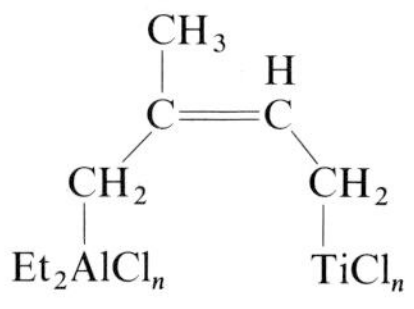

$$
\begin{array}{c}
\overset{\text{CH}_3}{\underset{|}{}} \quad \overset{\text{H}}{} \\
\text{C} = \text{C} \\
\underset{\text{CH}_2}{\diagup} \qquad \underset{\text{CH}_2}{\diagdown} \\
\underset{Et_2AlCl_n}{|} \qquad \underset{TiCl_n}{|}
\end{array}
$$

I

Fig. 16–6). The $TiCl_3$ acts to polarize the Al–C bond and thereby makes it active for polymerization.

B. GROWTH AT A TRANSITION METAL–CARBON BOND

According to these models, the growth bond involves only one metal atom, that is, the transition metal. In addition to the σ-type (localized) transition metal–carbon bond (II) previously suggested for olefin polymerizations, other

$$LnTi—CH_2—CH=CH—CH_2—Pn$$

II

structures have been considered, namely a stabilized σ bond (structure III) and a delocalized π-allyl bond (structure IV). There is a little confusion in

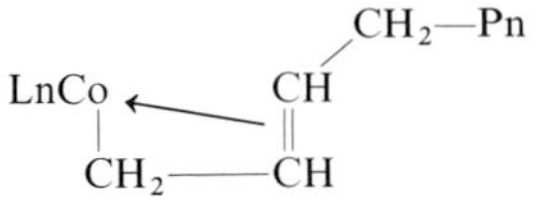

III 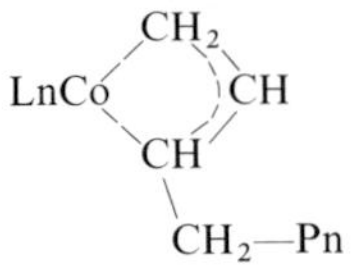   IV

the literature concerning the latter structures. The stabilized σ bond probably originated as an attempt by workers to show why cobalt-based catalysts were able to polymerize butadiene but not ethylene or propylene. In fact, these olefins were used to terminate chain growth in a butadiene polymerization (see Chapter 10). This apparent anomaly was explained by the ability of the double bond of the last added diene molecule to coordinate with the Co atom and thereby stabilize the σCo–C bond, as shown above. In the strictest sense, the σ bond retained its localized character.

The use of a π-allyl transition metal bond as the active center was accepted widely after it was shown that certain π-allyl compounds of transition metals were able to polymerize butadiene in the absence of base metal alkyl (for examples, see Chapter 12, Section III,B,2,c). As will be shown in Chapter 16, the mode of addition of dienes might be determined by which side of the π-allyl group the incoming diene becomes bonded.

Sometimes the stabilized σ- and the π-allyl transition metal bonds are considered to be virtually identical. The structures drawn for the π-allyl metal bond will also differ according to different authors. In addition to examples already cited, structures V and VI have also been used. An equili-

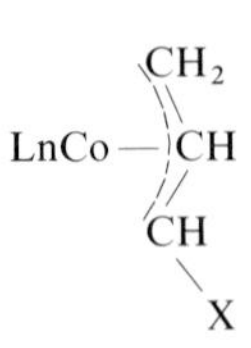

V 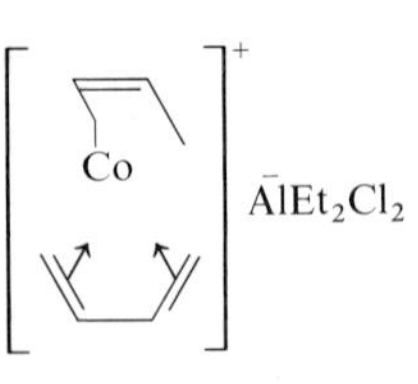   VI

brium between σ- and π-allyl metal bonds has also been proposed, and some believe that growth takes place only when the center has the σ bonding (10). The proposal has been made that this equilibrium can be shifted toward the σ-bonded structure by the addition of electron-donor ligands and toward the π-allyl structure by the influence of electron-acceptor ligands (11), as shown in Eq. 14-3.

$$M \longleftarrow \begin{matrix} CH_2 \\ \| \\ CH \\ // \\ CH \\ | \\ R \end{matrix} \quad \underset{\text{acceptor}}{\overset{\text{donor}}{\rightleftharpoons}} \quad M-CH_2-CH=CH-R$$

(14-3)

π-allyl bond σ bond

Cossee (12) suggests from molecular orbital considerations and qualitative kinetic data that pushing electrons into the complex, which decreases the ionization potential of the metal orbitals, results in a shift of the equilibrium toward the side of the σ form. He concludes that it would be more likely for elements like titanium, which has a low ionization potential, to form σ bonds and for transition metals like cobalt to form π-allyl bonds. Thus the σ- or π-allyl character of these centers may also vary according to the transition metal and the ligands attached to it.

Because the experimental data do not permit a clear delineation of these ideas, it is inevitable that the worker has a number of options to explain the rather large number of stereospecific paths possible. The true mechanism may involve aspects of all these models or hybrids of them.

III. Experimental Evidence

A. RADICAL AND CATIONIC MECHANISMS

As found in olefin polymerizations with Ziegler–Natta catalysts, overall activation energies are considerably lower than found for radical polymerizations, for example, <15 kcal/mole (13, 14) vs. about 25 kcal/mole. This suggests that ionic mechanisms rather than radical ones are responsible for chain growth. Cationic mechanisms of the conventional type, with BF_3, $AlCl_3$, etc., do not appear likely. Many of the most active catalysts have virtually no cationic character. So the fact that some catalysts do have acidic character is not in itself important. Furthermore, the obtained polymers have high contents of *cis*-1,4, *trans*-1,4, or 1,2 structures, which is a feature not characteristic of radical or cationic polymerizations of dienes (see Table 5–1). Electron donors can often be present without the destruction of centers; this also argues against cationic centers. Cationic polymerizations also lead to extensive gellation (cross-linking). While some gel is present in the Ziegler–Natta polymers, they can be prepared under controlled conditions without cross-linking.

We must look to growth at metal–carbon bonds as a more promising approach. But here, we cannot yet begin to differentiate mono- or bimetallic schemes, as the experimental data are insufficient. As we did for olefins, the discussion must be directed to answering whether it is the base metal or the transition metal that is the carrier of the polymer chain.

B. GROWTH AT A BASE METAL–CARBON BOND

It was concluded by some workers that because catalyst activity and stereoregulating ability were so intimately related to the structure of the metal alkyl, it meant that growth took place at the base metal–carbon bond. This was especially true for homogeneous catalysts. Supporting evidence is scarce. A few examples are cited.

Gippin (9) supported his mechanistic proposals with the observation that an aluminum alkyl pair lowest in Lewis acidity scale was non-catalytic with cobalt, such as $AlEt_3 + AlEt_2Cl \rightleftharpoons AlEt_2^+ \; AlEt_3Cl^-$ (8). Pairs that were high on the scale, $AlEtCl_2 + AlCl_3 \rightleftharpoons AlEtCl^+ \; AlCl_4^-$, produced a cationic polymerization marked by a fast rate and formation of cross-linked (gelled) polymer. In between were pairs of intermediate acidity, $AlEt_2Cl + AlEtCl_2 \rightleftharpoons AlEt_2^+ \; AlEtCl_3^-$, and these gave optimum rates.

Gippin also explained why small amounts of water or $AlCl_3$ activated the mixture of $AlEt_2Cl$ and $CoCl_2$. The reaction of $AlEt_2Cl$ and H_2O or $AlCl_3$ created the second metal alkyl necessary for his mechanism, $AlEt(OH)Cl$ or $AlEtCl_2$, respectively.

Gippin's mechanism would be hard to test because he stated that polymerization can take place via a carbonium ion or possibly via a carbanion mechanism. Hence, labeling experiments could not produce affirmative conclusions. Only if propagation took place at a conventional Al–C bond could we conclude from the labeling data (see the section on radioactive labeling) that this bond is not a carrier of the polymer chain.

Bressler and co-workers (15) observed that $\beta TiCl_3$ had a different kinetic activity for isoprene polymerization, depending on whether Al-*i*-Bu$_3$ or Al-*i*-Bu$_2$Cl was used to form the catalyst. The formed polymer had different molecular weight characteristics. They interpreted this to mean that the metal alkyl was an integral part of the center (as a bimetallic complex). Different adsorption capacities of the two metal alkyls on the $TiCl_3$ surface could also account for different rates, due to different numbers of active centers that were formed. These data, however, do not prove that the Al is the carrier of the polymer chain since alternate explanations could account for the results, including stabilization of the Ti–Pn growth center and variation in their number due to the two metal alkyls.

C. GROWTH AT A TRANSITION METAL–CARBON BOND

Most workers, including the author, accept the transition metal–carbon bond as the active center. The experimental evidence for this view will be presented presently. Whenever possible, we shall try to throw some light on whether σ or π bonds are in fact the active centers. The most cogent evidence is the ability of many metal alkyl-free catalysts to form active and highly stereoregulating catalysts. These appear to propagate by σ- and π-allyl-type growth centers, via Cr and Ni compounds, respectively.

1. Evidence Based on Metal Alkyl-Free Catalysts

Chapter 12 described a large number of metal alkyl-free catalysts, many of which were active and stereoregulating for butadiene. Of significant importance are the π-allyl derivatives of certain transition metals, including $(\pi$-allyl)NiX (X = Cl, Br, I) (16) and $(\pi$-allyl)$_3$Cr (17). Polybutadiene containing a high content of cis-1,4, trans-1,4, or 1,2 structures has been prepared, although it usually contained a very low molecular weight (about several thousand).

A growth process involving coordination of butadiene (BD) to the Ni atom, followed by insertion into the π-allyl bond, was suggested by Cooper (4), as shown in Eq. 14–4. π-Allyl groups derived from the π-allyl NiCl have

$$(14\text{--}4)$$

been identified in the polymer product [see Marconi (2, p. 265)]. Since π-allyl NiCl exists in the dimer state, only after coordination with butadiene (or a Lewis base) does it break up into the monomer state. It is important to note that butadiene can be polymerized with π-allyl Ni compounds without base metal alkyls being present.

Some workers assigned a greater ionic character to the Ni–Cl bond and, in fact, attributed the increased activity of the π-allyl Ni halide to the increased ionic character of the Ni–X bond when X = other halogens or ligands. Polymerizations were slow and increased in the order X = Cl < Br < I, which may parallel their ability to dissociate (16). Molecular weights were also low.

A more active catalyst and higher molecular weights were obtained when Lewis acids ($AlCl_3$, VCl_4, or $TiCl_4$) (18, 19) were also added. The increased activity was attributed to formation of ionic complexes of the π-allyl Ni chloride and these Lewis acids, such as with $AlCl_3$ when X = $AlCl_4^-$, D = an aromatic molecule (benzene), and the solvent was benzene, as shown in structure VII. The greater ionic character of the Ni–X bond was also

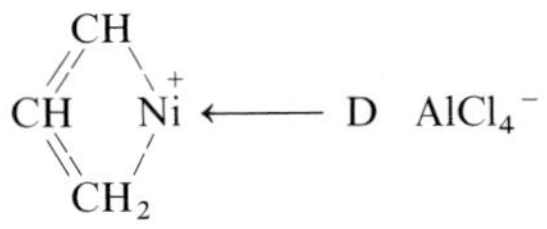

$$
\begin{array}{c}
CH \\
\diagup\!\!\diagup \quad \diagdown \\
CH \quad \overset{+}{Ni} \longleftarrow D \; AlCl_4^- \\
\diagdown\!\!\diagdown \quad \diagup \\
CH_2
\end{array}
$$

VII

suggested as the reason for the increased activity of these complexes. In other words, catalytic activity increased with increasing positive charge on the Ni atom (4).

This view has other support. Butadiene has been polymerized with the binary mixture bis(1,5-cyclooctadiene)nickel and a protonic acid (20). More active catalysts are produced as the strength of the acid is increased, as shown in the complex formed by the reaction shown in Eq. 14–5. Activity increased

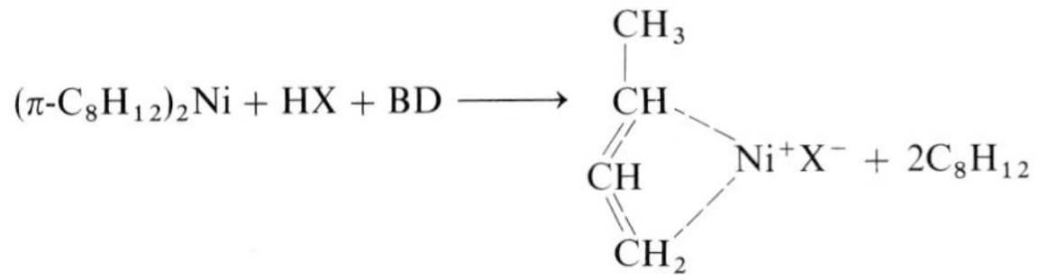

$$
(\pi\text{-}C_8H_{12})_2Ni + HX + BD \longrightarrow
\begin{array}{c}
CH_3 \\
| \\
CH \\
\diagup\!\!\diagup \quad \diagdown \\
CH \quad Ni^+X^- + 2C_8H_{12} \\
\diagdown\!\!\diagdown \quad \diagup \\
CH_2
\end{array}
\qquad (14\text{–}5)
$$

in the order: X = $CH_3CO_2^- < Cl_3CO_2^- < CF_3CO_2^-$.

Concomitant with the higher polymerization rates, the formed polymer had much higher molecular weights and often were equal to those measured for polymers obtained with Ziegler–Natta catalysts.

Cooper extended these views to Ziegler–Natta catalysts (1, 4). For the $Al_2Et_3Cl_3$–$CoCl_2$ catalyst, the following reactions leading to the active catalyst B and the polymer product D were proposed.

$$\text{(14-6)}$$

According to Cooper, the first added butadiene molecule is inserted into the Co–Et bond with the formation of a π-allyl terminal unit. The subsequent butadiene molecules are first coordinated to the Co(I) species and then inserted into the Co–π-allyl bond. This mechanism resembles the π-allyl-NiX-type in that X equals the $AlEtCl_3^-$ counteranion which allows a greater positive charge to develop on the Co atom.

According to Natta, Porri, and Carbonaro (21), another role of the $AlEt_2Cl$ in the Ziegler–Natta cobalt-based catalyst is to reduce Co(II) to Co(I), the latter reacting to form the active species. The active center is formed by the oxidative addition of butadiene to Co(I), as shown in Eq. 14–7.

$$\text{(14-7)}$$

The nature of the ligands (Ln) was specified in some cases (see Eq. 14–8). The formed radicals couple quickly so that growth occurs exclusively at the Co center. This view was based, in part, on the finding of Natta, Porri, and Carbonaro that a ternary mixture of $AlCl_3$, Al metal, and $CoCl_2$ in benzene produced a crystalline compound that alone was able to polymerize butadiene to 98% cis-1,4 polymer (21). A similar compound was synthesized directly from $Al(C_6H_5)Cl_2$, $Al(C_6H_5)_2Cl$, and $CoCl_2$ (22). Oxidative addition was suggested as the initiating step, as shown in Eq. 14–8, with

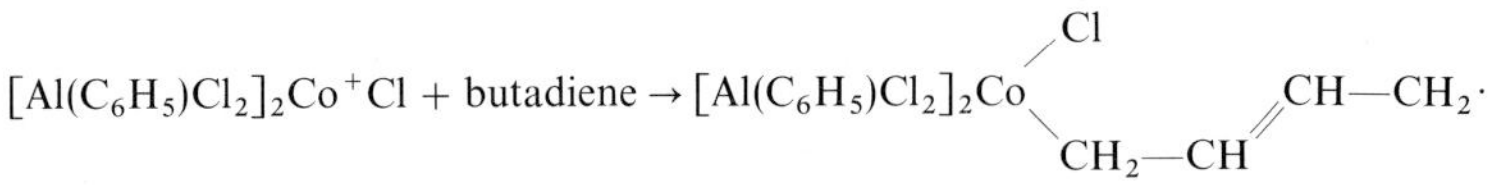

$$[Al(C_6H_5)Cl_2]_2Co^+Cl + butadiene \rightarrow [Al(C_6H_5)Cl_2]_2Co\begin{matrix} Cl \\ \diagdown \\ CH_2\!-\!CH \end{matrix}\begin{matrix} \\ \diagup \\ \diagdown \end{matrix}CH\!-\!CH_2\cdot \quad (14\text{–}8)$$

immediate destruction of the radical by dimerization. At that time, the polymer chain end was viewed as a stabilized σ bond, but more recently these authors consider it more as a π-allyl bond. Perhaps the differences are semantic.

The author suggests that the basic idea expressed by Cooper in Eq. 14–6 can be made to agree with the radiolabeling results if a small modification is made. Instead of complex A in Eq. 14–6, the complex A shown below

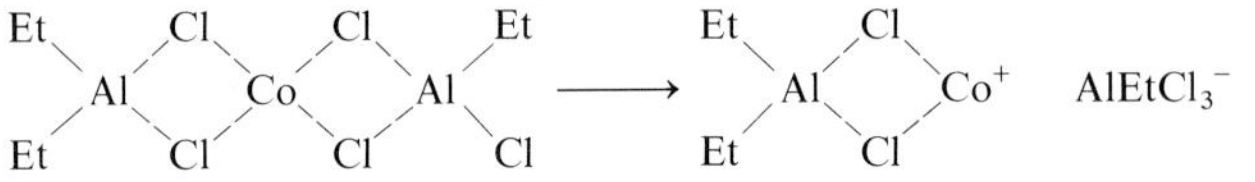

$$\begin{matrix} Et & & Cl & & Cl & & Et \\ \diagdown & \diagup & & \diagup & & \diagdown \\ & Al & & Co & & Al \\ \diagup & & \diagdown & & \diagdown & & \diagdown \\ Et & Cl & & Cl & & Cl \end{matrix} \longrightarrow \begin{matrix} Et & & Cl \\ \diagdown & \diagup & \\ & Al & Co^+ \\ \diagup & & \diagdown \\ Et & Cl \end{matrix} \quad AlEtCl_3^- \quad (14\text{–}9)$$

$$\qquad\qquad A \qquad\qquad\qquad\qquad B$$

forms and disproportionates to the polarized complex B. The oxidation state of Co is $+1$ and, following the idea of Natta, Porri, and Carbonaro, complex B oxidatively adds butadiene to form a divalent Co–C species, the active catalyst.

Natta, Giannini, Pino, and Cassata (23) cited spectroscopic evidence to show that the cobalt–carbonyl compound that caused the dimerization of butadiene to 3-methyl-1,4,6-heptatriene had a π-allyl structure as in VIII or IX.

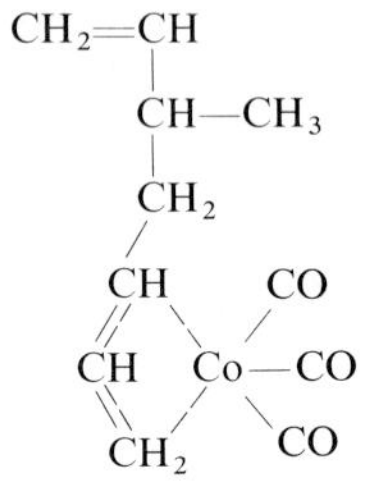

$$\begin{matrix} CH_2\!=\!CH \\ | \\ CH\!-\!CH_3 \\ | \\ CH_2 \\ \diagup \\ CH \quad CO \\ \diagdown \quad \diagup \\ CH \quad Co\!-\!CO \\ \diagdown \quad \diagdown \\ CH_2 \quad CO \end{matrix} \qquad or \qquad \begin{matrix} CH_2\!=\!CH \\ \diagdown \\ CH\!-\!CH_3 \\ | \\ CH \quad CO \\ \diagdown \quad \diagup \\ CH \quad Co\!-\!CO \\ \diagdown \quad \diagdown \\ CH \quad CO \\ | \\ CH_3 \end{matrix}$$

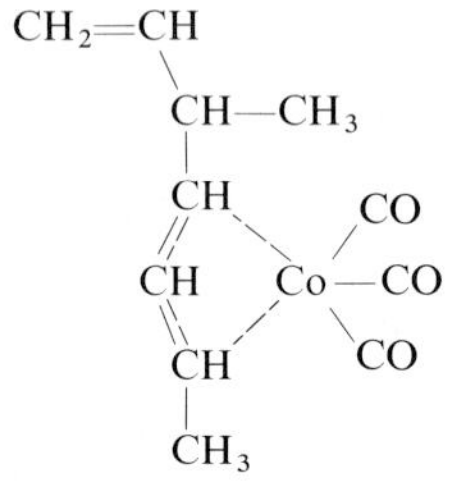

$$\qquad VIII \qquad\qquad\qquad\qquad IX$$

2. Evidence Based on Transfer Agents ZnEt₂ and H₂

That H_2 and $ZnEt_2$ are similarly effective transfer agents for diene polymerization suggests that, as in olefin polymerizations, transition metal–carbon bonds are the active centers (see Chapter 10). The experimental conditions under which these transfer agents are effective are fairly similar.

These transfer agents were effective for the Ziegler–Natta catalyst $AlEt_2Cl–CoCl_2$ and the metal alkyl-free catalyst π-allyl NiCl. Both of these catalysts are believed to involve π-allyl-transition metal-type bonds. The question is raised whether H_2 can react with the π-allyl transition metal bond or only if it is converted first to a σ carbon–transition metal bond or to a σ-stabilized bond. An unequivocal answer cannot be given.

One can argue that a Co–alkyl bond is unstable (such as Co–Et) and, therefore, as soon as the σCo–CH_2–CH=CH–Pn bond is formed, it decomposes before H_2 can participate. Thus, it might follow that π-allyl or stabilized σ-allyl transition metal–carbon bonds are actually split by hydrogen. However, if complexing of the Co center with H_2 would momentarily stabilize the Co–C bond, as shown in Eq. 14–10, then it may be possible

$$
\underset{\begin{array}{c}\\ CH_2\end{array}}{\overset{\begin{array}{c}CH-CH_2-Pn\\ \|\end{array}}{LnCo \longleftarrow CH}} + H_2 \longrightarrow LnCo \overset{\begin{array}{c}CH_2-CH\\ CH-CH_2-Pn\end{array}}{\underset{H}{\diagup}} \longrightarrow LnCoH + CH_3 - \overset{\begin{array}{c}CH-CH_2-Pn\\ \|\end{array}}{CH}
$$

$$(14\text{–}10)$$

for the σCo–C bond to be broken by H_2. The same sequence could have been written for the π-allyl–Co bond. Whether the metal–carbon has σ- or π-allyl character is not answered by the above.

3. Evidence Based on Homopolymerization and Copolymerization Studies

It was already recognized many years ago that some of the Ziegler–Natta catalysts were capable of homopolymerizing both monoolefins or dienes, especially Ti-, V-, Cr-, and Mo-based systems. These catalysts were capable of copolymerizing olefins and dienes.

Yet other catalysts, such as cobalt- and nickel-based ones, were active only for dienes, and when olefins were added, termination of chain growth occurred with the incorporation of some ethylene monomer into the chain (see Chapter 10). Labeling experiments suggested that one or two ethylene units were incorporated per chain. That occasionally more than one ethylene unit was added suggests a Co–C is stable long enough for another ethylene to add. But we cannot be sure if the formed σ-alkyl–Co bond is stabilized via the double bond, even though three methylenes insulated it from the Co atom, as seen in structure X. Such a stabilization would diminish as more

$$
\underset{\begin{array}{c}\\ CH_2-CH_2\end{array}}{\overset{\begin{array}{c}CH_2-Pn\\ HC\diagup\\ \diagdown CH\\ |\\ CH_2\end{array}}{LnCo \longleftarrow}}
$$

X

ethylenes were added. Hence long segments of polyethylene would not be formed. These homo- and copolymerization data suggest that both σ- and π-allyl-type transition metal bonds may be present. Catalysts based on Ti, V, Cr, or Mo may favor the σ-type, while π-allyl-type may prevail in Ni or Co catalysts. But copolymerization data that would allow us to differentiate this view more convincingly are not available. An answer was sought in labeling experiments with greater expectations.

4. Evidence Based on Radioactive Labeling

Labeling experiments with $Al(^{14}C_2H_5)_3$, $^{14}CO_2$, and tritiated methanol (CH_3O^3H) have been used as diagnostic tools to detect metal–carbon bond growth centers. On the basis of results obtained with olefins under similar conditions of polymerization, one would have expected the presence of ^{14}C and tritium in the formed polymer if the growth center consisted of a transition metal–carbon σ bond. This was not always the case, and the reader should be aware that contrary experimental results were obtained with apparently similar systems. Excellent detailed discussions of these experiments have been presented in review papers by Natta and Porri (3), Cooper (4), and Marconi (2).

For polymerizations of dienes with Ti, V, Cr, and Mo catalysts, the following expected results were obtained.

Termination of the polymerization by tritiated methanol produced polymers containing tritium, as expected from the reaction shown in Eq. 14–11. Natta and Porri disclosed as examples the polymerization of

$$\text{cat—Pn} + CH_3O^3H \longrightarrow \text{cat—OCH}_3 + {}^3H\text{–Pn} \qquad (14\text{–}11)$$

isoprene and pentadiene with the $AlEt_3$–$Ti(OR)_4$ catalyst (3, p. 653). Bressler and co-workers reported a similar finding for butadiene with the $Al\text{-}i\text{-}Bu_3$–TiI_4 catalyst (15).

Termination of the latter system with $^{14}CO_2$ resulted in ^{14}C in the polymer (3, p. 653), in accordance with the known reaction shown in Eq. 14–12.

$$\text{cat—Pn} + {}^{14}CO_2 \longrightarrow \text{cat—O—}\overset{\overset{\text{O}}{\|}}{C}\text{—Pn} \xrightarrow{H_2O} \text{catOH} + HO\overset{\overset{\text{O}}{\|}}{C}\text{—Pn} \quad (14\text{–}12)$$

$$(14) \qquad\qquad\qquad\qquad (14)$$

When labeled Al $(^{14}C_2H_5)_3$ alkyl was used, the polymer contained ^{14}C. This result can occur only if the C_2H_5 group is attached to the active metal center, as shown in Eq. 14–13. The cat–$^{14}C_2H_5$ bonds could have formed

$$\text{cat—}^{14}C_2H_5 + M \longrightarrow \text{cat—Pn—}^{14}C_2H_5 \xrightarrow{ROH} \text{catOR} + Pn^{14}C_2H_5 \quad (14\text{–}13)$$

during the reaction of $Al(^{14}C_2H_5)_3$ and the transition metal salt (either as $Al\text{–}^{14}C_2H_5$ if one accepts growth at this bond or as $Ti\text{–}^{14}C_2H_5$ centers formed by alkylation reactions). Since $Al-(^{14}C_2H_5)_3$ can function as a transfer agent for the polymerization, the $cat-^{14}C_2H_5$ bond can also result from chain-terminating reactions (see Chapter 10).

As examples of catalysts for butadiene and isoprene are cited the data of Natta and Porri (3, p. 653) and Cooper and co-workers (24) for the $AlEt_3$–VCl_3 system; for butadiene, the data obtained with the $Al(i-Bu)_3$–TiI_4 system (25); for 1,3-pentadiene and isoprene, the data obtained with the $AlEt_3$–$Ti(OR)_4$ catalyst (3, p. 653; 26).

These results show that, for these catalysts, the polymer chain is attached to the metal center. They do not, however, differentiate clearly between base metal and transition metal centers. If one accepts the growth center to be a transition metal–carbon bond, the combined copolymerization and labeling data argue strongly for a σ-type bonding in the Ti-, V-, Cr-, and Mo-based catalysts, as examined above.

When two of these diagnostic tests were applied to the polymerization of butadiene with the soluble $AlEt_2Cl$–Co salt catalyst, different results were obtained. However, different authors found contrary results in the case of termination by tritiated methanol (CH_3O^3H).

Tritium was not always found in the polymer when polymerizations were terminated with CH_3O^3H. For example, Childers (27) and Bressler (15) did not find tritium in the polymer when a butadiene polymerization was terminated with CH_3O^3H. Childers also found ^{14}C in the polymer if termination was made with a ^{14}C-containing butyl alcohol. These results suggested to him that a cationic mechanism was operating. In contrast, Cooper and co-workers found activity in the polymer (28). Natta and co-workers later showed that the presence of tritium was at least partly derived from secondary transfer reactions (dead polymer was exposed to the CH_3O^3H and found to contain tritium when worked up). Cooper showed that when the polymerization using an $AlEt_2\ ^{36}Cl$–Co catalyst was terminated with methanol, the polymer contained some ^{36}Cl (29). It was suggested that $H^{36}Cl$ was formed via $AlEt_2Cl + CH_3OH \rightarrow HCl$ and this $H^{36}Cl$ added to the double bonds of the dead polymer. This suggests one path by which tritium of the alcohol could get into the polymer without actually involving termination of chain growth.

These diagnostic tests may be of no utility if π-allyl-type bonded polymer chains do not react quantitatively with MeOH, as suggested by Natta and Porri (3). They cite the known fact that butadiene polymerizations with π-allyl Ni halides have been made in alcohol as solvent; thus π-allyl structures must be stable in alcohol. If the Co-based Ziegler–Natta catalysts described above have polymer attached by π-allyl bonds, one would expect similar insensitivity to alcohol. Hence, the CH_3O^3H method does not apply to this cobalt system.

No ^{14}C was found in the polymer when $Al^{14}Et_2Cl–Co(acac)_2$ was used to polymerize butadiene (21). Childers considered this along with his CH_3O^3H data (see above) as evidence that butadiene was polymerized by a cationic mechanism.

Natta and Porri suggested an alternate and more cogent explanation that significantly contrasts the $AlEt_2Cl–Co(acac)_2$ systems with those based on $AlEt_2Cl$ and the transition metal salts Ti, V, or Cr. Whereas, in the latter case for both olefin and diene polymerizations, the base metal alkyl acts to alkylate the transition metal center and perhaps to reduce it, in the cobalt catalyst it only reduces it to a lower valent state, from Co(II) to Co(I). In the first case, the center is generated by alkylation. In the second case, the center is generated by reaction of Co(I) and butadiene (see Eq. 14–7). Evidence that Co(I) is present in these catalysts has been reported (30, 31). If this scheme is valid, it represents a rather unique example of initiation in Ziegler–Natta catalysis. Natta and Porri earlier recognized this polymerization scheme to account for the activity of the $[(C_6H_5)AlCl_2]_2Co(I)Cl$ complex. For the Ziegler–Natta catalyst, they visualize, like Cooper, a cation–anion complex involving π-allyl transition metal bond centers, as shown in structure XI, where the aluminum alkyl in the active center also

XI

plays the role of a stabilizing ligand. Like other workers, they caution that this may be an oversimplification, and the real center may change structure during the growth step.

But $CoCl_2$ combined with nonreducing $AlCl_3$ in benzene was also shown to be active for stereospecific polymerization of butadiene (30, 32). Natta and Porri (3) suggested two reaction schemes that would generate Co–C bonds in this metal alkyl-free catalyst: (1) butadiene coordinates with Co of the $CoCl_2 \cdot 2AlCl_3$ complex and inserts into a Co–Cl bond, as shown in Eq. 14–14, or (2) $CoCl_2 \cdot 2$ $AlCl_3$ disproportionates to generate the Co(I)

$$CoCl_2 \cdot 2AlCl_3 + C_4H_6 \longrightarrow CoCl_2 \cdot 2AlCl_3 \cdot C_4H_6 \longrightarrow 2AlCl_3 \cdot Co \begin{matrix} Cl & CH_2 \\ \diagdown & \diagdown \\ & CH \\ \diagup & | \\ CH & \\ \diagdown & \\ & CH_2Cl \end{matrix} \quad (14\text{--}14)$$

$$CoCl_2 \cdot 2AlCl_3 \longrightarrow [Co^+Cl]^+[AlCl_3 \cdot AlCl_4]^- \quad (14\text{--}15)$$

containing complex, and the butadiene oxidatively adds to the Co(I).

5. Evidence Based on Electron Spin Resonance Analysis

Hiraki and co-workers (33) studied by ESR spectroscopy the polymerization mixtures from dienes such as butadiene, 1,3-pentadiene, and isoprene, using $(n\text{-BuO})_4\text{Ti-AlEt}_3$ catalysts. At $\text{Al}/\text{Ti} > 2.9$, which was required to polymerize the dienes, the butadiene polymerization mixture gave an ESR signal with a g value of 1.983 and with a hyperfine structure of about 19 components. From this and other considerations, they assigned this signal to the following growing end of polybutadiene, shown in XII. Analogous structures were proposed from the ESR data for the polymers

XII

of butadiene with the $\text{AlEt}_3\text{–Ti(acac)}_3$ catalyst, which required $\text{Al}/\text{Ti} = 100$ to be effective (due to lower ligand exchange).

IV. Conclusion

Experimental evidence supports the view that the transition metal atom is the carrier of the polymer chain and that it can be attached to the metal atom by either a σ-allyl or a π-allyl bond. The general view is that transition metals on the left side of the periodic table favor the σ-type bond (such as Ti and V), while π-allyl bonding occurs more readily with metals on the right side of the periodic table (Ni and Co). That both σ- and π-allyl bonds are present and in equilibrium has also been suggested with the likelihood that growth actually takes place at the σ-allyl bond. One of the roles that ligands can play is to alter this equilibrium.

It is well established in Chapter 12 that dienes can be polymerized to highly stereospecific products with some transition metal salts in the absence of added base metal alkyl, such as AlEt_2Cl. While the rates of polymerization are often low and the polymers tend to have low molecular weights, a few systems have been found that equal the typical Ziegler–Natta catalyst. A base metal alkyl is clearly not always necessary to form a highly stereo-regulating catalyst. When present, however, the base metal alkyl can alter the kinetic and stereochemical behavior of the catalyst. This suggests that it can play one or more major roles. One of these roles is the formation of the transition metal–carbon bond center by alkylation, and this is probably the case for Ti-, V-, and Cr-based catalysts. Another role is the reduction of the transition metal to a lower valent state species, which then oxidatively adds the diene molecule to form a π-allyl bond growth center. This occurs most likely with Ni- and Co-based catalysts. The base metal alkyl or a derivative can also function as a stabilizing ligand in a bimetallic-type complex. Soluble rather than heterogeneous catalysts are probably more affected by this role of the base metal alkyl. Because the crystal lattice in heterogeneous catalysts can act as a stabilizing ligand, the necessity of a bimetallic complex with a base metal alkyl is eliminated.

References

1. W. Cooper and G. Vaughan, *Prog. Polym. Sci.* **1**, 93–160 (1967).
2. W. Marconi, *in* "The Stereochemistry of Macromolecules" (A.D. Ketley, ed.), Chapter 5. Dekker, New York, 1967.
3. G. Natta and L. Porri, *in* Polymer Chemistry of Synthetic Elastomers by J. P. Kennedy and E. G. M. Törnqvist (eds.), **23**, pt. 2, Interscience, 1969, p. 597.
4. W. Cooper, *Ind. Eng. Chem., Prod. Res. Dev.* **9**, No. 4, 457–466 (1970).
5. M. Roha, *Adv. Polym. Sci.* **4**, 353–392 (1965).
6. F. Dawans and P. Teyssie, *Am. Chem. Soc., Div. Org. Coat. Plast. Chem., Pap.* **30**, No. 1, 208–219 (1970); P. Teyssie and F. Dawans, *Ind. Eng. Chem., Prod. Res. Dev.* **10**, 261–269 (1971).
7. B. A. Dolgoplosk, *Polym. Sci. USSR (Engl. Transl.)* **13**, 367–393 (1971); *Vysokomol. Soedin., Ser. A* **13**, 325–347 (1971).
8. W. M. Saltman, *J. Polym. Sci.* **46**, 375 (1960).
9. M. Gippin, *Ind. Eng. Chem., Prod. Res. Dev.* **4**, 160 (1965); *Rubber Chem. Technol.* **39**, 508 (1966).
10. P. Teyssie and F. Dawans, *Ind. Eng. Chem., Prod. Res. Dev.* **10**, 267 (1971).
11. For earlier references and comments, see G. Henrici-Olive and S. Olive, *Angew. Chem., Int. Ed. Engl.* **10**, 105 (1971).
12. P. Cossee, *in* "The Stereochemistry of Macromolecules" (A. D. Ketley, ed.), Chapter 3. Dekker, New York, 1967.
13. W. M. Saltman, W. E. Gibbs, and J. Lal, *J. Am. Chem. Soc.* **80**, 5615 (1958).
14. A. Mazzei, M. Araldi, W. Marconi, and M. De Malde, *J. Polym. Sci., Part A* **3**, 753 (1965).
15. L. S. Bressler, I. Ya. Poddubnyi, and V. N. Sokolov, *J. Polym. Sci., Part C* **16**, 4337 (1969).
16. L. Porri, G. Natta, and M. C. Gallazi, *J. Polym. Sci., Part C* **16**, 2525 (1967); L. Porri, M. Gallazi, and G. Vetulli, *ibid. Part B* **5**, 629 (1967).
17. G. Wilke, U.S. Patent 3,379,706, April 23, 1968, Studiengesellschaft M. B. H. Mülheim; G. Wilke, B. Bogdanović, P. Hardt, P. Heimbach, W. Keim, M. Kroner, W. Oberkirch, K. Tanaka, D. Walter, and H. Zimmermann, *Angew. Chem., Int. Ed. Engl.* **5**, 151 (1966).

18. E. I. Tinyakova, A. V. Alferov, T. G. Golenko, B. A. Dolgoplosk, I. A. Oreshkin, O. K. Sharaev, G. N. Chernenko, and G. N. Yakovlev, *J. Polym. Sci., Part C* **16**, 2625 (1967).
19. Many examples are cited in Dawans and Teyssie (6) and Dolgoplosk (7).
20. J. P. Durand, F. Dawans, and P. Teyssie, *J. Polym. Sci., Part B* **6**, 757 (1968); *Part A-1* **8**, 879 (1970).
21. G. Natta, L. Porri, and A. Carbonaro, *Atti Accad. Naz. Lincei., Cl. Sci. Fis., Mat. Nat., Rend.* [8] **29**, 491 (1960).
22. L. Porri and A. Carbonaro, *Makromol. Chem.* **60**, 236 (1963).
23. G. Natta, U. Giannini, P. Pino, and A. Cassata, *Chim. Ind. (London)* **47**, No. 5, 524 (1965).
24. W. Cooper, D. E. Eaves, G. D. T. Owen, and G. Vaughan, *J. Polym. Sci., Part C* **4**, 211 (1964).
25. J. F. Henderson, *J. Polym. Sci., Part C* **4**, 233 (1964).
26. G. Natta, L. Porri, and A. Carbonaro, *Makromol. Chem.* **77**, 126 (1964); G. Natta, L. Porri, A. Carbonaro, and G. Stoppi, ibid., p. 114.
27. C. W. Childers, *J. Am. Chem. Soc.* **85**, 229 (1963).
28. W. Cooper, D. E. Eaves, and G. Vaughan, *Makromol. Chem.* **67**, 229 (1963).
29. W. Cooper, D. Degler, D. E. Eaves, R. Hank, and G. Vaughan *in* Elastomer Stereospecific Polymerization. American Chemical Society, Advances in Chemistry Number 52, 1966, p. 2243.
30. H. Scott, Great Britain Patent 916,384, January 23, 1963, Goodrich-Gulf Chemicals, Inc.
31. C. E. Bawn, *Rubber Plast. Age* **46**, 510 (1965).
32. J. G. Balas, H. E. DeLaMare, and D. D. Schissler, *J. Polym. Sci., Part A* **3**, 2243 (1965).
33. K. Hiraki, T. Inoue, and H. Harai, *J. Polym. Sci., Part A-1* **8**, 2543 (1970).

15

Mechanisms for Stereochemical Control of α-Olefins

I. Introduction

This chapter and Chapters 16 and 17 will describe the various mechanistic proposals and, whenever possible, the corresponding experimental evidence that has been put forth in favor of specific stereochemical processes in the polymerization of α-olefins and diene monomers. While these two types of monomers share some common mechanistic features, the differences are great enough to warrant separate treatments. Isotactic and syndiotactic propagations using propylene as a monomer are discussed in this chapter. Conjugated and nonconjugated dienes are discussed next in Chapter 16. The special case of isotactic polymerizations involving stereoselective and stereoelective propagations of one of the enantiomers of a racemic mixture of α-olefins is presented in Chapter 17.

Propylene and higher α-olefins are polymerized in the presence of heterogeneous catalysts to polymers that can have low to very high isotacticity. Only propylene, however, has been polymerized to a syndiotactic polymer, and it can also be obtained in low to very high steric purity. Soluble catalysts are most effective, but a few percent of a partially syndiotactic polymer have been obtained with specific heterogeneous catalysts.

The growth step involves complexing of the olefin to the transition metal followed by a *cis*-insertion, or more accurately, by a *cis*-ligand migration. It is convincingly established that propylene adds to form a primary metal alkyl (primary addition) in an isotactic propagation. Recent experimental work suggests that a secondary metal alkyl (secondary addition) is formed in a syndiotactic propagation, but this needs independent confirmation.

The driving forces for both isotactic and syndiotactic are steric in nature. The asymmetric nature of each center in the heterogeneous catalyst forces the propylene to always add either in the d or in the l configuration, and isotactic chains are formed. This driving force is absent in the soluble catalyst that polymerizes propylene at $-78°C$. In this polymerization, steric interactions between the methyl groups of the last added and the incoming propylene force the propylene molecule to be inserted in opposite configurations after each growth step, and syndiotactic polypropylene is obtained.

II. Mode of Addition

In addition to establishing a driving force for the isotactic or syndiotactic propagation, attention must be given to the way the α-olefin adds to the metal–carbon bond. This section considers the importance of complexing, *cis*- vs. *trans*-addition and primary vs. secondary addition.

A. COMPLEXING VS. DIRECT INSERTION OF OLEFIN

Most workers have taken the view that the olefin molecule forms a complex with the transition metal atom of the active center just before it is inserted into the metal–carbon bond. The chemistry of transition metal compounds suggests strongly that this should occur. Yet direct experimental evidence is scarce. The possibility that a direct insertion of olefin without involving an identifiable complex was also suggested as an alternate path but without proof (1). By an identifiable complex is meant one as described by Cossee in his molecular orbital treatment (MO) (see Fig. 15–1 and Chapters 13 and 14).

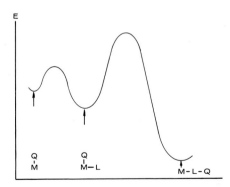

Fig. 15–1. Potential energy as a function of the reaction coordinate in a *cis*-ligand insertion preceded by complex formation; L = olefin and M = the center (4).

Henrici–Olive and Olive (2) suggested as evidence for complexation their observation that the reduction–decomposition of $(C_5H_5)_2TiClEt \cdot AlEtCl_2$ [Ti(IV) to Ti(III)] was increased in the presence of olefins that were not polymerizable by this catalyst. The rate of reduction was increased in the series 1-octene $\gg$ *cis*-2-butene > *trans*-2-octene. A reduction step involving two octahedral complexes was postulated (Chapter 13, structures VI, VII, and VIII). The data and mechanistic interpretation were presented in Chapter 10 (Section V) and in Chapter 13 (Section IV,B).

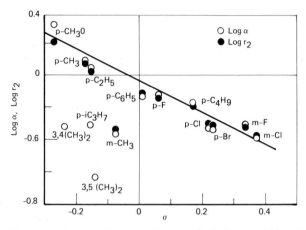

Fig. 15–2. Representation on a Hammett diagram of relative reactivities α in the homopolymerization, $\bigcirc$, and the reactivity ratios r_2 in the copolymerization, $\bullet$, of a set of nuclear substituted styrenes (reference monomer is styrene).

Other evidence comes from Danusso and Sianesi (3) and Cossee (4). Danusso and Sianesi investigated the homo- and copolymerization of a series of vinyl aromatic monomers that differed in the electron density at the double bond. Figure 15–2 summarizes these results in a plot of log α (relative reactivities in the homopolymerizations) and log r_2 (relative reactivities in the copolymerizations) vs. Hammet's σ values for each substituent. The monomers that had comparable steric requirements lined up on the diagram in a straight line with a negative slope. This indicated that there was a strong polar affect of the substituent on the rate of addition. Electron-donating substituents in the *para*-position gave higher rates. Polarization of the styrene monomers was suggested.

These authors proposed that once the complex was formed, electron movement 1 preceded electron movement 2, as shown in structure I.

$$\overset{\delta^+}{\text{Ar}-\text{CH}} \text{=}\!\!\text{=}\!\!\text{=}\!\!\text{=} \overset{\delta^-}{\text{CH}_2}$$

$$\begin{array}{cc} \uparrow 2 & \downarrow 1 \\ \text{P--CH}_2\text{------Cat} \\ - & + \end{array}$$

I

Danusso and Sianesi pointed out that reactivities of monomers in cationic and anionic coordinative polymerizations increased when the electron density at the double bond increased. In conventional anionic polymerizations the reverse was true, and in this case, process 2 could precede process 1 via the path shown in structure II, where R is an electron-attracting group.

$$\overset{\delta^+}{\text{CH}_2} \text{=}\!\!\text{=}\!\!\text{=}\!\!\text{=} \overset{\delta^-}{\text{CH}}-\text{Ar}$$

$$\begin{array}{cc} \uparrow 1 & \downarrow 2 \\ \text{P--CH------Cat} \\ |- & + \\ \text{R} \end{array}$$

II

An attractive reaction path was proposed by Cossee (4) on the basis of quantum-chemical considerations (see Chapter 14, Section II,B,1). Figure 15–1 shows the variation of potential energy as a function of the reaction coordinate in a *cis*-ligand insertion. One requirement of this MO treatment is complex formation between the olefin and the center.

B. PRIMARY VS. SECONDARY ADDITION

The α-olefin can potentially add in two ways, primary addition, as shown in Eq. 15–1, or secondary addition, as in Eq. 15–2.

$$\text{Ti--Pn} + \text{CH}_2\text{=CHCH}_3 \longrightarrow \text{Ti--CH}_2\text{--}\overset{\overset{\textstyle \text{CH}_3}{|}}{\text{CH}}\text{--Pn} \qquad (15\text{--}1)$$

$$\text{Ti--Pn} + \text{CH}_2\text{=CH--CH}_3 \longrightarrow \text{Ti--}\overset{\overset{\textstyle \text{CH}_3}{|}}{\text{CH}}\text{--CH}_2\text{--Pn} \qquad (15\text{--}2)$$

These two modes differ in the way the α-olefin becomes inserted, that is, the unsubstituted end becomes attached to Ti in a primary addition, while in the secondary addition, it is the substituted end that becomes attached to Ti.

The observation of vinylidene III (5) and vicinal end groups IV in

$$\overset{\overset{\textstyle \text{CH}_3}{|}}{\text{CH}_2\text{=C--Pn}} \qquad\qquad \overset{\overset{\textstyle \text{CH}_3}{|}}{\text{CH=CH--Pn}}$$

III   IV

isotactic polymerizations supported a primary mode of addition. A primary addition was subsequently assumed by most workers in describing syndio-tactic polymerizations but without similar evidence.

Only in recent years has the latter become recognized as the prevailing mode of addition in syndiotactic propagations. In 1969 and 1970, Takegami and Suzuki (6, 7) reported that, while primary addition occurred for the isotactic-specific catalysts $AlEt_3$–$TiCl_3$ and $AlEt_3$–VCl_4 (heterogeneous systems), secondary addition occurred when soluble, syndiotactic-specific catalysts were used such as $AlEt_2Cl$–VCl_4 or the same plus anisole. This conclusion was based on the insertion of one molecule of the olefin (cyclohexene, 1-pentene, or 4-methyl-1-pentene) into the V–C bond. With pentene and 4-methyl-1-pentene and $AlEt_2Cl$–VCl_4 catalyst, the amount of a secondary

alkyl, $M–\overset{\overset{\displaystyle R}{|}}{C}H–CH_2–Et$, exceeded the primary alkyl, $M–CH_2CHR–Et$, by four times or more at $-70°C$. When 1-pentene was reacted, n-heptane and 3-methylhexane were isolated and identified after hydrolysis of the reaction product. Per 5 mmole 1-pentene and 1 mmole VCl_4, up to 0.1 mmole of the secondary alkyl was formed. At higher temperatures, the primary metal alkyl resulted. The primary metal alkyl was said to have relatively lower activity at $-70°C$ but higher activity above $0°C$.

TABLE 15–1

Infrared Bands from Propylene and Ethylene Sequences of Different Lengths (8)

Band H	Group	Polyolefin structure
12.25	$-(CH_2-)_1$ (rocking vibration)	$-\overset{\overset{\displaystyle CH_3}{|}}{C}H-CH_2-\overset{\overset{\displaystyle CH_3}{|}}{C}H-CH_2-{}^a$
13.30	$-(CH_2-)_2$ (rocking vibration)	$-\overset{\overset{\displaystyle CH_3}{|}}{C}H-CH_2-CH_2-\overset{\overset{\displaystyle CH_3}{|}}{C}H-{}^b$
13.64	$-(CH_2-)_3$ (rocking vibration)	$-\overset{\overset{\displaystyle CH_3}{|}}{C}H-CH_2-CH_2-CH_2-\overset{\overset{\displaystyle CH_3}{|}}{C}H-CH_2-{}^c$
13.85	$(CH_2)_{n \geq 5}$ (rocking vibration)	$-CH_2-CH_2-CH_2-CH_2-CH_2-CH_2{}^d$

[a] Head-to-tail propylene sequences.

[b] Two head-to-head propylene or one ethylene inserted between two tail-to-tail propylene units.

[c] One ethylene unit between two head-to-tail propylene.

[d] Long ethylene units.

To show that this conclusion was valid for syndiospecific polymerization of propylene, Zambelli, Tosi, and Sacchi (8) characterized ethylene–propylene copolymers by infrared spectroscopy (both C_2H_4 and C_2D_4 were used). These spectra revealed bands originating from propylene and ethylene sequences of different lengths, as shown in Table 15–1.

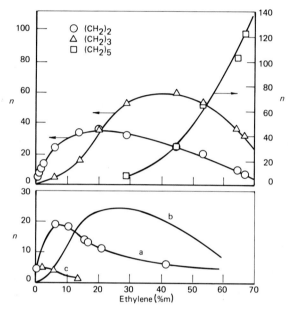

Fig. 15–3. (Upper) Distribution of methylene groups (expressed as the number n of CH_2's per 100 monomer units) over sequences of different length. (Lower) (a) Methylene groups in $(CH_2)_2$ sequences originated from two head-to-head propylene units (deduced from the spectra of C_2D_4–C_3H_6 copolymers); (b) methylene groups in $(CH_2)_2$ sequences originating by insertion of one ethylene unit between two tail-to-tail propylene units (obtained by subtracting the number of methylenes in $(CH_2)_2$ sequences of Type 1 (a) from the total number of methylenes in $(CH_2)_2$ sequences plotted in the upper part of the diagram); (c) vicinal methyls (8).

Both the catalyst systems and the reaction temperatures for the preparation of these copolymers were typically syndiospecific. A plot (Fig. 15–3) of the numbers of $(CH_2)_2$, $(CH_2)_3$, $(CH_2)_5$ and $-CH(CH_3)-(CH_2)_2-CH(CH_3)-$ groups (line b) in the copolymers and in the sample of syndiotactic polypropylene shows that: (1) both homo- and copolymer contain $(CH_2)_2$ and $-CH(CH_3)-(CH_2)_2-CH(CH_3)-$units; (2) the maximum number of $(CH_2)_2$ groups found in copolymers is much higher than twice that found in

TABLE 15–2

Mechanism of Propylene Insertion

Last unit of the growing chain	Catalytic Me–C bond	Newly formed catalytic Me–C bond
Propylene	Secondary + C_3H_6	Secondary
	Primary + C_3H_6	Secondary
Ethylene	Primary + C_3H_6	Primary

corresponding homopolymer; (3) at low ethylene concentrations, almost all ethylene units give rise to $-(CH_2)_2$ groups; and (4) the vicinal methyls do not appreciably decrease on introducing ethylene until $(CH_2)_3$ sequences become detectable. From a kinetic analysis of the obtained data, Zambelli, Tosi, and Sacchi also concluded that secondary addition prevailed in syndiotactic polymerizations. Table 15–2 shows that the preferred behavior of propylene in the insertion depends more on steric effects than on the polarity of the catalytic metal–carbon bond.

Supporting evidence has recently been reported by Mitani, Suzuki, Matsuo, and Takegami (8a), who examined isotactic, atactic, and syndiotactic polypropylenes by 220 MHz proton magnetic resonance and by infrared spectroscopy. They concluded that syndiotactic polypropylene has, in addition to the normal head-to-tail arrangements, some head-to-head and tail-to-tail arrangements of propylene units and that this was the origin of randomness of syndiotacticity. The earlier mechanism of Takegami was considered valid.

Other investigations bearing on the mode of addition are now briefly mentioned. G. Henrici–Olive and S. Olive (9) showed that by going from mixtures of high acid character ($TiCl_4$ + $AlEtCl_2$) to those of less acidity ($EtOTiCl_3$ + $AlEt_2Cl$), they could prepare oligomers of propylene that were linear and had high contents of head-to-tail structures instead of branched structures.

Reichert and Berthhold (10) concluded that 1,2-addition of monomer units occurs from kinetic studies involving α- and β,β'-deuterated styrene and the soluble catalyst $(C_5H_5)_2TiCl_2–AlEt_2Cl$. The rate of polymerization was greater for the α-deuterated styrene relative to the β,β'-deuterated styrene. They explained the isotopic effect by a termination reaction (Eq. 15–3)

$$\text{(15–3)}$$

whereby cleavage of a C–H bond of the growing chain occurs. Reichert and Schubert (11) showed that perdeuterioethylene-d_4 produced a higher molecular weight polymer than did normal ethylene when the homogeneous catalyst $(C_5H_5)_2TiCl_2/AlEtCl_2$ was used.

The occurrence of head-to-head additions in copolymerization is discussed in Chapter 20 [see Van Schooten et al. (24)].

The ability of these soluble vanadium-based catalysts to undergo a secondary addition in a syndiotactic propagation suggests a coordinated cationic mechanism is taking place. A coordinated anionic mechanism most easily explains a primary addition in an isotactic polymerization. The polarizations of bonds in the two centers are opposite, as seen in structures V (secondary addition) and VI (primary addition).

$$
\begin{array}{cc}
\underset{\delta^+}{\text{CH}_3} & \text{CH}_3 \\
\underset{\delta^-\ \ \ \delta^+}{\text{Ln}\overset{}{\text{V}}\!\!-\!\!\overset{}{\text{CH}}\!-\!\text{CH}_2\!-\!\text{Pn}} & \underset{\delta^+\ \ \ \delta^-}{\text{LnTi}\!-\!\text{CH}_2\!-\!\text{CH}\!-\!\text{Pn}} \\
\underset{\delta^+\ \ \ \ \delta^-}{\text{CH}_3\!-\!\text{CH}\!\!=\!\!\text{CH}_2} & \underset{\delta^-\ \ \ \delta^+}{\text{H}_2\text{C}\!\!=\!\!\text{CH}\!-\!\text{CH}_3} \\
\text{V} & \text{VI}
\end{array}
$$

Coordinated cationic mechanisms were previously discussed by Vandenberg and Danusso for polymerization of vinyl ethers and styrenes with Ziegler–Natta catalysts. Kennedy and Langer (11a) in 1964 proposed that a continuous spectrum of bonding may exist between the polymer end and the gegenion, covering the entire range between the extremes of anionic and cationic mechanisms. They even pointed out that the coordination step is influenced to a great extent by changing the transition metal component, citing V and Ti catalysts as examples for different stereochemical reactions. But the development of this idea to syndiotactic propagations was left to Takegami and Suzuki and Zambelli and their co-workers. Additional evidence based on other experimental approaches is still needed to establish this young idea.

C. CIS VS. TRANS ADDITION

That the addition of olefin molecules in both isotactic and syndiotactic propagations was cis was established by Natta (12), Miyazawa (13), and Zambelli (14, 15) and their co-workers using deuterium-labeled propylenes. Figure 15–4 shows the expected stereochemistry of the added deuterium-labeled propylene if the addition were cis or trans. *Cis*-monomers gave threo monomer units while *trans*-monomers gave erythro units when *cis*- and *trans*-1-*d*-propylene were homopolymerized. This showed that the addition was cis for isotactic-specific catalysts.

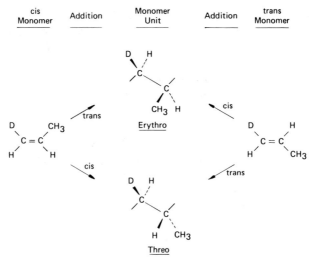

Fig. 15–4. Potential stereochemical structures from *cis*- and *trans*-additions (14, 15).

Zambelli and co-workers (14) found that the syndiotactic copolymers of perdeuteriopropylene and *trans*-1-*d*-propylene are gauche syndiotactic (structure VII) while copolymers of perdeuteriopropylene and *cis*-1-*d*-propylene are trans syndiotactic (structure VIII). The addition for syndio-

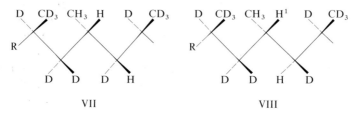

tactic-specific catalysts is thus also cis. Proton NMR was used to establish these structures.

The question of isotactic and syndiotactic propagation is now discussed.

III. Isotactic Propagation

The discovery by Natta and co-workers that some binary mixtures of metal alkyls and transition metal salts were able to polymerize α-olefins to isotactic polymers is just as remarkable as their ability to polymerize these olefins at all. It was also Natta who first recognized the importance of crystal surface as the primary driving force for the isotactic propagation of α-olefins.

A. PROPOSED MODELS

Since 1959, about a dozen or so proposals have been made to explain the driving force of the isotactic propagation. These are presented here according to the following themes: (1) active participation of metal alkyl molecule, (2) steric repulsions between the α-olefin and the ligands of the center, (3) active participation of the helix, and (4) other models.

1. Active Participation of the Metal Alkyl

Natta (16) in 1959 suggested that the active centers in the isotactic-specific heterogeneous catalyst were bimetallic complexes of metal alkyl molecules and transition metal atoms that were located on the surface of the crystal. Because these complexes had asymmetric character, each site could complex the α-olefin only in one or the other of two possible modes (Fig. 15–5). A polymer chain formed at a center would have prevailingly the α-olefin units in the same configuration.

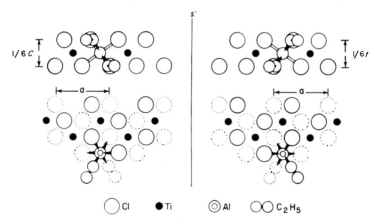

Fig. 15–5. Asymmetric bimetallic complex model, according to Natta (16). Schematic model of how epitactic adsorption of $AlCl(C_2H_5)_2$ on the border of the basal plane of a crystal of $TiCl_3$ may give rise to enantiomorphous active centers ($a = 6.12$ Å and $c = 17.50$ Å). Synthesis of isotactic polymers can take place since each macromolecule, derived from a single center, would contain α-olefin units having the same configuration. (Figure reproduced from *J. Inorg. Nucl. Chem.* through the courtesy of Pergamon Press.)

Fig. 15–6. Monomer–metal alkyl interaction model of isotactic placement. According to Rodriguez and Van Looy (17), the active sites are bimetallic complexes of aluminum alkyls and alkylated titanium atoms (a). The latter are located on the lateral faces and defects and contain two chlorine vacancies; (b) monomer complexation.

Rodriguez and Van Looy (17) proposed in 1965 that the active centers in aluminum alkyl–titanium chloride catalysts were bimetallic complexes of aluminum alkyls and alkylated titanium atoms, as shown in Eq. 15–4. From

(a)   (b)

$$(15\text{-}4)$$

their experimental data, they concluded that these titanium centers were present on lateral faces and defects. They proposed that each Ti had two missing chlorines (vacancies) (Fig. 15–6). Isotactic orientation was favored because the propylene molecule can complex with the exposed Ti center only in one configuration, that is, with its methyl group directed away from the alkyl groups of the aluminum atoms.

2. Steric Repulsions between the Olefin and Ligands of the Center

The basic idea of these proposals is that the preferred complexing mode minimized steric interactions between the pendent group of the α-olefin and the ligands of the active center to which it is complexed as well as adjacent atoms. The isotactic scheme had its beginning in the recognition by Natta in 1955 that the isotactic propagation required a specific surface of crystalline transition metal salts.

The idea that isotactic propagation is intimately linked to the geometry of an exposed Ti (chlorine vacancy) was worked out independently for both bimetallic complexes (18) and monoalkylated titaniums (19) as active centers. The exposed Ti is an important feature of these mechanisms in two ways: (1) the olefin can complex at the vacant octahedral position and thereby its energetics become more suitable for addition, and (2) the unique ligand environment can force or favor a complexing mode corresponding to one or the other configuration.

Figure 15–7 shows the scheme worked out by Boor (18) for bimetallic complex centers located at the edges of the $TiCl_3$ crystal.

A more appealing model, using the monoalkylated Ti centers, was developed by Cossee and Arlman (19), as shown in Fig. 15–8. According to

Fig. 15–7. Ligand–monomer interaction model of isotactic placement (18). Oblique view of $\alpha TiCl_3$ model illustrating orientation of α-olefin molecules coordinated to the exposed titaniums. Using a bimetallic complex model, Boor proposed that the ligand environment at the transition metal affected isotactic stereoregulation. It was suggested that at edge sites (IV, VI), the only possible complexing mode of the α-olefin with the transition metal was one where the α-substituent protruded out of the crystal lattice, while at more exposed sites (V and VII), both complexing modes may occur.

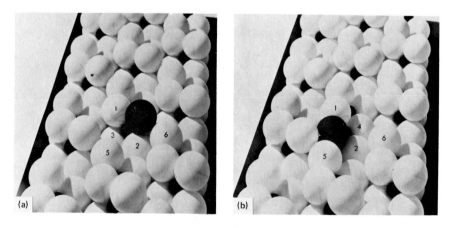

Fig. 15–8. Ligand–monomer interaction model of isotactic placement according to Arlman and Cossee (19).

this scheme, the alkylated Ti are located on the lateral faces of $TiCl_3$ (those formed by combining edges of the elementary layers). The large black sphere in Fig. 15–8 is a polymer chain and the small black sphere is the titanium atom to which it is attached. Polymerization takes place at site A. Complexing of the α-olefin at site A can occur only when the pendant side group is placed over the exposed Cl-1 ligand, and only one configuration is allowed for the polymer chain (isotactic growth). Sites A and B, however, do not possess equal abilities to orient the α-olefin. Site B, though, is converted to site A by migration of R (the growing polymer chain) by virtue of the steric crowding of R in the high ionic environment of the Cl ligands.

Allegra (20) rejected two features of the Cossee mechanism: (1) that the Ti–C and C–C were parallel in the complex (on the basis of excessive steric repulsions between CH_3 of the propylene and the methylene unit attached to Ti), and (2) back-migration of the growing polymer chain to its original octahedral orientation (on the basis of observed low activation energies, which were about 5 kcal/mole), which he considered too low to account for the required migration.

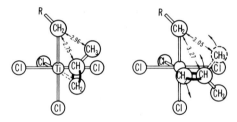

Fig. 15–9. Different modes of olefin π-complexation on an octahedral Ti atom, with the double bond parallel to a Ti–X bond. The mode of complexation assumed by Cossee is shown on the left. A *cis*- and *trans*-like mode of complexation are shown on the right. The former corresponds to the methyl group orientation drawn with a dashed line, the latter with a continuous line. Distances are given in Ångström units.

A different mechanism was proposed in which a *trans*-like mode of complexing took place (see Fig. 15–9). The *cis* mode was ruled out on the basis of steric repulsions.

The two arrangements of the olefin, corresponding to outward-like complexing on either Ti valency, are locally related by twofold axis, and the chirality of the olefin π-complex is the same. It is determined by the chirality of the coordination around the Ti atom. The symmetry of the system composed of the Ti atoms and of its nearest neighbors within the layer is described by the D_3 point group that allows the existence of chiral objects.

The optical configuration of any Ti atom is opposite to that of the three first neighboring Ti atoms.

The reaction path, according to Allegra's scheme, consists of complexing of the olefin in the outward *trans*-like arrangement, followed by a rotation of the complexed olefin to bring the carbon to a suitable orientation so that insertion can occur (Fig. 15–10). The outward *trans*-like orientation most efficiently decreases all steric repulsions involving the olefin side group, the bonded alkyl (polymer chain), and the chlorine ligands of the layer. After

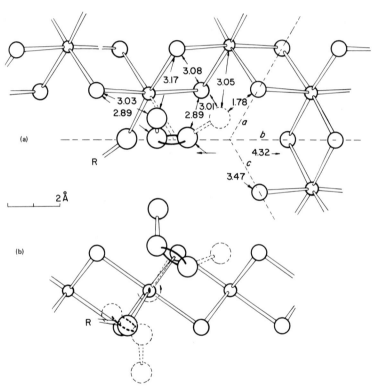

Fig. 15–10.   The scheme of the catalytic complex is normal to the average plane of the TiCl₃ layer (above, a) and parallel to the same plane (below, b). All the critical nonbonding distances between the complexed olefin and the layer atoms are reported (Å). In section a the outward and inward *trans*-like arrangements of the olefin are represented corresponding to the CH₃ group drawn with continuous lines and dashed lines, respectively. Three possible boundary lines are indicated with *a*, *b*, and *c*. In section b, in addition to the above, another complexed olefin with an outward *trans*-like arrangement is also shown (in dashed lines, at the lower left), representing the next step of complexation. A (local) twofold axis is shown to relate the two outward *trans*-like complexed olefins.

the addition, the Ti–R bond is in a new position, but a new olefin molecule can coordinate on the free Ti valency with the same chirality as before. This scheme requires the active center (the exposed Ti on which olefin complexing and chain addition occurs) to protrude above both neighboring layers.

Harrod (20a) also recognized that Ti(III) in the αTiCl$_3$ lattice has D_3 symmetry and suggested that the octahedral position occupied by the growing polymer chain and the vacancy would lead to identical configurations.

3. Active Participation of the Helix

Coover (21) speculated that the rate of placement preserving the symmetry of the helix was faster than the rate of the alternate placement. Thus, the driving force was due to preferential placement of the α-olefin as a result of interactions between the α-olefin and the helix (Fig. 15–11).

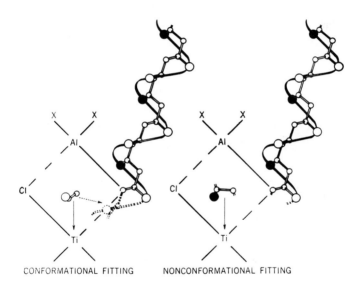

CONFORMATIONAL FITTING   NONCONFORMATIONAL FITTING

Fig. 15–11. Helical model conformational fitting in active site (21) (Coover).

Kissin, Chirkov, and Mezhikovsky (22, 23) proposed a model requiring two vacancies on a basal plane of the TiCl$_3$ crystal. Stereospecific control, that is, $k_{\text{isotactic}}$ vs. $k_{\text{syndiotactic}}$, depended on steric interactions of the olefin with the growing helical polymer chain, with neighboring transition metal atoms and with neighboring ligands (Fig. 15–12).

These workers rejected the Cossee–Arlman model (Fig. 15–8) on the grounds that the single vacancy was blocked by CH and CH$_3$ groups of the

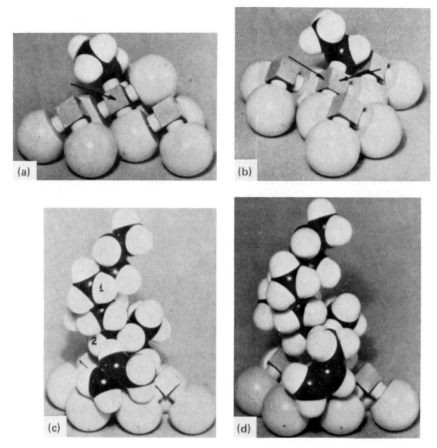

Fig. 15–12. Origin of isotactic and syndiotactic propagations as proposed by Kissin, Chirkov, and Mezhikovsky (22,23). (Upper) Models of potential active sites of polymerization connected with ethyl groups: (a) site with one vacancy; (b) site with two vacancies. (Lower) Scheme of propylene coordination on active site with two vacancies: (c) coordination leading to isotactic monomer addition; (d) coordination leading to syndiotactic monomer addition.

second propylene unit in the polymer chain (counting from Ti). Models suggested to them that this blocking could not be avoided by rotation of Ti–C or C–C bonds because both ends of the polymer chain were rigidly connected (one to Ti and the other being cocrystallized with adjacent chains). They considered the bimetallic complex model of Rodriguez and Van Looy to have its single vacancy blocked in the same way. To overcome this blocking, the authors suggested a $PnTiCl_3 \square \square$ site (Pn = polymer) on the basal plane that had two vacancies (two open octahedral positions). One of these vacancies also becomes blocked by the helical polymer chain, but the other remains available for coordination of olefin. Figure 15–12 shows their view of stereochemical control.

This model was developed in part from their kinetic studies of catalyst poisoning with methanol and water. They assumed that these reactants were adsorbed on the active centers which were thus inactivated. From their kinetic analysis, they concluded that the $AlEt_3–\alpha TiCl_3$ catalyst contained 3×10^{18} sites per square meter. This corresponds to about 35 Å^2 per active center. Thus, they concluded that the active centers must practically cover the entire surface of the $TiCl_3$ crystal. Since 95% of the surface is taken up by the basal planes, these sites must be located largely on this plane and not on the lateral planes, as suggested earlier by Rodriguez and Arlman.

4. Other Models

Other models for explaining stereoregulation have also been recently suggested.

Luisi and Mazo (24) have developed a statistical model that would predict the formation of isotactic, atactic, syndiotactic, and stereoblock polymer in terms of the asymmetry of the active center and the interactions of the incoming monomers, as well as those added last. Since these parameters are not measurable, predictions cannot yet be made with this model.

A rather unique approach has been taken by Corradini and co-workers (25) to devise an experimental basis for an asymmetric induction in the Ziegler catalyst. A study has been made of cis- and trans-π complexes of the square planar type, using olefin–Pt–dichloro-1-phenylethylamine. The observed asymmetric induction was attributed to a different steric interaction of the (R) and (S) complexed olefin and the cis-coordinated amine ligand.

B. EXPERIMENTAL EVIDENCE FOR AND AGAINST PROPOSED MECHANISMS

The detailed features of the above proposed mechanisms are beyond experimental confirmation at this time. What workers have attempted to do was to establish the general features, including the roles that the metal alkyl, the chlorine vacancy, or the helix play in causing the isotactic propagation.

1. Metal Alkyl

The literature is overwhelmed with papers that show that stereochemical control in Ziegler–Natta polymerizations is strongly influenced by the structure of the metal alkyl. This finding convinced many workers during the early period of Ziegler–Natta chronology (including this author) that the metal alkyl molecule was an integral part of the active center in the heterogeneous catalyst. This idea was, however, abandoned by many of the same workers (including the author) when they realized that the role of the alkyl might be merely to alter the surface of the $TiCl_3$ crystal and, in this way, generate sites of different isotactic-regulating ability.

The most convincing argument against the required participation of the metal alkyl comes from isotactic polymerizations with certain transition metal salts that do not contain added base metal alkyls, for example, a mixture of $TiCl_3$ (formed by reduction of $TiCl_4$ by H_2) and an amine or ball-milled $TiCl_2$. Chapter 12 and parts of Chapters 13 and 14 discuss various examples of metal alkyl-free catalysts in detail.

The ability of certain transition metal salts to behave as isotactic-specific catalysts cannot by itself be taken as evidence that isotactic propagations do not take place by sites such as those proposed by Natta (Fig. 15–5) or Rodriguez (Fig. 15–6). They do show, however, that an isotactic propagation does not require the presence of a metal alkyl molecule as an integral part of the active center in the heterogeneous catalyst.

Boor (26) suggested an additional argument against centers containing a complexed base metal alkyl as part of an alkylated Ti center as suggested by Rodriguez and co-workers. The active Ti center occupies about $16 \, Å^2$ (diameter $\simeq 4.5 \, Å$). The cross-sectional areas of polymer helices of polypropylene, poly-1-butene, and poly-4-methyl-1-pentene are about 35, 45, and $69 \, Å^2$, corresponding to diameters of about 7.0, 7.6, and $9.4 \, Å$, respectively. Steric repulsions between the metal alkyl and the helix would prevent close complexing of the aluminum alkyl and the titanium via a common chlorine bridge. This steric interaction would become more pronounced as the size of alkyl increased, contrary to experimental findings.

As already discussed in Chapter 13, Pasquon, Natta, and co-workers have accepted structures such as suggested by Fig. 15–6 as active centers. They recognized that isolated alkylated transition metal centers can polymerize α-olefins to isotactic polymer (for example, metal alkyl-free catalysts). But they held the view that, in the presence of Ziegler–Natta catalysts, these

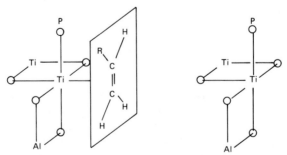

Fig. 15–13.   Schematic model of a stereospecific catalytic complex in the polymerization of α-olefins to isotactic polymers (27).

alkylated Ti (transition metal) centers became complexed with metal alkyl (27) (see also Fig. 15–13). The number, activity, and stereochemical behavior of these centers was said to be strongly influenced by this metal alkyl. In addition to their arguments discussed in Chapter 13, they stress evidence based on the stereochemical structure of the whole polymer and fractionated products from polymerizations having different metal alkyls (27). This is discussed next.

When whole polymers were fractionated by ethyl ether extraction and subsequently by boiling n-heptane extraction, the following data were obtained (see Tables 15–3, 15–4, 15–5, 15–6, and Fig. 15–14):

TABLE 15–3

Polypropylene Fractions Obtained by the $\gamma TiCl_3–Al(C_2H_5)_2I$ System[a]

Polymer temp. (°C)	Ether extract (%)	Heptane extract (%)	Octane extract (%)	Heptane residue (%)
84	1.5–3	7–10	60–80	87–90
66	1.5–2	3–5	15–50	94–96
46	~0.4	0.5–1	6–10	~99
27.5	0–0.4	0.5–1	6–10	~99

[a] The pressure is 500–2,000 mm Hg; 10–40 mmoles $\gamma TiCl_3$/liter; Al/Ti = 1:8.

TABLE 15–4

Characteristics of Polypropylene Fractions Obtained by the $\gamma TiCl_3–Al(C_2H_5)_2I$ System[a]

Polymer temp. (°C)	Heptane extract[b]		Octane extract[c]	
	Melting temp. (°C)	X-rays cryst. (%)	Melting temp. (°C)	X-rays cryst. (%)
84	149–155	70–75	~170	74–80
66	151–155	70–75	~168	74–80
46	145–150	not determined	165–170	74–80
27.5	150	63	170–172	74–80

[a] See Table 15–3 (27).
[b] $\bar{M}_v$ between <10,000 and 25,000.
[c] $\bar{M}_v$ between 100,000 and 200,000.

TABLE 15–5

Characteristics of Polypropylene Residues Obtained by the $\gamma TiCl_3$–$Al(C_2H_5)_2I$ System[a]

Polymer temp. (°C)	Heptane residue[b]		Octane residue[c]	
	Melting temp. (°C)	X-rays cryst. (%)	Melting temp. (°C)	X-rays cryst. (%)
84	170–172	75–80	173–174	~78
66	170–171	75–80	~172	75–80
46	172–173	75–80	173–174	75–80
27.5	173–175	75–80	174–176	75–80

[a] See Table 15–3 (27).

[b] $\bar{M}_v$ from 30,000 to 2,000,000 depending on the operating conditions.

[c] $\bar{M}_v$ from 400,000 to 2,000,000 depending on the operating conditions.

TABLE 15–6

Coefficients of Steric Irregularity in Polypropylene Fractions Obtained by the $\gamma TiCl_3$–$Al(C_2H_5)_2I$ System[a,b]

Ether extract	~0.2
Heptane extract	0.05–0.1
Octane extract	0.02–0.05
Heptane residue	0.005–0.03
Octane residue	0.000–0.02

[a] See Table 15–3 (27).

[b] The coefficients of steric irregularity evaluated for the different fractions are collected in Table 15–5. The data obtained were correlated one with another by the relationship (Fig. 15–16):

$$n = A + B/P_{C_3H_6}$$

where n = coefficient of steric irregularity; A, B = factors as function of temperature only; and $P_{C_3H_6}$ = partial pressure of propylene.

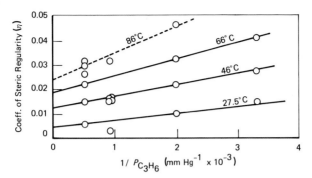

Fig. 15–14.   Coefficients of steric irregularity in crude polypropylenes obtained by $\gamma TiCl_3$– $Al(C_2H_5)_2$ at different temperatures vs. the reciprocal of the partial pressure of the monomer (27).

(a) The boiling ethyl ether extractable fractions consisted of amorphous molecules. Their stereoisomeric compositions differed according to the catalyst used. When listed according to increasing inversion of steric configuration of the ether fraction, the following order was obtained: $AlEt_2Br$–$TiBr_3$ < $AlEt_3$–$CrCl_3$ < $AlEt_3$–$\gamma TiCl_3$ < $AlEt_3$–VCl_3 < $BeEt_2$–$\beta TiCl_3$ < $GaEt_3$–VCl_3 < $GaEt_3$–$TiBr_3$.

(b) The boiling n-heptane extractable fractions generally consisted of block macromolecules mostly of the isotactic type. In addition, block macromolecules of segments of the isotactic, syndiotactic, and atactic type were present in this fraction.

(c) The presence of stereoblock polymer containing crystallizable syndiotactic segments in the heptane fraction was particularly high with some catalysts, $GaEt_3$ or $BeEt_2 + TiCl_3$, for example, but was low in other catalysts, especially those containing VCl_3.

(d) The insoluble polymer remaining from an extraction with boiling n-heptane consisted of completely isotactic polymer or one containing long isotactic blocks (Table 15–5).

These data suggested to these authors that the catalysts contain different types of catalytic complexes at which the atactic, isotactic, and syndiotactic macromolecules (or segments) are synthesized. The relative amounts of these complexes were said to be different in each of the above examined catalysts and depended both on the nature of the transition metal and the alkylating and complexing power of the organometallic alkyl used.

An alkylated titanium center bearing two vacancies, one of which is mostly complexed with a metal alkyl, was also suggested by Langer (27a) as shown in Fig. 15–15.

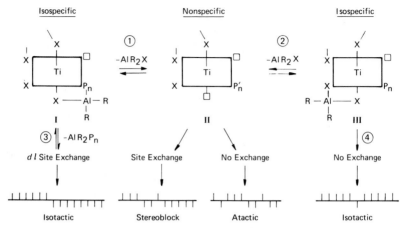

Fig. 15–15. Site structures and scheme proposed by Langer (11a) to account for tacticities.

Sites I and III in Fig. 15–15 differ in that the growing polymer chain Pn acts as a bridge to the aluminum alkyl in structure I but not in structure III. Isotactic polymer chains are formed at both centers. Because complex I is less stable than complex III, however, the metal alkyl can migrate to another site (step 3), carrying the growing polymer chain with it. If the new site forms chains of opposite configuration and this exchange continues to occur, then the polymer chain will consist of long segments of d and l configurations. Site III, because both bridge bonds are halogens, is said to be stronger, and the metal alkyl leaves less frequently. The polymer chain formed at this center would consist entirely of d or l configurations, with an occasional accidental inversion. Loss of AlR_2X from I or III (steps 1 and 3) gives II, a non-specific center, which produces only atactic polymer. But if exchange reactions occur between II and I or III, then stereoblock chains would be formed. In Langer's mechanism, the metal alkyl is an integral part of the isospecific center, and its structure determines the isospecificity of that center. Growth, however, takes place at a transition metal–carbon bond.

Langer (27a) also took a similar position that the metal alkyl, by virtue of complexation with the alkylated transition metal center, played an integral role in the stereochemical process (see Fig. 15–15). As evidence, he cited the experimental data that showed increased isotactic control whenever the metal alkyl was capable of forming more stable complexes (see structures I and III). He concluded that more isospecific centers were formed when: (1) the Lewis acidity of the metal alkyl was greater, for example, $BeR_2 > AlR_3 > MgR_2 > ZnR_2$ and in AlR_2X, $X = I > Br > Cl > F$ or

NR_2; (2) the metal alkyl had low steric bulk, that is, $AlEt_3 > Al(octyl)_3 >$ Al(hexadecyl); (3) the oxidation product had low solubility, that is, $EtZnCl =$ $Et_2Cd > ZnEt_2$ ($ZnCl_2 = EtCdCl \gg EtZnCl$); and (4) the metal alkyl contained an anchoring donor group, such as $Et_2Al-OCH_2CH_2-N(CH_3)_2 >$ $Et_2Al-O-CH_2-CH_3$ and base $\cdot MR_n$ complex $> MR_n$.

2. Ligand Environment of the Exposed Transition Metal Center

The basic feature of this proposal is that the transition metal center has a ligand missing and that the olefin becomes complexed to the center through this open octahedral position. The steric interactions between the pendant group of the α-olefin and the ligands of the center favor one of the two possible complexing modes, and isotactic chains are formed. Both theoretical and experimental evidence support this view.

The theoretical basis is due largely to the elegant and comprehensive work of Arlman (28–30) and Arlman and Cossee (19), who have evolved beautiful and convincing models for explaining the various stereochemical processes. This was done on the basis of consideration of the principles of the inorganic chemistry and the crystallography of crystalline titanium trichloride salts that were used in Ziegler–Natta catalysts. Figure 15–16 shows a photograph of a model of lateral ($10\overline{1}0$) faces containing one of the different possible arrangements of chlorine, that is, in a random distribution. One of these was used in describing isotactic propagation (Fig.

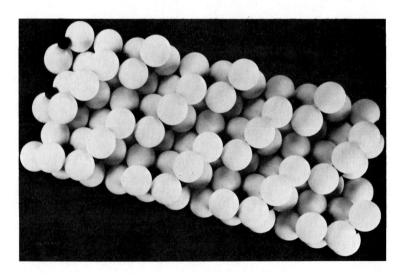

Fig. 15–16.   Structure of face ($10\overline{1}0$). Model shows one of the possible chlorine arrangements (28).

Fig. 15–17. The situation around a pair of Ti ions from Fig. 15–16; (a) square base parallel to *d* axis; (b) square base parallel to *a* axis; (c) after interchange of L and Cl vacancy of (a); and (d) after interchange of L and Cl vacancy of (b) (28).

15–17). Some of the important features of the exposed Ti (resulting from a missing chlorine) are the following:

(a) All potential active centers consist of a square base of four chloride ions around a titanium (III) ion, which is in an oblique position with respect to the crystal face (except in the case of the $10\overline{1}0$ face).

(b) The four Cl ions of a square base are not equivalent. Two are sterically blocked, (B), and two, (F and L), protrude from the crystal. One of the latter, (F), is bound to two Ti(III) ions, whereas the other, (L), is only bound to the Ti(III) of the active center. The ion, F, is nonreactive, but the ion, L, is reactive and might exchange its site with the Cl vacancy of the active center.

(c) The sites of the loosely bound Cl ion, (L), and of the Cl vacancy are not equivalent.

Arlman (30) also evaluated the anion polarization and lattice energy of some metal halides MX_2 and MX_3 with layer and chain structures.

Experimental evidence comes from studies involving addition of electron donors to a Ziegler–Natta polymerization (31). Chapter 9, Section II,B describes the nature of this diagnostic test in detail, and these results should be considered an integral part of this chapter. As an example, Et_3N added to a polymerization of propylene in the presence of $ZnEt_2 + TiCl_3$ causes a decrease in the rate of polymerization. Because Et_3N and $ZnEt_2$ do not form complexes, it is interpreted that Et_3N complexes and inactivates the exposed Ti centers. Table 9–3 shows other electron donors that caused inactivation of the catalyst. The ability of the donor to inactivate a center depends both on the ligand environment of the exposed Ti and the steric bulkiness of the donor. The donor effect can become complicated, and the reader is urged to read Chapter 9 and Section III,C,3 of this chapter for a fuller appreciation.

3. The Helix

Arguments in favor of the helix as a driving force were reported by Turner-Jones (32), who concluded from her copolymerization studies of branched and linear α-olefins that the crystal or helical structure of the growing chain can influence the acceptance or rejection of structurally different monomers.

The following observations were made. The comonomers 1-butene and 3-methyl-1-butene formed crystalline random copolymers, while the comonomers 1-butene and 4,4-dimethyl-1-pentene formed very block products, that is, chains very rich in the latter monomer were present even though its content in the copolymer was about 5%. The following explanation was given. In the randon copolymer, each olefin was interchangeable in the helix with minimal distortion and, for this reason, the polymer chain assumed the helix of the prevalent olefin. In this situation, the helix accepted both monomers favorably. This favorable situation did not exist for the 1-butene and 4,4-dimethyl-1-pentene comonomer pair, because neither monomer fits nicely into the helix of the other. Therefore, the formation of nonrandom (block) chains occurred since the last unit favored its own kind.

Hoeg and Liebman (33) rejected the helix model on the basis that solution polymerizations of propylene at high temperatures (above the melting point of polypropylene) still produced highly isotactic polymer. Isotactic polypropylene was made even at temperatures as high as 200°C (5). Infrared spectra of polypropylene solutions do not show absorption bands due to highly ordered helices. Isotactic poly-1-butene, which dissolved in hydrocarbon solvents at 60° to 90°C (about 40° to 60°C below its melting point) has also been made in solution. Steric effects might, however, favor the retention of some helical character in solutions of branched α-olefins, such as studied by Turner-Jones.

An additional argument against the model of Kissin and Chirkov (22, 23) is the unlikelihood of two chlorine ligands being removed from the same titanium. Lateral movement of neighboring chlorine ligands might occur to compensate for the double vacancy so as to distort the required feature of the required site.

C. STRUCTURAL VARIATION IN ACTIVE CENTERS

It is generally viewed that the Ziegler–Natta catalysts can have active centers of varying stereochemical-regulating ability. It is not clear, however, whether we are dealing with just a few types of sites or many types. Alternate causes of lower tacticity have also been suggested, namely, inversions of configuration caused by jumping of the growing polymer chain from one site to another. The possibility of inversions or change in complexing mode at the same site has been suggested if one of the ligands undergoes movement, is lost, or exchanged in the course of the propagation of a given chain or after several chains have grown at it. This could occur because a localized hot spot developed due to the heat of polymerization.

The variation in structure of sites is expected to be smaller for homogeneous in contrast to heterogeneous catalysts since the macro scale would allow a much wider range of structural variations.

Three types of diagnostic tests bear on this aspect of the center.

1. Polymer Fractionation

Whole polymers having low to very high crystallinity can be prepared by the proper choice of a catalyst. By a suitable extraction, the less crystalline

TABLE 15–7

Isotactic and Stereoblock Polypropylenes (34)

Boiling solvents		Melting point (°C)	X-ray crystallinity (%)
Extracting	Nonextracting		
	Trichloroethylene	176	75–85
	n-Octane	174–175	65–68
n-Octane	2-Ethylhexane	174–175	60–66
2-Ethylhexane	n-Heptane	168–170	52–64
n-Heptane	n-Hexane	147–159	41–54
n-Hexane	n-Pentane	110–135	25–37
n-Pentane	Diethyl ether	106–114	15–27

polymer can be removed so that the remaining residue has a crystallinity that is often as high as obtained with the most specific catalysts. If these partially crystalline polymers are extracted by a series of solvents and at different temperatures, it is possible to isolate fractions having decreasing melting points and crystallinities [see Tables 15–7 (34) and 15–3].

Pasquon (27) fractionated and then characterized the individual fractions of a polymer synthesized with the highly isotactic-specific catalyst $AlEt_2I + \gamma TiCl_3$ for melting point and X-ray crystallinity. The total crystallinity decreased as the temperature of polymerization was increased from 27° to 84°C.

Using a simplified equation (originally proposed by Flory for copolymers), he calculated the coefficients of steric irregularity in the polypropylene fractions (Table 15–6). According to Pasquon, steric inversions are more frequent at higher temperatures of polymerization. An inverse dependence was also shown for the pressure of propylene.

Other interpretations given to this data considered the catalyst to contain sites of varying stereochemical ability (18). The less exposed Ti centers produce more isotactic polymer chains, while chains of lower isotacticity are produced at relatively more exposed centers.

The latter idea was also used by other workers to explain their kinetic data. Schnecko, Lintz, and Kern (35) explained the variations in which the catalyst activity was lowered when oxygen was added to polymerization of ethylene, propylene, or 1-butene. Keii and co-workers (36) explained the formation of high and low isotactic polypropylene by the presence of stable and labile centers. Lovering and Wright (36a) rationalized the kinetics of polymerization and molecular weight distribution of polyisoprene formed with the VCl_3–$AlEt_3$ catalyst at a series of times and temperatures in terms of multiple sites.

The possibility that more than one of these factors is operating simultaneously is likely, but how much depends on the conditions of polymerization.

2. Copolymer Fractionation

It is well established that copolymer chains of ethylene and propylene synthesized with soluble and colloidally dispersed catalysts tend to have comonomer compositions that are uniform (random or partially alternating). Heterogeneous catalysts produce copolymer chains that have a nonuniform comonomer composition. (These two types of catalysts produce polypropylene of low and high isotacticity, respectively.) Table 15–8 shows the fractionation data.

Several explanations have been proposed. The soluble and colloidally soluble catalysts have sites that are relatively similar, and polymer chains are produced that also tend to be similar. In contrast, heterogeneous catalysts

TABLE 15–8

Fractionation of Ethylene–Propylene Copolymers by Successive Fractionation with Solvent (1)

Catalyst	Original copolymer C_3 (mole%)	Ether extract P (wt%)	Ether extract C_3 (mole%)	n-Hexane extract P (wt%)	n-Hexane extract C_3 (mole%)	n-Heptane extract P (wt%)	n-Heptane extract C_3 (mole%)
Heterogeneous							
$Al(C_6H_{13})_3–TiCl_3$	67.5	41.3	73.5	42.8	67.4	14.9	54.6
$Al(C_6H_{13})_3–VCl_3$	58.6	39.5	72.5	53.7	50.4	6.3	25.0
$Al(C_6H_{13})_3–VOCl_3$	37.9	28.5	61.0	59.0	32.0	12.2	18.5
$Al(C_6H_{13})_3–VOCl_3$	50.8	70.3	60.5	25.1	32.5	4.0	19.0
Homogeneous or colloidal							
$AlEt_2Cl–VO(OC_3H_5)_3$	32.5	38.5	33.0	61.5	32.0	0	—
$(C_6H_5)_4Sn–VCl_4–AlBr_3$	3.8	0	—	0	—	100	3.7

contain centers that have varying capacities to complex and polymerize each of the olefins, and these produce chains that have different comonomer compositions.

The presence of several kinds of sites has been reported by Suminoe and co-workers (37) in butadiene–isoprene polymerizations, by Murahashi and co-workers (38) in styrene–alkenyltrimethylsilane copolymerizations, and by Wiman and Rubin (39) in the copolymerization of ethylene and 1-butene with $AlR_3–TiX_4$, $AlEt_3–TiCl_4$, and $Cp_2TiCl_2–AlEt_2Cl$ catalysts, respectively. The results are in agreement with the results reported by Cozewith and Ver Strate, and described in Chapter 20, Section III.

3. Selective Site Removal by Electron Donors

Chapter 9 and Section II,C already described how active centers of different isotactic specificities are selectively removed by the addition of a small amount of Et_3N. The more exposed centers that yield polymer of lower isotacticity and molecular weight were preferentially complexed and deactivated. Polymerization at the remaining relatively less exposed centers produced polymer that had a higher isotacticity and molecular weight. Also, when bulkier amines were used in place of Et_3N, their addition led to increased conversions but polymer isotacticity was lower in every case (Fig. 15–18) (40). With very bulky amines, such as n-octyl$_3$N, or tribenzylamine, even the most exposed sites in the $ZnEt_2–TiCl_3$ catalyst are barely complexed or not at all.

While these results definitely show that active centers have different complexing abilities for electron donors, they do not define precisely the exact nature of the exposed Ti centers. Models suggested by Cossee and

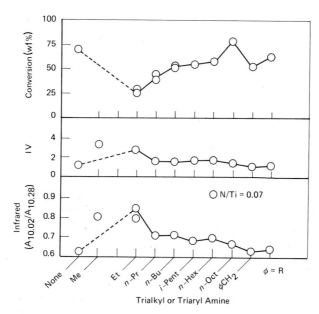

Fig. 15–18. Dependence of conversion and infrared ratio (index of stereospecificity) on amine bulkiness for $ZnEt_2$ and $\gamma Al_{0.3}TiCl_4$ catalyst (40). Materials: 8 oz bottle, 100 ml heptane, 20 hours at 50°C, 27 g propylene, 4.2 mmole $ZnEt_2$, and 1.3 mmole $\gamma Al_{0.3}TiCl_4$.

Arlman are possibilities. Less isotactic-specific centers (more exposed) might be present on edges (where planes meet), at corners, or crystalline defects. The ability of a site to form a long chain exclusively of one configuration may be diminished by small changes in ligand positions, so that it is not difficult to visualize a range of centers, some of which may differ only slightly.

IV. Syndiotactic Propagation

Synthesis of highly syndiotactic polypropylene was first disclosed in 1962 by Natta, Zambelli, and co-workers. While a crystalline surface is mandatory in the synthesis of highly isotactic polymer, soluble catalysts lead to the highest syndiotacticity. However, relatively few mechanistic studies have been published on syndiotactic polymerization.

A. GENERAL COMMENTS

The polymerization chemistry of the syndiotactic-specific catalyst was presented in Chapter 4, Section III,E.

The polymerization behavior of the syndiotactic-specific catalyst argues against classical radical or cationic mechanisms (41). The view that the active center is a metal–carbon bond is supported by the following experimental evidence: (1) when the metal alkyl was labeled with carbon-14, the radio-active carbon was found in the polymer; (2) when the polymerizations were terminated with tritiated alcohols, bound tritium was detected in the polymer (42); and (3) the molecular weight of the formed polymer was lowered by molecular hydrogen (42).

Evidence that complexing occurs is only suggestive (43). Zambelli and Tosi (15) proposed that the particular dependence of the polymerization rate on the monomer concentration suggests that coordination occurs. The form of such dependence suggested to them two successive rate-determining reactions: coordination and insertion (15, 43).

Zambelli, Tosi, and Sacchi (8) have questioned the data from terminations with tritium-labeled methanol (CH_3OT) on the grounds that such radio-activity was low and of doubtful significance. They found appreciable radioactivity in the syndiotactic polypropylene when ^{14}C- labeled methanol was used. In the absence of more convincing experiments, they suggested classifying the syndiospecific polymerization as a coordinated nonradical reaction. A $\overset{\delta^- \delta^+}{V-Pn}$ type center is implied (see Section II,B).

The valence state of the center is most likely to be (III) (44–47), even though active soluble catalysts were formed when the valence of the starting vanadium compound was (V), such as $VO(OEt)_2Cl$; (IV), such as VCl_4(41, 48); and (III), such as $V(acac)_3$ (48). Divalent vanadium salts (VCl_2) are not active (43). When mixtures of these higher valence vanadium salts and $AlEt_2Cl$ were allowed to warm above $-40°C$, precipitation occurred with the formation of an inactive catalyst containing divalent vanadium (43). Divalent vanadium (VCl_2) was not detected by electron spin resonance in the active catalysts (44). Lehr, on the basis of electron spin resonance, visible spectroscopy, magnetic susceptibility, and chemical titration, also concluded that the active vanadium center was trivalent. He differed with Zambelli in that his data suggested a complete reduction of VCl_4 to trivalent V within a short time and that the active V(III) center was derived from another V(III) species that was inactive.

B. PROPOSED MODELS AND EXPERIMENTAL EVIDENCE

Four different driving forces have been proposed to explain the syndiotactic propagation. These are now described.

1. Syndiotacticity due to Alternation of Growth Step between Two Octahedral Positions

While their mechanistic scheme was largely devoted to isotactic propagations, Arlman and Cossee (49) pointed out that syndiotactic propagation could occur under specific conditions. When the rate of migration of the growing polymer chain back to its starting octahedral position is lowered, it becomes possible for a new propylene to react before migration takes place. Because the propylene molecule complexed at the two octahedral positions would have opposite configurations, the continued growth would lead to alternation of *d* and *l* configuration (syndiotactic). The center in this polymerization would be an alkylated transition metal on a surface of crystal.

2. Syndiotacticity due to Steric Repulsion between Methyls of Complexed Propylene and Last Added Propylene

Zambelli and co-workers (41, 48), and later Boor and Youngman (42), favored a bimetallic complex of an alkylated trivalent vanadium chloride and AlEtCl$_2$ or AlEt$_2$Cl for the structure of the active center. The primary reason for having the metal alkyl present was to stabilize the very labile EtVCl$_2$ species, which has not been isolated in pure state. (In the heterogeneous catalysts, EtTiCl$_2$ species are an integral part of the TiCl$_3$ crystal and are presumed to be stabilized against decomposition by the nature of the crystal structure, that is, via chlorine bridges to other Ti metal centers.)

Both workers utilized the Cossee–Arlman octahedral-type structure, previously used for describing centers in the heterogeneous catalysts, to show how the syndio centers might look. Several different structures have been proposed, depending on whether AlEtCl$_2$ or AlEt$_2$Cl is the complexing alkyl, whether a donor is present, and whether one or two alkyls are coordinated to the center.

The sequence of steps leading to high molecular weight polymer with the syndio catalyst has been visualized within the framework of the Cossee–Arlman model proposed earlier for isotactic growth (Fig. 15–19). In the development of this model, a primary addition mode was assumed. Because an ionic surface of a solid crystal is absent, this model does not require

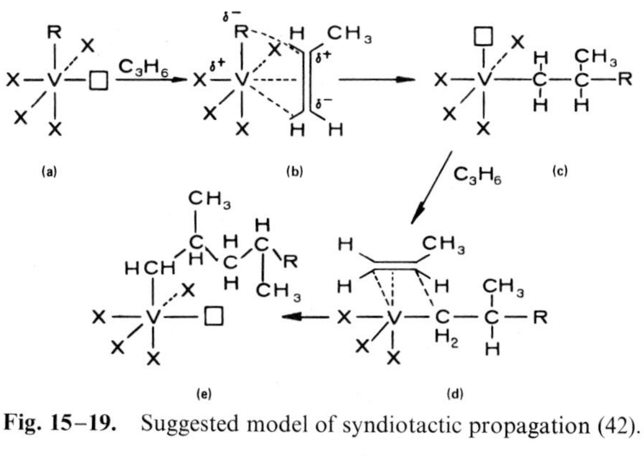

Fig. 15–19. Suggested model of syndiotactic propagation (42).

migration of the growing chain to its starting octahedral position after each insertion step.

The syndiotactic model, as does the isotactic model, attributes stereo-specificity to steric factors. According to the syndio model proposed by Boor and Youngman (42), propylene can become complexed to the vanadium center only if its methyl group does not face the methyl group of the last added propylene unit (Fig. 15–20). Steric compression would be too great if the methyl groups were on the same side (see C, Fig. 15–17). This model also requires that rotation about the V–R bond is hindered (it is difficult to pass the CH_3 of the last added propylene unit over the adjacent Cl ligand).

If this rotation barrier is overcome, such as by raising the temperature or attaching to the V center ligands smaller than Cl, a methylene unit could also face the vacancy at which propylene complexes, as shown in Eq. 15–5. In this situation (structure X), the complexed propylene can assume either of the two configurations, since the methyl of the last added propylene unit and the methyl of the complexed propylene are too far removed to interact. Propagation under these conditions, free rotation and subsequent unrestricted complexing of propylene, produced heterotactic polypropylene.

$$(15-5)$$

IX   X

This model has the following experimental support.

(i) Only at very low polymerization temperatures ($-78°C$) are highly syndiotactic polypropylenes synthesized. Increasing the temperatures to about $-40°C$ significantly lowers the syndio content. This temperature effect is explained in terms of increased rotation about the V–C bond at the higher temperature.

(ii) Even though 1-butene polymerizes extremely slowly at about $-50°C$ or not at all at $-78°C$, it copolymerizes readily with ethylene and to a lesser extent with propylene. It has been suggested that steric interactions between the incoming 1-butene molecules and those added last are too great, even if the ethyl groups are not on the same side in the complexing mode. Apparently, the methyl group located on the pendant ethyl group of the 1-butene added last interacts with the ethyl

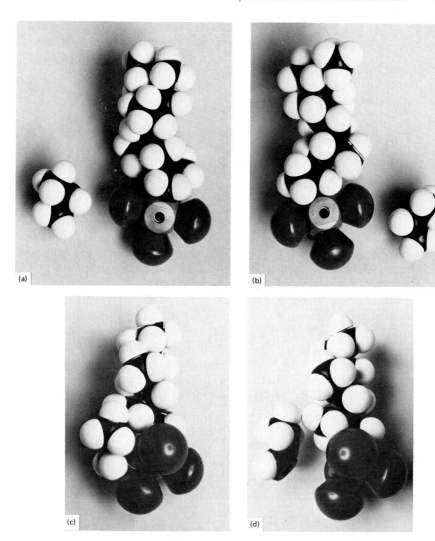

Fig. 15-20. (a) and (b) Front view of octahedral complex and two possible orientations of propylene molecules. The latter are shown after double bond opening. (c) Side view showing monomer orientation to minimize methyl–methyl repulsion and allow close approach to vanadium. (d) Side view showing monomer orientation with high methyl–methyl repulsion preventing close approach to vanadium.

group of the complexed 1-butene. This steric barrier, however, was alleviated by inserting one or more ethylene units between the 1-butene added last and the V center (that is, by copolymerization), as seen in Eq. 15–6.

$$
\begin{array}{c}
\text{Pn} \\
| \\
\overset{\displaystyle H}{\underset{\displaystyle CH_2}{\overset{\diagup}{\underset{\diagdown}{C}}}} \\
CH_2\!-\!CH_3 \\
| \\
CH_2 \\
| \\
CH_2\!-\!CH_2
\end{array}
\qquad (15\text{--}6)
$$

Copolymerization studies by Zambelli and co-workers (50) showed a tendency for alternation for the monomer pairs: ethylene–propylene, ethylene–1–butene, and propylene–1–butene. The authors suggested that the last unit of the growing chain end was sterically controlling.

Also, these workers found that when propylene (monomer 1) and butene-1 (monomer 2) were copolymerized, the following rate constants and reactivity ratios were found: (a) $K_{11} \simeq K_{12}$; $K_{12} \simeq K_{21}$; $K_{11} \gg K_{22}$ and (b) $r_1 \simeq 0.7$; $r_2 \simeq 0.7$; $r_1 r_2 \simeq 0.49$. Clearly, there is a lesser tendency to alternate for these two olefins in contrast to ethylene–propylene or ethylene–1-butene. Zambelli explained this by recognizing that two stable confirmations of 1-butene can be complexed, as seen in Eq. 15–7. Structure XII

$$
\begin{array}{ccc}
\underset{\text{XI}}{
\begin{array}{c}
\overset{H}{|}\ \overset{\nearrow CH_3}{} \\
C \\
\diagup\ \diagdown H \\
\diagdown C\!=\!C \diagdown
\end{array}}
&
\overset{\text{Rotation}}{\underset{\text{about C–Et bond}}{\xrightleftharpoons{\qquad}}}
&
\underset{\text{XII}}{
\begin{array}{c}
\overset{H}{|}\ \overset{\nearrow H}{} \\
C \\
\diagup\ \diagdown CH_3 \\
\diagdown C\!=\!C \diagdown
\end{array}}
\end{array}
\qquad (15\text{--}7)
$$

presents a smaller steric compression if one assumes that the V center and the unit added last are below the $\frac{1}{2}$ plane of this page. Rotation about the C–Et bond is easy. For the monomer unit at the growing chain end, however, a similar rotation of ethyl group may be forbidden, due to the bulky counterion and to steric hindrance of a penultimate 1-butene or propylene unit.

Zambelli and co-workers (50a) have supported this model with ^{13}C spectroscopic data obtained from examination of ethylene–propylene copolymers enriched with $[1\text{-}^{13}C]$ethylene. Both isotactic and syndiospecific catalysts were used to prepare these copolymers. Two types of configurations were envisioned for the propylene–ethylene units: meso, as in structure

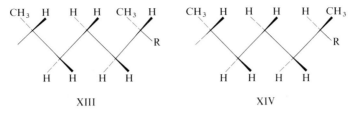

XIII   XIV

XIII, and racemic, as in structure XIV. In copolymers obtained with iso-tactic–specific catalysts, the ethylene units occurred only in meso-type situations, supporting the view that the mechanism of isotactic stereoregulation was not asymmetric induction involving only the last unit of the growing chain end. In contrast, the meso and racemic structures were present in comparable frequencies in copolymers prepared with syndiospecific catalysts. This agreed with the conclusion that steric control in the latter catalysts was exerted by the last unit of the growing chain end.

3. Syndiotacticity due to the Bimetallic Nature of the Center

Pasquon (27) and co-workers proposed that syndiotactic propagation was intimately linked to the structure of the bimetallic complex. Experimentally, it was observed that while $AlEt_2Cl–VCl_4$ is syndiospecific, $GaEt_3–VCl_4$ produces atactic polypropylene. The two metal alkyls have similar alkylating power, but $GaEt_3$ has a lower complexing ability (the latter is largely monomeric, while $AlEt_2Cl$ exists as dimer). They speculated that the $AlEt_2Cl–VEtCl_2$ complex is syndiospecific, while uncomplexed $VEtCl_2$ produced atactic polymer. In another paper, however, Pasquon (51) reported that the relative amount of syndiotactic segments in the heptane fractions (analyzed by ^1H-NMR spectroscopy) increased in the order: $AlEt_2I < AlEt_2Cl < AlEt_3 < BeEt_2 < GaEt_2$. $GaEt_3–\beta TiCl_3$ was found to produce the greatest amount of syndiotactic segments. The significance of this is not obvious.

4. Syndiotacticity due to Steric Repulsion between Olefin and the Combined Steric Front of the Last Added Propylene and Center

Takegami and Suzuki (6, 7) proposed steric repulsion as a driving force similar to that of Boor and Youngman, except that the secondary vanadium alkyl was used (see Section II,B of this chapter). By having the substituted end of the propylene directly attached to the vanadium center, steric repulsions between the last added propylene unit and the incoming propylene molecule were greater. Similarly, alternation occurs because the incoming propylene is complexed only from the less hindered side of the active center.

As in the previous mechanism (Fig. 15–17), the configuration of propylene being complexed alternates at each growth step. It follows that Zambelli's copolymerization data also support this model.

V. Steric Purity of Isotactic and Syndiotactic Polypropylenes

Flory and co-workers (52) have suggested that highly isotactic polypropylene would contain 5 to 10% of racemic dyads and that the figure of 2% asserted to be representative of typical isotactic polypropylene was based on a dubious interpretation of the NMR spectrum. It was generally concluded by the Natta school that steric imperfections were present in concentrations that were lower than the latter value. To appreciate the full scope of this apparent discrepancy, the reader is urged to read the original papers. Only brief comments are made here.

Zambelli, Segre, and their co-workers (53–56) investigated deuterated isotactic, syndiotactic, and atactic polypropylenes by NMR. Segre (56) studied the line width of high resolution NMR spectra of highly isotactic polypropylene. By virtue of being able to observe the ^{13}C side bands for the polymer having the unit $CD(CD_3)–CDH$, she concluded that since the side band was only 0.56% of the main signal, a steric purity of greater than 98.5% was present. Zambelli and co-workers (55) also reached this conclusion on the basis of NMR spectroscopy. Syndiotactic dyads isolated between blocks of isotactic dyads were said to be the main defect. A smaller content of imperfections of pairs of *dl* dyads was also found.

Differences of opinion were still held by the two groups at the time this was written, and these are presented in papers by Flory (57, 57a) and Zambelli and Tosi (15) and in reference cited therein.

Zambelli and Tosi (15) state that syndiotactic polypropylenes contain variable contents of isotactic dyads. The dyad distribution is not Bernoullian. Some of the less sterically pure syndiotactic polypropylenes also contain short isotactic stereoblocks. According to their findings, syndiotactic (as well as atactic) polypropylenes always contain up to several percent head-to-head and tail-to-tail units.

This is a subject that will receive considerable attention in the future. Proton NMR and ^{13}C-NMR spectroscopy, becoming increasingly more sophisticated, will continue to differentiate subtle structural differences. These findings will allow workers to more sharply focus on the mechanism of propagation with different catalysts. In essence, these will continue to become more powerful diagnostic tools to study and differentiate catalysts. It is clear from the Flory–Zambelli papers that, while a high level of detail is being sought regarding the stereochemical constitution of the polymer chain, the sophistication that theory and experiment have reached assures the solution of this problem in the near future.

VI. Conclusion

The syndiotactic-specific catalyst is able to polymerize only propylene to a highly syndiotactic polymer. In contrast, isotactic-specific catalysts are active and stereospecific for propylene and higher α-olefins. Both catalysts are active for the copolymerization of ethylene, propylene, and 1-butene. Both catalysts are similar in that the olefin first complexes to the transition metal atom and inserts by a *cis*-addition into the transition metal–carbon bond.

The catalysts are different in the direction in which the α-olefin adds and in the driving force for the stereochemical process. In isotactic polymerization, primary addition occurs, that is, the unsubstituted carbon of the olefin attaches to the transition metal. In contrast, recent evidence suggests that a secondary addition occurs in syndiotactic polymerizations, in which the substituted carbon of propylene becomes attached to the transition metal atom. The latter view still needs additional confirmation. Isotactic propagations are due to the asymmetric character of the center in a heterogeneous catalyst, which allows or favors the complexing of the α-olefin in only either the *d* or *l* configuration. Because this asymmetric feature is absent in soluble syndiotactic-specific catalysts, another driving force is permitted to occur, namely, steric interactions between the last added and incoming propylenes. This forces opposite configurations to add in any two successive insertions.

While differences may exist as to details of the structures of the isotactic- and syndiotactic-specific catalysts, the experimental data strongly support the mechanistic processes described above.

References

1. J. Boor, *Macromol. Rev.* **2**, 115–268 (1967); E. A. Youngman and J. Boor, *Macromol. Rev.* **2**, 33–69 (1967); J. Boor, *Ind. Eng. Chem., Prod. Res. Dev.* **9**, No. 4, 437–456 (1970).
2. G. Henrici-Olive and S. Olive, *J. Organomet. Chem.* **16**, 339 (1969).
3. F. Danusso, *Chim. Ind. (Milan)* **44**, 611 (1962); D. Sianesi, *ibid.* p. 474.
4. P. Cossee, *Rec. Trav. Chim. Pays-Bas* **T85**, No. 9–10, (1966).
5. P. Longi, G. Mazzanti, A. Roggero, and M. P. Lachi, *Makromol. Chem.* **61**, 63 (1963).
6. Y. Takegami and T. Suzuki, *Bull. Chem. Soc. Jpn.* **42**, No. 3, 848 (1969).
7. T. Suzuki and Y. Takegami, *Bull. Chem. Soc. Jpn.* **43**, 1484 (1970).
8. A. Zambelli, C. Tosi, and C. Sacchi, *Macromolecules* **5**, 649 (1972).
8a. K. Mitani, T. Suzuki, A. Matsuo, and Y. Takegami, *J. Polym. Sci., Polym. Chem. Ed.* **12**, 771 (1974).
9. G. Henrici-Olive and S. Olive, *J. Polym. Sci., Part B* **8**, 205 (1970).

10. K. H. Reichert and J. Berthhold, *Makromol. Chem.* **24**, 103 (1969).
11. K. H. Reichert and E. Schubert, *Makromol. Chem.* **123**, 58 (1969).
11a. J. P. Kennedy and A. W. Langer, *Fortschr. Hochpolym.-Forsch.* **3**, 508–580 (1964).
12. G. Natta, M. Farina, and M. Peraldo, *Chim. Ind. (Milan)* **42**, 255 (1960).
13. T. Miyazawa and T. Ideguchi, *J. Polym. Sci., Part B* **1**, 389 (1963); *Makromol. Chem.* **79**, 89 (1964).
14. A. Zambelli, M. G. Giongo, and G. Natta, *Makromol. Chem.* **112**, 183 (1968).
15. A. Zambelli and C. Tosi, *Adv. Polym. Sci.* **15**, 32–60 (1974).
16. G. Natta, *J. Inorg. Nucl. Chem.* **8**, 589 (1958).
17. L. A. M. Rodriguez and H. M. van Looy, *J. Polym. Sci., Part A-1* **4**, 1971 (1966).
18. J. Boor, *J. Polym. Sci., Part C* **1**, 237 (1963).
19. E. J. Arlman and P. Cossee, *J. Catal.* **3**, 99 (1964).
20. G. Allegra, *Makromol. Chem.* **145**, 235 (1971).
20a. Private communication to the author, May 1970.
21. H. W. Coover, Jr., *J. Polym. Sci., Part C* **4**, 1511 (1963).
22. Yu. V. Kissin and N. M. Chirkov, *Eur. Polym. J.* **6**, 525 (1970).
23. Yu. V. Kissin, S. M. Mezhikovsky, and N. M. Chirkov, *Eur. Polym. J.* **6**, 267 (1970).
24. P. L. Luisi and R. M. Mazo, *J. Polym. Sci., Part A-2* **7**, 775 (1969).
25. P. Corradini, G. Paiaro, and A. Panunzi, *J. Polym. Sci., Part C* **16**, 2905 (1967).
26. J. Boor, *Ind. Eng. Chem., Prod. Res. Dev.* **9**, No. 4, 437–456 (1970).
27. I. Pasquon, *Pure Appl. Chem.* **15**, 465 (1967).
27a. A. W. Langer, (Unpublished) *Lect., Bienn. Polym. Symp., 7th, 1974.*
28. E. J. Arlman, *J. Catal.* **3**, 89 (1964).
29. E. J. Arlman, *J. Catal.* **5**, 178 (1966).
30. E. J. Arlman, *Recl. Trav. Chim. Pays-Bas* **87**, 1217 (1968).
31. J. Boor, *J. Polym. Sci.* **6**25, 45 (1962); *Part C* **1**, 237 and 257 (1963). J. Boor and E. A. Youngman, *ibid. Part B* **2**, 265 (1964); J. Boor, *ibid.* **3**, 7 (1965); *Part A* **3**, 995 (1965); *Pap., 150th Meet., Am. Chem. Soc., 1965*; *Polym. Prepr., Am. Chem. Soc., Div. Polym. Chem.* **6**, No. 2, 890 (1965); E. A. Youngman and J. Boor, *J. Polym. Sci., Part B* **3**, 577 (1965); J. Boor, *ibid. Part A-1* **9**, 617 (1971).
32. A. Turner-Jones, *Polymer* **7**, 23 (1966).
33. D. F. Hoeg and S. Liebman, *Ind. Eng. Chem., Process Des. Dev.* **1**, 120 (1962).
34. G. Natta, *J. Polym. Sci.* **34**, 531 (1959).
35. H. Schnecko, W. Lintz, and W. Kern, *J. Polym. Sci., Part A-1* **5**, 205 (1967); see also H. Schnecko, W. Dost, and W. Kern, *Makromol. Chem.* **121**, 159 (1969).
36. T. Keii, K. Soga, S. Go, and A. Takahashi, *J. Polym. Sci., Part C* **23**, 453 (1968).
36a. E. G. Lovering and W. B. Wright, *J. Polym. Sci., Part A-1* **6**, 2221 (1968).
37. T. Suminoe, K. Sasaki, N. Yamazaki, and S. Kambara, *Kobunshi Kagaku* **21**, 9 (1964).
38. S. Murahashi, S. Nozakura, and M. Sumi, *Bull. Chem. Soc. Jpn.* **33**, 1170 (1960).
39. R. E. Wiman and I. D. Rubin, *Makromol. Chem.* **94**, 160 (1966).
40. J. Boor, *J. Polym. Sci., Part C* **1**, 257 (1963).
41. A. Zambelli, G. Natta, and I. Pasquon, *J. Polym. Sci., Part C* **4**, 411 (1963).
42. J. Boor and E. A. Youngman, *J. Polym. Sci., Part A-1* **4**, 1861 (1966).
43. A. Zambelli, I. Pasquon, T. Signori, and G. Natta, *Makromol. Chem.* **112**, 160 (1968).
44. G. Natta, A. Zambelli, G. Lanzi, I. Pasquon, E. R. Mognaschi, A. L. Segré, and P. Centola, *Makromol. Chem.* **81**, 161 (1965).
45. M. H. Lehr, *Macromolecules* **1**, 178 (1968).
46. C. J. Carman, *Macromolecules* **2**, 217 (1969).
47. G. Natta, P. Corradini, I. Pasquon, M. Pegoraro, and M. Peraldo, Australian Patent 61,150/60, December 8, 1960, Montecatini.

48. G. Natta, I. Pasquon, and A. Zambelli, *J. Am. Chem. Soc.* **84**, 1488 (1962).

49. E. J. Arlman and P. Cossee, *J. Catal.* **3**, 103 (1964).

50. A. Zambelli, A. Letz, G. Tosi, and I. Pasquon, *Makromol. Chem.* **115**, 73 (1968).

50a. A. Zambelli, G. Gatti, C. Sacchi, W. O. Crain, Jr., and J. D. Roberts, *Macromolecules* **4**, 475 (1971).

51. I. Pasquon, G. Natta, A. Zambelli, A. Marinangeli, and A. Surico, *J. Polym. Sci., Part C* **16**, 2501 (1967).

52. P. J. Flory, J. E. Mark, and A. Abe, *J. Polym. Sci., Part B* **3**, 973 (1965); *J. Am. Chem. Soc.* **88**, 639 (1966); P. J. Flory and J. D. Baldeschwieler, *ibid.* p. 2873.

53. A. Zambelli, L. Zetta, C. Sacchi, and C. Wolfsgrubber, *Macromolecules* **5**, 440 (1972).

54. A. Zambelli, "NMR Basic Principles and Progress," Vol. 4, p. 101. Springer-Verlag, Berlin and New York, 1971.

55. A. Zambelli, A. L. Segre, M. Farina, and G. Natta, *Makromol. Chem.* **110**, 1 (1967).

56. A. L. Segre, *Macromolecules* **1**, 93 (1968).

57. P. J. Flory, *Macromolecules* **3**, 613 (1970); Y. Fujiwara, *ibid.* **2**, 237 (1969).

57a. P. J. Flory, *J. Polym. Sci., Polym. Phys. Ed.* **11**, 621 (1973).

16

Mechanisms for Stereochemical Control of Conjugated and Nonconjugated Dienes

I. Introduction

Conjugated and nonconjugated dienes share one common polymerization feature with each other and with α-olefins: they can polymerize by 1,2-type addition involving only one of the double bonds. They differ in that conjugated dienes can also undergo 1,4-addition with formation of a new double bond, while nonconjugated dienes undergo a cycloaddition propagation with the formation of an in-chain ring at the expense of both double bonds.

In the polymerization of conjugated dienes, it has been proposed that the stereochemistry of the polymer is developed at the moment the diolefin becomes complexed to the center during the formation of the transition state, or after the diene has been inserted into the transition metal–carbon bond. For each of these time intervals, various driving forces have also been proposed, including steric, electronic, isomeric state of the diene, etc. It is not possible at this time to choose the correct mechanism; probably more than one mechanistic driving force can become operative, depending on the particular parameters involved.

In the polymerization of nonconjugated dienes, cycloaddition may involve one concerted step or a two-step process.

II. Conjugated Dienes

A. GENERAL COMMENTS

The reaction path and driving force for stereochemical polymerization of diene monomers is not well understood. A neat, unified picture is not possible at this time. Many elegant mechanisms have been proposed, and sophisticated experimental data have been offered in support of many of these. Relative to α-olefin polymerizations, these systems are more complicated and difficult to interpret and pin down mechanistically because of the many different stereochemical paths that are possible. The use of metal alkyl-free catalyst systems has been especially useful, and the reader is urged to refer to the appropriate sections of Chapter 12.

As in α-olefin polymerizations, the growth step occurs by insertion of diene molecules in transition metal–carbon bonds. These diene systems differ, however, in that both σ- and π-allyl transition metal bonds are potentially active in diene polymerizations, whereas only the σ-type bond was considered active in the propagation step involving olefins (see Chapter 13, Section III,B).

This chapter discusses three important aspects of the mechanism by which stereochemical processes occur: (1) at what stage of the growth step do the stereochemical processes occur? (Section III,B); (2) what are the driving forces that cause one or more structures to be formed? (Section III,C); and (3) when are mixtures of structures obtained, if more than one type of center is active? (Section III,D).

For views by other reviewers, consult Cooper *et al.* (1).

B. WHEN DOES THE STEREOCHEMICAL EVENT TAKE PLACE?

There are three intervals in the growth step that have been suggested as the time when stereoregulation occurs. (1) At the moment the diene becomes complexed to the transition metal. Coordination through both double bonds leads to *cis*-1,4 units, while coordination through one double bond favors the formation of 1,2- and *trans*-1,4-structures. (2) During the formation of the transition state. The diene is complexed through one double bond, but a 1,4-addition occurs as the complexed diene becomes inserted into the transition metal–carbon bond. (3) After the coordinated diene molecule has been inserted. It is attached to the transition metal by a σ- or π-allyl-type bond.

The next section describes the potential driving forces that have been proposed to explain the formation of different structures: *cis*-1,4, *trans*-1,4, 1,2, etc.

C. POTENTIAL DRIVING FORCES FOR ORIGIN OF STEREOREGULATION

Before these are presented, some relevant observations should be made. It is important to always keep in mind that a catalyst that gives one type of stereochemical control with one diene may give a completely different stereo-chemical control with another diene. For example, Natta, Porri, and Carbonaro (2) showed that the catalyst [AlEt$_3$ + Ti(OR)$_4$], which produces 3,4 amorphous polyisoprene or 1,2-syndiotactic polybutadiene, forms poly-1,3-pentadiene with a *cis*-1,4-isotactic structure.

The phase state of the catalyst does not necessarily determine the stereo-regulating ability of a catalyst in diene polymerizations. Both soluble and heterogeneous catalysts are stereoregulating. For example, both highly isotactic and syndiotactic polybutadiene have been synthesized with homo-geneous catalysts. This is in contrast to propylene polymerizations, where only a heterogeneous catalyst leads to a highly isotactic polymer and only a homogeneous catalyst yields a highly syndiotactic polymer.

Steric and electronic driving forces have been suggested to explain specific stereoregulating paths. It is possible that two or more forces may act co-operatively or independently, each with a coefficient of contribution. While in some examples it is easy to see why a particular path is taken, in others the changes are too subtle, especially since we lack detailed knowledge of the structure of the center.

The proposed schemes are now brought together according to the three time intervals cited in Section II,B.

1. Forces Operant at the Moment the Diene Becomes Coordinated

The following factors have been proposed to be partially or strongly influential.

a. Conformation of Diene Dissolved in Solvent.   Butadiene exists as an equilibrium mixture of trans and cis conformations, with the former pre-vailing. While this may be influential in polymerizations in which other driving forces may be weak (see Section II,C,1,c, below), in general this is not a dominant factor. Polybutadiene has been synthesized exclusively in one of the four possible structures. Similarly, isoprene, which exists mostly in cis conformation in solution, has been polymerized to products that had very high contents of the *trans*-1,4 or *cis*-1,4 structures.

b. Radius of the Transition Metal.   Furukawa (3) suggested that co-ordination of butadiene molecules to the transition metal atom in a *cis*-1,4 complexing mode prevailed when the radius of the metal was not favorable for coordination at two coordination positions of the atom. Ni and Co had this correct spatial arrangement, and orbital overlap favored coordination at the two positions.

c. Coordination through One or Two Double Bonds. To explain the formation of isotactic and syndiotactic *cis*-1,4-polypentadienes with AlR_3–$Ti(OR)_4$ and $AlEt_2Cl$–$Co(acac)_3$ catalysts, respectively, Natta and Porri (4) proposed the following mechanism. While 1,3-pentadiene complexes through only one double bond at the Ti center (Fig. 16–1), it coordinates at two double bonds at the Co center (Fig. 16–2). The subsequent reaction paths shown in structures a, b, and c in these figures lead to transition states whereby, in the Ti catalyst, the *cis*-1,4-inserted 1,3-pentadiene has the same stereochemical structure as the previous unit (hence isotactic), while in the Co catalyst, the *cis*-1,4-inserted 1,3-pentadiene has the opposite configuration (hence syndiotactic). In this polymerization, the mode of complexing and insertion in the transition state influences the configuration of the asymmetric carbon to which the methyl is attached. In the isotactic placement, the metal center before and after each growth has an identical structure, while in the syndiotactic placement, its structure is different, and it will then produce a configuration opposite to the last added unit.

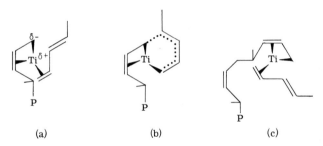

(a)   (b)   (c)

Fig. 16–1. Suggested scheme for the isotactic propagation of 1,3-pentadiene (4). Structure c is identical to structure a. Hence, the newly coordinated monomer of c will give, after insertion into the Ti–C bond, a monomeric unit isotactic with respect to the preceding one. (Figure reproduced from *J. Am. Chem. Soc.* through the courtesy of the Editor.)

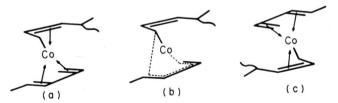

(a)   (b)   (c)

Fig. 16–2. Suggested scheme for the syndiotactic propagation of 1,3-pentadiene (4). Structures c and a are enantiomorphs having opposite configuration. Hence, the newly coordinated monomer of c will give, after insertion into the Co–C bond, a monomeric unit syndiotactic with respect to the preceding one. (Figure reproduced from *J. Am. Chem. Soc.* through the courtesy of the Editor.)

Porri, Di Corato, and Natta (5, 6), using cobalt-based catalysts containing $AlEt_2Cl$, $Al_2Et_3Cl_3$, or aluminum aloxane derivatives, were able to polymerize 1,3-pentadiene to polymers that prevailed in cis-1,4 syndiotactic, amorphous cis-1,4 and 1,2, and weakly crystalline 1,2-syndiotactic structures. They postulated that a cis-1,4 unit comes from coordination (in two steps) of monomer to Co through the two double bonds, as shown in Eq. 16–1.

$$\text{cat} + \quad \longrightarrow \quad \rightarrow \text{cat} \quad \overset{I}{\underset{II}{\nearrow \atop \searrow}} \quad \overset{\text{cat}}{\longrightarrow} \quad cis\text{-1,4 units}$$

$$\text{1,2 units}$$

$$(16-1)$$

Dolgoplosk (7) also suggested that the way butadiene complexes at the reaction centers plays a decisive role in determining whether cis-1,4- or trans-1,4-addition occur, as shown in Eq. 16–2. The syn and anti forms of the

$$
\begin{array}{c}
CH_2{=}CH \\
\quad | \\
\quad CH{=}CH_2 \\
PCH_2CH{=}CHCH_2MX_n
\end{array}
\xleftarrow{C_4H_6}
\begin{array}{c}
CH_2 \\
HC \quad\quad MX_n \\
CH \\
| \\
CH_2P
\end{array}
\xrightarrow{C_4H_6}
PCH_2CH{=}CHCH_2MX_n
$$

$$(16-2)$$

$$
\underset{\substack{syn \text{ configuration}\\(gives \; trans\text{-}1,4)}}{
\begin{array}{c}
CH_2 \\
HCH \quad MX_n \\
C \\
PCH_2CH{=}CHCH_2CH_2 \quad H
\end{array}
}
\xrightarrow{3}
\underset{\substack{anti \text{ configuration}\\(gives \; cis\text{-}1,4)}}{
\begin{array}{c}
CH_2 \\
HCH \quad MX_n \\
C \\
H \quad CH_2CH_2CH{=}CHCH_2P
\end{array}
}
$$

π-allyl complex were regarded as precursors of the trans and cis configurations of bonds, respectively. He suggested that a syn–anti isomerization (path 3) could be influenced by the presence of electron donors, and this was later discussed in the context of diene copolymerization studies (8). The decrease in the positive charge and the number of vacant sites on the transition metal atom as a result of coordination with electron donors in Ni and Ti systems led to increased content of trans-1,4 units. A similar effect was attributed to substituted dienes, that is, the alkyl substituent at the 2 position also decreased the positive charge on the transition metal. In both cases, the probability of the diene coordinating via one double bond (which would produce trans-1,4 units) was increased.

The importance of one vs. two double coordination, as suggested above, was disputed by Vasiliev and co-workers (9). On the basis of polymerization studies and ^{13}C-NMR spectroscopy, they concluded that the relative amount of *syn* and *anti* isomers in the 1,2-disubstituted adducts of bis(π-perdeuteriocrotyl nickel iodide) and 2-alkyl-1,3-butadienes depended on the nature of the substituent at carbon 2 of the allylic ligand. The ratio of *syn* to *anti* isomers was found to be approximately 3:1 for isoprene, 2:1 for 2-ethyl-1,3-butadiene and prevailingly *apto* for 2-isopropyl and 2-*tert*-butyl-1,3-butadiene. They suggested that formation of *syn* and *anti* complexes did not depend on the type of coordination of diene to the metal, since it is unlikely that, in passing from isoprene to 2-*tert*-butyl-1,3-butadiene, the coordination type is changed from mono- to bidentate.

d. Electron Density on Metal Center. Matsuzaki and Yasukawa (10–12) proposed that the electrostatic interactions between the nearly nonbonding electrons of a transition metal atom and a butadiene molecule or a growing end of polymeric chains play an important role in the stereochemical polymerization. As evidence, they cited the varying activities and stereo-regulating abilities of catalysts consisting of a transition metal acetate (Ti to Ni salts were examined), $AlEt_3$, and AlX_3 (X = Cl, Br, or I).

In their proposal, the transition atom was taken to be in a nearly neutral state, according to the electroneutrality principle. When the metal atom had a spherical electron distribution (as in the case for the V center), butadiene preferred to coordinate in its natural conformation (trans). In contrast, in Ni and Co complexes, the electron distribution was deformed from spherical symmetry, and high *cis*-1,4-polybutadiene was formed. But when the crystal field stabilization energies were very high [$Fe^0(d^8)$ or $Mn^0(d^7)$], the butadiene complexes were so stable that catalyst activity was low.

Why does an unsymmetrical electron distribution favor *cis*-1,4 coordination? The authors suggested that repulsion between the two terminal carbons C-1 and C-4 in butadiene are lowered by the screening effect of the d_{xy} electrons that are interposed between them (Fig. 16–3). According to the

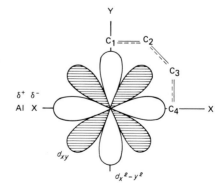

Fig. 16–3. Schematic illustration of the spatial distribution of d electrons in an intermediate complex (11).

crystal field theory, the energy level of d_{xy} orbital is lower than that of $d_{x^2-y^2}$ orbitals, and therefore the electrons prefer to go into the former in most complexes, except those of d^0, d^5, and d^{10} ions.

The effect of added bases was explained in terms of this model (see Fig. 16–5). Added bases diminished the *cis*-1,4-addition in the following order: $(C_6H_5)_3P$ > *o*-phenanthroline > aliphatic amine. This order parallels the order in which the energy of the lowest unoccupied orbital assumes lower values and, therefore, corresponds to electron affinity of the compound. Consequently, when the transition metal of the catalyst is complexed with these bases, back donation of $d\pi$ electrons (Fig. 16–4) from the Co atom to the base occurs in the same order. This leaves a lower electron density in the

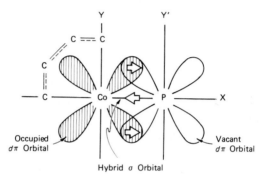

Fig. 16–4. Schematic illustration of the electron donor–acceptor interaction between a cobalt atom and a phosphine derivative (12).

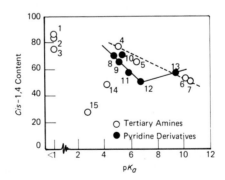

Fig. 16–5. Plot (12) of the *cis*-1,4 content of polybutadiene obtained with the Co(acac)$_3$–Et$_2$AlCl–base catalyst system against the pK_a value of the base for ●, pyridine derivatives and ○, tertiary amines: 1, TPA; 2, THF; 3, thiophene; 4, DMA; 5, DEA; 6, TEA; 7, NMP; 8, quinoline; 9, pyridine; 10, MQ; 11, MP; 12, DMP; 13, TMP; 14, BP; and 15, TPP.

$d\pi$ orbitals. $(C_6H_5)_3P$ is very effective as an acceptor, since P has vacant d orbitals.

An apparently contradictory view was taken by Dolgoplosk and co-workers (7, 13), who suggested that factors which increase the positive charge on the metal atom lead to significant increases in efficiency and stereo-chemical control in the direction of *cis*-1,4-addition. They cited the attach-ment of electron-withdrawing ligands to the metal as an example. While (π-crotyl)$_3$Cr produces 1,2-addition, *cis*-1,4-polybutadiene was favored with the chloride derivative. Also, addition of nucleophilic additives to π-crotyl NiCl led to increased formation of *trans*-1,4 structures at the expense of *cis*-1,4 structures. The reduction in the effective charge in the metal atom favored the coordination of diene, whereby only one double bond was used and the coordinated butadiene retained its natural *trans*-1,4 configuration. As they discussed in Section II,C,1,c, the ratio of *syn* to *anti* forms was influenced by the presence of donors.

The effect of electron-donating and electron-withdrawing compounds on the stereochemical behavior of Cr and Ni metal alkyl-free catalysts is shown clearly in Eqs. 12–11 and 12–12 and Table 12–7.

e. Number of Chlorine Vacancies. It has been well established in a num-ber of laboratories that the layer–lattice modifications (α, γ, and δ) TiCl$_3$ · xAlCl$_3$ (where $x = 0$ to about 0.5) promote polymerizations of propylene to isotactic structure and butadiene to *trans*-1,4 structure. In contrast, the chain–lattice modification βTiCl$_3$ · xAlCl$_3$ ($x = 0$ to 0.5), while it also gives isotactic polypropylene, induces the formation of butadiene to a mixture of *trans*-1,4 and *cis*-1,4 structures (see Chapter 5).

Arlman (14) concluded from theoretical considerations of crystallography of these TiCl$_3$ modifications that the layer α, γ, and δ forms have only one kind of active center (αI). In contrast, the β-chain modification has two kinds of centers, one with a single and the other with two chlorine vacancies (βI and βII) (Table 16–1). Because the αI and βI centers complex butadiene the same way, the *trans*-1,4 structure prevails. The βII center offers two coordination sites in the correct position for butadiene to be chemisorbed in the single cis conformation as a kind of bidentate ligand. According to the scheme, *cis*-1,4 polymer is formed at βII centers. Arlman supported his conclusions with the data of Natta: the polymerization of butadiene at 15°C with βTiCl$_3$–AlEt$_3$ produced a polymer that was separated by extraction methods into fractions having >90% *trans*-1,4 (about two-thirds of the product) and >90% *cis*-1,4 (about one third of the product) structures. Apparently, these βI and βII centers polymerized independently, and the growth of a single chain did not take place at more than one center.

TABLE 16–1

Crystal Structure of Catalyst and Molecular Structure of Polymers (14)

Crystal structure of catalyst	Active centers			Structure of polymers	
	Composition	Location	Characteristics	α-Olefins	1,3-Diolefins
Layer Lattice $TiCl_3 (\alpha, \gamma)$	$TiRCl_4 \square (\alpha I)$	Edges of elementary sheets	1. All Cl firmly bound 2. Sites of R and □ nonequivalent	Isotactic	trans-1,4
Chain Lattice $TiCl_3 (\beta)$	$TiRCl_4 \square (\beta I)$	Ends of elementary chains	1. One Cl loosely bound 2. R preferentially at outer site 3. Sites of Cl_L and □ nearly equivalent	Varying from atactic to isotactic	trans-1,4
	$TiRCl_3 \square_2 (\beta II)$	Ends of elementary chains	1. All Cl firmly bound 2. Sites of □'s equivalent	Atactic	cis-1,4

f. Arrangement of Atoms on the TiCl₃ Surface. Saltman (15) proposed that coordination on the crystal surface of βTiCl₃ favored 1,4-addition because of the unique spacing of the Al and Ti atoms to coordinate isoprene in the *cis*-1,4 complexing model (Fig. 16–6).

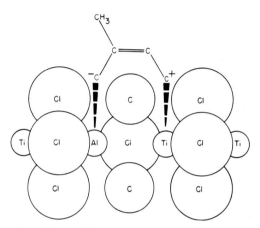

Fig. 16–6. Top view of the β TiCl₃ with an epitactically adsorbed AlR₂Cl molecule, and showing the approach from above of a polarized *cis*-isoprene monomer unit (15). Growth, according to Saltman, takes place by insertion of the coordinated isoprene into the Al–C bond. Growth continues until the new polymeric AlPnEtCl or Al(Pn)₂Cl is desorbed from the TiCl₃. Saltman proposed that because a range of sizes of lattice flaws exist on the TiCl₃ surface, butadiene (lacking a side methyl group) may enter the complex either *cis* or on its side in the *trans*-1,4 configuration; hence, the formed polymer contains a mixture of *cis*- and *trans*-1,4 structures. With αTiCl₃, the longer trans structure is highly favored by the much longer Ti–Ti distance (3.5 Å in the layer lattice vs. 2.9 Å for αTiCl₃) and polymers having a trans structure are formed.

g. Blocking of Coordination Sites. Natta and Porri (16) cited that two typically homogeneous catalysts that yield *cis*-1,4 polybutadiene in benzene (LiAlH₄ + TiI₄ and AlEt₂Cl + cobalt salt) can be made to produce exclusively *trans*-1,4 polymer if a small amount of a Lewis base is added. Cooper and co-workers (17) suggested that the base complexes and blocks one coordinate site and forces the diene to complex via one double bond. The view was taken that only when butadiene is complexed at two coordination sites can *cis*-1,4-addition occur; hence, *trans*-1,4-addition is more favored.

h. Cis–Trans Isomerism. Steric compression was cited as the reason why the cis isomer of 1,3-pentadiene did not polymerize in the presence of the

$AlEt_2Cl$–$CoCl_2$ catalyst, whereas the trans isomer did. Natta and co-workers (18) attributed this to the inability of cis isomer to coordinate by the cis mode. The trans isomer of 1,3-pentadiene produced a syndiotactic *cis*-1,4-polymer under similar conditions. In the formation of the isotactic *cis*-1,4-structure from the cis isomer, a rotation would have to occur at each addition step (18a).

In the presence of another catalyst AlR_3–$TiCl_4$, *cis*-1,3-pentadiene, because it could not assume a cis conformation, polymerized predominantly to a *trans*-1,4-polymer. In contrast, *trans*-1,3-pentadiene can assume both cis and trans conformations and yielded *cis*-1,4-polypentadienes. Highly *cis*-1,4-polyisoprene was formed with the same catalyst.

i. Steric Compression of Substituents.   Steric compression due to methyl groups has been cited as the reason why 4-methyl-1,3-pentadiene $[CH_2{=}CH{-}CH{=}C(CH_3)_2]$ preferentially undergoes 1,2-addition with catalysts made from $AlEt_3 + VCl_3$, VCl_4, $TiCl_4$, or $TiCl_3$ (19). These catalysts normally have a strong tendency to polymerize butadiene exclusively to the *trans*-1,4 structures.

2. During Formation of Transition State

In the presence of $AlEt_3$–$Ti(OR)_4$ catalyst, isoprene undergoes 3,4-addition, while 1,3-pentadiene polymerizes to *cis*-1,4 and 1,2 structures (2, 5, 20). This was attributed to inductive affects of the methyl substituent, which placed a different electron density on carbons in the monomer unit in the transition state.

For example, isoprene preferentially polymerizes by the path, shown in reaction 16–3. In comparison, 1,3-pentadiene polymerizes by the path shown

$$
\begin{array}{ccccc}
\hat{C}H_2 & & \hat{C}H_2 & & \hat{C}H_2 \\
\diagdown & & \diagdown & & \diagdown \\
cat \quad \diagdown & & cat \quad CH_2 & & cat \quad CH_2 \\
\diagdown CH_2 & \longrightarrow & \parallel & \longrightarrow & \diagup \\
HC \diagdown & & HC & & CH \\
\diagdown CH_2 & & \diagdown CH_2 & & \mid \\
C \diagup & & C \diagup & & C{=}CH_2 \\
\mid & & \mid & & \mid \\
CH_3 & & CH_3 & & CH_3
\end{array}
\qquad (16\text{–}3)
$$

Transition state

in Eq. 16–4. In the *cis*-1,4-addition of 1,3-pentadiene, the inductive effect of the methyl increased the electron density on the carbon undergoing bonding. In a similar situation with isoprene, the inductive effect of methyl would be

$$(16\text{–}4)$$

less effective since it would be on a carbon (marked *) that is not undergoing bonding in structure I. However, in the presence of cobalt-based catalysts

I

[AlEt$_2$Cl–Co(acac)$_3$] in heptane, 1,3-pentadiene preferentially undergoes 1,2-addition, suggesting that perhaps steric compression due to the methyl group exceeds the beneficial inductive effect described above in the titanium-based catalyst. Butadiene, with the same catalyst, exclusively undergoes cis-1,4-addition.

Takegami and co-workers (21) studied the insertion reaction of isoprene to the metal–ethyl bond of TiCl$_4$–AlEt$_3$ catalyst by operating with a high catalyst concentration. Structures II, III, and IV were identified. They concluded that II was an intermediate in the cis-1,4-polymerization, but structure

M—C—C=C—C—Et   C=C—C—C—M   C=C—C—C—Et
      | | | | |
      C   C   Et   C   M

     II   III   IV

IV isomerized to II under conditions of polymerization. Structure III was an intermediate in 3,4-polymerization.

3. Forces Operant after Diene Has Been Inserted

Several paths have been recognized, including rotation about the double bond, the direction of the σ-bonding, and the direction of the diene approach.

a. Rotation of Double Bond (22).   The double bond of the diene unit added last is excited when it coordinates with the metal atom to which the diene is attached. This excitation allows rotation about the bond, and cis–

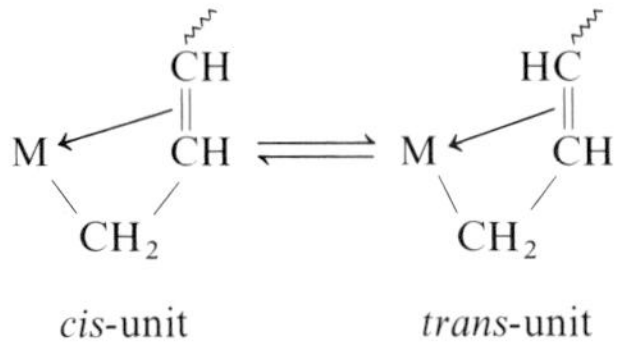

$$M \xleftarrow{} \overset{\overset{CH}{\|}}{\underset{CH_2}{CH}} \rightleftharpoons M \xleftarrow{} \overset{\overset{HC}{\|}}{\underset{CH_2}{CH}} \qquad (16\text{–}5)$$

cis-unit   trans-unit

trans isomerism can occur, as shown in Eq. 16–5. Unspecified factors can influence the cis–trans isomerism, including ligands, temperature, nature of metal, etc.

b. Direction of σ-Bonding.   Natta and Porri (20) suggested that 1,4- vs. 1,2-addition is decided by the nature of the ligands attached to the transition metal atom. They visualized the following reaction path for polymerization of butadiene (BD) with the $AlEt_2Cl$–$V(acac)_3$ and $AlEt_3$–$V(acac)_3$ catalysts to trans-1,4 and 1,2 structures, respectively. Both catalysts are soluble, but they differ in that the former contains chlorine ligands.

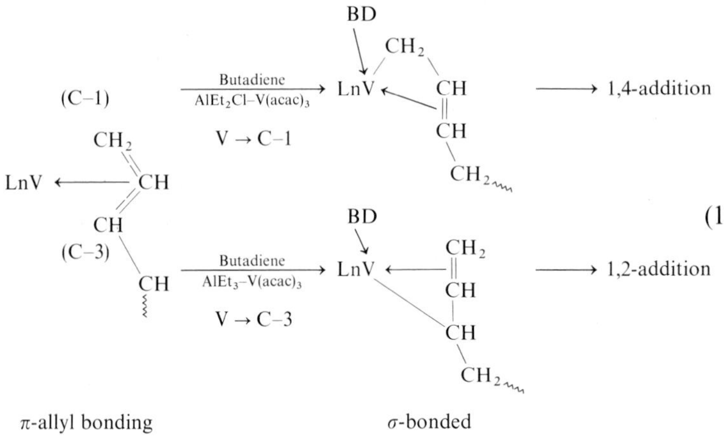

$$(16\text{–}6)$$

π-allyl bonding   σ-bonded

The mode of coordination was considered unimportant. It was assumed that the butadiene coordinated only through the vinyl group. In the halogen-containing catalyst when the next diene is coordinated, the π-allyl-bonded unit becomes σ-bonded to the vanadium at the 1-carbon, while in the other catalyst it becomes bonded at the 3-carbon. The detailed mechanistic influence of the ligand was not elucidated.

c. Direction of Diene Approach.   Marconi (23) considered the deciding factor to be the approach of the complexing diene. If the monomer coordinates with the Me—CH_2 bond of the complex and the 1 and 4 carbon atoms of the diene being involved (direction 1), all units of the forming chain will have a 1,4 (*cis* or *trans*) configuration, as shown in Eq. 16–7, where $X_n =$

$$(16-7)$$

Ln = ligands attached to the transition metal (Me) and ☐ = butadiene monomer being complexed. If the coordination of monomer in 1,4-configuration and its incorporation into the growing chain occur on the side of the substituent at the Me—CH bond (direction 2), 1,2 (or 3,4 units in isoprene) can be formed, as shown in Eq. 16–8.

$$(16-8)$$

The double bond of a penultimate diene unit appears to be in the electric field of the transition metal center. Klepikova and co-workers (24) identified a *syn*-π-crotyl structure by proton NMR studies when a deuterated butadiene was polymerized with both undeuterated and deuterated π-crotylnickel iodide to give a *trans*-1,4-polybutadiene, as shown in structure V. They concluded that the $-CH^e{=}CH^e-$ unit was still in the transition metal field, as

$$CD_2{-}CD_2{-}CH^e{=}CH^e{-}CD_2(CD_2CH^f{=}CH^f{-}CD_2)_n{-}CD_2{-}CD{=}CD{-}CD_3$$

V

evidenced by the difference in the chemical shifts of H^e and H^f protons ($\tau = 4.60$ vs. 4.50 ppm). The incorporation of one butadiene-d_4 molecule between the perdeuteriocrotyl group and nickel gave rise to a complex that showed two doublets at $\tau = 5.23$ (H^a) and 7.29 (H^d), corresponding to a *syn*-π-crotyl ligand shown above.

D. SINGLE OR MULTIPLE CENTERS

Are the various stereochemical structures, *cis*-1,4-, *trans*-1,4-, and 1,2- (or 3,4-) units, formed at the same or different centers? Very often, the whole product can be extracted by use of different solvents into fractions containing one of the structures in high purity. This suggests that the different structures were formed on different sites. Section II,C offers a good example.

III. Nonconjugated Dienes

A. GENERAL COMMENTS

Nonconjugated dienes can polymerize by a simple 1,2-addition (path a) and by cycloaddition (path b), as shown in Eq. 16–9. When polymerized by

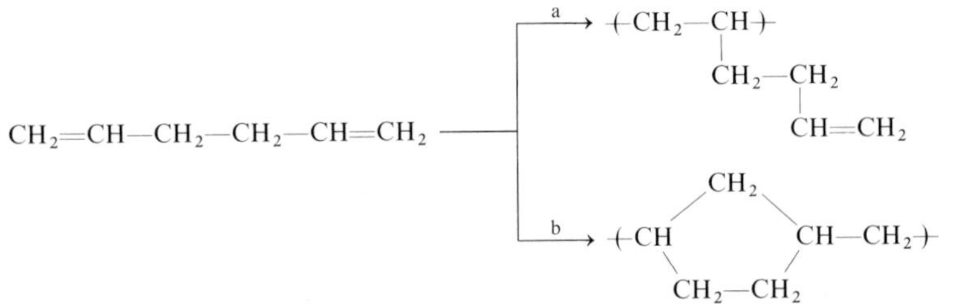

CH₂=CH—CH₂—CH₂—CH=CH₂
a
(CH₂—CH)
CH₂—CH₂
CH=CH₂
b
(CH
CH₂
CH—CH₂)
CH₂—CH₂

(16–9)

path a, the diene acts as a simple α-olefin. But why should cycloaddition take place? The reader should recall that cycloaddition also takes place readily in radical and cationic polymerizations for a variety of monomers that have two double bonds (structure VI) or two reactive functional groups (structure VII).

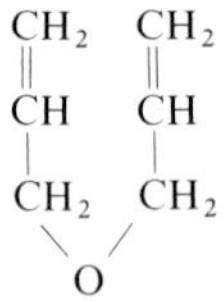

CH₂ CH₂
CH CH
CH₂ CH₂
O

VI

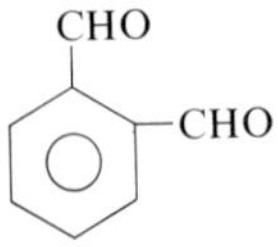

CHO
CHO

VII

B. EXPERIMENTAL FACTORS FAVORING 1,2- AND CYCLOADDITION

Chapter 19, Section XII describes the general polymerization chemistry of nonconjugated dienes, and it should be consulted for general background and references used in the discussion.

1. Factors Favoring 1,2-Addition

The factors that favor 1,2-addition include soluble, colloidal, or non-isotactic-specific catalysts, such as $AlEt_2Cl + V(acac)_3$; low polymerization temperatures ($-20°C$ has been investigated); and catalysts that appear to give random–alternating ethylene–propylene copolymers, for example, vanadium-based systems such as $AlEt_2Cl + VO(OEt)_2Cl$.

2. Factors Favoring Cycloaddition

The factors that favor cycloaddition, in contrast, include heterogeneous catalysts that are isotactic-specific, such as $AlEt_2Cl$ or $AlEt_3 + TiCl_3 \cdot xAlCl_3$ or $TiCl_4$; intermediate polymerization temperatures ($25°$ to $100°C$ have been examined); catalysts that appear to give random-block ethylene–propylene copolymers, such as titanium-based systems; a low concentration of nonconjugated diolefins; and a specific number of methylene units separating double bonds in the diolefin, for example, 1,5-hexadiene ($n = 2$) is most favorable.

C. PROPOSED MECHANISMS

In evolving a plausible mechanism to explain cycloaddition, two paths can be considered: the Marvel–Garrison two-step addition (25), in which consecutive insertion of the two double bonds takes place, as in Eq. 16–10;

$$(16–10)$$

or a concerted one-step addition. The author suggests that the diolefin in the transition state may involve both double bonds, as shown in structures

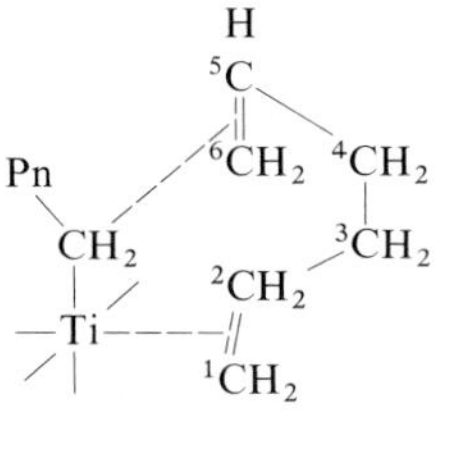

VIII

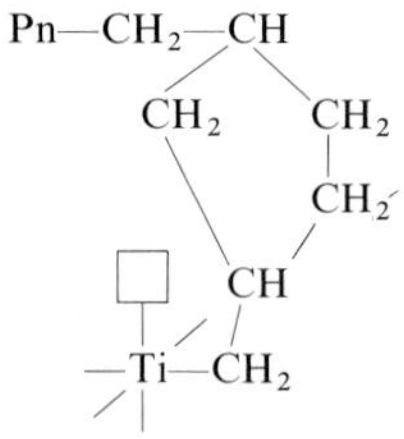

IX

VIII and IX, where Pn = a growing polymer chain and □ = an open octahedral center for complexing.

Indeed, molecular models suggest that, for 1,5-hexadiene, this concerted addition is possible because C–2 and C–6, as well as C–5 and the CH_2 attached to Ti, appear in a favorable position for bonding to occur via the resonance structure shown. Butler's proposal that cycloaddition of non-conjugated dienes is promoted by interactions of orbitals of the double bonds via lowering of the reaction path probably applies here, even though the sources of initiation are different. As far as the author is aware, no one has attempted a molecular orbital analysis (such as Cossee did for α-olefins and conjugated dienes) on this system.

TABLE 16–2

Percent Cyclization vs. Size of the Incipient Cycle for
Polymerization[a] of $CH_2{=}CH{-}(CH_2{-})_x CH{=}CH_2$ (25)

		Soluble polymer (C_6H_6)	
x	Size of incipient cycle	η_{inh}	% Cyclic units
4	7	0.08	25
5	8	0.14	9
6	9	0.35	6
7	10	0.23	10
8	11	0.14	11
9	12	0.10	11
10	13	0.11	—
11	14	0.06	15
12	15	0.25	15
14	17	0.10	4
18	21	0.13	8

[a] Catalyzed by triisobutylaluminum and titanium tetrachloride.

Which model is more plausible or operative? While attractive for 1,5-hexadiene, the concerted model is less attractive for higher homologs that polymerize in part by cycloaddition (Table 16–2). Models indicate considerable difficulty in arranging the necessary carbon atoms of the higher diolefin in a favorable position for insertion by a concerted path.

For higher monomers, a two-step mechanism is more plausible. According to the Marvel–Garrison mechanism, the second double bond becomes complexed at the center after the first double bond is inserted. Because one of the diolefin units is fixed to the metal center, there is an improved chance for complexing. In terms of the Cossee–Arlman model, the second double bond could complex at the newly formed octahedral position, and a second insertion would restore the growing chain to the original octahedral position without requiring migration of the polymer chain.

The Marvel–Garrison model is supported experimentally by the observation that cyclization is favored when the concentration of the diolefin is decreased. With fewer diolefins competing for the center, there is a better chance for the double bond of the newly added nonconjugated diene to complex and insert.

Other supporting evidence is also cited. Heterogeneous catalysts (isotactic-specific centers) favor cyclization because steric factors make complexing difficult, and again the terminal double bond has a better chance to complex and insert before a second diolefin can complex at the center. Nonisotactic-specific catalysts do not present this steric barrier, and complexing of incoming diolefin molecules is easier, allowing 1,2-addition to occur at a faster rate. In this situation, less time is available for the second double bond to complex and insert, so 1,2-addition prevails.

Lower polymerization temperatures favor 1,2-addition, because rotations that are necessary to bring the second double bond in the vicinity of the transition metal center vacancy are energetically less favored.

Neither cycloaddition nor 1,2-addition appears to be affected by the nature of the last added olefin unit. Copolymerizations with ethylene did not cause the catalyst to lose its ability to affect the mode of addition occurring preferentially during the homopolymerization of 1,5-hexadiene.

It would be mechanistically interesting to homopolymerize and copolymerize 1,5-hexadiene with the syndiotactic-specific catalyst.

IV. Conclusion

We are confronted by a rather large number of mechanistic schemes that have been proposed by different workers to explain the origin of stereoregulation in the polymerization of conjugated dienes. These proposals vary not only in the nature of the driving force for stereochemical control but also as to the time interval when this stereochemical event takes place. It is not possible at this time to decide which is the correct mechanism. While some of these proposals are cogently supported by some experimental data, most are still largely speculative. The author suggests that a single unique mechanistic scheme is not operating. Variations may exist among the different catalyst systems, both in the time interval when the stereo-chemical event occurs, as well as in the driving force that causes the stereo-chemical event. More than one driving force may be responsible for the stereochemical event, each characterized by a contributing coefficient. The most dramatic differences may exist between the σ- and π-allyl-type bonding systems. The value of the coefficient above would undoubtedly be affected by the electronic nature of the transition metal, by the steric and electronic nature of the ligands attached to the metal center, by the physical state of the catalysts, and by the structure of the metal alkyl. It is not unexpected nor surprising that a variation in one of these parameters would alter the stereochemical features of the catalyst, as has been reported by different workers. Examples are cited in Chapters 5 and 14.

In contrast, the mechanism by which nonconjugated dienes are polymerized appears to be better understood.

References

1. W. Cooper and G. Vaughan, *Prog. Polym. Sci.* **1**, 93–160 (1967); G. Natta and L. Porri, in "Polymer Chemistry of Synthetic Elastomers," (J. P. Kennedy and E. G. M. Törnqvist, eds.), Chap. 7A. **23**, Part 2, p. 597–678, Wiley Interscience, 1969. I. Pasquon, *Encycl. Polym. Sci. Technol.* **13**, 13–86 (1970); W. Cooper, *Ind. Eng. Chem., Prod. Res. Dev.* **9**, No. 4, 457–466 (1970); B. A. Dolgoplosk, *Polym. Sci. USSR (Engl. Transl.)* **13**, 367–393 (1971).
2. G. Natta, L. Porri, and A. Carbonaro, *Makromol. Chem.* **77**, 126 (1964).
3. J. Furukawa, *Bull. Inst. Chem. Res., Kyoto Univ.* **40**, 130 (1962).
4. G. Natta and L. Porri, *Abstr. Pap., 148th Meet., Am. Chem. Soc., 1964* p. 2x (1964).
5. L. Porri, A. Di Corato, and G. Natta, *Eur. Polym. J.* **5**, 1 (1969).
6. G. Natta and L. Porri, *Adv. Chem. Ser.* **52**, 24 (1960).
7. B. A. Dolgoplosk, *Polym. Sci. USSR (Engl. Transl.)* **13**, 367–393 (1971); *Vysokomol. Soedin., Ser, A* **13**, 325–347 (1971).
8. B. A. Dolgoplosk, S. I. Beilin, Yu. V. Korshak, K. L. Makovetsky, and E. I. Tinyakova, *J. Polym. Sci., Polym. Chem. Ed.* **11**, 2569 (1973).
9. V. A. Vasiliev, N. A. Kalinicheva, V. A. Kormer, M. I. Lobachi, and V. I. Klepikova, *J. Polym. Sci., Polym. Chem. Ed.* **11**, 2499 (1973).
10. K. Matsuzaki and T. Yasukawa, *J. Polym. Sci., Part B* **3**, 907 (1965).
11. K. Matsuzaki and T. Yasukawa, *J. Polym. Sci., Part A-1* **5**, 511 (1967).
12. K. Matsuzaki and T. Yasukawa, *J. Polym. Sci., Part A-1* **5**, 521 (1967).

13. B. A. Dolgoplosk, E. I. Tinyakova, P. A. Vinogradov, O. P. Parenago, and B. S. Turov, *J. Polym. Sci., Part C* **16**, 3685 (1968).
14. E. J. Arlman, *J. Catal.* **5**, 178 (1966).
15. W. M. Saltman, *J. Polym. Sci.* **46**, 375 (1960).
16. G. Natta and L. Porri, *in* "Polymer Chemistry of Synthetic Elastomers," (J. P. Kennedy and E. G. M. Törnqvist, eds.), Chap. 7A, **23**, Part 2, p. 660, Wiley Interscience, 1969.
17. W. Cooper, G. Degler, D. E. Eaves, R. Hank, and G. Vaughan, *Adv. Chem. Ser.* **52**, 46 (1966).
18. G. Natta, L. Porri, A. Carbonaro, F. Ciampelli, and G. Allegra, *Makromol. Chem.* **51**, 229 (1962).
18a. W. Cooper, *Ind. Eng. Chem., Prod. Res. Dev.* **9**, No. 4, 463 (1970).
19. L. Porri and M. C. Gallazzi, *Eur. Polym. J.* **2**, 189 (1966).
20. G. Natta and L. Porri, *in* "Polymer Chemistry of Synthetic Elastomers," (J. P. Kennedy and E. G. M. Törnqvist, eds.), Chap. 7A, **23**, Part 2, p. 662, Wiley Interscience, 1969.
21. Y. Takegami, T. Suzaki, and T. Okazaki, *Polym. Rep.* (*Jpn.*) **107**, 21 (1967).
22. This possibility was recognized independently by many workers.
23. W. Marconi, *in* "The Stereochemistry of Macromolecules" (A. D. Ketley, ed.), pp. 239–307. Dekker, New York, 1967.
24. V. I. Klepikova, G. P. Kondratenkov, V. A. Kormer, M. I. Lobachi, and L. A. Churlyaeva, *J. Polym. Sci., Polym. Lett. Ed.* **11**, 193 (1973).
25. C. S. Marvel and W. E. Garrison, *J. Am. Chem. Soc.* **81**, 4737 (1959).

17

Mechanisms for Stereochemical
Control of Stereoselective
and Stereoelective Isotactic
Propagations

I. Introduction

As described in Chapter 15, it is the nature of isotactic-specific centers to generate asymmetric carbons during the growth step when an α-olefin is polymerized. These asymmetric carbons become part of the backbone of the polymer chain (Fig. 3–3).

Additional stereochemical polymerizations with isotactic-specific catalysts were reported in 1962 by Pino and co-workers (1, 2), when the pendant group of the α-olefin had an asymmetric carbon (designated by *), as shown in structures I and II. These workers disclosed that centers of the catalyst

$$
\begin{array}{cc}
CH_2{=}CH & CH{=}CH \\
| & | \\
{*}CH{-}CH_3 & {*}CH{-}CH_3 \\
| & | \\
CH_2 & (CH_2)_3 \\
| & | \\
CH_3 & CH \\
 & {/}\;{\backslash} \\
 & CH_3\;\;CH_3 \\
\end{array}
$$

(R,S)-3-methyl-1-pentene   (R,S)-3,7-dimethyl-1-octene

I   II

had the remarkable ability to polymerize to a large extent selectively into single polymer chains only one or the other of the two antipodes in the racemic monomeric mixture. Two stereochemical polymerizations were

442

experimentally demonstrated: stereoselection [the monomeric (R) and (S) antipodes polymerize at the same rate]; and stereoelection (one of these antipodes polymerizes at a faster rate).

Various aspects of stereoselection and stereoelection have been described for α-olefins and other types of monomers in several earlier reviews (3–8). This chapter will first describe the experimental characteristics of these two stereochemical polymerizations and then compare the various mechanisms proposed to explain why they occur.

Studies of this novel propagation are extremely important, because it offers a unique and powerful diagnostic probe to investigate the nature of the active center and the mechanistic process.

II. Stereoselection

When simple, optically inactive Ziegler–Natta catalysts were used by Pino to polymerize racemic α-olefins [such as (R,S)-4-methyl-1-pentene], only an optically inactive isotactic polymer was obtained (1, 2). But when this product was fractionated by chromatography, it was separated into optically active fractions containing predominantly (R) or (S) antipode units. A column of insoluble optically active poly-(S)-3-methyl-1-pentene [prepared directly from optically active (S)-3-methyl-1-pentene] was used as the adsorbing material (Table 17–1).

TABLE 17–1

Chromatographic Resolution of Racemic Poly-4-methyl-1-hexene[a] with Crystalline (+)-Poly-(S)-3-methyl-1-pentene as the Supporting Medium[b,c]

Fraction	Eluent	T (°C)	Fraction weight (g)	$[\alpha]_D^{44}$ ($l = 2$) in benzene solution (deg.)	$[\alpha]_D^{44}$ (deg.)
1	1:1 Acetone diethyl ether mixture	25	0.0304	−0.175	−23.6
2	Diisopropyl ether	25	0.0170	−0.035	−13.6
3		56.2	0.0348	−0.055	−6.7
4	Diisopropyl ether	56.2	0.0466	−0.040	−4.1
5		56.2	0.0154	+0.015	+4.8
6	Benzene	64.5	0.0322	+0.185	+25.0
7	Benzene	77.1	0.0162	+0.155	+47.7

[a] A diethyl ether-insoluble–diisopropyl ether-soluble fraction having $[\eta] = 0.92$ at 120°C in tetralin.

[b] Boiling decalin-insoluble fraction.

[c] Weight of supporting polymer/weight of supported polymer = 30.4; 0.22 g supported polymer; 0.1986 g total polymer eluted.

Pino attributed this unique stereochemical process to the ability of the catalyst centers to predominantly select one or the other antipode of the racemic mixture, with the consequence that each single polymer chain contained a prevalence of one of these antipodes as shown in Eq. 17–1. In

$$
\begin{array}{l}
HC{=}CH_2 \\
\ \ | \\
\ \ CH_2 \\
\ \ | \\
*CHCH_3 \xrightarrow[TiCl_4]{Al\text{-}i\text{-}Bu_3} \\
\ \ | \\
\ \ CH_2 \\
\ \ | \\
\ \ CH_3
\end{array}
$$

Optically inactive polymer consisting of equal numbers of poly-(R) and poly-(S) chains and optically inactive unreacted monomer

Fractionation by adsorption chromatography on optically active poly-(S)-3-methyl-1-pentene →

Optically active polymer fractions having opposite signs of rotation   (17–1)

a stereoselective propagation, both monomeric antipodes are polymerized at the same rate but are selectively incorporated into separate chains at the individual centers.

These chromatographic separations confirmed their earlier conclusions that the polymerization rates of the two antipodes of the racemic mixture at the same active centers were sufficiently different to permit the formation of polymer chains consisting mostly of one or the other antipode (9). This conclusion was reached in 1960 on the basis of the observed structural identities of poly-(S)-3-methyl-1-hexene and the corresponding isotactic polymer made from racemic (R,S)-4-methyl-1-pentene. Both polymers had the same number of monomer units per turn of helix (3.5 or $\frac{7}{2}$) and the same identity period (2.0 Å). The X-ray patterns of these polymers showed identical positions and intensities of reflections. This identity was rationalized by concluding that the polymer obtained from racemic monomer consisted of polymer chains, each derived prevalently from one of the enantiomers, and that the crystals consisted of polymer chains formed from the same enantiomer.

In order to explain the relatively high optical rotations of isotactic polymer chains containing optically active monomer units, it has been suggested by several workers that these chains must contain substantial contents of helical sections prevailingly having a single screw sense determined by the absolute structure of the asymmetric carbons (10–13). Thus, the more isotactic was the polymer chain, the greater was the content of the prevailing antipode and the greater was the concentration of helical sections.

Supporting evidence that spiralization was responsible for the augmentation of optical activities came from recent copolymerization studies by Carlini and co-workers (14). They observed that the copolymer of (S)-4-methyl-1-hexene and 4-methyl-1-pentene, like poly-(S)-4-methyl-1-hexene, was spiraled in the left screw sense. The presence of 4-methyl-1-pentene in the copolymer did not destroy the stereoregularity or left screw sense of the poly-(S)-4-methyl-1-hexene chains. Conformational analysis of isotactic poly-4-methyl-1-pentene shows that the left- and right-handed helical conformations of the main chain have optical rotations of +240 and −240 degrees, respectively.

Pino and his co-workers (13) pointed out that the chromatographic separations of the polymers of racemic α-olefins to fractions having optical activity of opposite signs, can be taken only as a qualitative indication of the existance of polymer chains in which (R) or (S) antipode units prevail. While the minimum prevalence of (R) or (S) units in the polymer chains present in the fractions having negative or positive rotatory power can be evaluated, the efficiency of the separation process is not completely known. A strong dependence on the structural relationship between the polymer to be separated and the adsorbing support is indicated.

The mechanism of separation is only partially understood. Pino and co-workers observed that low molecular weight hydrocarbons could not be similarly resolved. A dependence on polymer molecular weight and isotacticity was shown. They speculated that a polymer chain having a helical conformation of a same screw sense was much more adsorbed than a polymer chain having a different conformation. Presumably, polymer chains predominantly consisting of either (R) or (S) antipode units have the same helical conformation.

Further experimental evidence for the stereoselective polymerization came from the laboratories of Pino, from Pino's former students, and other workers.

Pino, Ciardelli, and Montagnoli demonstrated that stereoselection was not due to statistical placement of the (R) and (S) antipode units (15). They observed that atactic poly-(R,S)-4-methyl-1-hexene, previously prepared by Ciardelli by hydrogenating poly-(R,S)-4-methyl-1-hexyne, could not be fractionated into optically active fractions by the same chromatographic methods. A soluble nonstereospecific $AlEt_3$–$Fe(acac)_3$ catalyst was used by Ciardelli and co-workers (16) to synthesize the hexyne polymer.

Ciardelli and co-workers (17, 18) found that when (R,S)-3,7-dimethyl-1-octene was copolymerized with (S)-3-methyl-1-pentene with the Zn-i-Bu_2–$TiCl_4$ catalyst, the (S)-3-methyl-1-pentene molecules were preferentially incorporated into the polymer chains that predominantly contained (S)-3,7-dimethyl-1-octene units. A fraction containing substantial poly-(R)-3,7-dimethyl-1-octene was isolated by solvent fractionations. This fraction had a high negative rotatory power, showing it to be derived from (R)-3,7-dimethyl-1-octene.

Additional copolymerization data by Goodman and Furuyama (19) confirm the view that the stereoselective driving force is not destroyed

TABLE 17-2

Comparison between Observed and Calculated Prevalence[a] of (S) or (R) Monomeric Units in the Fraction Having the Highest Absolute Value of the Rotatory Power Obtained in the Chromatographic Separations of Low Molecular Weight Polymers of Some Racemic α-Olefins (24)

	Polymer to be separated		Fraction having the highest absolute value of the rotatory power			Calculated prevalence[a] of (S) or (R) monomeric units in a random copolymer of the two antipodes (%)[c]
Type	Catalytic system[d,h]	Fraction	$\overline{DP}_n$	Wt%[e]	Observed prevalence[a] of (S) or (R) monomeric units[b]	
Poly-(R, S)-4-methyl-1-hexene	TiCl$_4$—AlR$_3$	Acetone-soluble	12[f]	7.6	11	56
	TiCl$_3$ "ARA"—AlR$_3$	Acetone-soluble	10[f]	12.9	5	52
	TiCl$_3$ "ARA"—AlR$_3$	Ethyl acetate-soluble–acetone-insoluble	100[g]	8.7	10	17
	TiCl$_4$—ZnR$_2$	Ethyl acetate-soluble–acetone-insoluble	50[g]	3.2	23	31
	VCl$_4$—AlR$_3$	Ethyl acetate-soluble–acetone-insoluble	100[g]	4.4	16	20
Poly-(R, S)-3,7-dimethyl-1-octene	TiCl$_4$—AlR$_3$	Acetone-soluble	7[f]	13.7	44	61
Poly-(R, S)-3-methyl-1-pentene	TiCl$_3$ "ARA"—AlR$_3$	Acetone-soluble	12[f]	5.1	61	60

[a] [(S) monomeric units − (R) monomeric units/((S) monomeric units + (R) monomeric units] × 100.

[b] Determined on the basis of the rotatory power of polymers obtained from α-olefins having different optical purity.

[c] On the basis of weight percent of the separated fraction, for a purely statistical copolymer of the two antipodes, monodispersed with respect to molecular weight, taking for $\overline{DP}$ of the statistical copolymer the $\overline{DP}_n$ of the polymer to be separated.

[d] R = i − C$_4$H$_9$.

[e] Calculated with respect to the total weight of the eluted polymer.

[f] Determined by cryoscopy in benzene solution.

[g] Determined in benzene at 25°C, using a Mechrolab apparatus Mod. 301 A.

[h] ARA = Aluminum Reduced–Activated.

TABLE 17-3

Comparison between Observed and Calculated Prevalence[a] of (S) Monomeric Units in the Fractions Having Positive Rotatory Power Obtained in the Chromatographic Separations of High Molecular Weight Polymers of Some Racemic α-Olefins in Fractions Having Optical Activity of Opposite Sign (24)

Polymer to be separated		$\bar{M}_v$[d]	Fraction in which (S) monomeric units are prevailing		Calculated prevalence[a] of (S) monomeric units in a random copolymer of the two antipodes (%)[c]	
Type	Fraction		Wt%[e]	Observed average (S) unit prevalence[a,b]	f	g
Poly-(R, S)-4-methyl-1-hexene[h]	Diethyl ether-soluble–ethyl acetate-insoluble	220,000	24.7	6.5	3	8
	Diisopropyl ether-soluble–diethyl ether-insoluble	95,000	32.1	5.5	3.5	10.5
	Isooctane-soluble–diisopropyl ether-insoluble	185,000	13.4	11	4	12
Poly-(R, S)-3,7-di-methyl-1-octene[i]	Diethyl ether-soluble–acetone-insoluble	35,000	68.4	25	<4	<14
	Diethyl ether-soluble–ethyl acetate-insoluble	180,000	41.4	42	2	7
Poly-(R, S)-3-methyl-1-pentene[h]	Diethyl ether-soluble–acetone-insoluble	6,600	24.8	16	13	—

[a] [(S) monomeric units − (R) monomeric units/(S) monomeric units + (R) monomeric units] × 100.

[b] Determined on the basis of optical activity of polymers obtained from monomer having different optical purity.

[c] On the basis of weight percent of the fraction for a purely statistical copolymer of the two antipodes, monodispersed with respect to molecular weight.

[d] Viscometric average molecular weight calculated by the equation $[\eta] = 1.93 \times 10^{-4} \bar{M}_v^{0.74}$ determined for polypropylene.

[e] With respect to the total weight of the eluted polymer.

[f] Calculated by assuming that the statistically monodispersed copolymer has a molecular weight equal to $\bar{M}_v$ of the polymer to be separated.

[g] Calculated by assuming that the statistically monodispersed copolymer has a molecular weight equal to 0.1 $\bar{M}_v$.

[h] Catalytic system: $TiCl_3$–$Al(i\text{-}C_4H_9)_3$.

[i] Catalytic system: $TiCl_4$–$Al(i\text{-}C_4H_9)_3$.

during copolymerization. They found that a copolymer of (R,S)-4-methyl-1-hexene and (R,S)-3-methyl-1-pentene (1 to 1 molar ratio and melting point of 204°C) had a helix identical to that of poly-(R,S)-4-methyl-1-hexene (a $\frac{7}{2}$ helix).

The position of the asymmetric carbon in the pendant group of the α-olefin is important (15, 19a,b). Stereoselectivity decreases sharply as the position changes from the α- to β-carbon and is virtually destroyed when the asymmetric carbon is in the γ-position (relative to the double bond).

Two papers clearly show that olefins bearing the asymmetric carbon in the γ-position do not undergo a stereoselective propagation.

Chiellini and Marchetti (19a) reported that the polymerization of racemic 5-methyl-1-heptene produced polymer chains, each of which contained both enantiomers in equal concentrations. They were not able to obtain optically active fractions by chromatographing the product on a column of poly-[(S)-3-methyl-1-pentene]. In control experiments, they synthesized isotactic poly-[(S)] and poly-[(R)-5-methyl-1-pentene] and demonstrated at least a partial separation of an artificial mixture of these under similar conditions. More recently, Carlini, Ciardelli, and Pini (19b) found that (S)-5-phenyl-1-heptene was polymerized to an amorphous polymer even though 5-phenyl-1-pentene gave a moderately crystalline product. One explanation suggested was a copolymerization of the (S) antipode with the (R) one, which was present up to 4 to 5% in the starting monomer. This copolymerization occurred because stereoselection is absent with α-olefins bearing the asymmetric carbon in the γ-position, and both antipodes entered the same chain.

Tables 17–2 and 17–3 compare the observed and calculated prevalence of (S) monomeric units in fractions obtained in the chromatographic separations of low and high molecular weight poly-α-olefins prepared with various conventional Ziegler–Natta catalysts.

III. Stereoelection

Pino and co-workers extended their investigation to Ziegler–Natta catalysts containing optically active metal alkyls with the objective of forming optically active polymers directly, that is, without chromatographic separations. Earlier, it was shown (as expected on the basis of then prevailing mechanistic schemes) that polymers prepared with optically active and inactive catalysts were identical if propylene (20) and styrene (21) were used as monomers. Backbone chain isomerism was definitely not affected by the use of optically active catalysts, such as tris[(S)-2-methylbutyl]aluminum + TiCl$_3$ (21).

Optically active catalysts became uniquely different from optically inactive catalysts only under certain conditions, and these are now discussed.

The position of the asymmetric carbon in the pendant group of the racemic olefin was found to be dominating. For example, optically inactive polymer was obtained from racemic 4-methyl-1-hexene with both the $Al(i\text{-}Bu)_3$–$TiCl_4$ catalyst and catalysts formed from $TiCl_4$ or $TiCl_3$ + tris[(S)-2-methylbutyl]aluminum, bis[(S)-2-methylbutyl]zinc, or (R)-2-methylbutyllithium (22). The small optical activity observed in the low molecular weight fractions was attributed to the asymmetric 2-methylbutyl group attached at the end of the polymer chain. This conclusion was supported by the observation that the rotatory power of the fractions decreased as the molecular weight of the polymer increased.

In comparison, if the α-olefin had the asymmetric carbon in the 3-position with respect to the double bond, for example in 3-methyl-1-pentene, then the polymer formed with the above optically active catalyst was optically active. Furthermore, the unreacted 3-methyl-1-pentene was also optically active. Both racemic (R,S)-3-methyl-1-pentene and (R,S)-3,7-dimethyl-1-octene were found effective by Pino, who designated this stereochemical process as stereoelection (Eq. 17–2). A stereoelective polymerization of a racemic α-olefin is a polymerization process in which one of the monomeric antipodes polymerizes at a higher rate than the other antipode.

$$
\begin{array}{ccc}
CH_2{=}CH & & \\
\quad | & \xrightarrow[\text{TiCl}_4]{\text{Zn}[(S)\text{-2-methyl-1-butyl}]_2} & \text{Optically active polymer} \\
*CH(CH_3) & & \\
\quad | & & + \qquad\qquad (17\text{–}2) \\
R & & \\
\end{array}
$$

(RS)-3-methyl-1-pentene
or
(RS)-3,7-dimethyl-1-octene

Optically active
unreacted monomer

The polymer and the unreacted monomer had opposite signs of rotation (Tables 17–4 and 17–5). The sign of rotation of the recovered monomer indicated that the enantiomer that had the same absolute structure as the optically active (S)-2-methyl-1-butyl groups present in the optically active catalyst was preferentially polymerized (Table 17–6). From the optical activity of the recovered monomers and conversions to polymer, Pino estimated that the (S) enantiomer was 1.2 to 1.7 times more active than the (R) enantiomer (Table 17–4). Later investigations showed the lower number to be more valid.

Pino found that stereoelection was not favored if the asymmetric carbon in the α-olefin was in the β-position. Also, if the asymmetric carbon of the metal alkyl was too far removed from the metal (in the γ- vs. β-position), stereoelection did not occur. Goodman and co-workers (22a) showed that the $LiAlX_4$–$TiCl_4$ catalyst did not polymerize (R,S)-3,7-dimethyl-1-octene

TABLE 17–4

Relative Polymerization Rate[a] of (S) and (R) Antipodes of Racemic α-Olefins in the Presence of (+)-Bis[(S)-2-methylbutyl]zinc/TiCl₄ Catalyst

Monomer	Run	Initial monomer (g)	Total polymer (g)	Conversion[b] (%)	Recovered monomer		Optical purity[c] (%)	Polymerized monomer optical purity[c] (%)	Relative polymerization rate[a]
					$[\alpha]_D^{25}$ (deg.)	$[\alpha]_D^{25}$ (deg.)			
3-Methyl-1-pentene	A	2.90	0.21	7.2	−0.07[d]	−0.75	2.0[f]	5.6[h]	1.7
3,7-Dimethyl-1-octene	B	3.50	0.51	14.6	∓0.20[e]	−0.28	1.7[g]	9.97[i]	1.2
	C	5.40	1.17	21.7	−0.36[e]	−0.52	3.2[g]	1.6[j]	1.3

[a] Moles of polymerized (S) monomer/moles of polymerized (R) monomer.
[b] Weight of the total unfractionated polymer/weight of initial monomer.
[c] Evaluated as optical purity of recovered monomer × (weight of recovered monomer/weight of polymerized monomer).
[d] l = 2 dm; c = 4.7 g/dl in methanol.
[e] l = 1 dm; homogeneous.
[f] $[\alpha]_D^{25}$ = 36.0° for the optically pure 3,7-dimethyl-1-pentene; (c = 4.65 g/dl in methanol).
[g] $[\alpha]_D^{25}$ = 16.3° for the optically pure 3,7-dimethyl-1-octene (homogeneous).
[h] ±20%
[i] ±10%
[j] ±5%

TABLE 17–5

Relationship between Absolute Structure of the Alkyl Group Present in the Catalyst and Absolute Structure of the Recovered Monomer and of the Optically Active Polymer

| | Absolute structure of the alkyl group present in the catalyst | Recovered monomer | | Polymer | |
| | | Rotation sign | Absolute structure | Rotation sign | Absolute configuration of the asymmetric C atom of the lateral chain |
Monomer					
3-Methyl-1-pentene	S	$-$	R	$+$	S
3,7-Dimethyl-1-octene	S	$-$	R	$+$	S

to an optically active polymer as did the ZnX_2–$TiCl_4$ catalyst, as shown in the tabulation below.

$LiAlX_4$	Stereoelection
$$X = (S)—CH_2—\overset{\displaystyle CH_3}{\underset{\displaystyle H^*}{C}}—CH_2—CH_3 \ (*\beta\text{-position})$$	yes
$$X = (R)—CH_2—CH_2—\overset{\displaystyle CH_3}{\underset{\displaystyle H^*}{C}}—(CH_2)_3—CH(CH_3)_2$$ $$(*\gamma\text{-position})$$	no

What seemed to be contradictory results were reported by Ciardelli and co-workers (23), who found that catalysts prepared from $TiCl_4$ and bis[(S)-3-methylpentyl]zinc had stereoelectivity comparable to that obtained with $TiCl_4$ and bis[(S)-2-methylbutyl]zinc. However, small amounts of (S)-3-methyl-1-pentene were detected in the purified nonpolymerized monomer. This suggested that this α-olefin was eliminated from the center by disproportionation [Ti-(S)-3-methyl-1-pentyl → TiH + (S)-3-methyl-1-pentene or by displacement with the olefin being polymerized] and then subsequently polymerized. The occurrence of a displacement reaction could account for the relatively low degree of stereoelectivity found in these polymerizations.

TABLE 17–6

Comparison between the Optical Activity of Different Fractions of Poly-4-methyl-1-hexene and Poly-3,7-dimethyl-1-octene (Run B) Obtained from the Racemic Monomer by $TiCl_4/(+)$-Bis[(S)-2-methylbutyl]zinc Catalyst

Fraction	Poly-4-methyl-1-hexene[e]				Poly-3,7-dimethyl-1-octene			
	(%)	$[\alpha]_D^a$ (deg.)	mp[b] (°C)	$[\eta]^c$ (dl/g)	(%)	$[\alpha]_D^{25}$ (deg.)	mp[b] (°C)	$[\eta]^c$ (dl/g)
Acetone-soluble	30.2	+15.2	n.d.[f]	n.d.[f]	49.2	+7.8	n.d.[f]	n.d.[f]
Acetone-insoluble–ethyl acetate soluble	20.0	+4.4	138–142	0.10	14.6	+12.6	145–150	0.22
Ethyl acetate insoluble–diethyl ether soluble	11.8	+4.1	196–199	0.21	8.0	+12.1[d]	n.d.[f]	0.28
Diethyl ether insoluble–diisopropyl ether soluble	5.1	+4.8	n.d.[f]	0.32	9.2	+17.5	198–202	n.d.[f]
Diisopropyl ether insoluble–isooctane soluble	32.9	+1.7	212–215	0.87	19.0	+18.0	242–248	0.91

[a] Determined at 20°C for the acetone soluble and acetone insoluble, ethyl acetate soluble fractions; at 50°C for other fractions.

[b] Determined by X-ray method.

[c] Determined at 120°C in tetralin.

[d] Determined at 50°C.

[e] P. Pino, F. Ciardelli, and G. P. Lorenzi, paper presented at International Symposium of Macromolecular Chemistry, Paris, July 1–6, 1963.

[f] Data not determined.

The observed optical activity in the polymers was shown not to be due to (S)-2-methylbutyl end groups (24). In contrast with the results obtained with 4-methyl-1-hexene, optical activity of the polymer fractions did not decrease as the molecular weight of the fractions increased. On the contrary, it was shown for poly-3,7-dimethyl-1-octene that optical activity increased as the molecular weight and isotactic content of the fraction increased (Table 17-6).

A small amount of stereoelection ($k_s/k_r \simeq 1.04$ to 1.12) was reported by Ciardelli (17, 18) to occur with optically inactive catalysts when the initial monomer mixture contained unequal concentrations of the antipodes, for example, $(S) > (R)$. They showed, however, that a displacement reaction involving the metal alkyl component and the monomers produced a small excess of optically active metal alkyl derived from the extra (S) enantiomer. This in effect produced the weak stereoelective catalyst.

When racemic (R,S)-3,7-dimethyl-1-octene and (S)-3-methyl-1-pentene were copolymerized, the nonpolymerized 3,7-dimethyl-1-octene had a slight negative optical activity (optical purity was 1 to 2%). Ciardelli and co-workers (17) suggested that the antipode of the racemic monomer having the same absolute configuration as the optically active comonomer was preferentially polymerized. One possible explanation for this is that, for steric reasons, the addition of (S)-3,7-dimethyl-1-octene to Ti-3-methyl-1-pentene was faster than if Ti-3,7-dimethyl-1-octene was the active center.

Ciardelli and co-workers (25, 26) also investigated the polymerization of racemic and optically active α-olefins (optical activity from 10 to 95%) with Al- and Zn-based–titanium chloride catalysts. Their data indicated that when the asymmetric carbon atom was in α-position with respect to the double bond, the polymerization was stereoelective. But the average rate constant for the polymerization of the antipode present in higher concentration was greater than the average rate constant of the other antipode. This was more pronounced for $Zn-i-Bu_2-TiCl_4$ as the catalyst, compared to $Al-i-Bu_3-TiCl_4$. When the asymmetric carbon was in the β-position, this was less evident, and for asymmetric carbon in the γ-position, the effect was absent or very small.

They also reported that, in the case of 4-methyl-1-hexene and 3,7-dimethyl-1-octene, the tacticity content seemed to increase if the monomer optical purity was larger than 50 to 70%. They suggested that an asymmetric induction by the monomer asymmetric carbon in determining the absolute configuration of the main chain asymmetric carbon atom seemed probable.

Carlini, Ciardelli, and co-workers (14) found that stereoelectivity remained practically constant or increased slightly with increasing conversions. It was affected by the nature of the transition metal halide and the nature of the alkyl group in the optically active organometallic compound of the catalyst.

TABLE 17-7

Degree of Separation^a Obtained in the Chromatography of Polymers Prepared by Polymerizing Racemic α-Olefins Having Different Structures with TiCl$_4$ or TiCl$_3$–Al(i-C$_4$H$_9$)$_3$ Catalytic Systems (24)

Monomer CH$_2$=CH—(CH$_2$)$_n$—CH—R | CH$_3$		Starting polymer	Degree of separation^a
n	R		
0	—(CH$_2$)$_3$—CH—CH$_3$ | CH$_3$	—^{b}	37.6
0	—C$_2$H$_5$	—^{c}	8.9
1	—C$_2$H$_5$	—^{b}	5.1
2	—C$_2$H$_5$	—^{c}	0

^{a} Calculated as

$$(2 \times 10^2/\text{Wt}) \sum_{i=1}^{m} ([\alpha]_i/[\alpha]^\circ)w_i$$

where m = number of eluted fractions having optical activity of opposite sign to that of the supporting polymer; Wt = total weight of eluted polymer: w_i = weight of the fraction having $[\alpha]_i$, and $[\alpha]^\circ$ = specific rotatory power measured at the same λ and temperature of $[\alpha]_i$ of the polymer of the corresponding monomer with high optical purity (>90%) having stereoregularity similar to that of the fraction to be separated.

^{b} Ethyl acetate-insoluble–diethyl ether-soluble fraction.
^{c} Acetone-insoluble–diethyl ether-soluble fraction.

Table 17–7 shows the degree of chromatographic separation as dependent on the structure of the pendant group bearing the asymmetric carbon (15). Table 17–8 shows that stereoselectivity and stereoelectivity depend closely on the structure of the pendant group (24).

IV. Proposed Mechanisms

Mechanisms have been proposed in which the stereoselective and stereoelective propagations take place at alkylated transition metal species or at complexes of these alkylated transition centers and the base metal alkyl molecules. Subsequent to the discovery of this unique isotactic propagation,

TABLE 17-8

Influence of Monomer Structure, Type of Organometallic Compound, and Transition Metal Compound Used in the Preparation of the Catalyst on the Stereoselectivity and Stereoelectivity in the Polymerization of Some Racemic α-Olefins (24)

Monomer

$$CH_2{=}CH{-}(CH_2)_n{-}\underset{\underset{\textstyle CH_3}{|}}{CH}{-}R$$

n	R	Stereoselectivity[a]			Stereoelectivity[b]		
		$TiCl_4{-}ZnR'_2$	$TiCl_n{-}AlR'^e_3$	$VCl_4{-}AlR'_3$	$TiCl_4{-}ZnR^*_2$	$TiCl_4{-}AlR^*_3$	$TiCl_4{-}BeR^*_2$
0	$-(CH_2)_3{-}\underset{\underset{\textstyle CH_3}{|}}{CH}{-}CH_3$	+[c]	++	n.d.	++	(+)[d]	~0
0	$-C_2H_5$	+[c]	++	n.d.	++	(+)[d]	n.d.
1	$-C_2H_5$	+	+	+	~+[e]	~+[d]	n.d.

[a] $R' = -CH_2{-}\underset{\underset{\textstyle CH_3}{|}}{CH}{-}CH_3$

[b] $R^* = -CH_2{-}\overset{\overset{\textstyle CH_3}{|}}{\underset{\underset{\textstyle H}{|}}{C^*}}{-}C_2H_5$

[c] Evidenced by stereoelective copolymerization.

[d] Extremely low.

[e] $n = 3, 4$.

Pino attributed it to the growth at Al–C bonds in Natta's mirror image bimetallic centers (1) (Fig. 15–5).

In 1965, Pino (3) formulated a generalized scheme using monometallic and bimetallic structures III and IV, respectively.

$$
\begin{array}{cc}
\text{III} & \text{IV}
\end{array}
$$

Structure III (monometallic):

```
            R*
            |
       CH₂—CH
            |
            Cl*
            /
Cl—Ti←—CH₂
  / \   ‖
 Y   Cl  CH
         |
         R*
```

Structure IV (bimetallic):

```
        R*
        |
ClₙTi←—CH₂----Me
        ‖      |
        CH----CH₂
        |      |
        R*    *CH—R*
               ⁞
```

III	IV
Monometallic center where $Y = Cl$ or alkyl group; R^* = pendant group containing asymmetric carbon	Bimetallic center where Me = base metal

Pino suggested that stereoselection could occur at structure III if this center was asymmetric. He recognized that if $Y = Cl$, asymmetry would exist by virtue of the different arrangement of $Y = Cl$ and the Cl trans to it in the crystal lattice. But a preference was shown for $Y = $ an alkyl group derived from the metal alkyl. In other words, the center was a doubly alkylated Ti, but polymerization occurred only at one of the alkylated positions. Stereoselection took place because each of these asymmetric centers could complex and polymerize only one of the monomeric antipodes.

If $Y = $ 2-methylbutyl in structure III, there exist two asymmetric carbons in the β-position with respect to the transition metal, one in the growing chain and the other one in the low molecular weight alkyl group. According to Pino and co-workers (ref. 3 and later papers), preferential activation of one monomeric antipode at the transition metal center and a stereoelective polymerization should occur only if the asymmetric induction by the two asymmetric carbons is cooperative.

According to the second hypothesis that Pino considered, the α-olefin is adsorbed on the transition metal at structure IV, and the polymer chain grows on the other metal forming the center. The asymmetry of this catalyst center was attributed to the presence of asymmetric alkyl groups (R^*) bound to the transition metal (see structure IV). In the alternate model, which he rejected, asymmetry arose from the special arrangement of two metal atoms and of the last carbon atom of the growing polymer chain on the surface of the solid catalyst (structures V and VI).

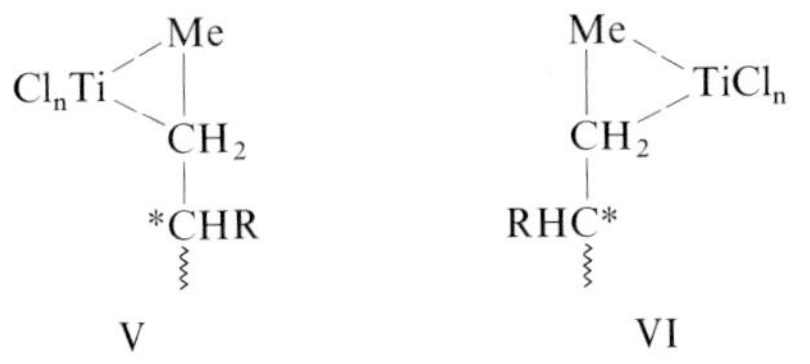

$$\underset{V}{\overset{Cl_nTi}{\underset{\underset{{}^*CHR}{\overset{|}{CH_2}}}{\overset{Me}{\diagup}}}}\qquad\qquad\underset{VI}{\overset{Me}{\underset{\underset{RHC^*}{\overset{|}{CH_2}}}{\diagdown TiCl_n}}}$$

Stereoelection at structure IV was attributed to the absolute configuration of R* (S)-2-methylbutyl bound to the transition metal at which the α-olefin is complexed. According to this view, the monomeric antipode that has the same configuration as R* is preferentially complexed and polymerized at this center.

The hypothesis that stereoelection occurs because reduction of $TiCl_4$ with an optically active metal alkyl produces an excess of one mirror image center was considered and rejected by Ciardelli, Pino, and co-workers (25). In structure III, Y would be Cl, according to this mechanism. A mechanism in which Y = the alkyl group containing the asymmetric carbon was still preferred on the basis of a definite relationship between the number of (S)-2-methylbutyl groups present as end groups of the polymer and stereoelectivity.

In 1967 and 1970, Boor (5, 6) proposed that stereoselection and stereo-election propagations can take place at alkylated titanium centers of the type proposed by Cossee and Arlman and the type proposed and rejected by Pino and co-workers (Y = Cl in structure III above.)

In the Cossee–Arlman type centers, the ligand environment is not symmetrical. Each Ti center will have a mirror image counterpart, and the α-olefin complexes in one of the two possible configurations at each of these sites. Only one complexing mode is permitted because Cl-1 presents a smaller steric repulsion than Cl-2, and the pendant group of the complexing α-olefin prefers to go over it rather than over Cl-2.

Figure 17–1 shows these centers; the Cl ligands are numbered identically to Fig. 15–8, which shows in model form the Cossee–Arlman centers. In general, Fig. 17–1 can fit other mechanistic models that have similar asymmetric character.

Fig. 17–1. Mirror image $RTiCl_4$ centers. R = alkyl group derived from a metal alkyl. Vacant octahedral position through which (R) or (S) enantiomer can potentially complex to Ti center.

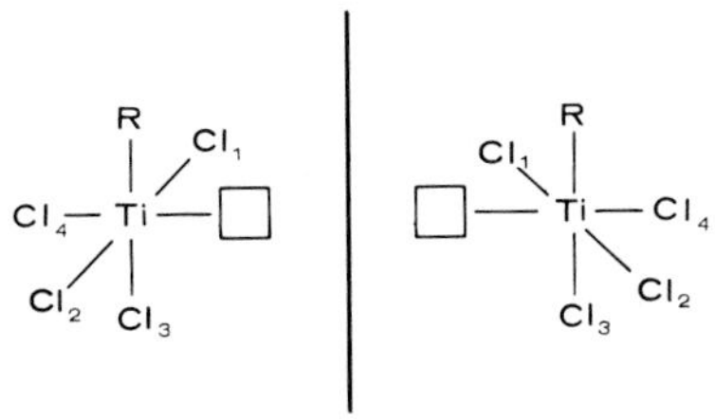

Both simple and branched α-olefins, propylene and 4-methyl-1-hexene, for example, are polymerized at these centers to isotactic chains, each of which contains prevailingly d or l units. Because the number of left-handed and right-handed centers is identical, the same number of d and l chains will be formed.

But because of their greater bulkiness, branched α-olefins will have a greater steric interaction with the Cl-1 ligand of the Ti center relative to propylene. Boor speculated that this steric interaction can explain stereoselection and stereoelection in the following way.

Stereoselection took place with an optically inactive catalyst because the (R) and (S) enantiomers (antipodes) were preferentially complexed and polymerized at the Mr–R and Ms–R centers, respectively, as shown in Fig. 17–1. The mirror image centers were arbitrarily designated by r and s to show whether they polymerize (R) or (S) enantiomer, respectively. As an illustration, when the optically inactive catalyst Al-i-Bu$_3$/TiCl$_4$ was used, equal numbers of Mr-i-Bu and Ms-i-Bu centers were formed. According to this proposed scheme, the (R) enantiomer complexed and predominantly polymerized at Mr-i-Bu centers at the same rate that the (S) antipode polymerized at the Ms-i-Bu centers, as shown in Eq. 17–3.

$$\text{Mr-}i\text{-Bu} + n(R) \text{ enantiomer} \xrightarrow{k_r}$$

$$\text{Mr-}[(R) \text{ enantiomer}]_n\text{-}i\text{-Bu}$$

$$\text{Ms-}i\text{-Bu} + n(S) \text{ enantiomer} \xrightarrow{k_s} \qquad (17\text{–}3)$$

$$\text{Ms-}[(S) \text{ enantiomer}]_n\text{-}i\text{-Bu}$$

where $k_r = k_s$ and M = Ti.

To explain these preferred propagations, it was suggested that the steric repulsion between Cl-1 and the pendant group of the α-olefin favors complexing only that enantiomer at a metal center, which places the H of the asymmetric carbon of the pendant group over the Cl-1 ligand. For this reason, the (R) enantiomer preferentially complexes at Mr–R centers. If the (S) enantiomer had complexed at the Mr–R center, the CH$_3$ of the asymmetric carbon of the pendant group would have to face Cl-1 (by facing was meant closest contact.) Boor proposed that this complexing mode α-olefin $(S) \rightarrow$ Mr–R presented a greater steric repulsion than the complexing mode α-olefin-$(R) \rightarrow$ Mr–R. The same reasoning was offered to explain why the (S) enantiomer complexes and polymerizes more easily at Ms–R sites than does the (R) enantiomer.

A mixture of isotactic polymer containing an equal number of chains consisting of either (R) or (S) enantiomer is formed because the concentrations of the mirror image centers, their activity, and the enantiomers are

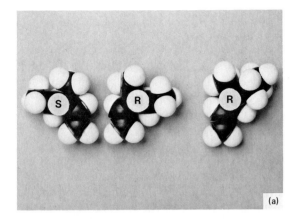

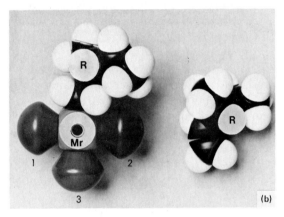

Fig. 17-2. Stereoselection of 3-methyl-1-pentene. (a) Models of (R,S)-3-methyl-1-pentene and (R)-4-methyl-1-hexene. (b) Models of Mr–Pn center and (R)-3-methyl-1-pentene. According to this mechanistic model, stereoselection occurs because the (R)-3-methyl-1-pentene and (S)-3-methyl-1-pentene enantiomers can polymerize only at Mr–Pn and Ms–Pn centers, respectively. Because [Mr–Pn] = [Ms–Pn], the number of chains consisting of the (R) enantiomer is same as the number of chains consisting of the (S) enantiomer, and the product is therefore optically inactive. The (R) enantiomer is preferentially complexed and polymerized at the Mr–Pn center, because the H of the asymmetric carbon of the α-substituent (which bears the letter R in these models) faces Cl-1 and, in this way, the steric repulsion between the branched substituent and Cl-1 is minimized (lower). For the same reason, the (S) enantiomer complexes preferentially at the Ms–Pn centers.

equal (Ms–R = Mr–R, $(S) = (R)$, and $k_s = k_r$). Figure 17–2 shows this scheme in model form.

Boor (27) supported his view by citing the low reactivity of branched α-olefins relative to linear monomers. Branched α-olefins become increasingly difficult to polymerize as the branching is brought closer to the double bond as the number and size of the attached groups increase. For the olefin series $CH_2{=}CH{-}R$, the following decreasing order was found: $R = H > CH_3 > CH_2CHR^1R^2 > CHR^1CH_2R^2 \gg CR^1R^2R^3$. The fully substituted carbon must be removed by at least three methylenes from the double bond before the olefin becomes reasonably active. For supporting detailed data, the reader is urged to read Chapter 19, Section III.

These results are easy to see from molecular models of these α-olefins and the center. In the complexed form (Mr–Pn + 3,3-dimethyl-1-butene), a methyl of this olefin would have to face Cl–1. Even when one or two methylenes are interposed between the double bond and the $(CH_3)_3C$ group, this CH_3 vs. Cl-1 steric interaction is present.

This alternate argument that steric repulsions between the incoming and last added α-olefin are the main barrier was not supported by copolymerization data. When 3,3-dimethyl-1-butene and ethylene were copolymerized (the former serves as reaction solvent), only 2 wt% of the branched α-olefin was found in the copolymer. Even this had an isomerized form, indicating it arose from a cationic polymerization.

The view that stereoselection arises from the asymmetric nature of the active center rather than from the structural features of the last added α-olefin unit has been supported by Pino, Ciardelli, and co-workers (3, 15, 28, 29). Very convincing evidence was recently reported that involved the copolymerization of racemic 3,7-dimethyl-1-octene and ethylene (28, 29). The product obtained was separated by chromatography into fractions having optical activity of the opposite sign. The degree of separation was found to be the same as observed for the isotactic homopolymer of racemic 3,7-dimethyl-1-octene.

To explain why stereoelection occurred when the optically active metal alkyl bis[(S)-2-methylbutyl]zinc was used, the proposal was made that the first additions of (S) and (R) enantiomers to Ms-(S)-2-methylbutyl and Mr-(S)-2-methylbutyl centers did not take place at the same rate, as indicated in the tabulation.

Initial center	Enantiomer	Relative rate of first addition
Ms-(S)-2-methylbutyl	(S)	Fast
Mr-(S)-2-methylbutyl	(R)	Slow

The same situation would hold had bis[(R)-2-methylbutyl]zinc been used instead, except now the first addition of (R) enantiomer to Mr-(R)-2-methyl-butyl center would be facile, while the (S) enantiomer would add only

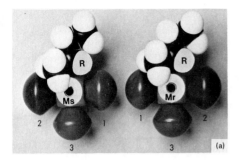
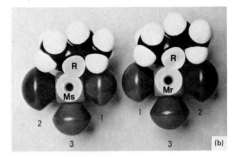

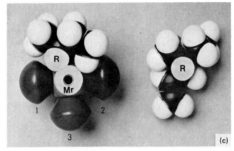

Fig. 17-3. Stereoelection of 3-methyl-1-pentene. (a) Mr–Pn and Ms–Pn centers with the CH_3 and the H facing the vacancy. (b) Same centers after slight rotations. (c) Model showing easy first addition of (R) of 3-methyl-1-pentene. Stereoelection takes place only when the metal alkyl is optically active, for example $Zn[(R)$-2-methyl-1-butyl$]_2$, while stereoselection (see Fig. 17-2) occurs with conventional metal alkyls such as Al-i-Bu$_3$. According to this model (see also Fig. 17-2), the (R) enantiomer polymerizes to a greater extent when it and the 2-methyl-1-butyl have the same configuration, because the first addition of the (R) enantiomer to the Mr-(R)-2-methyl-1-butyl center takes place more easily (see c) than the first addition of (S) enantiomer to the Ms-(R)-2-methyl-1-butyl center. The latter complexing mode encounters a greater steric repulsion.

slowly to Ms-(R)-2-methyl-1-butyl. Figure 17-3 shows this scheme in model form.

Figure 17-3 (a) shows the Ms-(R)-2-methylbutyl and Mr-(R)-2-methyl-butyl sites with the H and CH_3 groups on the side of the vacancy. Since steric interactions with CH_3 prevent complexing of the α-olefin at the vacancy, a rotation must take place to remove the CH_3 from the vicinity of the vacancy. This is easily done by a slight rotation about C_1 and C_2 but at the expense of bringing the methyl group of the ethyl over the *trans*-Cl (center). In a Ms-(R)-2-methylbutyl site, the methyl of the ethyl group comes over Cl-1, while in the Mr-(R)-2-methylbutyl site, it comes over the Cl-2 ligand of the center. The lower picture in Fig. 17-3 shows that the methyl that originally blocked the vacancy was completely removed from the vacancy and the Cl-1 ligand when the site was Mr-(R)-2-methylbutyl. At the same time, the methyl of the ethyl group was brought close to the Cl-2 ligand of the center but did not block the vacancy. Thus, the complexing of the (R)-3-methyl-1-pentene molecule was allowed.

This was not the case with the Ms-(R)-2-methylbutyl site. To remove the CH_3 that was on the side of the vacancy (Fig. 17–3, a), it was necessary to bring the methyl of the ethyl over Cl-1. This, however, was the ligand over which the substituent of the incoming α-olefin was placed. The increased steric encumbrance about Cl-1 made the complexation of the (S) enantiomer more difficult. For this reason, the first addition of (S) enantiomer (S)-3-methyl-1-pentene to a Ms-(R)-2-methylbutyl site was difficult. Once a first addition took place, however, the subsequent polymerizations were normal and took place at the same rate as additions of (R) enantiomer to Mr–Pn sites.

Boor suggested that the importance of a metal alkyl or of doubly alkylated Ti in promoting stereoselection could be ascertained by use of metal alkyl-free catalysts instead of the Ziegler–Natta catalysts. If stereoselection occurred, then the need of a base metal alkyl is removed. Since it would be unlikely that doubly alkylated Ti would be present in the metal alkyl-free systems (the number of sites is small), this finding would also weigh heavily against the doubly alkylated center mechanism.

References

1. P. Pino, F. Ciardelli, G. P. Lorenzi, and G. Natta, *J. Am. Chem. Soc.* **84**, 1487 (1962).
2. G. Natta, P. Pino, and G. P. Lorenzi, Italian Patent 10/3490, May 13, 1960.
3. P. Pino, *Adv. Polym. Sci.* **4**, 393–456 (1965).
4. F. Schué, *Rev. Gen. Caoutch. Plast., Ed. Plast.* **4**, 261–267 (1967).
5. J. Boor, *Macromol. Rev.* **2**, 115–268 (1967).
6. J. Boor, *Ind. Eng. Chem., Prod. Res. Dev.* **9**, No. 4, 437–456 (1970).
7. T. Tsuruta, *J. Polym. Sci., Part D* **6** pp. 180–246 (1972).
8. M. Sheldon, *Nuova Chim.* **46**, 47–56 (1970).
9. P. Pino, G. P. Lorenzi, and L. Lardicci, *Chim. Ind.* (*Milan*) **42**, 712 (1960).
10. P. Pino, G. P. Lorenzi, and L. Lardicci, *J. Polym. Sci.* **53**, Issue 158, 340 (1961).
11. P. Pino, F. Ciardelli, G. P. Lorenzi, and G. Montagnoli, *Makromol. Chem.* **61**, 207 (1963).
12. P. Pino, G. P. Lorenzi, D. P. Lardicci, and F. Ciardelli, *Vysokomol. Soedin.* **3**, 1597 (1961).
13. P. Pino, G. Montagnoli, F. Ciardelli, and E. Benedetti, *Makromol. Chem.* **93**, 158 (1966).
14. C. Carlini, H. Bano, and E. Chiellini, *J. Polym. Sci., Part A-1* **10**, 2803 (1972); C. Carlini, F. Ciardelli, and P. Pino, *Makromol. Chem.* **119**, 244 (1968); F. Ciardelli, H. Bano, and C. Carlini, *Chim. Ind.* (*Milan*) **52**, No. 1, 81 (1970).
15. P. Pino, F. Ciardelli, and G. Montagnoli, *J. Polym. Sci., Part C* **16**, 3265 (1967).
16. F. Ciardelli, E. Benedetti, and O. Pieroni, *Makromol. Chem.* **103**, 1 (1967).
17. F. Ciardelli, E. Benedetti, G. Montagnoli, L. Lucarini, and P. Pino, *Chem. Commun.* p. 285 (1965).

18. F. Ciardelli, C. Carlini, and G. Montagnoli, *Macromolecules* **2**, 296 (1969).
19. M. Goodman and T. Furuyama, private communication (1969).
19a. E. Chiellini and M. Marchetti, *Makromol. Chem.* **169**, 59 (1973).
19b. C. Carlini, F. Ciardelli, and D. Pini, *Makromol. Chem.* **174**, 15 (1974).
20. G. I. Fray and R. Robinson, *Tetrahedron* **18**, 261 (1962).
21. S. Murahashi, S. Nazakura, and S. Takeguchi, *Bull. Chem. Soc. Jpn.* **33**, 658 (1960).
22. P. Pino, F. Ciardelli, and G. P. Lorenzi, *J. Polym. Sci., Part C* **4**, 21 (1963).
22a. M. Goodman, K. J. Clark, M. A. Stake, and A. Abe, *Makromol. Chem.* **72**, 131 (1964).
23. F. Ciardelli, E. Benedetti, G. Montagnoli, L. Lucarini, and P. Pino, *Chem. Commun.* No. 13, p. 285 (1965).
24. P. Pino, F. Ciardelli, and G. P. Lorenzi, *Makromol. Chem.* **70**, 182 (1964).
25. F. Ciardelli, G. Montagnoli, D. Pini, O. Pieroni, C. Carlini, and E. Benedetti, *Makromol. Chem.* **147**, 53 (1971).
26. G. Montagnoli, D. Pini, A. Lucherini, F. Ciardelli, and P. Pino, *Macromolecules* **2**, No. 6, 684 (1969).
27. J. Boor, *Pure Appl. Chem.* **8**, 57 (1971); *Pap., Int. Congr. Pure Appl. Chem., 23rd, 1971* Vol. **8**, pp. 57–68 (1971).
28. F. Ciardelli, P. Locatelli, M. Marchetti, and A. Zambelli, *Makromol. Chem.* **175**, 923 (1974).
29. O. Pieroni, G. Stigliani, and F. Ciardelli, *Chim. Ind. (Milan)* **52**, No. 3, 289 (1970).

18

Kinetics

I. Introduction

The elucidation of the basic kinetic parameters that characterize the Ziegler–Natta polymerizations is of prime importance. Only then can the origin of the obtained polymerization rate curves and the nature of the homo- and copolymers that are formed in the presence of these catalysts be understood.

A unified, universally accepted kinetic picture has not yet emerged, in spite of many excellent experimental and theoretical studies that have been done on this elusive catalyst. A kinetic phenomena exists that is the result of a delicate interplay between a number of operational factors that can be controlled during the synthesis of the catalyst and its use in polymerization. Such factors include the choice of catalyst components, their absolute and relative concentrations, temperature of catalyst preparation and use, etc. Of course, these operational factors decide the nature and magnitude of basic kinetic parameters: structure, concentration and stability of active centers, lifetime of the growing polymer chains, the morphology of the catalyst–polymer particle in a heterogeneous polymerization, etc.

While the effects of operational factors have been fairly well identified, the evaluation of basic parameters has been subject to considerable interpretation and hence, debate. And therein lies the problem. Because of the experimental difficulties in measuring the value or contribution of individual parameters, their true effect in a polymerization is generally not well established. Often a worker assumes, without firm evidence, the contribution of a parameter to remain constant or negligible in order to simplify the kinetic treatment.

The problem is compounded by the fact that we are trying to understand not just one but many different Ziegler–Natta catalysts. What is learned from a study of one catalyst may not necessarily be applicable to the understanding of another. The reader should be aware that the reported kinetic data (here and in other reviews) has often been evaluated under these conditions and, for this reason, may differ from worker to worker.

This chapter first focuses briefly on the basic kinetic results: the nature of the polymerization rate curve and the nature of the polymerization product. In effect, the rest of the chapter attempts to elucidate these kinetic results. Section III discusses proposed kinetic schemes and models, while Sections IV and V describe the operational factors and basic parameters, respectively, that may account for the kinetic results.

For more detailed descriptions and kinetic analysis, the reader should consult an excellent book by Keii (1) and earlier reviews by Natta and Pasquon (2), Reich and Schindler (3), Boor (4), and Burfield, McKenzie, and Tait (5–5b). Keii's book is especially useful if comprehension and mathematical descriptions are sought for the different parts of the rate curves, etc. The review by Natta and Pasquon describes the early basic kinetics studies of that school and should be consulted for their earlier original papers.

II. Basic Kinetic Results

The practical objective of these kinetic studies is to enable us to understand and better control homopolymerizations, copolymerizations, and block polymerizations in the presence of Ziegler–Natta catalysts. Only the first will be described here since the kinetic features of copolymerizations and block polymerizations are discussed in Chapters 20 and 21, respectively.

With respect to homopolymerization, two important features are described in this section: the shape of the polymerization rate curve and the character of the polymer product, including molecular weight, stereoregularity, etc.

A. THE POLYMERIZATION RATE CURVE

A plot of the measured rate of polymerization of a monomer vs. the time of polymerization constitutes a typical polymerization rate curve. Rate curves having different shapes have been found for Ziegler–Natta polymerizations. Two types of rate curves are identified for discussion purposes: a constant-rate type and decaying-rate type.

Natta and Pasquon (2) recognized the constant-rate-type curve for the polymerization of propylene with the $AlEt_3-\alpha TiCl_3$ catalyst (Fig. 18–1). The rate of polymerization gradually increased to a constant value that was maintained for at least 27 hours. The initial period was designated as an "induction period."

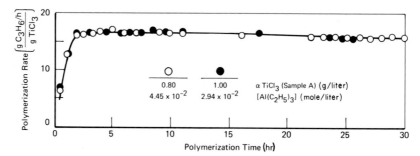

Fig. 18–1. Constant rate curve (2). The characteristic behavior of propylene polymerization rate is plotted vs. polymerization time. The data was obtained by operating at constant pressure ($P_{C_3H_6}$ = 1450 mm Hg, temp. = 70°C) with a catalytic system containing αTiCl$_3$ crystals having initial sizes between 1 and 10 μm (αTiCl$_3$ sample A).

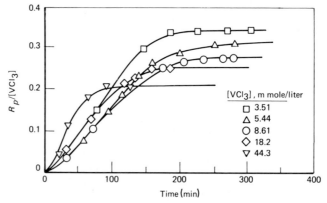

Fig. 18–2. Variation of rate of polymerization with time for different VCl$_3$ concentrations (5–5b). [4MP-1] = 2.00 mole/liter; [Al(i-Bu$_3$)] = 37.0 mmole/liter.

Later workers identified other examples of constant polymerization rate curves (5–5b). For example, Fig. 18–2 shows a constant rate curve for 4-methyl-1-pentene polymerized with Al-i-Bu$_3$–VCl$_3$. Note that the time period is about 4 to 5 hours.

A decaying-type polymerization rate curve was identified in a number of investigations (Table 18–1) (2, 5–11). Figure 18–3 shows an example for the polymerization of propylene with the AlEt$_3$–TiCl$_3$AA catalyst (6). As in the constant polymerization rate curve, there is an induction period before a maximum rate is reached, and then the rate continues to decline with time. Because, in most cases, the polymerizations were not carried out under similar conditions, direct comparisons are not possible. For example, care must be taken to note the times of polymerization, which vary from 1 to 2 hours up to 30 hours.

TABLE 18–1

Selected Examples of Constant and Decaying Polymerization Rate Curves

Catalyst	Monomer	Polymerization		Reference
		°C	hr	
Constant polymerization rate curves				
$AlEt_3$–$\alpha TiCl_3$	Propylene	70	30	2
Al-i-Bu_3–VCl_3	4-Methyl-1-pentene	30	5	5–5b
Decaying polymerization rate curves				
$AlEt_3$–$TiCl_3AA$	Propylene	44	2	6
$AlEt_2Cl$–VCl_4	Ethylene	30	2	7
$AlEt_2Cl$–$TiCl_3AA$	Propylene	50	6	8
$AlEt_3$–$\alpha TiCl_3$	Propylene	—	1–2	9
$AlEt_2Cl$–$\alpha TiCl_3$	Propylene			9
$AlEtCl_2$–$\alpha TiCl_3$-Donor	Propylene			9
$ZnEt_2$–$TiCl_3$	Propylene			10
$AlEt_3$–$VOCl_3$	Propylene			11

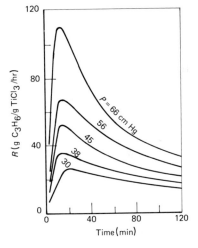

Fig. 18–3. Initial polymerization rate at 44°C; 2 g/liter of $TiCl_3$, [Al] = 15 mmole/liter (6).

Natta and Pasquon (2) identified another type of a polymerization rate curve that has features of both. When the same $\alpha TiCl_3$ was mechanically crushed just before the polymerization, the rate of polymerization increased quickly above the steady-state rate but immediately declined to the steady-state value (curves 1 and 2 in Fig. 18–4). This figure also shows the constant-type rate curve for comparison. Natta, Pasquon, and co-workers (2, 12) found that when $AlEt_2Cl$ was used in place of $AlEt_3$, a similar maximum in the curve was not obtained.

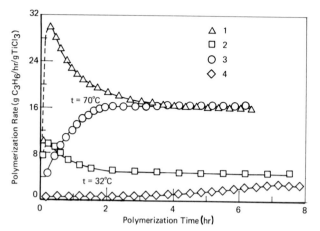

Fig. 18–4. Effect of previous physical treatments on a sample of $\alpha TiCl_3$ on the propylene polymerization rate at constant pressure and temperature (temp. = 70°C, $P_{C_3H_6}$ = 1,450 mm Hg). 1 and 2: Ground $\alpha TiCl_3$ (sample A) (sizes $\leqslant 2$ μm). 3 and 4: Unground $\alpha TiCl_3$ (sample A) (sizes within 1 to 10 μm) (2).

The three rate curves shown above are typical. But the actual rate vs. time relationship will vary considerably according to the catalyst–monomer system. Thus, induction periods (or the early part of the curve) can vary from several minutes to hours. Also, while a few systems hold a constant stationary rate for a long time, as shown in Fig. 18–1, others maintain a constant rate for only short periods before a decline is observed. Some catalyst–monomer systems show a faster decline in the polymerization rate than others.

What are the origins of the different shapes of these polymerization rate curves? Here we can offer only partial answers. For heterogeneous Ziegler–Natta catalysts, it is necessary to consider, in part, the rather complex changes that take place during the growth of the polymer particle from the catalyst particle (Chapter 8 is entirely devoted to this subject). An apparently simpler situation should exist if one is dealing with a soluble catalyst, that is, the active center is part of a simple molecular metal complex. But this may be deceptive because in the latter case bimolecular interactions become possible between reactive species, leading to termination of the center. The reader can easily generate the potential origins of the rate curves by considering the various stages of the polymerization. What are the suggestions in the literature?

1. The Induction Period

This part of the curve has also been described as the "build-up" or "adjustment" period. The initial increase in the rate curve may require only minutes, and at other times hours, depending on the catalyst–monomer system. A number of events can occur: (1) the metal alkyl and transition metal surface react to form active centers; (2) monomer diffuses from the gas cap to the liquid phase and finally to the active centers; (3) polymerization takes place; and (4) in a heterogeneous catalyst, the catalyst particle begins to break up, exposing new surfaces and allowing new centers to form, etc. The temperature in the vicinity of the center may become much higher than that of the reaction medium, and if heat transfer is insufficient, chemical reactions and physical changes characteristic of higher polymerization temperatures may occur to some extent.

Natta and Pasquon (2) demonstrated that the effect of mechanically grinding the αTiCl$_3$ was to decrease the induction period. Similarly, when unground αTiCl$_3$ was used, they found that the time required to reach three-fourths of the steady-state rate varied inversely with the value of the steady-state rate (see Fig. 18–5). In effect, the higher the steady-state rate, the less time was required to reach it. Factors such as higher temperature, higher monomer concentration, and higher rate of formation of active centers can be used to increase the value of steady-state rate. Apparently, the break up of catalyst into small fragments is an integral part of this induction

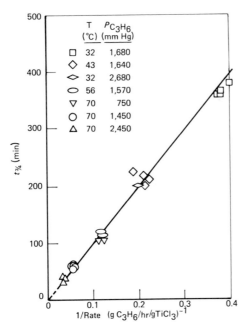

Fig. 18–5. Dependence of the $t_{3/4}$ index of the adjustment period on the reciprocal of the propylene polymerization rate in steady-state conditions (2). Tests performed with unground αTiCl$_3$ (sample A).

period. (The author feels that the maximum obtained with ground $\alpha TiCl_3$ may be due to higher temperatures at individual centers. The grinding process exposed extra centers and caused the localized temperatures to be considerably higher than present under stationary rate conditions because of the higher activity possessed by these more exposed centers.)

Shorter induction periods were observed with $TiCl_3$'s containing $AlCl_3$, such as $TiCl_3 \cdot xAlCl_3$. Possibly, the mode of break up of these $TiCl_3$ followed different paths, since their morphologies were different from the $\alpha TiCl_3$ type. There is strong evidence that the primary particles which make up the $TiCl_3 \cdot xAlCl_3$ particles are smaller than those which make up the $\alpha TiCl_3$ particle. They also have higher surface areas and are more active (see Chapter 8).

2. Constant-Rate Period

Only a few catalyst–monomer systems have been reported that give this shape curve. The $AlEt_3$–$\alpha TiCl_3$ system reported by Natta and co-workers (2) is a relatively low activity catalyst. Natta concluded that when the steady-state rate was reached (also called a stationary rate), the system achieved an equilibrium state, that is, the number of centers remainded constant. For the $AlEt_3$–$\alpha TiCl_3$, the steady rate R was maintained for over 27 hours. The value of R depended on the temperature of polymerization, monomer concentration, catalyst concentration, etc., and was altered by changing these reaction variables.

Figure 18–6 shows two polymerizations that were conducted at two temperatures (70° and 43°C) and then allowed to remain dormant (at 70° and 20°C, respectively) after each attained the steady-state rate. Upon continuation of both polymerizations at 70°C, identical steady-state rates were obtained, equal to the earlier polymerization done at 70°C. (It is somewhat

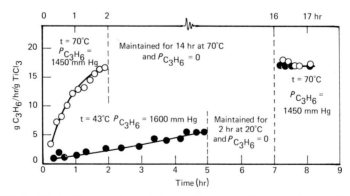

Fig. 18–6. Effect of the variations of the steady-state conditions on the propylene polymerization rate (2).

surprising that the rate of the 43°C run, after two hours dormancy, rose so quickly to the 70°C level; it took about $1\frac{3}{4}$ hours initially to go from a rate of 5 to 16 at 70°C.)

3. Decaying Period

Most of the catalyst–olefin systems are of the decaying type, in which the rate curve reaches a maximum and then continues to decline. The rate of decline varies according to the catalyst–monomer system. It appears that at increased time of polymerization, the decline in rate becomes smaller.

The decline in polymerization rate has been attributed to a number of factors: (1) a decrease in a number of active centers; (2) a lowering of activity of individual centers due to structural changes; and (3) a lowering of activity of individual centers due to a shortage of monomer in the vicinity of the center. This shortage is said to occur when the centers become encapsulated in polymer, and the monomer has difficulty diffusing from the liquid through the polymer to the centers.

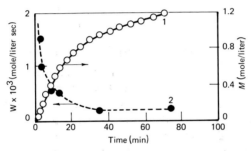

Fig. 18–7. Polymerization of butadiene in benzene solution at 20°C. $[C_4H_6] = 0.5$ mole/liter; $[TiI_2Cl_2] = 1$ mmole/liter; Al:Ti = 6. Curve 1, amount of polymerized monomer M; curve 2, rate of polymerization.

Bressler and co-workers (13) have shown for the polymerization of butadiene with the Al-i-Bu$_3$–TiCl$_2$I$_2$ catalyst that catalyst activity varied considerably during the period of polymerization. The rate of polymerization increases rapidly to a maximum and then decreases with time to a constant value. The final curve was described by a second-order equation (Fig. 18–7).

B. POLYMERIZATION PRODUCT

The "as obtained" polymer consists of a mixture of polymer chains that can be described basically by four parameters: (1) linearity (or degree of branching); (2) mode of coupling; (3) stereochemical structure; and (4) molecular weight. Not all polymer chains are alike, and variations in any of these parameters may occur. Because the properties of the polymer depend closely on these parameters, much effort has been made to elucidate their kinetic and chemical origin.

1. Linearity

Branching can occur in the synthesis of polyethylene if competing side reactions form low molecular weight olefins (such as dimers, trimers), which then copolymerize with ethylene (see Chapter 3, Section V). Dimerization reactions involving propylene or higher olefins do not occur under normal conditions of polymerization, and similar interfering copolymerizations are absent.

2. Mode of Coupling

The predominating coupling mode in polymerization of α-olefins is head-to-tail. Under certain conditions, however, head-to-head coupling can occur, such as in the synthesis of partially syndiotactic polypropylene (13a) and in copolymers of ethylene and propylene (14). The coupling mode in diene polymerization involves other possibilities: *cis*-1,4-, *trans*-1,4-,1,2-isotactic, and 1,2-syndiotactic structures (see Chapter 5).

3. Stereoregularity and Stereoregularity Distribution (SD)

When the "as prepared" whole polymer product contains more than one steric structure, it is necessary to establish whether this product is a mixture of polymer chains, each containing only one of the structures, or whether two or more structures are present in a single chain. If the former is true, then separation of the different structures can often be made by solvent fractionation. Similar attempts on the second type will lead to fractions having identical compositions.

By doing these fractionations with a series of solvents at increasing temperatures, fractions of different steric compositions can be isolated, and a stereoregularity distribution can be established for the catalyst–monomer systems. This was done for isotactic polypropylene; see Tables 15–3 to 15–6 (Natta) and Fig. 4–12 (Coover).

In the case of isotactic polypropylene, besides the main isotactic polypropylene fraction, the other fractions included stereoblock, syndiotactic, and atactic structures. In the case of diene polymers, the whole polymer can contain *cis*-1,4-, *trans*-1,4-, and 1,2-addition structures.

With this stereoregularity distribution data, one can begin to describe the catalyst system in terms of one or multiple centers, as well as to suggest if competing radical, cationic, or anionic initiations are present.

Chapters 4, 5, and 9 have already dealt with the problem of controlling stereoregularity in both α-olefin and diene polymerizations. The reader may recall that, in addition to the choice of catalyst (specific combination of metal alkyl and transition metal salt), many other factors can play a significant role in determining the level of stereoregularity, including concentration of components, temperature of polymerization, nature of the solid transition metal salt (crystal structure and composition), added electron donors, etc.

How can one control the distribution of the various stereoregular structures that are possible for a particular monomer system? One way is to select a catalyst that is known to produce the highest content of the desired structure. The other way involves the proper choice of polymerization variables for the catalyst–monomer system.

An illustration is given for isotactic polypropylene. Experience has shown that several catalyst systems produce isotactic polypropylene of the highest known isotactic content, such as $TiCl_3 \cdot xAlCl_3$ or $TiCl_3AA + AlEt_2Cl$ or $AlEtX_2$ combined with a donor ($X = Br, I, or Cl$). These polymers, when fractionated, would have only a small content of stereoblock and atactic polymer. The latter would be formed in higher fractions if $AlEt_3$ was used in place of $AlEt_2Cl$. Surprisingly, by the use of $AlEt_2F$ at comparable temperatures (about 60°C), a polymer would be formed of comparable crystallinity as obtained with $AlEt_3$, except that a significantly more syndiotactic structure would be present in the fraction that is soluble in hexane. There is a tendency to form more syndiotactic polymer by the use of soluble or colloidal catalysts, especially those based on vanadium. Lowering the polymerization temperature helps, too.

In diene polymerizations, higher contents of the desired microstructure can be obtained if classical initiation paths are avoided: radical, cationic, or anionic. The latter usually produces chains of low stereoregularity. Also, by careful synthesis of catalysts and the proper selection of polymerization conditions so as to obtain maximum content of one structure, one can decrease the relative contents of other structures. Unlike the case of α-olefin polymerizations, the physical state of the catalyst is not of paramount importance in deciding if a certain stereochemical process can occur, and polydienes having high contents of *cis*-1,4, *trans*-1,4, and 1,2 (3,4) structures have been made with both soluble and hetereogeneous catalysts.

4. Molecular Weight and Molecular Weight Distribution (MWD)

Relative to heterogeneous catalysts, soluble catalysts produce polymers consisting of chains whose molecular weights are not significantly different from one another. They have a narrow molecular weight distribution (MWD). While in recent years gel permeation chromatography has been used to measure MWD, most of the literature refers to the ratio $Q = \bar{M}_w/\bar{M}_n$ as a measure of dispersity of molecular weighs ($\bar{M}_w$ and $\bar{M}_n$ are the weight average and number average molecular weights, respectively).

For soluble Ziegler–Natta catalysts, Q values in the range of 2 to 4 have been reported for polyethylene. In contrast, heterogeneous catalysts produce polymers that have much wider MWD, for example, $Q > 4$ to 20 or higher.

Beyond recognizing that soluble catalysts lead to narrower MWD than obtained with heterogeneous catalysts, not much else of significant importance has been reported on the control of MWD. Workers have, however, recognized that changes in MWD take place during polymerization.

The reasons for the wide MWD in polymers prepared by heterogeneous catalysts are not well understood, and, for the present, these should be considered as proposals. A compilation is given here; they will be elaborated upon later in this chapter in the appropriate section. The cited reasons are: (1) not all of the active centers become active at the same time; (2) the active centers have a range of propagation rate constants (15); (3) the values of the propagation and termination rate constants change as the length of the polymer chain increases; (4) chain growth at a center is slowed down as the center becomes more and more encapsulated in the polymer mass. Olefin concentration at the center becomes decreasingly lower because the olefin finds it difficult to pass through the polymer mass; and (5) the lifetime of a growing chain increases as the center becomes more encapsulated because it is less susceptible to chain transfer by metal alkyl. The metal alkyl, being polar, finds it even more difficult to diffuse to the center through the polymer mass than the olefin.

III. Proposed Kinetic Models and Schemes

Chapters 13 and 14 discussed the origin and structure of the active center largely from the view of organic and inorganic chemistry. It was concluded that the metal in the active metal–carbon bond was the transition metal and not a base metal. Cases were recognized, however, whereby this alkylated transition metal center was part of a bimetallic complex involving both metals.

This section describes some of the kinetic approaches to this subject. Kinetic models are described that view the centers as bimetallic complexes or as alkylated transition metal complexes. The main equations describing the initiation, propagation, and chain termination steps are subsequently presented.

A. PROPOSED KINETIC MODELS

Eirich and Mark (16) first suggested in 1956 that adsorption kinetics probably played a significant role in Ziegler–Natta polymerizations and, therefore, may provide a tool to understand the observed rate laws. Subsequently, other workers applied the well-known Rideal and Langmuir–Hinshelwood-type rate laws in various forms. The latter were derived originally from two types of kinetic models describing the surface reactions of gases on a catalyst surface but later were extended to reactions taking place on surfaces of solids suspended in solvents such as polymerizations with heterogeneous catalysts.

Keii (1), Reich and Schindler (3), and Burfield et al. (5–5b) should be consulted for more rigorous derivations of these kinetic laws, their applications, and other literature references. This section was drawn largely from these sources, especially from Keii's book (1).

In the Rideal model, the rate-determining step is the surface reaction, for example, for propylene (M), as shown in Eq. 18–1.

$$(AlEt_3 \text{ dimer})_{ads} + (C_3H_6)_{sol} \xrightarrow{k_s} \text{polymer} \qquad (18-1)$$

The degree of coverage of $AlEt_3$ dimer on the surface is shown by Eq. 18–2,

$$\theta_A(AlEt_3 \text{ dimer}) = \frac{K_A[A]}{1 + K_A[A]} \qquad (18-2)$$

where [A] is the concentration of metal alkyl and K_A is the equilibrium constant for adsorption of A with the surface. The rate of surface reaction (R) is given by Eq. 18–3,

$$R = k_s P_B \theta_A \qquad (18-3)$$

where k_s = rate constant describing the reaction of adsorbed metal alkyl, P_B is a measure of concentration of monomer (M), and θ_A = degree of coverage of metal alkyl (A) on the surface.

This model was tested experimentally (16a) with the use of the expression shown in Eq. 18–4,

$$\frac{[A]}{R_\infty} = \frac{1}{k_s K_A[M]} + \frac{1}{k_s[M]}[A] \qquad (18-4)$$

where $R_\infty = k\theta_A[A]$ = rate of polymerization, $\theta_A = K_A[A]/1 + K_A[A]$ = fraction of surface covered by Al alkyl, [A] = concentration of aluminum alkyl, [M] = concentration of monomer, and k_s, K_A = constants. Figure 18–8 shows a plot of $[A]/R_\infty$ vs. [A] for ethylene and propylene polymerizations using AlR_3–$TiCl_3$ catalysts. A linear dependence was obtained at low concentrations. Figure 18–9 shows the applicability of the Rideal-type rate law to the data of Veisely and Tait et al. (5a,b).

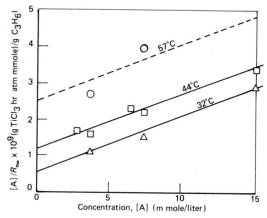

Fig. 18–8. Langmuir plot of stationary rate against $AlEt_3$ concentration. ($[TiCl_3]$ = 2 g/liter, 250 ml n-heptane).

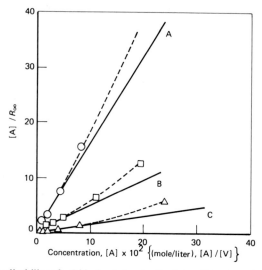

Fig. 18–9. Applicability of a Rideal rate law to the data of Vesely and Tait *et al.* (1).

In the Langmuir–Hinshelwood model, the rate-determining step is the surface reaction between an adsorbed $AlEt_3$ dimer (A) and an adsorbed olefin monomer (M). Both the $AlEt_3$ dimer and the olefin (for example, propylene) are said to be adsorbed on identical sites, that is, they compete for the same sites:

$$(AlEt_3 \text{ dimer}) + (C_3H_6) \longrightarrow \text{polymer}$$

The adsorption equations for the AlEt$_3$ dimer and olefin (propylene) are:

$$\theta_A(\text{AlEt}_3 \text{ dimer}) = \frac{K_A P_A}{1 + K_A P_A + K_M P_M} \tag{18-5}$$

$$\theta_M(\text{propylene}) = \frac{K_M P_M}{1 + K_A P_A + K_M P_M}$$

The rate of surface reaction (R) is given by the relation:

$$R = k_s \theta_A \theta_M \tag{18-6}$$

where k_s = the rate constant describing the reaction of the two adsorbed species, and θ_A, θ_M are the corresponding degrees of coverage of surface sites by molecules A and M, respectively.

The Langmuir–Hinshelwood model has been tested in the form:

$$\left(\frac{[A]}{R_\infty}\right)^{1/2} = \frac{1 + K_M[M]}{(k_s K_A K_M[M])^{1/2}} + \frac{K_A}{k_s K_M[M]}[A] \tag{18-7}$$

where R_∞ is the stationary (steady-state) rate of polymerization. Figure 18–10 shows a plot of $([A]/R_\infty)^{1/2}$ vs. [A].

The major difference between the two kinetic models is that in the Langmuir–Hinshelwood model both reactant species are adsorbed on the

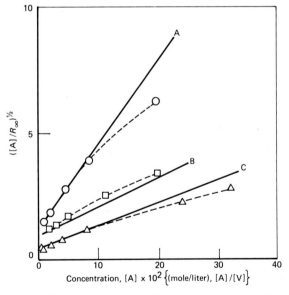

Fig. 18–10. Applicability of a Langmuir–Hinshelwood rate law to the data of Vesely (A and B) and Tait *et al.* (C) (1).

surface of the catalyst, while in the Rideal model, only one of the reactants is adsorbed while the other is in solution or in the gas phase.

Most workers who applied these models to Ziegler–Natta polymerizations considered adsorbed single or dimeric metal alkyl molecules to be the active centers. The precise structures of these active centers usually were not described.

Further application of these models is found in Section V,A.

B. OTHER MODELS

The Rideal and Langmuir–Hinshelwood models consider an adsorbed $AlEt_3$ dimer (or dissociated $AlEt_3$) to be the active center. Kinetic models have also been developed whereby the active center is an alkylated transition metal salt.

Burfield, McKenzie, and Tait (5, 5a) and Tait (5b) considered the rate to be as shown in Eq. 18–8,

$$R_p = k_p \theta_M C_0 \tag{18-8}$$

where k_p is the propagation rate constant with respect to adsorbed monomer; θ_M is the fraction of the surface covered by adsorbed monomer, that is, $(\theta_M = K_M[M])/(1 + K_M[M] + K_A[A])$; and C_0 is the concentration of active centers taken to be alkylated transition metal.

Keii (1) viewed Cossee's molecular orbital scheme in the light of kinetic theory, as shown in Eq. 18–9, where R is the growing polymer chain and PC

is the active polymerization center. In this mechanism, propylene is complexed directly at the centers bearing the growing polymer chain. The reaction can be represented as shown in Eq. 18–10,

and the rate expression as shown in Eq. 18–11 is obtained.

$$R_\infty = k_s[PC]\theta_M = k_s[PC]\left(\frac{K_M}{1 + K_M[M]}\right) \tag{18-11}$$

C. KINETIC SCHEMES

An addition polymerization, be it radical, cationic, or anionic, usually involves three different steps: initiation, propagation, and termination. The

Ziegler–Natta polymerization can also be described in this way, and the following reactions were proposed by Natta and co-workers (17) as possible for the three steps.

TABLE 18–2

Kinetic Scheme Proposed by Natta and Co-workers (17)^a

Initiation

1. $[\text{cat}]\text{H} + \text{CH}_2=\text{CH}-\text{CH}_3 \xrightarrow{k_{i,1}} [\text{cat}]\text{C}_3\text{H}_7$

2. $[\text{cat}]\text{C}_2\text{H}_5 + \text{CH}_2=\text{CH}-\text{CH}_3 \xrightarrow{k_{i,2}} [\text{cat}]\text{CH}_2-\underset{\underset{\text{CH}_3}{|}}{\text{CH}}-\text{C}_2\text{H}_5$

3. $[\text{cat}]\text{C}_3\text{H}_7 + \text{CH}_2=\text{CH}-\text{CH}_3 \xrightarrow{k_{i,3}} [\text{cat}]\text{CH}_2-\underset{\underset{\text{CH}_3}{|}}{\text{CH}}-\text{C}_3\text{H}_7$

Propagation

4. $[\text{cat}]\text{CH}_2-\underset{\underset{\text{CH}_3}{|}}{\text{CH}}\left(\text{CH}_2-\underset{\underset{\text{CH}_3}{|}}{\text{CH}}\right)_n R + \text{CH}_2=\text{CH}-\text{CH}_3 \xrightarrow{k_p} [\text{cat}]\text{CH}_2-\underset{\underset{\text{CH}_3}{|}}{\text{CH}}\left(\text{CH}_2-\underset{\underset{\text{CH}_3}{|}}{\text{CH}}\right)_{n+1} R$

5. $[\text{cat}]\text{CH}_2-\underset{\underset{\text{CH}_3}{|}}{\text{CH}}\left(\text{CH}_2-\underset{\underset{\text{CH}_3}{|}}{\text{CH}}\right)_{n+1} R + \text{CH}_2=\text{CH}_2-\text{CH}_3 \xrightarrow{k_p} [\text{cat}]\text{CH}_2-\underset{\underset{\text{CH}_3}{|}}{\text{CH}}\left(\text{CH}_2-\underset{\underset{\text{CH}_3}{|}}{\text{CH}}\right)_{n+2} R$

Chain Termination

6. $[\text{cat}]\text{CH}_2-\underset{\underset{\text{CH}_3}{|}}{\text{CH}}\left(\text{CH}-\text{CH}\right)_n R \xrightarrow{k_{t,1}} [\text{cat}]\text{H} + \text{CH}_2=\underset{\underset{\text{CH}_3}{|}}{\text{C}}\left(\text{CH}_2-\underset{\underset{\text{CH}_3}{|}}{\text{CH}}\right)_n R$ (Disproportionation)

7. $[\text{cat}]\text{CH}_2-\underset{\underset{\text{CH}_3}{|}}{\text{CH}}\left(\text{CH}-\text{CH}\right)_n R + \text{AlEt}_3 \xrightarrow{k_{t,2}} [\text{cat}]\text{Et} + \text{AlEt}_2-\text{CH}_2-\underset{\underset{\text{CH}_3}{|}}{\text{CH}}\left(\text{CH}_2-\underset{\underset{\text{CH}_3}{|}}{\text{CH}}\right)_n R$ (Transfer by AlEt_3)

8. $[\text{cat}]\text{CH}_2-\underset{\underset{\text{CH}_3}{|}}{\text{CH}}\left(\text{CH}-\text{CH}\right)_n R + \text{CH}_2=\text{CH}-\text{CH}_3 \xrightarrow{k_{t,3}}$ (Transfer by monomer)

$[\text{cat}]\text{C}_3\text{H}_7 + \text{CH}_2=\underset{\underset{\text{CH}_3}{|}}{\text{C}}\left(\text{CH}_2-\underset{\underset{\text{CH}_3}{|}}{\text{CH}}\right)_n R$

9. $[\text{cat}]\text{CH}_2-\underset{\underset{\text{CH}_3}{|}}{\text{CH}}\left(\text{CH}_2-\underset{\underset{\text{CH}_3}{|}}{\text{CH}}\right)_n R + \text{ZnEt}_2 \xrightarrow{k_{t,4}} [\text{cat}]\text{C}_2\text{H}_5 + \text{EtZnCH}_2-\underset{\underset{\text{CH}_3}{|}}{\text{CH}}\left(\text{CH}_2-\underset{\underset{\text{CH}_3}{|}}{\text{CH}}\right)_n R$ (Transfer by ZnEt_2)

10. $[\text{cat}]\text{CH}_2-\underset{\underset{\text{CH}_3}{|}}{\text{CH}}\left(\text{CH}_2-\underset{\underset{\text{CH}_3}{|}}{\text{CH}}\right)_n R + \text{H}_2 \xrightarrow{k_{t,5}} [\text{cat}]\text{H} + \text{CH}_3-\underset{\underset{\text{CH}_3}{|}}{\text{CH}}\left(\text{CH}_2-\underset{\underset{\text{CH}_3}{|}}{\text{CH}}\right)_n R$ (Transfer by H_2)

^{a} $[\text{cat}]\text{R}$ denotes active center where the metal could be a transition metal or a base metal.

The notation [cat]–R was used by Natta to designate the active center, taken to be a metal–carbon bond. R can be a polymer chain (R = Pn), a hydride (R = H), or an alkyl (R = methyl, ethyl, . . . , etc.), or any suitable moiety. As long as R is one of the above, the center is active or can become active. If, however, R becomes a halogen, sulfur, nitrogen, or oxygen, then the center is no longer active for the polymerization of olefins, dienes, etc.

In this scheme, the active metal could be a transition metal (now accepted by most workers to be the case) or a base metal (previously considered valid by many workers but now by only a few). The Natta kinetic scheme is shown in the following for the polymerization of propylene with the $AlEt_3$–$TiCl_3$ catalyst and with $ZnEt_2$ or H_2 also being present as transfer agents. Having these equations together will be helpful in the evaluation of the kinetic data. In all of these termination reactions (Eqs. 18–1 to 18–10), chain growth stops, but the [cat]–R center remains active for further polymerization.

Adventitious or added impurities may also react with the [cat]–Pn center and, in this case, termination of chain growth will be accompanied by inactivation of the center, as shown in Eq. 18–12.

$$[\text{cat}]-CH_2-\underset{\underset{CH_3}{|}}{CH}(CH_2-\underset{\underset{CH_3}{|}}{CH})_n R + R'OH \xrightarrow{k_t} [\text{cat}]-OR' + CH_3-\underset{\underset{CH_3}{|}}{CH}(CH_2-\underset{\underset{CH_3}{|}}{CH})_n R$$

$$\text{Dead center}$$

$$(18-12)$$

When the Ziegler–Natta catalyst contains multiple centers, the kinetic analysis gives only an average value. Chapter 10 describes other types of chain and site terminators.

IV. Operational Factors That Can Affect Kinetic Results

This section describes selected examples of some of the operational factors that can influence basic kinetic results. The reader should recognize that an individual catalyst–monomer system may be insensitive to, or only partially to strongly affected by a change in the operational factor, and this relationship is usually established through an experiment. Indeed, many catalyst–monomer systems show dramatic similarities, but, then, they may also show dramatic dissimilarities. Thus, a neat list showing the universal effect of an operational change on catalyst–monomer systems does not exist.

The operational factors described now are: (1) those related to catalyst: choice of catalyst components, stability, concentration of components, and order of mixing; (2) those related to polymerization: concentration of monomer, type of solvent, stirring speed, time and temperature of polymerization, yield of polymer, and competing reactions; and (3) those related to third components: transfer agents, reactants, and electron donors.

A. FACTORS RELATED TO THE CATALYST

1. Choice of the Catalyst

The selection of a metal alkyl and a transition metal salt to form the Ziegler–Natta catalyst is by far the most important factor in determining what kind of kinetic results will be obtained. Even if one operates in what appears to be a close family of components, such as aluminum alkyls and titanium chlorides, significant differences are observed between the various combinations (compare data of Table 18–4). Thus, the kinetic parameters obtained for $AlEt_3$–$\alpha TiCl_3$ will vary from those obtained with $AlEt_2Cl$–$\alpha TiCl_3$.

The similarities and differences will become apparent in this section and in Section V.

2. Stability of the Catalyst

When the transition metal salt component is in one of its higher valence states, such as VCl_4 or $TiCl_4$, then, upon addition of the reducing metal alkyl, a very unstable catalyst is formed (18–21a). Polymerization begins upon addition of the monomer, but the basic features of this polymerization can be masked by changes in the elemental composition, crystal structure, valence state, and relative concentrations of the reactants and products. These changes can take place throughout the entire duration of the polymerization. The common practice has been to stabilize the catalyst mixture for specified times and temperatures before admitting monomer or to use preformed transition metal salts that are already in a lower valence state.

The following examples are illustrative. Feldman and Perry (18), using an AlR_3–$TiCl_4$ catalyst for the ethylene polymerization, concluded from labeling experiments that stabilized catalysts (aged 15 minutes at 69°C) developed fewer active centers, C*, than unstabilized catalysts in the 40 minute polymerization period. They also reported that the number of centers decreased upon increasing the temperature of polymerization from 39° to 54°C, as shown in the tabulation.

Temperature (°C)	C*	k_p (liter/sec-mole)
39.4	10.9	52 ± 2.7
54.5	2.3	127 ± 16

Bier and co-workers (19) compared the activity of the VCl_4–$AlEt_2Cl$ and $TiCl_3$–$AlEt_2Cl$ catalyst for polymerization of ethylene during a 2 hour period. The former catalyst was twice as active initially, but within 20 to 30 minutes its activity declined below that of $TiCl_3$–$AlEt_2Cl$. It became negligible within an hour, while the $AlEt_2Cl$–$TiCl_3$ catalyst maintained a nearly constant rate. Bier and co-workers attributed the decline to the loss of centers as a result of the instability of the vanadium-based centers (reduction, etc.).

Some soluble and colloidally dispersed catalysts, such as $VO(OEt_3)_3$– $AlEt_2Cl$, when made at 0°C are active for only 5 to 10 minutes and become inactive if $AlEt_2Cl$ is replaced by $AlEt_3$ (20). Another example of an unstable vanadium-based catalyst is the syndiotactic-specific catalyst that must be used below -40°C (see Chapter 4).

A catalyst may be unstable toward reduction even when the transition metal salt is already in a lower state, such as VCl_3 or $TiCl_3$, if a powerful reducing agent is added as a cocatalyst (such as Li alkyl of even AlR_3 and ZnR_2 in higher concentrations) (21).

The active center may be reduced to a lower valent inactive center. Often, the addition of third components reactivates these catalyst centers [see Christman (21a) and Chapter 9, Section II,C,2,g].

3. Absolute and Relative Concentrations of Metal Alkyl and Transition Metal Salt

The kinetic behavior of a catalyst is sensitive to these factors and a direct dependence often exists.

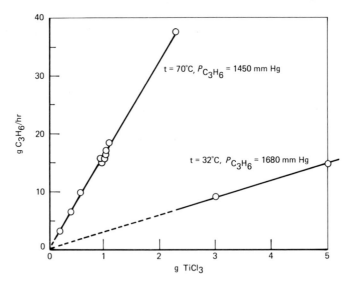

Fig. 18–11. Dependency of propylene polymerization rate in steady-state conditions on the amount of $\alpha TiCl_3$ (sample A) in the catalytic system (2).

a. Transition Metal Salts. Natta and Pasqon (2) first recognized that the rate of polymerization was directly proportional to the concentration of the transition metal salt when holding other experimental variables constant (Fig. 18–11). Other workers confirmed this dependency for other catalysts in agreement with the heterogeneous nature of the catalysis.

An apparent exception may occur if the polymerization becomes rate limiting due to secondary causes, such as slow mass transfer of olefin (22). This is discussed in Section IV,B,3 and V,F.

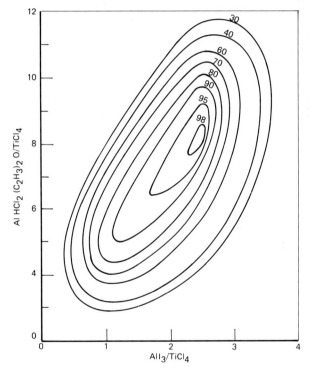

Fig. 18–12. Regions of equal conversions at variable ratios of $AlHCl_2 \cdot Et_2O/TiCl_4$ and $AlI_3/TiCl_4$. Conditions: 100 ml toluene; 15 g butadiene; 2.27×10^{-4} moles of $TiCl_4$; temp. $+5°C$; time = 2 hours.

Marconi and co-workers (23) extensively examined the polymerization of butadiene with ternary catalysts, consisting of $TiCl_4$, AlI_3, and aluminum hydride derivatives. *Cis*-1,4 content was increased from 60 to 94% on increasing the $AlI_3/TiCl_4$ ratio to 1. Figure 18–12 shows a contour-type representation of equal conversions at variable ratios of the reactants.

b. Metal Alkyl.   The polymerization rate is obviously at least partially dependent on the absolute concentration of metal alkyl, for the catalyst would be inactive if no alkyl was present. The extent of this dependency, however, varies at increasing concentrations, depending on the transition metal salt and on concentration and type of impurities present.

For the $AlEt_3–\alpha TiCl_3$ catalyst (propylene polymerization), Natta and Pasquon (2) found little difference in kinetic results when the $AlEt_3/TiCl_3$ ratio was varied from 1 to 8.5. A rate curve similar to Fig. 18–1 was obtained. At an $AlEt_3/\alpha TiCl_3$ molar ratio of 0.4, the rate curve initially rises to a maximum, then declines rapidly without obtaining a stationary value. They attributed the decrease in activity to the consumption of $AlEt_3$ and found the activity was restored upon addition of $AlEt_3$. For the molar ratio range cited, Natta assigned a zero reaction order with respect to aluminum concentration.

In general, other investigators confirmed these results for other Ziegler–Natta catalysts, except that in some cases the concentration of metal alkyl needed to achieve an optimum stationary rate was higher (as high as six). For some metal alkyls ($ZnEt_2$), the polymerization rate decreased at high concentrations, possibly due to side reactions (24a).

The mechanistic path by which metal alkyl molecules can decrease the polymerization rate is discussed in Section V,A.

The concentration of $AlEt_3$ relative to the transition metal salt is more important when $TiCl_4$ is used (25). At higher Al/Ti ratios, reduction to valences below III occurs even at room temperature (Fig. 18–13) (25–25a). The kinetic behavior of this catalyst is variable; it depends on temperature and time of polymerization.

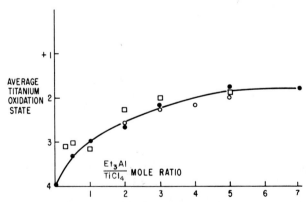

Fig. 18–13.   Average titanium oxidation state vs. Al/Ti mole ratio after 30 minutes at 25°C. □ , Cooper and Rose (25a); ○, Kern and Hurst (25); ●, data of Malatesta corrected for transfer (25).

Natta and Pasquon (2) found a dependency of intrinsic viscosity of polymer on $AlEt_3$ concentration and concluded that the metal alkyl took part in a transfer or termination process of the growing polymer chains (see Chapter 10). This conclusion was supported by their observation that the amount of aluminum bound in the polymer is higher in tests performed with higher $AlEt_3$ concentrations. (The polymer bound to Al was isolated by solvent extraction methods.)

4. Order of Mixing

There is an obvious advantage in saturating the reaction solvent with the monomer before the catalyst is added or before the catalyst components are brought into contact in the reaction vessel. Polymerization can begin immediately, and the time needed to saturate the solvent with the necessary supply of monomer is gained, since only the consumed monomer has to be dissolved. An exception to this may be a system with a low activity rate, in which monomer is slowly consumed and also replenished relatively easily.

Such is probably the case for the polymerization of propylene with the $AlEt_3$–$\alpha TiCl_3$ catalyst. Natta and Pasquon (2) found no dependence on the polymerization rate of the order in which the components were mixed.

Other workers (16a, 26) showed that the initial rate of propylene polymerization was affected by the way propylene and the catalyst components were mixed. Keii and co-workers (16a) found that both the initial and maximum rates were greater when $TiCl_3$ and propylene were premixed before addition of $AlEt_3$, rather than when $TiCl_3$ and $AlEt_3$ were premixed before addition of propylene.

The catalyst used by Keii ($AlEt_3 + TiCl_3AA$) was considerably more active than used by Natta ($AlEt_3 + \alpha TiCl_3$). If the reactor design and conditions of polymerization allow a rapid equilibration of olefin in the solvent, then perhaps highly active catalysts would show a lesser dependency on the order of mixing. Polymerization variables such as stirring speed and concentration of catalyst can be determining.

Gavrilova and co-workers (27) found that order of addition of components influenced the kinetics of polymerization of butadiene with the Al-i-Bu_2Cl–Co naphthenate catalyst.

B. FACTORS RELATED TO POLYMERIZATION

1. Concentration of the Monomer

The rate of polymerization under steady-state conditions is directly proportional to the concentration of the monomer. This was first reported by Natta and Pasquon (2) for the polymerization of propylene with the $AlEt_3$–$\alpha TiCl_3$ catalyst (Fig. 18–14). Ingberman and co-workers (28) later

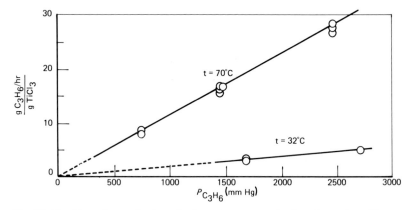

Fig. 18-14. Dependence of polymerization rate on concentration of propylene (2).

reported a linear increase in polymer yield with increasing pressures for the catalyst α- and γTiCl₃ and AlEt₂Cl for propylene polymerization.

2. The Solvent

Aliphatic and aromatic hydrocarbons have generally been used as solvents, preferably aliphatic for olefins and aromatic for dienes. Occasionally, aromatic and aliphatic chlorides have been used.

Keii found a higher initial polymerization rate in toluene relative to heptane when propylene was polymerized with $AlEt_3$–$TiCl_3AA$ catalyst (the stationary rates were the same in both solvents).

The rate of polymerization of propylene with $TiCl_3$–Al alkyl catalysts was greater in aromatic solvents. Ingberman and co-workers (28) attributed the increase not to high dielectric constants but rather to the greater ability of the aromatic relative to aliphatic solvent to cause desorption of poisons from the $TiCl_3$ surface.

Dyachkovskii and co-workers (29) reported a higher initial rate for ethylene with the $(C_5H_5)_2TiCl_2 \cdot AlMe_2Cl$ catalyst in toluene relative to heptane. This was attributed to a higher dielectric constant. When ethylene was polymerized with $(C_5H_5)_2TiCl_2$–$AlEt_2Cl$ in a polar solvent, such as ethyl chloride or 1,2-dichloroethane, the polymerization rate remained constant for a long time, while in nonpolar solvents, the polymerization rate was high initially but rapidly declined to zero (30).

Belov and co-workers (31) showed that the polymerization of ethylene with $(C_5H_5)_2TiCl_2$–$AlEt_2Cl$ produced a polymer that had a bimodal MWD in ethyl chloride solvent and a monomodal MWD if a hydrocarbon solvent was used. In ethyl chloride, increasing the relative concentration of one component brought the two peaks closer, while increasing the concentration of total catalyst and ethylene increased the width of the MWD.

TABLE 18–3

Effect of the Dielectric Constant Value of the Solvent on Butadiene Polymerization[a]

Solvent	ε	% Conversion	cis-1,4
Benzene	2.3	96.5	98
Toluene	2.4	89.4	98
o-Chlorotoluene	4.7	49.0	98
m-Chlorotoluene	5.5	31.6	97
Chlorobenzene	5.9	63.5	97
o-Dichlorobenzene	7.4	31.6	96
m-Dichlorobenzene	10.2	26.1	95

[a] Other aspects related to solvents are discussed in various parts of this book (33).

Gippin (32), in the polymerization of butadiene with the $AlEt_2Cl/H_2O/Co$ octoate catalyst, found a decrease in percent conversion but only little change in the cis-1,4 constant as the dielectric constant ε of the solvent was increased from 2.3 to 10.2, as shown in Table 18–3. Other aspects relating to solvents are discussed in various parts of this book (33).

3. Stirring Speed (Agitation)

It is definitely established that mass transfer of gaseous olefin to the reaction centers becomes an increasingly more important factor as the activity of the catalyst increases. Only by judicious choice of reactor and conditions of polymerization can this barrier be overcome.

A number of workers established the dependence of the polymerization rate on the degree of agitation. Lehmann and Gumboldt (26) reported a positive dependency of polymerization of propylene on the stirring speed. Berger and Grieveson (34), using a vibrating stirrer, showed a dependency existed over a range of $TiCl_3$ concentrations for the system $AlEt_2Cl-\gamma TiCl_3-$ ethylene. Stirring speeds of 250, 450, and 650 rpm were used. Boocock and Haward (22) showed that the rate of adsorption of ethylene in the solvent coincided with the maximum polymerization rate of ethylene when stirring speed was increased from 0 to 500 rpm. Keii and co-workers (35) demonstrated that different kinetic curves were obtained for propylene at each stirring speed and one could shift curves by changing the stirring speeds, which ranged from 300 to 600 rpm. In all of the above, the design of the reactor and the stirring mechanism is important in deciding the practical effect of stirring.

The above studies involved slurry-type polymerizations. In solution polymerizations, solution viscosities increased markedly, even at low conversions.

These viscous solutions become difficult to stir and mass transport of monomer undoubtedly becomes important.

4. Time of Polymerization

Chien (36) showed for the $(C_5H_5)_2TiCl_2-(CH_3)_2AlCl$ soluble system (ethylene) that $\bar{M}_w/\bar{M}_n$ values (MWD) decreased during the first 10 minutes and increased (2.2 to 4.5) during the next 40 minutes. Also, $\bar{M}_w/\bar{M}_n$ values increased as the temperature of polymerization decreased, as shown in the tabulation below.

Temperature (°C)	$Q = \bar{M}_w/\bar{M}_n$
60	2.4
45	3.6
30	4.5
15	5.4
0	6.1

This decrease was attributed to the increase in the rates of initiation and termination and the higher activation energy for the termination process. Broader molecular weight distributions were obtained at higher catalyst concentrations and lower Al/Ti ratios.

Bier and co-workers (37) found that $\bar{M}_w/\bar{M}_n$ values increased from 5 to about 9 during the first 30 minutes of polymerization and then decreased to about 4 during the next 3 hours or so for propylene polymerization $(AlEt_2Cl-TiCl_3 \cdot xAlCl_3)$. They attributed the narrowing of the MWD to the lack of formation of new sites after the initial period.

Bier and co-workers (37) reported that the molecular weights of the formed polymer increase throughout a 4 hour polymerization of propylene with both the Hoechst catalyst, as well as Natta's $AlEt_3-\alpha TiCl_3$ catalyst (see Chapter 21 for details). The Hoechst catalyst showed a higher rate of polymerization, and this was attributed by them to a greater number of active centers rather than differences in size of the k_p's for the two catalysts. In contrast, Natta and Pasquon (2) reported that they did not observe any effect of polymerization time on the molecular weight or steric composition of polymer, either after a few minutes of polymerization or after many hours.

5. Temperature of Polymerization

Temperature of polymerization is an important factor, especially if one operates in a wide range, 0° to 200°C. Obviously, much depends on the catalyst involved.

Novokshonova and co-workers (38) showed that increasing the temperature from 30° to 60°C changed the rate curve from the constant to the decay type for the $AlEt_3$–VCl_3 catalyzed polymerization of propylene.

From their kinetic analysis, Berger and Grieveson (34) proposed that the polymerization of ethylene with the $AlEt_2Cl$–$TiCl_3 \cdot xAlCl_3$ catalyst became mass transfer-limited under their operating conditions only above 60°C.

Combs and co-workers (39) showed the variation in stereoregularity distribution (SD) to be a function of the catalyst used, the temperature of polymerization, and the ratio of components. The narrowest SD values were obtained at lower temperatures of polymerization. They also showed that a narrower MWD was obtained by using highly stereospecific catalysts, high monomer concentrations, and high polymerization temperatures. Polypropylene molecular weights ($[\eta]$ dl/g) decreased as the temperature of polymerization increased (39a), as shown in the following tabulation.

Temperature (°C)	$[\eta]$ (dl/g)
30	9.5
45	6.4
60	4.5

Belea and co-workers (40) found the molecular weight of cis-1,4-polybutadiene decreased from 600×10^3 to 300×10^3 upon raising the temperature of polymerization from 5° to 30°C. The catalyst $CoCl_2 \cdot MeEtpyridine$–$AlEt_2Cl$ was used.

6. Limiting Production

For every reactor system and under specific operating conditions, there is a rate at which mass and heat transfer becomes rate limiting. Natta and co-workers (12) established this rate at 70°C and in 250 cm^3 solvent to be 20 g of polymerized propylene per hour for their reactor system. In their investigations, lower rates than this were always used.

Yoshimoto and co-workers (40a) found a first-order dependence on monomer concentration for the polymerization of butadiene with Ni carboxylate/BF_3 etherate/$AlEt_3$ catalyst for most of the polymerization period. Molecular weights (measured as $[\eta]$) increased at low conversions but appeared to remain the same at higher (about 60%) concentrations of butadiene. Chain transfer by monomer was postulated to be a major chain termination path.

7. Competing Reactions

Sometimes competing reactions make kinetic analysis difficult, especially if one is observing a loss in monomer content as a function of time. If part of the monomer undergoes a secondary reaction, then the evaluated polymerization rate will be too high. An example would be the polymerization of propylene with a catalyst that is acidic, and some of the propylene is consumed by alkylation of the aromatic solvent (see Chapter 22, Sections VII and VIII).

C. FACTORS RELATED TO THIRD COMPONENTS

The effect of adventitious impurities, transfer agents, and electron donors is considered.

1. Adventitious Impurities

Some of the differences in kinetic results can be attributed to impurities initially introduced to the catalyst–solvent, later in the monomer, or in the process of polymerization. Small amounts of water and oxygen can be introduced in spite of apparently meticulous procedures made to purify solvents. Commercial sources of metal alkyls can contain up to several percent of contaminant metal alkyls; $AlEt_3$, for example, contains some $AlEt_2H$. During handling, some $AlEt_2OEt$ can form by reaction of $AlEt_3$ with oxygen. It is recognized that some of the metal alkyl used in the polymerization actually acts as a scavenger. Natta and Pasquon (2) showed that, by operating with very pure solvents and reagents, the concentration of $AlEt_3$ could be lowered to a 10^{-3}–10^{-4} mole/liter level. Normally, concentrations of 1.4×10^{-2} mole/liter were used.

2. Transfer Agents

Both H_2 and $ZnEt_2$, the most studied transfer agents, can affect the kinetics (see Chapter 10).

3. Electron Donors

The effect that electron donors have on the kinetic and stereochemical behavior of a catalyst has been extensively studied. Addition of compounds such as ethyl acetate (41), oxygen (42), and water (32) can cause pronounced changes. For the many other examples, the reader is urged to read Chapter 9.

V. Growth of Polymer Chain and Particle

Only by understanding the basic kinetic parameters can one explain the origin of the shapes of the rate curves and the nature of the polymer product, as discussed in Section II. This section focuses on this aspect of the kinetic studies. As the reader will see, much controversy exists, and this is illustrated in the kinetic data of Table 18–4. Some of the reasons for this state of affairs will become obvious, and other reasons will remain obscure, awaiting more conclusive kinetic studies in the future. Yet, much valuable understanding has emerged that has considerably helped us to understand the elusive Ziegler–Natta catalyst.

A. CENTER FORMATION AND RATE-DETERMINING STEP

It has generally been accepted that the growth involves two steps: (1) the complexing of the monomer (M) at the metal–carbon center M–Pn, as shown in Eq. 18–13;

$$
\underset{\text{A} \qquad \text{M}}{\text{M–Pn} + \text{Monomer}} \underset{k_2}{\overset{k_1}{\rightleftarrows}} \underset{\text{MA}}{\overset{\text{Pn}}{\text{M} \longleftarrow \text{Monomer}}} \qquad (18\text{–}13)
$$

and (2) the insertion of monomer into the metal–carbon bond, shown in Eq. 18–14, where Pn is the growing polymer chain.

$$
\underset{\text{A}'}{\text{MA}} \xrightarrow{k_3} \text{M–Pn}' \qquad (18\text{–}14)
$$

For the AlR_3–$TiCl_3$ catalyst, M–Pn would be a Ti–Pn bond if the active center is an alkylated titanium chloride. This aspect has been discussed in detail in Chapter 13 and will be mentioned here only in the kinetic context.

Views have been expressed in favor of the above steps 1 or 2 as the rate-determining step. These conclusions were based on a number of factors, including personal interpretation of the structure of the active center and interpretation of kinetic measurements.

A common characteristic of all Ziegler–Natta catalysts is the direct dependence of the polymerization rate on concentrations of the transition metal salt and the monomer. This in itself, however, does not reveal whether k_1 (the complexation step) or k_3 (insertion step) is rate determining. Many workers argue that only a fraction of the total available centers are complexed with the monomer at any moment, because a first-order rate dependence is experimentally observed over a wide range of monomer concentration. If the rate-determining step was step 1, then $k_2 \gg k_1[\text{monomer}] > k_3$.

Cossee (43) has reached this conclusion in the following way: under steady-state conditions $d(\text{MA})/dt = 0$ and

$$
\frac{-d[\text{M}]}{dt} = \frac{k_1 k_3 [\text{C}][\text{M}]}{k_1[\text{M}] + k_2 + k_3} \qquad (18\text{–}15)
$$

where $\text{C} = [\text{MA}] + [\text{A}]$, and $[\text{A}]$ = all vacant sites = $\Sigma[\text{A}]$ (the active center is an alkylated transition metal). The observed dependence of the polymerization rate on monomer pressure requires either $k_2 \gg k_1[\text{M}]$ or $k_3 \gg k_1[\text{M}]$.

TABLE 18-4

Kinetic Features as Reported by Different Workers (4)

Investigator	Olefin	Catalyst	Polymerization						Dependence on conversion			Comments[e]
			Temp. (°C)	Time (hr.)	Conv.[a]	R_p[b]	C*[c]	k_p[d]	R_p	C*	k_p	
Natta and co-workers	Propylene	AlEt₃ + αTiCl₃	70	4	—	500	1×10^{-2}	5×10^4	Constant	Constant	Constant	1
Kohn and co-workers	Propylene	AlEt₂Cl—TiCl₃	70	11	—	440	0.4×10^{-2}	11×10^4	Constant	Constant	Constant	1
	Propylene	AlEt₂Cl–TiCl₃ (Al/TiCl₄)	50–90	6	~2	3000–1500	$2.5\text{–}14 \times 10^{-3}$	28×10^4	Declining	Declining	Constant	2
Bier and co-workers	Propylene	AlEt₃–αTiCl₃	50	3	40	3900–1300	$11\text{–}45 \times 10^{-3}$	$36\text{-}3 \times 10^4$	Declining	Incr.	Declining	3
	Propylene	AlEt₂Cl–TiCl₃ (AlEt₃/TiCl₄)	50	4	46	4400–2100	$23\text{–}70 \times 10^{-3}$	$19\text{-}3 \times 10^4$	Declining	Incr.	Declining	3
Chien	Propylene	AlEt₂Cl–αTiCl₃	50	50	1000	2700	112×10^{-3}	2.4×10^4	Constant	Constant	Constant	4
Coover	Propylene	AlEt₃–TiCl₃	70	0.5	—	3480	3×10^{-3}	116×10^4	Constant	Constant	Constant	5
Coover	Propylene	AlEt₂Cl–TiCl₃	70	1.2	—	960	1×10^{-3}	96×10^4	Constant	Constant	Constant	5
Coover	Propylene	1.0 AlEtCl₂ + 1.0 TiCl₃ + 0.6 hexamethyl phosphoramide	70	1.3	—	240	0.2×10^{-3}	120×10^4	Constant	Constant	Constant	5
Caunt	Propylene	AlEt₂Cl–TiCl₃	60	—	105	—	—	—	Declining	Constant	—	6
Youngman and Boor	Propylene	TiCl₃–Et₃N (Al/TiCl₄)	25	240	—	0.2–0.5	0.8×10^{-5}	$2.4\text{–}6.7 \times 10^4$	Declining	Incr.	Declining	7
Feldman and co-workers	Ethylene	Al(i-Bu)₃–TiCl₄ (stab. and unstab.)	35–62	—	—	—	$0.3\text{–}15 \times 10^{-2}$	—		Incr.		8
Fukui	Ethylene	AlEt₃–TiCl₃	—	—	—	—	—	—		Constant		9
Berger and Grieveson	Ethylene	AlEt₂Cl–γTiCl₃	40	17	2020	28400	1.5×10^{-2} to 0.6×10^{-2}	—	Constant up to 2020 g/TiCl₃ conversion	Constant	Constant	10
Grieveson	Ethylene	AlEt₂Cl–γTiCl₃	—	—	—	—	—	—	—	—	—	11

For row 11 (Grieveson), Comments:

$k_p \sim 80$ liters/mole-sec at 40°C;
$k_{sl}/k_p = 3.4 \times 10^{-6}$;
$k_s \approx 3 \times 10^{-4}$ liter/mole-sec
$k_{ml}/k_p \approx 1.8 \times 10^{-4}$;
$k_m \approx 1 \times 10^{-2}$ liter/mole-sec
$k_{H_2}/k_p \approx 3.7 \times 10^{-2}$;
$k_{H_2} \approx 3$ liters/mole-sec

^{a} Conversion, g polymer per g TiCl$_3$.

^{b} Rate, g polymer per mole TiCl$_3$ × hr × atm olefin.

^{c} Concentration of active centers, moles per mole TiCl$_3$.

^{d} Propagation rate constant, g polymer per mole center × hr × atm olefin.

^{e} Pertinent comments made by investigator:

1. Main lifetime of growing chains is on the order of several minutes; the number of sites was estimated to be about 1×10^{-2} mole per mole TiCl$_3$ (adsorption and kinetic methods); all sites are active when polymerization begins; proposes that chain transfer by metal alkyl terminates chain growth; initial polymerization rate (induction period) is greater than final steady rate for AlEt$_2$Cl–TiCl$_3$ but same as for AlEt$_3$–TiCl$_3$; site lifetimes could be maintained for long periods of time, i.e., >14 hr at 70°C. Reaction rate is dependent on [TiCl$_3$]$\cdot$ [propylene]$\cdot$ [propylene]' but independent of AlEt$_2$Cl concentration above a threshold value; established that the formed polymer chains are attached to aluminum atoms and to Zn atoms when ZnEt$_2$ was used as a transfer agent; propagation rates at sites producing slightly crystalline polymer are lower than at sites producing highly crystalline polymer.

2. Found that essentially all polymer chains were attached to a metal; concentration of metal–carbon bonds increased during polymerization (0–6 hr); viscosity molecular weight $\bar{M}_v$ increased slightly during first 2 hr and then remained constant during 2–4 hr; invokes bimetallic mechanism to explain kinetics.

3. Assumes that all metal–carbon bonds are active sites; concludes that chain termination by the metal alkyl does not take place at 30° to 70°C; polymer molecular weight increases with polymerization time up to 4 hr; molecular weight distribution maximized then declined as reaction time was increased; as polymerization temp. is increased, the number of sites (= metal–carbon bonds) increases; proposes that polymer presents a diffusion barrier to monomer and that concentration of monomer at sites decreases as conversion increases.

4. To get C*, extrapolate M–C bond values to $T = 0$; at low temperatures C* increases as AlEt$_2$Cl concentration increases; number average molecular weight increased steadily from 0.5×10^5 at about 1 hr to 5.5×10^5 at 50 hr; average life of growing chains is about 410 min; proposes that sites giving amorphous polymer are formed in increasing amounts at higher temperatures while sites giving crystalline polymer decrease in number; chain transfer by AlEt$_2$Cl proposed [α(AlEt$_2$Cl)$^{1/2}$]; no termination processes observed.

5. [C*] taken as equivalent to concentration of metal–carbon bonds at polymerization time when rate became constant, i.e., after about 10 min; the formation of metal–carbon bonds after the above time was attributed to chain transfer by metal alkyl; lifetime of growing chains at low conversions is about 1 min.

6. Metal–carbon bond concentration increases sharply up to 80 min and then slowly between 80 and 200 min; proposes that rate of chain transfer by metal alkyl highest during early stages of polymerization and that chain transfer process involves both monomer and aluminum alkyl, i.e., $R_{tr} = K_{tr}[C_3H_6][AlEtCl_2][C^*]$; presents evidence which invalidates extrapolation methods of counting [C*]. At higher conversions than 2 g polymer per g TiCl$_3$, author proposes diffusion barriers which limit kinetic features.

7. Proposes that (AlEt$_2$Cl)$_2$ $\xrightarrow{\text{TiCl}_3}$ Et$_3$Al (activator adsorbed on solid TiCl$_3$) + EtAlCl$_2$ (poison left in solution) takes place. AlEt$_3$ functions as activator while AlEtCl$_2$ poisons sites; the latter is removed by addition of NaCl or amine; chain growth times of 12 min proposed; proposes that catalyst–polymer bond weakest during addition of propylene, and thereby transfer by metal alkyl is facilitated; number average molecular weight increases with polymer conversion.

8. Assumed all metal–carbon bonds to be active sites; active centers are heterogeneous and polymer chain lifetimes range from 1 to 30 min (4 min = average lifetime); polymer molecular weight $\bar{M}_n$ increased throughout polymerization; as polymer molecular weight decreases the number of chains which are attached to a metal decreases.

9. Polymerization rates were directly proportional to [TiCl$_3$] and ethylene pressure and rose with AlEt$_3$ concentration. Latter not found by Natta for propylene polymerization with similar catalyst.

10. Rate independent of AlEt$_2$Cl above threshold concentration (~ 1 mmole AlEt$_2$Cl/liter); first-order dependence on ethylene below 50° polymerization temp.; initial rate (up to 60 min) was consistently higher or lower than final rate (1–17 hr).

11. Polymer chains attain near maximum molecular weight during initial 120 min of polymerization; molecular weight independent of TiCl$_3$ (1–7.9 mmoles/liter range) and of AlEt$_2$Cl (2–52 mmoles/liter range); concludes that no chain termination by AlEt$_2$Cl takes place; concludes that in absence of modifiers, chain termination by transfer with ethylene is the most important factor in determining polymer molecular weight; transfer with H$_2$ is 200 times more likely than transfer with ethylene; the growing chains have an average lifetime of 30 min but when hydrogen is present the time is only 7 min.

If every complexed monomer would be instantaneously inserted, then $k_3 \gg k_1[M]$. Cossee argues against this view on the grounds that the measured activation energy ($E_a = 11$ to 14 kcal/mole) is too high to account for putting a neutral olefin molecule into a vacant position. He prefers to attribute this higher E_a value to a rearrangement–insertion step.

In terms of the Cossee mechanism, the polymerization rate R_p is expressed in Eq. 18–16,

$$R_p = \frac{k_1 k_3}{k_2}[C][M] \qquad (18-16)$$

and the measured activation energy ΔE is

$$\Delta E = \Delta E_R - \Delta H_c \qquad (18-17)$$

where ΔE_R is the activation energy for the rearrangement and ΔH_c is heat of complex formation (assumed to be small) and therefore $\Delta E \simeq \Delta E_R$.

Schindler (44), on the other hand, argued in favor of the adsorption step being rate-limiting. This view was reached from the effect of inert gases such as N_2 or ethane on the polymerization of ethylene in the presence of catalysts such as $AlEt_2Cl/AlEt_2H + TiCl_4$. He observed that the rate of polymerization was depressed more than would be proportional to the decrease in the partial pressure of ethylene. Schindler concluded that these "inert" gases are adsorbed on the active centers and depress activity. Relative to ethylene, nitrogen and ethane are less strongly held. In a true adsorption–desorption equilibrium, the number of centers complexed by nitrogen or ethane relative to ethylene would be expected to be relatively smaller. But under the conditions of a very fast propagation of the complexed ethylene and a slow rate of adsorption, the number of active sites blocked by adsorbed ethane or N_2 was considered to be much higher than by ethylene, even though the latter is more strongly adsorbed.

Many kinetic studies have been reported that attempted to describe the polymerization in terms of a Rideal or Langmuir–Hinshelwood model (see Section III,B). Langmuir–Hinshelwood rate laws were favored by Vesely (45) for propylene polymerization with the $AlEt_3$–$TiCl_3$ catalyst and by Reich and Schindler (3) for ethylene polymerization using $AlEt_2Cl$–$TiCl_4$. Schnecko and co-workers (46) and Keii and co-workers (6) reported that the Rideal rate law was more applicable.

The diagnostic test is to make the appropriate plots of R_p vs. metal alkyl concentration, as given by Eqs. 18–4 and 18–7 for the Rideal and Langmuir–Hinshelwood rate laws, respectively. The interpretation of the experimental data is not always clear-cut and sometimes depends on the conditions of the experiment.

Figure 18–8 shows a plot of $[A]/R_\infty$ vs. $[A]$ for the polymerization of propylene with the $AlEt_3$–$TiCl_3AA$ catalyst, according to the Rideal rate law (see page 34 of ref. 1).

Keii (1) has plotted the data of Vesely and Tait according to both rate laws (Figs. 18–9 and 18–10). Apparently, at low $AlEt_3$ concentrations and at specific monomer concentrations, the data can be explained within the framework of both kinetic models.

One characteristic feature of the Langmuir–Hinshelwood rate law is that a maximum should be obtained in the polymerization rate curve as the concentration of metal alkyl is increased. The reason for this is that the metal alkyl and the monomer compete for adsorption with the available centers. When the concentration of AlR_3 is high, these centers complex more favorably with the polar AlR_3 and become less available for complexing with the hydrocarbon monomer. In contrast, according to the Rideal model, the monomer comes directly from the solvent or the gas phase. In the latter case, R_p will be independent of AlR_3 concentration at high concentration levels.

TABLE 18–5

Some Metal Alkyl Concentrations Employed in Ziegler–Natta Systems (5–5b)

System	Effect of increasing [A]	Range of [A] employed (mmoles/liter)
$TiCl_3$/$AlEt_3$	Maximum produced	10–200
Propylene	Independent	29–45
	Independent	5–30
	Independent	2–35

Burfield, McKenzie, and Tait (5) pointed out that the apparent observations of Schnecko (46), Keii (6), and Natta (47) that the rate was independent of Al alkyl concentration for the polymerization of propylene with the $AlEt_3$–$TiCl_3$ catalyst was due to the insufficient concentration range used for the AlR_3 (Table 18–5). They also point out that the value of θ_A depends on the actual concentration of the metal alkyl and not on the ratio of catalyst components. They stressed that Vesely obtained a maximum rate at a concentration of 50 mmoles/liter of $AlEt_3$, a value higher than used by the other workers. The fact that all of these workers used Al/Ti ratios in the same range was not adequate.

Burfield, McKenzie, and Tait (5–5b) have recently re-examined these earlier kinetic views and data. On the basis of kinetic data derived from studies of polymerization of 4-methyl-1-pentene with AlR_3–VCl_3 catalyst, they developed kinetic interpretations in which the propagation occurred with an active center formed by interaction of metal alkyl and the transition metal halide rather than with a metal alkyl molecule adsorbed onto the $TiCl_3$ or VCl_3 surface.

As evidence for this view, they argued that if propagation were to occur with the adsorbed metal alkyl, then the active center concentration (C_0) would be proportional to the fraction of the surface covered by the adsorbed monomeric alkyl (θ_A). They presented data to show that this was not true (Table 18–6). They also reported that C_0 can vary under conditions of constant θ_A.

TABLE 18–6

Variation in the Active Center Concentration with θ_A (5–5b)[a]

Metal alkyl	$C_0 \times 10^4$ (mole/mole VCl_3)	θ_A	$C_0/\theta_A \times 10^3$
Al-i-Bu$_3$	3.78	0.125	3.00
AlEt$_3$	6.10	0.445	1.37
Al(n-Bu)$_3$	3.30	0.090	3.65
Al(n-Hex)$_3$	2.30	0.110	2.10

[a] $[AlR_3] = 37.0$ mmole/liter; $[4\text{-MP-1}] = 2.0$ mole/liter; $[VCl_3] = 18.5$ mmole/liter; solvent = benzene; temp. = 30°C.

The rate expression that they used was:

$$R_p = k_p \theta_M C_0 \tag{18-18}$$

where k_p is the propagation rate constant with respect to adsorbed monomer and θ_M is the fraction of the surface covered by adsorbed monomer. The adsorption of monomer and metal alkyl onto the catalyst surface was considered to be described by Langmuir–Hinshelwood isotherms (see Section III,B).

In summary, while helpful in explaining the polymerization process, the kinetic studies have not enabled us to elucidate structures of the active centers, and we must seek other types of measurements.

B. TYPES, CONCENTRATION, AND DECAY OF ACTIVE CENTERS

Measurements aimed at understanding both the type of centers and their concentration at various stages of a polymerization have received considerable attention. Data of this type would permit elucidation of the kinetic results as described in Section II with respect to the polymerization rate curve and the polymerization product.

The problem is that a variety of methods have been used to measure active centers, each of which suffers from some ambiguity. Hence, compilations such as Table 18–1 do not necessarily show true comparisons of kinetic features characteristic of different Ziegler–Natta catalysts, and one should not accept these without some reservation.

The large variation in reported active centers (C*) can be attributed to the different methods of measurements.

1. Some workers (7) assume that all metal–carbon bonds present in the reactor at the time of termination are active centers. Chain transfer by metal alkyl is rejected (see Bier papers in Chapter 21).

2. Other workers (2, 8, 18) accept chain transfer, extrapolate metal–carbon bond concentrations to zero time, and take the extrapolated value to be the site number.

3. Some workers (2) accept the total adsorbed metal alkyl as a measurement of active centers.

4. Some (48) calculate C* from rate of transfer reaction, the assumption being made that the rate of chain transfer is proportional to the number of active sites, $R_{tr} = k_{tr}[M][A][C^*]$.

1. Types of Centers

Natta and co-workers (11) reported that 65% of the centers present formed 93% of the total polymer that had isotactic structure; the remaining 35% of the centers accounted for only 7% of the polymer which had atactic structure. They concluded that isotactic sites were seven times more active than atactic sites. Feldman and Perry (18), for the AlR_3–$TiCl_4$ catalyst, also concluded that the k_p of an atactic-forming center was 10% of the isotactic-forming center. To account for their kinetic data and the observed molecular weight distribution in ethylene–propylene copolymerization, Meshkova and co-workers (49) concluded that two types of activated reaction centers were present. Grieveson (15) attributed wide molecular weight distribution to unequal activities of reaction centers.

2. Concentration Centers

The following methods have been used to measure concentration of active centers. Natta and co-workers (2, 11) used kinetic and adsorption techniques to determine the number of active centers in the $AlEt_2Cl$–$\alpha TiCl_3$ and $AlEt_3$–$\alpha TiCl_3$ catalysts, as shown in Table 18–7.

TABLE 18–7

Number of Active Centers in AlEt$_2$Cl–αTiCl$_3$ and AlEt$_3$–αTiCl$_3$ Catalysts (2, 11)

| | Concentration of active center mole/mole TiCl$_3$ | | |
| | | Kinetic method | |
Catalyst	Adsorption method	Atactic	Isotactic
αTiCl$_3$–AlEt$_3$	1×10^{-2}	—	—
αTiCl$_3$–AlEt$_2$Cl	0.3×10^{-2}	0.2×10^{-2}	0.4×10^{-2}

Chien (50) calculated [C*] values from ^{131}I$_2$ in polymer samples resulting from the reaction of labeled ^{131}I$_2$ with metal alkyl complexes. A maximum was shown after about 20 minutes polymerization, followed by a sharp decrease for 280 minutes.

Bier (51) concluded from his studies that the number of active centers does not stay constant throughout the polymerization, and its value is influenced by the temperature, by the polymerization period, or possibly by the nature of the aluminum alkyl and its concentration. Decrease in R_p (even though the number of centers increased) was attributed to the lower activity of centers, due to limited diffusion of monomer and metal alkyl from solution to the centers. He believed chain transfer or chain termination reactions were insignificant and concluded that the polymer chains had living character even though their activity was diminished at increased conversion.

Coover and co-workers (9, 48) concluded from their kinetic studies that the number of active centers was much smaller than determined previously by Natta, Chien, and Kohn. The following tabulation summarizes the active site data for catalyst systems at 70°C.

Investigator	Method	Active sites (moles/mole TiCl$_3 \times 10^3$)
Natta	^{14}C	6
Chien	^{14}C	80–120
Kohn	^{3}H	28
Coover	^{3}H	1

This discrepancy was attributed to the way the kinetic data were interpreted rather than to its generation (all of the experimental methods were similar). Coover and co-workers, unlike the other workers, did not extrapolate to zero time to obtain the concentration of centers. Their studies indicated that the rate of chain transfer during the early changes of the polymerization was much higher than at later stages, and their evaluation of site number took this into account.

Coover and co-workers made measurements of active sites for three catalysts (Table 18–8). According to their calculations, the absolute propagation rate was ten times greater than earlier estimates, and the lifetime of the growing polymer chain at low conversion was very short, about 1 minute. Soga (52), however, criticized the Coover method in that R and transfer constant R_{tr} were not constant during the initial 50 to 60 minute period.

TABLE 18–8

Active Site Measurements (9, 48)

Catalyst	$R_p \times 10^4$ (mole/liter sec)	$R_{tr} \times 10^7$ (mole/liter sec)	$C^* \times 10^5$ (mole/liter)	k_p (liter/mole sec)	τ (sec)	Active sites (mole/mole TiCl₃)
AlEt₃–αTiCl₃	22.6	26	∼10	48	38	0.003
AlEt₂Cl–αTiCl₃	5.45	3.2	3	40	94	0.001
AlEtCl₂–αTiCl₃	1.86	1.3	0.8	50	62	0.0002

Tanaka (53) calculated the active sites for the $AlEt_2Cl–TiCl_3AA$ catalyst and found $C^* = 1.0 \times 10^{-2}$ mole/mole $TiCl_3$ ($R_p = 2.52$ mole propylene per minute per mole $TiCl_3$ at 70°C and 2 atm propylene pressure).

Jung and Schnecko (54) found that even with a single, heterogeneous catalyst ($AlEt_2Cl + TiCl_3$), marked differences in polymerization behavior and kinetic data were obtained for the three olefins ethylene, propylene, and 1-butene. Their kinetic analysis was based on tritium end-group determinations. Graphic or computational extrapolations of the measured end-group content to the start of the polymerization was made to differentiate true active centers from those metal–carbon bonds formed by transfer reactions with metal alkyl molecules. The number of centers, C^*, was different for the three monomers: $C^*_{C_2}:C^*_{C_3}:C^*_{C_4} = 2:1.5:1$ (initial) and 3.6:1.7:1 (final). They concluded that the different activity of the catalyst toward ethylene, propylene, and 1-butene at the beginning of the polymerization was attributable to different k_p values and to different active center concentrations [C^*]. The change in R_p during polymerization was explained on the basis of a change in the number of active centers. The broad molecular weight distribution was attributed to different k_p values for different active centers, depending on their location on the catalyst surface. Speculations were also made on how the physical state of the polymer being formed affects kinetic behavior of the catalyst.

3. Decay of Centers

Many workers speculated that the decline in the rate of polymerization with many of the olefin–catalyst systems was due to the loss of active centers. The paths by which these centers became inactive varied, and this section describes the different views expressed.

Kohn and co-workers (8) studied the polymerization of propylene with the $AlEt_2Cl–TiCl_3AA$ catalyst. They observed a decline in R_p during a 6 hour polymerization time, even though the concentration of metal–carbon bonds increased significantly during the same period. They postulated that not all of the metal–carbon bonds were active, as previously suggested by Feldman and Perry, and that decay of R_p was due to the rearrangement that occurred in the bimetallic complex center, as shown in Eq. 18–19.

$$
\begin{array}{cc}
\underset{\substack{\text{Cl} \diagdown \quad \diagup \text{Cl} \diagdown \quad \diagup \text{R} \\ \text{Ti} \qquad \text{Al} \\ \text{Cl} \diagup \quad \diagdown \text{Pn} \diagup \quad \diagdown \text{Cl}}}{\text{Active growing center}}
&
\rightleftharpoons
\quad
\underset{\substack{\text{Cl} \diagdown \quad \diagup \text{Cl} \diagdown \quad \diagup \text{R} \\ \text{Ti} \qquad \text{Al} \\ \text{Cl} \diagup \quad \diagdown \text{Cl} \diagup \quad \diagdown \text{Pn}}}{\text{Inactive center}}
\end{array}
\qquad (18\text{–}19)
$$

$$
\Updownarrow
$$

$$
\underset{\substack{\text{Active site but} \\ \text{Pn not growing}}}{\substack{\text{Cl} \diagdown \quad \diagup \text{Cl} \diagdown \quad \diagup \text{Cl} \\ \text{Ti} \qquad \text{Al} \\ \text{Cl} \diagup \quad \diagdown \text{R} \diagup \quad \diagdown \text{Pn}}}
$$

Caunt (55) examined the $AlEt_2Cl–TiCl_3$ ($TiCl_3 \cdot xAlCl_3$) catalyst for the polymerization of propylene at 50°C. He postulated that site formation is preceded by the surface reaction shown in Eq. 18–20. The $AlEtCl_2$ acts as

$$
\underset{\text{Inactive system}}{(AlEt_2Cl)_2 + TiCl_3 \text{ surface}} \longrightarrow \underset{\text{Active centers}}{AlEt_3 \text{ on } TiCl_3 \text{ surface} + AlEtCl_2 \text{ in solution}} \qquad (18\text{–}20)
$$

a poison and accounts for the decrease in activity. Upon addition of electron donors, the $AlEtCl_2$ poisoning effect is negated, and higher activities were obtained. He attributed decreasing R_p values to the accumulation of inhibitor and exhaustion of activator.

Chien (56) showed that the concentration of centers [C*] was five times higher at 50°C than at 90°C for the $AlEt_2Cl–\alpha TiCl_3$ catalyst when used for

propylene polymerization. He suggested that changes in surface characteristics and surface area were responsible for the variation with temperature.

Schnecko and co-workers (46) concluded that the valence state of the
catalyst was not as critical (over reduction) as was the geometric nature of
the surface in determining whether or not a maximum occurred in R_p vs. Al/Ti
curve; for example, $AlEt_3$–$TiCl_4$ has maxima while $AlEt_2Cl$–$TiCl_4$ does not.

Soga and co-workers (57) suggested from their studies that a decrease in
the polymerization rate was due to loss of centers and not to a change in
the nature of centers. They (58) attributed the decline in rate to the deactivation of the centers which they presumed to be a bimetallic complex func-
tioning by a Rideal-type mechanism.

Keii and co-workers (59) attributed the rate decline to an irreversible
deactivation of the polymerization centers localized on unstable surface
sites due to a spontaneous process and partially to a propylene (or polymer)
attack. Such deactivation was possibly attributed to a structural change in
the $TiCl_3$ crystal. Polymer coverage of centers was also suggested to account
for the loss of activity of some centers. Ambroz and co-workers (60) similarly
expressed that, for the $AlEt_3$–$TiCl_3$ catalyst in propylene polymerization,
the solid phase becomes depleted in chloride, and this loss contributes to
decomposition of active centers.

Smith and Zelmer (61) suggested a second-order deactivation of active
species with time on the basis of kinetic measurement for the polymerization of ethylene with $AlEt_3$–$TiCl_4$ catalyst.

Natta and co-workers (62) studied the polymerization of propylene at
subatmospheric pressures and found the rate of initiation to be lower than
the rate of propagation. They suggested that cat–H, cat–Et, and cat–Pr are
in a lower activation state than complexes of cat–Pn.

C. TYPE, CONCENTRATION, AND DECAY
OF POLYMER CHAINS

Polymer chain lifetimes ranging from a few minutes to hours (or days)
have been reported, as shown in Table 18–9 (62a).

Natta and Pasquon (2) suggested that the polymer chain remains alive
for about 10 minutes before termination by chain transfer by the metal
alkyl occurs. Propylene was polymerized with the $AlEt_3$–$\alpha TiCl_3$.

In contrast, Bier and co-workers (63) concluded that polypropylene chains
remained alive throughout the polymerizations and that termination of
chain growth did not take place by chain transfer with metal alkyl. They
proposed monomolecular disproportionation reaction to be the main route
to terminate chain growth and this depended on temperature. A plot of η_r

TABLE 18-9

Average Chain Growth Lifetimes Reported for Ziegler Catalysts (62a)

Monomer	Catalyst system[a]	Temp. (°C)	Average chain lifetime	Author
C_2H_4, C_3H_6, C_4H_8	Hoechst $TiCl_3$/Al alkyl	15–70	Several hours	Bier and Lehmann
C_2H_4	$TiCl_4$/AlR_3	68	1–30 minutes	Feldman and Perry
C_2H_4	$TiCp_2Cl_2$/$Al(CH_3)_2Cl$	0–30	~24 minutes	Chien
C_2H_4	$\gamma TiCl_3$/$Al(C_2H_5)_2Cl$	40	30 minutes	Grieveson
$C_2H_4 + H_2$	$\gamma TiCl_3$/$Al(C_2H_5)_2Cl$	40	~7 minutes	Grieveson
C_2H_4	$TiCl_4$, $TiCl_3$/$Al(C_2H_5)_3$	50	Limited dependence of η on reaction time	Schnecko et al.
C_2H_4	Hoechst $TiCl_3$/Al alkyl	30	>15 hours	Bier et al.
C_3H_6	$TiCl_3$/$Al(C_2H_5)_2Cl$	60	~12 minutes	Caunt
C_3H_6	Hoechst $TiCl_3$/$Al(C_2H_5)_2Cl$	50	30–72 minutes	Bier (recalculated by Caunt)
C_3H_6	$TiCl_3$/$Al(C_2H_5)_3$	70	5–12 minutes	Natta and Pasquon
C_3H_6	$TiCl_3$/$Al(C_2H_5)_2Cl$	50	410 minutes	Chien
C_3H_6	Stauffer $TiCl_3$/Al/$Al(C_2H_5)_2Cl$	70	33–40 minutes	Tanaka and Morikawa
C_3H_6	$TiCl_3$/$Al(C_2H_5)Cl_2$/HMPT	70	Initially 1 minute	Coover et al.
C_3H_6	$TiCl_3$/$Al(C_2H_5)_2Cl$	40	Essentially no termination	Ingberman et al.
C_3H_6	$TiCl_3$/$Al(C_2H_5)_3$ or $Al(C_4H_9)_2H$	56	Very limited change of viscosity number with time	Schnecko et al.

[a] HMPT, Hexamethylphosphoric triamide $O:P[N(CH_3)_2]_3$; Cp, cyclopentadienyl.

vs. $AlR_3/TiCl_3$ ratio showed no change in the range 2 to 11. Bier accepted the Al–C bond of the bimetallic complex as the active center (for examples, see Figs. 21–3 through 21–7).

Chien (56) found a chain lifetime of about 410 minutes for $AlEt_2Cl$–$\alpha TiCl_3$–propylene, which is more or less comparable to Bier's values. Figure 4 of Chien (56) shows that the molecular weight increases from 0.4×10^5 (2 hours) to 5.5×10^5 (50 hours) with polymerization time. Chain transfer by $AlEt_2Cl$ accounts for the termination of chain growth, k_t at $50°C = 6.54 \times 10^{-5}$ mole$^{-1/2}$ sec$^{-1/2}$. The rate of polymerization R_p was constant throughout the 50 hour polymerization.

Tanaka (53) concluded from his kinetic data that the average lifetime of the growing chain of polypropylene ($AlEt_2Cl$–$TiCl_3$ catalyst) was about 33 to 40 minutes. With increasing conversions (0 to 8 hours), the average lifetime of growing chains increased from 10 to 30 minutes. Also, as the polymerization temperature was decreased from 70° to 50° to 30°C, average chain lifetimes increased from 28 to 48 to 226 minutes. Tanaka calculated these chain lifetimes from changes of the number average molecular weights from measurements of molecular weight distributions of polypropylene by a turbidimetric method. Previous workers had ignored changes in MWD in their calculations. The lifetime of the growing polymer chains was not affected by increasing the concentration of $AlEt_2Cl$ and the pressure of monomer.

Feldman and Perry (18) suggested chain lifetimes on the order of minutes (about 4 to 10 minutes). They concluded that after 40 minutes, each site formed about 7 moles of polymer.

Molecular weights of polypropylene samples synthesized in the presence of the metal alkyl-free catalyst $TiCl_3 \cdot xAlCl_3 + n\text{-}Bu_3N$ increased throughout the 144 hour duration of the polymerization (64), as shown in the tabulation below.

Time (hr)	$[\eta]$ (dl/g)
1	3
10	8
144	18

A large number of the Ti–Pn growth centers apparently can survive in the absence of the metal alkyl, but the lifetime of the polymer chains was not calculated.

For ethylene polymerization with the soluble catalyst $(C_5H_5)_2TiCl_2$–$(CH_3)_2AlCl$, Chien (50) showed an inverse relationship between polymer molecular weight and catalyst concentration. This suggested a bimolecular termination mechanism. The values of k_t in this system were several orders of magnitude lower than found in radical polymerizations; even so, polymers having low molecular weights were obtained.

Kissin (65) proposed that mechanical termination of polymer chains can occur. Even in the absence of monomer, the polymer chain connected to the center continues to be drawn from the catalyst surface by neighboring chains making up a common crystallite. The rate of mechanical termination would be proportional to the polymerization rate and probably then indistinguishable from chain termination to monomer.

D. ACTIVATION ENERGY

Generally, overall activation energy values in the range 8 to 17 kcal/mole have been reported for different catalyst–monomer systems (Table 18–10) (18, 34, 50, 56, 66–71). There is, however, some confusion about the meaning of these values.

TABLE 18–10

Activation Energies Reported for Different Catalyst–Olefin Systems

Catalyst	Olefin	E_a (kcal/mole)	Ref.
Al-i-Bu$_3$–TiCl$_3$	Ethylene	10	66
AlEt$_3$–(sec-BuO)$_4$Ti	Ethylene	13	67
Al-i-Bu$_2$H–(C$_5$H$_5$)$_2$TiCl$_2$	Ethylene	12.2	50
Al-i-Bu$_3$–TiCl$_4$	Isoprene	14.4	68
AlEt$_3$–TiCl$_3$	Propylene	10.5	69
AlBu$_2$H–TiCl$_4$ ⎫ Al-i-Bu$_3$–TiCl$_4$ ⎬	Ethylene	11	18
AlEt$_2$Cl–αTiCl$_3$	Propylene	13.0	56
AlEt$_3$–TiCl$_3$	Styrene	8.1	70
AlEt$_2$Cl–TiCl$_3$(AlR)	Ethylene	17.3	34
AlEt$_3$–TiCl$_3$	3-Methyl-1-butene	0.10	71

Some have suggested that these values are in agreement with a diffusion-controlled mechanism and do not reveal anything about the growth step. Interpretations can also be cloudy if the polymerization is mass-transfer limited.

At best, they describe the overall polymerization process, but one should not relate them to the propagation step without definite knowledge of what is happening to the number and types of centers as the polymerization temperature is increased.

TABLE 18-11

Absolute Rate Constant in Ziegler–Natta Catalyst Systems

Monomer	Catalyst system	Temp. (°C)	k_p (liter/mole-sec)	Ref.
Propylene	$AlEt_2Cl-TiCl_3AA$	70	5.6	53
Propylene	$AlEt_3-TiCl_3$	70	2.5	47
Propylene	$AlEt_2Cl-TiCl_3AA$	50	1.2	37
Propylene	$AlEt_2Cl-TiCl_3$	50	0.426	56
		70	1.15	56
		90	3.50	56
Ethylene	$Al-i-Bu_3-TiCl_4$	54.5	127	18
Ethylene	$Al(CH_3)_2Cl-(C_5H_5)TiCl_2$	30	13.6	56

TABLE 18-12

Kinetic Constants at Various Temperatures for $AlEt_2Cl-TiCl_3$–Propylene System (56)

Temp. (°C)	k_p (liter-mole^{-1}-sec^{-1})	k_{tr_1} (mole$^{-1/2}$-liter$^{1/2}$-sec^{-1})	k_{tr_2} (liter-mole^{-1}-sec^{-1})
30	0.1 ± 0.2	—	—
50	0.426 ± 0.018	6.54×10^{-5}	9.54×10^{-4}
70	1.15 ± 0.20	3.50×10^{-4}	1.90×10^{-3}
90	3.50 ± 0.12	1.58×10^{-3}	3.60×10^{-3}

E. SPECIFIC RATE CONSTANTS

Tanaka and Morikawa (53) collected k_p values for a number of systems and suggested that differences may be attributed to the method of calculating the number of centers (Table 18-11). Chien (56) reported constants at various temperatures for the $AlEt_2Cl-TiCl_3$–propylene system, as shown in Table 18-12.

F. THE POLYMER AS A BARRIER TO THE DIFFUSION OF THE MONOMER AND METAL ALKYL (72)

In Section V,B, the decline in the rate of polymerization was attributed by many workers to loss of active centers with increasing time of polymerization (or conversion). Other workers have attributed this decline in rate to the formed polymer (9, 18, 19, 33, 73, 74).

Coover (9) noted that, for the $AlEt_2Cl-\alpha TiCl_3$ catalyst, the rate of formation of metal–polymer bonds decreased markedly at a yield of polymer near

2 g/g $TiCl_3$. The inherent viscosity of the polymer being formed began to increase, and the MWD broadened. They attributed this effect to the occlusion of active sites by relatively large amounts of polymer. This would also reduce the rate at which the polar metal alkyl can reach the site and affect chain transfer. The polymerization rate was not diminished, suggesting that the hydrocarbon monomer was still able to reach the centers easily, and only after higher conversions were obtained was its access to them limited.

Lehmann and Gumboldt (26) also attributed the decline in R_p (despite the apparent increase in concentration of centers taken to be M–Pn bonds) to be a diffusional barrier of polymer being formed.

Singh and Merrill (75) explained the broad MWD in polyethylene by the consideration of the diffusion of the monomer and the movement of catalytic sites in a growing polymer chain. Various mathematical models were considered to show that a broad MWD can be obtained for Thiele parameters greater than 6.

Begley (76), on the other hand, concluded that a loss of centers rather than diffusion barriers accounted for the decline in R_p. According to his calculations, particle size of the polymer–catalyst would have to be 10 to 100 times the size of the catalyst before diffusion time would be significant. Lipman and Norrish (77) studied the vapor-phase polymerization of ethylene with $AlMe_3$–$TiCl_4$ and also concluded that diffusion of monomer through a polymer layer was not the rate-controlling process.

Schmeal and Street (78, 79) described four mathematical models in which the polymer particles were presumed to expand as new polymer was formed. According to these models, the active centers are dispersed throughout the polymer matrix, and the monomer must diffuse through the polymer to reach the active centers. In the base model, particle growth was ignored, and centers were assumed to be dispersed in a spherical matrix of polymer. In the solid core model, polymer was assumed to accumulate about a solid catalyst core. In one expansion model, catalyst centers were assumed to be moving outward in a polymer matrix with a negligible velocity. In the other expansion model (flow model), centers were assumed to move outward unhindered by the expanding polymer matrix. Figure 18–15 compares changes in MWD (Q values) as the Thiele modulus $\alpha = R(kl/D)^{1/2}$ increases from zero to 60. The Thiele modulus is a ratio characteristic of the diffusion time to reaction time (a measure of the importance of diffusion to reaction). Broad MWD were predicted by the authors in cases of diffusion control (large α) for those models in which centers were not accessible to monomer, that is, in all but the solid core model. They concluded that polymerization rates also declined toward an asymptotic value as the particles expanded in diffusion-controlled cases. Most of the observable decline in rate would have occurred by the time

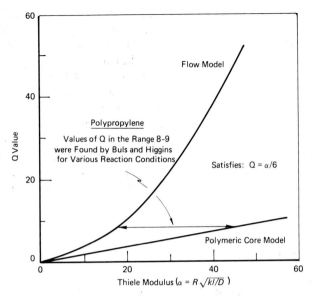

Fig. 18–15. Q value of the molecular weight distribution for polymer produced on a particle as a function of the Thiele modulus predicted by the expansion models (78, 79).

the particle radius had increased to about three times its original value. Any further decline was attributed largely to deactivation of centers.

Buls and Higgins (80) concluded from their kinetic studies that the rate decline in the polymerization of propylene with $AlEt_2Cl$–$TiCl_3$ occurred mostly during the first half hour of reaction. The rate decline was minimized at higher temperatures (70° vs. 40°C) and they suggested that the final rate was controlled by the rate of diffusion of the monomer through the polymer. Under their conditions, the rate was proportional to the first power of $TiCl_3$ concentration and the one-half power of the $AlEt_2Cl$ concentration. Variations were noted from use of different preparations of $TiCl_3$, and the final rate responded differently to changing reaction conditions than did the initial rate. The polymerization was initially first-order with respect to propylene concentration. But, following a rapid rate decline during the initial period, the reaction became functionally second-order. Hydrogen followed a first-order consumption and the rate of polymerization was increased in its presence.

Crabtree and co-workers (81) made an allowance for an observed first-order decay in intrinsic catalyst activity and proceeded to study the effect of catalyst encapsulation. A model was used in which the primary particles (subparticles with about 100 to 500 Å in diameter) were encapsulated by the formed polymer. Ethylene was polymerized in a slurry reactor using a highly active Ziegler–Natta catalyst in the presence of hydrogen. They found agreement between theoretical predictions and experimental findings. A positive conclusion was reached that the polymerization was diffusion-controlled throughout the major part of the catalyst lifetime. The combined effect of this diffusion barrier and the gradual loss of centers determined the observed molecular weight and molecular weight distribution of the polyethylene. Molecular weight distribution was predicted and found to be wider during the early stages but narrowed with increasing conversions and reaction times; for example, the value decreased from about 15 to 8 as the yield of polymer increased from about 0.3 to 10 kg per gram of catalyst.

Guttman and Guillet (82), from their examination of propylene polymerizations, also suggested that monomer diffusion through the formed polymer can be rate-determining.

VI. Conclusion

In spite of many excellent studies, the basic kinetic parameters are not well understood. This is partly due to the complex nature of the many different forms of the Ziegler–Natta catalysts and partly due to the lack of a suitable diagnostic tool for counting the numbers of active centers under variable conditions. The situation, however, is not hopeless. The basic kinetic results are fairly well defined experimentally (Section II), and their control in the laboratory is reasonably understood (Section IV). Indeed, the explanations that have been put forth are cogent. What we seem to be lacking is the ability to assign the relative contributions of these explanations for the various catalysts under the conditions of polymerization. The problem is also compounded because what we learn kinetically from one catalyst may not be applicable to another.

The author feels that once adequate site counting methods are found (which will not be subject to controversy themselves), we shall take a great step forward in overcoming these obstacles.

References

1. T. Keii, "Kinetics of Ziegler-Natta Polymerization." Chapman and Hall, 1972.
2. G. Natta and I. Pasquon, *Adv. Catal.* **11**, 1–65 (1959).
3. L. Reich and A. Schindler, "Polymerization by Organometallic Compounds," Chapters III-1 and IV. Wiley (Interscience), New York, 1966.
4. J. Boor, *Macromol. Rev.* **2**, 115–268 (1967).
5. D. R. Burfield, I. D. McKenzie, and P. J. Tait, *Polymer* **13**, 321 (1972).

5a. I. D. McKenzie, P. J. Tait, and D. R. Burfield, *Polymer* **13**, 307 (1972); D. R. Burfield and P. J. Tait, *ibid.* p. 315; D. R. Burfield, P. J. Tait, and I. D. McKenzie, *ibid.* p. 321.

5b. P. J. Tait, *Polym. Prepr., Am. Chem. Soc., Div. Polym. Chem.* **15**, 292 (1974).

6. T. Keii, K. Soga, and N. Saiki, *J. Polym. Sci., Part C* **16**, 1507 (1967).

7. G. Bier, A. Gumboldt, and G. Schleitzer, *Makromol. Chem.* **58**, 43 (1962).

8. E. Kohn, H. J. L. Schuurmans, J. V. Cavender, and R. A. Mendelson, *J. Polym. Sci.* **58**, 681 (1962).

9. H. W. Coover, Jr., *J. Polym. Sci., Part C* **4**, 1511 (1964).

10. K. Soga and T. Keii, *J. Polym. Sci., Part A-1* **4**, 2429 (1966).

11. G. Natta, G. Mazzanti, D. DeLuca, U. Giannini, and F. Bandini, *Makromol. Chem.* **71**, 54 (1964).

12. G. Natta, I. Pasquon, G. Pajaro, and E. Giachetti, *Chim. Ind. (Milan)* **40**, 556 (1958).

13. L. S. Bressler, V. A. Grechanovskii, A. Muzhai, and I. Poddubnyi, *Polym. Sci. USSR (Engl. Transl.)* **11**, 1325 (1969).

13a. See Chapter 15, Section II for discussion and references; A. Zambelli and C. Tosi, *Adv. Polym. Sci.* **15**, 32–60 (1974).

14. J. van Schooten, E. W. Duck, and R. Berkenbosch, *Polymer* **2**, 351 (1961).

15. B. M. Grieveson, *Makromol. Chem.* **84**, 93 (1965).

16. F. Eirich and H. Mark, *J. Colloid Sci.* **11**, 748 (1956).

16a. T. Keii, K. Soga, and N. Saiki, *J. Polym. Sci., Part C* **16**, 1507 (1967).

17. G. Natta, I. Pasquon, J. Svab, and A. Zambelli, *Chim. Ind. (Milan)* **44**, 621 (1962).

18. C. F. Feldman and E. Perry, *J. Polym. Sci.* **46**, 217 (1960).

19. G. Bier, A. Gumboldt, and G. Schleitzer, *Makromol. Chem.* **58**, 43 (1962).

20. C. A. Lukach and H. M. Spurlin, *in* "Copolymerization" (G. E. Ham, ed.), Chapter IVA. Wiley (Interscience), New York, 1964.

21. A. Zambelli, I. Pasquon, A. Marinangeli, G. Lanzi, and E. R. Mognaschi, *Chim. Ind. (Milan)* **46**, 1464 (1964).

21a. D. L. Christman, *J. Polym. Sci., Part A-1* **10**, 471 (1972).

22. G. Boocock and R. N. Haward, *SCI Monogr.* **20**, 3 (1966).

23. W. Marconi, A. Mazzei, M. Araldi, and M. De Malde, *J. Polym. Sci., Part A* **3**, 735 (1965).

24. J. Boor, *J. Polym. Sci., Part C* **1**, 237 (1963).

25. R. J. Kern and H. G. Hurst, *J. Polym. Sci.* **44**, 272 (1960).

25a. M. L. Cooper and J. B. Rose, *J. Chem. Soc.* p. 795 (1959).

26. G. Lehmann and A. Gumboldt, *Makromol. Chem.* **70**, 23 (1964).

27. L. V. Gavrilova, V. A. Grechanovskii, Y. Kropacheva, and B. Dolgoplosk, *Polym. Sci. USSR (Engl. Transl.)* **10**, 2324 (1968).

28. A. K. Ingberman, I. J. Levine, and R. J. Turbett, *J. Polym. Sci., Part A-1* **4**, 2781 (1966).

29. F. S. Dyachkovskii, A. K. Shilova, and A. E. Shilov, *J. Polym. Sci., Part C* **16**, 2333 (1967).

30. E. A. Fushman, V. J. Tsvetkova, and N. M. Chirkov, *Dokl. Akad. Nauk SSSR* **164**, 1085 (1965).

31. G. P. Belov, A. P. Lisitskaya, N. M. Chirkov, and V. I. Tsvetkova, *Vysokomol. Soedin., Ser. A* **9**, No. 6, 1269 (1967).

32. M. Gippin, *Rubber Chem. Technol.* **39**, 508 (1966).

33. See Section IVB in Chapter 3 and Section II in Chapter 7.

34. M. N. Berger and B. M. Grieveson, *Makromol. Chem.* **83**, 80 (1965).

35. T. Keii, M. Taira, and T. Takagi, *Can. J. Chem.* **41**, 206 (1962).

36. J. C. W. Chien, *J. Polym. Sci., Ser. A* **1**, 1839 (1963).

37. G. Bier, W. Hoffman, G. Lehmann, and G. Seydel, *Makromol. Chem.* **58**, 1 (1962).

38. L. A. Novokshonova, G. P. Berseneva, V. I. Tsvetkova, and N. M. Chirkov, *Polym. Sci. USSR (Engl. Transl.)* **9**, 631 (1967).

39. R. L. Combs, D. F. Slonaker, F. B. Joyner, and H. W. Coover, Jr., *J. Polym. Sci., Part A-1* **5**, 215 (1967).
39a. L. A. Novokshonova, V. I. Tsvetkova, and N. M. Chirkov, *Izv. Akad. Nauk SSSR, Ord. Khim. Nauk* p. 1176 (1963).
40. I. Belea, V. Dimonie, E. Melega, P. Blenke, and V. Provinceanu, *Rev. Roum. Chim.* **14**, 473 (1969).
40a. T. Yoshimoto, K. Komatsu, R. Sakata, K. Yamaoto, Y. Takeuchi, A. Onishi, and K. Ueda, *Makromol. Chem.* **135**, 61 (1970).
41. K. Vesely, J. Ambroz, R. Vilim, and O. Hamrik, *J. Polym. Sci.* **55**, 25 (1961).
42. G. P. Belov, L. N. Raspopov, A. P. Lisitskaya, V. I. Tsvetkova, and N. M. Chirkov, *Polym. Sci. USSR (Engl. Transl.)* **8**, 1728 (1966); S. M. Mezhikovskii, Yu. V. Kissin, and N. M. Chirkov, *Vysokomol. Soedin., Ser. A* **10**, 2231 (1968).
43. P. Cossee, *in* "The Stereochemistry of Macromolecules" (A. D. Ketley, ed.), Chapter 3, p. 155. Dekker, New York, 1967.
44. L. Reich and A. Schindler, "Polymerization by Organometallic Compounds," p. 328, and earlier references cited therein. Wiley (Interscience), New York, 1966; *Makromol. Chem.* **70**, 94 (1964).
45. K. Vesely, *Pure Appl. Chem.* **4**, 407 (1962).
46. H. Schnecko, M. Reinmoller, K. Weirauch, and W. Kern, *J. Polym. Sci., Part C* **4**, 71 (1963).
47. G. Natta, *J. Polym. Sci.* **34**, 21 (1959).
48. H. W. Coover, Jr., J. E. Guillet, R. L. Combs, and F. B. Joyner, *J. Polym. Sci., Part A-1* **4**, 2583 (1966).
49. I. N. Meshkova, I. L. Dubnikova, E. I. Vizer, and N. M. Chirkov, *Vysokomol. Soedin., Ser. B* **11**, 486 (1969).
50. J. C. W. Chien, *J. Am. Chem. Soc.* **81**, 86 (1959).
51. G. Bier, *Makromol. Chem.* **70**, 44 (1964).
52. K. Soga, *J. Polym. Sci., Part B* **5**, 347 (1967).
53. S. Tanaka and H. Morikawa, *J. Polym. Sci., Part A* **3**, 3147 (1965).
54. K. A. Jung and H. Schnecko, *Makromol. Chem.* **154**, 227 (1972).
55. A. D. Caunt, *J. Polym. Sci., Part C* **4**, 49 (1963).
56. J. C. W. Chien, *J. Polym. Sci., Part A* **1**, 425 (1963).
57. K. Soga, T. Keii, and Y. Murayama, *J. Polym. Sci., Part B* **4**, 199 (1966).
58. K. Soga, *Shokubai* **11**, 67 (1969).
59. T. Keii, K. Soga, and N. Saiki, *J. Polym. Sci., Part C* **16**, 1507 (1967).
60. J. Ambroz, P. Osecky, J. Mejzlik, and O. Hamrik, *J. Polym. Sci., Part C* **16**, *423* (1967).
61. W. E. Smith and R. G. Zelmer, *J. Polym. Sci., Part A* **1**, 2587 (1963).
62. G. Natta, I. Pasquon, J. Svab, and A. Zambelli, *Chim. Ind. (Milan)* **44**, 621 (1962).
62a. M. N. Berger, G. Boocock, and R. N. Haward, *Adv. Catal.* **19**, 227 (1969).
63. G. Bier, A. Gumboldt, and G. Lehmann, *Plast. Inst., Trans. J.* **28**, 98 (1960).
64. J. Boor, *Macromol. Rev.* **2**, 237 (1967).
65. Yu. V. Kissin, *Vysokomol. Soedin., Ser A* **11**, 1569 (1969).
66. E. J. Badin, *J. Am. Chem. Soc.* **80**, 6545 (1958).
67. C. E. H. Bawn and R. Symcox, *J. Polym. Sci.* **34**, 139 (1959).
68. W. M. Saltman, W. E. Gibbs, and J. Lal, *J. Am. Chem. Soc.* **80**, 5615 (1958).
69. G. Natta, *Angew. Chem.* **7**, 213 (1957).
70. F. D. Otto and G. Parravano, *J. Polym. Sci., Part A* **2**, 5131 (1964).
71. M. P. Berdnikova, Yu. V. Kissin, and N. M. Chirkov, *Vysokomol. Soedin.* **5**, 63 (1963).
72. The reader might find it useful to reread Chapter 8, Sections III and IV, which describe the microstructure of the catalyst and polymer particles.

73. A. S. Hoffman, B. A. Fries, and P. C. Condit, *J. Polym. Sci., Part C* **4**, 109 (1964).
74. K. Fukui, T. Schimidzu, T. Yagi, S. Fukumoto, T. Kagiya, and S. Yuosa, *J. Polym. Sci.* **55**, 321 (1961).
75. D. Singh and R. P. Merrill, *Macromolecules* **4**, 599 (1971).
76. J. W. Begley, *J. Polym. Sci., Part A-1* **4**, 319 (1966).
77. R. D. A. Lipman and R. G. W. Norrish, *Proc. R. Soc. London, Ser. A* **275**, 310 (1963).
78. W. R. Schmeal and J. R. Street, *AIChE J.* **17**, No. 5, 1188 (1971).
79. W. R. Schmeal and J. R. Street, *J. Polym. Sci., Polym. Phys. Ed.* **10**, 2173 (1972).
80. V. W. Buls and T. L. Higgins, *J. Polym. Sci., Polym. Chem. Ed.* **11**, 925 (1973).
81. J. R. Crabtree, F. N. Grimsby, A. J. Nummelin, and J. M. Sketchley, *J. Appl. Polym. Sci.* **17**, 939 (1973).
82. J. Y. Guttman and J. E. Guillet, *Macromolecules* **1**, 461 (1968).

Polymerization of Monomers

I. Introduction

Ziegler–Natta catalysts are best suited for the polymerization of olefins and dienes. In the case of olefins, ethylene, propylene, and higher α-olefins were found most active. Similarly, simple dienes such as butadiene, isoprene, and 1,3-pentadiene were most active. Commercial developments were largely based on these simple monomers. Yet many other types of monomers have been found active with different forms of Ziegler–Natta catalysts, such as allenes, acetylenes, polar vinyl, and related monomers. The mechanisms by which some of these polymerize are not always those by which simple olefins or dienes polymerize, because the Ziegler–Natta catalysts can provide a source of conventional initiation: cationic, anionic, and radical, as well as novel initiation or catalysis (see Chapter 22). An example of the latter is the metathesis-type polymerization of cyclopentene.

This chapter describes the many different monomers that have been polymerized with Ziegler–Natta catalysts. Their presentation is according to the structure type of the monomer. Some of the more common monomers are not discussed in great detail in this chapter since their polymerization was already described in earlier chapters, including ethylene, propylene, butadiene, and isoprene in Chapters 4, 5, 13, 14, 15, 16, and 17.

II. Ethylene

Because it poses the least steric hindrance, ethylene is the most easily polymerized olefin. Furthermore, because it does not have the potential of forming asymmetric carbons, stereoisomerism is not possible. In formulating

512

catalysts, one has a very wide lattitude in the choice of the metal alkyls and transition metal salts, as well as modifications on this catalyst. All kinds of Ziegler–Natta catalysts have been successfully used to form highly linear polyethylene, including soluble (1–3), colloidal (4), and heterogeneous (5–7). Many metal alkyl-free catalysts have also been found active for the polymerization of ethylene (see Chapter 12). In this respect, ethylene is unique because, as will be described shortly, propylene and higher α-olefins are polymerized with fewer types of catalysts.

In practice, the polyethylene chains formed by Ziegler–Natta catalysts are not perfectly linear. This is due to oligomerization of ethylene to low molecular weight α-olefins, which then copolymerize with ethylene to form a partially branched polyethylene. Relative to unbranched polyethylene, the latter has a lower crystallinity and melting point (8).

Some catalysts that are active for ethylene are not active for the homopolymerization of propylene and higher α-olefins, such as Cp_2TiCl_2–$AlEt_2Cl$ (1–3). All catalysts that are found active for propylene or higher α-olefins are also active for ethylene.

III. 1-Olefins

The importance of 1-olefins lies in their ability to form chains with asymmetric carbons. By a proper choice of components that make up the Ziegler–Natta catalyst, one can synthesize polymers of variable isotacticity. The attachment of a substituent to the double bond, however, significantly alters the reactivity of the bond at the Ziegler–Natta center. A sufficiently high number of different linear, branched, cyclic, and aromatic olefins have been investigated so that some general conclusions can be made about their reactivity. The combined literature data clearly show that monomer reactivity decreases drastically as the branching in the pendant group (R in CH_2=CHR) moves close to the double bond. Examples are now presented.

A. R IS LINEAR

The largest decrease in activity occurs when going from ethylene (R = H) to propylene (R = CH_3); ethylene was reported to be about ten (9) to 50 to 100 (9a) times more active, depending on the catalyst system. In contrast, propylene is about three times more active than 1-butene (R = Et) (10). The decrease in rate becomes progressively less when increasing the size of the linear pendant group, and only small changes in polymerization activity are reported when going from 1-pentene to 1-dodecene (11).

TABLE 19-1

Branched vs. Linear Olefins Polymerized (13)

	k (60°C)		
Catalyst	1-Heptene	4-Methyl-1-hexene	3-Methyl-1-pentene
Al-i-Bu$_3$–γTiCl$_3$	>617	109	43
AlEt$_2$Cl–TiCl$_3$AA	—	60[a]	17[a]
Al-i-Bu$_3$–TiCl$_3$HA	—	47	10
AlEt$_3$–VCl$_3$	>69	1.38[a]	0.84[a]

[a] Average value from two or more runs.

B. BRANCHED VS. LINEAR R

When the pendant group becomes branched, a large change in activity was observed by Campbell and Haven (12); 1-hexene was 9.7 times more active than 4-methyl-1-hexene with the LiAlR$_4$–TiCl$_4$ catalyst. Boor confirmed this finding for several olefin-catalyst systems (13), as shown in Table 19-1. The relative reactivities of different 1-olefins were compared through an arbitrary rate constant, k. This was done for 1-olefins, which were re-examined with the above four catalysts but also for selected literature data. Only the k values for olefins examined by the same worker should be compared; see below.

C. POSITION OF BRANCH IN R

A significant decrease in olefin activity takes place when the branch in R moves closer to the double bond; compare 3-methyl-1-pentene in Table 19-1 (branch at carbon-3) and 4-methyl-1-hexene (branch at carbon-4). When the branch was moved away from the double bond to carbon-5 or carbon-6, however, the olefin activities increased; k values of 1.0, 1.8, and 15.8 were evaluated for 4-methyl-1-hexene, 5-methyl-1-hexene, 6-methyl-1-heptene, respectively, for polymerizations reported earlier by Campbell and Haven (12). These authors also found a large increase in activity for the olefin series CH$_2$=CH—(CH$_2$)$_n$-phenyl as the value of n increased, as shown in the tabulation below.

n	k
1	0.27
2	4.7
3	6.8

TABLE 19–2

One vs. Two Branched Carbons (13)

	k (60°C)		
Catalyst	3,4-Dimethyl-1-pentene	3-Methyl-1-pentene	4-Methyl-1-hexene
Al-i-Bu$_3$–γTiCl$_3$	0.7	43	109
AlEt$_2$Cl–TiCl$_3$AA	0.25^a	17^a	60^a
Al-i-Bu$_3$–TiCl$_3$HA	0.21	10	47
AlEt$_3$–VCl$_3$	0.030^a	0.84^a	1.38^a

^{a} Average value from two or more runs.

D. NUMBER OF BRANCHES IN R

The olefin activity drastically decreased when both carbons-3 and -4 were substituted with methyls (13), as shown in Table 19–2.

E. SIZE OF THE SUBSTITUENT AT THE BRANCH IN R

When 3-methyl-1-pentene and 3-ethyl-1-pentene were compared, the former was considerably more active (13), as shown in the tabulation below.

	k (60°C)	
Catalyst	3-Methyl-1-pentene	3-Ethyl-1-pentene
Al-i-Bu$_3$–γTiCl$_3$	43	1.3
AlEt$_2$Cl–TiCl$_3$AA	17	0.54
Al-i-Bu$_3$–TiCl$_3$AA	10	0.68
AlEt$_3$–VCl$_3$	0.84	0.05

F. R IN CYCLIC

When the 3-alkyl is folded back by making it part of a ring structure, its contributions to steric repulsion become lower, as shown by the data of Hewitt (14) and Holler (15) below.

	k (50°C)	
Catalyst	Vinylcyclohexane	3-Ethyl-1-pentene
AlEt$_2$Cl–γTiCl$_3$	0.84	0.10
AlEt$_2$Cl–TiCl$_3$ARA	1.0	0.25

TABLE 19-3

Size of Cyclic Substituent Varied

| Researcher | R | | Conditions |
	Vinylcyclo-hexane	Vinylcyclo-pentane	
Campbell-Haven (12)	0.64	0.07	LiAl(decyl)$_4$–TiCl$_4$/25°C
Overberger *et al.* (16)	5.7	2.0	Al-*i*-Bu$_3$–TiCl$_3$/68°C

Apparently, the cyclic R poses a lower steric hindrance if it can assume certain conformations. Both Campbell and Haven (14) and Overberger and co-workers (16) observed that vinylcyclohexane was more active than vinyl-cyclopentane, as shown in the tabulation in Section III,E (see Table 19–3). The differences were attributed to the ability of the cyclohexane ring to form ring conformations which minimized its steric repulsions with Cl of the RTiCl$_4$ □ center (13). The cyclopentane substituent, being more planar, presents a greater steric repulsion (see Chapter 17 for the mechanistic significance in stereoselection).

G. AROMATIC VS. CYCLIC R

Styrene is less active than vinylcyclohexane (13). Two factors may be responsible. The aromatic ring of styrene may complex with the center and block it. Alternately, because the cyclohexane ring of vinylcyclohexane can assume several conformations, its complexing is facilitated, as discussed in the previous section.

H. R HAS A FULLY SUBSTITUTED CARBON

Dunham and co-workers (17) compared the reactivities of several fully branched α-olefins with the AlEt$_3$–VCl$_3$ catalyst. Only when the fully sub-stituted carbon was removed by more than two methylenes was the α-olefin highly active, as shown in structure I.

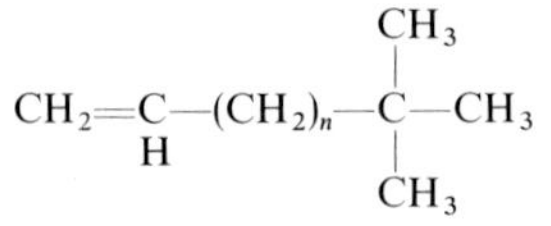

$$CH_2{=}C{-}(CH_2)_n{-}\underset{\underset{CH_3}{|}}{\overset{\overset{CH_3}{|}}{C}}{-}CH_3$$
$$\underset{H}{|}$$

I

n:	0	1	2	4
% conversion:	0.2	0.4	1.7	57

Other workers (14, 15, 18), using VCl$_3$ and TiCl$_3$ catalysts, were not able to polymerize 3,3-dimethyl-1-butene ($n = 0$) to isotactic polymer (Dunham and co-workers reported a melting point of 260°C for the product obtained when $n = 0$). Boor (13) concluded from ^{1}H-NMR analysis that the small amount of solid material formed in his polymerization more resembled an isomerized polymer in accordance with Kennedy's cationic isomerization–polymerization mechanism involving a 3-2 methide shift (19,20). Braun and Heimes (18) reported that 3,3,3-triphenyl-1-propene was not polymerized.

The sensitivity of olefin activity to the full substitution of carbon-4 ($n = 1$) is made more apparent in the comparison of its activity with those of other branched α-olefins. Relative to 4,4-dimethyl-1-butene, the olefins 3-methyl-1-pentene and vinylcyclohexane are up to ten times more active with the AlEt$_2$Cl–γTiCl$_3$ catalyst (13). It has about the same activities as 3-ethyl-1-pentene and 3,4-dimethyl-1-pentene (13). It may at first seem surprising that 3-ethyl-1-pentene should have the same activity as 4,4-dimethyl-1-pentene, which has a methylene adjacent to the double bond. Apparently, the presence of a fully methyl substituted carbon even in the 4-position hinders complexing and polymerization at the active centers. Even a bulky group such as phenyl in the 3-position has the same rate-depressing ability as do two methyls in the 4-positions (12), as shown in structures II and III.

$$\underset{H}{\overset{\phi}{CH_2{=}CH{-}C{-}CH_2{-}CH_3}} \simeq \underset{CH_3}{\overset{CH_3}{CH_2{=}CH{-}CH_2{-}C{-}CH_3}}$$

$$k = 0.38 \qquad\qquad k = 0.46$$

$$\text{II} \qquad\qquad\qquad \text{III}$$

Braun and Heimes (18) found that 4,4-dimethyl-1-pentene was more reactive than styrene for the catalyst AlR$_3$–TiCl$_4$ (or TiCl$_3$).

Many other papers have been published that are in general agreement with the primary conclusion that the reactivity of a double bond decreases most drastically when the 3- and 4-carbons contain branching and the size of the alkyl substituent increases (see Table 19–4) (16, 21–32).

I. R IS A SUBSTITUTED AROMATIC NUCLEUS

Using the Al-i-Bu$_3$–TiCl$_4$ catalyst and temperature of 60°C, Danusso and co-workers (33, 34) investigated the polymerizability of a large number of vinyl aromatic monomers to form isotactic polymers. Table 19–5 shows

TABLE 19–4

Other Branched α-Olefins That Have Been Investigated

R in $CH_2{=}CH{-}R$	Ref.
$-CH_2-\overset{S'}{\bigcirc}$ (cyclohexyl)	16, 21, 22
$-CH_2-\overset{S}{\triangle\!\!\!\bigcirc}$ (cyclopentyl)	16, 23
$\triangleleft$ (cyclopropyl)	24, 25
$+CH_2+_n C_6H_5$ $n = 1, 2, 3$	26, 27
$-CH_2-\overset{\overset{H}{\vert}}{C}\overset{CH_3}{\underset{CH_3}{\diagdown}}$	28, 29
(tolyl, CH_3 in 2-, 3-, or 4-position)	30
$-CH(CH_3){-}(CH_2)_3{-}CH(CH_3)_2$	31, 32
$-CH_2{-}CH(CH_3){-}CH_2{-}CH_3$	31
$-CH(CH_3)CH_2CH_3$	31

TABLE 19–5[a]

Relative Reactivity, α, in Stereospecific Homopolymerization and Reactivity Ratios r_2 in Stereospecific Copolymerization for Different Aromatic Vinyl Monomers [Relative to Styrene, at 60°C; Catalyst Made from $Al(i\text{-}C_4H_9)_3$ and $TiCl_4$] (33, 34)

Monomer	α	r_2	Monomer	α	r_2
o-Methylstyrene	0.10	0.13	p-Fluorostyrene	0.74	0.72
m-Methylstyrene	0.43	0.47	m-Chlorostyrene	0.40	0.43
p-Methylstyrene	1.20	1.17	p-Chlorostyrene	0.47	0.50
2,4-Dimethylstyrene	0.10	0.13	p-Bromostyrene	0.45	0.49
2,5-Dimethylstyrene	0.04	0.03	p-Methoxystyrene	2.0–2.2	1.6
2,4,6-Trimethylstyrene	0	0	1-Vinyldiphenyl	0.73	0.76
p-Ethylstyrene	1.10	1.05	2-Vinylnaphthalene	0.67	—
p-Cyclohexylstyrene	0.50	0.48	3,4-Dimethylstyrene	0.48	—
o-Fluorostyrene	0.20	0.18	3,5-Dimethylstyrene	0.23	—
m-Fluorostyrene	0.50	0.47			

[a] The ratio of the polymerization rate constant for the monomer under study to the corresponding rate constant for styrene, chosen as a reference monomer, was determined for different monomers by suitable measurements of polymerization rates. These ratios have been defined as "relative reactivities," α, in homopolymerization. For convenience, a set of r_2 values referring to 60°C are listed in the second column.

the relative reactivity α in stereospecific homopolymerization and the reactivity ratio r_2 in copolymerizations for different aromatic vinyl monomers.

In a Hammet plot (Fig. 15–1), the data for most of the monomers examined fell on a straight line which had a negative slope. Danusso interpreted this to mean that the addition reaction was sensitive to purely polar effects exerted by the substituent on the vinyl group and was favored by electron-donating substituents. The monomers having a relatively low reactivity suffered from steric hindrance, mainly from substitution at the *ortho* position to the vinyl group or, in some cases, at the *meta* and *para* positions (35).

Sianesi and co-workers (36) investigated the polymerization of several sterically hindered vinyl aromatic monomers: 1-vinylnaphthalene, 2-vinylnaphthalene, 1-vinyl-4-chloronaphthalene, 1,2,3,4-tetrahydro-6-vinylnaphthalene, 4-vinyldiphenyl, *p*-cyclohexylstyrene, 9-vinylanthracene, and 9-vinylphenanthrene. The catalyst components were $AlEt_3$ or Al-*i*-Bu_3 mixed with $TiCl_4$ or $TiCl_3$, and polymerizations were done at 40° to 70°C for 7 hours. Of the monomers examined, all polymerized except 9-vinylanthracene. Conversions were low, about 10 to 40%. Only the polymer from 1-vinylnaphthalene was crystalline, as shown by X-ray measurements (melting point of 360°C); the other products were amorphous. All had high molecular weights (intrinsic viscosities = 0.7 to 3.4 dl/g).

Heller and Miller (37), using infrared and NMR measurements, reported that polymers from 1- and 2-vinylnaphthalene and 4-vinyldiphenyl were at least 90% isotactic. The isotactic fractions were isolated by extraction of the whole polymer with 2-butanone. $AlEt_3$–$TiCl_3$ was used as catalyst for the polymerizations.

While 3-vinylpyrene was polymerized with the $AlEt_3$–$TiCl_4$ catalyst to a polymer with a melting point of 214°C, Tanikawa and co-workers concluded from supplementary experiments that the mechanism was cationic (38).

Hatada and co-workers (39) concluded that a bromine atom in the *para* position in *p*-bromostyrene prevented isotactic polymerization when this monomer was polymerized with the Ziegler–Natta catalyst, $AlEt_3$–$TiCl_3$. The poly-*p*-bromostyrene product was reduced to polystyrene ($LiAlH_4$ or LiR plus hydrolysis), and the latter was compared with authentic polystyrene samples synthesized by anionic and Ziegler–Natta polymerizations. In contrast, Saunders (40) reported the polymerization of *p*-*tert*-butylstyrene to an isotactic crystalline polymer (melting point of 290° to 300°C) with mixtures of $AlEt_3$ + $TiCl_4$, VCl_3, or $TiCl_3$.

Styrene was investigated extensively by Kapur and co-workers with a variety of Ziegler–Natta catalysts, including VCl_4–$AlEt_3$ (41), $ZrCl_4$ or $ZrCl_3$–Al-*i*-Bu_3 (42), $VOCl_3$–$AlEt_3$ or Al-*i*-Bu_3 (43), $VOCl_3$–$AlEt_2Br$ (44), $VOCl_3$–Li-*i*-C_5H_{11} (45), $TiCl_4$–Li-*i*-C_5H_{11} (46), and $ZrCl_4$–$ZrCl_3$–$AlEt_3$

(47). Other workers prepared active catalysts by mixing $TiCl_4$ and n-BuLi (48) or $TiCl_4$ and $AlEt_3$ (49). The stereospecific polymerization of styrene in the presence of BuLi–metal salt catalysts was reported (50) where the metal salt was nickel naphthenate, vanadium naphthenate, cobalt octanoate, vanadium acetylacetonate, cobalt(II) acetylacetonate, nickel(II) acetylacetonate, nickel benzoate, and nickel carbonyl and the temperature of polymerization was below 40°C. This is a rather surprising finding, because α-olefins such as propylene are not polymerized with these Ni- or Co-based catalysts. One must suspect that a radical mechanism might have taken place.

IV. 1,1-Disubstituted Ethylene

No examples are reported that show that a 1,1-disubstituted olefin has been polymerized at the Ziegler–Natta center. When polymerization does take place, it is by a cationic mechanism; that is, the Ziegler–Natta catalyst serves as a source of a cationic initiator (51–53).

Topchiev and co-workers (52) first reported the polymerization of isobutylene to polymers (MW $= 7 \times 10^3$ to 10×10^3) with the $AlEt_3$–$TiCl_4$ catalyst. They tentatively assigned the rearranged structure in IV to the polymer compared to normal 1,2-addition in structure V.

$$\begin{array}{cc} \underset{\text{IV}}{\overset{\displaystyle \overset{\text{CH}_3}{|}\ \overset{\text{CH}_3}{|}}{-\text{CH}_2-\text{CH}-\text{CH}-}} & \underset{\text{V}}{\overset{\displaystyle \overset{\text{CH}_3}{\underset{|}{|}}}{-\text{CH}_2-\underset{\underset{\text{CH}_3}{|}}{\text{C}}-}} \end{array}$$

When Bacskai and Lapporte (53) repeated these earlier experiments, they found the structure to be V, normally found for cationic catalysis. Ziegler–Natta catalysts that had less cationic character (such as $AlEt_3 + \alpha TiCl_3$) were inactive. More recently, Korneev and co-workers (51a), using mostly $VOCl_3$-based catalysts, reported that the polymer contains some of the isomerized structure IV. The relative ratios of structures IV and V could be influenced by the choice of metal alkyl [$AlEt_3$ and $AlEt-(NSi-Me_3)_2$] and ratio of Al/V. A kinetic study was made by Hamada and Gary (51).

Alpha-methylstyrene was also polymerized via cationic mechanism in the presence of a Ziegler–Natta catalyst (54, 55). Variable catalyst activities were found when $TiCl_4$ was combined with a different metal alkyl ($AlMe_3$, $AlEt_3$, $AlEt_2Cl$, $AlEtCl_2$, etc.). The intermediate complex $TiCl_3^+$ $AlEt_3Cl^-$ was suggested as the active initiator. Acidic aluminum alkyls (such as

AlEt$_2$Cl or AlEtCl$_2$) were also active initiators. While AlEt$_3$ alone was not active, hydrolysis with H$_2$O produced an active catalyst (56).

Diisopropenylbenzene was polymerized to a linear polyindane (no unsaturation was found in product) by a cationic mechanism using LiBu–TiCl$_4$HCl (57).

Pinazzi and Brossas (58) reported that isopolyisoprene was formed in a practically pure state from methylene cyclobutane by the polymerization

$$
\begin{array}{c}
\text{[cyclobutane ring]} \diagup^{CH_2} \\
\end{array}
+ \text{AlEt}_2\text{Cl} + \text{acetylacetone} + \text{acetylacetonate of V, Cr, Co, Mn, or Zr} \longrightarrow
$$

$$
\begin{array}{c}
CH_2 \\
\parallel \\
-C-(CH_2)_3- \quad (19\text{--}1)
\end{array}
$$

shown in Eq. 19–1. A cationic polymerization gave ring retention. In later work, these workers and Clouet (58a) re-examined the polymerization in the presence of the catalyst AlEt$_3$–TiCl$_4$–(C$_6$H$_5$)$_3$P. The structure and the product were confirmed by ozonolysis, and the corresponding polyketone was identified by NMR spectroscopy, shown in structure VI.

$$
\begin{array}{c}
-C-(CH_2)_3- \\
\parallel \\
O
\end{array}
$$

VI

V. 1,2-Disubstituted Ethylenes

The Natta school (59, 60) demonstrated that while certain internal olefins such as 2-butene could be copolymerized with ethylene, they could not be homopolymerized, as shown in Eq. 19–2. The inability of 2-olefins to homopolymerize was attributed to high steric repulsions.

$$
\text{RCH}{=}\text{CH}{-}\text{R} \xrightarrow[\text{polymerization}]{\text{no}} {+}\text{CH}{-}\text{CH}{+} \qquad (19\text{--}2)
$$
$$
 \underset{R \quad\ R}{\mid\qquad\mid}
$$

Later, Jezl (61), Natta (62), Symcox (63), Iwamoto (64), and Shimuzo and Otsu (65) demonstrated the polymerizability of 2-butene with Ziegler–Natta catalysts but by a rather circuitous path. The polymer that they formed was poly-1-butene, however, not poly-2-butene. Neither were copolymers of 1-butene and 2-butene formed in measurable amount. 1-Pentene and 1-hexene behaved similarly (65, 66).

These findings suggested that the catalyst first isomerized the 2-olefin (2-butene) to 1-butene, and the 1-butene was polymerized exclusively to the poly-1-butene (66). Indeed, when the unreacted butene mixture was analyzed by gas chromatography, it was found to contain the three isomers: 1-butene (5.1%), cis-2-butene (22.5%), and trans-2-butene (72.4%) (66). The values in parentheses are the equilibrium percentages of each isomer. No isobutene was detected, ruling out skeletal isomerization.

An excellent review by Kennedy (67) elaborates upon this subject in considerable detail up to 1970. There has been speculation that isomerization–polymerizations of this type may find significant commercial applications. On one hand, internal (nonpolymerizable) olefins could be converted to useful polymers. In addition, coupling isomerization–polymerization catalysts with metathesis catalysts (see Chapter 22) would yield a variety of routes to homopolymers and copolymers (67).

Because the equilibrium favors internal olefins, the concentration of the 1-olefin is low, and only low yields of the poly-1-olefins are obtained. Otsu and co-workers showed that the rate of polymerization of 2-butene was directly dependent on the rate of isomerization (68). Some catalysts were more active. For example, $AlEt_3–TiCl_3$ polymerized 2-butene and 2-pentene while $AlEt_3–VCl_3$ did not (68, 69). The addition of other metal salts such as nickel dimethylglyoxime or $Fe(acac)_3$ increased the polymerization ability of both catalysts. While the Ni and Fe salts were able to isomerize both 2-butene and 1-butene, they were not polymerization catalysts. This suggested to these authors that the isomerization–polymerization mechanism involved two types of sites. One of these centers isomerized the 2-olefin to the polymerizable 1-olefin, and the second center (which was a Ziegler–Natta center) polymerized the 1-olefin to a poly-1-olefin (66).

The importance of choosing the right catalyst and temperature of polymerization is illustrated in the following comparison. Natta and co-workers, using $AlEt_2Cl–V(acac)_3$ as catalyst at $-30°C$, were successful in copolymerizing 2-butene and ethylene without isomerization (59). In contrast, Chauvin and co-workers (70), using the $AlEt_2Cl–NiCl_2–Ti(OC_4H_9)_4$ catalyst at $+50°C$, were able to synthesize a copolymer of 1-butene and ethylene when 2-butene and ethylene were reacted. More recently, Otsu, Aoki, and Nishimura (71) reported that they detected a 1,2-dimethyl structure in products synthesized at $0°C$ with VCl_4-based but not with $TiCl_4$-based catalysts. [The other components were $AlEt_3$ and Ni (dimethylglyoxime)$_2$.] They speculated that some 2-butene copolymerized because the activity of 1-butene with the VCl_4 catalyst was relatively lower.

Isomerization side reactions have been detected when 1-olefins have been polymerized with Ziegler–Natta catalysts (72–74). Kleiner and co-workers found that, during the polymerization of vinylcyclohexane, isomerization occurred, forming unreactive isomers with an internal double bond as well as transformation of monomer into ethylcyclohexane (72). A detailed mechanistic path was offered. Aubrey and Barnatt (73) detected *trans-n*-octa-2-decene when *n*-octa-1-decene was polymerized with AlEt$_3$–TiCl$_4$ to high molecular weight products. Yerasova and co-workers (74) reported that, during the polymerization of 1-butene with the catalyst AlEt$_3$ + CrCl$_3$, some 1-butene was isomerized to the *cis*- and *trans*-isomers.

VI. Cycloolefins

Many but not all of the cycloolefins from cyclobutene to cyclododecene have been successfully polymerized. A wide range of Ziegler–Natta catalysts were explored, with different results being obtained with different combinations of the catalyst components. Both practical and theoretical motivations have prompted much research with this type of monomer.

The practical interest lay in the polymerization of cyclopentene at high rates by a ring opening path to a linear, high *cis*- or *trans*-poly-1,5-pentenamer (see Eq. 2–4). The properties of the trans product were sufficiently attractive to spur commercial development as a new elastomer (see Chapter 3, Section V,B).

This monomer–catalyst system is theoretically important because of the different modes of polymerization that are possible. Also of mechanistic importance is the finding that some of these cycloolefins will copolymerize with ethylene, even though they do not homopolymerize.

Comprehensive reviews that are addressed to a wider scope than possible here have been published by Marconi (75), Natta and Dall'Asta (76), Dall'Asta and Matroni (77), Dall'Asta (78), and Calderon (79). A symposium on this subject was held in 1972 (80).

Copolymerizations of cycloolefins and other monomers are discussed in Chapter 20, Section IV,E.

A. POLYMERIZATION PATHS

It will be advantageous at this point to outline some of the polymerization modes that have been established for cyclic and bicyclic olefins. The exact mechanistic paths by which these products may be obtained will be discussed near the end of this section.

The scheme for monocyclic olefins (such as cyclobutene) is shown in Eq. 19–3.

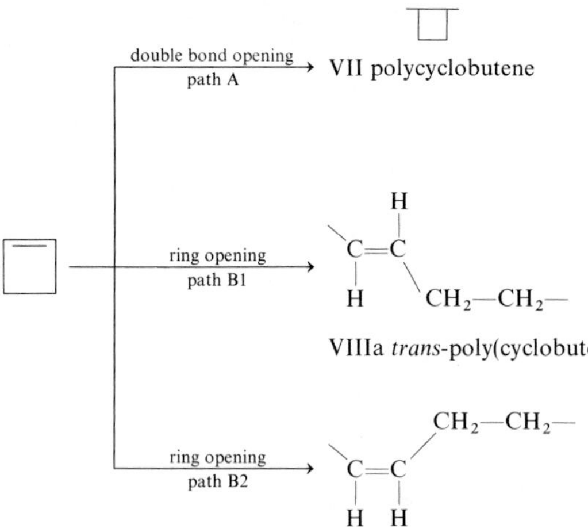

VII polycyclobutene

VIIIa *trans*-poly(cyclobutenamer)   (19–3)

VIIIb *cis*-poly(cyclobutenamer)

When polymerization occurs by double bond opening, the product can have two structures, erythro-diisotactic polycyclobutene, shown in VIIa, and erythro-disyndiotactic polycyclobutene, in VIIb.

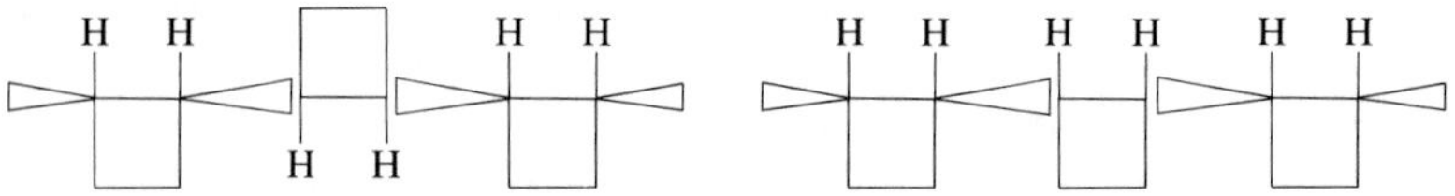

VIIa   VIIb

The scheme for bicyclic olefins (norbornene and bicyclo[2.2.2]-2-octene) is shown in Eqs. 19–4 and 19–5.

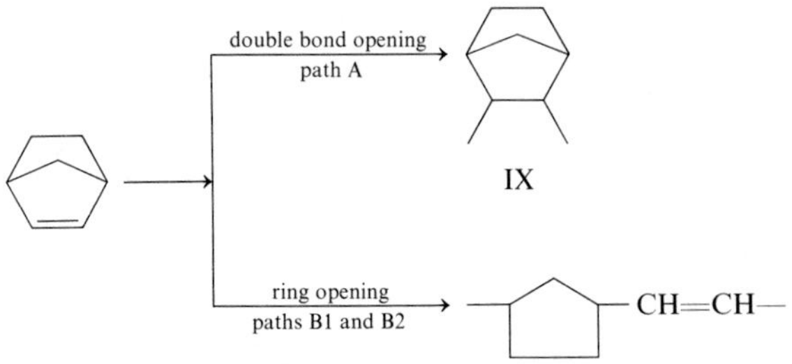

IX   (19–4)

Xa *cis* and/or Xb *trans*

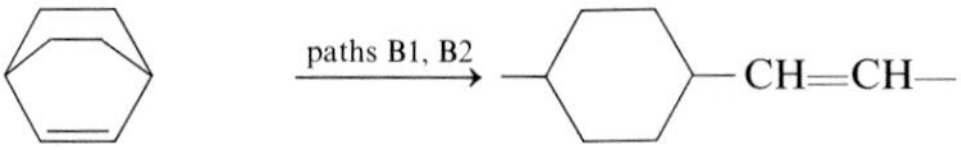

bicyclo[2.2.2]-2-octene   Xia *cis* and/or XIb *trans*   (19–5)

B. CATALYSTS

Table 19–6 collects selected examples of different cycloolefins that have been polymerized with various catalysts (76, 81–91). Salient features of these polymerizations are presented here.

The activity of the cycloolefin and the polymerization mode depends closely on: (1) the ring size of the cycloolefin; (2) the degree of alkyl substitution; (3) whether the ring is part of a mono- or bicyclic system; and (4) the nature of the Ziegler–Natta catalyst. The relative percentages of polymerization modes reported were usually made on the whole polymer. In only a few cases, the product was fractionated to establish if the structural units were in the same or different chains.

Cyclopropene polymerized spontaneously above $-80°C$ by path A (92).

Cyclobutene is very highly active and polymerized easily with different Ziegler–Natta catalysts by double bond opening (path A) or ring opening [with both paths B1 for trans and B2 for cis, or combinations of them (see Eq. 19–3)]. Cr- and V-based catalysts predominantly induce polymerization through the double bond (path A), while Ti- and Mo-based catalysts favor ring opening polymerizations to *cis*- and *trans*-polybutenamers. Dall'Asta (93) showed that 3-methyl-1-cyclobutene was predominantly polymerized with ring retention with mixtures of VCl_4 or $VOCl_3$ and a variety of metal alkyls, such as $AlEt_3$, $GaEt_3$, $BeEt_2$, or $MgEt_3$. When LiBu was used, a ring opening polymerization occurred. It should be noted that cyclobutene has also been polymerized with metal alkyl-free catalysts, including Ni, Rh, or Ru salts in water or ethanol solvents (94, 95).

Both bicyclo[4.2.0]octa-7-ene and bicyclo[3.2.0]hepta-2,6-diene were polymerized by Dall'Asta and Matroni (96) with Ziegler–Natta catalysts but only through the cyclobutene group. Homopolymerization occurred by opening of the double bond as well as by ring opening; for example, 98% 1,2 addition with $AlEt_3 + VCl_4$ and 85% *trans*-addition with $AlEt_3 + TiCl_4$ was noted (both were done at $-70°C$).

Cyclopentene has been most extensively investigated (97–97d). As already cited in Chapter 3, the *trans*-product from the ring opening polymerization has already achieved commercial status as a new elastomer (97a). Cyclopentene was easily polymerized exclusively by paths B1 and B2 when catalysts containing Mo and W were used, but very slowly (about 1% yield) by combinations of paths A, B1, and B2 when catalysts containing V were used (Table 19–6).

Emphasis was placed on the tungsten-based catalysts, and many modifications have emerged. For example, catalysts for the production of the *trans*-isomer were modified by presence of an organic epoxide (97c) or a halogenated alcohol (97d).

TABLE 19–6

Polymerization of Cycloolefins

Olefin and catalyst system	Percent polymerization path			Ref.	Comments
	Double bond opening A	Ring opening B(*trans*)	Ring opening B(*cis*)		
Cyclobutene					
Cr(acac)$_3$ or V(acac)$_3$ + AlEt$_2$Cl	100	0	0	81	Erythrodiisotactic structure VIIa
VCl$_4$ + AlEt$_3$	Predominant	—	—	81	Erythrodiisotactic structure VIIa
TiCl$_4$ + AlEt$_3$	5	65	30	81, 82	Ratio VIIa/VIIb inverted by use of heptane vs. toluene.
V(acac)$_3$ or VO(OBu)$_3$ + AlEt$_2$Cl	Predominant	—	—	81, 82, 76	Erythrodisyndiotactic structure VIIb.
MoCl$_3$ + AlEt$_3$	10	30	60	81	—
WCl$_6$ + AlEt$_3$	30	40	30	81	—
Cyclopentene					
MoCl$_5$ + AlEt$_3$	0	0	100	83–85	
WCl$_6$ + AlEt$_3$	0	100	0	83–85	
TiCl$_4$ + ZrCl$_4$ + AlEt$_3$	0	100	0	83–85	

VCl$_4$ + AlMe$_2$Cl	50–80	30–50	Combined	86
TiCl$_3$AA + AlMe$_3$	50–80	30–50	Combined	86
Cyclohexene	Inactive with all catalysts tested			
Cycloheptene				
MoCl$_5$ + AlEt$_3$	0	93	7	85
WCl$_6$ + AlEt$_2$Cl	—	91	—	—
Cyclooctene (*cis*)				
WCl$_6$ + AlEt$_3$	0	85	15	85
Cyclodecene (*cis*)				
WCl$_6$ + AlEt$_2$Cl (with dicumyl peroxide)	0	85	15	88
Cyclododecene (*cis–trans*, 1:2)				
WCl$_6$ + AlEt$_2$Cl	0	94	6	85
Norbornene				
LiAlH$_4$–TiCl$_4$				
Al/Ti = 95	Predominant	—	—	89–90
Al/Ti = 2	—	Predominant	—	89–90
AlR$_3$–MoCl$_5$ (Al/Mo = 3)	—	—	—	91

Cyclohexene has not been polymerized; this has been attributed to the high ring stability of the normal twisted chain conformation that it takes. Ofstead [see ref. 25 of Calderon (79)] showed that if the cyclohexene ring is part of a bicyclic system (Eq. 19–5), then polymerization occurs by ring opening when Calderon's $AlEtCl_2$–WCl_6–EtOH catalyst was used. Calderon (79) suggested that polymerization occurred because the cyclohexene was forced into a higher energy boat conformation.

Higher cycloolefins have been polymerized with relatively more ease when tungsten-containing catalysts were used, and the polymerization occurred chiefly by ring opening (path B1) to give *trans*-polycycloalkenamers. These higher cycloolefins can have the double bond in a *cis*- or a *trans*-configuration. The geometric configuration of the double bond, however, is not necessarily conserved during the polymerization.

Calderon, Ofstead, and Judy (98) showed that properly substituted cycloolefins can provide unique syntheses of alternating copolymers or terpolymers, as shown in Eq. 19–6.

$$\longrightarrow \quad [(CH_2-CH=CH-CH_2)(CH_2-CH_2)(\underset{\underset{X}{|}}{CH}-CH_2)]_n \quad (19\text{–}6)$$

Unit = butadiene–ethylene–vinyl

The cycloolefin 1,1-dimethyl-1-silacyclopent-3-ene was reported by Lammens and co-workers (98a) to undergo a ring opening polymerization, as shown in Eq. 19–7.

$$\longrightarrow \quad (CH=CH-CH_2-\underset{\underset{CH_3}{|}}{\overset{\overset{CH_3}{|}}{Si}}-CH_2) \quad (19\text{–}7)$$

C. MECHANISMS

The mechanistic paths by which cycloolefins are polymerized in the presence of the Ziegler–Natta catalysts have been satisfactorily elucidated only for several cycloolefin–catalyst combinations.

Dall'Asta (78) speculated that path A occurred by 1,2-addition of the internal double bond and paths B1 and B2 involved scission of the α and β bonds to produce the *trans*- and *cis*-polyalkenamer. Similar reaction paths can be envisaged for polymerizations of norbornene and bicyclo[2.2.2]-2-octene (Eqs. 19–4 and 19–5).

The polymerization scheme in Eq. 19–8 and 19–9 was suggested by Calderon (79), in which ring opening occurs through the double bond rather than through scission of the α or β bonds. This mechanism has been designated as metathesis or transalkylidenation as is outlined below.

Initiation:

$$RAlCl_2 + [WCl_6 + ROH] + 2\ CH{=}CH{-}(CH_2)_n \longrightarrow \quad (19\text{--}8)$$

Propagation:

$$(19\text{--}9)$$

The repetition of this growth step as well as of macrocyclization of only larger rings formed during the polymerization would constitute propagation. In the absence of impurities, a true termination step is absent.

From a historical point of view, it should be mentioned that metathesis reactions were discovered and examined earlier. Euleterio (99) first described the polymerization of cyclopentene with a hydrogen-reduced molybdena on alumina catalyst in 1963, but this finding did not attract scientific or industrial attention. Banks (100) reported the disproportionation of acyclic olefins with molybdena catalysts, a process that was expressed mechanistically in 1967 by Bradshaw and co-workers (101), as shown in Eq. 19–10. Calderon

$$\begin{array}{c} C{=}C{-}C{-}C \\ C{=}C{-}C{-}C \end{array} \rightleftharpoons \begin{array}{c} C{-}C{-}C{-}C \\ C{-}C{-}C{-}C \end{array} \rightleftharpoons \begin{array}{c} C \quad C{-}C{-}C \\ C \quad C{-}C{-}C \end{array} \qquad (19\text{--}10)$$

and co-workers (102) proved this mechanistic scheme, using deuterated 2-butene and the WCl_6–$AlEt_2Cl$–EtOH catalyst, and postulated that ring opening polymerization of cyclopentene was a special case of the disproportionation (metathesis) reaction. When the double bond becomes part of a cyclic olefin, polymerization, instead of disproportionation, occurs. Chain growth occurs via macrocyclic rings of increasing size. Alkema and Van Helden (103) oligomerized cyclooctene with an Al_2O_3–MoO_3–CoO catalyst to multiunsaturated olefins. When these multiunsaturated fractions were fractionated and hydrogenated, they obtained 24-, 32-, 40-, . . . , 72-membered cycloalkanes (multiples of cyclooctene). In 1968, Wasserman and co-workers (104) isolated some oligomers from a low molecular weight fraction of polyoctenamer obtained with Calderon's catalyst. Scott and co-workers (105) obtained analogously up to 120-membered cyclic oligomers.

More recently, Dall'Asta and Matroni (106, 107) ozonized [1-^{14}C]-cyclopentene/cyclooctene random copolymer and examined the radioactive products by gas chromatography from decomposition of the ozonides. From the product analysis, they concluded that polymerization by double bond opening had occurred.

The metathesis scheme has been mainly demonstrated for tungsten-containing catalysts. One should ask if any of the cycloolefins are polymerized by the accepted olefin mechanism. Simply, the latter assumes the olefin is complexed at a transition metal center that bears an alkyl. The complexing weakens the transition metal bond and, subsequently, the olefin becomes inserted into the transition metal–carbon bond.

It appears that this mechanism is operative with certain catalyst–olefin combinations. For example, Natta and co-workers have established that *cis*- or *trans*-2-butene and several cycloolefins such as cyclopentene are copolymerized to alternating copolymers when ethylene is added in low concentrations. When cyclopentene is a monomer, it is incorporated with complete ring retention. Polymerization exclusively occurs by 1,2-addition. This happens when catalysts containing V or Ti but not W are used.

That catalysts containing Ti and W are significantly different is illustrated when 1,5-cyclooctadiene is polymerized with each type. Calderon (108) showed ring opening occurs with tungsten catalysts, while Marvel (109) using a titanium catalyst, showed a product with no unsaturation, indicating cycloaddition to yield the structure shown in XII. Whether both types of

XII

mechanisms can operate with each metal type under suitable reaction conditions (ligands, temperature, oxidation state, solvent, etc.) is not known at this time. While most of the higher cycloolefins polymerize exclusively or (almost exclusively) by ring opening, cyclobutene ($AlEt_3$–WCl_6 catalyst) gives 40% cyclobutane retention (path A) and 60% of ring-opened products (30% cis and 30% trans) quantitatively. While cyclopentene only sluggishly polymerizes (1% yield) with V catalysts, cyclobutene is considerably more active.

VII. Bicyclo[n.1.0]alkanes

A series of bicyclo[n.1.0]alkanes was polymerized by Pinazzi and co-workers (110) in the presence of Ziegler–Natta catalysts such as $AlEt_3$–$TiCl_4$ and $AlEt_2Cl$–$TiCl_4$ as shown in Eq. 19–11, where $n = 3$, 4, or 5. The polym-

$$\qquad\qquad\qquad\qquad\qquad\qquad\qquad\qquad\qquad (19\text{--}11)$$

erizations were done in hexane at temperatures between 80° to 120°C and for durations between 24 to 72 hours. Low yields (typically 10 to 30%) were formed, and the polymers had molecular weights below 1,500. A polymerization mechanism was proposed in which one of the cyclopropane bonds not common to both rings was inserted into the Ti–C bond of the active center. The same monomers were polymerized with cationic initiators, but the products had structures that suggested a hydride shift–rearrangement had occurred.

Earlier, Yamada and co-workers (110a) polymerized bicyclo[4.1.0]heptane and phenylcyclopropane with $AlEt_3$–$TiCl_4$, but the degree of polymerization was low, that is, DP = 4 and 2, respectively.

A review by Pinazzi and co-workers (110b) compares this type of polymerization with both Ziegler–Natta and cationic systems.

VIII. Polar Monomers

While many polar monomers have been reported to be polymerized by Ziegler–Natta catalysts, only a few have been shown to be polymerized at the same centers (transition metal–carbon bonds) that polymerize ethylene and α-olefins. For most of the polar monomers, the Ziegler–Natta catalyst serves as another source (however, sometimes unique) of cationic, radical, or anionic centers. Some of the polar monomers are said to be polymerized by mechanisms involving coordination–anionic or coordination–cationic-type centers, which differ from the Ziegler–Natta centers. This distinction will be made later.

The more closely the polar compound resembles an α-olefin, the greater is the opportunity for it to polymerize at the same centers that are active for olefins. In practice, this can be done by: (1) insulating the double bond from the heteroatom by methylene units; (2) by increasing the steric encumbrance about the heteroatom; (3) by lowering the electron-donating character of the heteroatom (by choosing a particular heteroatom or by attaching electron-attracting group to it); and (4) by simply choosing catalyst components that do not react with the heteroatom. Additional assistance can be obtained by lowering the reactivity of the metal alkyl components toward the heteroatom by complexing it with a suitable donor. In accordance with these criteria, it follows that monomer XIII is more suitable than monomer XIV.

$$CH_2{=}CH{-}(CH_2)_5{-}N(i\text{-}Bu)_2 \qquad CH_2{=}CH{-}CH_2NH_2$$

$$\text{XIII} \qquad\qquad\qquad \text{XIV}$$

The polymerization of the polar monomers is now described according to the heteroatom: silicon, nitrogen, halogen, oxygen, and miscellaneous.

A. SILICON MONOMERS

Natta, Mazzanti, Bassi, and co-workers (111, 112) have polymerized several silane monomers to polymers in an isotactic configuration, as shown in Eq. 19–12. The polymer product had a melting point above 350°C. Other

$$(CH_3)_3Si{-}CH_2{-}CH{=}CH_2 \xrightarrow[70°C,\ 20\ hours]{AlEt_3\ +\ TiCl_4} \left(CH_2{-}\underset{\underset{Si(CH_3)_3}{\overset{|}{CH_2}}}{\overset{|}{CH}} \right)_n \qquad (19\text{-}12)$$

silicon-containing polymers that were also synthesized included poly-allylsilane and poly-5-trimethylsilyl-1-pentene: all of these polymers showed an identity period along the fiber axis of about 6.5 Å and possessed three monomer units per turn of the helix. The above suggests that allylsilanes behave similarly to α-olefins and are also polymerized at the Ziegler–Natta sites (transition metal–carbon bonds).

Homopolymers (113) and copolymers (114) (with propylene) of dimethyl-diallylsilane and methylphenyldiallylsilane were also reported. Bogomol'nyi

(113) reported that cycloaddition occurred, leading predominantly to the structural unit shown in XV.

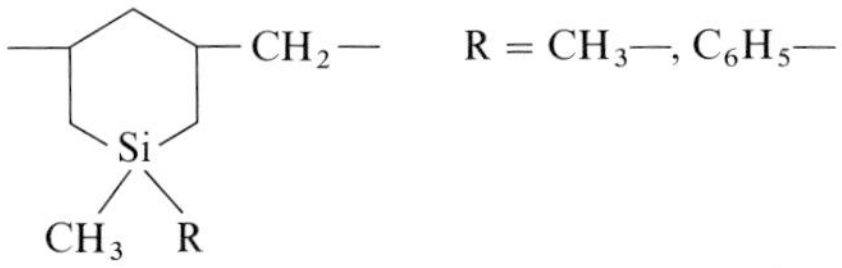

R = CH$_3$—, C$_6$H$_5$—

XV

Carbonaro, Greco, and Bassi (115) showed that vinylethyl silane could be homopolymerized in low yields with the AlEt$_3$–VCl$_3$ catalyst at ambient temperatures. The product had a low molecular weight ([η] > 0.3 dl/g) and a low melting point (65° to 75°C), but X-ray measurements showed it to have isotactic crystallinity (identity period = 6.50 ± 0.05 Å).

Longi, Greco, and Rossi (116) found that if allylsilane was copolymerized (AlEt$_2$Cl–TiCl$_3$ARA catalyst) with propylene or 1-hexene, the copolymers contained 0.4 wt% Si. The propylene copolymer still had a high melting point (about 163°C). These findings show that a silicon attached directly to a double bond lowers its activity but does not completely deactivate it.

B. NITROGEN MONOMERS

Several different vinyl monomers containing aliphatic and heterocyclic nitrogens have been polymerized with Ziegler–Natta catalysts. If the criteria suggested above were followed, the formed polymers had isotactic structures.

Heller and co-workers (117) polymerized carbazole-containing monomers,

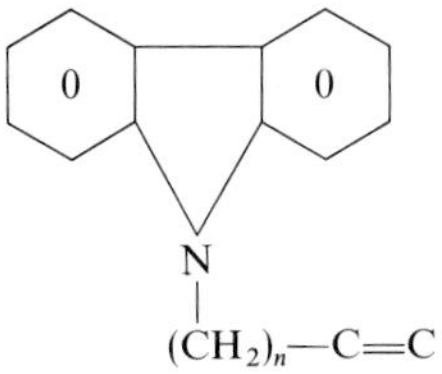

XVI

shown in structure XVI. Polymerization activity increased as n = 1 to 2 to 5. The polymers were isotactic-crystalline, suggesting that polymerizations took place at Ziegler–Natta centers. When vinyl carbazole was polymerized, both crystalline (118, 119) and amorphous (120) (n-BuLi–TiCl$_4$ and AlEt$_3$–TiCl$_4$ catalysts, respectively) polymers were claimed. The interpretation of data of vinyl carbazole is difficult, since this monomer polymerizes so easily with different types of catalysts.

Giannini and co-workers (121) synthesized high molecular weight isotactic polymers from the Type 1 and 2 monomers in structures XVII and XVIII using $AlEt_2Cl$–$TiCl_3AA$ as the catalyst.

Type 1

$$CH_2{=}CH{-}(CH_2)_n{-}N\begin{array}{c}R\\\diagdown\\R\end{array}$$

XVII

XVIIa: $n = 2$, $R = CH\begin{array}{c}CH_3\\\diagup\\\diagdown\\CH_3\end{array}$

XVIIb: $n = 3$, $R = CH_3$, C_2H_5, or $CH\begin{array}{c}CH_3\\\diagup\\\diagdown\\CH_3\end{array}$

XVIIc: $n = 3$ or 5, $R = CH\begin{array}{c}CH_3\\\diagup\\\diagdown\\CH_3\end{array}$

Type 2

$$CH_2{=}CH{-}(CH_2)_n{-}X{-}Si(CH_3)_3$$

XVIII

$n = 3$, $X =$ oxygen or $N{-}CH\begin{array}{c}CH_3\\\diagup\\\diagdown\\CH_3\end{array}$

$n = 9$, $X =$ oxygen

This elegant and comprehensive work strengthened the criteria cited above for polymerizability of polar monomers at Ziegler–Natta sites.

Pomogailo and co-workers (122) first complexed the transition metal component with some of the polar monomer (2-vinylpyridine) before combining it with $Al\text{-}i\text{-}Bu_2Cl$ for copolymerization of ethylene and the 2-vinylpyridine. A copolymer containing 0.1 to 0.9% nitrogen was made.

Gehrke and co-workers (123) polymerized 2-methyl-5-vinylpyridine with catalyst $AlR_2X + TiCl_4$, $VOCl_3$, or $TiCl_3$. A high molecular weight polymer (melting point about $178°$ to $210°C$) was formed. Radical polymerizations gave a low viscosity (0.336 vs. 2.82 dl/g) and a melting point of $179°C$.

Copolymers of 4-methyl-1-pentene and undecylenamide (5 to 8% of latter in copolymer) were made by Clark (124) if he first complexed the metal alkyl component $(AlEt_2Cl)$ with the polar monomer. Also, undecylenol and undecylenic acid were copolymerized.

Acrylonitrile and methacrylonitrile have been copolymerized with Ziegler–Natta catalysts. No evidence is presented that these are polymerized by the Ziegler–Natta centers. It is known that these monomers can be polymerized by Al–N, Mg–N (125), or Ti–N (126) catalysts to stereoregular polymers. It is probable that monomer–catalyst interactions produce similar active centers.

A few examples are given. Simionescu and co-workers (127) reported the copolymerizations (binary and ternary) of such monomers as acrylonitrile with methylmethacrylate, styrene, α-methylstyrene and epichlorohydrin, and styrene and epichlorohydrin. The $(C_5H_5)_2TiCl_2$–$AlEt_3$ catalyst was used. Dixit, Deshpande, and Kapur reported that acrylonitrile was polymerized with the $AlEt_3$–VCl_4 catalyst (128). Yamamoto and co-workers (129) used mixtures of $AlEt_2Et$ and $Fe(C_2H_5)_2(dipy)_2$ or $Co(C_2H_5)_2(dipy)_2$ (where dipy is dipyridyl) to polymerize acrylonitrile or methacrylonitrile to high molecular weight polymers (MW about 600,000). Even monomers such as succinonitrile have been reported to be polymerized (130).

C. HALOGEN MONOMERS

Only a few halogen-containing monomers have been polymerized or co-polymerized at Ziegler–Natta centers. Bacskai (131) reported that the copolymerizations of 4-methyl-1-pentene or propylene and ω-halo-α-olefins (6-chloro-1-hexene) were favorable if the $AlEt_3$ component was complexed with pyridine to lower its reactivity with the halogenated monomer. In this way, copolymers of the halogenated monomer were synthesized with the $AlEt_3$–$TiCl_3$ catalyst.

Clark and Powell (132) reported that the homo- and copolymerizations of ω-halo-α-olefins were enhanced when the size of the halogen was increased (I > Br > Cl), its distance from the polymerizing bond was increased, and when the halogen was attached to a primary rather than secondary or tertiary carbon. Halogenated olefins such as 11-iodoundecene were polymerized with the $AlEt_2Cl$–$TiCl_3$ catalyst.

Overberger and Khattab (133) showed that fluorine-containing 1-butenes and 1-pentenes could be copolymerized with the Al-i-Bu_3–VCl_3 and other Ziegler–Natta catalysts, as shown in structure XIXa,b,c, and d. Their homo-

$$CH_2\!\!=\!\!CH\!\!-\!\!CH_2\!\!-\!\!CF_3 \qquad\qquad CH_2\!\!=\!\!CH\!\!-\!\!\underset{\overset{|}{CF_3}}{CH}\!\!-\!\!CH_3$$

XIXa   XIXb

$$CH_2\!\!=\!\!CH\!\!-\!\!CH_2\!\!-\!\!CH_2\!\!-\!\!CF_3 \qquad CH_2\!\!=\!\!CH\!\!-\!\!CH_2\!\!-\!\!\underset{\overset{|}{CF_3}}{CH}\!\!-\!\!CH_3$$

XIXc   XIXd

polymerization and copolymerization data showed that the trifluorobutenes were less reactive than the trifluoropentenes. This was attributed to the inductive affect of the trifluoromethyl group, which was apparently most overwhelming in the 1-butenes (structures XIXa and XIXb) and less in the 1-pentene (structure XIXc and XIXd). Consequently, the electron-deficient vinyl group of the fluorobutenes did not allow these monomers to coordinate with the active sites of the catalyst. Comparisons of nuclear magnetic resonance spectra of these fluoroolefins and their hydrocarbon α-olefin analogs also indicated an electron-deficiency of the vinyl group of the fluorobutenes. When corresponding organic molecules containing the tri-fluoromethyl group were added to the polymerizations of the hydrocarbon α-olefin analogs, the rate of polymerization was not similarly retarded.

The monomer, $CF_3-CH=CH_2$, was polymerized to an amorphous polymer by means of the $AlEt_2F-VO(acac)_2$ catalyst in methylene chloride or dimethyl formamide solvent (134). Perfluoropropylene was also polym-erized (135). Allyl fluoride was polymerized with the $AIR(OR)Cl-V(acac)_3$ catalyst, but the mechanism was radical (136).

Several chlorine-containing olefins have been polymerized, including allyl chloride, vinylidene chloride, chloroprene, and vinyl chloride.

Allyl chloride was polymerized by Matkovskii and co-workers (137) with a mixture of $VOCl_3$ and Al-*i*-BuCl or Al-*i*-Bu$_3$. A partially dehydro-chlorinated polymer was obtained, shown in structure XX, that had 19% Cl vs. 46% calculated for allyl chloride.

$$\left[\left(-CH_2-CH-\right)_x\left(-CH_2-C=\right)_y\right]_n$$

with CH_2 side group and Cl on the first unit, CH_2 double-bonded on the second unit.

XX

Vinyl chloride has probably been the most investigated of polar monomers by Ziegler–Natta catalysis (138–149). The preponderance of evidence is that the mechanistic path is radical for most of the catalysts investigated. When vinyl chloride and other polar monomers, such as vinyl acetate, were copolymerized with the Al-*i*-Bu$_3$–VOCl$_3$ catalyst, the copolymer composi-tion was characteristic of radical polymerizations (138). Radical initiation was suggested for the catalysts AlR_3-TiCl_3–alkyl halide (140, 141), $AlEtClOEt-VO(C_5H_7O_2)_2$ (139), Al-*i*-Bu$_2$Cl-Ti(OC$_4$H$_9$)$_4$ (142), Al-*i*-Bu$_3$–VOCl$_3$–ethyl acetate (144), Al-*i*-Bu$_2$Cl-Ti(OBu$_4$)–ethyl acetate (145), $AlEt_3$–Ti(OBu)$_4$–epichlorohydrin (146), and $AlEtCl_2$–Ti(OBu)$_4$ (149).

With many of the aforementioned catalysts, some dehydrochlorination occurs. Higashi and co-workers (143), however, found that when an $AlEt_2OCH_2CH_2X$ (X = halogen or NR$_2$) catalyst was used along with $VOCl_3$, high yields of polyvinyl chloride were formed without elimination of HCl. Some workers, however, have suggested that some coordinative or ionic mechanistic paths take place when vinyl chloride was homopolymerized in presence of donors (149). Misono and co-workers (147) reported that, depending on the molar ratio r of Al/V in the $VO(OEt)_3-AlR_3$ (or AlEt$_2$Cl or AlEtCl$_2$) catalyst, a random copolymer ($r < 1.5$) or almost exclusively polyethylene ($r > 1.5$) was formed. Also, it might be mentioned that, when vinylidene and ethylene were copolymerized with the $AlEt_2OEt$–$TiCl_4$–donor system, the polymerization rate and copolymer composition were influenced by both the Al/Ti and donor/Ti ratios, and the product was said to vary from a statistical copolymer to a block copolymer (148).

Haszeldine, Hyde, and Tait (148a) homopolymerized vinyl chloride and vinyl fluoride with the ternary catalyst Al-i-Bu_3–$VOCl_3$–tetrahydrofuran. On the basis of kinetic and copolymerization studies, they concluded that polymerization mechanism was not free radical but resembled that encountered in conventional Ziegler–Natta polymerizations. Differences were noted in the thermal properties of the polyvinyl chloride prepared above and by radical initiation. The role of the tetrahydrofuran (THF) donor was to restrict the reduction of $VOCl_3$ to $VOCl_2$. In fact, the complex $VOCl_2 \cdot$ THF was isolated and, when combined with Al-i-Bu_3 and some THF, it formed the catalyst comparable to Al-i-Bu_3–$VOCl_3$–THF. They proposed the active center to be similar to that shown in structure XXI.

XXI

Mejzlik and co-workers (148b) interpreted their copolymerization studies of propylene and vinyl chloride with the proposal that copolymerization proceeds via an ionic–coordination mechanism only if the monomer charge is rich in propylene. When the charge was rich in vinyl chloride, a radical mechanism was operative.

D. OXYGEN-CONTAINING MONOMERS

Vinyl monomers such as vinyl ethers, acrylic acid, and acrylic esters have been reported to be polymerized.

Vandenberg (150) polymerized methyl, ethyl, n-propyl, isopropyl, n-butyl, isobutyl, $tert$-butyl, and neopentyl vinyl ethers with a catalyst made by combining a complex of Al-i-Bu_3 and tetrahydrofuran with $VCl_3 \cdot AlCl_3$ (made by reducing $3VCl_4$ with $1AlEt_3$). While only the vinyl ethyl ether of the n-alkylvinyl ethers gave crystalline polymer (0.06% conversion), the polymerization of various branched alkyl vinyl ethers gave low conversions of highly crystalline polymers. A cationic insertion mechanism was proposed.

Natta (151) showed that the reactivity of alkyl vinyl ethers increased on going from $AlEt_2Cl$ to $AlEtCl_2$ to $AlCl_3$, when each was combined with Cp_2TiCl_2. Ethylene polymerization decreased simultaneously.

Isotactic polymers of vinyl ethers have been prepared at low temperatures in the presence of soluble acidic metal alkyl by Natta and co-workers (152) and BF_3–etherate by Schildknecht (153) (see Chapter 2, Section III for Schildknecht's 1947–1949 contributions).

Yoh and co-workers (154) reported the polymerization of isobutylvinyl ether to crystalline polymer at 30°C with the Al-i-Bu$_3$–VCl$_3$ · LiCl catalyst. The VCl$_3$ · LiCl composition was made by reduction of 1VCl$_4$ with 1 or 2 moles n-BuLi. In other experiments, they showed that isotactic polymer could be obtained even if Al-i-Bu$_3$ was not added (155). In contrast, Yuki and Hatada (156) found that benzylvinyl ether was polymerized to a non-crystalline product (0° to 60°C, with AlEt$_3$–TiCl$_4$), which was partly debenzylated and complexed with a Ti derivative. When they attempted a copolymerization of styrene and butylvinyl ether with the Al-i-Bu$_2$Cl–Cp$_2$TiCl$_2$ catalyst, only a polyether was formed (157).

Chiellini and Nocci (158) showed by use of solvent fractionation and careful analysis of the obtained fractions that copolymerization of 4-methyl-1-pentene and alkyl methacrylates or alkyl acrylates did not take place with a Ziegler–Natta catalyst. This paper casts doubt that these polar monomers polymerize at a transition metal–carbon bond as do olefins. Methyl methacrylate has been polymerized with Ziegler–Natta catalysts such as AlEt$_3$–VOCl$_3$ (159), AlEt$_3$–Cr(C$_5$H$_7$O$_2$) (160), Fe and Co alkyl complexes (161), (C$_2$H$_5$)$_4$Al$_2$SO$_4$–TiCl$_3$AA (162), AlEt$_3$–Ti(O-i-C$_3$H$_7$)$_4$ (163), transition metal–π-allyl complexes (Cr) (164), and AlEt$_2$Cl–VCl$_4$ (165). The mechanistic paths for most of these have not been elucidated. Matzuzaki and co-workers (165) showed their product was equivalent to that obtained by radical means. Abe, Imai, and Matsumoto (163) reported that high syndiotacticities are possible with their catalyst. As for vinyl chloride and related vinyl monomers, the question of head-to-tail additions vs. stereoisomerism must be considered. Since acrylate monomers are polymerizable by metal alkyls, especially by their reaction products with functional organic molecules, mechanistic paths such as for polymerization of epoxides, aldehydes, etc., may be possible.

Hopkins and Miller (166) reported the polymerization of $tert$-butylacrylate to a stereoregular, crystalline polymer with the n-BuLi–TiCl$_4$ catalyst. Unfortunately, control syntheses with LiBu were not run to see if highly crystalline polymer can be formed in the absence of TiCl$_4$. Methyl methacrylate and other acrylates are readily polymerizable with lithium alkyls.

E. MISCELLANEOUS MONOMERS

Otsu and co-workers (167) polymerized several epoxide monomers in low conversions (propylene oxide, epichlorohydrin, styrene oxide, and phenyl glycidyl ether) with a variety of Ziegler–Natta catalysts, with $AlEt_3$–$TiCl_3$ giving the highest yields, molecular weights, and crystallinities. In general, however, the products had low molecular weights.

An isotactic polymer in 92% yield was reported by Aylward (168) when *n*-propyl-*p*-vinyl benzene sulfonate was polymerized with $AlEt_3$–$TiCl_4$ in toluene at 50°C for 4 hours. A melting point of about 349°C was found.

Marktscheffel and co-workers (169) converted bis(β-chloroethyl)vinyl phosphate in low conversions to a crystalline but not fusible polymer (it was, however, soluble in hot dimethylformamide).

A crystalline polymer with a fiber repeat distance of 4.95 Å was synthesized at −78°C by Kobayanshi and Sumitomo (170) from β-cyanopropion-aldehyde with combinations of $TiCl_4$ or $TiCl_3$ and $AlEt_3$, $AlEt_2Cl$, or $AlEtCl_2$.

Heterocyclic compounds such as α-methylfuran, furan, thiophene, and dihydropyran were polymerized with $AlEt_3$–$TiCl_4$ (171). Sometimes, the metal alkyl component is active under certain conditions in the absence of the transition metal salt; for example, benzofuran was polymerized in the presence of aluminum chloroalkyls and a Lewis base (172).

IX. Allenes

Allene has been polymerized by a variety of Ziegler–Natta catalysts (173–178), as well as catalysts containing only transition metal salts and organic molecules. The polymer can predominantly contain one or several structural units, as shown in XXII, depending on the catalyst and conditions of polymerization.

The polymerization of allene with Ziegler–Natta catalysts was first reported by Baker (173), who found that the polymer products contained vinylidene, as shown in structure XXIIa; vinyl, XXIIc; and *cis*-unsaturation, XXIId. Al-*i*-Bu$_3$, in combination with TiCl$_4$, VOCl$_3$, CoCl$_2$, FeCl$_3$, Co(NH$_3$)$_6$Cl, or NiCl$_2$ was used as a catalyst. Baker proposed that the mechanistic path involved the attack of anion on the terminal rather than the central carbon atom of allene, as shown in Eq. 19–13, where M is the metal

$$\overset{\delta^-\ \delta^+}{\text{R–M}} + \text{C}{=}\text{C}{=}\text{C} \longrightarrow \text{RCH}_2-\overset{\overset{\displaystyle CH_2}{\|}}{\text{C}}-\text{M} \qquad (19{-}13)$$

of the active center.

Polyallene products that had a predominance of the 1,2 (structure XXIIa) or 1,2,2,1 structure (XXIIb) were prepared more recently (174, 175). Havinga and Schors (174) examined VOCl$_3$ in combination with several metal alkyls and found the order of activity to be: AlR$_3$ > AlR$_2$X > AlRX$_2$ where R = Et, *i*-Bu, and X = halogen. A polymer with melting point of 115° to 125°C, almost 65°C greater than formed under Baker's conditions, was formed. They also showed the following order of polymerization activity: CH$_2$= C=CH$_2$ > R—CH=C=CH$_2$ > R$_1$R$_2$C=C=CH$_2$ > R$_1$CH=C= CHR$_2$ > R$_1$R$_2$C=C=CR$_3$R$_4$. They suggested that coordination of these allenes is favored at the internal double bond due to an increased electron density at this bond. Steric factors may, however, override this driving force.

Van den Enk and Van der Ploeg (175) proposed that the ratio of 1,2 (structure XXIIa) to 1,2,2,1 (XXIIb) structural units in the polymer could be controlled by the strengths of electron-donating and electron-withdrawing ligands attached to the transition metal center. They cited the earlier finding of Otsuka that, on polymerizing allene with π-allylnickel halide complexes, the proportion of 1,2,2,1 polymer increased in the order: I > Br > Cl, the ratios 1,2/1,2,2,1 being, respectively, 79/21 for I, 87/13 for Br, and 90/10 for Cl. This indicated that ligands with a strong electron-withdrawing affect (due to a greater overlap) gave rise to a greater degree of 1,2,2,1 addition. Thus, when Ni(II) acetylacetonate (stabilized by complexing with allene) is alkylated with Al-*i*-Bu$_3$, the *i*-Bu ligand increases the formation of 1,2,2,1 structure.

Poly(2,3-dimethylenebutylene) (structure XXIIb) was also synthesized by Shier (176) with Pd(NO$_3$)$_2$·2H$_2$O complexed with (C$_6$H$_5$)$_3$P in glacial acetic acid–methanol solvent. Also, highly crystalline 1,2 polymers (melting point of 125°C) were prepared by Otsuka and co-workers (177,178) with RhCl(CO)$_2$·(C$_6$H$_5$)$_3$P and [Co(CO)$_3$(C$_6$H$_5$)$_3$P]$_2$.

X. Acetylenes

Since Natta and co-workers (179) first reported the polymerization of acetylene with binary mixtures of metal alkyls and transition metal salts, many acetylene compounds have also been polymerized (180–215). Benes, Janic, and Peska reviewed this subject in 1970 (189). It has not been established whether or not acetylenes polymerize at Ziegler–Natta centers in the same way that olefins polymerize or whether other mechanisms are taking place. The experimental observations are more difficult to interpret, because the acetylenic monomers can undergo linear or cyclopolymerizations, as well as be trimerized to aromatic compounds. Because the terminal acetylene, such as phenylacetylene, contains an acidic hydrogen, the metal alkyl component can react with one or more of the monomer molecules (180), as shown in Eq. 19–14. The alkylating metal alkyl thus need not be that initially added.

$$C_6H_5C \equiv CH + AlEt_3 \longrightarrow C_6H_5C \equiv C—AlEt_2 + \text{ethane} \qquad (19\text{–}14)$$

Terminal acetylenes, $RC \equiv CH$, have been polymerized when R is H (179, 181–184), CH_3 (183), Et (185), isopropenyl (186, 190), butyl (187), octyl (187), phenoxy (188), decyl (189), vinyl (203), phenyl (182, 187, 191–193), β-naphthyl (194), and 9-anthracenyl (204).

Highly dispersed or soluble catalysts have been preferred: $TiCl_4$ or $Ti(OR)_4 + AlEt_2Cl$ or $AlEt_3$ (184, 190) and transition metal chelates + $AlEt_3$ where the transition metal is Co, Ni, V, or Fe (191, 196–198). Some workers modified these catalysts by the addition of electron-donors (199). Berlin and co-workers (199) reported that soluble catalysts containing $AlEt_3 + Cp_2TiCl_2$ or $Cp_2Ti(OC_4H_9)_2$ did not homopolymerize phenyl acetylene (ethylene was previously polymerized with $AlEt_3–Cp_2TiCl_2$). Unlike α-olefins, acetylenes are easily polymerized with aluminum alkyl-group VIII salt mixtures (200–202). Acetylenes, similarly to olefins, were shown to be polymerized when no metal alkyl is added. Kormer and co-workers (203) showed that vinyl acetylene polymerized in the presence of π-crotyl-NiI exclusively through a triple bond.

Both high and low molecular weight linear polymers were formed, depending on the catalyst and structure of R in $RC \equiv CH$.

Usually, aliphatic acetylenes can be converted to high molecular weight polymers. Trepka and Sonnenfeld (201) found inherent viscosity values in the range 2 to 6 dl/g for polymers from 1-butyne, 1-hexyne, and 1-dodecyne when $Al\text{-}i\text{-}Bu_3 + Fe(III)$ naphthenate catalyst was used. In comparison, Michel (204) showed that 9-ethylnylanthracene in the presence of $Al\text{-}i\text{-}Bu_3–TiCl_4$ catalyst gave polymers that had molecular weights of about 1,000. The heterogeneous catalyst $AlEt_3–TiCl_3$ produced a polymer from phenyl acetylene that had a molecular weight of about 5,000 (199). The polymer from β-ethynylnaphthalene ($Al\text{-}i\text{-}Bu_3–TiCl_4$ catalyst) had a molecular weight of about 2,091 (DP = 14) (184).

Highly linear polymers consisting of long sequences of trans—CH= CH—units have been prepared from acetylene (179, 181). The crude product was fractionated into a black insoluble residue and an acetone-soluble fraction that was analyzed.

Ito and co-workers (181a), using the $AlEt_3$–Ti $(O\text{-}n\text{-}Bu)_4$ catalyst (Al/Ti = 3 to 4), synthesized an all *cis*-polymer at temperatures lower than $-78°C$ and an all *trans*-polymer at temperatures higher than $150°C$. They developed a method of polymerizing the acetylene monomer directly into thin films.

The higher aliphatic acetylenes were converted to an amorphous product with this catalyst. In comparison, the light brown to red products made from 9-ethynylanthracene had a crystalline melting point near $255°C$ (endotherm in DTA thermogram) (194).

Aso and co-workers (205) showed that, in addition to linear polymerization, some cycloaddition occurred when *o*-diethynylbenzene was polymerized with binary mixtures of $TiCl_4$ or $Ti(acac)_3$ and aluminum alkyls, as shown in Eq. 19–15. Natta (207) examined poly-1-hexyne by X-ray and

$$(19\text{-}15)$$

found no crystallinity (the polymer lacked planarity).

In 1948, Reppe and Schweckendiek (207a) reported that acetylenes can be trimerized to aromatic compounds in the presence of nickel carbonyl-phosphine complexes. Furlani and co-workers (206) showed an intimate dependence existed between monomer structure and catalyst composition in determining to what degree linear polymerization and trimerization occurred for the systems shown in Eqs. 19–16 and 19–17. For example,

$$(19\text{-}16)$$

$$(19\text{-}17)$$

trimerization of phenylacetylene gave triphenylbenzenes.

Shirakawa and Ikeda (206a) proposed the scheme in Eqs. 19–18 and 19–19 for oligomerization of acetylene to ethyl benzene with the catalyst $AlEt_2Cl + Ti(acac)_3$.

$$\text{Ti(acac)}_3 + \text{TiCl}_3 \rightarrow \underset{\substack{\text{active}\\\text{center}}}{\text{cat–Et}} \xrightarrow{\text{HC}\equiv\text{CH}} \underset{\text{cat}}{\overset{H}{>}}C=C\underset{Et}{\overset{H}{<}} \longrightarrow \tag{19-18}$$

$$\underset{\text{cat}}{\overset{H}{>}}C=C\overset{H}{\underset{\underset{\underset{Et\quad H}{\diagdown}}{\overset{\|}{C}}}{\diagdown}}\text{—H} \longrightarrow \underset{\text{cat}}{\overset{H}{>}}C=C\overset{H}{\underset{\underset{\underset{H}{>}C=C\overset{}{\underset{H}{<}}}{\overset{\|}{C}}}{\diagdown}} \longrightarrow \text{cat–H} + \text{Et—}\left\langle\bigcirc\right\rangle \tag{19-19}$$

A similar sequence involving cat-H instead of cat-Et was said to give benzene. This study used an equimolar mixture of acetylene and acetylene-d_2.

When one of the monomers contains two ethynyls, there is the possibility that trimerization will lead to a branched polyphenylene structure. Several examples are cited here.

Bracke (208) reported that the polyphenylenes formed by copolymerizing phenylacetylene and *o*-diethynylbenzene were highly branched but were still soluble in chlorobenzene (MW $\simeq$ 70,000). Chaulk and Gilbert (209) also reported the synthesis of soluble copolymers having the polyphenylene structure (MW $\simeq$ 10,000), when phenylacetylene and *m*-diethynylbenzene were copolymerized with the $AlEt_2Cl–TiCl_4$ catalyst.

Nicolescu and Angelescu (184, 210, 211) investigated several catalyst systems for the polymerization of acetylenes with the objective of gaining a better understanding of the catalyst. Their results led them to conclude that the velocity of polymerization was directly related to the presence of complexes, which were detected by visible and UV spectroscopy and by conductivity measurements. Mixtures of $AlEt_3$ and $VO(C_5H_7O_2)_2$, $Cr(C_5H_7O_2)_3$, and $Co(C_5H_7O_2)$ (described as soluble catalysts) were used.

Luttinger and Colthup (212, 213) oligomerized monosubstituted acetylenes mostly to dimers and trimers in the presence of a mixture of sodium borohydride plus a salt or complex of a group VIII metal, such as $NiCl_2$ (acetylene, with the same catalysts, was polymerized to a linear, trans, high molecular weight polyacetylene).

XI. Conjugated Cyclic and Acyclic Dienes

This section describes some of the cyclic and acyclic conjugated dienes that have been investigated with Ziegler–Natta catalysts. A brief description of higher conjugated dienes will also be presented.

A. CYCLIC DIENES

Cyclopentadiene (214, 215) and 1,3 cyclohexadiene (216, 217) have been polymerized with Ziegler–Natta catalysts by several additional modes.

Aso and co-workers (215) reported the polymer from cyclopentadiene contained 1,4- and 1,2-structural units, as shown in Eq. 19–20. They noted,

$$(19{-}20)$$

1,4 unit   1,2 unit

however, that a product with the same structure was made by cationic initiation.

Marvel, Frey, and co-workers (216, 217) found that the polymerization of 1,3-cyclohexadiene with Ziegler–Natta catalysts (Al-i-Bu$_3$ + TiCl$_3$ or VCl$_3$) led to a regular 1,4-structure (see Eq. 19–21) with a melting point of 160° to

$$(19{-}21)$$

180°C (products from cationic initiations contained a mixture of 1,2 and 1,4 structures). The regular structure was dehydrogenated to produce a p-polyphenyl. Other catalysts that were effective for polymerization of 1,3-cyclohexadiene were C$_6$H$_5$MgBr–TiCl$_4$, LiAlH$_4$–TiCl$_4$, and n-BuLi–TiCl$_4$ (218).

Mabuchi, Saegusa, and Furukawa (219) polymerized 1-methylene 2-cyclohexene in the presence of Ziegler–Natta catalysts (AlEt$_3$ + TiCl$_4$ or VCl$_4$) and cationic initiators (BF$_3$·Et$_2$O). According to IR and NMR analyses, polymers from both systems were similar and largely contained the 1,4-addition unit, as shown in Eq. 19–22.

$$(19{-}22)$$

B. ACYCLIC DIENES

Chapters 3, 13, 15, 16, and 17 have already described the polymerization chemistry of three important dienes, namely, butadiene, isoprene, and 1,3-pentadiene in detail. Most of the diene experimental work has centered about these three dienes, because the monomers are relatively inexpensive, the polymers have attractive useful properties, and they have suitable structures to study stereochemical processes. No further elaboration is necessary here, and only other dienes will be described.

The mono- and disubstituted derivatives of 1,3-butadiene, shown in structure XXIIIa through f, have also been investigated.

$$CH_2\!=\!CH\!-\!CH\!=\!CH\phi$$

1-phenyl-1,3-butadiene

XXIIIa

$$CH_2\!=\!\overset{\overset{\displaystyle CH_3}{|}}{C}\!-\!CH\!=\!CH\!-\!CH_3$$

2-methyl-1,3-pentadiene

XXIIIb

$$CH_2\!=\!\overset{\overset{\displaystyle \phi}{|}}{C}\!-\!CH\!=\!CH_2$$

2-phenyl-1,3-butadiene

XXIIIc

$$CH_2\!=\!CH\!-\!CH\!=\!\overset{\overset{\displaystyle CH_3}{|}}{C}\!-\!CH_3$$

4-methyl-1,3-pentadiene

XXIIId

$$CH_2\!=\!\overset{\overset{\displaystyle Pr}{|}}{C}\!-\!CH\!=\!CH_2$$

2-n-propyl-1,3-butadiene

XXIIIe

$$CH_2\!=\!\overset{\overset{\displaystyle \phi}{|}}{C}\!-\!\overset{\overset{\displaystyle \phi}{|}}{C}\!=\!CH_2$$

2,3-diphenyl-1,3-butadiene

XXIIIf

Polymers having melting points in the range of 105° to 120°C were made from trans-1-phenyl-1,3-butadiene with the Al-i-Bu$_3$–TiCl$_4$ catalyst (220).

Marconi, Mazzei, and co-workers (221) found that amorphous, low molecular weight homopolymers of 2-phenylbutadiene having prevailingly cis-1,4 structure were prepared with ternary catalysts consisting of aluminum hydrides, AlI$_3$ and TiCl$_4$. Vessel and Stille (222), using Al-i-Bu$_3$ and TiCl$_4$, also prepared low molecular weight products that had a high degree of crystallinity (80 to 95% cis-1,4 structure if a 3:1 Al/Ti ratio was used). Under the same conditions, 2,3-diphenylbutadiene gave only 1 to 5% conversion of a low molecular weight material.

Marconi, Mazzei, and co-workers (223) found that the stereochemical composition of polymers of 2-n-propyl-1,2-butadiene could be controlled by choice of catalyst and closely resembled the behavior of isoprene. For example, 1,4-cis structures prevailed when AlEt$_3$–TiCl$_4$ or AlHCl$_2$·Et$_2$O–TiCl$_4$ systems were used; 3,4- and trans-1,4 structures were obtained when the catalysts AlEt$_3$–Ti(i-PrO)$_4$ and AlEt$_3$–VCl$_3$ were used, respectively.

Cuzin and co-workers (224) polymerized trans-2-methyl-1,3-pentadiene to a crystalline polymer (1,4-cis structure) with the AlEt$_3$–TiCl$_4$ catalyst and to a crystalline polymer (1,4-trans structure) with the ternary catalyst AlR$_3$–VCl$_3$–Ti(OR)$_4$. Competing cationic polymerizations were recognized.

Porri and Gallazzi (225) found that 4-methyl-1,3-pentadiene formed crystalline (1,2-isotactic) polymers with heterogeneous catalysts ($AlEt_3$ + $TiCl_4$, $TiCl_3$, or VCl_3). Only amorphous products, however, were formed with homogeneous (perhaps colloidal) systems, such as $AlEt_3$ + $Ti(OR)_4$ or $VO(OR)_3$. The 1,2-isotactic polymer crystallizes in two modifications (226). Catalysts consisting of AlE_2Cl + $V(acac)_3$ or $AlEt_3$ + $Co(acac)_3$ were inactive (225). The fact that 1,2-addition prevailed was attributed to steric repulsions due to terminal methyl groups.

Murahashi and co-workers (226a) polymerized 2,4-hexadiene to a crystalline product (melting point $69°$ to $78°C$) and $[\eta] = 0.07$ dl/g) in the presence of several Ziegler–Natta catalysts ($Al\text{-}i\text{-}Bu_3$ + $TiCl_4$, and $AlEt_2Cl$ + $Ti(acac)_3$ or $Co(acac)_3$). The prevailing structures were trans-1,4 units with a small number of 1,2 units being present. X-ray studies suggested that the polymer was not syndiotactic but was either erythrodiisotactic or threodiisotactic.

C. HIGHER CONJUGATED DIENES

A number of dienes that contain more than two double bonds were investigated with Ziegler–Natta catalysts. These monomers are interesting in that 1,2-, 1,4-, and 1,6-addition modes are possible, as shown in Eq. 19–23.

$$
CH_2=CH-CH=CH-CH=CH_2 \quad
\begin{cases}
\xrightarrow{\text{1,2 addition}} & -CH_2-CH- \\
& \qquad\quad\ CH=CH-CH=CH_2 \\
\xrightarrow{\text{1,4 addition}} & -CH_2-CH=CH-CH- \quad (19\text{–}23) \\
& \qquad\qquad\qquad\quad CH=CH_2 \\
\xrightarrow{\text{1,6 addition}} & -CH_2-CH=CH-CH=CH-CH_2-
\end{cases}
$$

1,3,5-hexatriene

Bell (227) investigated the monomers 1,3,5-hexatriene, 1,3,5-heptatriene, and 2,4,6-octatriene with different catalysts at low ($-30°$ to $+25°C$) temperatures. The 1,3,5-hexatriene monomer preferentially polymerized by 1,6-addition. A 75% crystalline product (exclusive trans,trans-diene units were indicated by infrared analysis) was synthesized with $Al\text{-}i\text{-}Bu_3$–$VOCl_3$ at $-30°C$. An amorphous polymer (mixture of trans,trans- and cis, trans-diene units were present) was synthesized with the $AlEt_2Cl \cdot$ anisole–$VOCl_3$–$Ti(OC_6H_5)_4$ catalyst at $-25°C$. The other trienes were also polymerized with the $Al\text{-}i\text{-}Bu_3$–$VOCl_3$ (Al/V = 4) catalyst, but the polymers were amorphous. All three dienes polymerized with LiBu to give a polymer containing 1,2, 1,4, and 1,6 units.

Marvel and co-workers (228) concluded from a comparison of products obtained by cationic initiation ($BF_3 \cdot Et_2O$ or $TiCl_4$) and by Ziegler–Natta catalyst (Al-i-Bu_3 + VCl_3 or $TiCl_3$) that Ziegler–Natta catalyst provides a cationic initiator for alloocimene, as shown in structure **XXIV**. Polymeri-

$$CH_3\!-\!\underset{\underset{CH_3}{|}}{C}\!=\!CH\!-\!CH\!=\!CH\!-\!\underset{\underset{CH_3}{|}}{C}\!=\!CH\!-\!CH_3$$

XXIV

zation in both cases occurred at the 6,7 bond by 1,2-addition (40 to 50%) and at the 4,7 bond by 1,4-addition (50 to 60%). Marvel and Rogers (229) also polymerized 1,9,12-octadecatriene (Al-i-Bu_3 + $TiCl_4$), but the product was cross-linked.

Other trienes that were polymerized with Ziegler–Natta catalysts are 1,3,8-nonatriene (230) and myrcene (231). In presence of $AlEt_3$–$TiCl_4$, the *trans*-isomer of 1,3,8-nonatriene cyclopolymerized to a product that was only partially soluble in benzene and that showed only a trans internal double bond (no terminal unsaturation). The polymerization mode in Eq. 19–24 was suggested. In the case of myrcene, the polymer contained two

$$(19\text{–}24)$$

internal double bonds (Al-i-Bu_3–VCl_3 or $TiCl_3$ catalysts), and the path in Eq. 19–25 was suggested.

$$CH_2\!=\!\underset{\underset{(CH_2)_2}{|}}{C}\!-\!CH\!=\!CH_2 \longrightarrow \text{---}\!(CH_2\!-\!\underset{\underset{(CH_2)_2}{|}}{C}\!=\!CH\!-\!CH_2)_n$$

$$\underset{\overset{|}{CH}}{\overset{}{}} \qquad \underset{\overset{|}{CH}}{\overset{}{}}$$

$$\underset{CH_3 \quad CH_3}{\overset{\parallel}{C}} \qquad \underset{CH_3 \quad CH_3}{\overset{\parallel}{C}} \qquad (19\text{–}25)$$

XII. Nonconjugated Acyclic and Cyclic Diolefins

A variety of acyclic and cyclic nonconjugated diolefins have been reported to be homopolymerized with heterogeneous and homogeneous Ziegler–Natta catalysts, e.g. structures **XXVa** and **XXVb** and **XXVI**.

$$CH_2{=}CH{-}(CH_2)_n{-}CH{=}CH_2 \qquad CH_2{=}\overset{\overset{\displaystyle R}{|}}{C}{-}(CH_2)_n{-}\overset{\overset{\displaystyle R}{|}}{C}{=}CH_2$$

XXVa	XXVb	XXVI
$n = 1, 2, \ldots, 20$	R = H, alkyl, or phenyl	

As with conjugated dienes, these diolefins can be polymerized by different types of initiators and catalysts. Hence, it does not necessarily follow that all of these reported polymerizations took place at transition metal–carbon bonds. As illustrated with many examples in Chapter 22, the Ziegler–Natta catalyst can serve as a source of cationic, anionic, or radical species. In many of the above polymerizations, the products have very low molecular weight, characteristic of products obtained by polymerizing the same diolefin with cationic initiators. In the absence of more definitive evidence, the reader should be cautious in accepting any mechanism.

A good diagnostic test is a copolymerization involving ethylene (or propylene); these do not homopolymerize to high molecular weight polymers with cationic initiators. Furthermore, the products from the copolymerizations should be subjected to a fractionation to rule out the possibility of concurrent homopolymerizations by several mechanisms producing a mixture of homopolymers.

A. ACYCLIC MONOMERS

We have convincing evidence that 1,5-hexadiene polymerizes at transition metal–carbon centers. This diolefin is to the class of nonconjugated olefins what butadiene is to the conjugated diolefins.

Marvel and Stille first reported its polymerization with Ziegler–Natta catalysts (232); in addition to the expected 1,2-addition product (structure XXVII), they also reported a cyclopentane-enchained unit was present, as shown in Eq. 19–26. In fact, structure XXVIII prevailed for this catalyst.

$$(19{-}26)$$

XXVII	XXVIII

Possible mechanistic paths are described in Chapter 16. The uniqueness and versatility of the Ziegler–Natta centers is again demonstrated.

How do the conditions of polymerization, type of catalyst, and structure of the diolefin affect the relative formations of structures XXVII and XXVIII?

Heterogeneous catalysts appear to favor cyclization (232), while homogeneous catalysts favor 1,2-addition (233) (Table 19–7).

TABLE 19–7

1,2-Addition for Heterogeneous vs. Homogeneous Catalysts

Catalyst	Phase	Hexadiene polymerized with 1,2 enchainment (%)
$AlEt_2Cl + TiCl_4$	Heterogeneous $(+25)$	< 10
$Al(C_6H_{13})_3 + VCl_4$	Heterogeneous $(+25)$	< 10
$AlEt_2Cl + V(acac)_3$	— $(+25)$	$30–35$
$Al\text{-}i\text{-}Bu_3 + VCl_4$	(-20)	< 10
$AlEt_2Cl + VCl_4$	(-20)	$35–40$
$AlEt_2Cl + V(acac)_3$	Homogeneous (-20)	$75–80$
$AlEt_2Cl + VO(OEt)_3$ or $VO(OEt)_2Cl$	Homogeneous (-20)	$75–80$

Lower temperatures favor the formation of homogeneous catalysts. In addition to the vanadium salts above, homogeneous catalysts were formed at $-20°C$ by combining $AlEt_2Cl + VCl_3 \cdot 3THF$ and $VCl_3 \cdot 2(CH_3)_3N$. Marvel and Garrison found that cyclization also increased as the concentration of the diolefin increased (234).

Homopolymers of 1,5-hexadiene, in which cyclopentane prevails, are powderlike, while the 1,2-addition product has elastomeric properties. Makowskii and co-workers (234a) characterized several preparations of poly-1,5-hexadiene. The polymer was crystalline, had a high tensile strength (about 5,420 psi), a high melting point (138°C), a high density (1.22 g/cm^3), and was flexible. The structures of the extracted and residue fractions were similar, showing that they differed only in molecular weight and not in composition. Since not all of the diene units were not cyclized, the polymer could be considered as an interpolymer rather than a homopolymer. The polymer resembled high-density polyethylene in many of its properties.

According to Trifan and co-workers (235), 1,4-pentadiene does not undergo an analogous cyclization polymerization, as indicated in Eq. 19–27,

$$CH_2{=}CH{-}CH_2{-}CH{=}CH_2 \xrightarrow[\text{Reaction}]{\text{No}} \left(CH_2{-}CH \underset{CH_2}{\overset{CH_2}{\diagup\diagdown}} CH \right)_n \quad (19\text{–}27)$$

but rather forms the structure, shown in **XXIX**, by a combination of 1,2-

XXIX

and cycloaddition. They cited as evidence that the CH_2 stretching frequencies were characteristic of unstrained 6-membered rings rather than part of highly strained 4-membered rings.

Some researchers have copolymerized 1,5-hexadiene with ethylene, propylene, and higher olefins to true copolymers (233, 234a, 236). According to Valvassori and co-workers (233), the ethylene–hexadiene copolymers, obtained with soluble catalysts, are amorphous, have elastomeric character, are completely soluble in boiling heptane, and are unsaturated. The relative amount of monomeric units of hexadiene polymerized in 1,2-units was of the same order as observed in the homopolymer prepared with the same catalyst. Makowskii and co-workers (234a) characterized a series of ethylene–1,5-hexadiene copolymers containing 15 to 93 mole % ethylene, which were prepared with the $AlEt_3$–$TiCl_3 \cdot xAlCl_3$ catalyst. Their X-ray diffraction analysis showed the presence of two phases in the copolymer. One of these phases was rich in polyethylene and the other was rich in poly-1,5-hexadiene.

Livshits and co-workers (236) have found that, during the copolymerization of 1,5-hexadiene with ethylene and propylene in the presence of aluminum–vanadium catalysts, cyclization prevailed over 1,2-addition. The ring content in these copolymers varied with the particular catalyst used and decreased in the order: $AlEt_3Cl_3$–$VOCl_3$ > Al-i-Bu_2Cl–VCl_4 > $AlEt_2$Cl–$V(acac)_3$. The latter forms a homogeneous catalyst; this agrees with the conclusions of Valvassori that 1,2-addition is favored by homogeneous catalysts. These workers also found that cyclization decreased in the order: 1,5-hexadiene (84 to 87%) > 1,5-heptadiene (78 to 85%) > 2-methyl-1,5-hexadiene (61 to 78%) > 2,5-dimethyl-1,2-hexadiene (62%).

Larger nonconjugated diolefins ($n = 3$ to 12 and 16) were polymerized at $-70°C$ by Marvel and Garrison (234). The polymers obtained possessed unreacted vinyl groups, internal unsaturation, some simple cyclic units, and, in the case of poly-1,7-octadiene, considerable vinylidene unsaturation. The molecular weights tended to be low; inherent viscosities were usually about 0.1 dl/g. The estimated maximum yield of the smallest possible ring size was 25% for the 6–7 ring and ranged from 6 to 15% for ring sizes C_8 to C_{21} rings. When poly-1,6-heptadiene was dehydrogenated with $KClO_4$, aromatic *meta*-substituted adsorption maxima were observed in the infrared spectrum and an aromatic adsorption maximum in the ultraviolet region of the spectrum (232).

Romanov and co-workers (236a) proposed the structure shown in **XXX** for the polymer of 1,5-heptadiene.

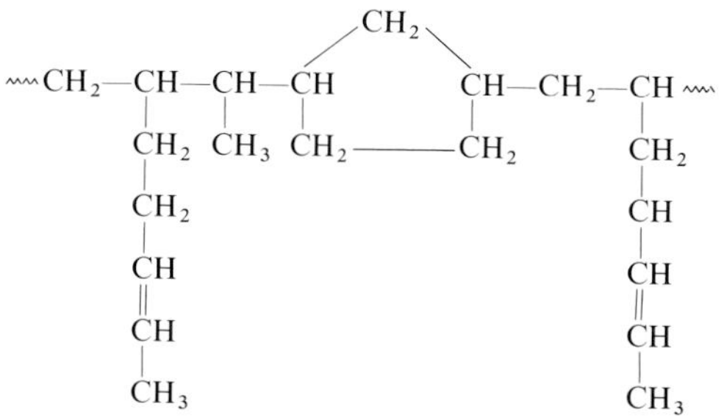

XXX

It is uncertain whether 1,5-disubstituted nonconjugated diolefins (structure **XXVb**) have also been polymerized at transition metal–carbon centers. Marek and co-workers (237) showed that 2,5-dimethyl-1,5-hexadiene was polymerized by a cyclization mechanism in the presence of BF_3 or $TiCl_3$ to give a structure as shown in **XXXI**.

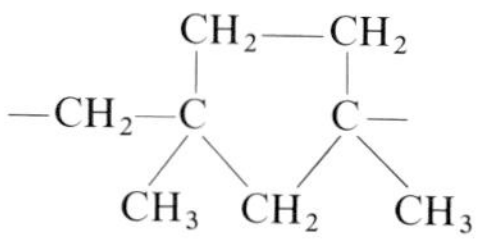

XXXI

It is probable that 2,5-diphenyl-1,5-hexadiene also polymerized by a cationic mechanism in the presence of Al-i-Bu$_3$–TiCl$_4$ catalyst. Marvel and Gall (238) and Field (239) showed this monomer to polymerize by cationic, anionic, and radical initiation. The use of 5,7-dimethyl-1,6-octadiene as comonomer in terpolymers with ethylene and propylene was reported (240).

B. CYCLIC MONOMERS

Both 1,2-divinylcyclohexane (241) and 1,3-divinylcyclohexane (242) were cyclopolymerized by Ziegler–Natta catalysts, as shown in Eqs. 19–28 and 19–29. Whereas 1,2-divinylcyclohexane was polymerized by cationic initiators, 1,3-divinylcyclohexane was inactive in the presence of BF_3-etherate and common peroxide catalysts, indicating growth at transition metal centers. Earlier, Valvassori and co-workers (243) polymerized 1,2-divinylcyclobutane.

$$(19\text{--}28)$$

$$MW = 1100$$

$$(19\text{--}29)$$

Some cyclic nonconjugated diolefins have also been cyclopolymerized to variable degrees.

Butler and Miles (244) used the Al-i-Bu$_3$–TiCl$_4$ catalyst at ambient temperatures to polymerize 4-vinylcyclohexene to a product that was 45% soluble in chloroform. This fraction was essentially saturated, had a melting temperature of 91°C, and an intrinsic viscosity of 0.04 dl/g. They interpreted their results to mean that double bonds interacted to provide a lower energy pathway to polymer through cyclization. In comparison, Sadykh–Zade and co-workers (244a) reported a softening point of 240° to 250°C for a polymer of 4-vinylcyclohexene made with the Al-i-Bu$_2$Cl–TiCl$_4$ catalyst. The nature of the mechanism is not clear because this monomer has also been polymerized by cationic initiators, but most likely a prevalence of cationic polymerization is suggested.

Monomeric cis, cis-1-methyl-1,5-cyclooctadiene has been cyclopolymerized in the presence of Al-i-Bu$_3$–TiCl$_4$ (245), as shown in Eq. 19–30. In a

$$(19\text{--}30)$$

terpolymerization with ethylene and propylene in the presence of aluminum alkyl–vanadium salt catalysts (AlEt$_2$Cl or Al$_2$Et$_3$Cl$_3$ + VOCl$_3$ or VCl$_4$), however, this diolefin copolymerized by opening of the nonsubstituted double bond. A similar reaction path was found for 1,5-cyclooctadiene and 2-methylene-5-norbornene (246), as shown in Eq. 19–31, where a is the path

$$(19-31)$$

taken during homopolymerization and b is the path taken during terpolymerization with ethylene and propylene.

Six different structural units have been identified by Dall'Asta and co-workers (247) when dicyclopentadiene was polymerized with catalysts containing transition metals from groups IV, V, VI, or VIII, as shown in variations of structure XXXII.

Catalyst	Prevalent structure (%)
Al-i-Bu$_3$–TiCl$_4$	XXXIIa (85)
AlEt$_2$Cl–MoO$_2$(acac)$_2$	XXXIIc (70)
AlEt$_2$Cl–MoCl$_5$	XXXIIe (85)

The structure of the product was sensitive to the choice of catalyst, relative concentrations, and temperature of polymerization.

While 1,4-dimethylenecyclohexane has been polymerized with both Ziegler–Natta (Al-i-Bu$_3$ + TiCl$_4$) and cationic (BF$_3$) systems (248), as shown in Eq. 19–32, 1,2-dimethylenecyclohexane failed to polymerize with

$$CH_2 = \underbrace{} = CH_2 \longrightarrow \left(CH_2 - \underbrace{} \right)_n \qquad (19\text{–}32)$$

Ziegler–Natta systems (249).

As indicated by NMR analysis, polymerization of 1-methylene-4-vinylcyclohexane with the AlEt$_3$–TiCl$_4$ catalyst gave a product that contained mostly cyclic units (250), as shown in Eq. 19–33.

$$CH_2 = \underbrace{} - CH = CH_2 \longrightarrow \left(CH_2 - \underbrace{} \right)_n \qquad (19\text{–}33)$$

When an internal and a vinyl bond are both present, polymerization through the vinyl bond occurs preferentially (251), as indicated in Eq. 19–34.

$$\underset{\text{CH}=\text{CH}_2}{\bigcirc} \xrightarrow{\text{AlEt}_3 + \text{TiCl}_3} \left(\underset{H}{\overset{}{C}} - CH_2 \right)_n \qquad (19\text{–}34)$$

In this sense, the polymer behaves as styrene and forms an isotactic polymer of high crystallinity. Partial cationic polymerization can occur, producing the cycloaddition product in small yields (244).

Norbornadiene has been polymerized by ring opening to a product that contains at least partly *trans*-unsaturation (252), as shown in Eq. 19–35.

$$\bigtriangleup\!\!\!\bigtriangledown \longrightarrow \left(CH = CH - \Box \right)_n \qquad (19\text{–}35)$$

Many other cyclic dienes have been polymerized with Ziegler–Natta catalysts, including mixtures of α- and β-pinenes (253), D-limonene (254), and bicyclo[2.2.1]hept-2-ene-5-butylene (255).

References

1. D. S. Breslow and N. R. Newburg, *J. Am. Chem. Soc.* **79**, 5073 (1957).
2. G. Natta, P. Pino, G. Mazzanti, and R. Lanzo, *Chim. Ind.* (*Milan*) **39**, 1032 (1957).

3. W. L. Carrick, *J. Am. Chem. Soc.* **80**, 6455 (1958).

4. M. N. Berger and K. Fletcher, *Polymer* **2**, 441 (1961).

5. K. Ziegler, E. Holzkamp, H. Breil, and H. Martin, *Angew. Chem.* **67**, 541 (1955).

6. A. Delbouille and R. Speltinckx, U.S. Patent 3,454,547, July 8, 1969, Solvay & Cie.

7. W. L. Carrick, G. L. Karapinka, and R. J. Turbett, U.S. Patent 3,324,095, June 6, 1967, Union Carbide Corporation.

8. See Chapter 3, Section VB.

9. This comparison is made for a heterogeneous catalyst that is active for both olefins, such as AlEt + TiCl$_3$ or TiCl$_4$; see G. Bier, *Angew. Chem.* **73**, No. 6, 186 (1961).

9a. T. G. Heggs, *in* "Block Copolymers" (D. C. Allport and W. H. Janes, eds.), p. 138. Wiley, New York, 1973.

10. J. Boor, *J. Polym. Sci., Part C* **1**, 237 (1963).

11. The conclusion was made by comparison of results reported by different workers.

12. T. W. Campbell and A. C. Haven, *J. Appl. Polym. Sci.* **1**, 73 (1959).

13. J. Boor, *Spec. Lect., Int. Congr. Pure Appl. Chem., 23rd, 1971*, Butterworths, London, **8**, 57 (1971).

14. W. A. Hewitt, see Ref. 29 of Boor (13).

15. H. Holler, see ref. 30 of Boor (13).

16. C. G. Overberger, P. E. Borchert, and A. Katchman, *J. Polym. Sci.* **44**, 491 (1960).

17. K. R. Dunham, J. Vandenbergh, J. W. H. Faber, and L. E. Contois, *J. Polym. Sci., Part A* **1**, 751 (1963).

18. D. Braun and H. Heimes, *Angew. Makromol. Chem.* **3**, 122 (1968).

19. J. P. Kennedy, G. G. Wanless, and J. J. Elliott, *Polym. Prepr., Am. Chem. Soc., Div. Polym. Chem.* **5**, No. 2, 676 (1964).

20. V. V. Mal'tsev, N. A. Plate, T. Azimov, and V. A. Kargin, *Polym. Sci. USSR (Engl. Transl.)* **11**, 248 (1970).

21. A. V. Topchiev, E. L. Fantolova, and L. V. Osipova, *Dokl. Akad. Nauk SSSR* **147**, No. 4, 857 (1962).

22. G. F. D'Alelio, A. B. Finestone, L. Taft, and T. J. Miranda, *J. Polym. Sci.* **45**, 83 (1960).

23. A. V. Topchiev, L. V. Osipova, and E. L. Fantalova, *Dokl. Akad. Nauk SSSR* **147**, 1090 (1962).

24. G. Natta, D. Sianesi, D. Moreno, I. W. Bassi, and G. Caporiccio, *Atti Accad. Naz. Linei, Cl. Sci. Fis., Mat. Nat., Rend.* [8] **28**, 552 (1960).

25. C. G. Overberger and G. W. Halek, *J. Polym. Sci., Part A-1* **8**, 359 (1970).

26. G. Pregaglia and M. Binaghi, *Gazz. Chim. Ital.* **90**, 1554 (1960).

27. F. J. Golemba, J. E. Guillet, and S. C. Nyburg, *J. Polym. Sci., Part A-1* **6**, 1341 (1968).

28. W. R. Watt, *J. Polym. Sci.* **45**, 509 (1960).

29. Y. Atarashi, *Kobunshi Kagaku* **21**, 264 (1964).

30. C. G. Overberger and J. E. Mulvaney, *J. Am. Chem. Soc.* **81**, 4697 (1959).

31. P. Pino, *Adv. Polym. Sci.* **4**, 393–456 (1965); earlier references can be found in this review.

32. M. Goodman, K. J. Clark, M. A. Stake, and A. Abe, *Makromol. Chem.* **72**, 131 (1964).

33. F. Danusso and D. Sianesi, *Chim. Ind. (Milan)* **44**, No. 5, 474 (1962).

34. F. Danusso, *Chim. Ind. (Milan)* **44**, No. 6, 611 (1962).

35. G. Natta, F. Danusso, D. Sianesi, and A. Machi, *Chim. Ind. (Milan)* **41**, No. 10, 968 (1959).

36. D. Sianesi, A. Machi, and F. Danusso, *Chim. Ind. (Milan)* **41**, No. 10, 964 (1959).

37. J. Heller and D. B. Miller, *J. Polym. Sci., Part A-1* **5**, 2323 (1967).

38. K. Tanikawa, S. Kusabayashi, H. Kirata, and H. Mikawa, *J. Polym. Sci., Part B* **6**, 275 (1968).

39. K. Hatada, T. Niki, S. Nozakura, and S. Murahashi, *Bull. Chem. Soc. Jpn.* **35**, 1686 (1962).
40. F. L. Saunders, *J. Polym. Sci., Part A-1* **5**, 2187 (1967).
41. L. C. Anand, S. S. Dixit, and S. L. Kapur, *J. Polym. Sci., Part A-1* **6**, 909 (1968).
42. S. L. Malhotra, A. B. Deshpande, and S. L. Kapur, *J. Polym. Sci., Part C* **22**, 1 (1968).
43. L. C. Anand, A. B. Deshpande, and S. L. Kapur, *J. Polym. Sci., Part A-1* **5**, 2079 (1967).
44. L. C. Anand, A. B. Deshpande, and S. L. Kapur, *J. Polym. Sci., Part A-1* **5**, 665 (1967).
45. L. C. Anand, A. B. Deshpande, and S. L. Kapur, *Makromol. Chem.* **117**, 224 (1968).
46. A. B. Deshpande, R. V. Subramanian, and S. L. Kapur, *Makromol. Chem.* **98**, 90 (1966).
47. S. L. Malhotra, A. B. Deshpande, and S. L. Kapur, *J. Polym. Sci., Part A-1* **6**, 193 (1968).
48. K. C. Tsou, J. F. Megee, and A. Malatesta, *J. Polym. Sci.* **58**, 299 (1962).
49. R. J. Kern, H. G. Hurst, and W. R. Richard, *J. Polym. Sci.* **45**, 195 (1960).
50. S. Anzai, K. Irako, A. Onishi, and J. Furukawa, *Kogyo Kagaku Zasshi* **72**, 2076 (1969).
51. M. M. Hamada and J. H. Gary, *Polym. Prep., Am. Chem. Soc., Div. Polym. Chem.* **9**, 413 (1968).
51a. N. N. Korneev, S. K. Goryunovich, and I. F. Leshecheva, *Sov. Plast. (Engl. Transl.)* p. 7 (1968).
52. A. V. Topchiev, B. A. Krentsel, N. F. Bogomolova, and Y. Ya. Gol'dfarb, *Dokl. Chem. Technol. (Engl. Transl.)* **106**, 659 (1957); *Dokl. Akad. Nauk SSSR* **111**, 121 (1956).
53. R. Bacskai and S. J. Lapporte, *J. Polym. Sci., Part A* **1**, 2225 (1963).
54. Y. Sakurada, *J. Polym. Sci., Part A* **1**, 2407 (1963).
55. Y. Sakurada, K. Imai, and M. Matsumoto, *Kobunshi Kagaku* **20**, 422 (1963).
56. T. Saegusa, H. Inai, and J. Furukawa, *Makromol. Chem.* **79**, 207 (1964).
57. A. A. D'Onofrio, *J. Appl. Polym. Sci.* **8**, 521 (1964).
58. C. P. Pinazzi and J. Brossas, *Kinet. Mech. Polyreactions, Int. Symp. Macromol. Chem., Prep. 1969* Vol. 2, p. 285 (1969); *Makromol. Chem.* **122**, 105 (1969); **147**, 15 (1971).
58a. C. P. Pinazzi, J. Brossas, and G. Clouet, *Makromol. Chem.* **148**, 81 (1971).
59. G. Natta, G. Dall'Asta, G. Mazzanti, and F. Ciampelli, *Kolloid Z & Z. Polym.* **182**, 50 (1962); G. Natta, G. Dall'Asta, G. Mazzanti, F. Ciampelli, I. Pasquon, and A. Valvassori, *Makromol. Chem.* **54**, 95 (1962).
60. G. Natta, *Experientia* **19**, No. 12, 609 (1963).
61. J. L. Jezl, U.S. Patent 2,956, 989, October 18, 1960, Sun Oil.
62. G. Natta, G. Dall'Asta, G. Mazzanti, I. Pasquon, A. Valvassori, and A. Zambelli, *J. Am. Chem. Soc.* **83**, 3343 (1961).
63. R. O. Symcox, *J. Polym. Sci., Part B* **2**, 947 (1964).
64. A. Iwamoto and S. Yaguchi, *Bull. Chem. Soc. Jpn.* **40**, 159 (1967); Preprint *Annu. Meet. Polym. Sci. of Vapor, 13th*, 453 (1964); A. Shimizu, T. Otsu, and M. Imoto, *J. Polym. Sci., Part B* **3**, 453 (1965).
65. A. Shimizu, T. Otsu, and M. Imoto, *J. Polym. Sci., Part B* **3**, 449 (1965).
66. T. Otsu, A. Shimizu, and M. Imoto, *J. Polym. Sci., Part A-1* **4**, 1579 (1966).
67. J. P. Kennedy and T. Otsu, *Adv. Polym. Sci.* **7**, 369–385 (1970).
68. T. Otsu, A. Shimizu, and M. Imoto, *J. Polym. Sci., Part A-1* **7**, 3111 (1969).
69. T. Otsu, A. Shimizu, K. Itakura, and M. Imoto, *Makromol. Chem.* **123**, 289 (1969).
70. Y. Chauvin and G. Lefebvre, Great Britain Patent 1,027,758, April 27, 1966, Celanese Corporation.
71. T. Otsu, S. Aoki, and M. Nishimura, *Makromol. Chem.* **128**, 272 (1969).
72. V. I. Kleiner, B. A. Krentsel, and L. L. Stotskaya, *Eur. Polym. J.* **7**, 1677 (1971).
73. D. W. Aubrey and A. Barnatt, *J. Polym. Sci., Part A-1* **5**, 1191 (1967).
74. Ye. L. Yerasova, B. A. Krentsel, N. A. Pokatilo, and A. V. Topchiev, *Polym. Sci. USSR (Engl. Transl.)* **4**, 558 (1963).

75. W. Marconi, *in* "The Stereochemistry of Macromolecules" (A. D. Ketley, ed.), Chapter 5, Dekker, New York, 1967.

76. G. Natta and G. Dall'Asta *in* "Polymer Chemistry of Synthetic Elastomer," Vol. 2 (J. P. Kennedy and E. G. M. Törnqvist, eds.), pp. 703–725, Wiley Interscience, New York, (1968).

77. G. Dall'Asta and G. Matroni, *Angew. Makromol. Chem.* **16/17**, 51–74 (1971).

78. G. Dall'Asta, *Makromol. Chem.* **154**, 1–19 (1972).

79. N. Calderon, *J. Macromol. Sci., Rev. Macromol. Chem.* **C 7**, 105–159 (1972).

80. *Pap., 164th Meet., Am. Chem. Soc., 1972* pp. 874–897 (1972); *Polym. Prepr., Am. Chem. Soc., Div. Polym. Chem.* **13**, No. 2 (1972).

81. G. Natta, G. Dall'Asta, G. Mazzanti, and G. Matroni, *Makromol. Chem.* **69**, 163 (1963).

82. G. Dall'Asta, G. Mazzanti, G. Natta, and L. Porri, *Makromol. Chem.* **56**, 224 (1962).

83. G. Dall'Asta, G. Natta, and G. Mazzanti, *Pap., Macromol. Colloq., 1964*; *Chem. Eng. News* **42**, 41 (Mar. 16, 1964).

84. G. Natta, G. Dall'Asta, and G. Mazzanti, *Angew. Chem.* **76**, 765 (1964); U.S. Patent 3,549,607, December 22, 1970, Montecatini Edison S.p. A.

85. G. Natta, G. Dall'Asta, I. Bassi, and G. Carella, *Makromol. Chem.* **91**, 87 (1966).

86. J. Boor, E. A. Youngman, and M. Dimbat, *Makromol. Chem.* **90**, 26 (1966).

87. G. Natta, G. Dall'Asta, I. W. Bassi, and G. Carella, *Makromol. Chem.* **91**, 87 (1966).

88. G. Dall'Asta and R. Manetti, *Eur. Polym. J.* **4**, 145 (1968).

89. A. W. Anderson and N. G. Merckling, U. S. Patent 2,721, 189, E. I. duPont de Nemours & Company.

90. I. M. Robinson and W. L. Truett, *J. Am. Chem. Soc.* **82**, 2337 (1960).

91. T. Tsujino, T. Saegusa, and J. Furukawa, *Makromol. Chem.* **85**, 71 (1965); **78**, 231 (1964).

92. K. B. Wiberg and W. J. Bartley, *J. Am. Chem. Soc.* **82**, 6375 (1960).

93. G. Dall'Asta, *J. Polym. Sci., Part A-1* **6**, 2397 (1968).

94. G. Natta, G. Dall'Asta, and L. Porri, *Makromol. Chem.* **81**, 253 (1965).

95. G. Natta, G. Dall'Asta, and G. Matroni, *J. Polym. Sci., Part B* **2**, 349 (1964).

96. G. Dall'Asta and G. Matroni, *J. Polym. Sci., Part A-1* **6**, 2405 (1968).

97. G. Dall'Asta, *Chim. Ind. (Milan)* **46**, No. 12, 1525 (1964).

97a. *Chem. Eng. News* **50**, 12, (July 24, 1972); **42**, 42 (April 6, 1964).

97b. G. Natta, G. Dall'Asta, and G. Mazzanti, *Angew. Chem., Int. Ed. Engl.* **3**, 723–729 (1964); G. Dall'Asta, *Corsi Semin. Chim.* **8**, 89–91 (1968); G. Natta, G. Dall'Asta, I. W. Bassi, and G. Carella, *Makromol. Chem.* **91**, 87 (1966).

97c. G. Pampus and J. Witte, U.S. Patent 3,632,849, January 4, 1972, Farbenfabriken Bayer Aktien.

97d. J. Witte, N. Schon, and G. Pampus, U.S. Patent 3,631,010, December 28, 1971, Farbenfabriken Bayer Aktien.

98. N. Calderon, E. A. Ofstead, and W. A. Judy, *J. Polym. Sci., Part A-1* **5**, 2209 (1967).

98a. H. Lammens, G. Sartori, J. Siffert, and N. Sprecher, *J. Polym. Sci., Part B* **9**, 341 (1971).

99. H. Souza Euleterio, U.S. Patent 3,074,918, January 22, 1963, E. I. du Pont de Nemours and Company.

100. For references, see review by R. L. Banks and G. C. Bailey, *Ind. Eng. Chem., Prod. Res. Dev.* **3**, 170 (1964); also Belgian Patents 620,440, to R. L. Banks, January 21, 1963, and 642,916 to M. A. Albright, July 23, 1964, Phillips Petroleum Company.

101. C. P. C. Bradshaw, E. J. Howman, and L. Turner, *J. Catal.* **7**, 269 (1967).

102. N. Calderon, E. A. Ofstead, J. P. Ward, W. A. Judy, and K. W. Scott, *J. Am. Chem. Soc.* **90**, 4133 (1968).

103. H. J. Alkema and R. Van Helden, Great Britain Patent 1,118,517, May 1967, Shell Oil Company.

104. E. Wasserman, D. A. Ben Efraim, and R. Wolovsky, *J. Am. Chem. Soc.* **90**, 3286 (1968).
105. K. Scott, N. Calderon, E. A. Ofstead, W. A. Judy, and J. P. Ward, *Adv. Chem. Ser.* **91**, 399 (1969).
106. G. Dall'Asta and G. Matroni, *Eur. Polym. J.* **7**, 707 (1971).
107. G. Dall'Asta, G. Matroni, and L. Motta, *J. Polym. Sci., Part A-1* **10**, 1601 (1972).
108. N. Calderon, E. A. Ofstead, and W. A. Judy, *J. Polym. Sci., Part A-1* **5**, 2209 (1967).
109. B. Reichel, C. S. Marvel, and R. Z. Greely, *J. Polym. Sci., Part A* **1**, 2835 (1963).
110. C. P. Pinazzi, J. Brossas, J. C. Brosse, and A. Pleurdeau, *Makromol. Chem.* **144**, 155 (1971).
110a. A. Yamada, M. Yanagita, and M. Suzuki, *Inst. Phys. Chem. Res., Tokyo Rep.* **37**, 429 (1961).
110b. C. P. Pinazzi, J. C. Brosse, A. Pleurdeau, J. Brossas, G. Legeay, and J. Cattiaux, *Adv. Polym. Sci.* **10**, 141–174 (1973).
111. G. Natta, G. Mazzanti, P. Longi, and F. Bernardini, *J. Polym. Sci.* **31**, 181 (1958); *Angew. Chem.* **70**, 597 (1958).
112. I. W. Bassi, G. Natta, and P. Corradini, *Angew. Chem.* **70**, 597 (1958).
113. V. Ia. Bogomol'nyi, *Vysokomol. Soedin.* **1**, 1469 (1959); *Polym. Sci. USSR (Engl. Transl.)* **2**, 6 (1961).
114. N. S. Nametkin, A. V. Topchiev, S. G. Durgar'ian, and I. M. Tolchinskii, *Vysokomol. Soedin.* **1**, 1739 (1959); *Polym. Sci. USSR (Engl. Transl.)* **2**, 133 (1961).
115. A. Carbonaro, A. Greco, and I. W. Bassi, *Eur. Polym. J.* **4**, 445 (1968).
116. P. Longi, F. Greco, and U. Rossi, *Makromol. Chem.* **116**, 113 (1968).
117. J. Heller, D. J. Lyman, and W. A. Hewett, *Makromol. Chem.* **73**, 48 (1964).
118. O. F. Solomon, M. Dimonie, K. Ambrozh, and M. Tomescu, *J. Polym. Sci.* **52**, 205 (1961).
119. A. Kimura, S. Yoshimoto, Y. Akana, H. Hirata, S. Kusabayashi, H. Mikawa, and N. Kasai, *J. Polym. Sci., Part A-2* **8**, 643 (1970).
120. J. Heller, D. O. Tieszen, and D. B. Parkinson, *J. Polym. Sci., Part A* **1**, 125 (1963).
121. U. Giannini, G. Bruckner, E. Pellino, and A. Cassata, *J. Polym. Sci., Part B* **5**, 527 (1967); *Part C* **22**, 157 (1968).
122. A. D. Pomogailo, P. E. Matkovskii, V. P. Konovalov, G. A. Beikhol'd, and I. D. Leonov, *Dokl. Akad. Nauk SSSR* **184**, 1364 (1969).
123. K. Gehrke, D. Richter, and P. Hiep, *Plaste Kautsch.* **18**, 325 (1971).
124. K. J. Clark, U.S. Patent 3,492,277, January 27, 1970, Imperial Chemical Industries, Ltd.
125. Y. Yoh and Y. Kotake, *Macromolecules* **3**, No. 3, 337 (1970).
126. A. D. Jenkins, M. F. Lappert, and R. C. Srivastava, *Eur. Polym. J.* **7**, 289 (1971).
127. Cr. I. Simionescu, I. Benedek, and N. Asandei, *Eur. Polym. J.* **7**, 1549 (1971); earlier references of group cited in this paper.
128. S. S. Dixit, A. B. Deshpande, and L. S. Kapur, *J. Polym. Sci., Part A-1* **9**, 1167 (1971).
129. A. Yamamoto, T. Shimizu, and S. Ikeda, *Makromol. Chem.* **136**, 297 (1970).
130. D. Woehrle and G. Manecke, *Makromol. Chem.* **138**, 283 (1970).
131. R. Bacskai, *J. Polym. Sci., Part A* **3**, 2491 (1965).
132. K. J. Clark and T. Powell, *Polymer* **6**, 531 (1965).
133. C. G. Overberger and G. Khattab, *J. Polym. Sci., Part A-1* **7**, 217 (1969).
134. D. Sianesi and G. Caporiccio, *Makromol. Chem.* **81**, 264 (1965).
135. D. Sianesi and G. Caporiccio, *Makromol. Chem.* **60**, 213 (1963).
136. D. Sianesi and G. Caporiccio, *J. Polym. Sci., Part A-1* **6**, 335 (1968).
137. P. Ye. Matkovskii, G. A. Beikhol'd, I. D. Leonov, A. D. Pomogailo, and N. M. Chirkov, *Polym. Sci. USSR (Engl. Transl.)* **12**, 2281 (1970).
138. U. Giannini and S. Cesca, *Chim. Ind. (Milan)* **44**, 371 (1962).
139. W. P. Baker, Jr., *J. Polym. Sci.* **42**, 578 (1960).

140. K. S. Minsker, Yu. A. Sangalov, and G. A. Razuvaev, *J. Polym. Sci., Part C* **16**, 1489 (1967).

141. G. A. Razuvaev, Yu. A. Sangalov, K. S. Minsker, and N. Y. Kovaleva, *Polym. Sci. USSR (Engl. Transl.)* **7**, 597 (1965).

142. G. P. Budanova and V. V. Mazurek, *Polym. Sci. USSR (Engl. Transl.)* **9**, 2703 (1967).

143. H. Higashi, K. Watabe, and S. Namikawa, *J. Polym. Sci., Part B* **5**, 1125 (1967).

144. V. V. Mazurek, G. T. Nesterchuk, and A. V. Merhur'eva, *Polym. Sci. USSR (Engl. Transl.)* **11**, 693 (1970).

145. G. P. Budanova, V. V. Mazurek, *Polym. Sci. USSR (Engl. Transl.)* **12**, 1201 (1971).

146. Y. Susuki and M. Saito, *J. Polym. Sci., Part A-1* **9**, 3639 (1971).

147. A. Misono, Y. Uchida, K. Yanada, and T. Saiki, *Bull. Chem. Soc. Jpn.* **41**, 2995 (1968).

148. J. Ulbricht and M. Arnold, *Plaste Kautsch.* **18**, 250 (1971).

148a. R. N. Haszeldine, T. G. Hyde, and P. J. Tait, *Polymer* **14**, 215, 221, and 224 (1973).

148b. J. Mejzlik, M. Navratil, and L. Vilimova, *Collect. Czech. Chem. Commun.* **38**, 3457 (1973).

149. N. Yamazaki and S. Kambara, *J. Polym. Sci., Part C* **22**, 75 (1968).

150. E. J. Vandenberg, *J. Polym. Sci., Part C* **1**, 207 (1963).

151. G. Natta, *Chim. Ind. (Milan)* **42**, 1207 (1960).

152. G. Natta, G. Dall'Asta, G. Mazzanti, U. Giannini, and S. Cesca, *Angew. Chem.* **71**, 205 (1959).

153. C. E. Schildknecht, S. T. Gross, H. R. Davidson, J. M. Lambert, and A. O. Zoss, *Ind. Eng. Chem.* **40**, 2140 (1948); C. E. Schildknecht, S. T. Gross, and A. O. Zoss, *ibid.* **41**, 1998 (1949); C. E. Schildknecht, A. O. Zoss, and F. Grosser, *ibid.* p. 2891.

154. Y. Yoh, H. Yuki, and S. Murahashi, *J. Polym. Sci., Part A-1* **8**, 2775 and 3311 (1970).

155. Y. Yoh, K. Harada, H. Yuki, and S. Murahashi, *J. Polym. Sci., Part A-1* **9**, 1089 (1971).

156. H. Yuki and K. Hatada, *Polym. J.* **1**, No. 3, 271 (1970).

157. A. Kogerman, H. Martinson, T. Evseev, and A. Kongas, *Eesti NSV Tead. Akad. Toim., Keem., Geol.* **18**, 232 (1969).

158. E. Chiellini and R. Nocci, *J. Polym. Sci., Polym. Chem. Ed.* **11**, 493 (1973).

159. S. L. Kapur, *Macromol. Prepr. Int. Congr. Pure Appl. Chem.*, 23rd, *1971* Vol. 1, p. 129 (1971).

160. A. B. Deshpande, S. M. Kale, and S. L. Kapur, *J. Polym. Sci., Part A-1* **1**, 195 (1972); *J. Polym. Sci.* **11**, 1307 (1973).

161. A. Yamamoto, T. Shimizu, and S. Ikeda, *Makromol. Chem.* **136**, 297 (1970).

162. K. Matsumura and O. Fukumoto, *J. Polym. Sci., Part A-1* **9**, 471 (1971).

163. H. Abe, K. Imai, and M. Matsumoto, *J. Polym. Sci., Part B* **4**, 589 (1966).

164. D. G. H. Ballard, W. H. Janes, and T. Medinger, *J. Chem. Soc. B* **10**, 1168 (1968).

165. K. Matzuzaki, T. Uryu, A. Ishida, and M. Takeuchi, *J. Polym. Sci., Part A-1* **5**, 2167 (1967).

166. E. A. H. Hopkins and M. L. Miller, *Polymer* **4**, 75 (1963).

167. T. Otsu, A. Akimoto, and S. Aoki, *J. Appl. Polym. Sci.* **12**, 1477 (1968).

168. N. N. Aylward, *J. Polym. Sci., Part B* **8**, 377 (1970).

169. F. Marktscheffel, A. F. Turbak, and Z. W. Wilchinsky, *J. Polym. Sci., Part A-1* **7**, 2433 (1966).

170. K. Kobayanshi and H. Sumitomo, *J. Polym. Sci., Part A-1* **7**, 925 (1969).

171. A. V. Topchiev, Yu. Ya. Gol'dfarb, and B. A. Krentsel, *Vysokomol. Soedin.* **3**, 870 (1961); *Polym. Sci. USSR (Engl. Transl.)* **3**, 696 (1962).

172. G. Natta, M. Farina, M. Peraldo, and G. Bressan, *Chim. Ind. (Milan)* **43**, 161 (1961).

173. W. P. Baker, *J. Polym. Sci., Part A* **1**, 655 (1963).

174. R. Havinga and A. Schors, *J. Macromol. Sci., Chem.* **2**, 1, 21, and 31 (1968).

175. J. E. Van den Enk and H. J. Van der Ploeg, *J. Polym. Sci., Part A-1* **9**, 2395 (1971).
176. G. D. Shier, U.S. Patent 3,442,883, May 6, 1969, Dow Chemical Company.
177. S. Otsuka and A. Nakamura, *J. Polym. Sci., Part B* **5**, 973 (1967); U.S. Patent 3,536,692, October 27, 1970, Japan Synthetic Rubber Company.
178. S. Otsuka, A. Nakamura, S. Ueda, and H. Minamida, *Polym. Rep.* **137**, 26 (1967).
179. G. Natta, G. Mazzanti, and P. Corradini, *Atti Accad. Naz. Lincei, Cl. Sci. Fis., Mat. Nat., Rend.* [8] **25**, 3 (1958).
180. G. I. Bantsyrer, M. I. Cherkashin, and A. A. Berlin, *Izv. Akad. Nauk SSSR, Ser. Khim.* **5**, 2705 (1968).
181. E. Lombardi and L. Guiffré, *Atti Accad. Naz. Lincei, Cl. Sci. Fis., Mat. Nat., Rend.* [8] **25**, 70 (1958).
181a. T. Ito, H. Shirakawa, and S. Ikeda, *J. Polym. Sci., Polym. Chem. Ed.* **12**, 11 (1974).
182. G. Champetier and M. Martynoff, *Bull. Soc. Chim. Fr.* p 2083 (1961).
183. W. H. Watson, Jr., W. C. McMordie, Jr., and L. G. Lands, *J. Polym. Sci.* **55**, 137 (1961).
184. I. V. Nicolescu and E. M. Angelescu, *J. Polym. Sci., Part A* **3**, 1227 (1965).
185. A. Valvassori, G. Sartori, V. Turba, and F. Ciampelli, *Makromol. Chem.* **61**, 256 (1963).
186. A. Furlani, G. Moretti, and A. Guerriere, *J. Polym. Sci., Part B* **5**, 527 (1967).
187. V. O. Reikhsfeld and K. L. Makovetskii, *Dokl. Akad. Nauk SSSR* **155**, 414 (1964).
188. A. A. Berlin, M. I. Cherkashin, I. P. Chernysheva, Y. G. Aseev, Ye. I. Barkanand, and P. A. Kisilitsa, *Polym. Sci. USSR* (*Engl. Transl.*) **9**, 2075 (1967).
189. M. J. Benes, M. Janic, and J. Peska, *Chem. Listy* **64**, 1094 (1970).
190. A. Furlani, E. Cervone, and G. Moretti, *Ric. Sci.* **33**, Part IIA, 619 (1963).
191. H. Noguchi and S. Kambara, *J. Polym. Sci., Part B* **1**, 553 (1963).
192. A. A. Berlin, M. I. Cherkashin, Yu. G. Aseev, and I. M. Shcherbakova, *Vysokomol. Soedin.* **6**, 1773 (1964).
193. K. Higashiura, S. Yokomichi, and M. Oiwa, *Kogyo Kagaku Zasshi* **66**, 379 (1963).
194. R. H. Wiley and J. Y. Lee, *J. Macromol. Sci., Chem.* **5**, 507 (1971).
195. A. Valvassori, G. Sartori, and F. Ciampelli, *Chim. Ind.* (*Milan*) **44**, 1095 (1962).
196. F. K. Shmidt, V. G. Lipovich, and I. V. Kalechits, *Kinet. Katal.* **11**, 251 (1970).
197. G. Natta, L. Porri, P. Corradini, and D. Moreno, *Atti Accad. Naz. Lincei, Cl. Sci. Fis., Mat. Nat., Rend.* **20**, 560 (1956).
198. G. Natta, L. Porri, G. Zanini, and L. Fiore, *Chim. Ind.* (*Milan*) **41**, 526 (1959).
199. A. A. Berlin, M. I. Cherkashin, P. P. Kisilitsa, and O. N. Pirogov, *Polym. Sci. USSR* (*Engl. Transl.*) **9**, 2069 (1967).
200. R. J. Kern, *J. Polym. Sci., Part A-1* **7**, 621 (1969).
201. W. J. Trepka and R. J. Sonnenfeld, *J. Polym. Sci., Part A-1* **8**, 2721 (1970).
202. G. A. Chukhadshyan, Zh. I. Abramyan, and V. G. Grigoryan, *Arm. Khim. Zh.* **23**, 608, (1970).
203. V. Kormer, L. Churlyaeva, and J. L. Yufa, *Vysokomol. Soedin., Ser. B* **12**, 4834 (1970).
204. R. H. Michel, *J. Polym. Sci., Part A-1* **5**, 920 (1967).
205. C. Aso, T. Kunitake, and K. Saiko, *Makromol. Chem.* **151**, 265 (1972).
206. A. Furlani, G. Moretti, and A. Guerrieri, *J. Polym. Sci., Part B* **5**, 523 (1967).
206a. H. Shirakawa and S. Ikeda, *J. Polym. Sci., Polym. Chem. Ed.* **12**, 1929 (1974).
207. G. Natta, *Gazz. Chim. Ital.* **89**, 465 (1959).
207a. W. Reppe and W. J. Schweckendiek, *Justus Liebigs Ann. Chem.* **560**, 104 (1948).
208. W. Bracke, *J. Polym. Sci. Part A-1* **10**, 2097 (1972).
209. A. J. Chaulk and A. R. Gilbert, *J. Polym. Sci., Part A-1* **10**, 2033 (1972).
210. E. A. Angelescu and I. V. Nicolescu, *J. Polym. Sci., Part C* **22**, 203 (1968).
211. I. V. Nicolescu and E. A. Angelescu, *J. Polym. Sci., Part A-1* **4**, 2963 (1966); **3**, 1227 (1965).

212. L. B. Luttinger, *Org. Chem.* **27**, 1591 (1962).

213. L. B. Luttinger and E. C. Colthup, *Org. Chem.* **27**, 3752 (1962).

214. Belgian Patent 547,699, November 9, 1956, Imperial Chemical Industries, Ltd.

215. C. Aso, T. Kunitake, and Y. Ishimoto, *J. Polym. Sci., Part A-1* **6**, 1163 (1968).

216. C. S. Marvel and G. E. Hartzell, *J. Am. Chem. Soc.* **81**, 448 (1959).

217. D. A. Frey, M. Hasegawa, and C. S. Marvel, *J. Polym. Sci., Part A* **1**, 2057 (1963).

218. G. Lefebvre and F. Dawans, *J. Polym. Sci., Part A* **2**, 3277 (1964).

219. K. Mabuchi, T. Saegusa, and J. Furukawa, *Makromol. Chem.* **81**, 112 (1965).

220. A. D. Aliev, A. V. Arbatskii, M. V. Shishkina, and B. A. Krentsel, *Dokl. Akad. Nauk SSSR* **153**, No. 2, 333 (1963).

221. W. Marconi, A. Mazzei, G. Lugi, and M. Bruzzone, *J. Polym. Sci., Part C* **16**, 805 (1967).

222. E. D. Vessel, Polymerization of Phenyl Substituted Butadienes by Metal Alkyl Coordinate Catalysts, Ph. D. dissertation, State Univ. Iowa, Iowa City, 1960. Univ. Microfilms No. 60-5702.

223. W. Marconi, A. Mazzei, S. Cucinella, M. Cesari, and E. Pauluzzi, *J. Polym. Sci., Part A* **3**, 123 (1965).

224. D. Cuzin, Y. Chauvin, and G. Lefebvre, *Eur. Polym. J.* **5**, 283 (1969).

225. L. Porri and M. C. Gallazzi, *Eur. Polym. J.* **2**, 189 (1966).

226. G. Natta, P. Corradini, I. W. Bassi, and G. Fagherazzi, *Eur. Polym. J.* **4**, 297 (1968).

226a. S. Murahashi, M. Kamachi, and N. Wakabayashi, *J. Polym. Sci., Part B* **7**, 135 (1969).

227. V. Bell, *J. Polym. Sci., Part A* **2**, 5291 (1964).

228. C. S. Marvel, P. E. Kiener, and E. D. Vessel, *J. Am. Chem. Soc.* **81**, 4694 (1959).

229. C. S. Marvel and J. R. Rogers, *J. Polym. Sci.* **49**, 335 (1961).

230. G. B. Butler and T. W. Brooks, *Polym. Prep., Am. Chem. Soc., Div. Polym. Chem.* **3**, 168 (1962).

231. C. S. Marvel and C. C. L. Hwa, *J. Polym. Sci.* **45**, 25 (1960).

232. C. S. Marvel and J. K. Stille, *J. Am. Chem. Soc.* **80**, 1740 (1958).

233. A. Valvassori, P. Sartori, and F. Ciampelli, *Chim. Ind.* (*Milan*) **44**, 1095 (1962).

234. C. S. Marvel and W. E. Garrison, Jr., *J. Am. Chem. Soc.* **81**, 4737 (1959).

234a. H. S. Makowskii, B. K. C. Shim, and Z. E. Wilchinsky, *J. Polym. Sci., Part A* **2**, 1549 and 4973 (1964).

235. D. S. Trifan, R. A. Shelden, and J. J. Hoglen, *J. Polym. Sci., Part A-1* **6**, 1605 (1968).

236. I. A. Livshits, L. M. Korobova, I. Ya. Poddubnyi, V. N. Sokolov, N. N. Marasanova, and V. N. Zyabina, *Vysokomol. Soedin., Ser. A* **12**, 1794 (1970).

236a. L. M. Romanov, A. P. Verkhoturova, Y. V. Kissin, and G. V. Rakova, *Polym. Sci. USSR* (*Engl. Transl.*) **4**, 1424 (1963).

237. M. Marek, M. Roosova, and D. Doskocilova, *J. Polym. Sci., Part C* **16**, 971 (1967).

238. C. S. Marvel and E. J. Gall, *J. Org. Chem.* **25**, 1784 (1960).

239. N. D. Field, *J. Org. Chem.* **25**, 1006 (1960).

240. J. M. Wilber, Jr. and C. S. Marvel, *J. Polym. Sci., Part A* **2**, 4415 (1964).

241. C. Aso, T. Kunitake, and H. Uchio, U.S. Patent 3,642,757, February 15, 1972, Asahi Kasei Kogyo Kabushiki Kaisha.

242. G. C. Corfield and A. Crawshaw, *J. Macromol. Sci., Chem.* **5**, 21 (1971).

243. A. Valvassori, P. Sartori, V. Turba, and F. Ciampelli, *Makromol. Chem.* **61**, 256 (1963).

244. G. B. Butler and M. L. Miles, *J. Polym. Sci., Part A* **3**, 1609 (1965).

244a. S. I. Sadykh-Zade, T. I. Mamedov, and A. T. Abbasova, *Sint. Prevrashch. Monomernykh Soedin.* p. 105 (1967).

245. A. Valvassori, P. Sartori, V. Turba, and M. P. Lachi, *J. Polym. Sci., Part C* **16**, 23 (1967).

246. P. Sartori, A. Valvassori, V. Turba, and M. P. Lachi, *Chim. Ind.* (*Milan*) **45**, 1529 (1963).

247. G. Dall'Asta, G. Matroni, R. Manetti, and C. Tosi, *Makromol. Chem.* **130**, 153 (1969).

248. L. E. Ball and H. J. Harwood, *Polym. Prepr., Am. Chem. Soc., Div. Polym. Chem.* **1**, 59 (1961).
249. A. T. Blomquist and D. T. Longone, *J. Am. Chem. Soc.* **79**, 3916 (1957).
250. G. M. Butler, M. L. Miles, and W. S. Brey, *J. Polym. Sci., Part A* **3**, 723 (1965).
251. W. Marconi, S. Cesca, and G. Della Fortuna, *J. Polym. Sci., Part B* **2**, 301 (1964).
252. P. Sartori, A. Valvassori, and S. Faina, *Atti Accad. Naz. Lincei, Cl. Sci. Fis., Mat. Nat., Rend.* [8] **34**, 565 (1963).
253. M. A. Achon, M. I. Garcia-Bonon, and J. L. Mateo, *Makromol. Chem.* **26**, 175 (1958).
254. M. Modena, R. B. Bates, and C. S. Marvel, *J. Polym. Sci., Part A* **3**, 949 (1965).
255. H. G. G. Dekking, *J. Polym. Sci.* **55**, 525 (1961).

20

Copolymerizations

I. Introduction

Random and alternating-type copolymerizations have been demonstrated with Ziegler–Natta catalysts. The properties of these copolymers depend largely on the relative contents and placement of comonomers, molecular weight, and mode of addition. The architecture of a copolymer can be controlled experimentally to a large degree by the following factors: operating conditions, chemical composition and physical state of catalyst, the physical state of the copolymer being formed, and the structure of the comonomers. The most important practical copolymer is made from ethylene and propylene. Many other binary monomer pairs have been polymerized with the formation of amorphous or crystalline copolymers. Titanium- and vanadium-based catalysts have been used to synthesize copolymers that have a prevailingly random, block, or alternating structure.

When two or more monomers are present in a copolymerization mixture, each must compete with the others for the available centers. The structure, comonomer composition, and molecular weight of the formed copolymer is the consequence of this competition. To a large extent, the behavior of an olefin or a diolefin monomer in a copolymerization can be predicted from homopolymerization data obtained under similar conditions. Yet, there are surprises. The primary purpose of this chapter is to describe the experimental factors that determine the architecture of the formed copolymer and to relate some of these surprises.

Many of the reported copolymerization studies contributed much to the understanding of the Ziegler–Natta catalyst. Industrial efforts were largely directed toward finding novel and more efficient catalysts for the synthesis of the desired copolymer. The practical interest lay in the synthesis of inexpensive elastomers, such as an ethylene–propylene copolymer. Section IV, which briefly describes copolymerizations of selected comonomer pairs, indicates the scope of these copolymerization investigations.

Extensive general and specific reviews have been written by Lukach and Spurlin (1) in 1964; by Bier and Lehmann (1a) in 1964; by Crespi, Valvassori, and Sartori (1b) in 1964; by Natta, Crespi, Valvassori, and Sartori (2) in 1963; by Pasquon, Valvassori, and Sartori in 1967 (3); by Boor (4) in 1967; by Natta, Valvassori, and Sartori (5, 5a) in 1969; and by Baldwin and Ver Strate (5b) in 1972. Recent reviews by Tosi and Ciampelli and by Kissin on structural determinations are also highly recommended (see Chapter 3 for refs.).

II. Copolymerization Equation and Nomenclature

The reactivity ratios r_1 and r_2 are most frequently used to describe the ability of added comonomers to add to the copolymer chains. For a mathematical development of the copolymerization equation and its experimental use, the reader is referred to Lukach and Spurlin (1), Bier and Lehmann (1a), and Crespi et al. (1b).

The copolymerization equation has the form shown in Eq. 20–1, where

$$\frac{d[M_1]}{d[M_2]} = \frac{[M_1]}{[M_2]} \frac{(r_1[M_1] + [M_2])}{([M_1] + r_2[M_2])} \qquad (20\text{–}1)$$

$r_1 = k_{11}/k_{12}$ and $r_2 = k_{22}/k_{21}$, and the copolymer composition $d[m_1]/d[m_2] = m_1/m_2$ is the molar ratio of the two monomer units in the copolymer, and $[M_1]$ and $[M_2]$ are instantaneous concentrations of the two monomers in the feed. The k's are the characteristic propagation rate constants, and the subscripts associated with each k refer to the nature of the last added unit and the monomer being added, respectively. As an example, k_{12} is the characteristic propagation rate constant of adding M_2 to a chain end that has M_1 as the last added monomer unit. Applied to a Ziegler–Natta polymerization for an $AlEt_2Cl$–$VOCl_3$ catalyst in which the chain ends would be V–M_1–Pn and V–M_2–Pn, r_1 measures the relative abilities of monomers M_1 and M_2 to add to the V–M_1–Pn centers and r_2 measures the relative abilities of monomers M_1 and M_2 to add to the V–M_2–Pn centers (Pn is the polymer chain). The rate of reaction of each of these additions is the product of the characteristic rate constant, the concentration of the corresponding chain end centers, and the concentration of the monomer being added; for example, the rate of M_1 adding to $V–M_1–Pn = k_{11}[V–M_1Pn][M_1]$ (the ligand environment of vanadium is not described above).

The product of r_1 and r_2 is usually between zero and one. When $r_1r_2 \simeq 1$, the copolymerization is random; as the r_1r_2 product approaches zero, there is an increasing tendency toward alternation. In a random copolymerization, the last added monomer unit does not influence the next addition. In an alternating copolymerization, there is an effect. When $r_1r_2 > 1$, there is a tendency for the comonomers to form long segments or perhaps undergo homopolymerization.

Also, even though the copolymerization is random, the formed copolymer may have a block structure as a result of changes in the concentration of monomer near the center (see Section III,A,4).

The literature (especially patents) lacks a uniform nomenclature to describe the structure of the synthesized copolymers. The following nomenclature will be used in this text; illustrations are for copolymers made from ethylene (E) and propylene (P).

> *Random*: EEPEPEPPPEP ...
> *Block*: EEPEPPPPPEEPEEEEEP ...
> *Alternating*: EPEPEPEPEPEPEPEP ...
> *Tapered*: PPPPPEPPP ... EPPEPEE ... EEPEEEEE

Block polymers will be considered in the next chapter. The author prefers to consider the products described in the next chapter as "block polymers" rather than "block copolymers." Frequently, the patents describe products as true block polymers when, in fact, they are copolymers with variable block content, length and distribution.

III. Experimental Control over the Architecture of Copolymer Chains

The properties of a copolymer are dependent to a large extent on several structural features of the copolymer chains: (a) the relative content of comonomer units in a polymer chain, the way that the comonomer units are distributed in the chain, and variation in the comonomer composition of different chains; (b) molecular weight and molecular weight distribution; and (c) relative content of normal head-to-tail addition vs. head-to-head and tail-to-tail additions.

A. CONTROL OVER THE COMONOMER COMPOSITION
IN THE COPOLYMER

Because the elastomeric properties of the copolymer are so closely dependent on comonomer composition, considerable work was done to elucidate
this aspect of copolymerization (1–5a, 6). Five influencing factors have been
reported that allow control over the composition of the copolymer: (1) operating conditions, (2) chemical composition of catalyst, (3) physical state
of catalyst, (4) physical state of the copolymer being formed, and (5) structure of comonomers. Some of these are closely interdependent.

1. Operating Conditions

The strictest adherence to operating conditions is required if copolymers
of a desired monomer composition are to be synthesized. The dissolved
monomer must be equilibrated with the gas phase of known composition
to keep the monomer concentration constant throughout the copolymerization. In a process in which the reacting olefins are continuously passed
through the solvent and unreacted olefins are vented, the rate of flow of
olefins must be higher than the rate of reaction. To ensure saturation of the
reacting olefins at all times, it is necessary to maintain high stirring rates,
low solution viscosities, and low reaction rates; otherwise, the kinetics will
become dominated by mass transfer barriers. Two examples are cited.

Cozewith and Ver Strate (6) found that the very high initial rates in an
ethylene–propylene copolymerization resulted from a single batch addition
of catalyst, and this caused deviations from saturation. To counteract this,
they added the catalyst components continuously to well-stirred reactor
containing solvent saturated with the olefin mixture. Under these conditions,
they found that the copolymer composition and the copolymerization rate
were unchanged for the catalysts $VO(acac)_2Cl$–$AlEt_2$ and $VOCl_3$–Al-i-Bu_3
in the time intervals 75 to 115 and 70 to 160 minutes, respectively.

Suminoe and co-workers (7) observed that, if the catalyst $AlEt_3$–VCl_4 was
prepared in the presence of propylene, the ethylene–propylene copolymer
was richer in propylene than if the catalyst was prepared in the presence
of both olefins. When the catalyst was prepared in the presence of only
ethylene or both olefins, the corresponding copolymers had the same monomer composition.

2. Chemical Composition of the Catalyst

The copolymer composition depends on the structure of the transition
metal component and sometimes on the metal alkyl component.

Carrick and co-workers (8) first reported that, for an ethylene–propylene
copolymerization, the relative reactivities of the olefins were determined

exclusively by the choice of the transition metal salt and were almost independent of the metal alkyl. For example, the relative reactivities did not change significantly when VCl_4 was combined with $ZnEt_2$, Al-i-Bu$_3$, $Zn(C_4H_9)_2$, and CH_3TiCl_3. In contrast, when Al-i-Bu$_3$ was used as the metal alkyl component, the reactivity of propylene increased in the order: $HfCl_4 < ZrCl_4 < TiCl_4 < VOCl_3 < VCl_4$ (see Fig. 13–6). They noted that this order followed increasing metal electronegativities.

Later investigations by others showed that this was not a general characteristic of Ziegler–Natta catalysts. Junghanns and co-workers (9) found a decrease in the relative reactivity of ethylene when $AlEt_3$ or $AlEt_2Cl$ was replaced by $AlEtCl_2$ or $Al_2Et_3Cl_3$ (VCl_4 and $VOCl_3$ were coreactants). This was not confirmed by Ichikawa (10), however, who examined $VOCl_3$ in combination with $AlEt_2Cl$ and $AlEtCl_2$. Junghanns (9) reported that with increasing contents of chlorine in the catalyst, the copolymer became increasingly amorphous when the ethylene content of the copolymer was in the 65 to 75 wt % range.

A dependence on valence of the transition metal was also found. In the series $VOCl_3$, VCl_4, and VCl_3, as well as $TiCl_4$ and $TiCl_3$, the relative reactivity of ethylene decreased as the valence of V and Ti decreased [see Natta et al. (5, p. 687)]. The relative reactivity of ethylene was higher for titanium-based catalysts relative to the corresponding vanadium-based catalysts (8).

Cozewith and Ver Strate (6) found considerable variation in the relative monomer reactivities when different combinations of metal alkyls and vanadium salts were used. But they did not see an apparent relationship between reactivities and the structure of the catalyst. For example, the catalysts VCl_4–$AlEt_3$ and VCl_4–Al-i-Bu$_3$ gave similar reactivity ratios but the catalysts $VOCl_3$–$AlEt_2Cl$ and $VOCl_3$–Al-i-Bu$_3$ did not. Also, whereas binary combinations of $VOCl_3$ with $AlEt_2Cl$ or $Al_2Et_3Cl_3$ gave different results, no differences were observed when VCl_4 was combined with $AlEt_2Cl$ or $Al_2Et_3Cl_3$. They concluded that the determination of reactivity ratios alone was of little value in assessing the chemistry of the catalyst.

Fractionations of the copolymers by Cozewith and Ver Strate revealed variable compositional homogeneity, depending on the catalyst used. Some copolymers showed no more than $\pm 3\%$ variation about the mean and showed no tendency to higher ethylene contents as molecular weights of the fractions increased. These copolymers had narrow molecular weight distributions ($Q = \bar{M}_w/\bar{M}_n$ approached 2). The $\pm 3\%$ variation was attributed to operational causes. Other copolymers showed compositional spreads greater than 10% about the mean (Figs. 20–1 and 20–2). In addition, the molecular weight distributions were multimodal and broad, with Q values often exceeding 10 (Fig. 20–3).

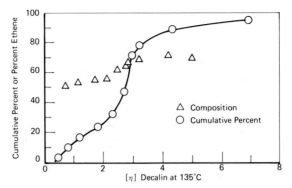

Fig. 20–1. Fractionation data; VCl$_4$–Et$_2$AlCl catalyst, prepared in a continuous-flow stirred reactor (6).

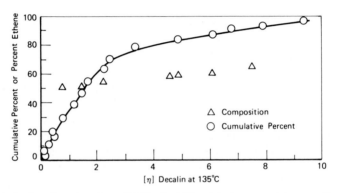

Fig. 20–2. Fractionation data; VOCl$_3$–Et$_2$AlCl catalyst, prepared in a continuous-flow stirred reactor (6).

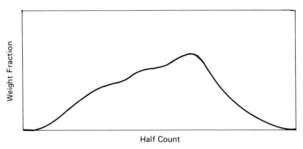

Fig. 20–3. Gel permeation chromatogram of the polymer prepared as in Fig. 20–2 in trichlorobenzene at 135°C (6).

These workers attributed the large compositional spreads to the presence of multiple catalyst centers. When only one type of center is present in the homogeneous catalyst, the compositional spread is absent and narrow Q values are found. Apparently, these "apparently homogeneous" catalysts that gave wider compositional spreads were colloidal or soluble molecular complexes formed that had different structures.

Cozewith and Ver Strate redetermined the reactivity ratios for a number of published vanadium catalysts (Table 20–1) and compared their values with literature data (Table 20–2). Some disagreement was found, especially in the $r_1 r_2$ products. Whereas most of the early studies led to $r_1 r_2$ products equal to approximately unity for most of the catalysts, their reevaluated $r_1 r_2$ products lay entirely in the range of 0.13 to 0.45. The significance of this discrepancy is that the earlier data supported the view that the co-polymerization was largely random, whereas the Cozewith–Ver Strate data suggest that the last added olefin unit influences the next addition. The author is inclined to agree with the latter view, provided certain copolymerization conditions are met. Two experiments are cited: (1) most of these vanadium catalysts favor syndiotactic polymerization if used to polymerize propylene at low temperatures, and (2) copolymerize ethylene with internal acyclic and cyclic olefins, the latter not being homopolymerizable alone. Syndiotactic polymerization occurs because the last added propylene sterically interacts with the incoming propylene (see Chapter 15). Acyclic and

TABLE 20–1

Reactivity Ratios in Ethylene–Propylene Copolymerizations (6)

Catalyst	Cocatalyst	$r_{C_2H_4}$	95% Confidence interval	$r_{C_3H_8}$	95% Confidence interval	$r_{C_2H_4} r_{C_3H_8}$	Active species present[d]
VCl$_4$	AlEt$_3$[a]	10.3	12.6–8.6	0.025	0.030–0.021	0.25 ± 0.11	
	Al-i-Bu$_3$[a]	11.0	11.5–10.7	0.028	0.030–0.027	0.30 ± 0.07	
	AlEt$_2$Cl	5.9	5.6–4.5	0.029	0.031–0.029	0.14 ± 0.02	2
	Al$_2$Et$_3$Cl$_3$[a]	9.1	9.6–8.9	0.031	0.033–0.030	0.28 ± 0.02	1
VOCl$_3$	Al-i-Bu$_2$Cl	20.3	21.8–19.2	0.022	0.024–0.021	0.45 ± 0.04	
	AlEt$_2$Cl	12.1	13.4–11.1	0.018	0.021–0.017	0.22 ± 0.04	3
	Al$_2$Et$_3$Cl$_3$	10.1	10.8–9.7	0.025	0.027–0.023	0.25 ± 0.03	1
VO(OBu)$_3$[b]	AlEt$_2$Cl	16.8	18.4–15.8	0.019	0.022–0.017	0.32 ± 0.06	2
VO(OBu)$_3$[c]	AlEt$_2$Cl	19.8	21.0–19.1	0.012	0.014–0.011	0.24 ± 0.04	
VO(acac)$_2$	AlEt$_2$Cl	11.7	12.5–11.0	0.011	0.013–0.010	0.13 ± 0.03	2
VO(acac)$_2$Cl	AlEt$_2$Cl	16.4	17.0–15.9	0.018	0.019–0.017	0.30 ± 0.02	2
VO(acac)Cl$_2$	AlEt$_2$Cl	16.5	17.4–15.8	0.012	0.013–0.011	0.20 ± 0.03	2

[a] J. Van de Castle, personal communication.
[b] Heptane solvent.
[c] Chlorobenzene solvent, 30°C.
[d] Where more than one species is indicated, the numbers represent a lower limit.

TABLE 20-2

Results of Previous Ethylene–Propylene Reactivity Ratio Determinations with Soluble Catalysts (6)

Catalyst	Cocatalyst	Solvent	Temp (°C)	$r_{C_2H_4}$	$r_{C_3H_6}$	$r_{C_2H_4}r_{C_3H_6}$	Ref.[a]
VCl$_4$	AlEt$_3$					1.0	1
	AlEt$_2$Cl					0.26	1
		Heptane	−10	13.7	0.021	0.29	9
		Heptane	21	3.0	0.073	0.23	19
	Al-i-Bu$_2$Cl			20.0	0.023	0.46	20
VOCl$_3$	AlEt$_2$Cl					0.37	1
	Al-i-Bu$_2$Cl	Heptane	30	16.8	0.052	0.87	19
				35.3	0.027	0.95	20
		Heptane	30	14.8	0.037	0.55	21
V(acac)$_3$	AlEt$_2$Cl					0.35	1
	Al-i-Bu$_2$Cl			16.0	0.04	0.64	20
VOCl$_2$(OEt)	Al-i-Bu$_2$Cl	Heptane	30	16.8	0.055	0.93	21
VOCl(OEt)$_2$	Al-i-Bu$_2$Cl	Heptane	30	18.9	0.069	1.06	21
VO(OBu)$_3$	Al-i-Bu$_2$Cl	Heptane	30	22.0	0.046	1.01	21
VO(OEt)$_3$	Al-i-Bu$_2$Cl	Heptane	30	15.0	0.070	1.04	21
	AlEt$_2$Cl	Chlorobenzene	30	26.0	0.039	1.02	11
VO(OR)$_3$	AlR$_2$Cl	Chlorobenzene	30	24.4	0.041	1	11

[a] These numbers refer to references cited in ref. 6.

cyclic olefins do not homopolymerize because steric interactions are too great. Similarly, 1-butene does not undergo syndiotactic polymerization. Furukawa's copolymerization studies of propylene and butadiene add further support (see Section IV,F).

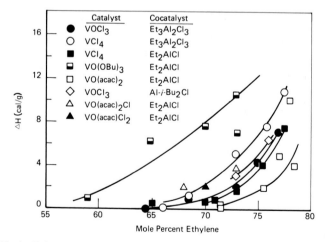

Fig. 20-4. Polymer composition and crystallinity as detected by heat of fusion (6).

According to Cozewith and Ver Strate (6), the major perturbation caused by the multiple species is the compositional spread in the polymer. This leads to an enrichment in longer sequences of ethylene over that present in a homogeneous polymer of the same ethylene–propylene composition. Thus, the heterogeneous polymers appear too crystalline (relative to those prepared with true homogeneous catalysts, which contain only one type center) for their composition and reactivity ratios (Fig. 20–4).

If multiple centers are present, reactivity ratios are a function of the mode of catalyst addition, and reactivity ratios determined by the normal copolymerization equation do not correctly predict the sequence distribution of the copolymer. Differences in catalyst preparation and use may then produce catalysts with different contents of multiple centers, and this may account in large part for the anomolous $r_1 r_2$ products in Tables 20–1 and 20–2.

TABLE 20–3

Distribution Function of Different Lengths in Ethylene–Propylene Copolymers Having Different Compositions Prepared with the $Al(C_6H_{13})_3$–VCl_4 Catalyst (11)

Mole % of m_x in the copolymer	$\%P(m_x)_n{}^a$										
	$n = 1^b$	$n = 2$	$n = 3$	$n = 4$	$n = 5$	$n = 6$	$n = 7$	$n = 8$	$n = 9$	$n = (l)$–14^b	$n \geqslant 15$
85	15.80	13.20	11.20	9.430	7.941	6.687	5.630	4.741	3.992	(10) 12.271	9.006
75	27.07	19.74	14.40	10.50	7.658	5.584	4.073	2.970	2.166	(10) 4.631	1.204
60	44.47	24.69	13.71	7.614	4.228	2.347	1.303	0.724	—	(9) 0.762	0.142
50	55.89	24.65	10.87	4.796	2.116	0.933	0.412	—	—	(8) 0.324	0.003
40	66.70	22.21	7.398	2.464	0.820	0.273	—	—	—	(7) 0.136	<0.001
25	81.20	15.27	2.871	0.540	0.101	—	—	—	—	(6) 0.023	<0.001
15	89.53	9.377	0.982	0.103	—	—	—	—	—	(5) 0.012	<0.001

$^a \%P(m_x)_n$ = the percentage of sequences of ethylene (or of propylene) containing n monomer units; n = the number of monomer units present in each sequence.

^{b} Note that a range is given; the first number of the range (l) being set in parentheses in the table.

Earlier, in 1960, Natta and co-workers (11) calculated the sequence distribution for ethylene–propylene copolymers of various compositions (Table 20–3) prepared in the presence of the catalyst system $Al(C_6H_{13})_3$–VCl_4. According to these workers, the sequence distribution did not differ significantly from the random distribution for which $r_1 r_2 = 1$.

The conventional copolymerization equations proposed by Alfrey and Goldfinger were used to make these calculations. The following conclusions were made: (1) As the content of the olefin in the copolymer increases, the probability of longer sequences of that olefin increases; (2) The largest dispersion of sequence lengths is found for copolymers that contain equimolar amounts of the two olefins; (3) Only in copolymers containing in excess of 80 mole % ethylene are found crystallizable ethylene sequences (≥ 15 units); and (4) The copolymerizations are random, since $r_1 r_2$ products are near unity.

3. Physical State of Catalyst

As already mentioned, variability in site structure can play an important role in determining the composition of the copolymer. Soluble and certain colloidal catalysts, in which sites are all nearly identical, produce copolymers that have a random or moderately alternating distribution of comonomer units. Catalysts that have been found most suitable are based on ortho-vandate esters, $VOCl_3$ or VCl_4, and aluminum alkyl chlorides.

Block copolymers have been prepared with heterogeneous catalysts that contain multiple centers. If these centers have ligand environments that determine the ease with which an olefin is complexed at the center, then the composition of the copolymer chain formed at this site will be different from that formed at another site which has a different ligand environment. The heterogeneous catalyst such as $AlEt_3$–$TiCl_3$ contains mostly isotactic-regulating centers, but a small number of syndiotactic-regulating and non-regulating centers are also present.

When random (or random-alternating) copolymers are solvent frac-tionated, the fractions and the parent copolymer have similar compositions. In addition, the random copolymer has a narrow molecular weight dis-tribution, for example, $Q = \bar{M}_w / \bar{M}_n = 2.2$ was obtained for the ethylene–propylene copolymer formed with $\phi_4 Sn$–$AlBr_3$–VCl_4 catalyst (12). The fractions from a solvent fractionation of block copolymers have a variable composition (see also Table 15–8).

Valvassori and co-workers (13) observed that, in an ethylene–propylene copolymerization with heterogeneous catalysts, increasing the ethylene con-centration caused an increase in the copolymerization rate beyond that expected from the greater reactivity of ethylene. They speculated that sites existed at which ethylene could homopolymerize or copolymerize with propylene but at which propylene could not homopolymerize.

4. Physical State of the Copolymer

Lukach and Spurlin (1) speculated that a diffusion barrier, which they called "the gel effect," was responsible for the origin of blockiness in ethylene–propylene copolymers (see Section II for definition). According to their proposal, the following situation exists in the copolymer gel during the copolymerization. As the center of gel is approached, the concentration of propylene increases relative to the concentration of ethylene because ethylene is consumed more readily at all sites and the diffusion barrier prevents immediate replenishment of ethylene in the locality of the sites.

Consequently, sites closer to the center of gel tend to produce chains containing longer segments of propylene units.

5. Structure of the Comonomers

As will be illustrated in the next section, copolymers can be formed from combinations of monomers of different structure. In general, the more active the monomer is for homopolymerization with the catalyst being used, the more active is its reactivity in a copolymerization, that is, ethylene > propylene > 1-butene > 3-methyl-1-butene. A suitable catalyst must be selected for certain combinations; for ethylene and butadiene copolymerization, titanium- or vanadium-based catalysts may suffice but not cobalt-based catalysts. Ethylene is not polymerized with the latter and, in fact, acts as a transfer agent.

The disruption of long isotactic sequences of α-olefin units by other monomer units sometimes, but not always, lowers the ability of the resulting copolymer to crystallize. This is true when ethylene and propylene are copolymerized. Both monomers produce crystalline homopolymers, but the copolymer is elastomeric (see Section IV). This is not true when the two olefins have similar structures. For example, for 3-methyl-1-butene and 4-methyl-1-butene, the copolymer can be crystalline since either monomer can fit into the crystal lattice structure of the other homopolymer (14). Another example was discussed in Chapter 15, Section III,B,3, namely, 1-butene/3-methyl-1-butene. In this case, the formed copolymer had a random structure. Block copolymers were, however, formed with other dissimilar comonomers, such as 1-butene/4,4-dimethyl-1-pentene.

Dall'Asta, Natta, and co-workers discovered that certain monomers that do not polymerize or polymerize only sluggishly can, under certain conditions, be copolymerized with ethylene (15–18). These conditions are: (1) ethylene is added very slowly to the "inactive" or "sluggishly" active monomer that is present in excess concentrations (sometimes as its own solvent); (2) soluble or colloidal catalysts are used; and (3) the copolymerization is done at specific (low) temperatures. Following these rules, Natta and co-workers copolymerized ethylene with cis- and trans-2-butene, cyclopentene, cyclohexene, cycloheptene, and cyclooctene. In some cases, 1:1 alternating crystalline copolymers were formed. Natta, Zambelli, and co-workers were able to copolymerize 1-butene with ethylene or propylene, even though it would not homopolymerize with the syndiotactic-regulating catalyst (19).

These results can be explained by the proposal that ethylene units minimize steric interactions between the incoming and last added α-olefin units; that is, they act as spacers. The $r_1 r_2$ products are low, in agreement with the view that the copolymerization is not random.

TABLE 20–4

Reactivity Ratios of Ethylene(C_2), Propylene(C_3), and 1-Butene(C_4) (5, 20)

Catalyst	C_2–C_4			C_3–C_4			C_2–C_3		
	r_1	r_2	r_1r_2	r_1	r_2	r_1r_2	r_1	r_2	r_1r_2
VCl_4–$Al(C_6H_{13})_3$	29.60	0.019	0.56	4.39	0.227	1.0	7.08	0.088	0.62
VCl_3–$Al(C_6H_{13})_3$	26.96	0.043	1.16	4.04	0.252	1.0	5.61	0.145	0.81

Table 20–4 shows the reactivity ratio data for three combinations of olefin pairs. The reactivity ratios shown for the propylene/1-butene pair were determined by Pajaro, using radioactive 1-butene (20). Values that are in good agreement, however, were calculated from the reactivity ratios of the ethylene/propylene and ethylene/1-butene copolymerizations (5).

B. CONTROL OVER MOLECULAR WEIGHT AND MOLECULAR WEIGHT DISTRIBUTION

As in homopolymerizations, transfer agents such as hydrogen and diethylzinc can be used to decrease molecular weights in a copolymerization. In addition, variations can be made in operational procedures, relative concentrations of catalyst components and comonomer units, etc., with concomitant changes in molecular weight of the formed copolymer. It is important to recognize that these factors will vary according to the particular catalyst used and, hence, one should be wary of generalities.

Some factors appear to hold, at least for most systems (5). For example, copolymer molecular weights increase when: (1) the ethylene/propylene ratio is increased in the reaction solvent; (2) the concentration of catalyst is decreased; (3) the polymerization temperature is lowered; and (4) the concentration of monomers is decreased. Other factors depend on the particular catalysts used [Natta *et al.* (5, pp. 687–689) lists many examples]. For example, aging does not affect copolymer molecular weight when $Al(C_6H_{13})_3$–VCl_4 is used (5), but molecular weight increases when $AlEt_2Cl$–$V(acac)_3$ and $VOCl_3$–$AlRCl_2(Al_2Et_3Cl_3)$ systems are aged (9). Also, increasing copolymerization time increases molecular weight with certain catalysts (21): $VOCl_3$–$AlRCl_2(Al_2Et_3Cl_3)$ (9, 22), $VOCl_2(OR)$–$AlRCl_2$, or VCl_4–AlR_2Cl (22), but negligible changes take place when $Al(C_6H_{13})_3$–VCl_4 is used (23). Copolymer molecular weights can be varied according to the choice of aluminum alkyl structure, as well as the ratio of Al/V when VCl_4 or $VOCl_3$ are used as the transition metal component. Molecular weights increase in the order: $AlEt_2Cl > Al_2Et_3Cl_3 > AlEtCl_2 > AlEt_3$ (9).

For the catalysts $AlEt_2Cl-V(acac)_3$ (21) or $Al(C_6H_{13})_3-VCl_4$, aged at 60°C for 30 minutes and used at 25°C (5), molecular weight decreased when Al/V ratio increased.

It is important to note that these changes hold only for the ratios examined. In all cases, intrinsic viscosity values were determined as indices of molecular weight. The variation in molecular weight distribution according to the catalyst used was already mentioned (see Fig. 20–3).

C. MODE OF ADDITION

In Chapter 15, it was strongly indicated that, in an isotactic homopolymerization, primary addition was favored, whereas in syndiospecific homopolymerization, secondary addition prevailed. Heterogeneous titanium-based and soluble vanadium-based catalysts were used to prepare the isotactic and syndiotactic polymers, respectively.

Because random distribution of ethylene and propylene units was sought in the formed copolymers, vanadium-based catalysts have been preferred in copolymerization synthesis. Van Schooten and co-workers (24) examined these random copolymers and established the presence of some head-to-head orientation of propylene units by infrared spectroscopy, in addition to the normal head-to-tail units. These copolymers contained between 46 and 61% propylene.

More recently, Zambelli and co-workers (24a) have found that atactic and highly syndiotactic polypropylenes always contained up to several percent of head-to-head and tail-to-tail arranged units. Highly isotactic polypropylene (after extraction) was free of such irregularities. Both head-to-head and tail-to-tail structures were identified by the following infrared bands: by a band at 752 cm^{-1} due to rocking vibration of the two methylene units

$$-CH-CH_2-CH_2-CH- \\ \quad | \qquad\qquad\qquad | \\ \quad CH_3 \qquad\qquad\quad CH_3$$

and by a band at 1,153 cm^{-1} due to vibration of methyl in the unit

$$-CH-CH- \\ \quad | \quad\; | \\ \quad CH_3\; CH_3$$

Zambelli's investigation led him to the conclusion that, if the last added unit was ethylene, the next added propylene molecule became attached by a secondary addition, whereas if the last added unit was propylene, primary addition was favored.

The irregularities cited above can be explained by the various modes of addition of propylene in the copolymerization.

IV. Copolymerizations of Selected Comonomer Pairs

A. ETHYLENE–PROPYLENE (INCLUDING DIENE TERPOLYMERS)

The copolymerization of ethylene and propylene has been extensively investigated for two reasons. First, following the wide examinations of ethylene and propylene homopolymerizations, it was advantageous to examine these same olefins to establish the features of copolymerization with the Ziegler–Natta catalysts. Secondly, amorphous ethylene/propylene copolymers have desirable elastomeric properties. If successfully developed, this copolymer has the potential of a large volume, inexpensive elastomer.

The commercial applications of these elastomers has already been described in Chapter 3. One of these applications is in tire manufacture using conventional methods. A serious deficiency of the copolymer was that it was 100% saturated and could not be cured by commercial processes. It lacked a reactive center (double bonds) that could serve as a loci for cross-linking. The practical introduction of reactive groups into ethylene/propylene copolymers was, therefore, a serious objective in many industrial laboratories. The future of this elastomer was virtually dependent on a successful method. Most laboratories concentrated on the introduction of a small amount of unsaturation by copolymerization of ethylene and propylene with a nonconjugated diene monomer. It was later shown that the unsaturation should not be in the main chain, because oxidative stability (excellent for the parent copolymer) would be decreased. Many nonconjugated dienes, which polymerized only through one of the double bonds, were most applicable, such as 1,4-hexadiene and ethylidene norbornene. Conjugated dienes, however, have also been used (24b).

References 1–3, 5, 5b, and 24c review this copolymerization. For comprehensive original papers on the synthesis and characterization of these terpolymers, Arrighetti *et al.* (24b) and Duck and Cooper (24d) are valuable. Various methods of characterization of these co- and terpolymers have also been disclosed, including infrared, NMR, and pyrolysis–gas chromatography [Seno *et al.* (24e) contains many original references]. The papers of Bucci and Simonazzi (24f) and Ciampelli and Valvassori (24g) disclosed infrared spectroscopic methods for determining the number of methylene sequences and the distribution of propylene units in the copolymer chain. Perhaps the most revealing information about the detailed chain structure will continue to come from ^{13}C-NMR spectroscopy (see Chapter 3).

TABLE 20–5

Selected Reactivity Ratios Reported for Ethylene/Propylene and Several Catalyst Types

Catalyst[a]	r_1 (M_1)[b]	r_2 (M_2)[b]
$TiCl_3-Al(C_6H_{13})_3$ (H)	15.72	0.110
$TiCl_4-Al(C_6H_{13})_3$ (CD)	33.36	0.032
$VCl_3-Al(C_6H_{13})_3$ (H)	5.61	0.145
$VCl_4-Al(C_6H_{13})_3$ (CD)	7.08	0.088
$HfCl_4-Al-i-Bu_3$ (H)	76	—
$ZrCl_4-Al-i-Bu_3$ (H)	61	—
$VCl_4-AlBr_3 + Sn(C_6H_5)_4$ (S)	16	—
$VO(OR)_xCl_{3-x}-AlR_2Cl$ or $AlRCl_2$	17–28	—
$TiCl_2-Al(C_6H_{13})_3$ (H)	15.72	0.110

[a] H = heterogeneous; CD = colloidal; S = soluble.
[b] Monomer 1, ethylene; monomer 2, propylene.

Tables 20–1, 20–2, and 20–5 collect some of the catalyst systems and corresponding reactivity ratios. In all cases, the reactivity ratio of ethylene was greater than for propylene. That the most active monomer in homopolymerizations has the greater reactivity ratio is a general finding for all olefin pairs.

B. ETHYLENE–1-OLEFINS (α-OLEFIN $\geq$ 1-BUTENE)

Natta, Mazzanti, Valvassori, and Pajaro (25) synthesized amorphous copolymers (free of homopolymers) from ethylene and butene using $VCl_4-Al(C_6H_{13})_3$ and $VCl_3-Al(C_6H_{13})_3$ catalysts. Table 20–4 contains the obtained reactivity ratios.

Lanovskaya and co-workers (26) examined the effect that second monomers (α-methylstyrene, isobutylene, styrene, butadiene, and isoprene) have on the polymerization of ethylene ($AlEt_3-\alpha TiCl_3$ catalyst). They found a decrease in the reaction rate, but the molecular weight was unchanged.

Agakishieva and co-workers (27) investigated the copolymerization of ethylene with 3-methylbutene in the presence of $Al-i-Bu_2Cl-V(acac)_3$ catalyst. The yield of copolymer declined with increasing reaction temperatures.

Seidov and co-workers (28) extensively studied the copolymerization of ethylene-1-butene in the presence of $VCl_4-Al-i-Bu_2Cl$ catalyst. Reactivity ratios and monomer unit sequential distribution were calculated.

Cote and Gregorian (29) reported the synthesis of an ethylene–1-butene copolymer having density in the range of 0.920 to 0.945 g/cm^3 and 2 to 10 wt % 1-butene units in the presence of a catalyst consisting of AlR_3 and a mixture of ammonium silicofluoride and chromium oxide. This copolymer was said to be capable of biaxial orientation at 98° to 125°C temperature without cross-linking.

C. 1-OLEFIN–1-OLEFIN

Mazzanti, Valvassori, Sartori, and Pajaro (30) evaluated reactivity ratios for propylene and 1-butene which are shown in Table 20–4. One can see that the reactivity ratio values for propylene and 1-butene differ less from each other than values of ethylene–propylene or ethylene–1-butene. This is obviously due to these two monomers having more similar homopolymerization rates than the ethylene-based pairs (due to a more similar steric compression). On the basis of reactivity ratios derived from copolymerizations of ethylene–propylene and ethylene–1-butene comonomer pairs, they calculated reactivity ratios for the propylene–1-butene pair (Table 20–4). They checked very closely with experimental values obtained with the same catalysts. For this comonomer–catalyst system, it was concluded that the rate at which a given monomer adds to growing chains does not depend upon the terminal monomer unit (the copolymerization was random).

Ketley (31) investigated the copolymerization of 3-methyl-1-butene and butene with Al-i-Bu$_3$–TiCl$_3$. Fractionation produced non-homogeneous fractions, indicating sites capable of forming random as well as block copolymers. Reactivity ratios $r_1 = 8.5$ and $r_2 = 0.013$ (M$_1$ = 1-butene and M$_2$ = 3-methyl-1-butene) were determined; the product $r_1 r_2 = 0.11$ indicates considerable deviation from a random copolymerization. This suggested that the addition of the incoming monomer is limited when the chain end is the bulky 3-methyl-1-butene molecule (steric compression).

Coover and co-workers (32) reported a block copolymer structure when propylene and 1-butene were copolymerized with AlEtCl$_2$–TiCl$_3$ donor-type catalysts. Copolymers containing 3 to 80% 1-butene had two DTA melting points, which they attributed to polypropylene and poly-1-butene crystallinity. Multiblock structure was supported by observations: $r_1 r_2 > 1$, and absence of homopolymers by fractionation studies. The 1-butene segments had modification I crystallinity, and Gianotti and Capizzi (33) found that, contrary to what takes place with pure poly-1-butene, the 1-butene segments of the copolymer crystallize from the melt into a low melting modification I.

A crystalline terpolymer containing 2% 1-hexene, 16% 3-methyl-1-pentene, and 91% 4-methyl-1-pentene was prepared by Hambling and Woodhead (34) using AlEt$_2$Cl–TiCl$_3$ as a catalyst.

The cocrystallizing behavior shown by copolymers of 1-butene and a variety of linear and branched olefins (AlEt$_2$Cl–TiCl$_3$ catalyst) was extensively investigated by Turner–Jones (35) using X-ray methods. The effect of copolymerization on the spontaneous crystal–crystal transformation (modification II to modification I) was also studied. Ethylene, propylene, or pentene units markedly accelerated this transformation and in insufficient amounts caused direct crystallization of 1-butene units from melt to modification I, while linear α-olefins with C > 5 and branched comonomers retarded the transformation.

D. COPOLYMERS CONTAINING VINYL AROMATIC MONOMERS

Styrene and substituted styrenes are unique monomers, because they are polymerizable by many types of mechanistic paths, including thermal initiation. In the absence of proper control experiments, copolymerizations of styrene and other monomers can lead to deceptive conclusions, and one should be aware of this complication in assessing the literature.

Relative to aliphatic (linear and most branched) α-olefins (36, 37), styrene is less active in a homopolymerization in the presence of Ziegler–Natta catalysts. This manifests itself in a copolymerization; sufficient amounts of styrene must be added in the comonomer mixture if styrene is to be incorporated in the copolymer chain.

Table 20–6 collects some examples of investigated comonomer pairs where the comonomer is not a styrene derivative (38–43).

TABLE 20–6

Copolymerizations with Styrene and α-Olefins

M_1	M_2	Catalyst	r_1	r_2	$r_1 r_2$	Ref.
Styrene	1-Heptene	VCl_3–Al-i-Bu_3	0.47	1.34	0.63	(38)
Styrene	1-Heptene	$VOCl_3$–Al-i-Bu_3	0.43	0.95	0.41	(38)
Styrene	1-Octadecene	VCl_3–Al-i-Bu_3	1.94	0.75	1.46	(38)
Styrene	1-Octadecene	$VOCl_3$–Al-i-Bu_3	1.32	0.87	1.15	(38)
Styrene	4-Methyl-1-pentene	VCl_3–Al-i-Bu_3	0.55	1.23	0.68	(38)
Styrene	4-Methyl-1-pentene	$VOCl_3$–Al-i-Bu_3	0.49	1.15	0.56	(38)
4-Methyl-1-pentene	Styrene	$TiCl_3$–Al-i-Bu_3	3.92	0.98	3.84	(39)
Styrene	1-Hexene	$\alpha TiCl_3$–$AlEt_3$	0.19	9.75	1.85	(40)
Styrene	1-Heptene	$\alpha TiCl_3$–$AlEt_3$	0.61	5.70	3.48	(40)
Styrene	4-Methyl-1-hexene	$\alpha TiCl_3$–$AlEt_3$	1.80	1.3	2.34	(40)
4-Vinylcyclohexene	p-Cl-styrene	$TiCl_4$–$AlEt_3$	1.45	0	0	(41)
Propylene	Styrene	$TiCl_4$–$AlEt_3$	62	0.93	57	(42)

Baker and Tait (38) concluded from their characterization data that the copolymer from styrene and various olefins (1-heptene, 1-octadecene, and 4-methyl-1-pentene) had a block structure. These copolymers showed a high degree of X-ray crystallinity.

A maximum of 19 mole % styrene in the copolymer was obtained under best conditions when ethylene and styrene were copolymerized with $AlEt_3$–$\alpha TiCl_3$ (44). Copolymers of styrene with isoprene (45) and butadiene (46) were also reported.

That some Ziegler–Natta catalysts can polymerize styrene monomers by free radical mechanisms was reported by Amass and co-workers (47, 48). Styrene–p-methylstyrene were polymerized with a mixture of $(C_5H_5)_2TiCl_2$ and $AlEt_3$. The mechanistic interpretation of a comprehensive study by Danusso and Sianesi (36, 37) on some thirty vinyl aromatic monomers is discussed in Chapter 19. Table 19–5 should also be consulted.

E. OLEFIN–INTERNAL OLEFIN

When ethylene at low concentrations is copolymerized with certain co-monomers, there is a tendency for the formation of alternating copolymers. These regular copolymers can be crystalline if sufficient alternation occurred. Sometimes, when the crude polymer is solvent fractionated, more regular crystalline fractions are isolated.

Two types of alternating copolymers have been disclosed: ethylene–2-butene, and ethylene–cycloolefin.

Acyclic 2-olefins do not homopolymerize in the absence of isomerization (see Chapter 19, Section V). When ethylene is present, however, the removal of steric compression at the transition metal–carbon center allows 2-butene to add (49). Thus, while two or more 2-butene units are unlikely in the polymer chain, alternation of ethylene and 2-butene makes possible the growth of 2-butene. Long segments of ethylene units are prevented by keeping the concentration of ethylene low and using the 2-olefin as a reaction solvent.

Similarly, cyclopentene, cyclohexene, cycloheptene, and cyclo-cis-octene do not homopolymerize extensively, if at all. With ethylene, however, the monomers, especially cyclopentene and cycloheptene, can be made to yield alternating copolymers that are crystalline (15–18).

F. OLEFIN–DIENE

Alternating copolymers of ethylene and butadiene were prepared at $-30°C$ with VCl_4–AlR_3–AlR_2Cl–anisole catalysts (mixed at $-78°C$). Solvent fractionation produced cuts that had 28 to 50 mole % ethylene, and the content of *trans*-1,4 structure was greater than 90% (50).

Suminoe and co-workers (7) showed that conventional Ziegler–Natta titanium-based catalysts produced block copolymers of butadiene and propylene. Furukawa and co-workers (51–54) and Kawasiki and co-workers (55) showed that, with the use of vanadium-based catalysts, highly alternating copolymers of butadiene and propylene could be prepared. Such catalysts as VCl_4–$AlEt_3$, $VOCl_3$–$AlEt_2Cl$, and $VO(acac)_2$–$AlEt_3/AlEt_2Cl$ were used (54). The microstructure of the diene was *trans*-1,4. Furukawa and co-workers (52) prepared random copolymers of butadiene and propylene with the $AlEt_3$–$TiCl_4$ in the presence of phosgene. The microstructure was 68.7% *cis*-1,4, 27.5% *trans*-1,2, and 3.8% vinyl. Both copolymers were evaluated as potential new synthetic rubbers. In more recent work, Furukawa and co-workers (53) have reported the synthesis of alternating copolymers of butadiene with several α-olefins and of isoprene with propylene in the presence of the catalyst $VO(acac)_2$, $AlEt_3$, and $AlEt_2Cl$. The alternating copolymerization ability of the olefins decreased in the order: propylene > 1-butene > 4-methyl-1-pentene > 3-methyl-1-butene.

G. NONCONJUGATED DIENE–OLEFIN

Valvassori, Sartori, and Ciampelli (56) prepared amorphous copolymers of 1,5-hexadiene and ethylene in which the diene was polymerized mainly (70 to 80%) in the 1,2-enchainment. Control of 1,2- vs. cycloaddition was affected by variation of temperature ($AlEt_2Cl$–vanadium salts were used as catalysts). Di Pietro and Di Edwardo (57) prepared crystalline copolymers of 3-methyl-1-butene and the dienes: 2,4-dimethyl-2,7-octadiene and 2,6-dimethyl-2,7-octadiene. Makowskii and co-workers (58) described the synthesis of crystalline block copolymers of 1,5-hexadiene and ethylene in the presence of the catalyst $AlEt_2Cl$–$TiCl_3$.

H. COPOLYMERS OF DIENES

Because practical elastomeric properties were found in the homopolymers of dienes, there was little incentive to extensively investigate the corresponding copolymers. True copolymers of butadiene and 2-phenylbutadiene, piperylene, or isoprene have been synthesized, as shown in Table 20–7 (59–65). While, in some cases, the microstructure of each type of diene unit is that expected from homopolymerization with that catalyst, in other cases (59, 60), it appears that the microstructure of the incoming diene unit is affected by the mixture of the last added diene (64, 65).

TABLE 20-7

Copolymers of Butadiene and Substituted Diene

Substituted diene	Ref.	Catalyst	Reactivity ratios		Comment
			r_1	r_2	
2-Phenyl-1,3-butadiene	(59, 60)	$AlHCl_2 \cdot Et_2O–AlI_3–TiCl_4$ (S)[a]	1.85–1.90	2.58–2.6	Both dienes have a *cis*-1,4 structure in copolymer
4-Methyl-1,3-butadiene (Piperylene)	(61)	$AlEt_3–VCl_3$ (H)	?	?	*trans*-1,4 structure
	(61, 62)	$AlEt_2Cl–V(acac)_3$ (S)	?	?	BD units have *trans*- and piperylene mixed-1,4 and 1,2 structure
2-Methyl-1,3-butadiene (isoprene)	(63)	$AlEt_3–TiCl_4$ (H)	1.0	1.0	Structure of the homopolymers and copolymers are the same
	(64, 65)	$AlEt_2Cl–CoCl_2 \cdot EtOH$ (S)	2.3	1.15	Copolymer microstructure depends on copolymer composition.
	(64, 65)	$AlEt_2Cl–Co(acac)_2$ (S)	1.0	1.0	

[a] H = heterogeneous catalyst; S = soluble catalyst.

Random copolymers having an all *trans* structure and having 30 to 45% 1,3-pentadiene were synthesized by Carbonaro and co-workers (65a) from butadiene and 1,3-pentadiene with $Al_2Et_3Cl_3$–$V(acac)_3$-type catalysts. These copolymers were amorphous at room temperature when left undisturbed but crystallized upon stretching. Cured tread stocks were said to have improved technological properties relative to other synthetic diene-based elastomers, including high tack, tear strength, and abrasion resistance.

Dolgoplosk and co-workers have collected [Table III in Dolgoplosk *et al.* (65b)] reactivity ratios for a number of binary diene mixtures obtained with Ti and Ni catalysts.

I. POLAR MONOMER-CONTAINING COPOLYMERS

Hopkins and Miller (66) reported that copolymerization of ethylene and *tert*-butyl acrylate in the presence of *n*-BuLi–TiCl$_4$ at room temperatures produced a mixture of a homopolymer of *tert*-butyl acrylate, and a block copolymer of ethylene and *tert*-butyl acrylate, but no polyethylene. They suggested that the *tert*-butyl acrylate can add to growing chains of polyethylene but not the reverse. The reviewer suggests that *tert*-butyl acrylate could have added to Ti–C bonds to generate Ti–O bonds which then became the true propagating centers for the *tert*-butyl acrylate.

Bacskai (67) reported that ω-haloolefins, such as 6-chloro-1-hexene, could be copolymerized with α-olefins, like 4-methyl-1-pentene, in good yields to high molecular weight products if the aluminum alkyl was complexed with a Lewis base. In this way, the decomposition of the ω-halo-olefin by the aluminum alkyl (producing hydrogen halide) was suppressed or diminished. Copolymers of this type, after reaction with amines, could be dyed by acid dyes.

Matsummura and Fukumoto (68) reported that propylene could be copolymerized with the reaction product of acrylic acid and an aluminum alkyl, as shown in Eq. 20–2. The whole product contained 2% acrylic acid.

$$AlEt_2Cl + CH{=}CH{-}CO_2H \longrightarrow EtClAl{-}O\overset{\overset{\displaystyle O}{\|}}{C}{-}CH{=}CH_2 \qquad (20\text{–}2)$$

The possibility that the copolymerization involved radical–ion mechanisms has to be considered. Copolymers of styrene and 2-methyl-5-vinylpyridine were reported by Sakurada (69). The isotactic crystallinity of polystyrene decreased as the content of the polar monomer increased.

V. Conclusion

The versatility of the Ziegler–Natta catalyst is again demonstrated but now in its ability to copolymerize olefins and dienes. Many new polymeric products have been synthesized that could not have been prepared by other means. The copolymers of ethylene and propylene are very important in the family of commercial products. In addition, copolymerization studies have served as another powerful diagnostic tool to study the nature of the active center and how it polymerizes. The delicateness of the Ziegler–Natta catalyst has also manifested itself in copolymerization studies. Undoubtedly, some of the anomolies that have been encountered by different workers are due to subtle differences in catalyst preparation and use.

With the continued use of improved instrumentation and techniques, such as ^{13}C- and ^{1}H-NMR spectroscopy, gas chromatography–pyrolysis, IR, copolymerization studies will continue to be used to probe the detailed mechanistic aspects of the Ziegler–Natta catalyst. New insight will be achieved with respect to the microstructure of the polymer chain, which will more definitively reveal the structure of the active center and its polymerization behavior. There undoubtedly will occur a reevaluation of the copolymerization results that are reported in this chapter.

References

1. C. A. Lukach and H. M. Spurlin, *in* "Copolymerization" (G. E. Ham, ed.), Chapter IVA. Wiley (Interscience), New York, 1964.
1a. G. Bier and G. Lehmann, *in* "Copolymerization" (G. E. Ham, ed.), Chapter IVB. Wiley (Interscience), New York, 1964.
1b. G. Crespi, A. Valvassori, and G. Sartori, *in* "Copolymerization" (G. E. Ham, ed.), Chapters IVB and IVC. Wiley (Interscience), New York, 1964.
2. G. Natta, G. Crespi, A. Valvassori, and G. Sartori, *Rubber Chem. Technol.* **36**, 1583 (1963); see also G. Natta, *SPE Trans.* **3**, 99–110 (1963).
3. I. Pasquon, A. Valvassori, and G. Sartori, *in* "The Stereochemistry of Macromolecules" (A. D. Ketley, ed.), Chapter 4. Dekker, New York, 1967.
4. J. Boor, *Macromol. Rev.* **2**, 115–268 (1967).
5. G. Natta, A. Valvassori, and G. Sartori *in* "Polymer Chemistry of Synthetic Elastomers" (J. P. Kennedy and E. G. M. Törnqvist, eds.), pp. 679–702. **23**, Part II, Wiley (Interscience) New York (1969).
5a. G. Natta and L. Porri *in* "Polymer Chemistry of Synthetic Elastomers" (J. P. Kennedy and E. G. M. Törnqvist, eds.), pp. 597–678. **23**, Part II, Wiley (Interscience) New York (1969).
5b. F. P. Baldwin and G. Ver Strate, *Rubber Chem. Technol.* **45**, No. 3, 709–881 (1972).
6. C. Cozewith and G. Ver Strate, *Macromolecules* **4**, 482 (1971).
7. T. Suminoe, N. Yamazaki, and S. Kambara, *Chem. High Polym.* **20**, 461 (1963).
8. W. L. Carrick, F. J. Karol, G. L. Karapinka, and J. J. Smith, *J. Am. Chem. Soc.* **82**, 1502 (1960).
9. E. Junghanns, A. Gumboldt, and G. Bier, *Makromol. Chem.* **58**, 18 (1962).
10. M. Ichikawa, *J. Chem. Soc. Jpn., Ind. Chem. Sect.* **68**, 535 (1965); *Int. Chem. Eng.* **5**, 724 (1965).
11. G. Natta, G. Mazzanti, A. Valvassori, G. Sartori, and D. Morero, *Chim. Ind. (Milan)* **42**, 125 (1960).

12. G. W. Phillips and W. L. Carrick, *J. Am. Chem. Soc.* **84**, 920 (1962); *J. Polym. Sci.* **59**, 401 (1962).

13. A. Valvassori, G. Sartori, G. Mazzanti, and G. Pajaro, *Makromol. Chem.* **61**, 46 (1963).

14. E. P. Reding and E. R. Walter, *J. Polym. Sci.* **37**, 55 (1955).

15. G. Dall'Asta, G. Mazzanti, G. Natta, and L. Porri, *Makromol. Chem.* **56**, 224 (1962).

16. G. Natta, G. Dall'Asta, G. Mazzanti, and G. Motroni, *Makromol. Chem.* **69**, 163 (1963).

17. G. Dall'Asta, G. Natta, and G. Mazzanti, *Macromol. Colloq.*, *1964*; *Chem. Eng. News* **42**, 41 (1964).

18. G. Natta, G. Dall'Asta, and G. Mazzanti, *Angew. Chem.* **76**, 765 (1964).

19. A. Zambelli, G. M. Giongo, and G. Natta, *Makromol. Chem.* **112**, 183 (1968); A. Zambelli, A. Lèty, G. Tosi, and I. Pasquon, **115**, 73 (1968); A. Zambelli, I. Pasquon, R. Signori and G. Natta, **112**, 160 (1968).

20. G. Pajaro, *Energ. Nucl. (Milan)* **6**, 273 (1959).

21. G. Natta, G. Mazzanti, A. Valvassori, G. Sartori, and O. Fiumani, *J. Polym. Sci.* **51**, 911 (1960).

22. G. Bier, A. Gumboldt, and G. Schleitzer, *Makromol. Chem.* **58**, 43 (1962).

23. G. Natta, G. Mazzanti, A. Valvassori, G. Sartori, and A. Barbagallo, *J. Polym. Sci.* **51**, 429 (1961).

24. J. Van Schooten and S. Mostert, *Polymer* **4**, 135 (1963); J. Van Schooten, E. W. Duck, and R. Berkenbosch, *ibid.* **2**, No. 3, 357 (1961).

24a. Original references and review papers are given in Chapter 15 (Syndiotactic Polypropylene).

24b. S. Arrighetti, G. Bertolini, S. Cesca, G. Ghetti, and A. Roggero, *Macromol. Prepr.*, *Int. Congr. Pure Appl. Chem.*, *23rd*, *1971* Vol. II, 732 (1971).

24c. E. J. Vandenberg and B. C. Repka, *in* "Polymer Processes" (C. Schildknecht, ed.). Wiley (Interscience), New York, 1975.

24d. E. W. Duck and W. Cooper, International Symposium on Macromolecular Chemistry, Macromolecular Preprints. *Int. Union Pure Appl. Chem.*, *23rd*, Vol. II, p. 722 (1971).

24e. H. Seno, S. Tsuge, and T. Takeuchi, *Makromol. Chem.* **161**, 195 (1972).

24f. G. Bucci and T. Simonazzi, *J. Polym. Sci.*, *Part C* **7**, 203 (1964).

24g. F. Ciampelli and A. Valvassori, *J. Polym. Sci.*, *Part C* **16**, 377 (1967).

25. G. Natta, G. Mazzanti, A. Valvassori, and G. Pajaro, *Chim. Ind. (Milan)* **41**, 764 (1959).

26. L. M. Lanovskaya, A. R. Gantmakher, and S. S. Medvedev, *Vysokomol. Soedin.* **2**, 1391 (1960).

27. M. Ya. Agakishieva, N. M. Seidov, and T. A. Kuliev, *Azerb. Khim. Zh.* **6**, 111 (1969).

28. N. M. Seidov, M. A. Dalin, and S. M. Kyazimov, *Azerb. Khim. Zh.* No. 1, p. 68 (1968).

29. J. A. Cote and R. S. Gregorian, U.S. Patent 3,509,116, April 28, 1970, W. R. Grace & Company.

30. G. Mazzanti, A. Valvassori, G. Sartori, and G. Pajaro, *Chim. Ind. (Milan)* **42**, 468 (1960).

31. A. D. Ketley, *J. Polym. Sci.*, *Part B* **1**, 121 (1963).

32. H. W. Coover, Jr., R. L. McConnell, F. B. Joyner, D. F. Slonaker, and R. L. Combs, *J. Polym. Sci.*, *Part A-1* **4**, 2563 (1966).

33. G. Gianotti and A. Capizzi, *Makromol. Chem.* **124**, 152 (1969).

34. J. K. Hambling and D. A. Woodhead, U.S. Patent 3,635,921, January 18, 1972, British Petroleum Company.

35. A. Turner-Jones, *Polymer* **7**, 23 (1966).

36. F. Danusso and D. Sianesi, *Chim. Ind. (Milan)* **44**, 474 (1962).

37. F. Danusso, *Chim. Ind. (Milan)* **44**, 611 (1962).

38. B. Baker and P. J. T. Tait, *Polymer* **8**, 225 (1967).

39. Yu. V. Kissin, Yu. Ya. Gol'dfarb, B. A. Krentsel, and H. Uyliem, *Eur. Polym. J.* **8**, 487 (1972).

40. I. H. Anderson, G. M. Burnett, and W. C. Geddes, *Eur. Polym. J.* **3**, 161 (1967).
41. W. Kawai and S. Katsuta, *J. Polym. Sci., Part A-1* **8**, 2421 (1970).
42. R. P. Chernovskaya, V. P. Levedev, K. S. Minsker, and G. A. Razuvaev, *Vysokomol. Soedin.* **6**, 1313 (1964).
43. G. A. Razuvaev, K. S. Minsker, and J. Z. Shapiro, *Vysokomol. Soedin.* **4**, 1833 (1962).
44. K. Gehrke, A. Bledzki, B. Schmidt, and J. Ulbricht, *Plast Kautsch.* **18**, 87 (1971).
45. V. P. Shatalov, B. I. Mihant'ev, and F. P. Afanason, *Sb. Nauchn. Rab. Aspir. Voronezh. Gos. Univ.* No. 4, p. 148 (1968).
46. M. A. Mardanov, V. M. Akhmedov, L. I. Zakharakin, and A. A. Khanmetov, *Zh. Org. Khim.* **6**, 1765 (1970).
47. A. J. Amass, J. N. Hay, and J. C. Robb, *Br. Polym. J.* **1**, 282 (1969).
48. A. J. Amass, J. N. Hay, and J. C. Robb, *Br. Polym. J.* **1**, 277 (1969).
49. G. Natta, *J. Am. Chem. Soc.* **83**, 3343 (1961).
50. G. Natta, A. Zambelli, I. Pasquon, and F. Ciampelli, *Makromol. Chem.* **79**, 161 (1964); G. Natta, A. Zambelli, and I. Pasquon, U.S. Patent 3,407,185, October 22, 1968, Montecatini Edison, S.p.A.
51. J. Furukawa, *Rubber Chem. Technol.* **45**, 1532 (1972).
52. J. Furukawa, E. Kobayashi, and K. Haga, *J. Polym. Sci., Polym. Chem. Ed.* **11**, 629 (1973).
53. J. Furukawa, S. Tsuruki, and J. Kiji, *J. Polym. Sci., Polym. Chem. Ed.* **11**, 2999 (1973).
54. J. Furukawa, Int. Sym. on Macromolecular Chemistry, Macromolecular Preprints, *Int. Union Pure Appl. Chem., 23rd*, Vol. II, p. 695 (1971).
55. A. Kawasiki, I. Maruyama, M. Taniguchi, R. Hirai, and J. Furukawa, *J. Polym. Sci., Part B* **7**, 613 (1969).
56. A. Valvassori, G. Sartori, and F. Ciampelli, *Chim. Ind. (Milan)* **44**, 1095 (1962).
57. J. Di Pietro and A. Di Edwardo, *Makromol. Chem.* **98**, 275 (1966).
58. H. S. Makowskii, B. K. C. Shim, and Z. W. Wilchinsky, *J. Polym. Sci., Part A* **2**, 1549 (1964).
59. W. Marconi, A. Mazzei, M. Araldi, and M. Bruzzone, U.S. Patent 3,417,065, December 17, 1968.
60. W. Marconi, A. Mazzei, G. Lugli, and M. Bruzzone, *J. Polym. Sci., Part C* **16**, 805 (1967).
61. G. Natta, L. Porri, A. Carbonaro, and G. Lugli, *Makromol Chem.* **53**, 52 (1962).
62. L. Porri, A. Carbonaro, and F. Ciampelli, *Makromol. Chem.* **61**, 90 (1963).
63. L. S. Bresler, B. A. Dolgoplosk, M. F. Kolechkova, and E. N. Kopacheva, *Vysokomolekul. Soedin.* **5**, 357–62 (1963).
64. T. Suminoe, K. Sasaki, N. Yamazaki, and S. Kambara, *Chem. High Polym.* **21**, No. 225, 9 (1964).
65. I. Pasquon, L. Porri, A. Zambelli, and F. Ciampelli, *Chim. Ind. (Milan)* **43**, 509 (1961).
65a. A. Carbonaro, V. Zamboni, G. Novajra, and G. Dall'Asta, *Rubber Chem. Technol.* **46**, 1274 (1973).
65b. B. A. Dolgoplosk, S. J. Beilin, Yu. V. Korshak, K. L. Makovetskii, and E. I. Tinyakova, *J. Polym. Sci., Polym. Chem. Ed.* **11**, 2569 (1973).
66. E. A. H. Hopkins and M. L. Miller, *Polymer* **5**, 432 (1964).
67. R. Bacskai, *J. Polym. Sci., Part A* **3**, 2491 (1965).
68. K. Matsummura and O. Fukumoto, *J. Polym. Sci., Part A-1* **9**, 471 (1971).
69. Y. Sakurada, *Kobunshi Kagaku* **18**, 496 (1961).

21

Block Polymerizations

I. Introduction

Many patents and publications claim the synthesis of block polymers from ethylene and propylene, P–E, P–EP, or $(EP)_x$ where P, E, and EP are propylene, ethylene, and ethylene–propylene block segments, respectively. Direct evidence establishing that these products were obtained in high purity is nonexistent. Most probably, the claimed block polymer was present in small concentrations, and the major products were homopolymer and copolymer chains. The known kinetic features of heterogeneous Ziegler–Natta catalysts suggest that it is unlikely that neat block polymers can be synthesized in high concentrations.

The basis for synthesizing block polymers is that, if it is possible for a polymer chain to remain alive and grow for a long time, it should be possible to form long segments of two or more olefin molecules in the same polymer chain. Each olefin or a mixture of olefins that is to form a segment must be added sequentially. The last added olefin must be entirely consumed or removed if pure blocks are to be synthesized.

It is important to differentiate between a "live" polymer chain and a "live" center. Once the polymer chain becomes detached from the transition metal center, the chain is considered to be dead. The center, however, may continue to grow more polymer chains, and two or more chains can be formed at a center.

The first example of "living polymers" was disclosed in 1956 by Szwarc (1), who used sodium naphthenate as an anionic initiator for polymerizing styrene and isoprene. Since then, rather sophisticated tailored block polymers have been synthesized, usually with lithium alkyls as the preferred anionic initiators, for example, S–B–S, S–I–S, or S–I, where S, I, and B are styrene, isoprene, and butadiene segments. The lithium alkyl-initiated polymerizations are unique in that it is possible to control both the number and length of segments in the polymer chain for the above monomers.

Tailored block polymers of this type have not, however, been successfully synthesized from olefins with the help of Ziegler–Natta catalysts, in spite of much effort from industrial laboratories. This chapter describes the experimental research effort made in this area and speculates about the products that were actually made. Comprehensive reviews by Bier and Lehmann in 1964 (2) and by Heggs in 1973 (3) offer many details about the patent literature, which is beyond the scope of this chapter. Heggs has also reviewed the physical properties of the reported block polymers.

The types of block polymers sought via Ziegler–Natta polymerizations usually had the structures: P–E, P–EP, (EP)–(EP)′, P–EP–P, and multi-segment chains such as P–E–P–E, P–(EP)–P–(EP), or (EP)′–(EP)–(EP)′–EP, where E and P are long segments of monomers, ethylene, or propylene, respectively. The notation EP denotes a copolymer segment of ethylene and propylene.* These block polymers should not be confused with block copolymers as defined in Chapter 20; admittedly, some structures may fall into either classification.

II. Pioneering Work

The first scientific publications came from the Natta school at the time it was investigating the kinetic features of the Ziegler–Natta catalyst. Since the published work of this group was solely influential in provoking research into this area of Ziegler polymerization, our story of block copolymers properly begins with Natta's early publications on the kinetics of propylene polymerization (3a). These early studies, however, indicated a gloomy outlook for the synthesis of block copolymers. They concluded that the average lifetime of a growing chain was less than one minute for polymerizations with the $AlEt_3$–$\alpha TiCl_3$ catalyst at 70°C, a time too short to be practically used for the synthesis of blocks. Soon afterward, they revised their estimate of chain lifetime to several minutes or more at 70°C (Fig. 21–1) (4). The prospects for the synthesis of block copolymers was increased even further when they recognized that chain lifetimes were longer at lower temperatures of polymerization, about 15° to 18°C.

---

* This notation conforms with that used for anionic block polymers. Another acceptable notation, b_P–b_E, where b is a block segment and P and E are propylene and ethylene units, respectively, was used by Bier and co-workers. For a complete discussion on the accepted nomenclature in block polymers, see Heggs (3).

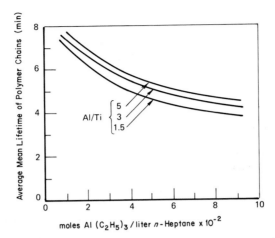

Fig. 21–1. Average mean lifetime of the chains of polypropylene growing on the active centers of a catalytic system: $\alpha TiCl_3Al(C_2H_5)_3$, temp. 70°C ($p_{C_3H_6} = 950$ mm Hg) (4).

Natta and co-workers believed that chain growth was limited by chain termination and by transfer reactions, which occurred during the polymerization. Viscosity values were used as indicators of changes in polymer molecular weight and, because they observed these viscosity values to remain virtually unchanged, they concluded that chain lifetimes were small. Transfer reactions involving propylene, $AlEt_3$, and soluble titanium-compound molecules were suggested.

In a typical synthesis using the $AlEt_3$–$\alpha TiCl_3$ catalyst in toluene solvent at 15° to 18°C, propylene (P) and ethylene (E) were each admitted five to seven times for a five minute duration. The reactor was purged with nitrogen each time to remove unreacted olefin. Propylene and ethylene were admitted at low partial pressures (200 and 100 to 200 mm Hg, respectively). Yields were between 1.1 and 5.1 g and the ratio of ethylene to propylene in the product was between 1.36 and 1.53. The products, if actually synthesized according to this scheme, would have the structure shown in I.

$$(P-E)_{5-7}$$

I

Block polymer structures were suggested on the basis of kinetic data and extraction data on the products (see Table 21–1).

While practical on a laboratory scale, this procedure could not be used commercially for three reasons: (1) very low polymerization rates; (2) too much propylene had to be removed by nitrogen purging; and (3) too little time was available for changing monomers. If all Ziegler–Natta catalysts behaved similarly, then there would be little hope of commercial production of block polymers.

TABLE 21-1

Comparison between the Fractionations of an Ethylene–Propylene Heteroblock Co-polymer; Ethylene and Propylene Homopolymers Obtained with the Same Catalytic System: [αTiCl₃ + Al(C₂H₅)₃]; and a Random Copolymer

| | % By weight of fraction extractable with different boiling solvents | | | |
Polymer	Ether	n-Heptane	n-Octane	Residual
Heteroblock copolymer,				
$C_2:C_3 = 1:1$ mole	0.9	17	30.5	51.6
Random copolymer, 1:1 mole	48.9	51.1	—	—
Polyethylene	traces	4.6	65.0	30.4
Polypropylene	14.4	6.5	22.8	56.6

Research investigations at the Hoechst Laboratory suggested this was not always the case (5–9). The apparent success of the Hoechst group lay in their recognition that lifetimes of growing polymer chains were longer than Natta calculated from the viscosity data, and of their use of another Ziegler–Natta catalyst, namely AlEt₂Cl + a TiCl₃. The Hoechst TiCl₃ was made by reduction of TiCl₄ with aluminum alkyls or aluminum metal, while αTiCl₃ is prepared by reduction with H₂ at high temperatures. Chain lifetimes of hours or more were reported for this catalyst. (As described in Chapter 4, the same catalyst produces highly isotactic polypropylene.) In their hands, polymer chain lifetimes appeared to be even longer for the AlEt₃–αTiCl₃ catalyst that Natta had previously used (4).

As pointed out by Bier and Lehmann (2), an increase in the viscosity value during polymerization cannot be automatically taken as an increase in the number average molecular weight. Knowledge of number average molecular weights is essential for determination of the kinetics of the polymerization. In addition, only approximate weight average molecular weights can be evaluated from viscosity numbers. If the $Q = \bar{M}_w/\bar{M}_n$ ratio changes during the polymerization, then viscosity values can lead to wrong interpretations. An increasing Q value can result in a constant viscosity value, even though polymer molecules continue to grow. Or, a decrease in Q values would result in increasing viscosity values without continual growth of polymer chains.

In order to demonstrate that the increasing viscosity values that they observed during their polymerization of propylene actually corresponded to increasing molecular weight resulting from continued growth of polymer chains, Bier, Gumboldt, and Lehmann (5) investigated the effect of a transfer agent on these viscosity values (Fig. 21–2). Polymerization was interrupted and hydrogen was introduced. Before propylene was again introduced, the

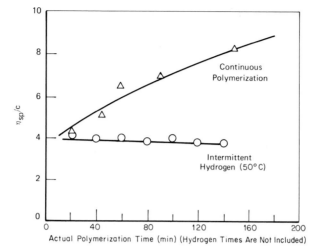

Fig. 21–2. Polymerization of propylene at 50°C with and without interruption; introduction of hydrogen during the interruptions (5).

hydrogen was displaced by nitrogen. Each time, a sample of the product was withdrawn for a viscosity measurement. Bier and co-workers interpreted their results to mean that the hydrogen interrupted the growth of living polymer chains. This diagnostic test, however, did not unequivocally establish that all polymer chain ends remained alive when hydrogen was absent.

Homopolymerizations were carried out for ethylene (Figs. 21–3 and 21–4), propylene (Figs. 21–5 and 21–6), and 1-butene (Fig. 21–7) for

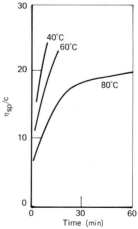

Fig. 21–3. Polymerization of ethylene. Increase of η_{sp}/c with polymerization time at different temperatures (5).

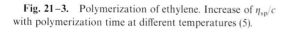

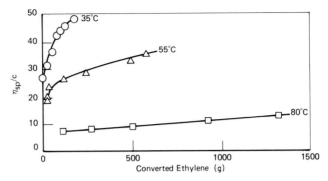

Fig. 21–4. Polymerization of ethylene. Increase of η_{sp}/c with conversion at different temperatures (5).

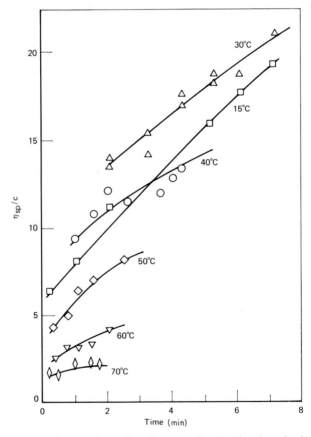

Fig. 21–5. Polymerization of propylene. Increase of η_{sp}/c with polymerization at different temperatures (5).

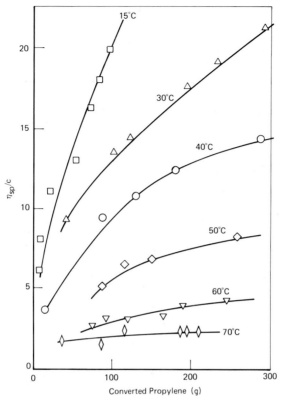

Fig. 21-6. Polymerization of propylene. Increase of η_{sp}/c with conversion at different temperatures (5).

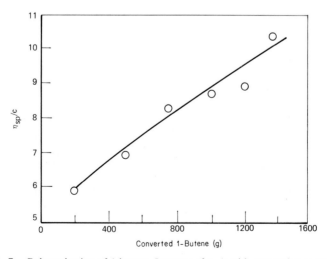

Fig. 21-7. Polymerization of 1-butene. Increase of η_{sp}/c with conversion at 30°C (5).

593

specific times or conversions. In all of these polymerizations, they observed an increase of viscosity values as conversion increased. The greatest increase took place at lower polymerization temperatures. Their data suggested that at about 70° to 80°C, chain termination became significant, and an increase in the η_{sp}/c values was absent with increasing conversions.

Bier and co-workers concluded from their kinetic studies that, "Macromolecules that have once begun to grow continue doing so until polymerization is stopped, and that transfer reactions, if any, occur only to a limited extent." This conclusion was based on viscometric-determined molecular weights and the assumption that $Q = \bar{M}_w/\bar{M}_n$ values are in the range of 2 to 15.

In so far as this catalyst is concerned, the validity of this conclusion determined the sophistication of the block copolymers that could be synthesized. Bier and co-workers apparently believed it to be essentially valid at that time, because they claimed synthesis of block copolymers having six to 40 blocks (Table 21–2).

TABLE 21–2

η_{sp}/c Value[a] as a Function of Experiment Time in the Block Copolymerization of Ethylene and Propylene (5)

Number of blocks	Experimental interval (min)	Solid product yield (g)	η_{sp}/c[b]
6	90	316	1.4
12	180	800	2.3
24	360	1613	2.9
40	600	2700	4.0

[a] Belgian Patent 591,623, June 7, 1960 (German priority June 6, 1959, Farbwerke Hoechst AG.).

[b] 0.1 g of polymer in 100 ml decahydronaphthalene, measured at 135°C.

The polymer products that were described by Bier had plasticlike properties, resembling isotactic polypropylene, isotactic poly-1-butene, or polyethylene. The synthesis of block copolymers that had properties more resembling rubbers were reported by Kontos and co-workers at United States Rubber shortly following the Bier disclosure (10, 11).

Block copolymers having structures such as shown in II, where relative

$$[(EP)-E]_n$$

$$[(EP)-(EP)']_n$$

II

ratios of E and P were different in (EP) and (EP)′ blocks, were claimed by Kontos to be synthesized by sequential additions of the appropriate olefin or olefin mixture. These products were amorphous but crystallizable or semicrystalline, depending on the structure. They were designated as *plastic rubbers*. Two catalysts, $AlEt_2Cl$ or $AlEtCl_2 + VCl_4$ or $VOCl_3$, and $LiAlR_4 + TiCl_4$, were used, and the polymerization temperatures were kept low, near 25° to 35°C.

As did Bier, Kontos concluded that block copolymerization had taken place on the basis of his observations that polymer viscosity values significantly increased with increasing conversions of polymer. In his procedure, Kontos added either one olefin or a mixture of olefins for each block and suggested that unreacted olefin be vented under reduced pressures or by purging with nitrogen if clean blocks were desired.

During this period, other groups apparently were also active in this area, but their findings did not become known until later when the corresponding patents were issued. These are now described in the next section.

III. Other Studies Aimed at Block Polymers

Heggs (3) has collected, in table form, virtually all of the patent literature on block polymers that has emerged from academic and industrial laboratories worldwide. The reader who is interested in detailed descriptions of catalysts, processes, and characterization studies will find Hegg's reviews [Chapters 4 and 8D in Heggs (3)] very worthwhile.

Why such a wide interest in developing commercial synthesis of block polymers? While the workers had a variety of specific improvements in mind, the most important one was to improve the performance of isotactic polypropylene at low temperatures (0°C and below). Isotactic polypropylene, because it has a glass transition temperature near 0°C, becomes brittle when refrigerated. Upon impact, it is very susceptible to breaking, and thus the application to consumer products is limited to temperatures above 0°C.

This deficiency was overcome, in part, by blending isotactic polypropylene with 10 to 20% of a rubber that had a low glass transition temperature, such as $T_g \simeq -60°C$. A variety of rubbers were investigated, including polyisobutylene, polyisoprene, polybutadiene, and ethylene–propylene copolymers. The toughening of polypropylene becomes greater as the concentration of the added rubber is increased. In the case of ethylene–propylene rubbers, random copolymers having about 40 to 60% propylene have been preferred. These gains in toughness, however, are made at a loss of tensile strength and modulus.

Apparently, workers speculated that incompatibility (which would lead to undesired segregation of the polymers) would be decreased if the rubber copolymer molecules were bonded to the isotactic polypropylene chains as block copolymers, for example as P–EP. Others speculated that, by forming simple P–E block polymers or $(P-E)_x$ ($x > 2$) multiblock polymers, similar improvements would be produced. Table 21–3 describes selected examples from the corresponding patents (12–37). Many variations of the examples cited have been described, which depend on specific detailed experimental procedures. Indeed, very detailed procedures were often used that simply cannot be described here. The reader must resort to individual patents to make a proper assessment of each patent. While many of the examples in Table 21–3 may be commercially important, selection was made to show the variety of structures, rather than applications.

Column 2 in Table 21–3 notes the types of block polymers that were claimed, implied, or suggested by the sequential addition followed in the patents. Only in several cases did the authors voluntarily acknowledge that their product contained block polymers as only one of the components of a mixture. For the most part, the notations are written only to show the experimental sequential addition of olefins that was followed in the patent

TABLE 21–3

Selected Examples of Reported Block Polymerizations[a]

Expected or claimed structure[b]	Ref.	Catalyst[c]	Polymerization	Comments[d]
$(E-P)_n$	12	$AlEt_2Cl + TiCl_3$	Slurry process at 50°C	Toughened polymer claimed
P–E–P and $(P-E)_n$	13	$AlEt_2Cl + TiCl_3$	Slurry process at 50°C	Up to 10% E in product
$P-PE-(PE)'$	14	$AlEt_2Cl + TiCl_3$	Second block contains more propylene than third block	Up to 5% E in product
P–E and P–EP	15	$AlEt_2Cl + TiCl_3$	Combined slurry and vapor phase process	Toughened polymers claimed
P–E, P–EP	16, 17	$AlEt_2Cl + TiCl_3$	Slurry polymerization	Toughened polymers claimed
B–S and B–BD	18	$AlEt_2Cl + TiCl_3$	Two stage polymerization (50° and 90°C)	Products used as oil-soluble viscosity index improvers
P–E	19	$AlEt_2Cl + TiCl_3$	Slurry polymerization at 70°C	Up to 4% E in product
P–EP and $(P-E)_3P$	20	$AlEt_2Cl$ or $AlEtCl_2$ + $Si(OEt)_4 + TiCl_3$	—	—
P–E	21	$AlEt_2Cl + TiCl_3$	Slurry polymerization at 70°C	Some propylene remaining after 1st stage; up to 6% E

TABLE 21-3 (*Continued*)

Expected or claimed structure[b]	Ref.	Catalyst[c]	Polymerization	Comments[d]
(P–E)$_n$	22	Al$_2$Et$_4$SO$_4$ + TiCl$_3$ and modifiers	Slurry polymerization at 50°C	Up to 7% E present in product
(EP–E)$_n$EP (EP–E–EP–P)$_n$–EP	23, 24	Al(C$_{12}$H$_{25}$)$_3$ + VOCl$_3$	—	Stereoblock elastomers
P–EP	25	AlEt$_2$Cl + TiCl$_3$	Slurry polymerization at 60°C	Up to 10% E; higher impact strengths claimed
P–EP–P	26	AlEt$_2$Cl + TiCl$_3$	Low temp. polymerization ~ 0°C	Fractionated material has improved properties
(P–E)$_{10}$	27	AlEt$_2$Cl + TiCl$_3$	N$_2$ purges between segments	Mixtures of homopolymers and copolymers reported
P–EB	28	AlEt$_2$Cl + TiCl$_3$	Combined slurry and vapor phase	Up to 20% E in product
P–E	29	AlEt$_3$ or AlEt$_2$Cl + TiCl$_3$	Continuous 2-stage polymerization	Up to 10% E in product
P–E	30	Modifier + LiAlH$_4$ + TiCl$_3$(1) or AlEt$_2$Cl + TiCl$_3$(2)	Catalysts 1 and 2 were used at 150° and 80°C, respectively	Polyallomers containing up to 5% E; toughened
(P–B)$_n$	31	AlEt$_3$ + TiCl$_3$	Vapor phase polymerization	Almost equal concentration of olefins
EP–E	32	AlEt$_2$Cl + VOCl$_3$	Solution polymerization	Crystallizable plastomers claimed
P–EP	33	AlEt$_2$Cl + TiCl$_3$ + modifiers	Slurry polymerization	Mixture of block polymer and by-products described
E–B	34	Al-i-Bu$_3$ + TiCl$_4$	Slurry polymerization at 90°C	Product was blended with PE to improve its stress cracking resistance
P–EP	35	AlEt$_2$Cl + TiCl$_3$ + Ti(OC$_8$)$_4$	50°–60°C	Higher impact strengths and lower brittleness temp.
P–PE–(PE)′	36	AlEt$_2$Cl + TiCl$_3$	Continuous 3 stage polymerization	Improved clarity and impact strength
PE–(PE)′ where P > E in PE and E >> P in (PE)′	37	AlEt$_3$ + TiCl$_3$	Slurry polymerization	—

[a] In most cases, the described polymerizations were very detailed and required precise procedures which had to be followed.

[b] Segments: E = polyethylene; P = polypropylene; B = poly(1-butene); S = polystyrene; BD = polybutadiene.

[c] TiCl$_3$ was, in most cases, either Al- or Al-alkyl reduced having the approximate composition TiCl$_3$ · xAlCl$_3$ where x = 0.1 to 0.4.

[d] Toughened polypropylenes had improved resistance to impact at low and ambient temperatures (see Chapter 3, Section V and Chapter 20, Section IV,A).

example. If the synthesis was based on kinetic background, this was not generally revealed in the patents.

IV. Conclusion

Have block copolymers that were claimed in Table 21–3 actually been synthesized? If so, in what purity were they obtained? It would be fruitless to attempt an analysis of each patent with the goal of answering these questions. It is, however, instructive to discuss some of the barriers that are inherent in heterogeneous Ziegler–Natta catalyses toward synthesizing block polymers and then to attempt to reach a general conclusion.

Ideally, if we wish to synthesize tailored block polymers from ethylene, propylene, etc., we would want to control the following in the polymer chain: (1) the length (molecular weight) and molecular weight distribution of each segment; (2) the order of segments of specific olefin; (3) the number of segments in the monomer chain; (4) the composition and order of olefins in each segment; and (5) the linearity, stereochemistry, and microstructure of the olefins and diolefins in each segment.

Suppose we wished to synthesize a simple P–E block polymer with the above five controls. The following conditions must take place: (1) all of the Ti centers must become active simultaneously (centers must not become active during the polymerization; (2) all of the Ti centers must stay alive during the polymerization (the propylene polymer chain must remain attached to the Ti centers); (3) all of the Ti centers must have equal activities for polymerizing propylene and ethylene; (4) all of the Ti centers must be equally accessible to the available olefin molecules; (5) the cross-over propagation rates must be high for Ti–propylene centers to be efficiently converted into Ti–polyethylene centers; and (6) the centers must be isotactic-specific for propylene polymerization and polymerize all ethylene molecules to a linear polymer.

Suppose we would want to synthesize a P–EP–P block polymer where EP is a random copolymer segment of ethylene and propylene. In addition to the above six requirements, the following should prevail: (1) the Ti–ethylene/propylene copolymer centers must consume ethylene and propylene at the same rate throughout the copolymerization; (2) the Ti center must be highly isotactic-specific when polypropylene segments are synthesized and should give a random ethylene–propylene segment for the center segment; (3) the activities of individual centers must remain unchanged throughout the entire polymerization; (4) the ethylene–propylene copolymer must remain attached to the Ti center throughout the copolymerization step;

and (5) the cross-over propagation rate must be high for Ti-propylene–ethylene centers to be converted to Ti–polypropylene centers.

Such a specific control is not possible with known heterogeneous Ziegler–Natta catalysts. A number of barriers must be overcome: (1) uneven initiation and uneven growth, leading to wide molecular weight distributions; (2) transfer reactions by metal alkyl (also by H_2, if added); (3) apparent ability of some centers to copolymerize ethylene and propylene but not to homopolymerize propylene; (4) tendency of catalysts that make highly isotactic-crystalline polypropylene to make a blocky EP, while those which produce only partially isotactic-crystalline polypropylene make a random EP copolymer; and (5) diffusion or transport of monomer from the gas cap to active centers, leading to monomer concentration gradients at the active center.

What kinds of block copolymers can one then synthesize from ethylene and α-olefins? Certainly, the rather sophisticated ABA-type (described as "confectionary" by one author) block polymers made by lithium alkyl initiation cannot yet be made with presently known heterogeneous Ziegler–Natta catalysts. However, kinetic studies by Natta, Bier, Kontos, and others established that some chains remain alive for minutes, hours, and perhaps days. Some block polymer chains should therefore be formed with certain catalysts, such as $AlEt_2Cl + TiCl_3$. The chance of forming block polymers decreases when H_2 is added, since chain growth is terminated. Polymerizations at high temperatures, usually above 70°C, lead to increased chain termination by disproportionation and decreased opportunity for block copolymerization.

In only a few cases were fractionations properly done to establish the compositions of these reported block polymers. In two examples (38, 39), a product from a reported block polymerization containing about 15 to 16% ethylene was found to contain between 70 and 80% isotactic polypropylene. In one of these examples, Slonaker, Combs, and Coover (38) isolated a fraction that had a melting point of isotactic polypropylene, was soluble even at low temperatures, but which contained 50% ethylene. It is possible that this product had the structure P–EP but other species can be envisaged.

The reported block polymers from propylene and ethylene, collected in Table 21–1, are most likely mixtures of largely isotactic polypropylene, some nearly pure polyethylene, and either random, block, or tapered co-polymers, depending on the exact method of synthesis. They are, at best, present in low concentrations, perhaps in the zero to 20% range. The observed differences in the performance of the product, relative to an artificial mixture of homopolymer and a copolymer, could be due to the fact that the latter does not contain the many structures that are uniquely synthesized in a block polymerization. The reader will more appreciate this point if he takes time to read a few of these patents, which describe detailed and precise recipes for the synthesis of the improved product.

References

1. M. Szwarc, *Nature (London)* **178**, 1168 (1956).
2. G. Bier and G. Lehmann, *in* "Copolymerization" (G. E. Ham, ed.), Chapter IV B. Wiley (Interscience), New York, 1964.
3. T. G. Heggs, *in* "Block Copolymers" (D. C. Allport and W. H. Janes, eds.), Chapters 4 and 8D. Wiley, New York, 1973.
4. G. Natta, *J. Polym. Sci.* **34**, 531 (1959).
5. G. Bier, A. Gumboldt, and G. Lehmann, *Plast. Inst., Trans. J.* **28**, 98 (1960).
6. G. Bier, G. Lehmann, and H. J. Leugering, *Makromol. Chem.* **44/46**, 347 (1961).
7. G. Bier, *Kunststoffe* **48**, 354 (1958).
8. G. Bier, W. Hoffman, G. Lehmann, and G. Seydel, *Makromol. Chem.* **58**, 1 (1962).
9. G. Bier, *Angew. Chem.* **73**, 186 (1961).
10. E. G. Kontos, E. K. Esterbrook, and R. D. Gilbert, Belgian Patent 609,239; *J. Polym. Sci.* **61**, 69–82 (1962).
11. E. G. Kontos, E. K. Esterbrook, and R. D. Gilbert, *J. Polym. Sci.* **61**, 69 (1962).
12. Great Britain Patent 986,189, March 17, 1965, Farbwerke Hoechst A.G.
13. J. A. Bond, M. Clark, and L. S. Rayner, Great Britain Patent 970,478, September 23, 1964, Imperial Chemical Industries, Ltd.
14. Mitsui Toatsu Kagaku K.K., Japan Patent 20621/69, September 4, 1969, see Derwent 35711Q.
15. Great Britain Patent 1,006,469, October 6, 1965, Montecatini.
16. A. A. Harban, U.S. Patent 3,345,431, October 3, 1967, Phillips Petroleum Company.
17. S. Renaudo, U.S. Patent 3,358,056, December 12, 1967, Phillips Petroleum Company.
18. T. H. Shephard, U.S. Patent 3,509,056, April 28, 1970, Princeton Chemical Research.
19. Great Britain Patent 1,074,383, July 5, 1967, Rexall Drug and Chemical Company.
20. J. A. Casey, U.S. Patent 3,256,237, June 14, 1966, Sun Oil Company.
21. K. M. Khelghatian, L. D. Hague, and J. L. Jezl, Great Britain Patent 994,416, June 10, 1965, Sun Oil Company.
22. Japanese Patent 17981/68, July 30, 1968, Toyo Rayon Company, see Derwent 10,613Q.
23. E. G. Kontos, U.S. Patent 3,378,606, April 16, 1968, Uniroyal, Inc.
24. Great Britain Patent 957,070, May 6, 1964, United States Rubber.
25. G. A. Short, U.S. Patent 3,354,239, November 21, 1967, Shell Oil Company.
26. H. L. Hassell, U.S. Patent 3,378,608 April 16, 1968, Shell Oil Company.
27. R. Holzer and K. Mehnert, U.S. Patent 3,262,992, July 26, 1966.
28. D. E. Hostetler, French Patent 1,352,024, February 7, 1964, W. R. Grace and Company.
29. A. M. Jones, J. A. Planchard, and R. A. Speed, U.S. Patent 3,378,607, April 16, 1968, Esso Research and Engineering Company.
30. H. J. Hagemeyer, Jr. and M. B. Edwards, U.S. Patent 3,529,037, September 15, 1970.
31. Great Britain Patent 1,045,221, October 12, 1966, E. I. duPont de Nemours and Company.
32. Great Britain Patent 1,175,670, December 23, 1969, Dunlop Company.
33. Canadian Patent 802,307, December 24, 1968, Avisun Corporation.
34. L. D. Hoblit and C. P. Strange, U.S. Patent 3,660,530, May 2, 1972, Dow Chemical Company.

35. I. Leibson and D. H. Getz, U.S. Patent 3,338,986, August 29, 1967, Rexall Drug and Chemical Company.
36. M. Sennari and T. Nakajima, U.S. Patent 3,670,053, June 13, 1972, Mitsubishi Petroleum Company.
37. Belgian Patent 687,748, March 16, 1967, International Synthetic Rubber Company.
38. D. F. Slonaker, R. L. Combs, and H. W. Coover, *J. Macromol. Sci., Chem.* **1**, 539 (1967).
39. Great Britain Patent 1,099,853, January 18, 1968, Mitsubishi Petrochemical Company, Ltd.; A. Schneider, K. M. Khelghatian, L. D. Hague, and J. L. Jezl, Great Britain Patent 889,230, January 25, 1960, Sun Oil Company.

22

Other Uses of the Ziegler–Natta Catalyst

I. Introduction

It is clear that for many of these reactions, especially radical and cationic polymerizations, mechanisms other than those described in Chapters 13 through 17 for Ziegler–Natta polymerizations are operating, i.e., one of the components or the product or by-product of both components acts as a source of radicals, cations, or anions. Yet, many of the reactions, isomerization, hydrogenation, and oligomerization, appear to share common mechanistic features that occur in the polymerizations of olefins and dienes with Ziegler–Natta and metal alkyl-free catalysts.

II. Radical and Cationic Polymerizations

Some vinyl and epoxide monomers have been polymerized with specific Ziegler–Natta catalysts. Paradoxically, most of these monomers are poisons, even when added to Ziegler–Natta catalysts in small concentrations. For example, vinyl chloride has been polymerized to high molecular weight products in the presence of the $AlEt_2OEt-VO(O_2C_5H_7)_2$ catalyst (1). But the mechanism was free radical. Natta (2) showed that the polymerizability of vinyl ethers increased and that of ethylene decreased when the metal alkyl component (combined with Cp_2TiCl_2) was changed from $AlEt_2Cl$ to $AlEtCl_2$ to $AlCl_3$.

Epoxide monomers, such as propylene oxide, epichlorohydrin, and styrene oxide, were polymerized with mixtures of metal alkyls ($AlEt_3$, $AlEt_2Cl$, or

AlEtCl$_2$) and transition metal salts (VCl$_4$, VCl$_3$, TiCl$_4$, or TiCl$_3$) (3). The products had low molecular weights, indicating a cationic mechanism.

Gaylord and co-workers (4–6) noted that the polymerization of isoprene with AlEtCl$_2$ to give cyclopolyisoprene products was markedly accelerated by the addition of TiCl$_4$. The active species were suggested to be cations formed by the dissociation of the reaction product of AlEtCl$_2$ and TiCl$_4$.

Aluminum alkyl halides act as cationic initiators even in the absence of transition metal salts. Tinyakova (6a) describes the polymerization of isoprene, isobutylene, butadiene, propylene, and styrene in the presence of AlEt$_2$Cl at $-78°$ to $+20°$C. The addition of water or alkyl halides often increased activities. Attempts to polymerize 3-methyl-1-butene and 3,3-dimethyl-1-butene with the syndiotactic-specific catalyst AlEt$_2$Cl-VCl$_4$ were not successful (6b). It appeared that the catalyst functioned as a cationic initiator, and an isomerization–polymerization had occurred in which a 3,2 hydride or a 3,2 methide shift took place, as described earlier by Kennedy and co-workers (6c) for cationic initiation.

III. Metathesis of Acyclic and Cyclic Olefins

The metal center of certain Ziegler–Natta catalysts has the ability to bring two unsaturated pairs of carbon atoms together to form a four-center intermediate, which dissociates by breaking either set of opposite bonds (metathesis), as shown in Eq. 22–1. When the starting olefins are acyclic, new

$$(22-1)$$

olefins are formed, while cycloolefins are converted to high molecular weight products. Obviously, adding acyclic olefins to the latter polymerization will produce lower molecular weight products.

The most important example of polymerization was disclosed in 1964 by Natta, Dall'Asta, and Mazzanti (7), who synthesized 100% *cis*-1,5-polypentenamer with MoCl$_5$ + AlEt$_3$ and 100% *trans*-1,5-polypentenamer with WCl$_6$ + AlEt$_3$ (see Chapter 19). A homogeneous catalyst system for olefin metathesis was disclosed in 1967 by Calderon, Chan, and Scott (8), who combined AlEtCl$_2$, WCl$_6$, and ethanol. Al-i-Bu$_3$ + ReCl$_4$ was effective, producing 94% cis-1,4 product (8a). Another homogeneous catalyst that appeared to be versatile was described recently by Doyle (8b), namely AlMeCl$_2$ + R$_4$N$^+$[M(CO)$_5$X]$^-$, where R is an alkyl, M is a group VI metal, and X is a halide or carboxylate.

Metathesis opens the door to synthesis of novel and otherwise hard to get polymers. For example, Furukawa and Mizoe (9) prepared a 1:2 copolymer of ethylene and butadiene by ring opening of $cis,trans$-cyclodeca-1,5-diene, as shown in Eq. 22–2. MoCl$_5$ was more effective than WCl$_6$ and AlEtCl$_2$ >

$$\longrightarrow \ [(CH_2{-}CH_2){-}(CH_2{-}CH{=}CH{-}CH_2)_2{-}]_n \quad (22\text{–}2)$$

AlEt$_2$Cl > AlEt$_3$. Eq. 22–3 is another example (9a).

$$\xrightarrow[\text{WCl}_5]{\text{AlEtCl}_2} \ {-}[CH_2{-}CH{=}CH{-}CH_2{-}CH_2{-}CH{-}CH{-}CH_2]_n \quad (22\text{–}3)$$

50% conversion   MW ≃ 2500 to 6500

Another novel use of metathesis was reported by Ast and Hummel (9b), who degraded polybutadiene into C$_{8-16}$ hydrocarbons by metathesis with 2-pentene in the presence of WCl$_6$ and EtAlCl$_2$. For other examples, see Chapter 19, Section VI.

Much work has been done to enhance the scope of these catalyses. The reader is referred to ref. 10 for leading references and for reviews of various aspects of this catalyst. It should be noted that metathesis of olefins by non-Ziegler–Natta catalysts was discovered independently by other workers, namely Bailey and Banks of Phillips Petroleum Company and Euleterio of Du Pont. Bailey (11) reviews the earlier work on olefin disproportionation. A more recent review presents theoretical views on the mechanism (12) and indicates that olefin metathesis was first disclosed by E. F. Peters $et\ al.$ in 1957.

IV. Oligomerization of Olefins and Acetylenes

Many examples of oligomerization of ethylene, α-olefins, and dienes have been reported. Only a few selected examples are described here.

Wilke (13) reported the cyclotrimerization of butadiene to form $trans,trans,trans$-1,5,9-cyclododecatriene in the presence of a Ziegler–Natta catalyst at 80° to 130°C, as shown in Eq. 22–4. In the presence of ϕ_3P and

$$3\ CH_2{=}CH{-}CH{=}CH_2 + Ni\,(O_2CR)_2 + AlEt_3 \longrightarrow \qquad\qquad (22\text{--}4)$$

chloride ion, cyclodimerization to 1,5-cyclooctadiene and 1,4-polymer took place.

Misono and co-workers (14) showed that propiolonitrile was trimerized to the aromatic product trimellitonitrile, as shown in Eq. 22–5.

$$3\ HC{\equiv}C{-}CN + Co\,(acac)_3 + AlEt_3 \longrightarrow \qquad\qquad (22\text{--}5)$$

Furlani and co-workers (15) trimerized phenylacetylene and isopropenyl-acetylene with $TiCl_4$–$AlEt_2Cl$ catalyst to a mixture of polymer and tri-isopropenyl benzenes (1,2,4- and 1,3,5-isomers) for the latter monomer but exclusively to 1,2,4- and 1,3,5-triphenylbenzene for the former monomer. The composition of the product was dependent on the reaction variables, e.g. Al/Ti ratios (for other examples, see Chapter 19, Section X).

Onsager, Wang, and Blindheim (16) reported the dimerization of ethylene, propylene, and butene in the presence of a mixture of π-tetramethylcyclo-butadiene–nickel dihalides, alkylaluminum halides, and triarylphosphine (a homogeneous system). Higher α-olefins were obtained, including 1-butene from ethylene, C_6 olefins from ethylene and 1-butene, etc.

Isoprene was predominantly dimerized in the presence of $AlEt_2Cl$, dimethyl sulfoxide, and titanium di(p-toluene sulfonate) dichloride (17), as shown in Eq. 22–6.

$$CH_2{=}C{-}CH{=}CH_2 \xrightarrow{\;90\%\;} \qquad + \quad CH_3{-}\overset{\displaystyle CH_3}{\underset{}{C}}{=}CH{-}CH_2{-}CH{=}CH{-}\overset{\displaystyle CH_3}{\underset{}{C}}{=}CH_2$$
$$\underset{\displaystyle CH_3}{|}$$

$$(22\text{--}6)$$

Other examples include: (1) butadiene was converted to 3-methyl-1,4,6-heptatriene in high yield with LiBu and Co(acac)$_2$ (18); (2) butadiene and ethylene were converted to a mixture of mostly hexenes and pentenes in the presence of Ni(acac)$_2$, silica, and AlEt$_3$ (19); (3) propylene was oligomerized in the presence of ethylaluminum sesquichloride, NiBr$_2$, and phosphoric amides (20); (4) ethylene and propylene were treated with AlH$_2$NR$_2$ and

Ni(acac)$_2$ or Ti(OR)$_4$ to promote dimerization (21); and (5) C$_6$–C$_{18}$ olefins were made in a one-step process using aluminum alkyls and titanium compounds, whereby additives increased the yield of straight chain olefins and suppressed polymerization (22).

Cannell (23) reported the synthesis of vinylcyclobutane when ethylene and 1,3-butadiene were catalyzed with organotitanium compounds alone or with the Ziegler–Natta catalyst Ti(O-n-Bu)$_4$ + AlEt$_3$.

Kuhlein and Clauss (24) reported the oligomerization of ethylene at $-70°$C with a number of methyltitanium (IV) compounds, including CH$_3$Ti(O-i-Pr)Cl$_2$, C$_3$ to C$_{33}$ olefins were obtained with CH$_3$TiCl$_3$. No aluminum alkyl was present. Ethylene has been dimerized to 1-butene (89% yield) with only Ni(C$_5$H$_5$)$_2$ (25).

Langer reported that a mixture of AlEtCl$_2$ + TiCl$_4$ oligomerizes ethylene in high selectivity at 25° to 70°C to linear α-olefins (26, 27). These olefins have a number average molecular weight from about 70 to more than 300. The catalyst functions only if alkylation, but not precipitation, is allowed to take place. The process was said to be under development for commercialization. The product olefins can be used as raw materials for the synthesis of high molecular weight alcohols.

Bestian and Clauss (27a) extensively studied the oligomerization of ethylene to low molecular weight olefins in the presence of AlMeCl$_2$–CH$_3$TiCl$_3$ and Al$_2$Me$_3$Cl$_3$–TiCl$_4$ in CH$_2$Cl$_2$ solvent and at low temperatures. The composition of the olefin products could be controlled by varying the temperatures at which CH$_3$TiCl$_3$ forms in the latter catalyst.

Henrici–Olive and Olive (27b) reported that, by systematically changing reaction conditions and ligands on titanium, they were able to prepare C$_6$–C$_{40}$ olefins having predominantly vinylic end groups. (C$_2$H$_5$O)$_3$TiCl$_3$ + AlEtCl$_2$ was used as catalyst. Oligomerization of 1-butene by a cationic mechanism occurred in the presence of acid products formed from (C$_5$H$_5$)$_2$TiCl$_2$ + AlEt$_2$Cl catalyst. When benzene was present, alkylation also occurred (27c). A comprehensive review of this subject by these authors describes the theoretical and experimental features of this subject (27c).

V. Isomerization of Olefins

Certain Ziegler–Natta catalysts can cause isomerization between α-olefins and internal olefins. Yerasova and co-workers (28) established the equilibrium: 1-butene $\rightleftharpoons$ 2-butene (cis + $trans$-isomers) by exposing 1-butene to the AlEt$_3$–CrCl$_3$ catalyst. Aubrey and Barnatt (29) showed the equilibrium: 1-octadecene $\rightleftharpoons$ 2-octadecene when the 1-olefin was treated with AlEt$_3$–

TiCl$_4$. The most interesting examples arise when 2-olefins are exposed to specific Ziegler–Natta catalysts and a polymer forms that has the structure of a 1-olefin. These examples are described in Chapter 19, Section V.

Skinner and co-workers (30) concluded that the *cis*-double bond of the hexa-1-*cis*-4-diene termonomer (ethylene and propylene were other co-monomers) was isomerized to the *trans*-configuration at the polymerization stage but not before. The catalyst Al$_2$Et$_3$Cl$_3$–VOCl$_3$ was used for the terpolymerization. They did not, however, rule out that the formed polymer did not isomerize.

A crystalline polyethylene and an amorphous ethylene–propylene co-polymer were synthesized from propylene with the ternary system VCl$_4$, Fe(acac)$_3$, and AlEt$_3$ in benzene, 1,2-dichloroethane, or chlorobenzene as the solvent (30a). A 1,3-isomerization–polymerization was postulated.

VI. Hydrogenation of Olefins and Unsaturated Polymers

A review by Lapporte (31) describes the main features of this hydrogenation. A typical catalyst is prepared by mixing a metal alkyl (AlEt$_3$ or LiBu) and an organic salt of the transition metal Ni, Co, Fe, or Cr. The ligands attached to the transition metal can be acetylacetonate, 2-ethyl hexanoate, etc. Hydrogenation proceeds from mild to stringent conditions, depending on the substrate to be hydrogenated. In general, the following order of transition metals leads to decreasing activities: Ni > Co > Fe > Cr. When alkenes are hydrogenated, the following decreasing order of reactivity is observed: monosubstituted > unsymmetrical disubstituted > cyclic > symmetrical disubstituted. Some isomerization takes place when H$_2$ is present or absent.

Other unsaturated compounds that can be hydrogenated are dienes, styrene, some functionally substituted olefins (such as maleic anhydride), alkynes, and both unsaturated and aromatic polymers. Aromatic-containing polymers, polystyrene for example, require higher temperatures. Block and homopolymers containing unsaturation have been polymerized with these types of catalysts (32, 33).

Witt and Hogan (34) hydrogenated a terminal vinyl unsaturation in highly linear polyethylene with AlEt$_3$–Ni octoate. They found that the terminal vinyl unsaturation of the polymer had to be isomerized to some internal position before hydrogenation occurred. The hydrogenated polyethylene did not show an increase in melt viscosity when subjected to extreme thermal treatments. Shikata and co-workers (35) hydrogenated cyclohexene to cyclohexane in the presence of $(\pi\text{-}C_5H_5)_2TiCl_2\text{--}n\text{-BuLi}$.

The chemical and physical evidence suggests that Ni(II) is reduced to a diamagnetic $3d^{10}$ Ni(0), probably via unstable EtNi and H–Ni species, as shown in Eq. 22–7. The isomerization reaction may occur during the re-

$$AlEt_3 + (RCO_2)_2Ni \longrightarrow L\!-\!\overset{\displaystyle L}{\underset{\displaystyle L}{\overset{|}{\underset{|}{Ni}}}}\!-\!Et \rightleftharpoons L\!-\!\overset{\displaystyle L}{\underset{\displaystyle L}{\overset{|}{\underset{|}{Ni}}}}\!-\!H + \overset{C}{\underset{C}{\|}} \qquad (22\text{--}7)$$

versible addition of olefins to the Ni–H species (31). Klinedinst and Boudart (36) showed high-spin Fe(II) was formed when Fe(acac)$_3$ and AlEt$_3$ react at $-83°C$. When warmed to room temperature, an active homogeneous hydrogenation catalyst was formed, but no metallic iron particles (even as small as 17 Å) were detected by Mossbauer spectroscopy.

VII. Alkylation of Aromatic Nuclei by Unsaturated Systems

When a polymerization of propylene is carried out in an aromatic solvent, some Ziegler–Natta catalysts can lead to an alkylation side reaction, as shown in Eq. 22–8 (37). Soga and co-workers showed that even in the

$$CH_3\!-\!CH\!=\!CH_2 \xrightarrow[\text{benzene}]{AlEt_2Cl + TiCl_3AA} \text{polypropylene} + \text{isopropylbenzene isomers} \qquad (22\text{--}8)$$

absence of aluminum alkyl, alkylation occurred and was the main reaction. Apparently, different centers are responsible for the alkylation and polymerization.

A cross-linked copolymer of benzene and butadiene was made with Et$_3$AlCl$_3$ and TiCl$_4$ (38).

VIII. Examples of Other Competing Side Reactions

Ketley and Moyer (39) reported the presence of crystalline polyethylene in the product made by polymerization of 3-methyl-1-butene with AlEt$_3$–TiCl$_3$. Reactions with ^{14}C-labeled AlEt$_3$ showed the polyethylene arose from ethylene displaced from AlEt$_3$–TiCl$_3$. When AlEt$_3$ was replaced by Al-i-Bu$_3$, no polyethylene was formed. The relative order in which monomers displace ethylene from AlEt$_3$ is: 3-methyl-1-butene = vinylcyclohexene > 3,3-dimethylheptene = 3,3-dimethyl-1-butene > 1-butene = 1-pentene, 1-hexene, 1-heptene, and 1-octene.

Atarashi (40) showed that when the polymerization is carried out above 60°C, mainly polyethylene formed, but at lower temperatures, copolymerization with 3-methyl-1-butene occurs. Pino and co-workers (41) showed that partial isomerization of 3-methyl-1-pentene occurred during polymerization of 3-methyl-1-pentene with Al-i-Bu_3–$TiCl_4$. Ho and co-workers (42) found that the polymerization of 4-methyl-1-pentene with Al-i-Bu_3–$TiCl_3$ gave about 90% polymer, but isomers of the monomer were also formed. The distribution products from the reaction of $AlEt_3$ and 1-octene were 12.1% $Al(n$-octyl$)_3$, 2.7% octene, 10.5% $Al(3$-methylnonyl$)_3$, 44% 2-ethyl-1-octene, 6.5% $Al($hexadecyl$)_3$, and 24.3% hexadecene (43). Tokuzumi (44) reported that 1-butene was formed during the polymerization of ethylene with $TiCl_3$/ $Ti(OEt)_4$ + $AlEt_3$, and that this 1-butene copolymerized randomly with ethylene.

References

1. W. P. Baker, Jr., *J. Polym. Sci.* **42**, 578 (1960).
2. G. Natta, *Chim. Ind.* (*Milan*) **42**, 1207 (1960).
3. T. Otsu, A. Akimoto, and S. Aoki, *J. Appl. Polym. Sci.* **12**, 1477 (1968).
4. N. G. Gaylord, I. Kossler, B. Matyska, and K. Mach, *J. Polym. Sci., Part A-1* **6**, 125 (1968).
5. V. Stepan, H. Vodehnal, I. Kossler, and N. G. Gaylord, *J. Polym. Sci., Part A-1* **5**, 503 (1967).
6. N. G. Gaylord, *J. Polym. Sci., Part D* **4**, 183–244 (1970).
6a. E. I. Tinyakova, T. G. Zhuravleva, T. N. Kurungina, and N. S. Kirikova, *Dokl. Akad. Nauk SSSR* **144**, No. 3, 592 (1962).
6b. J. Boor and E. A. Youngman, *J. Polym. Sci., Part A-1* **4**, 1861 (1966).
6c. J. P. Kennedy, G. G. Wanless, and J. J. Elliott, *Polym. Prepr., Am. Chem. Soc., Div. Polym. Chem.* **5**, No. 2, 676 (1964).
7. G. Natta, G. Dall'Asta, and G. Mazzanti, *Angew. Chem.* **76**, 765 (1964).
8. N. Calderon, H. Y. Chan, and K. W. Scott, *Tetrahedron Lett.* p. 3327 (1967).
8a. P. Gunther, W. Oberkirch, and G. Pampus, U.S. Patent 3,580,892, May 25, 1971, Farbenfabriken Bayer Aktiengesellschaft.
8b. G. Doyle, *J. Catal.* **30**, 118 (1973).
9. J. Furukawa and Y. Mizoe, *J. Polym. Sci., Part B* **11**, 263 (1973).
9a. C. P. Pinazzi, J. Cattiaux, J. C. Soutif, and J. C. Brosse, *J. Polym. Sci., Part B* **11**, 1 (1973).
9b. W. Ast and K. Hummel, *Naturwissenschaften* **57**, 545 (1970).
10. Symposium on Polymerization and Related Reactions by Metathesis, *164th Am. Chem. Soc., Div. Polym. Chem., Polym. Prepr.* **13**, No. 2, 874–902 (1972).
11. G. C. Bailey, *Catal. Rev.* **3**, 37–60 (1969).
12. F. D. Mango, *Coord. Chem. Rev.* **15**, 109 (1975).
13. G. Wilke, *Angew. Chem., Int. Ed. Engl.* **2**, 105 (1963).
14. A. Misono, H. Noguchi, and S. Noda, *J. Polym. Sci., Part B* **4**, 985 (1966).
15. A. Furlani, G. Moretti, and A. Guerrieri, *J. Polym. Sci., Part B* **5**, 523 (1967).
16. O. T. Onsager, H. Wang, and U. Blindheim, *Helv. Chim. Acta* **52**, Fasc. 1, 187–250 (1969).
17. J. Itakura and H. Tanaka, *Makromol. Chem.* **123**, 274 (1969).
18. E. A. Zuech, U.S. Patent 3,624,175, November 30, 1971, Phillips Petroleum Company (see also U.S. Patent 3,541,176).
19. T. Hill, U.S. Patent 3,663,451, May 16, 1972, B. P. Chemicals, Ltd.

20. W. Herwig, U.S. Patent 3,507,930, April 21, 1970, Farbwerke Hoechst.
21. S. Cesca, W. Marconi, and M. Santostasi, *J. Polym. Sci., Part B* **7**, 547 (1969).
22. K. Izumi, *Chem. Eng. (N.Y.)* **77**, 71 (1970).
23. L. G. Cannell, *J. Am. Chem. Soc.* **94**, 6867 (1972).
24. K. Kuhlein and K. Clauss, *Makromol. Chem.* **155**, 145 (1972).
25. M. Tsutsui and T. Koyano, *J. Polym. Sci., Part A-1* **5**, 681 (1967).
26. A. W. Langer and H. T. White, U.S. Patent 3,441,630, April 29, 1969, Esso Research and Engineering Company.
27. A. W. Langer, Jr., *J. Macromol. Sci., Chem.* **4**, 775 (1970); *Chem. & Eng News* **50** (No. 15), 16, (1972).
27a. H. Bestian, K. Clauss, H. Jensen, and E. Prinz, *Angew. Chem., Int. Ed. Engl.* **2**, No. 1, 32 (1963); H. Bestian and K. Clauss, *ibid. p.* 704.
27b. G. Henrici-Olive and S. Olive, *Angew. Chem., Int. Ed. Engl.* **9**, 243 (1970).
27c. G. Henrici-Olive and S. Olive, *Adv. Polym. Sci.* **15**, 1–30 (1974).
28. Ye. L. Yerasova, B. A. Krentsel, N. A. Pokatilo, and A. V. Topchiev, *Polym. Sci. USSR (Engl. Transl.)* **4**, 558 (1963).
29. D. W. Aubrey and A. Barnatt, *J. Polym. Sci., Part A-1* **5**, 1191 (1967).
30. G. A. Skinner, M. Viney, and S. R. Wallis, *Polymer* **13**, 242 (1972).
30a. S. Yuguchi and M. Iwamoto, *J. Polym. Sci., Part B* **2**, 1035 (1964).
31. S. J. Lapporte, *Ann. N.Y. Acad. Sci.* **158**, 510 (1969).
32. O. Johnson, U.S. Patent 3,415,759, December 10, 1968, Shell Oil Company.
33. T. Yoshimoto, S. Kaneko, T. Narumiya, and H. Yoshi, U.S. Patent 3,625,927, December 7, 1971, Bridgestone Tire Co., Ltd.
34. D. R. Witt and J. P. Hogan, *J. Polym. Sci., Part A-1* **8**, 2689 (1970).
35. K. Shikata, K. Nishino, K. Azuma, and Y. Takegami, *Kogyo Kagaku Zasshi* **68**, 358 (1965).
36. K. A. Klinedinst and M. Boudart, *J. Catal.* **28**, 322 (1973).
37. K. Soga, T. Keii, and A. Takahashi, *J. Polym. Sci., Part B* **3**, 1075 (1965).
38. R. G. Hay, L. F. Meyer, and C. M. Selwitz, U.S. Patent 3,509,119, April 28, 1970, Gulf Research and Development Company.
39. A. D. Ketley and J. D. Moyer, *J. Polym. Sci., Part A* **1**, 2467 (1963).
40. Y. Atarashi, *J. Polym. Sci., Part A-1* **8**, 3359 (1970).
41. P. Pino, G. P. Lorenz, and L. Lardicci, *Chim. Ind. (Milan)* **42**, 712 (1960).
42. W. Ho, Yu. Ya. Gol'dfarb, and B. A. Krentsel, *Izv. Akad. Nauk SSSR, Ser. Khim.* No. 6, p. 1391 (1970).
43. E. Perry and H. A. Ory, *J. Org. Chem.* **25**, 1685 (1960).
44. T. Tokuzumi, *Chem. High Polym.* **25**, No. 283 (1968); see *Kobunshi Kagaku* **25**, 721–725 (1968).

23

Final Comments and Outlook

The Ziegler–Natta catalyst has offered many challenges to those scientists and engineers who had an opportunity to work with it. There was the opportunity in catalyst research aimed at more effective systems that could eventually be used in commercial plants. There was the opportunity to study the fundamental aspects of the physical chemistry of these new stereochemical structures. Characterization and evaluation studies were made to assess these new materials for practical applications. Others directly investigated the potential use in plastics, fibers, elastomers, etc. New processes for future plants were designed. Continued studies of the mechanisms of polymerizations were made, not only to achieve a better understanding for science's sake but also to develop more effective catalyst–process systems. The fields of catalysis and polymer science have profited greatly. This book relates many of these accomplishments.

But the commercial plants are already built and are producing large quantities of polymers. Literally thousands upon thousands of journal papers and patents have been published on various aspects of Ziegler–Natta catalysts and polymerizations. What is left to do for the future investigator?

Certainly, there is no need for "another example" of a catalyst or a kinetic analysis. This book shows that enough examples are known. Yet the worker, while he cannot ignore the past work, cannot allow himself to be overwhelmed by the large volume of patents and journal papers and feel that nothing significant can be done. True, the Ziegler–Natta catalyst has joined ranks with cationic, radical, and anionic initiators as a major means of initiating polymerizations. It is true that the major waves of discovery for these systems have probably gone by. But work is being continued, and new significant findings are being made. And it will be the same for Ziegler–Natta catalysts. The investigator should not feel restricted by the present conclusions if his intuitive feelings suggest a different course. Many of the

611

accepted views concerning Ziegler–Natta catalysis are not unequivocally proven; some are merely accepted. Novel and convincing experiments are still needed.

Polymer producers will continue to build new plants that will require more efficient processes. New grades of plastics and elastomers will be required for novel applications. These are the "built-in" problems that will have to be solved.

The solution of unsolved mechanistic problems will significantly contribute not only to a better understanding of the Ziegler–Natta catalyst but also in its more effective application. Powerful analytical tools, such as ^{13}C-NMR, pyrolysis-GLC and GPC, coupled favorably with conventional infrared, X-ray, and differential thermal calorimetry methods, have already allowed giant leaps toward the understanding of the detailed structure of the polymer chains. If these studies unequivocally define the nature of inversions in the polymer chains, then we must reexamine the earlier mechanistic proposals, make the appropriate changes, or invent completely new ones. For example, an inversion leading to structural units I or II would have considerable mechanistic significance. Structure I would require the polymer

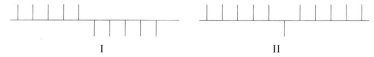

chain to be formed at two sites that had opposite stereochemical abilities or at the same site, that during the polymerization, underwent a change in its stereochemical control ability. In structure II, the inversion could be accidental, because the configurations of the remaining unit are the same as those of the early block. Definitive knowledge of polymer chain structure at this level will have an important bearing on the nature of the active center and the nature of the propagation step.

There is a great need for a site counting method that will instantaneously measure all centers, whether they are in an active or inactive state. Active or potentially active transition metal species, be the active ligand a hydride, an alkyl, a low molecular weight chain, or a high molecular weight chain, must be accounted for. With such a diagnostic tool on hand, the worker can more definitively probe the problems of rate decline and differentiate between the contributions of loss of centers vs. diffusion barriers with increased yield of polymer. Should this data allow better tailoring of catalysts, it might permit direct and fine control over molecular weight distribution for the first time.

It is unfortunate that, in spite of the magnitude of factual information on diene catalysts, we cannot make unequivocal assignments of the driving forces leading to the different microstructures. This is a situation that will undoubtedly be resolved in the future. The continued use of metal alkyl-free catalysts as simpler models and perhaps the subjection of both types of catalysts to the same diagnostic probes should both be fruitful.

As far as novel forms of the Ziegler–Natta catalysts, the author believes that the following would be significant accomplishments: (1) a highly active soluble or a supported catalyst that would be highly isotactic-specific. The latter would be applicable for high-yield, nondeashing processes; (2) a soluble catalyst that would have "living" character comparable to that of the lithium alkyls. This would permit synthesis of block polymers from olefins and dienes; and (3) probably, new metal alkyl-free catalysts will be found and developed that will offer distinct advantages in commercial processes.

Supplementary Bibliography

Following is a bibliography from 1972–October 1978, which updates the bibliographies at the end of each chapter. This bibliography was acquired using the Lockheed Information Services Dialog system to search the on-line Chemical Abstracts data bases. We wish to thank Lockheed Information Services and Chemical Abstracts, who were kind enough to give us permission to print this.

We also express our gratitude to Shirley Thompson and Jean Richards of the Shell Development Library for their excellent assistance in acquiring and assembling the update.

How to read the bibliography:

Chemical abstracts identification:
    080 = volume number
        20 = issue number
            108933X = abstract number
CA/080/20/108933X   Journal: J. Polym. Sci., Polym. Chem. Ed. ⎫ Source
Year of Publication ⟶ Publ: 74   Series: 12   Issue: 1   Pages: 11-20 ⎭
    Title of article ⟶ Simultaneous polymerization and formation of polyacetylene film on the surface of a concentrated soluble Ziegler-type catalyst solution.

CA08910075662P   Journal: Japan.   Kokai   Publ: 780427
Pages: 7 pp.
   Polymerization catalysts for ethylene
   Patent No: 78 47484

CA08910075661N   Journal: Japan.   Kokai   Publ: 780427
Pages: 8 pp.
   Polymerization catalysts for ethylene
   Patent No: 78 47483

CA08910075660M   Journal: Japan.   Kokai   Publ: 780427
Pages: 6 pp.
   Polymerization catalysts for olefins
   Patent No: 78 47481

CA08910075496N   Journal: Yuki Gosei Kagaku Kyokaishi
Publ:. 78'   Series: 36   Issue: 3   Pages: 237-41
Language: Japan
   Reinvestigation of the three-component catalysts system
   EtAlCl2-TiCl3-((CH3)2N)3PO for the stereospecific polymerizat-
   ion of propylene

CA08908066088D   Journal: Japan.   Kokai   Publ: 780204
Pages: 6 pp.
   Titanium trichloride alpha olefin polymerization catalyst
   Patent No: 78 12796

CA08908060753V   Journal: Kompleks. Metallorgan. Katalizatory
Polimerizatsii Olefinov   Publ: 77   Issue: 6   Pages:
98-103   Language: Russ
   Catalytic activity and stereospecificities of the effect of
   complex vanadium catalysts in the polymerization of isoprene

CA08908060515U   Journal: U.S.S.R.   Publ: 780530
   Polypropylene
   Patent No: 608810

CA08908060263K   Journal: Ger.   Offen.   Publ: 780601
Pages: 30 pp.
   Titanium trichloride polymerization catalyst
   Patent No: 2752536

CA08908060260G   Journal: Japan.   Kokai   Publ: 780329
Pages: 21 pp.
   Catalysts for manufacture of highly crystalline polyolefins
   Patent No: 78 33289

CA08908060255J   Journal: Brit.   Publ: 770901   Pages: 4
pp.
   The production of supported Ziegler catalyst components
   Patent No: 1484254

CA08908060110H   Journal: Kompleks. Metallorgan. Katalizatory
Polimerizatsii Olefinov   Publ: 77   Issue: 6   Pages:
132-5   Language: Russ
   Side reactions during the formation of high-molecular-weight
polypropylene on a titanium trichloride-0.3 aluminum chloride
(TiCl3.0.3AlCl3)-diethylaluminum chloride catalytic system

CA08906044425J   Journal: Japan.   Kokai   Publ: 780227
Pages: 14 pp.
   Polymerization catalysts for ethylene
   Patent No: 78 21286

CA08906044424H   Journal: Japan.   Kokai   Publ: 780303
Pages: 8 pp.
   Polymerization catalysts for olefins
   Patent No: 78 23383

CA08906044422F   Journal: Japan.   Kokai   Publ: 780111
Pages: 14 pp.
   Polymerization catalysts for propylene
   Patent No: 78 02583

CA08904025089M   Journal: Japan.   Kokai   Publ: 780217
Pages: 5 pp.
   Polymerization catalysts for .alpha.-olefins
   Patent No: 78 17684

CA08704023791U   Journal: Osaka Kogyo Gijutsu Shikensho Kiho
Publ: 76   Series: 27   Issue: 4   Pages: 224-31
Language: Japan
   Addition effect of n-tributyltin hydride to Ziegler catalyst
on the 1-butene polymerization

CA08904024918U   Journal: J. Polym. Sci.. Polym. Lett. Ed.
Publ: 78   Series: 16   Issue: 6   Pages: 303-8
  Observations   of   the   fundamental   particles   in   a
titanium-trichloride-based polymerization catalyst by electron
microscopy

CA08902006834J   Journal: Japan.   Kokai   Publ: 780322
Pages: 5 pp.
  .alpha.-Olefin polymerization catalyst and production of the
polymer
  Patent No: 78 30493

CA08902006833H   Journal: Fr.   Demande   Publ: 770930
Pages: 11 pp.
  Ziegler catalyst for the polymerization of olefins
  Patent No: 2342993

CA08902006629W   Journal: Kobunshi   Ronbunshu   Publ: 78
Series: 35   Issue: 3   Pages: 157-60   Language: Japan
  Effect   of   hexamethylphosphoric   triamide   on   the
polymerization   of   propylene   with   diethylaluminum
chloride-titanium trichloride catalyst

CA08826198372F   Journal: J.   Chem.   Phys.   Publ: 78
Series: 68'   Issue: 5   Pages: 2337-51
  Theoretical   study   on   a   reaction   pathway   of
Ziegler-Natta-type catalysis

CA08826191805R   Journal: Braz.   Pedido PI   Publ: 771101
Pages: 58 pp.
  Solid catalyst for the polymerization of olefins and the
production of olefinic polymers with high crystallinity
  Patent No: 77 01480

CA08826191803P   Journal: Ger.   Offen.   Publ: 780413
Pages: 48 pp.
  Catalyst components for olefin polymerization
  Patent No: 2744559

CA08826191801M   Journal: Ger.   Offen.   Publ: 780330
Pages: 29 pp.
  Stereoregular polymers or copolymers of olefins
  Patent No: 2743415

CA08826191799S   Journal: Japan.   Kokai   Publ: 780111
Pages: 8 pp.
  Polymerization catalysts for olefins
  Patent No: 78 02580

CA08826191797Q   Journal: Japan.   Kokai   Publ: 780126
Pages: 6 pp.
  Polymerization catalysts for olefins
  Patent No: 78 08685

CA08826191791H   Journal: Japan.   Kokai   Publ: 771027
Pages: 8 pp.
  Activated catalysts for polymerization of propylene with
other olefins
  Patent No: 77127994

CA08826191590S   Journal: Dokl. Bolg. Akad. Nauk   Publ: 77
Series: 30   Issue: 12   Pages: 1721-4
  On obtaining high-molecular-weight polyethylene by means   of
a thallium trichloride-modified Ziegler catalyst

CA08826191585U   Journal: Vysokomol. Soedin.. Ser. A   Publ:
78   Series: 20   Issue: 4   Pages: 927-35   Language:
Russ
  Molecular-weight distribution of isotactic polypropylene
prepared under the conditions of "quasiliving" polymerization

CA08826191557M   Journal: J.   Chem.   Soc..   Faraday Trans.   1
Publ: 77   Series: 73   Issue: 11   Pages: 1738-44
  Kinetic studies of olefin polymerization on a supported
Ziegler catalyst

CA08826191546G   Journal: Izv. Sev.-Kavk. Nauchn. Tsentra
Vyssh. Shk.. Ser. Estestv. Nauk   Publ: 77   Series: 5
Issue: 3   Pages: 62-3   Language: Russ
  Polymerization of propylene in the presence of hydrogen at
different temperatures

CA08824170795X   Journal: Ger.   Offen.   Publ: 780330
Pages: 35 pp.
  Polymers or copolymers of olefins with at least 3 carbon
atoms. and catalysts for their manufacture
  Patent No: 2743366

CA08824170794W   Journal: Ger.   Offen.   Publ: 780330
Pages: 89 pp.
  Polymerization catalysts
  Patent No: 2742586

CA08824170789Y   Journal: Japan.   Kokai   Publ: 780111
Pages: 7 pp.
  Polymerization catalysts for olefins
  Patent No: 78 02584

CA08822153294R   Journal: Ger.   Offen.   Publ: 780302
Pages: 31 pp.
  Olefin polymers or copolymers
  Patent No: 2738627

CA08822153285P   Journal: Ger.   Offen.   Publ: 780223
Pages: 40 pp.
  Polyolefin manufacture
  Patent No: 2737362

CA08822153278P   Journal: Japan.   Kokai   Publ: 771214
Pages: 20 pp.
  Polymerization of ethylene or .alpha.-olefins
  Patent No: 77150491

CA08822153275K   Journal: Ger.   Offen.   Publ: 780119
Pages: 21 pp.
  Violet titanium trichloride and its use
  Patent No: 2731241

CA08822153273H   Journal: Japan.   Kokai   Publ: 771128
Pages: 5 pp.
  Polymerization catalysts for olefins
  Patent No: 77142790

CA08820137212G   Journal: Ger.   Offen.   Publ: 780309
Pages: 36 pp.
  .alpha.-Olefin polymers or copolymers
  Patent No: 2739608

CA08820137197F   Journal: Japan.   Kokai   Publ: 771021
Pages: 11 pp.
  Activation of catalysts for polymerization of propylene
  Patent No: 77125591

CA08820137196E   Journal: Belg.   Publ: 770404   Pages: 8
pp.
  Titanium trichloride particles useful in the   stereospecific
polymerization of .alpha.-olefins
  Patent No: 846911

CA08820137195D   Journal: Ger.   Offen.   Publ: 770929
Pages: 54 pp.
  Catalyst for manufacturing polyolefins
  Patent No: 2711300

CA08818122456R   Journal: U.S.S.R.   Publ: 780205
  Preparation of high-molecular-weight cis-1.4-polyisoprene
  Patent No: 438278

CA08818122037E   Journal: Vysokomol. Soedin.. Ser. B   Publ:
78   Series: 20   Issue: 1   Pages: 73-7   Language: Russ
  Degree of ordering of the supramolecular structure and
strength of polyethylene synthesized on supported Ziegler
catalysts

CA08818121937M   Journal: Japan.   Kokai   Publ: 771214
Pages: 24 pp.
  Polymerization   catalyst   component   for   ethylene   or
.alpha.-olefin polymerization
  Patent No: 77150396

CA08702006669T   Journal: Japan.   Kokai   Publ: 770308
Pages: 9 pp.
  Polymerization catalysts for propylene
  Patent No: 77 30888

CA08818121933G   Journal: Japan.   Kokai   Publ: 771128
Pages: 5 pp.
  Polyolefin catalyst raw material or component
  Patent No: 77142691

CA08818121931E   Journal: Braz.   Pedido PI   Publ: 770906
Pages: 32 pp.
  Polymerization or copolymerization of .alpha.-olefins.   and
catalyst for the process
  Patent No: 76 01381

CA08818121924E   Journal: Ger.   Offen.   Publ: 771229
Pages: 56 pp.
  Titanium trichloride composition
  Patent No: 2716847

CA08818121920A   Journal: Japan.   Kokai   Publ: 771027
Pages: 12 pp.
  Catalysts for polymerization of ethylene in high yield
  Patent No: 77127990

CA08816105856D   Journal: J. Polym. Sci.. Polym. Chem. Ed.
Publ: 77   Series: 15   Issue: 12   Pages: 2893-2907
  Polymerization of acetylenic derivatives. XXVII. Synthesis
and properties of isomeric poly(N-ethynylcarbazole)

CA08816105838Z   Journal: Vysokomol. Soedin.. Ser. B   Publ:
77   Series: 19   Issue: 9   Pages: 674-6   Language: Russ
  Study of the characteristics of the copolymerization of
butadiene with its linear dimers in the presence of
coordination systems

CA08814090272J   Journal: Japan.   Publ: 771003   Pages: 6
pp.
  Catalysts for polymerization of .alpha.-olefins
  Patent No: 77 39075

CA08814090271H   Journal: Japan.   Kokai   Publ: 771001
Pages: 5 pp.
  Polymerization catalysts for propylene
  Patent No: 77117294

CA08814090270G   Journal: Japan.   Kokai   Publ: 770711
Pages: 10 pp.
  Polyolefins with improved stereoregularity
  Patent No: 77 82988

CA08814090269P   Journal: Ger.   Offen.   Publ: 780119
Pages: 12 pp.
  Purification   of   olefin   polymers   manufactured   using
Ziegler-Natta catalyst
  Patent No: 2729236

CA08812074742G   Journal: Ger.   Offen.   Publ: 771222
Pages: 28 pp.
  Treatment   of   transition   metal   compounds.   olefin
polymerization catalysts and the production of .alpha.-olefin
polymers and copolymers
  Patent No: 2727674

CA08812074741F   Journal: Ger.   Offen.   Publ: 771215
Pages: 18 pp.
  Ziegler catalysts and their use
  Patent No: 2723477

CA08812074619X   Journal: Conv. Ital. Sci. Macromol.. (Atti).
3rd   Publ: 77   Pages: 62-6   Language: Ital
  Examination of the interaction of atoms not linked to the
catalytic site in the stereospecific polymerization of
propylene

CA08810062760Y   Journal: Japan.   Kokai   Publ: 770913
Pages: 13 pp.
  Polymerization catalysts for .alpha.-olefins
  Patent No: 77109586

CA08810062759E   Journal: Japan.   Kokai   Publ: 770913
Pages: 9 pp.
  Catalysts for polymerization of propylene
  Patent No: 77109585

CA08414090791V   Journal: U.S.   Publ: 751209   Pages: 6
pp.
  Control of polymer particle size in olefin polymerization
  Patent No: 3925338

CA08808051394F   Journal: Ger.   Offen.   Publ: 771208
Pages: 30 pp.
  Polymers or copolymers of olefins with at least 3 carbon
atoms
  Patent No: 2724971

CA08808051393E   Journal: Ger.   Offen.   Publ: 771208
Pages: 12 pp.
  Ziegler catalyst
  Patent No: 2721263

CA08806038339X   Journal: Ger.   Offen.   Publ: 771124
Pages: 29 pp.
  Catalysts for the polymerization of olefins
  Patent No: 2722386

CA08806038333R   Journal: Japan.   Kokai   Publ: 770817
Pages: 7 pp.
  Polymerization catalysts for propylene
  Patent No: 77 98076

CA08804023635S   Journal: Japan.   Kokai   Publ: 770715
Pages: 11 pp.
  Polymerization catalysts for manufacturing stereoregular
propylene homopolymers
  Patent No: 77 85280

CA08802007700D   Journal: Ger.   Offen.   Publ: 771103
Pages: 9 pp.
  Poly(.alpha.-olefins)
  Patent No: 2616260

CA08802007697H   Journal: Hung.   Teljes   Publ: 770628
Pages: 9 pp.
  Polymerization and copolymerization of .alpha.-olefins
  Patent No: 13433

CA08802007694E   Journal: Japan.   Kokai   Publ: 770721
Pages: 9 pp.
  Polymerization catalysts for propylene
  Patent No: 77 87490

CA08802007692C   Journal: Japan.   Kokai   Publ: 770726
Pages: 6 pp.
  Polymerization catalysts for propylene
  Patent No: 77 89190

CA08802007690A   Journal: Japan.   Kokai   Publ: 770726
Pages: 6 pp.
  Catalysts for polymerization of olefins
  Patent No: 77 89189

CA08702006665P   Journal: Ger.   Offen.   Publ: 770414
Pages: 29 pp.
  Transition metal catalyst components for   Ziegler   catalysts
for olefin polymerization
  Patent No: 2543235

CA08526193139P   Journal: J. Appl. Polym.   Sci.   Publ: 76
Series: 20   Issue: 10   Pages: 2779-90
  Elementary steps in Ziegler-Natta catalysis

CA08522160636V   Journal: v   sb.,   Khimiya i Fiz.-khimiya
Vysokomolekul.   Soedinenii   Publ: 75   Pages: 74-82
Language: Russ
  Mechanism   of   polymerization on   modified   Ziegler-Natta
catalysts

CA08522160645X   Journal: Polim.   Protsessy,   Appar.-Tekhnol.
Oformlenie Mat.   Model.   Publ: 74   Pages: 136-41
Language: Russ
  Generalized   mathematical   model   of   the   kinetics   of   the
polymerization of ethylene with catalysts on supports

CA08522160548T   Journal: Compr.   Chem.   Kinet.   Publ: 76
Series: 15,   Pages: 133-257
  Kinetics of polymerization initiated   by   Ziegler-Natta   and
related catalysts

CA08426181413Q   Journal: Kauch. Rezina   Publ: 76   Issue:
3   Pages: 13-16   Language: Russ
  Synthesis,   structure,   and   properties   of   alternating
copolymers   of   butadiene   with   propylene,   a   new   type   of
all-purpose rubber

CA08726202358E   Journal: Fr.   Demande   Publ: 770304
Pages: 14 pp.
  Catalytic   elements   useful   for   the   stereospecific
polymerization of .alpha.-olefins
  Patent No: 2320309

CA08726202173R   Journal: Polymer   Publ: 77   Series: 18
Issue: 2   Pages: 179-84
  Polymerization of propylene by the Ziegler catalysts systems
.alpha.-titanium   trichloride-triethylaluminum   and   supported
titanium tetrachloride-triethylaluminum

CA08724185289P   Journal: Ger.   Offen.   Publ: 770908
Pages: 35 pp.
  Catalyst and method for polymerization of .alpha.-olefins
  Patent No: 2708588

CA08722168635F   Journal: Ger.   Offen.   Publ: 770915
Pages: 21 pp.
  Titanium   chloride   catalysts   for   stereoregulated
polymerization of .alpha.-olefins
  Patent No: 2709857

CA08722168631B   Journal: Ger.   Offen.   Publ: 770908
Pages: 16 pp.
  Ziegler catalysts and their use
  Patent No: 2708647

CA08722168489M   Journal: Mater.   Plast.   Elastomeri   Publ:
77   Issue: July-Aug.   Pages: 417-21   Language: Ital
  New catalysts for polypropylene

CA08720157615Y   Journal: Dokl.   Akad.   Nauk BSSR   Publ: 77
Series: 21   Issue: 7   Pages: 604-7
  Structure of coordinated-complex binary catalysts

CA08720152743D   Journal: Ger.   Offen.   Publ: 770908
Pages: 9 pp.
  Polyethylene
  Patent No: 2608863

CA08720152738F   Journal: Japan.   Kokai   Publ: 770312
Pages: 8 pp.
   Polymerization catalysts for olefins
   Patent No: 77 32985

CA08720152732Z   Journal: Neth.   Appl.   Publ: 770223
Pages: 9 pp.
   Supported Ziegler catalyst for the polymerization of olefins
   Patent No: 76 09030

CA08720152721V   Journal: Brit.   Publ: 770504   Pages: 8
pp. Division of Brit. 1.472.541.
   Gas-phase polymerization of halide monomers
   Patent No: 1472542

CA087181366420   Journal: Ger.   Offen.   Publ: 770804
Pages: 26 pp.
   Isotactic polypropylene
   Patent No: 2602182

CA08717133654J   Journal: Chem.   Phys.   Lett.   Publ: 77
Series: 49   Issue: 1   Pages: 8-12
   Restricted Hartree-Fock calculations on the   interaction   of
an olefin with a titanium compound

CA08716118438Y   Journal: U.S.S.R.   Publ: 770725
   Polypropylene
   Patent No: 565918

CA08716118428V   Journal: Japan.   Kokai   Publ: 770613
Pages: 4 pp.
   Highly   active   .alpha.-olefin   polymerization   catalyst
component
   Patent No: 77 70997

CA08424165261H   Journal: Makromol.   Chem.   Publ: 76
Series: 177   Issue: 3   Pages: 763-75
   Study of the mechanism of propagation and transfer reactions
in the polymerization of olefins by   Ziegler-Natta   catalysts,
2.   The influence of polymerization temperature on the kinetic
characteristics of propagation

CA08716118206W   Journal: Plaste Kautsch.   Publ: 77
Series: 24   Issue: 8   Pages: 540-4   Language: Ger
   Contributions to the homo- and copolymerization of styrene
and butadiene with a complex-coordinative initiator system
based on titanium and iodine.   II

CA08714102853U   Journal: Ger.   Offen.   Publ: 770721
Pages: 30 pp.
   Catalyst and method for polymerizing olefins
   Patent No: 2701647

CA08712085532U   Journal: Ger.   Offen.   Publ: 770714
Pages: 39 pp.
   Reformed titanium trichloride catalyst and its use in the
polymerization of .alpha.-olefins
   Patent No: 2657124

CA08712085527W   Journal: Japan.   Kokai   Publ: 770531
Pages: 10 pp.
   Polymerization catalysts for ethylene
   Patent No: 77 65592

CA08712085314Z   Journal: Yuki Gosei Kagaku Kyokai Shi
Publ: 76   Series: 34   Issue: 10   Pages: 726-9
Language: Japan
   Preparation and isolation of hexamethylphosphoric triamide
complexes with ethylaluminum chlorides (EtnAlCl3-n) and
abilities for propylene polymerization of the catalysts
composed of those complexes and titanium trichloride

CA08710068905A   Journal: Ger.   Offen.   Publ: 770623
Pages: 30 pp.
   Catalysts for polypropylene production
   Patent No: 2656055

CA08710068896Y   Journal: Ger.   Offen.   Publ: 770616
Pages: 22 pp.
   Titanium chloride catalysts. especially for the
polymerization of .alpha.-olefins
   Patent No: 2655278

CA08710068893V   Journal: Ger.   Offen.   Publ: 770602
Pages: 41 pp.
  Catalysts for the polymerization of .alpha.-olefins
  Patent No: 2643143

CA08708053876T   Journal: Japan.   Kokai   Publ: 770425
Pages: 8 pp.
  Modified Ziegler-Natta catalysts for polymerization   of
ethylene
  Patent No: 77 51485

CA08708053874R   Journal: Ger.   Offen.   Publ: 770608
Pages: 15 pp.
  Polymerization of olefins
  Patent No: 2654940

CA08706039903T   Journal: Fiz.-Khim.   Osn.   Sint.   Pererab.
Polim.   Publ: 76   Series: 1.   Pages: 3-6   Language:
Russ
  Polymerization of ethylene in the presence of phenoxy(or
naphthoxy) titanium trichlorides and aluminum alkyls

CA08704023971C   Journal: Japan.   Kokai   Publ: 770415
Pages: 6 pp.
  Titanium trichloride catalyst
  Patent No: 77 47594

CA08704023969H   Journal: Ger.   Offen.   Publ: 770428
Pages: 34 pp.
  Titanium trichloride for polymerization catalysts
  Patent No: 2645530

CA08704023965D   Journal: Ger.   Offen.   Publ: 770414
Pages: 17 pp.
  Transition metal catalyst components for Ziegler catalysts
for olefin polymerization
  Patent No: 2543219

CA08424165260G   Journal: Makromol.   Chem.   Publ: 76
Series: 177   Issue: 3   Pages: 747-61
  Study of the mechanism of propagation and transfer reactions
in the polymerization of olefins by Ziegler-Natta catalysts,
1. Determination of the number of propagation centers and the
rate constant

CA08626190674U   Journal: S.   African   Publ: 760928
Pages: 27 pp.
  Catalyst component supported on high performance carrier
  Patent No: 75 07382

CA08626190669W   Journal: Ger.   Offen.   Publ: 770407
Pages: 19 pp.
  Olefin polymers
  Patent No: 2543272

CA08626190527Y   Journal: Vestsi Akad. Navuk BSSR, Ser. Khim.
Navuk   Publ: 77   Issue: 2   Pages: 14-16   Language:
Russ
  Polymerization   of   ethylene   on   homogeneous   complex
coordination catalysts

CA08624172173A   Journal: U.S.   Publ: 770405   Pages: 5
pp.
  Olefin polymerization catalyst
  Patent No: 4016344

CA08624172170X   Journal: Japan.   Kokai   Publ: 770222
Pages: 4 pp.
  Polymerization catalysts for propylene
  Patent No: 77 23594

CA08624172169D   Journal: Belg.   Publ: 760903   Pages: 8
pp.
  Catalysts and methods for manufacturing highly stereoregular
polymers and copolymers from .alpha.-olefins
  Patent No: 839131

CA08624171947U   Journal: Vysokomol. Soedin., Ser. A   Publ:
77   Series: 19   Issue: 4   Pages: 813-18   Language:
Russ
  Polymerization of vinylphosphonic acid esters on complex
catalysts

CA08622156233Q   Journal: Japan.   Kokai   Publ: 770128
Pages: 8 pp.
  Highly crystalline .alpha.-olefin polymers
  Patent No: 77 11287

CA08622156224N   Journal: Japan.   Publ: 761007   Pages: 5
pp.
  Manufacture of stereoregular polyolefins
  Patent No: 76 36308

CA08622156064K   Journal: Macromol.   Synth.   Publ: 77
Series: 6,   Pages: 39-43
  Isotactic polystyrene

CA08622156046F   Journal: Deposited   Doc.   Publ: 73
Issue: VINITI 6929-73,   Pages: 20 pp.   Language: Russ
  Effect of the nature of organoaluminum compounds on the
reduction rate and catalytic activity of complexes based on
bis(cyclopentadienyl)ethyltitanium chloride

CA08620141404X   Journal: Ger.   Offen.   Publ: 770303
Pages: 12 pp.
  Copolymerization of olefin monomers for the production of
copolymers with elastomeric properties
  Patent No: 2636930

CA08620140942J   Journal: Japan.   Kokai   Publ: 770112
Pages: 10 pp.
  Impact-resistant propylene copolymers with improved
moldability
  Patent No: 77 03684

CA08620140703G   Journal: Ger.   Offen.   Publ: 770303
Pages: 13 pp.
  Ziegler support catalyst for the polymerization of olefins
  Patent No: 2637527

CA08620140689G   Journal: Japan.   Kokai   Publ: 761217
Pages: 5 pp.
  Stereoregular 1-butene polymers
  Patent No: 76147566

CA08620140539H   Journal: J. Polym. Sci., Polym. Chem. Ed.
Publ: 77   Series: 15   Issue: 3   Pages: 767-70
  Stereoelective polymerization of racemic .alpha.-olefins by
chirally modified Ziegler-Natta catalysts

CA08618122051P   Journal: Ger. Offen.   Publ: 770224
Pages: 22 pp.
  Catalyst components for polymerization of olefins
  Patent No: 2635163

CAC8618122047S   Journal: Ger. Offen.   Publ: 770210
Pages: 15 pp.
  Ziegler catalyst placed on a support
  Patent No: 2632730

CA08618121862S   Journal: Nippon Kagaku Kaishi   Publ: 77
Issue: 2   Pages: 279-83   Language: Japan
  Retardation of drop in polymerization of isoprene catalyzed
by titanium tetrachloride-triisobutylaluminum

CA08618121838P   Journal: Bull. Chem. Soc. Jpn.   Publ: 76
Series: 49   Issue: 12   Pages: 3530-9
  Structure of Ziegler-Natta catalysts for the polymerization
of isoprene

CA08616107243J   Journal: Japan. Kokai   Publ: 761216
Pages: 6 pp.
  Olefin polymers with high degree of stereoregularity
  Patent No: 76146587

CA08614090899C   Journal: U.S.S.R.   Publ: 761125
  Polyalkenamers
  Patent No: 468505

CA08614090610B   Journal: U.S.   Publ: 761207   Pages: 4
pp.
  Modified   titanium   component   for   catalysts   of   the
Ziegler-Natta type
  Patent No: 3992320

CA08614090371Z   Journal: Plast. Massy   Publ: 77   Issue:
1   Pages: 19-20   Language: Russ
  Modification of the titanium trichloride-diethylaluminum
chloride catalytic system during polymerization of propylene

CA08614090319P   Journal: Plast. Massy   Publ: 76   Issue:
12   Pages: 22-3   Language: Russ
  Organometallic catalysts for polymerization of ethylene on
modified organic supports

CA08614090318N   Journal: Plast. Massy   Publ: 76   Issue:
12   Pages: 18-19   Language: Russ
  Polymerization of ethylene using a triethylaluminum +
triethyl vanadate + titanium tetrachloride (Al(C2H5)3 +
VO(OC2H5)3 + TiCl4) catalytic system

CA08612074052R   Journal: Dokl. Akad. Nauk SSSR   Publ: 76
Series: 230   Issue: 3   Pages: 602-4   Language: Russ
  Stereospecificity of the action of benzyl derivatives of
titanium(IV) during polymerization of isoprene and the nature
of the active center

CA08612073203D   Journal: React. Kinet. Catal. Lett.   Publ:
76   Series: 5   Issue: 4   Pages: 429-34
  Determination of the rate constants for chain transfer with
the monomer and hydrogen in Ziegler-Natta polymerization

CA08612073187B   Journal: Plast. Massy   Publ: 76   Issue:
12   Pages: 24-6   Language: Russ
  Effect of the structure of an organoaluminum catalyst on
polymerization of .alpha.-olefins

CA08610055954V   Journal: Def. Publ., U. S. Pat. Off.   T
Publ: 760803   Pages: 79 pp.
  Olefin polymerization
  Patent No: 949005

CA08610055750A   Journal: Res. Discl.   Publ: 76   Series:
151,   Pages: 28-31
  Transition metal compound

CA08608044111S   Journal: Eur.   Polym.   J.   Publ: 76
Series: 12   Issue: 10   Pages: 727-8
  Stereospecificity of catalytic systems M trichloride-trieth-
ylaluminum in propylene polymerization reactions

CA08608044096R   Journal: Dokl. Akad.   Nauk SSSR   Publ: 76
Series: 231   Issue: 2   Pages: 413-15 (Chem.   Tech.)
Language: Russ
  Control of the molecular-mass distribution of polyethylene
synthesized on supported Ziegler catalysts

CA08606030321Y   Journal: Ger.   Offen.   Publ: 761118
Pages: 11 pp.
  Modified titanium components for catalysts of the
Ziegler-Natta type for polymerizing .alpha.-monoolefins
  Patent No: 2519582

CA08606030319D   Journal: Ger.   Offen.   Publ: 761111
Pages: 14 pp.
  Catalytic complexes for stereospecific polymerization of
.alpha.-olefins
  Patent No: 2618452

CA08606030158A   Journal: Hwahak Kwa Hwahak Kongop   Publ:
76 Series: 19   Issue: 4   Pages: 215-17   Language:
Korean
  Effects of a third substance on olefinic polymerization by
the Ziegler-Natta catalyst.   3.   Effect of moisture on
ethylene polymerization

CA08606030084Y   Journal: Chemsa   Publ: 76   Series: 2
Issue: 9   Pages: 164-5
  High density polyethylene by the FWH process

CA08604017096J   Journal: J.   Mol.   Catal.   Publ: 76
Series: 1   Issue: 6   Pages: 395-410
  The bonding of ethylene to titanium halides: an
SCF-X.alpha.-scattered-wave study of clusters relevant to
Ziegler-Natta catalysts

CA08604017029Q   Journal: Plast. Massy   Publ: 76   Issue:
10   Pages: 8-9   Language: Russ
  Monitoring the melt index of polypropylene in the
polymerization process

CA08424165222W   Journal: Kobunshi   Publ: 76   Series: 25
Issue: 4   Pages: 260-62   Language: Japan
  Recent development in coordinate polymerization

CA08422151004Q   Journal: Appl. Polym. Symp.   Publ: 75
Series: 26   Issue: Polym. Polycondensat   Pages: 1-11
  Second generation Ziegler polyethylene processes

CA08412074630M   Journal: J. Polym. Sci., Polym. Symp.
Publ: 75   Series: 51   Issue: Int. Symp. Macromol. Honor
Professor Herman F. Mark   Pages: 187-96
  Recent advances in Ziegler polyolefin processes

CA08404017866U   Journal: Makromol. Chem.   Publ: 75
Series: 176   Issue: 11   Pages: 3353-70
  Radiochemical study of the Ziegler-Natta catalyzed
polymerization of styrene

CA08404017755G   Journal: Chem. Technol.   Publ: 75
Series: 5   Issue: 11   Pages: 688-92
  Ziegler-Natta polymerization. Model, mechanism, and
kinetics

CA08402005427P   Journal: Hwahak Kwa Hwahak Kongop   Publ:
75   Series: 18   Issue: 2   Pages: 76-9   Language:
Korean
  Effects of third substances on olefinic polymerization by
the Ziegler-Natta catalyst. 1. Effect of alcohol on ethylene
polymerization

CA08526193661C   Journal: U. S. Reissue   Publ: 761012
Pages: 5 pp. Reissue of U.S. 3,876,600.
   Deactivating and removing aluminum and titanium contaminant
from Ziegler-Natta polymerization mixtures
   Patent No: 29004

CA08526193342Z   Journal: Ger.   Offen.   Publ: 761007
Pages: 21 pp.
   Catalyst for olefin polymerization
   Patent No: 2503718

CA08526193335Z   Journal: Ital.   Publ: 670715   Pages: 25
pp.
   Titanium amidic derivatives and their use in the
stereospecific polymerization and copolymerization of olefins
and diolefins
   Patent No: 778386

CA08526193140G   Journal: Acta Cient. Venez., Supl.   Publ:
73   Series: 24   Issue: 2   Pages: 12-20   Language: Span
   The mechanism of Ziegler-catalysis

CA08524178042H   Journal: Angew. Chem.   Publ: 76   Series:
88   Issue: 20   Pages: 689-90   Language: Ger
   Halogen-free soluble Ziegler catalysts for ethylene
polymerization. Control of molecular weight by the choice of
the reaction temperature

CA08522161595T   Journal: U.S.   Publ: 760817   Pages: 7
pp.
   Polymers of nonconjugated 1,4-dienes
   Patent No: 3975336

CA08522160877Z   Journal: Ger.   Offen.   Publ: 760826
Pages: 22 pp.
   Highly stereoregular polymers or copolymers from
.alpha.-olefins and a catalyst
   Patent No: 2605922

CA08522160876Y   Journal: Japan.   Publ: 760313   Pages: 7
pp.
  Catalysts for polymerization of butadiene
  Patent No: 76 08152

CA08522160867W   Journal: Ger.   Offen.   Publ: 760624
Pages: 36 pp.
  Improved polymerization catalysts
  Patent No: 2555165

CA08522160627T   Journal: Ionic Polym.: Unsolved Probl.,
Jpn.-U. S. Semin. Polym. Synth., 1st   Publ: 76   Pages:
259-70
  Monomer-isomerization   polymerization   of   some   branched
internal olefins with a Ziegler-Natta catalyst

CA08520143816M   Journal: Ger.   Offen.   Publ: 760610
Pages: 32 pp.
  Polymerization of ethylene
  Patent No: 2552845

CA08520143791Z   Journal: U.S.   Publ: 760713   Pages: 4
pp.
  4-Methyl-1-pentene homopolymers
  Patent No: 3969335

CA08520143790Y   Journal: U.S.   Publ: 760713   Pages: 3
pp.
  4-Methyl-1-pentene copolymers
  Patent No: 3969333

CA08520143609W   Journal: J. Polym. Sci., Polym. Chem. Ed.
Publ: 76   Series: 14   Issue: 8   Pages: 2083-5
  Monomer-isomerization polymerization. XV. Monomer-isomerization polymerization of 4-phenyl-2-butene with Ziegler-Natta
catalyst

CA08520143602P   Journal: J. Polym. Sci., Polym. Chem. Ed.
Publ: 76   Series: 14   Issue: 8   Pages: 1915-32
  Supported catalysts for stereospecific polymerization of
propylene

CA08520143532R   Journal: Chem. Econ. Eng. Rev.   Publ: 76
Series: 8   Issue: 6   Pages: 49-54
  Trends in development of polypropylene catalyst production

CA08518131264S   Journal: Neth.   Appl.   Publ: 760210
Pages: 22 pp.
  Modification of the titanium trichloride component of a
Ziegler catalyst
  Patent No: 74 10605

CA08518124696F   Journal: Neth.   Appl.   Publ: 760305
Pages: 25 pp.
  Stereoregular alkene polymers
  Patent No: 75 10394

CA08518124690Z   Journal: Japan.   Kokai   Publ: 760717
Pages: 6 pp.
  Catalysts for crystalline polyolefins
  Patent No: 76 81889

CA08518124686C   Journal: Can.   Publ: 760420   Pages: 9
pp.
  Process and catalyst for the polymerization of
alpha-monoolefins having three or more carbon atoms per
molecule
  Patent No: 987848

CA08518124684A   Journal: U. S. Reissue   Publ: 760608
Pages: 20 pp.   Reissue of U.S. 3,766,153 and Division of U.S.
3,714,133.
  Process for preparing an alternating copolymer of butadiene
and .alpha.-olefin and a novel alternating copolymer of
butadiene and .alpha.-olefin containing a cis-configuration
butadiene unit
  Patent No: 28850

CA08518124677A   Journal: Japan.   Kokai   Publ: 760709
Pages: 10 pp.
  Modified Ziegler catalysts for manufacture of polyethylene
having relatively broad molecular weight distributions
  Patent No: 76 79194

CA08518124449C   Journal: Dokl. Akad. Nauk SSSR   Publ: **76**
Series: 229   Issue: 3   Pages: 667-70 (Phys.   Chem.)
Language: Russ
  Effect   of radical size in aluminum trialkyls on the
ion-coordination polymerization of dienes

CA08516109798X   Journal: Ger.   Publ: 760408   **Pages: 8**
pp.
  High-molecular-weight copolymers of .alpha.-olefins   **and**
diolefins using modified organometallic mixed catalysts
  Patent No: 1495698

CA08516109063D   Journal: Plast. Massy   Publ: 76   **Issue:**
7   Pages: 27-30   Language: Russ
  Characteristics of ethylene polymerization on   nonfixed
Ziegler-Natta catalysts

CA08516109057E   Journal: Dokl. Akad. Nauk SSSR   Publ: **76**
Series: 229   Issue: 1   Pages: 119-21 (Phys.   Chem.)
Language: Russ
  Relative growth rate of a chain in cis- and trans-regulating
systems

CA08516109039A   Journal: J. Chem. Soc., Faraday Trans. **1**
Publ: 76   Series: 72   Issue: 7   Pages: 1685-93
  Infrared studies of a Ziegler catalyst supported   **on**
magnesium oxide and titanium dioxide

CA08514094744A   Journal: Dokl. Akad. Nauk SSSR   Publ: **76**
Series: 228   Issue: 4   Pages: 892-5 (Phys.   Chem.)
Language: Russ
  Organooxy- and organosiliconoxytitanium trichlorides **as**
catalysts of the "Ziegler" polymerization of ethylene

CA08512086532Q   Journal: Z. Naturforsch., B: Anorg. Chem.,
Org. Chem.   Publ: 76   Series: 31B   Issue: 6   **Pages:**
892-3   Language: Ger
  Surface compounds of transition metals, XVII.   **Oxygen**
sensitivity of coordinatively unsaturated surface compounds

CA08512078446Z   Journal: Khim. Vysokomol. Soedin. Neftekhim.
    Publ: 75   Pages: 73   Language: Russ
    Study of the polymerization of vinylphosphonates on complex
organometallic catalysts

CA08510063694E   Journal: Japan.   Kokai   Publ: 760124
Pages: 9 pp.
    Stereoregular polymerization catalysts of .alpha.-olefins
    Patent No: 76 09184

CA08510063424S   Journal: Khim. Vysokomol. Soedin. Neftekhim.
    Publ: 75   Pages: 67-8   Language: Russ
    Polymerization   of   .alpha.-olefins   in   the   presence   of
catalysts with a "microsupport"

CA08508047163A   Journal: Jpn.-USSR Polym.   Symp.,   (Proc.),
2nd   Publ: 76   Pages: 161-8
    Gas phase polymerization of ethylene and propylene with
titanium trichloride and alkylaluminum

CA08508047151V   Journal: Coord. Polym.   Publ: 75   Pages:
305-25
    Supported Ziegler-Natta catalysts

CA08508047141S   Journal: Polym. J.   Publ: 76   Series: 8
Issue: 2   Pages: 139-49
    The   polymerization of   acetylenic   derivatives.   XXV.
Synthesis and properties of isomeric poly(.beta.-ethynylnapht-
halene)

CA08506033709W   Journal: Ger.   Offen.   Publ: 760415
Pages: 12 pp.
    Modified   titanium   components   for   catalysts   of   the
Ziegler-Natta type
    Patent No: 2441541

CA08506033498B   Journal: Coord. Polym.   Publ: 75   Pages:
155-97
    A   Kinetic   model   for   heterogeneous   Ziegler-Natta
polymerization

CA08506033497A   Journal: Coord. Polym.   Publ: 75   **Pages:** 135-53
Ethylene polymerization with the catalysts of one and two component systems based on titanium trichloride complex

CA08506033492V   Journal: Makromol.   Chem.   Publ: **76** Series: 177   Issue: 5   Pages: 1465-76
Stereospecific polymerization of cis,cis-1,4-dideuterio-1,3- -butadiene to trans-1,4 or cis-1,4 polymers

CA08506033468S   Journal: Nippon Kagaku Kaishi   Publ: **76** Issue: 4   Pages: 554-9   Language: Japan
Kinetics of ethylene polymerization with dichlorobis(.eta.-- cyclopentadienyl)titanium(IV)-dichloroethylaluminum catalyst

CA08506033453H   Journal: Eur.   Polym.   J.   Publ: **75** Series: 11   Issue: 12   Pages: 829-32
Nature of active centers and stereospecificity of tetravalent titanium benzyl derivatives in polymerization of butadiene

CA08504025783A   Journal: J.   Chem.   Soc.,   Dalton Trans. Publ: 74   Issue: 22   Pages: 2390-5
Electron spin resonance studies of Ziegler-type catalysts. Part 1.   Characterization of a vanadium-aluminum complex obtained on mixing dichlorobis(.eta.-cyclopentadienyl)vanadium with ethylaluminum dichloride

CA08504021966Q   Journal: Coord. Polym.   Publ: 75   **Pages:** 263-89
A kinetic approach to elucidate the mechanism of Ziegler-Natta polymerization

CA08504021965P   Journal: Coord. Polym.   Publ: 75   **Pages:** 73-89
The number of active sites for the polymerization of ethylene, propylene and 1-butene by Ziegler-Natta catalyst

CA08504021896S   Journal: Polymer   Publ: 76   Series: 17
Issue: 2   Pages: 130-6
  Ziegler-Natta catalysis. 7. The settling period

CA08502006391E   Journal: Ger.   Offen.   Publ: 760226
Pages: 20 pp.
  Reducing agent and its use in preparing catalysts for olefin
polymerization
  Patent No: 2535956

CA08502006149G   Journal: Zh. Obshch. Khim.   Publ: 76
Series: 46   Issue: 4   Pages: 746-9   Language: Russ
  Polarographic determination of titanium(IV), titanium(III),
and titanium(II) during butadiene polymerization in catalytic
systems of alkyl orthotitanates and triisobutylaluminum

CA08502006141Y   Journal: Nippon Kagaku Kaishi   Publ: 76
Issue: 2   Pages: 216-20   Language: Japan
  Effect of hydrogen on propylene polymerization with titanium
trichloride-triethylaluminum

CA08426180901K   Journal: Belg.   Publ: 750903   Pages: 21
pp.
  Catalyst useful for the polymerization of olefins
  Patent No: 826220

CA08424165961M   Journal: Japan.   Kokai   Publ: 750926
Pages: 7 pp.
  Trans-1,4-polyisoprenes
  Patent No: 75122586

CA08422151907M   Journal: Can.   Publ: 751118   Pages: 22
pp.
  Butadiene polymers and preparation thereof
  Patent No: 978299

CA08422151271Z   Journal: Ital.   Publ: 720817   Pages: 13
pp.
  Catalysts for polymerization of olefins
  Patent No: 926445

CA08420136968P   Journal: Japan.   Kokai   Publ: 751121
Pages: 12 pp.
  Polymerization catalysts to give cis-1,4-rich polyisoprenes
  Patent No: 75145486

CA08420136325B   Journal: Neth.   Appl.   Publ: 750903
Pages: 28 pp.
  Catalyst for the production of polyalkenes
  Patent No: 75 02172

CA08418122430V   Journal: Acta Cient.   Venez.   Publ: 75
Series: 26   Issue: 1   Pages: 34-9 ,   Language: Span
  Polymerization kinetics of propylene using Ziegler-Natta
catalysts

CA08416106848G   Journal: Def. Publ., U. S. Pat. Off. T
Publ: 750506   Pages: 16 pp.
  Coordination polymerization catalyst
  Patent No: 934005

CA08416106336P   Journal: Belg.   Publ: 750418   Pages: 38
pp.
  Ziegler-Natta catalysts
  Patent No: 821260

CA08414091373X   Journal: U.S.   Publ: 751223   Pages: 4
pp.
  Polymers of 2-ethyl-1,3-butadiene and method of their
preparation
  Patent No: 3928301

CA08414090806D   Journal: U.S.   Publ: 751216   Pages: 5
pp.
  Catalysts for producing high trans-polybutadiene
  Patent No: 3926933

CA08326206571J   Journal: Polym. Prepr., Am. Chem. Soc., Div.
Polym. Chem.   Publ: 74   Series: 15   Issue: 1   Pages:
359-63
  Ziegler-Natta catalyst. Looking back

CA08414090639B   Journal: J.   Polym.   Sci.,   Polym.   Symp.
Publ: 75   Series: 52   Issue: Contrib.   Stud.   Friends
Professor Champetier Occas. His 70th Birthday   Pages: 107-17
    Language: Fr
  Study   of   the   catalytic   activity   of   violet   titanium
trichloride in the high-temperature polymerization of ethylene

CA08412074876W   Journal: Ger.   Offen.   Publ: 751016
Pages: 28 pp.
  Catalyst for the polymerization of olefins
  Patent No: 2515211

CA08410060248F   Journal: S.   African   Publ: 750619
Pages: 24 pp.
  Polymerization catalyst composition
  Patent No: 74 04748

CA08410060227Y   Journal: Ger.   Offen.   Publ: 750130
Pages: 15 pp.
  Ziegler-Natta   catalysts   for   aging-resistant   norbornene
polymers
  Patent No: 2421838

CA08410060082X   Journal: J. Appl. Polym. Sci.   Publ: 76
Series: 20   Issue: 1   Pages: 55-61
  Molecular weight and molecular weight distribution control
in   HDPE   with   trichlorooxovanadium-based   Ziegler-Natta
catalysts

CA08410060036K   Journal: Bull. Chem. Soc. Jpn.   Publ: 75
Series: 48   Issue: 9   Pages: 2470-2
  Monomer-isomerization polymerization. XIV. Monomer-isomer-
ization   polymerization   of   4-methyl-2-pentene   with
Ziegler-Natta catalyst

CA08408044975K   Journal: Brit.   Publ: 750917   Pages: 5
pp.
  Polymerizing olefins using a modified Ziegler-Natta catalyst
  Patent No: 1406282

CA08408044803C   Journal: Vysokomol. Soedin., Ser. A   Publ:
75   Series: 17   Issue: 10   Pages: 2163-7
Language: Russ
  Structure of copolymers of 4-methyl-1-pentene and 1-hexene

CA08408044767U   Journal: Kinet.   Katal.   Publ: **75**
Series: 16   Issue: 5   Pages: 1358 ,   Language: Russ
   Effect of reduction conditions for titanium tetrachloride by
organoaluminum compounds on the activity and stereospecificity
of titanium trichloride during polymerization of propylene

CA08406031599R   Journal: Mech.   Hydrocarbon React.,   **Symp.**
Publ: 75   Pages: 487-95
   Coordination mechanism of insertion reactions into the
metal-carbon bond during catalytic polymerization of olefins

CA08404018089E   Journal: Ger.   Offen.   Publ: **750925**
Pages: 22 pp.
   Catalyst for polymerizing .alpha.-olefins
   Patent No: 2413261

CA08404018085A   Journal: Ger.   Offen.   Publ: **750911**
Pages: 29 pp.
   Polymerization catalyst
   Patent No: 2409726

CA08404017858T   Journal: J. Polym. Sci., Polym. Chem. **Ed.**
Publ: 75   Series: 13   Issue: 11   Pages: 2491-7
   Molecular weight distribution and stereoregularity of
polypropylenes produced with vanadium tetrachloride-chlorodie-
thylaluminum catalyst

CA08402005675T   Journal: Ger.   Offen.   Publ: **750807**
Pages: 31 pp.
   Olefin polymerization catalysts
   Patent No: 2503880

CA08402005480A   Journal: Hwahak Kwa Hwahak Kongop   Publ:
75   Series: 18   Issue: 3   Pages: 134-7   Language:
Korean
   Effects of third substances on olefinic polymerization by
Ziegler-Natta catalyst.   2.   Effect of oxygen on ethylene
polymerization

CA08402005465Z   Journal: Polym. Prepr., Am. Chem. Soc., Div.
Polym. Chem.   Publ: 74   Series: 15   Issue: 1   Pages:
368-72
   Chain transfer in Ziegler type polymerization of ethylene

CA08402005463X   Journal: Polym. Prepr., Am. Chem. Soc., Div.
Polym. Chem.   Publ: 74   Series: 15   Issue: 1   Pages:
292-7
   Kinetic model for heterogeneous Ziegler-Natta polymerization

CA08326206951H   Journal: Plast. Massy   Publ: 75   Issue:
9   Pages: 42-3   Language: Russ
   Frost-resistant polypropylene-polyethylene composition

CA08326206826W   Journal: Ger.   Offen.   Publ: 750828
Pages: 11 pp.
   Homopolymers of 4-methylpent-1-ene
   Patent No: 2406900

CA08326206819W   Journal: Ger.   Offen.   Publ: 750807
Pages: 34 pp.
   Polymerization of .alpha.-olefins
   Patent No: 2504036

CA08326206653N   Journal: Nippon Kagaku Kaishi   Publ: 75
Issue: 10   Pages: 1664-8   Language: Japan
   Kinetic study of propylene polymerization with titanium
trichloride - aluminum ethyl X (X = chloride, bromide, iodide)

CA08326206635H   Journal: Vysokomol. Soedin., Ser. A   Publ:
75   Series: 17   Issue: 9   Pages: 1932-7   Language:
Russ
   Nature of active centers and stereospecificity of the effect
of benzyl derivatives of titanium(IV) during polymerization of
butadiene

CA08326206631D   Journal: J. Macromol. Sci., Chem.   Publ:
75   Series: A9   Issue: 7   Pages: 1245-54
   Monomer-isomerization polymerization. XII. Monomer-isomerization polymerization of 2-heptene and 3-heptene with
Ziegler-Natta catalyst

CA08326206612Y   Journal: Angew. Makromol. Chem.   Publ: **75**
Series: 44   Issue: 1   Pages: 31-46   Language: Fr
   Stereospecific   polymerization   of   propylene   using
.gamma.-titanium   trichloride   catalyst   prepared   from
.beta.-titanium trichloride

CA08324194090M   Journal: Ger.   Offen.   Publ: 750807
Pages: 37 pp.
   Olefin polymerization catalysts
   Patent No: 2503881

CA08324194078P   Journal: Japan.   Publ: 741015   **Pages: 3**
pp.
   Copolymerizing .alpha.-olefins
   Patent No: 74 38027

CA08324193915X   Journal: Makromol.   Chem.   Publ: **75**
Series: 176   Issue: 9   Pages: 2765-70
   Stereoregularity of polypropylene obtained with   **different**
isospecific catalyst systems

CA08324193805M   Journal: Izv. Akad.   Nauk SSSR,   Ser.   Khim.
Publ: 75   Issue: 7   Pages: 1615-17   Language: Russ
   Effect of the chemical nature of the substituent in titanium
tetrachloride derivatives on   the   Ziegler   polymerization   of
butadiene

CA08324193799N   Journal: Plaste Kautsch.   Publ: **75**
Series: 22   Issue: 3   Pages: 233-9   Language: Ger
   Butadiene   polymerization   with   a   coordinating-complex
initiator system based on titanium and iodine.   I

CA08322179959H   Journal: U.S.   Publ: 750812   **Pages: 8**
pp.
   Polymerization of ethylene
   Patent No: 3899477

CA08326206827X   Journal: Japan.   Kokai   Publ: 750805
Pages: 7 pp.
   Polymerization catalysts for olefins
   Patent No: 75 98585

CA08322179954C   Journal: Japan.   Kokai   Publ: 750705
Pages: 5 pp.
  Polymerization catalysts for .alpha.-olefin
  Patent No: 75 83284

CA08322179953B   Journal: Ger.   Offen.   Publ: 750717
Pages: 9 pp.
  Modified titanium component for Ziegler-Natta catalysts
  Patent No: 2400190

CA08322179949E   Journal: Ger.   Offen.   Publ: 750710
Pages: 28 pp.
  Polyolefins
  Patent No: 2461187

CA08322179945A   Journal: Ger.   Offen.   Publ: 750703
Pages: 37 pp.
  Olefin polymers
  Patent No: 2363697

CA08322179943Y   Journal: Japan.   Kokai   Publ: 750619
Pages: 6 pp.
  Catalysts for olefin polymerization
  Patent No: 75 74593

CA08322179942X   Journal: U.S.   Publ: 750624   Pages: 8
pp.
  Alpha titanium trichloride particles
  Patent No: 3891746

CA08322179712X   Journal: Dokl.   Resp.   Nauchno-Tekh.   Konf.
Neftekhim., 3rd   Publ: 74   Series: 2,   Pages: 44-54
Language: Russ
  Effect of organosilicon additives on polymerization of
ethylene in the presence of the vanadium(IV)
chloride-triisobutylaluminum catalytic system

CA08322179670G   Journal: J. Macromol. Sci.,   Chem.   Publ:
75   Series: A9   Issue: 6   Pages: 899-910
  Monomer-isomerization polymerization of some branched
internal olefins with Ziegler-Natta catalyst

CA08318148069P   Journal: Ger.   Offen.   Publ: **750703**
Pages: 22 pp.
  Catalysts for polymerization of ethylene
  Patent No: 2363696

CA08318148068N   Journal: Ger.   Offen.   Publ: **750703**
Pages: 29 pp.
  Catalysts for polymerization of ethylene
  Patent No: 2363695

CA08318148058J   Journal: Neth.   Appl.   Publ: **740328**
Pages: 19 pp.
  Stereoregular polymerization of .alpha.-alkenes
  Patent No: 73 12901

CA08318147830T   Journal: Makromol.   Chem.   Publ: **75**
Series: 176   Issue: 7   Pages: 2159-61
  Role of surface halogen ligands of titanium metals sites in
stereospecific polymerization with Ziegler-Natta catalyst

CA08318147829Z   Journal: Makromol.   Chem.   Publ: **75**
Series: 176   Issue: 7   Pages: 1959-68
  Role   of   organometallic   cocatalysts   in   catalytic
Ziegler-Natta systems

CA08318147821R   Journal: J. Polym. Sci., Polym. Chem. Ed.
Publ: 75   Series: 13   Issue: 7   Pages: 1601-5
  Monomer-isomerization polymerization. XIII. Monomer-isome-
rization   polymerization   of   1,4-cyclohexadiene   with
Ziegler-Natta catalyst

CA08318147795K   Journal: Polymer   Publ: **75**   Series: **16**
Issue: 5   Pages: 384-6
  Active center determination in donor-modified Ziegler-Natta
polymerization

CA08324194868W   Journal: U.S.S.R.   Publ: **750725**
  Control of continuous solution polymerization of conjugated
dienes on Ziegler-Natta catalysts
  Patent No: 478018

CA08316132313C   Journal: Brit.   Publ: 741120   Pages: 13
pp. Addn. to Brit. 1,374,579.
   Vinyl polymerization
   Patent No: 1374967

CA08316132312B   Journal: Japan.   Kokai   Publ: 740812
Pages: 12 pp.
   Improved Ziegler-Natta type catalyst
   Patent No: 74 83781

CA08316132013E   Journal: Vysokomol. Soedin., Ser. A   Publ:
75   Series: 17   Issue: 1   Pages: 46-53   Language: Russ
   Determination of the concentration of active centers and
rate   constants   of   elemental   acts   during   ethylene
polymerization   on   supported   titanium   tetrachloride-   and
triethylaluminum-based catalysts

CA08314115760D   Journal: Brit.   Publ: 750219   Pages: 5
pp.
   Trans-1,4-polydienes
   Patent No: 1384163

CA08314115487V   Journal: J. Polym. Sci., Polym. Chem. Ed.
Publ: 75   Series: 13   Issue: 5   Pages: 1071-81
   Enhancement of stereospecificity in isoprene polymerization
with   alkylaluminum-titanium chloride catalysts through use of
carbon disulfide

CA08314115434A   Journal: Ger.   Offen.   Publ: 750619
Pages: 17 pp.
   Polymerization catalyst
   Patent No: 2457968

CA08314115433Z   Journal: Brit.   Publ: 741030   Pages: 4
pp.
   Polymerization of alkenes
   Patent No: 1372440

CA08314115026U   Journal: Rev.   Roum.   Chim.   Publ: 74
Series: 19   Issue: 11   Pages: 1695-707
   Theoretical investigation of Ziegler-type catalysis.   II.
Species related to solid titanium trichloride

CA08314115023R   Journal: Makromol.   Chem.   Publ: **75**
Series: 176   Issue: 2   Pages: 507-9
   Catalytic regulation for isotactic orientation in propylene
polymerization with Ziegler-Natta catalyst

CA08314115011K   Journal: Eur.   Polym.   J.   Publ: **74**
Series: 10   Issue: 6   Pages: 481-8
   Ziegler-type catalysts for the polymerization of ethylene
and   propylene.   II.   Polymerization of ethylene with
trichlorooxovanadium-trichlorotriethyldialuminum (chlorinated
activator) catalysts

CA08312098190Y   Journal: Japan.   Publ: 750318   **Pages: 5**
pp.
   Polymerization of .alpha.-olefins
   Patent No: 75 06877

CA08310080017N   Journal: Japan.   Kokai   Publ: **750421**
Pages: 7 pp.
   Stereoregular polyolefins
   Patent No: 75 44273

CA08310080008K   Journal: Japan.   Kokai   Publ: **750327**
Pages: 6 pp.
   Stereoregular polymerization of olefins
   Patent No: 75 30983

CA08310079992V   Journal: Japan.   Kokai   Publ: **740314**
Pages: 8 pp.
   Poly (.alpha.-olefin)
   Patent No: 74 28689

CA08310079728P   Journal: Izv. Akad. Nauk SSSR, Ser. Khim.
Publ: 75   Issue: 5   Pages: 1116-20   Language: Russ
   Transformation   of   3-methyl-1-butene   on   complex
organometallic catalysts

CA08302010995F   Journal: Plast. Massy   Publ: 75   Issue:
2   Pages: 29-30   Language: Russ
   Properties of branched polyethylene prepared with Ziegler
catalysts on supports

CA08308061249Y   Journal: Japan.   Kokai   Publ: 750313
Pages: 7 pp.
  Controlling molecular weight distributions of copolymers
  Patent No: 75 23493

CA08308061239V   Journal: Japan.   Kokai   Publ: 750217
Pages: 5 pp.
  cis-1,4-Polyisoprenes
  Patent No: 75 14789

CA08308059882F   Journal: Japan.   Kokai   Publ: 740314
Pages: 7 pp.
  Poly(.alpha.-olefin)
  Patent No: 74 28690

CA08308059870A   Journal: Ger.   Offen.   Publ: 740502
Pages: 25 pp.
  Catalysts for stereoregular propene polymerization
  Patent No: 2347577

CA08308059869G   Journal: Ger.   Offen.   Publ: 740321
Pages: 12 pp.
  Polyolefins by stereospecific polymerization
  Patent No: 2345515

CA08308059807K   Journal: Ger.   Offen.   Publ: 740711
Pages: 29 pp.
  Polymerization of 2-olefins with recycling of the solvent
  Patent No: 2362281

CA08308059555B   Journal: Tr.   Kazan.   Khim.-Tekhnol.   Inst.
Publ: 74   Series: 54,   Pages: 82-6   Language: Russ
  Copolymerization of ethylene with bis(.beta.-chloroethyl)
vinylphosphonate on the .alpha.-titanium trichloride-diethylaluminum chloride system

CA08308059431H   Journal: Nippon Kagaku Kaishi   Publ: 74
Issue: 10   Pages: 2017-19   Language: Japan
  Effect of triethylaluminum or ethylaluminum dichloride on
the   catalyst   system   diethylaluminum   chloride-titanium
trichloride in the stereospecific polymerization of 1-butene

CA08304028843Q   Journal: Japan.   Kokai   Publ: 750114
Pages: 12 pp.
   Highly crystalline propylene copolymers
   Patent No: 75 03188

CA08302012562M   Journal: ACS   Symp.   Ser.   Publ: 75
Series: 10   Issue: Cellul.   Technol.   Res.,   Symp.,   1974
Pages: 147-59
   Nascent polyethylene-cellulose composite

CA08302011199M   Journal: Japan.   Publ: 741223   Pages: 12
pp.
   Titanium trichloride composition
   Patent No: 74 48637

CA08302010962T   Journal: J. Polym. Sci., Polym. Lett. Ed.
Publ: 75   Series: 13   Issue: 1   Pages: 1-9
   Pertinence of the scrambling behavior of ligands on
transition-metal centers to Ziegler-Natta catalyst activities

CA08302010959X   Journal: Vysokomol. Soedin., Ser. A   Publ:
75   Series: 17   Issue: 2   Pages: 309-17   Language:
Russ
   Nature of active centers of Ziegler-Natta catalytic systems

CA08302010934K   Journal: Polimery (Warsaw)   Publ: 74
Series: 19   Issue: 11   Pages: 530-3   Language: Pol
   Polymerization of allyl chloride in the presence of
aluminum-titanium catalysts

CA08226171745U   Journal: Japan.   Kokai   Publ: 750127
Pages: 6 pp.
   Stereospecific polymerization catalyst for .alpha.-olefin
   Patent No: 75 07793

CA08226171514T   Journal: Eur.   Polym.   J.   Publ: 75
Series: 11   Issue: 3   Pages: 247-51   Language: Fr
   Effect of temperature and monomer concentration on kinetics
of Ziegler-Natta catalyzed polymerization of propene

CA08224157574F　　Journal: U.S.　　Publ: 750211　　Pages: 8
pp.
　Butadiene polymerization catalyst
　Patent No: 3865749

CA08222145815X　　Journal: U.S.　　Publ: 750107　　Pages: 6
pp.
　Simplified process for manufacture of catalyst component
　Patent No: 3859231

CA08222140812J　　Journal:　Ger.　Offen.　　Publ:　750116
Pages: 33 pp.
　Modified Ziegler catalysts
　Patent No: 2428979

CA08222140582J　　Journal: Tr. Kazan. Khim.-Tekhnol. Inst.
Publ: 74　　Series: 54,　　Pages: 78-82　　Language: Russ
　Polymeriazation of bis(.beta.-chloroethyl) vinylphosphonate
on complex organometallic catalysts

CA08220125848G　　Journal:　Ger.　Offen.　　Publ:　741219
Pages: 23 pp.
　Polymerization of .alpha.-olefins
　Patent No: 2329641

CA08216098852V　　Journal: Japan.　　Publ: 740902　　Pages: 6
pp.
　Polymerizing .alpha.-olefin
　Patent No: 74 32670

CA08216098503G　　Journal: Dokl. Akad. Nauk B. SSR　　Publ: 74
　　　Series:　18　　Issue:　11　　Pages:　1008-10　　Language:
Russ
　Nucleophilic ligands and ethylene polymerization on
Ziegler-Natta catalysts

CA08214086883T　　Journal:　Ger.　Offen.　　Publ:　741107
Pages: 14 pp.
　Partially crystalline 1-butene polymers
　Patent No: 2318901

CA08214086868S   Journal: U.S.S.R.   Publ: 740815
  Trans-1,4-Polydienes
  Patent No: 418047

CA08212073681A   Journal: Japan.   Publ: 740612   Pages: 7
pp.
  Copolymerization of olefin hydrocarbons
  Patent No: 74 22954

CA08210058558X   Journal: Ger.   Offen.   Publ: 740905
Pages: 28 pp.
  Modified Ziegler catalyst
  Patent No: 2407095

CA08210058553S   Journal: Ger.   Offen.   Publ: 740725
Pages: 6 pp.
  Titanium trichloride-aluminum trichloride component   for
Ziegler-Natta catalysts
  Patent No: 2301136

CA08210058448M   Journal: Plast. Massy   Publ: 74   Issue:
5   Pages: 22-4   Language: Russ
  Polymerization of ethylene on   modified   Ziegler-Natta
catalysts

CA08208044080Y   Journal: Fr.   Demande   Publ: 740419
Pages: 15 pp.
  Stereospecific polymerization of .alpha.-olefins
  Patent No: 2200290

CA08208044077C   Journal: Japan.   Publ: 740426   Pages: 6
pp.
  Polymerizing .alpha.-olefin
  Patent No: 74 17034

CA08208043850N   Journal: Angew. Makromol. Chem.   Publ: 74
Series: 39   Issue: 1   Pages: 131-74   Language: Ger
  Elementary processes of Ziegler-Natta   catalysis.   I.
Oligomerization kinetics in the plug flow reactor

CA08208043849U   Journal: Dokl. Akad. Nauk SSSR   Publ: 74
Series: 218   Issue: 2   Pages: 353-5 ((Chem))   Language:
Russ
   Effect   of   orthotitanate   structure   on   ion-coordination
polymerization of butadiene

CA08208043831G   Journal: J. Organomet. Chem.   Publ: 74
Series: 77   Issue: 2   Pages: 231-40
   Titanium reduction in a soluble Ziegler-Natta catalyst

CA08206031739V   Journal: Japan.   Kokai   Publ: 740518
Pages: 5 pp.
   Stereoregular .alpha.-olefin polymers
   Patent No: 74 51375

CA08204017507M   Journal: Japan.   Publ: 740419   Pages: 6
pp.
   Titanium trichloride composite
   Patent No: 74 16040

CA08204017233U   Journal: Vysokomol. Soedin., Ser. A   Publ:
74   Series: 16   Issue: 9   Pages: 1972-9   Language:
Russ
   Polymerization of ethylene in the dicyclopentadienylethylti-
tanium chloride-ethylaluminum dichloride system

CA08126170276J   Journal: Japan.   Kokai   Publ: 740518
Pages: 6 pp.
   Stereoregular .alpha.-olefin polymers
   Patent No: 74 51377

CA08126169985H   Journal: J. Polym. Sci., Polym. Chem. Ed.
Publ: 74   Series: 12   Issue: 8   Pages: 1703-16
   Polymerization   of   vinyl   chloride   by   use   of   modified
Ziegler-Natta catalysts. I. Overall kinetic features

CA08124154280A   Journal: U.S.   Publ: 740528   Pages: 3
pp.
   Polymerization of conjugated diolefins with   catalysts
containing carbon oxysulfide
   Patent No: 3813374

CA08124152675J   Journal: Makromol.   Chem.   Publ: 74
Series: 175   Issue: 3   Pages: 923-33
  Stereoselective copolymerization of racemic .alpha.-olefins
with ethylene by isospecific Ziegler-Natta catalyst

CA08122136776N   Journal: Japan.   Kokai   Publ: 740518
Pages: 4 pp.
  Stereoregular .alpha.-olefin polymers
  Patent No: 74 51376

CA08122136764G   Journal: U.S.   Publ: 740604   Pages: 7
pp.
  Polymerizing .alpha.-olefins
  Patent No: 3814743

CA08120121449J   Journal: Japan.   Publ: 731226   Pages: 7
pp.
  Highly crystalline olefin polymer
  Patent No: 73 44674

CA08120121216F   Journal: J. Polym. Sci., Polym. Chem. Ed.
Publ: 74   Series: 12   Issue: 4   Pages: 771-84
  Stereochemical regulation of polypropylenes prepared with
various Ziegler-Natta Catalysts

CA08120121142D   Journal: Dokl. Akad. Nauk SSSR   Publ: 74
Series: 215   Issue: 3   Pages: 590-4 (Chem)   Language:
Russ
  Polymerization of propylene at lowered temperatures in the
presence of the homogeneous complex vanadium
tetrachloride-diisobutylaluminum chloride-carbon tetrachloride
system

CA08120121134C   Journal: Polym. Prepr., Amer. Chem. Soc.,
Div. Polym. Chem.   Publ: 72   Series: 13   Issue: 2
Pages: 1097-102
  Enhancement of stereospecificity in isoprene polymerization
with aluminum alkyl-titanium chloride catalysts through use of
carbon disulfide

Index

A

Acetylenes
oligomerization, 605
polymerization, 540–543
Acrylates, polymerization, 538
Acrylic acid, derivatives, copolymers, 583
Acrylonitrile, polymerization, 534, 535
Alkylation, of aromatic nuclei by unsaturated systems, 608
Allene, polymerization, 539
Alloocimene, polymerization, 547
Allyl chloride, polymerization, 536
Allyl fluoride, polymerization, 536
Aluminum alkyls
catalysts, 33, 80, 83
with cyclopentadienyltitanium chloride, in mechanistic studies, 115–117
with preformed titanium chlorides, 111–115
with vandium chloride, 117–124
complexing ability, 104–106
effect of alkyl group size, 86
of alkyl replacement, 87
with titanium chlorides, 109–111
Aluminum-titanium catalyst
azulene as third component, 222–224
for ethylene polymerization, 155
inorganic halides as third components, 215
organic compounds as third components, 215–217
Aluminum-vanadium catalyst, see Vanadium-aluminum catalyst
Amines, as donors for Ziegler-Natta catalysts, 220, 221
Arene catalysts, for polymerization, 302
Aufbau reaction for olefin synthesis, 22, 23
Azulene, third component in aluminum-titanium catalyst, 222–224, 226

B

Ball-milling, of transition metal salt catalysts, 302–304
Beryllium alkyls, catalysts, 33, 84, 86
Bicyclo[n.1.0]alkanes, polymerization, 531
Bicyclo[3.2.0]hepta-2,6-diene, polymerization, 525
Bicyclo[4.1.0]heptane, polymerization, 531
Bicyclo[2.2.1]hept-2-ene-5-butylene, polymerization, 554
Bicyclo[6.1.0]non-4-ene, metathesis by ring opening, 604
Bicyclo[2.2.2]-2-octene, polymerization, 524
Bicyclo[4.2.0]oct-7-ene, polymerization, 525
Bis(β-chloroethyl)vinyl phosphate, polymerization, 539
Block polymerization, see Polymerization
Boron catalysts, 83
p-Bromostyrene, polymerization, 518, 519
Butadiene
block polymerization, 588
catalysts, 130, 141
copolymerization, 581–583
with ethylene, 580
with propylene, 581
copolymer with ethylene, by metathesis, 604
cyclotrimerization, 604
dimerization, 605
polymerization, 4, 27, 54
with allyl-containing catalyst, 295–302
decaying period, 471
effect of hydrochloric acid, 234
effect of molecular hydrogen, 256
effect of water, 230
mechanism, 426, 428, 429
1-Butene